Turbulence

Turbulence

An Introduction for Scientists and Engineers

P.A. Davidson

University of Cambridge

OXFORD
UNIVERSITY PRESS

Great Clarendon Street, Oxford OX2 6DP

Oxford University Press is a department of the University of Oxford.
It furthers the University's objective of excellence in research, scholarship,
and education by publishing worldwide in

Oxford New York

Auckland Bangkok Buenos Aires Cape Town Chennai
Dar es Salaam Delhi Hong Kong Istanbul
Karachi Kolkata Kuala Lumpur Madrid Melbourne Mexico City Mumbai
Nairobi São Paulo Shanghai Taipei Tokyo Toronto

Oxford is a registered trade mark of Oxford University Press
in the UK and in certain other countries

Published in the United States
by Oxford University Press Inc., New York

First published 2004

Reprinted 2004

A catalogue record for this title is available from the British Library

Library of Congress Cataloging in Publication Data

(Data available)

ISBN 019852948 1 (Hbk)
ISBN 019852949 X (Pbk)

10 9 8 7 6 5 4 3 2

Typeset by Newgen Imaging Systems (P) Ltd., Chennai, India
Printed in China

For Henri and Mars

Preface

Turbulence is all around us. The air flowing in and out of our lungs is turbulent, as is the natural convection in the room in which you sit. Glance outside; the wind which gusts down the street is turbulent, and it is turbulence that disperses the pollutants, which belch from the rear of motor cars, saving us from asphyxiation. Turbulence controls the drag on cars, aeroplanes, and bridges, and it dictates the weather through its influence on large-scale atmospheric and oceanic flows. The liquid core of the earth is turbulent, and it is this turbulence that maintains the terrestrial magnetic field against the natural forces of decay. Even solar flares are a manifestation of turbulence, since they are triggered by vigorous motion on the surface of the sun. It is hard not to be intrigued by a subject which pervades so many aspects of our lives.

Yet curiosity can so readily give way to despair when the budding enthusiast embarks on serious study. The mathematical description of turbulence is complex and forbidding, reflecting the profound difficulties inherent in describing three-dimensional, chaotic processes.

This is a textbook and not a research monograph. Our principle aim is to bridge the gap between the elementary, heuristic accounts of turbulence to be found in undergraduate texts, and the more rigorous, if daunting, accounts given in the many excellent monographs on the subject. Throughout we seek to combine the maximum of physical insight with the minimum of mathematical detail.

Turbulence holds a unique place in the field of classical mechanics. Despite the fact that the governing equations have been known since 1845, there is still surprisingly little we can predict with relative certainty. The situation is reminiscent of the state of electromagnetism before it was transformed by Faraday and Maxwell. A myriad of tentative theories have been assembled, often centred around particular experiments, but there is not much in the way of a coherent theoretical framework.[1] The subject tends to consist of an uneasy mix of semi-empirical laws and deterministic but highly simplified cartoons,

[1] One difference between turbulence and nineteenth century electromagnetism is that the latter was eventually refined into a coherent theory, whereas it is unlikely that this will ever occur in turbulence.

bolstered by the occasional rigorous theoretical result. Of course, such a situation tends to encourage the formation of distinct camps, each with its own doctrines and beliefs. Engineers, mathematicians, and physicists tend to view turbulence in rather different ways, and even within each discipline there are many disparate groups. Occasionally religious wars break out between the different camps. Some groups emphasize the role of coherent vortices, while others downplay the importance of such structures and advocate the use of purely statistical methods of attack. Some believe in the formalism of fractals or chaos theory, others do not. Some follow the suggestion of von Neumann and try to unlock the mysteries of turbulence through massive computer simulations, others believe that this is not possible. Many engineers promote the use of semi-empirical models of turbulence; most mathematicians find that this is not to their taste. The debate is often vigorous and exciting and has exercised some of the finest twentieth century minds, such as L.D. Landau and G.I. Taylor. Any would-be author embarking on a turbulence book must carefully pick his way through this minefield, resigned to the fact that not everyone will be content with the outcome. But this is no excuse for not trying; turbulence is of immense importance in physics and engineering, and despite the enormous difficulties of the subject, significant advances have been made.

Roughly speaking, texts on turbulence fall into one of two categories. There are those that focus on the turbulence itself and address such questions as: where does turbulence come from, what are its universal features, to what extent is it deterministic? On the other hand, we have texts whose primary concern is the influence of turbulence on practical processes, such as drag, mixing, heat transfer, and combustion. Here the main objective is to parameterize the influence of turbulence on these processes. The word *modelling* appears frequently in such texts. Applied mathematicians and physicists tend to be concerned with the former category, while engineers are necessarily interested in the latter. Both are important, challenging subjects.

On balance, this text leans slightly towards the first of these categories. The intention is to provide some insight into the physics of turbulence and to introduce the mathematical apparatus which is commonly used to dissect turbulent phenomena. Practical applications, alas, take a back seat. Evidently such a strategy will not be to everyone's taste. Nevertheless, it seems natural when confronted with such a difficult subject, whose pioneers adopted both rigorous and heuristic means of attack, to step back from the practical applications and try and describe, as simply as possible, those aspects of the subject which are now thought to be reasonably well understood.

Our choice of material has been guided by the observation that the history of turbulence has, on occasions, been one of heroic initiatives which promised much yet delivered little. So we have applied the filter of time and chosen to emphasize those theories, both rigorous and heuristic, which look like they might be a permanent feature of the turbulence landscape. There is little attempt to document the latest controversies, or those findings whose significance is still unclear. We begin, in Chapters 1–5, with a fairly traditional introduction to the subject. The topics covered include: the origins of turbulence, boundary layers, the log-law for heat and momentum, free-shear flows, turbulent heat transfer, grid turbulence, Richardson's energy cascade, Kolmogorov's theory of the small scales, turbulent diffusion, the closure problem, simple closure models, and so on. Mathematics is kept to a minimum and we presuppose only an elementary knowledge of fluid mechanics and statistics. (Those statistical ideas which are required, are introduced as and when they are needed in the text.) Chapters 1–5 may be appropriate as background material for an advanced undergraduate or introductory postgraduate course on turbulence.

Next, in Chapters 6–8, we tackle the somewhat refined, yet fundamental, problem of homogeneous turbulence. That is, we imagine a fluid that is vigorously stirred and then left to itself. What can we say about the evolution of such a complex system? Our discussion of homogeneous turbulence differs from that given in most texts in that we work mostly in real space (rather than Fourier space) and we pay as much attention to the behaviour of the large, energy-containing eddies, as we do to the small-scale structures.

Perhaps it is worth explaining why we have taken an unconventional approach to homogeneous turbulence, starting with our slight reluctance to embrace Fourier space. The Fourier transform is conventionally used in turbulence because it makes certain mathematical manipulations easier and because it provides a simple (though crude) means of differentiating between large and small-scale processes. However, it is important to bear in mind that the introduction of the Fourier transform produces no new information; it simply represents a transfer of information from real space to Fourier space. Moreover, there are other ways of differentiating between large and small scales, methods that do not involve the complexities of Fourier space. Given that turbulence consists of eddies (blobs of vorticity) and not waves, it is natural to ask why we must invoke the Fourier transform at all. Consider, for example, grid turbulence. We might picture this as an evolving vorticity field in which vorticity is stripped off the bars of the grid and then mixed to form a seething tangle of vortex tubes and sheets. It is hard to picture a Fourier mode being stripped off the bars of the grid! It is the view of this author that, by and large, it is preferable to work in real space, where the relationship between mathematical representation and physical reality is, perhaps, a little clearer.

The second distinguishing feature of Chapters 6–8 is that equal emphasis is given to both large and small scales. This is a deliberate attempt to redress the current bias towards small scales in monographs on homogeneous turbulence. Of course, it is easy to see how such an imbalance developed. The spectacular success of Kolmogorov's theory of the small eddies has spurred a vast literature devoted to verifying (or picking holes in) this theory. Certainly it cannot be denied that Kolmogorov's laws represent one of the milestones of turbulence theory. However there have been other success stories too. In particular, the work of Landau, Batchelor, and Saffman on the large-scale structure of homogeneous turbulence stands out as a shining example of what can be achieved through careful, physically motivated analysis. So perhaps it is time to redress the balance, and it is with this in mind that we devote part of Chapter 6 to the dynamics of the large-scale eddies. Chapters 6–8 may be suitable as background material for an advanced postgraduate course on turbulence, or act as a reference source for professional researchers.

The final section of the book, Chapters 9 and 10, covers certain special topics rarely discussed in introductory texts. The motivation here is the observation that many geophysical and astrophysical flows are dominated by the effects of body forces, such as buoyancy, Coriolis and Lorentz forces. Moreover, certain large-scale flows are approximately two-dimensional and this has led to a concerted investigation of two-dimensional turbulence over the last few years. We touch on the influence of body forces in Chapter 9 and two-dimensional turbulence in Chapter 10.

There is no royal route to turbulence. Our understanding of it is limited and what little we do know is achieved through detailed and difficult calculation. Nevertheless, it is hoped that this book provides an introduction which is not too arduous and which allows the reader to retain at least some of that initial sense of enthusiasm and wonder.

It is a pleasure to acknowledge the assistance of many friends and colleagues. Alan Bailey, Kate Graham, and Teresa Cronin all helped in the preparation of the manuscript, Jean Delery of ONERA supplied copies of Henri Werle's beautiful photographs, while the drawing of the cigarette plume and the copy of Leonardo's sketch are the work of Fiona Davidson. I am grateful to Julian Hunt, Marcel Lesieur, Keith Moffatt, and Tim Nickels for many interesting discussions on turbulence, and to Alison Jones and Anita Petrie at OUP for their patience and professionalism. In addition, several useful suggestions were made by Ferit Boysan, Jack Herring, Jon Morrison, Mike Proctor, Mark Saville, Christos Vassilicos, and John Young. Finally, I would like to thank Stephen Davidson who painstakingly read the entire manuscript, exposing the many inconsistencies in the original text.

P.A. Davidson
Cambridge, 2003

It remains to call attention to the chief outstanding difficulty of our subject.

HORACE LAMB 1895

PART I

The classical picture of turbulence

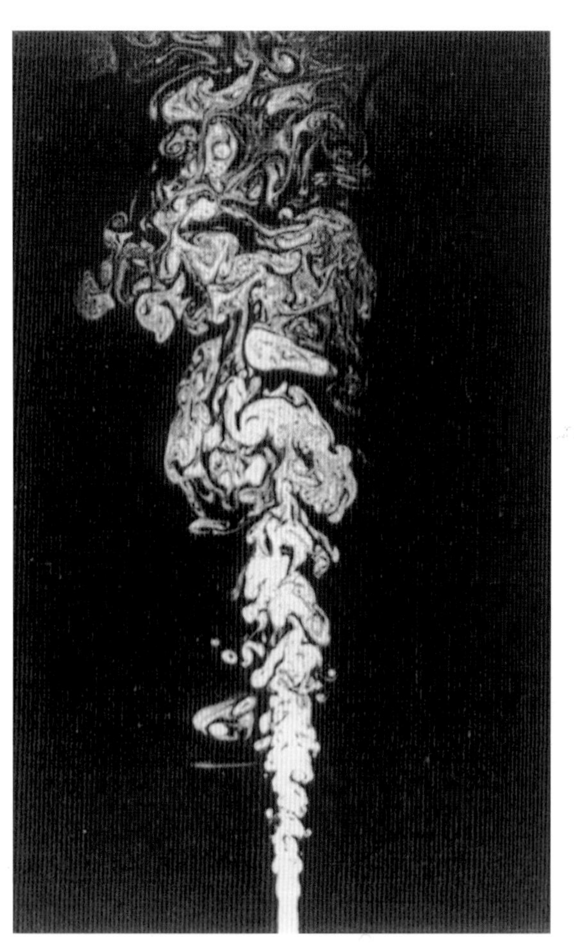

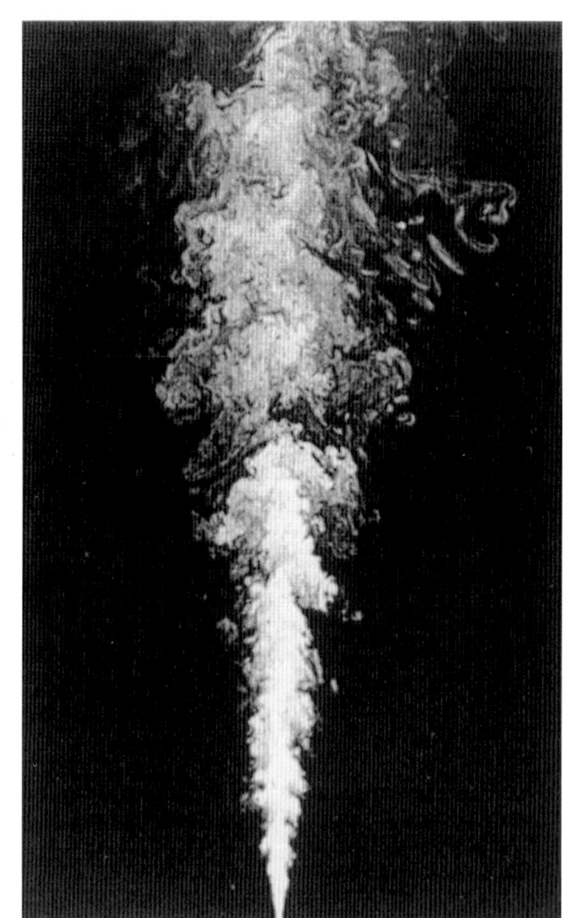

CHAPTER 1

The ubiquitous nature of turbulence

> Vether it's worth goin' through so much, to learn so little, as the charity-boy said ven he got to the end of the alphabet, is a matter o' taste.
>
> Charles Dickens, *Pickwick Papers*

The study of turbulence is not easy, requiring a firm grasp of applied mathematics and considerable physical insight into the dynamics of fluids. Worse still, even after the various theoretical hypotheses have been absorbed, there are relatively few situations in which we can make definite predictions!

For example, perhaps the simplest (and oldest) problem in turbulence concerns the decay of energy in a cloud of turbulence. That is, we stir up a fluid and then leave it to itself. The turbulence decays because of viscous dissipation and a natural question to ask is: how fast does the kinetic energy decline? Theoretical physicists and applied mathematicians have been trying to answer this question for over half a century and still they cannot agree. At times, one is tempted to side with Weller the elder in Pickwick Papers.

Nevertheless, there *are* certain predictions which can be made based on a variety of physical arguments. This is important because turbulent motion is the natural state of most fluids. As you read this book the air flowing up and down your larynx is turbulent, as is the natural convection in the room in which you sit. Glance outside: if you are lucky you will see leaves rustling in the turbulent wind. If you are unlucky you will see pollutants belching out from the rear of motor cars, and it is turbulent convection in the street which disperses the pollutants, saving the pedestrians from an unfortunate fate.

Engineers need to know how to calculate the aerodynamic drag on planes, cars, and buildings. In all three cases the flow will certainly be turbulent. At a larger scale, motion in the oceans and in the atmosphere is turbulent, so weather forecasters and oceanographers study turbulence. Turbulence is also important in geophysics, since it is turbulent convection in the core of the earth which maintains the earth's magnetic field despite the natural forces of decay. Even

Turbulent jets at different Reynolds numbers, with the higher value of Re on the right. Note that the edges of the jets are highly convoluted. (See the discussion in Section 4.3.1.) Note also that the small scales become finer as the Reynolds number increases. [Courtesy of Physics of Fluids & P E Dimotakis, California Institute of Technology.]

astrophysicists study turbulence, since it controls phenomena such as solar flares, sun-spots, and the 22-year solar cycle.

It is the ubiquitous nature of turbulence, from the eruption of solar flares to the rustling of leaves, which makes the subject both important and intriguing. The purpose of this chapter is to give some indication as to just how widespread turbulence really is. We start by describing some simple laboratory experiments.

1.1 The experiments of Taylor and Bénard

It is an empirical observation that the motion of a very viscous or slow moving fluid tends to be smooth and regular. We call this laminar flow. However, if the fluid viscosity is not too high, or the characteristic speed is moderate to large, then the movement of the fluid becomes irregular and chaotic, that is, turbulent. The transition from laminar to turbulent motion is nicely illustrated in a number of simple experiments, of which the most famous are probably those of Reynolds, Bénard, and Taylor.

In 1923 Taylor described a remarkably simple, yet thought-provoking experiment. Suppose that we have two concentric cylinders and that the annular gap between the cylinders is filled with a liquid, say water. The inner cylinder is made to rotate while the outer one remains stationary. At low rotation rates the fluid within the gap does what you would expect: it also rotates, being dragged around by the inner cylinder. At higher rotation rates, however, something unexpected happens. At a certain critical speed, toroidal vortices suddenly appear, superimposed on the primary circular motion (this is shown schematically in Figure 1.1(a)). These axisymmetric structures are called *Taylor vortices*, for an obvious reason, and they arise because of an instability of the basic rotary flow. The net motion of a fluid particle is now helical, confined to a toroidal surface.

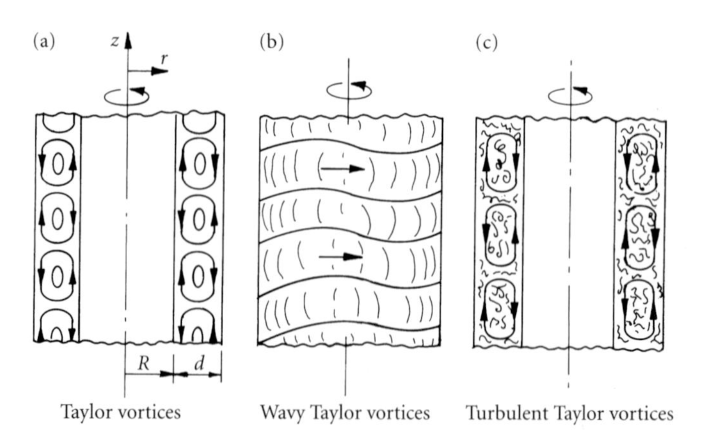

Figure 1.1 Flow between concentric cylinders. As the rotation rate of the inner cylinder increases, the flow becomes progressively more complex until eventually turbulence sets in.

The reason for the sudden appearance of Taylor vortices is related to the centrifugal force. This tends to drive the rotating fluid radially outward. Below the critical speed this force is balanced by a radial pressure gradient and there is no radial motion. However, the centrifugal force is greatest at points where the rotation rate is highest (near the inner cylinder) and so there is always a tendency for fluid at the inner surface to move outward, displacing the outer layers of fluid. This tendency is held in check by pressure and viscous forces. At the critical rotation rate, however, the viscous forces are no longer able to suppress radial disturbances and the flow becomes unstable to the slightest perturbation. Of course, all of the rapidly rotating fluid cannot move uniformly outward because the outer fluid is in the way. Thus the flow breaks up into bands (or cells) as shown in Figure 1.1(a). This is known as the Rayleigh instability.[1]

Now suppose we increase the rotation rate a little more, say by 25%. The Taylor vortices themselves become unstable and so-called *wavy Taylor vortices* appear. These have the appearance of non-axisymmetric toroidal vortices which migrate around the inner cylinder (Figure 1.1(b)). Note that, although complex, this flow is still laminar (non-chaotic).

Suppose we now increase the speed of the inner cylinder yet further. More complex, unsteady structures start to emerge (so-called modulated wavy Taylor vortices) until eventually, when the speed of the inner cylinder is sufficiently high, the flow becomes fully turbulent. In this final state the time-averaged flow pattern resembles that of the steady Taylor vortices shown in Figure 1.1(a), although the cells are a little larger. However, superimposed on this mean flow, we find a chaotic component of motion, so that individual fluid particles are no-longer confined to toroidal surfaces. Rather, as they are swept around by the mean flow, there is a constant jostling for position, almost as if the particles are in a state of Brownian motion (Figure 1.1(c)). A typical measurement of, say, $u_z(t)$ would look something like that shown in Figure 1.2. (Here $\bar{u}$ is the time-averaged value of u_z.)

In fact, it is not just the rotation rate, Ω, which determines which regime prevails. The viscosity of the fluid, ν, the annular gap, d, the radius of the inner cylinder, R, and the length of the apparatus, L, are also important. We can construct three independent dimensionless groups from Ω, ν, d, R, and L, say,

$$\mathrm{Ta} = \frac{\Omega^2 d^3 R}{\nu^2}, \qquad \frac{d}{R}, \qquad \frac{L}{R}.$$

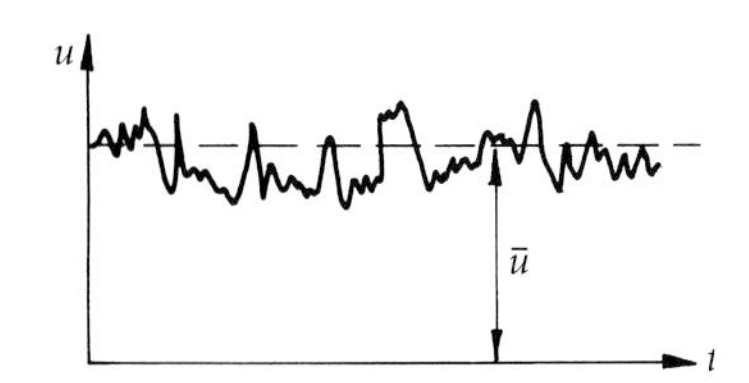

Figure 1.2 Variations of the axial component of velocity with time at some typical location in the annulus. When the flow is turbulent there is a mean component of motion plus a random component.

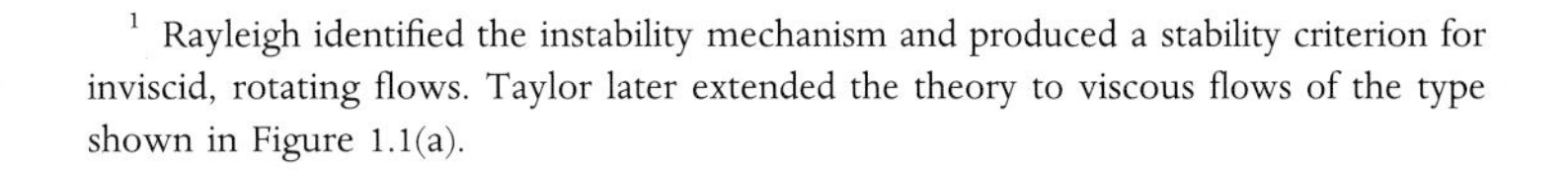
[1] Rayleigh identified the instability mechanism and produced a stability criterion for inviscid, rotating flows. Taylor later extended the theory to viscous flows of the type shown in Figure 1.1(a).

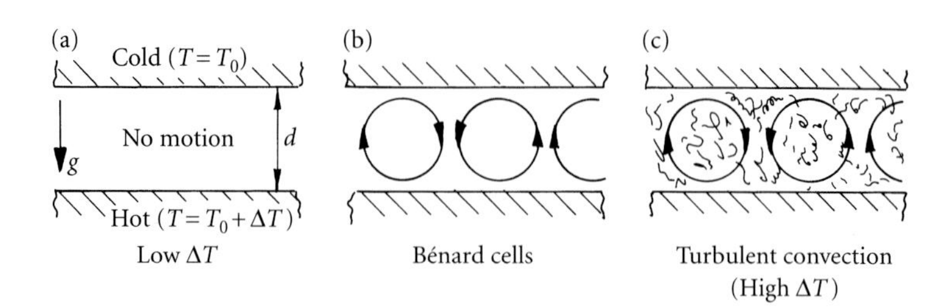

Figure 1.3 Fluid is held between two flat, parallel plates. The lower plate is heated. At low values of ΔT the fluid is quiescent. As ΔT is increased natural convection sets in, first in the form of regular convection cells and then in the form of turbulent flow.

The first of these groups, Ta, is called the Taylor number. (Different authors use slightly different definitions of Ta.) When the apparatus is very long, $L \gg R$, and the gap very narrow, $d \ll R$, it happens to be the value of Ta, and only Ta, which determines the onset of Taylor vortices. In fact, the critical value of Ta turns out to be 1.70×10^3 and the axial wave number of the vortices is $k = 2\pi/\lambda = 3.12/d$, λ being the wavelength.

Let us now consider a quite different experiment, often called Bénard convection or Rayleigh–Bénard convection. Suppose that a fluid is held between two large, flat, parallel plates, as shown in Figure 1.3(a). The lower plate is maintained at temperature $T = T_0 + \Delta T$ and the upper plate at temperature T_0. At low values of ΔT the fluid remains stagnant and heat passes from one plate to the other by molecular conduction. Of course there is an upward buoyancy force which is greatest near the lower plate and tends to drive the fluid upward. However, at low values of ΔT this force is exactly balanced by a vertical pressure gradient. We now slowly increase ΔT. At a critical value of ΔT the fluid suddenly starts to convect as shown in Figure 1.3(b).

The flow consists of hot rising fluid and cold falling regions. This takes the form of regular cells, called Bénard cells, which are reminiscent of Taylor vortices.[2] The rising fluid near the top of the layer loses its heat to the upper plate by thermal conduction, whereupon it starts to fall. As this cold fluid approaches the lower plate it starts to heat up and sooner or later it reverses direction and starts to rise again. Thus we have a cycle in which potential energy is continuously released as light fluid rises and dense fluid falls. Under steady conditions the rate of working of the buoyancy force is exactly balanced by viscous dissipation within the fluid.

The transition from the quiescent state to an array of steady convection cells is, of course, triggered by an instability associated with the buoyancy force. If we perturb the quiescent state, allowing hot fluid to rise and cold fluid to fall, then potential energy will be released and, if the viscous forces are not too excessive, the fluid will accelerate

[2] In plan view the convection cells can take a variety of forms, depending on the value of $(\Delta T)/(\Delta T)_{\mathrm{CRIT}}$ and on the shape of the container. Two-dimensional rolls and hexagons are both common.

giving rise to an instability. On the other hand, if the viscous forces are large then viscous dissipation can destroy all of the potential energy released by the perturbation. In this case the quiescent configuration is stable. Thus we expect the quiescent state to be stable if ΔT is small and ν large, and unstable if ΔT is large and ν is small.

It is possible to set up a variational problem where one looks for the form of perturbation which maximizes the rate of release of potential energy and minimizes the viscous dissipation. On the assumption that the observed convection cells have just such a shape this yields the critical values of ΔT and ν at which instability first sets in. It turns out that the instability criterion is

$$\text{Ra} = \frac{g\beta\Delta T d^3}{\nu\alpha} \geq 1.70 \times 10^3$$

where Ra (a dimensionless parameter) is known as the Rayleigh number, β is the expansion coefficient of the fluid, d is the height of the gap and α the thermal diffusivity. The wave number (2π/wavelength) of the convection cells is $k = 3.12/d$.

Of course we have seen these numbers before. The critical value of the Taylor number for narrow annuli is

$$\text{Ta} = \frac{\Omega^2 d^3 R}{\nu^2} = 1.70 \times 10^3$$

and the wave number of the Taylor cells at the onset of instability is $k = 3.12/d$. At first sight this coincidence seems remarkable. However, it turns out that there is an analogy between axisymmetric flow with swirl (in the narrow-gap approximation) and natural convection. The analogue of the buoyancy force is the centrifugal force and the angular momentum, $\Gamma = ru_\theta$, in a rotating axisymmetric flow is convected and diffused just like the temperature, T, in Bénard's experiment.[3] In fact, Rayleigh first derived his famous criterion for the instability of a rotating, inviscid fluid through a consideration of the analogy between the buoyancy and centrifugal forces.

Now suppose Ra is slowly increased above its critical value. Then there is a point at which the Bénard cells themselves become unstable and more complex, unsteady structures start to appear. Eventually, when Ra is large enough, we reach a state of turbulent convection.

It seems that the general picture which emerges from both Taylor's and Bénard's experiments is the following. For high viscosities the basic configuration is stable. As the viscosity is reduced we soon reach a point where the basic equilibrium is unstable to infinitesimal perturbations and the system bifurcates (changes) to a more complex state, consisting of a steady laminar flow in the form of regular cells.

[3] Actually it turns out that there is a 0.7% difference in the critical value Ta and Ra, so the analogy is very nearly, but not exactly, perfect.

As the viscosity is reduced even further the new flow itself becomes unstable and we arrive at a more complex motion. Subsequent reductions in viscosity give rise to progressively more complex flows until eventually, for sufficiently small ν, the flow is fully turbulent. At this point the flow field consists of a mean (time-averaged) component plus a random, chaotic motion. This sequence of steps is called the *transition to turbulence*.

1.2 Flow over a cylinder

This kind of behaviour, in which a fluid passes through a sequence of flow regimes of increasing complexity as ν is reduced, is also seen in external flows. Consider, for example, flow past a cylinder, as shown in Figure 1.4. An inverse measure of ν is given by the (dimensionless) *Reynolds number*, $\mathrm{Re} = ud/\nu$, where u is the upstream speed of the fluid and d the diameter of the cylinder. At high values of ν we have a steady, symmetric flow pattern. As Re approaches unity the upstream–downstream symmetry is lost and in the range $\mathrm{Re} \sim 5$–40 we find steady vortices attached to the rear of the cylinder. When Re reaches a

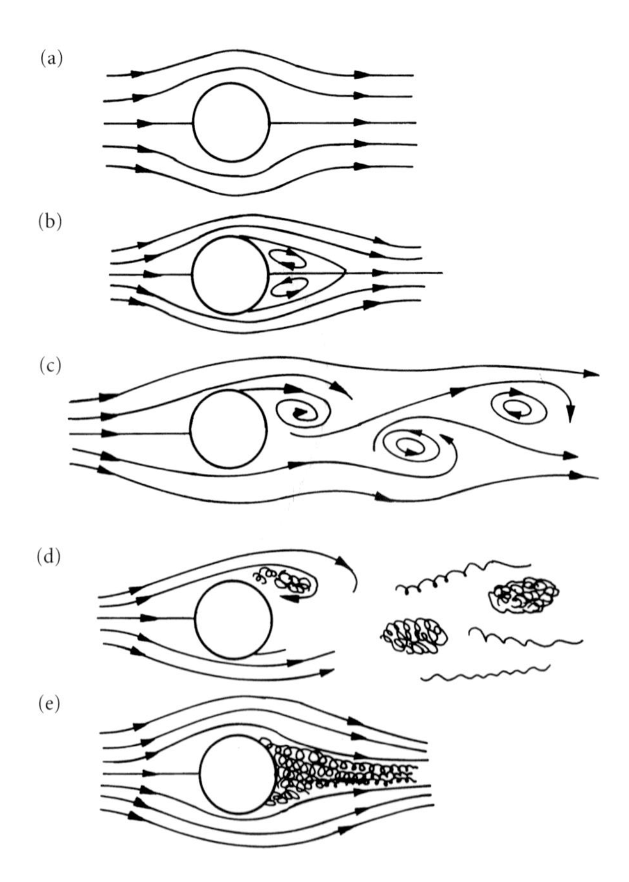

Figure 1.4 Flow behind a cylinder: (a) $\mathrm{Re} < 1$; (b) $5 < \mathrm{Re} < 40$; (c) $100 < \mathrm{Re} < 200$; (d) $\mathrm{Re} \sim 10^4$; and (e) $\mathrm{Re} \sim 10^6$.

value of around 40 an instability is observed in the form of an oscillation of the wake, and by the time we reach Re $\sim$ 100 the vortices start to peel off from the rear of the cylinder in a regular, periodic manner. The flow is still laminar but we have now lost the top–bottom symmetry. This is, of course, the famous Karman vortex street. The laminar Karman street persists up to Re $\sim$ 200 at which point three-dimensional instabilities develop. By Re $\sim$ 400 low levels of turbulence start to appear within the vortices. Nevertheless, the periodicity of the vortex shedding remains quite robust. Eventually, for high values of Re, say 10^5, the turbulence spreads out of the vortices and we obtain a fully turbulent wake. Within this wake, however, one can still detect coherent vortex shedding. Thus we have the same sort of pattern of behaviour as seen in Taylor's experiment. As ν is reduced the flow becomes more and more complex until, eventually, turbulence sets in.

This example is important since it illustrates the key role played by Re in determining the state of a flow. In general, Re $= ul/\nu$ represents the ratio of inertial to viscous forces in a fluid, provided that l is chosen appropriately.[4] Thus, when Re is large, the viscous forces, and hence viscous dissipation, is small. Such flows are prone to instabilities and turbulence, as evidenced by our example of flow over a cylinder.

1.3 Reynolds' experiment

The transition from laminar to turbulent flow, and the important role played by Re in this transition, was first pointed out by Reynolds in 1883. Reynolds was concerned with flow along a straight, smooth pipe which, despite having a particularly simple geometry, turns out to be rather more subtle and complex than either Taylor's or Bénard's experiments.

In his now famous paper, Reynolds clearly distinguished between the two possible flow regimes (laminar and turbulent) and argued that the parameter which controlled the transition from one regime to another had to be Re $= ud/\nu$, where d is the pipe diameter and u the mean flow down the pipe. He also noted that the critical value of Re at which turbulence first appears is very sensitive to disturbances at the entrance to the pipe. Indeed, he suggested that the instability which initiates the turbulence might require a perturbation of a certain magnitude, for a given value of Re, for the unstable motion to take

[4] The inertial forces are of order u^2/l_u where l_u is a length-scale typical of the streamline pattern, say the curvature of the streamlines, while the viscous forces are of order $\nu u/l_v^2$, where l_v is a length typical of cross-stream gradients in velocity (see Chapter 2).

root and turbulence to set in. For example, when the inlet disturbances were minimized, he found the laminar flow to be stable up to Re $\sim$ 13,000, whereas turbulence typically appears at Re $\sim$ 2000 if no particular effort is taken to minimize the disturbances. We now know that Reynolds was right. The current view is that fully developed laminar pipe flow is stable to *infinitesimal* disturbances for all values of Re, no matter how large, and indeed recent experiments (with very special inlet conditions) have achieved laminar flow for values of Re up to 90,000. It is the size of the disturbance, and the type of inlet, which matters.

Reynolds also examined what happens when turbulence is artificially created in the pipe. He was particularly interested in whether or not there is a value of Re below which the turbulence dies out. It turns out that there is, and that this corresponds to Re $\sim$ 2000.

The modern view is the following. The inlet conditions are very important. When we have a simple, straight inlet, as shown in Figure 1.5, and Re exceeds 10^4, turbulence tends to appear first in the annular boundary layer near the inlet. The turbulence initially takes the form of small, localized patches of chaotic motion. These 'turbulent spots' then spread and merge until a slug of turbulence fills the pipe (Figure 1.5). For lower values of Re, on the other hand, the boundary layer near the inlet is thought to be stable to small disturbances. Thus the perturbations which initiate transition in the range 2000–10^4 must be present at the pipe inlet, or else represent a finite-amplitude instability of the boundary layer near the inlet. In any event, whatever the origin of the turbulence, it appears that transition starts with a series of intermittent turbulent slugs passing down the pipe (Figure 1.5). Provided Re exceeds $\sim$2000, these slugs tend to grow in length at the expense of the non-turbulent fluid between them, eventually merging to form fully developed turbulence. When Re is less than $\sim$2000, on the other hand, any turbulent slugs which are generated near the inlet simply decay.

1.4 Common themes

The examples above suggest that there are at least two types of transition to turbulence. There are flows in which turbulent motion appears first in small patches. Provided Re is large enough, the

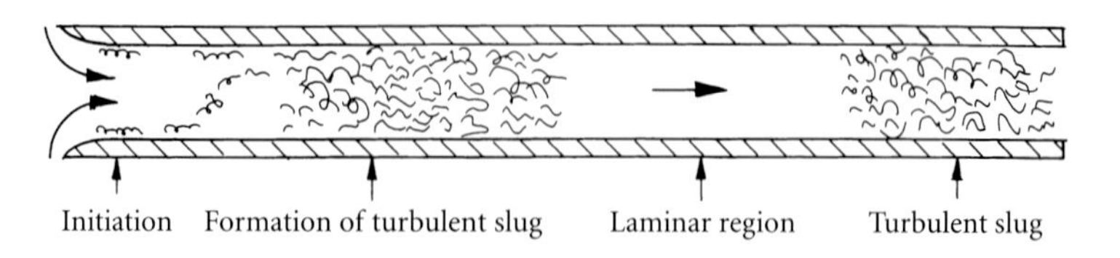

Figure 1.5 Turbulent slugs near the inlet of a pipe.

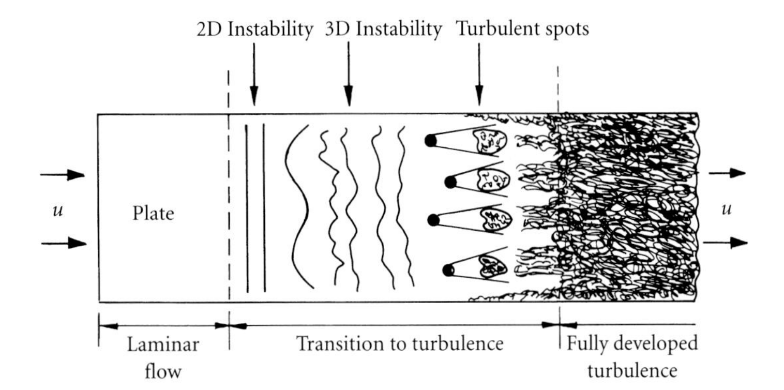

Figure 1.6 Transition to turbulence in a boundary layer.

turbulent patches grow and merge until fully developed turbulence is established. This is typical of transition to turbulence in boundary layers (Figure 1.6) and in pipes. The key point here is that, in the transition region, the turbulence is somewhat intermittent, being interspersed by quiescent, laminar regions.

Then there are flows in which, when a certain threshold is exceeded, chaos develops uniformly throughout the fluid. This might start out as a simple instability of the mean flow, leading to a more complicated laminar motion, which in turn becomes unstable and breaks up into even more complex structures. (See Figures 1.1 and 1.3.) A sequence of such instabilities leads eventually to random, chaotic motion, that is, turbulence.

A familiar example of this second type of transition occurs in the buoyant plume from a cigarette. As the plume rises the fluid accelerates and Re becomes larger. Sooner or later an instability sets in and the laminar plume starts to exhibit a complex, three-dimensional structure. Shortly thereafter the plume becomes fully turbulent (Figure 1.7).

The essential point is that fluid motion is almost always inherently unstable, and that incipient instabilities are suppressed only if the viscous dissipation is high enough. However, virtually all fluids have an extremely low viscosity. This is true of water, air, blood, the molten metal in the core of the earth, and the atmosphere of the sun. The implication is that turbulence is the natural state of things and indeed this is the case. Consider, for example, Bénard convection in a layer of water one inch deep. The initial instability sets in at a temperature difference of $\Delta T \sim 0.01\,^\circ\mathrm{C}$ and the final transition to turbulence occurs at $\Delta T \sim 0.1\,^\circ\mathrm{C}$, which is not a huge temperature difference! Alternatively, we might consider Taylor's experiment. If $R = 10$ cm, the annular gap is 1 cm and the fluid is water, then the flow will become unstable when the peripheral speed of the inner cylinder exceeds a mere 1 cm/s. Evidently, laminar flow is the exception and not the rule in nature and technology.

Figure 1.7 A schematic representation of a cigarette plume (after Corrsin 1961).

So far we have carefully avoided giving a formal definition of turbulence. In fact, it turns out to be rather difficult, and possibly not very useful, to provide one.[5] It is better simply to note that when ν is made small enough all flows develop a random, chaotic component of motion. We group these complex flows together, call them turbulent, and then note some of their common characteristics. They are:

(1) the velocity field fluctuates randomly in time and is highly disordered in space, exhibiting a wide range of length scales;

[5] Nevertheless, we provide a definition in Section 2.4 of Chapter 2!

(2) the velocity field is unpredictable in the sense that a minute change to the initial conditions will produce a large change to the subsequent motion.

To illustrate the second point consider the following experiment. Suppose we tow an initially stationary cylinder through a quiescent fluid at a fixed speed creating a turbulent wake. We do the experiment one hundred times and on each occasion we measure the velocity as a function of time at a point a fixed distance downstream of the cylinder, say $\mathbf{x}_0$ in a frame of reference travelling with the cylinder. Despite the nominally identical conditions we find that the function $\mathbf{u}(\mathbf{x}_0, t)$ is quite different on each occasion. This is because there will always be some minute difference in the way the experiment is carried out and it is in the nature of turbulence that these differences are amplified. It is striking that, although the governing equations of incompressible flow are quite simple (essentially Newton's second law applied to a continuum), the exact details of $\mathbf{u}(\mathbf{x}, t)$ appear to be, to all intents and purposes, random and unpredictable.

Now suppose that we do something different. We measure $\mathbf{u}(\mathbf{x}_0, t)$ for some considerable period of time and then calculate the time-average of the signal, $\bar{\mathbf{u}}(\mathbf{x}_0)$. We do this a second time and then a third. Each time we obtain the same value for $\bar{\mathbf{u}}(\mathbf{x}_0)$, as indicated in Figure 1.8. The same thing happens if we calculate $\overline{\mathbf{u}^2}$. Evidently, although $\mathbf{u}(\mathbf{x}, t)$ appears to be random and unpredictable, its statistical properties are not. This is the first hint that any theory of turbulence has to be a statistical one, and we shall return to this point time and again in the chapters which follow.

In summary, then, the statistical properties of a turbulent flow are uniquely determined by the boundary conditions and the initial conditions. If a sequence of nominally identical experiments are carried out,

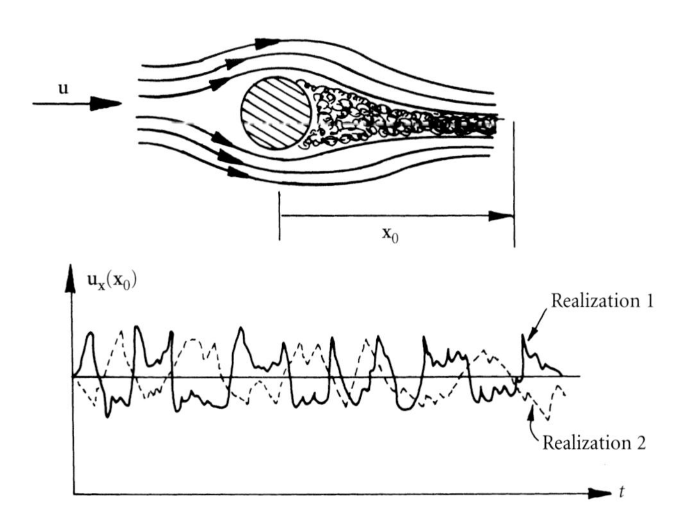

Figure 1.8 A cylinder is towed through a quiescent fluid and $\mathbf{u}_x$ is measured at location $\mathbf{x}_0$. The records of $\mathbf{u}_x(t)$ in two nominally identical realizations of the experiment are quite different.

then the statistical properties of the flow will not change from one experiment to the next, even though the details of the individual realizations will be different. To be more precise: the inevitable minute variations in the different realizations of the experiments produce only minute variations in the statistics. In principle, $\mathbf{u}(t)$ is also uniquely determined by the boundary conditions and initial conditions. However, in any laboratory experiment the initial conditions can be controlled only to within a certain level of accuracy and no matter how hard we try there will always be infinitesimal variations between experiments. It is in the nature of turbulence to amplify these variations so that, no matter how small the perturbations in the initial conditions, the resulting trace of $\mathbf{u}(t)$ eventually looks completely different from one realization to the next. It should be emphasized that this is not because we are poor experimentalists and that if only we could control conditions better the variability in $\mathbf{u}(t)$ would disappear. Almost *any* change in the initial conditions, no matter how small, eventually leads to an order one change in $\mathbf{u}(t)$. This extreme sensitivity to initial conditions, long known to experimentalists, is now recognized as the hallmark of mathematical chaos, and is exhibited by a wide range of non-linear systems (see Chapter 3, Section 3.1).[6]

1.5 The ubiquitous nature of turbulence

Of course turbulence is not restricted to the laboratory. It influences many aspects of our lives, operating at many scales, from the vast to the small. Let us take a moment to discuss some of the more common examples of turbulence, starting with large-scale phenomena.

Perhaps the most spectacular of the large-scale manifestations of turbulence is a solar flare. Solar flares are associated with so-called prominences: those vast, arch-like structures which can be seen at the surface of the sun during a solar eclipse (Figure 1.9(a)). Prominences extend from the chromosphere (the lower atmosphere of the sun) up into the corona, and contain relatively cool chromospheric gas, perhaps 300 times cooler than the surrounding coronal gas. They are huge, typically ten times the size of the earth. Prominences are immersed in, and surrounded by, magnetic flux tubes which arch up from the photosphere (the surface of the sun), criss-crossing the prominence. The magnetic flux tubes hold the prominence in place (Figure 1.9(b)). Those overlying the prominence (the so-called magnetic arcade) push down on it, while those lying below provide a magnetic cushion. Some prominences are stable, long-lived structures, surviving many

[6] We shall see in Chapter 3 that two initial conditions separated by a small amount, ε, typically lead to solutions which diverge at a rate $\varepsilon \exp(\lambda t)$ where λ is a constant.

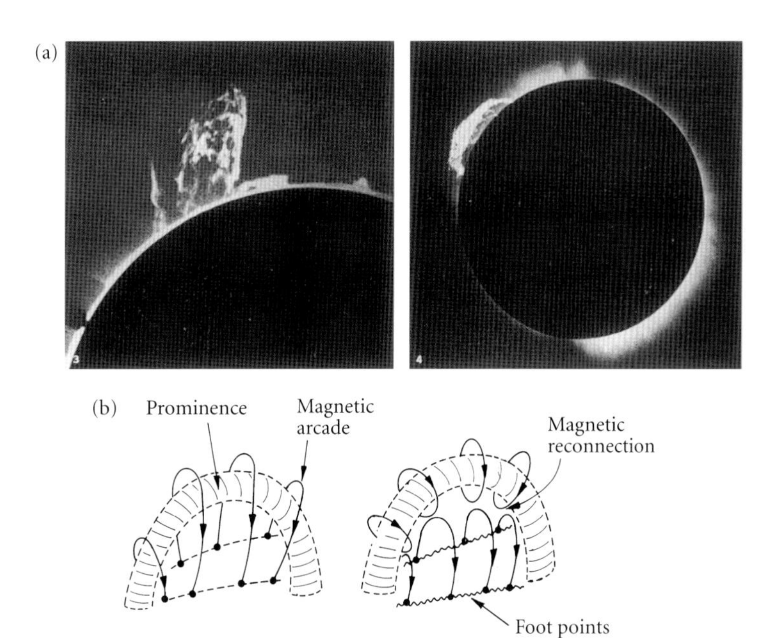

Figure 1.9 (a) Solar prominences (Encyclopaedia Britannica 1926). (b) A cartoon of a solar flare. Turbulent motion on the surface of the sun causes the footpoints of the magnetic flux tubes to jostle for position. The flux tubes then become entangled leading to magnetic reconnection, a loss of equilibrium, and finally to a solar flare.

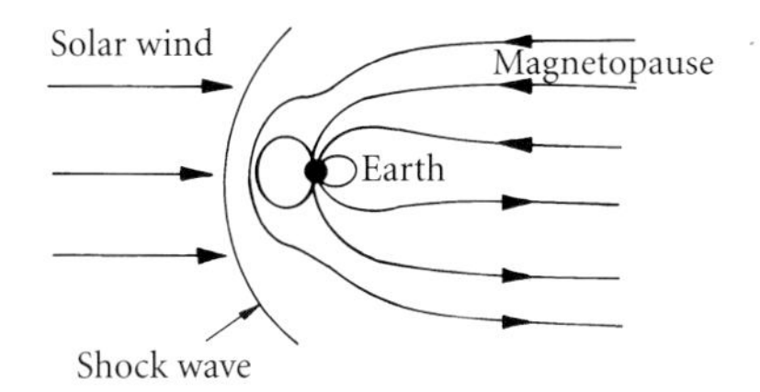

Figure 1.10 The solar wind causes turbulence in the magnetosphere.

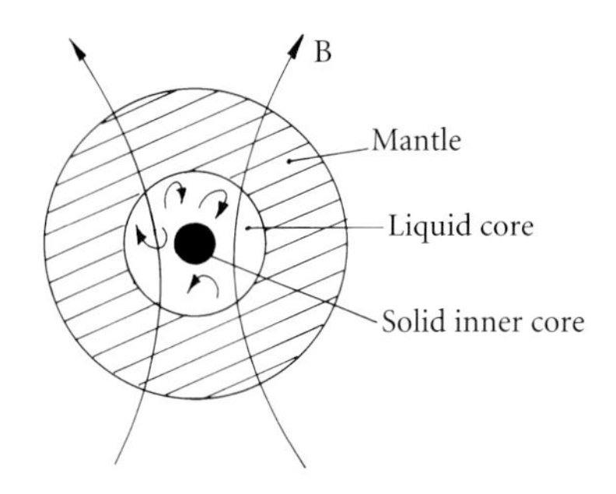

Figure 1.11 Turbulent convection in the core of the earth maintains the earth's magnetic field.

weeks. Others erupt explosively, releasing prodigious amounts of energy ($\sim 10^{25}$ J) in a matter of hours. Turbulence plays several key roles in these explosions. The first point is that the surface of the sun is highly turbulent and so there is a constant jostling of the foot points of the magnetic flux tubes. As a result, the flux tubes in the corona can become entangled, leading to so-called magnetic reconnection in which flux tubes are severed and recombine. (This severing of the flux tubes happens through the local action of turbulence in the corona.) In the process the equilibrium of the prominence is lost and vast amounts of mass and energy are propelled out into space.

This sudden release of mass and energy enhances the solar wind which spirals radially outward from the sun. Thus the mass released by a solar flare sweeps through the solar system and one or two days after a flare erupts the earth is buffeted by magnetic storms which generate turbulence in the magnetosphere. The earth is shielded from the solar wind by its magnetic field which deflects the charged particles in the wind around the earth (Figure 1.10). So, without the earth's magnetic field we would all be in bad shape.

But why does the earth, and indeed many of the planets, possess a magnetic field? The interiors of the planets do not contain any magnetic material, and the planets themselves are, by and large, too old for their magnetic fields to be some relic of the primordial field trapped within them at their birth. Such a field would long ago have decayed. The answer, it seems, is turbulence. It is now generally agreed that the source of these magnetic fields is turbulent convection within the core of the planets (Figure 1.11). This acts like a dynamo, converting

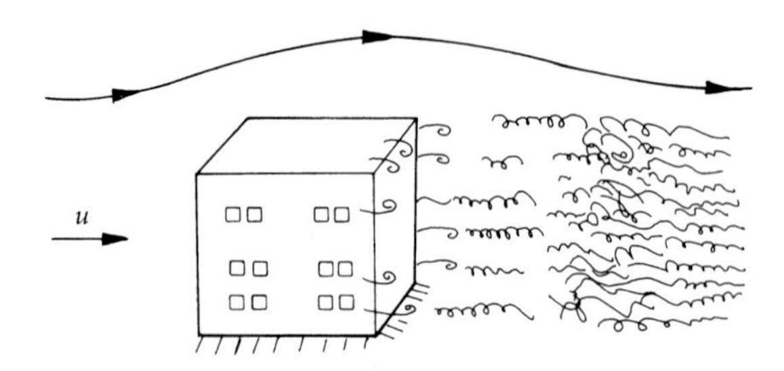

Figure 1.12 Flow around a building gives rise to a turbulent wake. (See also Plate 2.)

mechanical energy into magnetic energy. So both astrophysicists and geophysicists have to contend with the effects of turbulence.

Closer to home, large-scale oceanic and atmospheric flows are turbulent, so the art of weather forecasting is one of predicting the short-term evolution of a highly complex turbulent flow. Indeed, the need for accurate weather forecasting has provided much of the impetus for the development of numerical methods in turbulence research.

Of course, engineers of every sort have to contend with turbulence. The aerodynamic drag on a plane or a car is controlled by turbulent boundary layers (Plate 1) and indeed one of the major stumbling blocks to better wing design is our lack of understanding of turbulence. On the other hand, structural engineers have to worry about turbulent wind loading and the effects of turbulent wakes from high buildings (Figure 1.12), while the designers of internal combustion engines rely on the turbulent mixing of the fuel and gases to maximize efficiency. Turbulence is also crucial in the environmental arena as city planners have to model the turbulent dispersion of pollutants from chimney stacks (Plate 2) and car engines, while architects need to predict how natural convection influences the temperature distribution within a building. Even steel makers have to worry about turbulence since excess turbulence in an ingot casting can cause a deterioration in the metallurgical structure of the ingot.

Evidently, there is a clear need not just for a qualitative understanding of turbulence, but also for the ability to make quantitative predictions. Such predictions are, however, extremely difficult. Consider, for instance, the simple case of a cigarette plume (Figure 1.7). The smoke twists and turns in a chaotic manner, continually evolving and never repeating itself. How can we make definite predictions about such a motion? It is no accident that nearly all theories of turbulence are statistical theories. For instance, in the case of a cigarette plume some theory might try to estimate the time-averaged concentration of smoke at a particular location, or perhaps the time-averaged width of the plume at a particular height. No theory will ever be able to predict the exact concentration at a particular location and at a particular time. So the science of turbulence is largely about making statistical predictions about the chaotic solutions of a non-linear partial differential equation (the Navier–Stokes equation).

1.6 Different scales in a turbulent flow: a glimpse at the energy cascade of Kolmogorov and Richardson

One gets a similar impression when making a drawing of a rising cumulus from a fixed point; the details change before the sketch can be completed. We realise that big whirls have little whirls that feed on their velocity, and little whirls have lesser whirls and so on to viscosity. (L.F. Richardson 1922)

Let us now try to develop a few elementary theoretical ideas. We have already seen that a turbulent flow which is steady-on-average contains, in general, a mean flow plus a random, fluctuating component of motion (Figure 1.2). In the case of Taylor's experiment, for example, the mean flow is the array of (turbulent) Taylor cells while in Reynolds' experiment there is a mean axial flow in the pipe. Let us denote time-averages by an overbar. Then at each location in a steady-on-average flow we have

$$\mathbf{u}(\mathbf{x}, t) = \bar{\mathbf{u}}(\mathbf{x}) + \mathbf{u}'(\mathbf{x}, t)$$

where $\mathbf{u}'$ is the random component of motion. The difference between $\mathbf{u}$ and $\bar{\mathbf{u}}$ is illustrated in Figure 1.13 which shows the instantaneous and time-averaged flow over a sphere of diameter d. Note that, although the flow is turbulent, $\bar{\mathbf{u}}(\mathbf{x})$ is a smooth, ordered function of position. Note also that $\mathbf{u}$, and hence $\mathbf{u}'$, is highly disordered in space.

We shall see shortly that, at any instant, $\mathbf{u}'$ consists of a random collection of eddies (vortices). The largest of these eddies have a size

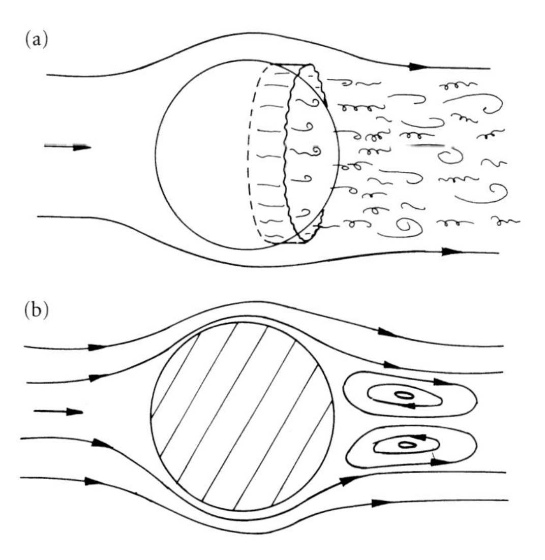

Figure 1.13 Schematic representation of flow over a sphere at $\mathrm{Re} = 2 \times 10^4$: (a) snapshot of the flow as illustrated by dye injected into the boundary layer; (b) time-averaged flow pattern as seen in a time-lapse photograph. See also Plate 4 for the actual flow at $\mathrm{Re} = 2 \times 10^4$ and 2×10^5.

comparable with the characteristic geometric length scale of the mean flow (d in Figure 1.13). However, most of the eddies are much smaller than this and indeed there are usually eddies which are very small indeed. For example, if the mean eddy size is, say, 1 cm, we can usually find eddies of size 0.1 mm, or even smaller. The size of the smallest eddies depends on the Reynolds number of the turbulence, as we shall see. For the present purposes, however, the details are not important. The key point to grasp is that, at any instant, there is a broad spectrum of eddy sizes within fully developed turbulence. This is nicely illustrated in Leonardo's famous sketch of water falling into a pool. (See Plate 3.)

Now the rate at which kinetic energy is dissipated in a fluid is $\varepsilon = 2\nu S_{ij}S_{ij}$ per unit mass (consult Chapter 2). Here S_{ij} is the strain rate tensor, $S_{ij} = \frac{1}{2}(\partial u_i/\partial x_j + \partial u_j/\partial x_i)$. Thus, dissipation is particularly pronounced in regions where the instantaneous gradient in velocity, and hence the shear stress, is large. This suggests that the dissipation of mechanical energy within a turbulent flow is concentrated in the smallest eddies, and this turns out to be true.

Let us now consider turbulence where $\text{Re} = ul/\nu$, is large. Here u is a typical value of $|\mathbf{u}'|$ and l is the typical size of the large-scale turbulent eddies. The observation that there exists a broad spectrum of eddy sizes, and that dissipation is associated predominantly with the smallest eddies, led Richardson to introduce the concept of the *energy cascade* for high-Re turbulence.

The idea is the following. The largest eddies, which are created by instabilities in the mean flow, are themselves subject to inertial instabilities and rapidly break-up or evolve into yet smaller vortices.[7] In fact, the lifespan of a typical eddy is rather short, of the order of its so-called turn-over time, l/u. Of course, the smaller eddies are themselves unstable and they, in turn, pass their energy onto even smaller structures and so on. Thus, at each instant, there is a continual *cascade* of energy from the large scale down to the small (Figure 1.14, Plate 3). Crucially, viscosity plays no part in this cascade. That is, since $\text{Re} = ul/\nu$ is large, the viscous stresses acting on the large eddies are negligible. This is also true of their offspring. The whole process is essentially driven by inertial forces. The cascade comes to a halt, however, when the eddy size becomes so small that Re, based on the size of the smallest eddies, is of order unity. At this point the viscous forces become significant and dissipation starts to become important. Thus we have a picture of large-scale eddies being continually created by the mean flow, and these eddies then breaking up through a sequence of inviscid instabilities into finer and finer structures. Energy

[7] The term 'break-up' is used rather loosely here to mean that energy is progressively transferred from large eddies to smaller ones.

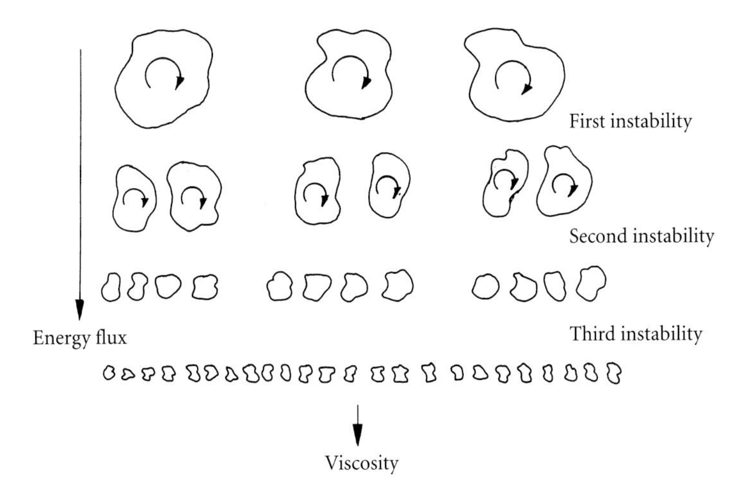

Figure 1.14 A schematic representation of the energy cascade (after Frisch 1995). See also Leonardo's sketch—Plate 3.

is destroyed only in the final stages of this process when the structures are so fine that Re, based on the small-scale structures, is of order unity. In this sense viscosity plays a rather passive role, mopping up whatever energy cascades down from above.

Let us now see if we can determine the smallest scale in a turbulent flow. Let u and v represent typical velocities associated with the largest and smallest eddies respectively. Also, let l and η be the length scales of the largest and smallest structures. Now we know that most eddies break-up on a timescale of their turn-over time (all of the experimental evidence confirms that this is so), and so the rate at which energy (per unit mass) is passed down the energy cascade from the largest eddies is,

$$\Pi \sim u^2/(l/u) = u^3/l.$$

When conditions are statistically steady this must match exactly the rate of dissipation of energy at the smallest scales. If it did not, then there would be an accumulation of energy at some intermediate scale, and we exclude this possibility because we want the statistical structure of the turbulence to be the same from one moment to the next. The rate of dissipation of energy at the smallest scales is,

$$\varepsilon \sim \nu S_{ij} S_{ij}$$

where S_{ij} is the rate of strain associated with the smallest eddies, $S_{ij} \sim \mathrm{v}/\eta$. This yields

$$\varepsilon \sim \nu(\mathrm{v}^2/\eta^2).$$

Since the dissipation of turbulent energy, ε, must match the rate at which energy enters the cascade, Π, we have,

$$u^3/l \sim \nu(\mathrm{v}^2/\eta^2). \tag{1.1}$$

However, we also know that Re based on v and η is of order unity

$$\mathrm{v}\eta/\nu \sim 1. \tag{1.2}$$

Combining these expressions we find

$$\eta \sim l\mathrm{Re}^{-3/4} \qquad \text{or} \qquad \eta \sim (\nu^3/\varepsilon)^{1/4} \tag{1.3}$$

$$\mathrm{v} \sim u\mathrm{Re}^{-1/4} \qquad \text{or} \qquad \mathrm{v} \sim (\nu\varepsilon)^{1/4} \tag{1.4}$$

where Re is based on the large-scale eddies, $\mathrm{Re} = ul/\nu$. In a typical wind tunnel experiment we might have $\mathrm{Re} \sim 10^3$ and $l \sim 1$ cm. The estimate above suggests $\eta \sim 0.06$ mm, so much of the energy in this flow is dissipated in eddies which are less than a millimetre in size! Evidently the smallest scales in a turbulent flow have a very fine structure. Moreover, the higher the Reynolds number, the finer the small-scale structures. (This is illustrated in Figure 1.15 which shows two nominally similar flows at different values of Re.) The scales η and v are called the *Kolmogorov microscales* of turbulence while l is called the *integral scale*.

Now the arguments leading up to (1.1)–(1.4) are more than a little heuristic. What, for example, do we mean by an eddy (is it spherical, tubular, or even sheet-like) or by the phrase 'eddy break-up'? Moreover, how do we know that there really is a cascade process in which eddies 'break up' through a sequence of instabilities, creating intermediate structures all the way from l down to η? Despite these reservations, estimates (1.1) to (1.4) all turn out to conform remarkably well to the experimental data. Indeed, they represent some of the more useful results in turbulence theory!

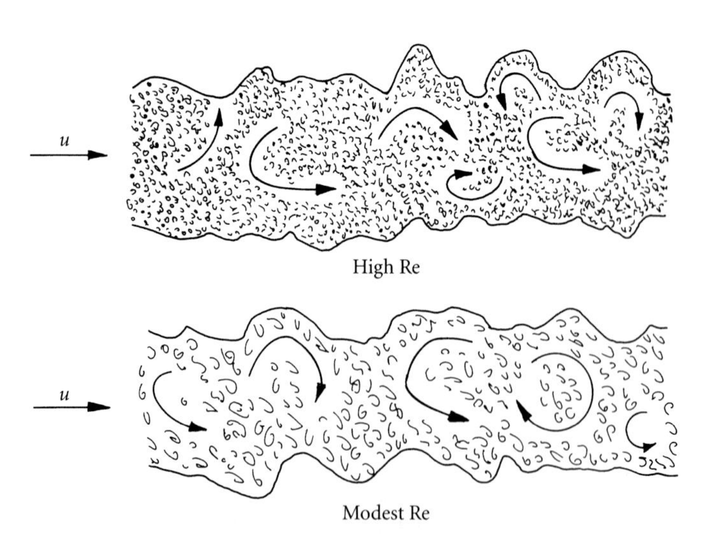

Figure 1.15 The influence of Re on the smallest scales in a turbulent wake. Note that the smallest eddies are much smaller in the high-Re flow. (After Tennekes and Lumley 1972.)

1.7 The closure problem of turbulence

Had I been present at the Creation, I would have given some useful hints for a better ordering of the universe. (Alfonso the Wise of Castile ~ 1260)

Figure 1.8 tells us something interesting about turbulence. It seems that, despite the random nature of $\mathbf{u}(\mathbf{x}, t)$, the statistical properties of the velocity field, such as $\bar{\mathbf{u}}(\mathbf{x})$ and $\overline{\mathbf{u}^2}(\mathbf{x})$, are quite reproducible. This suggests that any predictive theory of turbulence should work with statistical quantities, and indeed this is the basis of most theoretical approaches. The problem then arises as to how to derive dynamical equations for these statistical quantities. The starting point is invariably Newton's second law. When applied to a lump of fluid this takes the form of the Navier–Stokes equation (see Chapter 2, Section 2.1),

$$\rho \frac{\partial \mathbf{u}}{\partial t} = -\rho(\mathbf{u} \cdot \nabla)\mathbf{u} - \nabla p + \rho\nu\nabla^2\mathbf{u}. \tag{1.5}$$

The terms on the right of (1.5) represent inertial, pressure, and viscous forces respectively (p is the fluid pressure and ν the viscosity). The details of (1.5) are, for the moment, unimportant. It is necessary only to recognize that $\mathbf{u}$ obeys an equation of the form,

$$\frac{\partial \mathbf{u}}{\partial t} = F_1(\mathbf{u}, p).$$

Now, in an incompressible fluid $\mathbf{u}$ is solenoidal, $\nabla \cdot \mathbf{u} = 0$, and so an equation for p may be obtained by taking the divergence of (1.5):

$$\nabla^2(p/\rho) = -\nabla \cdot (\mathbf{u} \cdot \nabla\mathbf{u}).$$

In an infinite domain this may be inverted using the Biot–Savart law (see Chapter 2, Section 2.2) to give

$$p(\mathbf{x}) = \frac{\rho}{4\pi}\int \frac{[\nabla \cdot (\mathbf{u} \cdot \nabla\mathbf{u})]'}{|\mathbf{x} - \mathbf{x}'|}\, d\mathbf{x}'$$

and so p is uniquely determined by the instantaneous distribution of velocity. It follows that (1.5) may be rewritten in the symbolic form

$$\frac{\partial \mathbf{u}}{\partial t} = F_2(\mathbf{u}). \tag{1.6}$$

This is a perfectly deterministic equation and so, for given initial conditions, we can integrate forward in time to find $\mathbf{u}(\mathbf{x}, t)$. In practice this would have to be done numerically, and because $\mathbf{u}$ is so convoluted and chaotic, such an integration requires a vast computer, even for very simple geometries. Nevertheless, we can, in principle, integrate (1.6) to determine $\mathbf{u}(\mathbf{x}, t)$.

However, for many purposes it is not one realization of $\mathbf{u}(\mathbf{x}, t)$ which is important, but rather the statistical properties of $\mathbf{u}$, such as $\bar{\mathbf{u}}$

and $\overline{(\mathbf{u}')^2}$. After all, the details of **u** are chaotic and vary from one realization of a flow to the next, while the statistical properties of **u** seem to be well behaved and perfectly reproducible. So the focus of most 'turbulence theories' is not the determination of $\mathbf{u}(\mathbf{x}, t)$ itself, but rather its statistical properties. There is an analogy here to the statistical theories of, say, a gas. We are less interested in the motion of every gas molecule than in its statistical consequences, such as pressure.

It is natural, therefore, to try and develop a theory in which $\bar{\mathbf{u}}$, $\overline{(\mathbf{u}')^2}$ etc play the central role, rather than $\mathbf{u}(\mathbf{x}, t)$. This, in turn, requires that we have a set of dynamical equations for these statistical quantities which is in some ways analogous to (1.6). Actually, it turns out to be possible to manipulate (1.6) into a hierarchy of statistical equations of the form,

$$\frac{\partial}{\partial t}[\text{certain statistical properties of } \mathbf{u}] \\ = F\,(\text{other statistical properties of } \mathbf{u}). \tag{1.7}$$

We shall see how this is achieved in Chapter 4 where we shall discover something extraordinary about system (1.7). It turns out that this system of equations is *not* closed, in the sense that, no matter how many manipulations we perform, there are always more statistical unknowns than equations relating them. This is known as the *closure problem of turbulence*, and it arises because of the non-linear term on the right of (1.5). In fact, this closure problem is a common characteristic of non-linear dynamical systems, as illustrated by Example 1.1.

It is not possible to overstate the importance of the closure problem. It has haunted the subject from its very beginnings and we are no closer to circumventing this difficulty than we were when Reynolds first performed his famous pipe flow experiment. In effect, the closure problem tells us that there are no rigorous, statistical theories of turbulence! Those theories which do exist invariably invoke additional heuristic hypotheses. Of course, these stand or fall on the basis of their success in explaining the experimental evidence. This gloomy conclusion has led to the oft quoted phrase: 'turbulence is the last unsolved problem of classical physics'.[8]

[8] There have been a multitude of attempts to plug this gap by introducing additional, ad hoc, equations. Typically these relate certain statistical quantities to each other, or else propose relationships between the mean flow and the state of the turbulence. Either way, these additional hypotheses are empirical in nature, their plausibility resting on certain experimental evidence. The resulting closed set of equations are referred to as a *turbulence closure model*. Unfortunately, these models tend to work only for a narrow class of problems. Indeed, in many ways we may regard turbulence closure models as nothing more than a highly sophisticated exercise in interpolating between experimental data sets. This depressing thought has led some to give up all hope of constructing a theory (or theories) of turbulence. The whimsical advice of W.C. Fields comes to mind: 'If at first you don't succeed, try, try again. Then quit. No use being a damn fool about it.'

It seems that nature (God?) has a nice sense of irony. On the one hand we have a physical quantity, **u**, which behaves in a random fashion, yet is governed by a simple, deterministic equation. On the other hand the statistical properties of **u** appear to be well-behaved and reproducible, yet we know of no closed set of equations which described them!

Example 1.1 Consider a system governed by the equation $du/dt = -u^2$. Suppose that u is given a random value between 1 and 2 at $t = 0$ and we observe the subsequent trace of $u(t)$. We repeat the experiment 1000 times where each time $u(0) = u_0$ is chosen randomly from the range 1–2. We are interested in the expected value of u at any instant t, that is, the value of u at time t obtained by averaging over a large number of experiments. Let us denote this average by $\langle u \rangle(t)$, where $\langle \sim \rangle$ means 'averaged over many experiments'. Of course, our equation may be solved explicitly for u and hence we find that $\langle u^{-1} \rangle = \langle (u_0^{-1} + t) \rangle$. Suppose, however, that we did not spot that our equation has an exact solution. Instead we try to find an evolution equation for $<u>$ which we can solve explicitly. Simply taking the average of our governing equation will not work since we end up with $d\langle u \rangle / dt = -\langle u^2 \rangle$, which introduces the new unknown $\langle u^2 \rangle$. Of course, we can find an evolution equation for $\langle u^2 \rangle$ by multiplying our governing equation by u. However, this involves $\langle u^3 \rangle$, and so we still have a problem. Show that, if we try to establish a hierarchy of evolution equations for the variables $\langle u^n \rangle$ then there are always more unknowns than equations. This is reminiscent of the closure problem of turbulence.

1.8 Is there a 'theory of turbulence'?

At one time it was thought that there might be some kind of 'universal theory of turbulence', valid under a wide range of circumstances. That is to say, theoreticians hoped that, just as in the kinetic theory of gasses, one might be able to average out the apparently random motion of individual fluid lumps (atoms in the case of kinetic theory) and produce a non-random, macroscopic, statistical model. In the case of turbulence such a model might, perhaps, predict the rate of energy transfer between the mean flow and the turbulence, the distribution of energy across the different eddy sizes, and the average rate of dispersal of a pollutant by turbulent mixing. It would not, of course, predict the detailed evolution of every eddy in the flow.

Unfortunately, after a century of concerted effort, involving engineers, physicists, and mathematicians, no such theory has emerged. Rather, we have ended up with a multitude of theories: one relevant to, say, boundary layers, another to stratified flows, yet an other to magnetohydrodynamic turbulence, and so on. Worse, each theory invokes non-rigorous hypotheses, usually based on experimental observations, in order to make useful predictions. It is now generally agreed that there is no coherent 'theory of turbulence'. There are many problems and many theories.[9]

Just occasionally, however, we get lucky and the odd universal feature of turbulence emerges. The most commonly quoted example is Kolmogorov's theory of the very small eddies, whose statistical properties appear to be nearly (but not completely) universal, that is, the same for jets, wakes, boundary layers, etc. Another example is the behaviour of a turbulent shear flow very close to a smooth wall where, again, certain near-universal statistical properties are found. However, such universal laws are the exception and not the rule.

Of course, this is a profoundly unsatisfactory state of affairs for, say, the astrophysicist or structural engineer to whom the effects of turbulence are usually just a small part of a bigger picture. Such researchers usually want simple models which parameterize the effects of turbulence and allow them to focus on the more interesting problems at hand, such as quantifying how a star forms or a bridge oscillates in a high wind. So when the structural engineer or the physicist turns to his colleague working in fluid mechanics and asks for a 'turbulence model', he or she is usually met with a wry smile, a few tentative equations, and a long string of caveats!

1.9 The interaction of theory, computation, and experiment

The danger from computers is not that they will eventually get as smart as men, but we will meanwhile agree to meet them halfway. (Bernard Avishai)

Given the difficulties involved in developing rigorous statistical models of turbulence, and the rapid development of computer power, it might be argued that the way forward is to rely on numerical simulation of turbulence. That is, given a large enough computer, we can readily integrate (1.6) forward in time for a given set of initial conditions. So it would seem that the mathematician or physicist interested in the fundamental structure of turbulence can perform 'numerical experiments', while the engineer who needs answers to a particular problem can, in principle, simulate the flow in question on the computer. This is not entirely a pipe-dream. Researchers have already performed numerical simulations which capture not only the

[9] It seems likely that Horace Lamb saw the way the wind was blowing at an early stage if a quote, attributed to Lamb by S. Goldstein, is correct. Lamb is reputed to have said: 'I am an old man now, and when I die and go to heaven there are two matters on which I hope for enlightenment. One is quantum electrodynamics and the other is the turbulent motion of fluids. About the former I am rather optimistic' (1932).

mean flow, but every turbulent eddy, right down to the Kolmogorov microscale.

There is a catch however. Suppose, for example, that we wanted to simulate the effects of a strong wind on a tall chimney, say 100 m high (Plate 2). We might be interested in the drag forces exerted on the structure, or perhaps in the turbulence generated in the wake of the chimney. To fix thoughts, suppose that the integral scale of the turbulence in the wake is $l = 0.3$ m, and that the typical velocity of a large-scale eddy is $u = 2$ m/s. The Re for the turbulence is then $\text{Re} = ul/\nu \sim 0.5 \times 10^5$ and from (1.3) the Kolmogorov microscale is $\eta \sim 0.1$ mm. It is extraordinary that much of the dissipation of energy surrounding this 100 m tall chimney occurs in turbulent structures which are only 1 mm or so in size! Moreover, the turn-over time of the smallest eddies is very rapid, $\eta/\text{v} \sim 10^{-3}$ s. Now suppose we wished to simulate this flow for 2 min and that the speed and direction of the wind changes during this time. Then we would have to compute flow structures (in the mean flow and the turbulence) whose dimensions varied from 0.1 mm to 100 m and whose characteristic timescales span the range 10^{-3}–100 s! Such a monumental calculation is well beyond the capacity of any computer which exists today, or is likely to exist for quite some time.

The most which can be achieved at present is to perform numerical simulations at rather modest values of Re (so that η is not too small) and in *extremely* simple geometries. The geometry favoured by those investigating the fundamental structure of turbulence, say a cloud of turbulence slowly decaying under the influence of viscosity, is the so-called *periodic cube*. This is a cubic domain which has a special property: whatever is happening at one face of the cube happens at the opposite face. Of course, a freely evolving cloud of turbulence is, in reality, anything but periodic, and so this artificial periodicity must be enforced as a rather strange kind of boundary condition on the calculation. Periodic cubes are popular because they lend themselves to particularly efficient (fast) numerical algorithms for solving the Navier–Stokes equations. Nevertheless, even in this particularly simple geometry it is difficult to achieve values of Re (based on the integral scale) much above 10^3–10^4. (Most flows of interest to the engineer are in the range 10^5–10^8.)

Periodic cubes lend themselves to efficient simulations. However, there is a price to pay. The forced periodicity of the turbulence is quite unphysical. It ensures that the turbulence is statistically anisotropic at the scale of the box (consult Chapter 7). It also enforces unphysical, long-range statistical correlations on the scale L_{BOX} (see Figure 1.16). (What is happening on one side of the box is perfectly correlated to events at the other side.) To obtain results which are representative of *real* turbulence, and are not polluted by the presence of the artificial

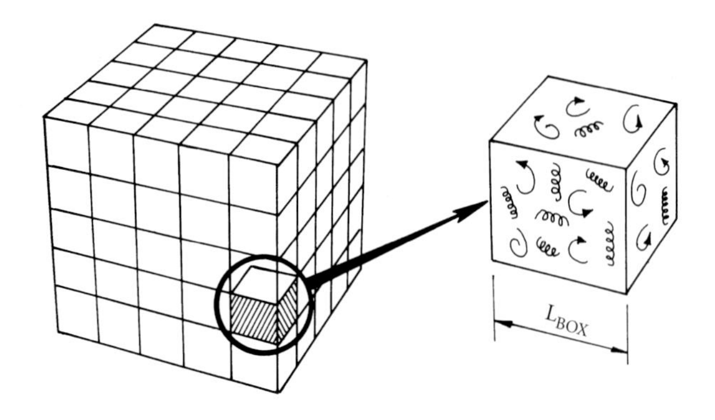

Figure 1.16 Flow in a periodic cube. Space is divided up into an infinite number of cubes and it is arranged for the flow in each box to be identical at any instant. We focus attention on just one cube and study the evolution of the turbulence within it. The dynamics in any one cube is influence by the pressure field set up by the surrounding cubes and so is not representative of real turbulence. However, if L_{BOX} is much greater than the eddy size then the hope is that the influence of the surrounding cubes is small.

boundary conditions, it is necessary to have $L_{BOX} \gg l$, where l is the integral scale (size of the large eddies). This is particularly problematic if information is required about the large eddies in the turbulence, as these are the most sensitive to the enforced periodicity. In such cases one needs, say, $L_{BOX}/l \sim 20$–50 (see Section 7.2). (The problem is less severe if one is interested in the small eddies, where $L_{BOX} \sim 6l$ may be sufficient.) Unfortunately, limitations in computing power means that it is very difficult to achieve simultaneously both $L_{BOX}/l \sim 50$ and $\text{Re} \sim 10^3$. One is reminded of Corrsin's quip, 'The foregoing estimate (of computing power) is enough to suggest the use of analogue instead of digital computation; in particular, how about an analogue consisting of a tank of water?' (1961). Great advances have been made since 1961, but Corrsin's words still carry a certain resonance.

It seems that, despite great strides in computational fluid dynamics (CFD), we are still a long way from performing *direct numerical simulations* (DNS) of flows of direct interest to the engineer. The most that we can achieve are simulation at modest Re and in extremely simple geometries (e.g. periodic cubes). These simulations provide valuable insights into the structure of turbulence, thus bolstering our understanding of the phenomena. However, they cannot be used to answer practical questions such as 'how quickly will the pollutant belching out of a chimney disperse?', or 'will my chimney fall over in a high wind?' To answer such questions engineers typically resort to experiment or to semi-empirical models of turbulence, such as the so-called k–ε model. Sometimes these models work, and sometimes they do not. To understand why and when these models fail requires a sound understanding of turbulence and this provides one of the motivations for studying turbulence at a fundamental level. The other motivation, of course, is natural curiosity.

Exercises

1.1 Show that there is an exact analogy between two-dimensional buoyancy-driven flow in a thin, horizontal layer and axisymmetric flow with swirl in a narrow annulus. Now show that Rayleigh's stability criterion for inviscid, swirling flow follows directly from the observation that heavy fluid overlying light fluid is unstable. (Rayleigh's theorem says that a flow is unstable if $(u_\theta r)^2$ decreases with radius, u_θ being the azimuthal velocity in (r, θ, z) coordinates.)

1.2 Estimate the Re at which transition occurs in the cigarette plume of Figure 1.7. (Note that most flows of interest to the engineer are in the range $\text{Re} = 10^5$–10^8.)

1.3 The rate, Π, at which energy is passed down the energy cascade must be independent of the size of the eddies in statistically steady turbulence, otherwise energy would accumulate at some intermediate scale. Let $(\Delta \text{v})^2$ be the kinetic energy of a typical eddy of size r, where $\eta \ll r \ll l$. (Here η is the Kolmogorov microscale and l is the integral scale.) On the assumption that eddies of size r break-up on a timescale of their turn-over-time, $r/(\Delta \text{v})$, show that

$$(\Delta \text{v})^2 \sim \Pi^{2/3} r^{2/3} \sim \varepsilon^{2/3} r^{2/3}.$$

Now show that this scaling is compatible with the Kolmogorov microscales in the sense that v and η satisfy the relationship above.

1.4 In a volcanic eruption a turbulent plume is created in which the integral scale of the turbulence is $l \sim 10$ m and a typical turbulent velocity is 20 m/s. If the viscosity of the gas is 10^{-5} m^2/s, estimate the size of the smallest eddies in the plume. Compare this with the mean-free-path length for air.

1.5 In wind tunnel turbulence generated by a grid the kinetic energy per unit mass is found to decay approximately at a rate $u^2 \sim t^{-10/7}$. On the assumption that the large eddies break-up on a time-scale of their turn-over time show that $u^2 l^5$ is approximately constant during the decay. [$I \sim u^2 l^5$ is called Loitsyansky's invariant.]

Suggested reading

Corrsin, S. (1961) Turbulent Flow. *Am. Sci.*, **49**(3). (A mere 24 pages crammed with physical insight.)

Tennekes, H. and Lumley, J.L. (1972) *A First Course in Turbulence*. MIT Press. (Still one of the best guides for the beginner. Consult chapter 1 for an excellent introduction to turbulence.)

Tritton, D.J. (1988) *Physical Fluid Dynamics*. Clarendon Press. (A beautiful book written from the perspective of a physicist. The experiments of Reynolds, Bénard, and Taylor are discussed in chapters 2, 17, and 22.)

CHAPTER 2

The equations of fluid mechanics

> In reflecting on the principles according to which motion of a fluid ought to be calculated when account is taken of the tangential force, and consequently the pressure not supposed the same in all directions, I was led to construct the theory explained in this paper . . . I afterwards found that Poisson had written a memoir on the same subject, and on referring to it I found that he had arrived at the same equations. The method which he employed was however so different from mine that I feel justified in laying the latter before this Society . . . The same equations have also been obtained by Navier in the case of an incompressible fluid, but his principles differ from mine still more than do Poisson's.
>
> Stokes (1845)

It is a tribute to Stokes that the modern derivation of the viscous equations of motion is virtually identical to that set out by him one and a half centuries ago.

This chapter is intended for those who have had only a limited exposure to fluid mechanics. Our aim is to derive and discuss those laws of fluid mechanics which are particularly important for an understanding of turbulence. We shall restrict ourselves to fluids which are incompressible and Newtonian. (See equation (2.4) and associated footnote.) The physical principles at our disposal are:

(1) Newton's second law applied to a continuum;
(2) a constitutive law, called Newton's law of viscosity, which relates shear stresses in a fluid to the rate of distortion of fluid elements; and
(3) the conservation of mass (i.e. what flows in must flow out).

When we put these all together we obtain a simple partial differential equation (the Navier–Stokes equation) which governs the motion of nearly all fluids. This equation is deceptively simple. It looks no more complex than a wave equation or a diffusion equation. Yet we know that the diffusion equation always leads to simple (almost boring) solutions. The Navier–Stokes equation, on the other hand, embodies such rich and complex phenomena as instabilities and turbulence. Clearly there must be something special about the Navier–Stokes equation. There is. It turns out that this equation is non-linear in the sense that the dependent variable, $\mathbf{u}(\mathbf{x}, t)$, appears in a quadratic form.

It is this seemingly innocent non-linearity which leads to such unpredictable phenomena as solar flares and tornadoes.

2.1 The Navier–Stokes equation

2.1.1 *Newton's second law applied to a fluid*

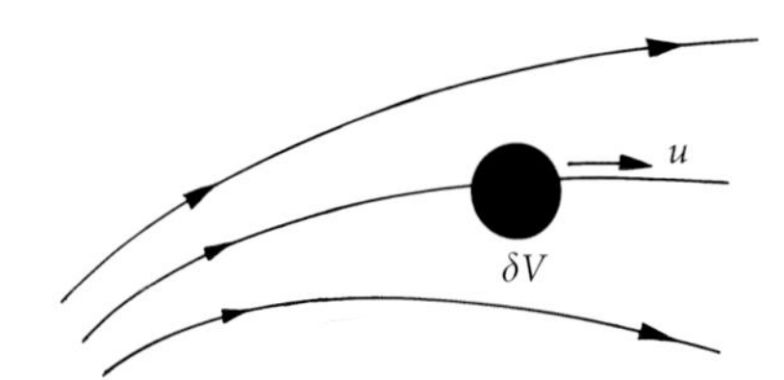

Figure 2.1 A blob of fluid moving with velocity **u**.

Let us apply Newton's second law to a lump of fluid of volume δV as it moves through a flow field (Figure 2.1). We have,

$$(\rho\delta V)\frac{D\mathbf{u}}{Dt} = -(\nabla p)\delta V + \text{viscous forces.} \tag{2.1}$$

In words this says that the mass of the fluid element, $\rho\delta V$, times its acceleration, $D\mathbf{u}/Dt$, is equal to the net pressure force acting on the lump plus any viscous forces arising from viscous stresses. The fact that the net pressure force can be written as $-(\Delta p)\delta V$ follows from Gauss' theorem in the form

$$\oint_S (-p)d\mathbf{S} = \int_{\delta V} (-\nabla p)dV = -(\nabla p)\delta V$$

where the surface integral on the left is the sum of all the elementary pressure forces, $-pd\mathbf{S}$, acting on the surface S of the blob of fluid.

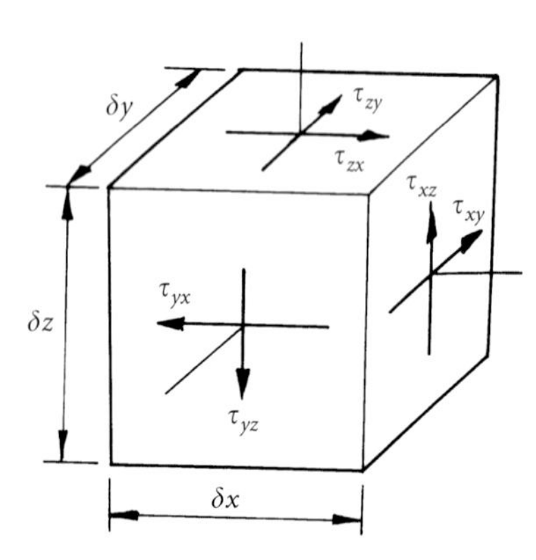

Figure 2.2 Viscous stresses acting on a rectangular element.

The task now is to evaluate the viscous term in (2.1). Suppose the fluid lump is instantaneously in the form of a rectangular element with edges dx, dy, dz as shown in Figure 2.2. The stresses arising from the presence of viscosity comprise both shear stresses, τ_{xy}, τ_{yz}, etc., as well as normal stresses, τ_{xx}, τ_{yy}, and τ_{zz}, which supplement the normal stresses due to pressure. These viscous stresses can influence the trajectory of a fluid lump because any imbalance in stress will lead to a net viscous force acting on the fluid element. For example, a difference between τ_{zx} at the top and the bottom of the element will produce a net force in the x-direction of magnitude

$$f_x = (\delta\tau_{zx})\delta A = \left[\left(\frac{\partial\tau_{zx}}{\partial z}\right)dz\right]dxdy.$$

Gathering all such terms we find the net viscous force in the x-direction is

$$f_x = \left[\frac{\partial\tau_{xx}}{\partial x} + \frac{\partial\tau_{yx}}{\partial y} + \frac{\partial\tau_{zx}}{\partial z}\right]\delta V.$$

We rewrite this in the abbreviated form,

$$f_x = \frac{\partial\tau_{jx}}{\partial x_j}\delta V$$

where it is understood that there is a summation over the repeated index j. (Those unfamiliar with tensor notation may wish to consult

Appendix I.) Similar expressions can be found for f_y and f_z and it is evident that the net viscous force in the *ith* direction acting on the rectangular element is,

$$f_i = \frac{\partial \tau_{ji}}{\partial x_j} \delta V.$$

Thus, (2.1) becomes

$$\rho \frac{D\mathbf{u}}{Dt} = -\nabla p + \frac{\partial \tau_{ji}}{\partial x_j}. \tag{2.2}$$

This is as far as Newton's second law will take us. If we are to progress we need additional information. There are two more principles available to us. First, we have conservation of mass which may be expressed as $\nabla \cdot (\rho \mathbf{u}) = -\partial \rho / \partial t$. Since we are treating ρ as a constant this reduces to the so-called continuity equation

$$\nabla \cdot \mathbf{u} = 0. \tag{2.3}$$

Next we need a constitutive law relating τ_{ij} to the rate of deformation of fluid elements. Most fluids obey Newton's law of viscosity, which says,[1]

$$\tau_{ij} = \rho \nu \left\{ \frac{\partial u_i}{\partial x_j} + \frac{\partial u_j}{\partial x_i} \right\}. \tag{2.4}$$

To understand where this comes from consider the simple shear flow $\mathbf{u} = (u_x(y),0,0)$ shown in Figure 2.3(a). Here the fluid elements slide over each other and one measure of the rate of sliding is the angular distortion rate, $d\gamma/dt$, of an initially rectangular element. In this simple flow a shear stress, τ_{yx}, is required to cause the relative sliding of the fluid layers. Moreover, it is reasonable to suppose that τ_{yx} is directly proportional to the rate of sliding $d\gamma/dt$, and we define the constant of proportionality to be the absolute viscosity $\mu = \rho\nu$. It follows that: $\tau_{yx} = \rho\nu(d\gamma/dt)$. However, it is clear from the diagram that $d\gamma/dt = \partial u_x/\partial y$ and so for this simple flow we might anticipate,

$$\tau_{yx} = \rho \nu \frac{\partial u_x}{\partial y}.$$

Of course, this is just a special case of (2.4). This kind of argument may be generalized in an obvious way. For simplicity we restrict ourselves to two-dimensional motion. Consider the element of fluid shown in Figure 2.3(b). In a time δt it experiences an angular distortion of,

$$\delta\gamma = \delta\gamma_1 + \delta\gamma_2 = \left(\frac{\partial u_x}{\partial y} + \frac{\partial u_y}{\partial x} \right) \delta t$$

[1] We shall use kinematic viscosity, ν, rather than absolute viscosity, $\mu = \rho\nu$, throughout. Fluids which obey (2.4) are called Newtonian fluids.

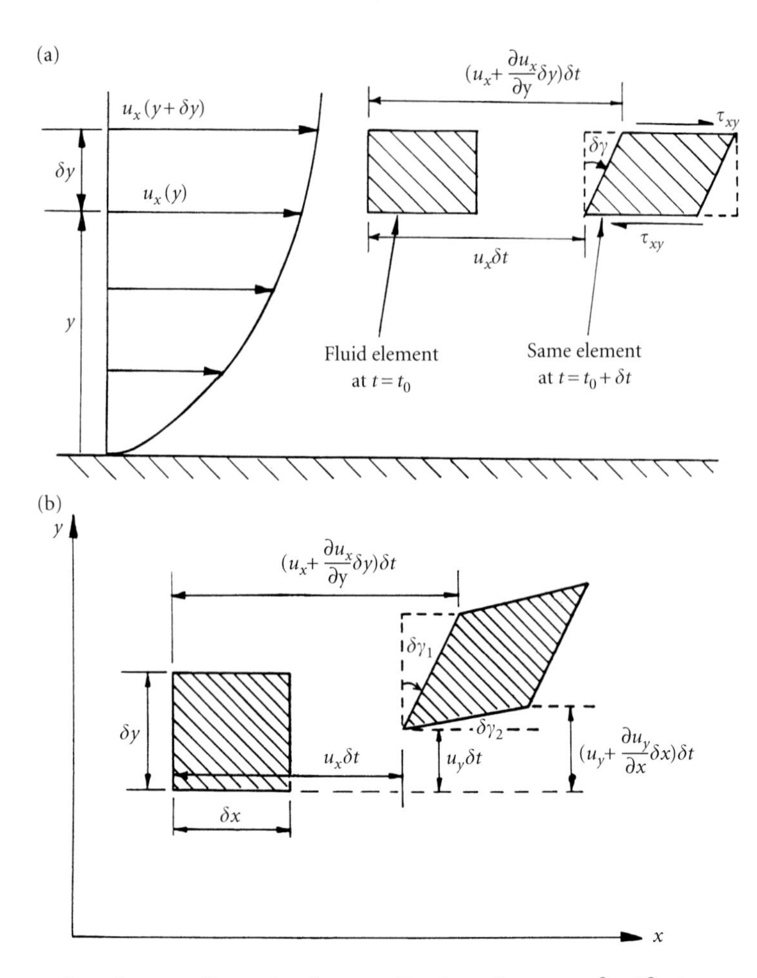

Figure 2.3 (a) Distortion of an element in a parallel shear flow. (b) Distortion of a fluid element in two-dimensions.

and so the two-dimensional generalization of $\tau_{yx} = \rho\nu\partial u_x/\partial y$ is

$$\tau_{xy} = \tau_{yx} = \rho\nu\left(\frac{\partial u_x}{\partial y} + \frac{\partial u_y}{\partial x}\right).$$

Equation (2.4) is simply the three-dimensional counterpart of this.

It is conventional to introduce the term *strain-rate tensor* for the quantity

$$S_{ij} = \frac{1}{2}\left[\frac{\partial u_i}{\partial x_j} + \frac{\partial u_j}{\partial x_i}\right] \tag{2.5}$$

and so the most compact form of Newton's law of viscosity is,

$$\tau_{ij} = 2\rho\nu S_{ij}.$$

In any event, substituting (2.4) into our equation of motion (2.2) yields, after a little work,

$$\frac{D\mathbf{u}}{Dt} = -\nabla\left(\frac{p}{\rho}\right) + \nu\nabla^2\mathbf{u}. \tag{2.6}$$

This is the Navier–Stokes equation. The boundary condition on $\mathbf{u}$ corresponding to (2.6) is that $\mathbf{u} = \mathbf{0}$ on any stationary solid surface, that is, fluids 'stick' to surfaces. This is known as the 'no-slip' condition. Sometimes it is convenient, though rarely realistic, to imagine a fluid with zero viscosity. These 'perfect fluids' (which do not exist!) are governed by the so-called Euler equation,

$$\frac{D\mathbf{u}}{Dt} = -\nabla\left(\frac{p}{\rho}\right). \tag{2.7}$$

The boundary condition appropriate to this equation is not $\mathbf{u} = \mathbf{0}$, but rather $\mathbf{u} \cdot d\mathbf{S} = 0$ at an impermeable surface, that is, there in no mass flux through the surface.

2.1.2 *The convective derivative*

So far we have focused on dynamical issues. We now switch to kinematics. We have written the acceleration of a fluid lump as $D\mathbf{u}/Dt$. The special symbol $D(\cdot)/Dt$ was first introduced by Stokes and means *the rate of change of a quantity associated with a given fluid element*. It is called the *convective derivative* and should not be confused with $\partial(\cdot)/\partial t$ which is, of course, the rate of change of a quantity at a fixed point in space. For example, DT/Dt is the rate of change of temperature, $T(\mathbf{x}, t)$, of a fluid lump as it moves around, whereas $\partial T/\partial t$ is the rate of change of temperature at some fixed point in space through which a succession of particles will pass. Thus the acceleration of a fluid element is $D\mathbf{u}/Dt$ and not $\partial\mathbf{u}/\partial t$.

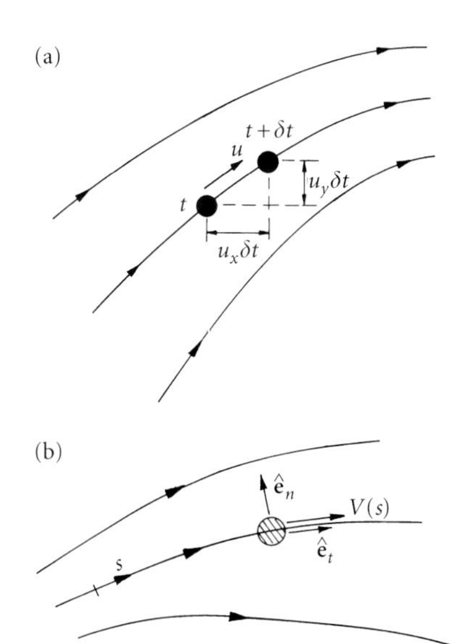

Figure 2.4 (a) Rate of change of temperature of a fluid element as it moves through a flow field. (b) Acceleration of a fluid lump in a steady flow.

An expression for $D(\cdot)/Dt$ may be obtained using the chain rule. Perhaps it is easier to develop the ideas in terms of scalar fields and so let us consider, for the moment, the temperature field, $T(\mathbf{x}, t)$. (See Figure 2.4(a).) The change in T due to small variations in x, y, z, and t is, $\delta T = (\partial T/\partial t)\delta t + (\partial T/\partial x)\delta x + \cdots$ Since we are interested in changes in T following a fluid particle we have $\delta x = u_x \delta t$ etc. and so

$$\frac{DT}{Dt} = \frac{\partial T}{\partial t} + u_x\frac{\partial T}{\partial x} + u_y\frac{\partial T}{\partial y} + u_z\frac{\partial T}{\partial z} = \frac{\partial T}{\partial t} + (\mathbf{u}\cdot\nabla)T. \tag{2.8}$$

The same expression applies to each component of any vector field, $\mathbf{A}(\mathbf{x}, t)$, and so we have

$$\frac{D\mathbf{A}}{Dt} = \frac{\partial \mathbf{A}}{\partial t} + (\mathbf{u}\cdot\nabla)\mathbf{A}. \tag{2.9}$$

Substituting $\mathbf{u}$ for $\mathbf{A}$ gives us an explicit expression for the acceleration of a fluid particle,

$$\frac{D\mathbf{u}}{Dt} = \frac{\partial \mathbf{u}}{\partial t} + (\mathbf{u}\cdot\nabla)\mathbf{u} \tag{2.10}$$

and so the Navier–Stokes equation may be rewritten as,

$$\frac{D\mathbf{u}}{Dt} = \frac{\partial \mathbf{u}}{\partial t} + (\mathbf{u} \cdot \nabla)\mathbf{u} = -\nabla\left(\frac{p}{\rho}\right) + \nu\nabla^2\mathbf{u}. \tag{2.11}$$

The crucial point to note here is that (2.11) contains a non-linear (quadratic) term in $\mathbf{u}$. It is this which leads to most of the complex and rich phenomena of fluid mechanics and in particular to turbulence.

In order to better understand why $(\mathbf{u} \cdot \nabla)\mathbf{u}$ appears on the left-hand side of (2.11) it is convenient to restrict ourselves to steady flows, that is, flows in which $\mathbf{u}$ is a function only of position. In such flows $\partial\mathbf{u}/\partial t = \mathbf{0}$, the shape of the streamlines is fixed for all t, and the streamlines represent particle trajectories for individual fluid 'lumps'. Consider one such lump moving along a streamline as shown in Figure 2.4(b). Let s be the curvilinear coordinate measured along the streamline and $V(s)$ be the speed $|\mathbf{u}|$. Since the streamline represents a particle trajectory we have, from elementary mechanics,

$$(\text{acceleration of lump}) = V\frac{dV}{ds}\hat{\mathbf{e}}_t - \frac{V^2}{R}\hat{\mathbf{e}}_n. \tag{2.12}$$

Here R is the radius of curvature of the streamline and $\hat{\mathbf{e}}_t$ and $\hat{\mathbf{e}}_n$ are unit vectors as shown in Figure 2.4(b).

Compare this with (2.10). Since $\partial\mathbf{u}/\partial t = \mathbf{0}$, this would have us believe that the acceleration of the lump is $(\mathbf{u} \cdot \nabla)\mathbf{u}$. In fact, if we rewrite $(\mathbf{u} \cdot \nabla)\mathbf{u}$ in curvilinear coordinates we find, after a little algebra

$$(\mathbf{u} \cdot \nabla)\mathbf{u} = V\frac{dV}{ds}\hat{\mathbf{e}}_t - \frac{V^2}{R}\hat{\mathbf{e}}_n \tag{2.13}$$

and so (2.12) and (2.10) do indeed give the same result. The physical meaning of $(\mathbf{u} \cdot \nabla)\mathbf{u}$ is now clear. Even in steady flows individual fluid lumps experience an acceleration because, as they slide along a streamline, they pass through a succession of points at which $\mathbf{u}$ is, in general, different. Thus, as we follow the fluid element, $\mathbf{u}$ changes and its rate of change (in a steady flow) is $(\mathbf{u} \cdot \nabla)\mathbf{u}$.

2.1.3 *Integral versions of the momentum equation*

Now (2.11) is applied sometimes in differential form and sometimes in integral form. If we have a fixed volume (a so-called control volume) then the integral of (2.11) throughout V yields, with the help of Gauss' theorem,

$$\frac{\partial}{\partial t}\int_V \rho u_i dV = -\oint_S u_i(\rho\mathbf{u} \cdot d\mathbf{S}) - \oint_S p d\mathbf{S} + (\text{viscous term}). \tag{2.14}$$

This can be interpreted in terms of a linear momentum budget for V. Since $\rho\mathbf{u}\cdot d\mathbf{S}$ is the rate of flow of mass through surface element $d\mathbf{S}$, the first integral on the right represents the net flux of linear momentum out of V. Thus (2.14) expresses the fact that the total linear momentum in V can change because momentum is transported across the bounding surface S, or else because pressure forces or viscous forces act on the boundary. Note that, since $u_i = \nabla\cdot(\mathbf{u}x_i)$, the net linear momentum in any closed domain is necessarily zero. ($\nabla\cdot(\mathbf{u}x_i)$ integrates to zero because of Gauss' theorem and the boundary condition $\mathbf{u}\cdot d\mathbf{S} = 0$.)

A second integral equation can be derived from (2.11). First we expand the convective derivative of $\mathbf{x}\times\mathbf{u}$ to give $\mathbf{x}\times D\mathbf{u}/Dt$ and then use (2.11) to rewrite the acceleration, $D\mathbf{u}/Dt$, in terms of pressure and viscous forces:

$$\begin{aligned}\rho\frac{D}{Dt}(\mathbf{x}\times\mathbf{u}) &= \rho\mathbf{x}\times\frac{D\mathbf{u}}{Dt} + \rho\frac{D\mathbf{x}}{Dt}\times\mathbf{u} = \rho\mathbf{x}\times\frac{D\mathbf{u}}{Dt}\\ &= \nabla\times(p\mathbf{x}) + (\text{viscous term}).\end{aligned}$$

Here we have taken advantage of the fact that $D\mathbf{x}/Dt = \mathbf{u}$ and $-\mathbf{x}\times\nabla p = \nabla\times(p\mathbf{x})$. Clearly this is an angular momentum equation for the fluid. It integrates to give,

$$\begin{aligned}\frac{\partial}{\partial t}\int_V \rho(\mathbf{x}\times\mathbf{u})_i dV = &-\oint_S (\mathbf{x}\times\mathbf{u})_i(\rho\mathbf{u}\cdot d\mathbf{S})\\ &-\oint_S \mathbf{x}\times(pd\mathbf{S}) + (\text{viscous term}).\end{aligned}\tag{2.15}$$

This has a similar interpretation to (2.14). That is to say, the angular momentum in V may change because angular momentum is transported across the bounding surface S or because of the action of a viscous or pressure torque acting on S.

2.1.4 *The rate of dissipation of energy in a viscous fluid*

We close Section (2.1) with a discussion of energy. In particular, we quantify the rate at which mechanical energy is converted into heat by friction. Let us start by calculating the rate of working of the viscous stresses in a Newtonian fluid. Suppose we have a volume V of fluid whose boundary, S, is subject to the viscous stresses $\tau_{ij} = 2\rho\nu S_{ij}$. Then the rate of working of these stresses on the fluid is

$$\dot{W} = \oint u_i(\tau_{ij}dS_j).$$

That is to say, the ith component of the viscous force acting on surface element $d\mathbf{S}$ is $\tau_{ij}dS_j$, and so the rate of working of this force is $u_i(\tau_{ij}dS_j)$.

From Gauss's theorem we may rewrite $\dot{W}$ as

$$\dot{W} = \int \frac{\partial}{\partial x_j}[u_i \tau_{ij}]dV$$

and so we conclude that the rate of working of τ_{ij} per unit volume is

$$\frac{\partial}{\partial x_j}[u_i \tau_{ij}] = \frac{\partial \tau_{ij}}{\partial x_j} u_i + \tau_{ij} \frac{\partial u_i}{\partial x_j}.$$

(Actually, this equation may be obtained directly from a consideration of the work done on a small cube by τ_{ij}.) It turns out that the two terms on the right of this expression represent quite different effects, as we now show. First we note that the net viscous force per unit volume acting on the fluid is $f_i = \partial \tau_{ij} / \partial x_j$, and so the first term on the right is $f_i u_i$. The second term, on the other hand, can be rewritten as,

$$\tau_{ij} \frac{\partial u_i}{\partial x_j} = \frac{1}{2}\left[\tau_{ij} + \tau_{ji}\right] \frac{\partial u_i}{\partial x_j} = \frac{1}{2}\left[\tau_{ij} \frac{\partial u_i}{\partial x_j} + \tau_{ij} \frac{\partial u_j}{\partial x_i}\right] = \tau_{ij} S_{ij}$$

since $\tau_{ij} = \tau_{ji}$. Thus the rate of working of τ_{ij} on the fluid is

$$\frac{\partial [u_i \tau_{ij}]}{\partial x_j} = f_i u_i + \tau_{ij} S_{ij}.$$

The two contributions on the right both represent changes in the energy of the fluid, as they must. However, they correspond to two rather different processes. The first term is the rate of working of the net viscous force acting on a fluid element. This necessarily represents the rate of change of mechanical energy of the fluid. The second term must, therefore, correspond to the rate of change of internal energy (per unit volume) of the fluid. Thus we conclude that the rate of increase of internal energy per unit mass is,

$$\varepsilon = \frac{\tau_{ij} S_{ij}}{\rho} = 2\nu S_{ij} S_{ij}.$$

In the absence of work being done by the boundaries $\int \rho \varepsilon dV$ must represent the rate of loss of mechanical energy to heat as a consequence of viscous dissipation.

We can arrive at the same conclusion via a slightly different route. If we take the product of (2.6) with $\mathbf{u}$, and note that $D(\cdot)/Dt$ obeys the usual rules of differentiation, we have

$$\mathbf{u} \cdot \frac{D\mathbf{u}}{Dt} = \frac{D}{Dt}\left(\frac{u^2}{2}\right) = -\nabla \cdot \left[\frac{p}{\rho}\mathbf{u}\right] + \nu \mathbf{u} \cdot (\nabla^2 \mathbf{u}).$$

Evidently, we have the makings of a kinetic energy equation. Noting that,

$$\nu \mathbf{u} \cdot (\nabla^2 \mathbf{u}) = u_i \frac{\partial}{\partial x_j}[\tau_{ij}/\rho] = \frac{\partial}{\partial x_j}[u_i \tau_{ij}/\rho] - 2\nu S_{ij} S_{ij}$$

our energy equation becomes,

$$\frac{\partial(u^2/2)}{\partial t} = -\nabla\cdot[(u^2/2)\mathbf{u}] - \nabla\cdot[(p/\rho)\mathbf{u}] + \frac{\partial}{\partial x_j}[u_i\tau_{ij}/\rho] - 2\nu S_{ij}S_{ij}. \tag{2.16}$$

This gives, when integrated over an arbitrary, fixed volume V,

$$\begin{aligned}\frac{d}{dt}\int_V (u^2/2)dV = &-(\text{rate at which kinetic energy is convected across the boundary})\\ &+(\text{rate at which the pressure forces do work on the boundary})\\ &+(\text{rate at which the viscous forces do work on the boundary})\\ &-\int_V 2\nu S_{ij}S_{ij}dV.\end{aligned}$$

Conservation of energy tells us that the final term on the right must represent the rate of loss of mechanical energy to heat. Since this equation may be applied to a small volume δV it follows that the rate of dissipation of mechanical energy per unit mass is simply,

$$\varepsilon = 2\nu S_{ij}S_{ij} \tag{2.17}$$

as anticipated above. It is conventional in some texts to rewrite (2.16) in a slightly different form. That is, we note that

$$\nu\mathbf{u}\cdot(\nabla^2\mathbf{u}) = -\nu(\nabla\times\mathbf{u})^2 + \nabla\cdot[\nu\mathbf{u}\times(\nabla\times\mathbf{u})]$$

and so (2.16) becomes,

$$\frac{\partial}{\partial t}\left(\frac{u^2}{2}\right) = -\nabla\cdot[(u^2/2 + p/\rho)\mathbf{u} + \nu(\nabla\times\mathbf{u})\times\mathbf{u}] - \nu(\nabla\times\mathbf{u})^2. \tag{2.18}$$

This second form is less fundamental than (2.16) but possibly more useful when dealing with a closed domain, V. In such cases we find,

$$\frac{d}{dt}\int\left(\frac{u^2}{2}\right)dV = -\nu\int(\nabla\times\mathbf{u})^2 dV. \tag{2.19}$$

Thus, for a closed domain with stationary boundaries, the total rate of dissipation of mechanical energy is,

$$\int\varepsilon dV = 2\nu\int S_{ij}S_{ij}dV = \nu\int(\nabla\times\mathbf{u})^2 dV. \tag{2.20}$$

The quantity $\boldsymbol{\omega} = \nabla \times \mathbf{u}$ is called the vorticity field and $\boldsymbol{\omega}^2/2$ is called the *enstrophy*. We shall return to the concept of vorticity shortly.

2.2 Relating pressure to velocity

Let us summarize the governing equations. We have Newton's second law applied to a viscous fluid,

$$\frac{D\mathbf{u}}{Dt} = -\nabla(p/\rho) + \nu\nabla^2\mathbf{u} \tag{2.21}$$

plus conservation of mass,

$$\nabla \cdot \mathbf{u} = 0. \tag{2.22}$$

We might anticipate that (2.21) and (2.22), subject to the boundary condition $\mathbf{u} = \mathbf{0}$, represents a closed system. If we look only at (2.21) it is not obvious that this is so since it contains two fields, velocity and pressure. However, we may exploit the solenoidal nature of $\mathbf{u}$ to obtain a direct relationship between $\mathbf{u}$ and p. Taking the divergence of (2.21) yields,

$$\nabla^2(p/\rho) = -\nabla \cdot (\mathbf{u} \cdot \nabla\mathbf{u}).$$

For infinite domains this may be inverted using the Biot–Savart law to give,

$$p(\mathbf{x}) = \frac{\rho}{4\pi}\int \frac{[\nabla \cdot (\mathbf{u} \cdot \nabla\mathbf{u})]'}{|\mathbf{x} - \mathbf{x}'|}d\mathbf{x}'. \tag{2.23}$$

Thus, in principle, we can rewrite (2.21) as an evolution equation involving only $\mathbf{u}$, and so our system is indeed closed.

The important feature of (2.23), from the point of view of turbulence, is that p is non-local, in the sense that an eddy at $\mathbf{x}'$ induces a pressure field which is felt everywhere in space. Of course, this is a manifestation of the propagation of information by pressure waves which, in an incompressible fluid, travel infinitely fast.

The fact that p is non-local has profound implications for the behaviour of turbulent flows. An eddy which evolves in space at one location, say $\mathbf{x}$, sends out pressure waves, the distribution of which are dictated by (2.23). These, in turn, induce far-field pressure forces, $-\nabla p$, which churn up the fluid at large distances from the eddy. Thus every part of a turbulent flow feels every other part and this means that eddies which are spatially remote can interact with each other.

Another consequence of (2.23) is that it makes little sense to think of velocity fields being localized in space. If, at $t = 0$, we specify that $\mathbf{u}$ is non-zero only in some small region, then, for $t > 0$, a pressure field is established throughout all space and this, in turn, induces motion at all points via the pressure force ∇p. So, even if $\mathbf{u}$ starts out as localized, it will not stay that way for long. Yet we have been talking about turbulent 'eddies' as if they were structures with a definite size! Clearly we have to refine our views a little. The first step is to introduce the vorticity field, $\boldsymbol{\omega} = \nabla \times \mathbf{u}$. We shall see that, although $\mathbf{u}$ is never localized in space, $\boldsymbol{\omega}$ can be. Moreover, while linear momentum can be instantaneously redistributed throughout space by the pressure field, vorticity can only spread through a fluid in an incremental fashion, either by diffusion or else by material transport (advection). Without doubt, it is the vorticity field, and not $\mathbf{u}$, which is the more fundamental.

2.3 Vorticity dynamics

2.3.1 *Vorticity and angular momentum*

We now explore the properties of the vorticity field defined by $\boldsymbol{\omega} = \nabla \times \mathbf{u}$. The reason why much attention is given to $\boldsymbol{\omega}$ is that it is governed by an evolution equation which is much simpler than the Navier–Stokes equation. Unlike $\mathbf{u}$, $\boldsymbol{\omega}$ cannot be created nor destroyed within the interior of a fluid and it is transported throughout the flow field by familiar processes such as advection and diffusion. Also, localized distributions of $\boldsymbol{\omega}$ remain localized, which is not the case for the velocity field. Thus, when we talk of an 'eddy' in a turbulent flow we really mean a blob of vorticity and its associated rotational and irrotational motion.

Let us try to endow $\boldsymbol{\omega}$ with some physical meaning. Stokes did not use the term vorticity. Rather, he referred to $\boldsymbol{\omega}/2$ as the *angular velocity of the fluid*. This is, perhaps, a better name, as we now show. Consider a small element of fluid in a two-dimensional flow, $\mathbf{u}(x, y) = (u_x, u_y, 0)$, $\boldsymbol{\omega} = (0, 0, \omega_z)$. Suppose that the element is instantaneously circular with radius r and bounding curve C. Then, from Stokes' theorem,

$$\omega_z \pi r^2 = \oint_C \mathbf{u} \cdot d\mathbf{l}. \tag{2.24}$$

Let the element have an angular velocity of Ω, defined as the average rate of rotation of two mutually perpendicular lines embedded in the element. Then we might anticipate, from (2.24), that

$$\omega_z \pi r^2 = \oint \mathbf{u} \cdot d\mathbf{l} = (\Omega r) 2\pi r$$

from which

$$\Omega = \omega_z / 2. \tag{2.25}$$

The rationale for Stokes' terminology is now clear. In fact, (2.25) is readily confirmed by exact analysis. Consider Figure 2.3(b). The

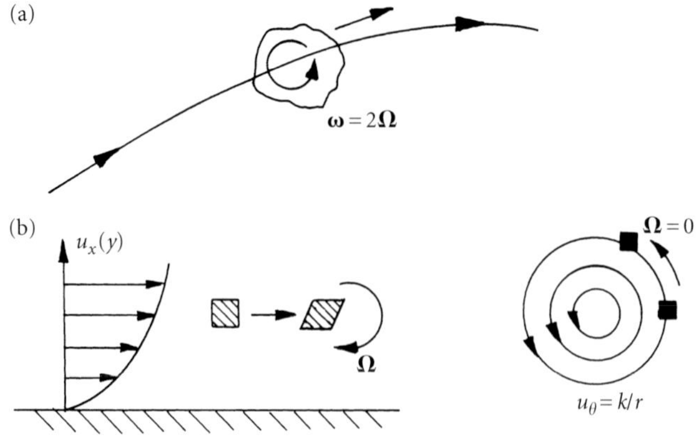

Figure 2.5 The physical interpretation of vorticity. (a) The vorticity at **x** is twice the angular velocity of a fluid blob instantaneously passing through **x**. (b) Vorticity has nothing at all to do with global rotation. Fluid elements in a rectilinear shear flow have vorticity, while those in a free vortex, $u_\theta = k/r$, do not.

anti-clockwise rotation rate of line elements dx and dy are $\partial u_y/\partial x$ and $-\partial u_x/\partial y$ respectively. According to our definition of Ω this gives $\Omega = (\partial u_y/\partial x - \partial u_x/\partial y)/2$ and so (2.25) follows directly from the definition $\boldsymbol{\omega} = \nabla \times \mathbf{u}$.

We note in passing that Figure 2.3(b) illustrates the three effects of a planar velocity field on a lump of fluid: it can move the lump from place to place, rotate it at a rate $\Omega = (\partial u_y/\partial x - \partial u_x/\partial y)/2$, and distort it (strain it) at a rate S_{ij}. (This strain has three components: an angular distortion rate of $S_{xy} = (\partial u_y/\partial x + \partial u_x/\partial y)/2$ and two normal strain rates, S_{xx} and S_{yy}.) In short, **u** can translate, rotate, and distort a fluid element.

These results generalize to three dimensions: the rate of straining of a fluid element is S_{ij} and the vorticity field $\boldsymbol{\omega}(\mathbf{x})$ is twice the average angular velocity of a spherical blob of fluid instantaneously located at **x**, $\boldsymbol{\omega} = 2\boldsymbol{\Omega}$. Thus $\boldsymbol{\omega}$ gives some measure of the local rotation, or spin, of fluid elements (Figure 2.5(a)). It is crucial, however, to note that $\boldsymbol{\omega}$ has nothing at all to do with the global rotation of a fluid. For example, the shear flow $\mathbf{u} = (u_x(y), 0, 0)$ possesses vorticity, yet the streamlines are straight and parallel. On the other hand, the flow $\mathbf{u}(r) = (0, k/r, 0)$ in (r, θ, z) coordinates has no vorticity (except at $r = 0$), yet the streamlines are circular. This is illustrated in Figure 2.5(b).

Note that the velocity gradients, $\partial u_i/\partial x_j$, at any one point can always be decomposed into a combination of strain and vorticity:

$$\frac{\partial u_i}{\partial x_j} = \frac{1}{2}\left(\frac{\partial u_i}{\partial x_j} + \frac{\partial u_j}{\partial x_i}\right) + \frac{1}{2}\left(\frac{\partial u_i}{\partial x_j} - \frac{\partial u_j}{\partial x_i}\right) = S_{ij} - \frac{1}{2}\varepsilon_{ijk}\omega_k$$

where ε_{ijk} is the Levi–Civita symbol (see Appendix I). Although S_{ij} and $\boldsymbol{\omega}$ represent very different processes, one measuring the rate of distortion of fluid elements and the other representing their rate of rotation, S_{ij} and $\boldsymbol{\omega}$ are not independent. This is exemplified by the relationship of each to the Laplacian:

$$\nabla^2 u_i = 2\frac{\partial S_{ij}}{\partial x_j} = -[\nabla \times \boldsymbol{\omega}]_i.$$

Thus gradients in the strain field are related to gradients in vorticity. Note, however, that a uniform strain field can exist without vorticity while a uniform distribution of vorticity can exist without strain.

A question which often arises in turbulence is: do the local velocity gradients contribute most to the rate-of-stain tensor or to the vorticity field? In this respect, it is useful to consider the quantity Q defined by

$$Q = -\frac{1}{2}\frac{\partial u_i}{\partial x_j}\frac{\partial u_j}{\partial x_i} = -\frac{1}{2}\left(S_{ij}S_{ij} - \frac{1}{2}\boldsymbol{\omega}^2\right)$$

which is an *invariant* of the matrix $\partial u_i/\partial x_j$: invariant in the sense that its value does not depend on the orientation of the coordinate system. This is often rewritten in normalized form

$$\Lambda = \frac{S_{ij}S_{ij} - 1/2(\boldsymbol{\omega}^2)}{S_{ij}S_{ij} + 1/2(\boldsymbol{\omega}^2)}.$$

Positive values of Λ tend to be associated with a flow in which there is a large amount of strain, while negative values of Λ suggest a flow dominated by vorticity. Another invariant of the matrix $\partial u_i/\partial x_j$ is

$$R = \frac{1}{3}\left(S_{ij}S_{jk}S_{ki} + \frac{3}{4}\omega_i\omega_j S_{ij}\right).$$

The two invariants, Q and R, are sometimes used to classify the local structure of a flow field. (This is discussed in section 3.6 of Chapter 5.)

Example 2.1 Consider the simple shear flow $u_x(y) = 2Sy$, $S =$ constant. It can be decomposed into the two velocity fields $\mathbf{u}_1 = (Sy, Sx, 0)$ and $\mathbf{u}_2 = (Sy, -Sx, 0)$. Show that the first represents irrotational strain (with no vorticity) and the second rigid-body rotation (with no strain). Sketch $\mathbf{u}_1$ and calculate the orientation of the principal axes of strain.

Let us now introduce some dynamics. Since $\boldsymbol{\omega} = 2\boldsymbol{\Omega}$, it follows that the angular momentum, $\mathbf{H}$, of a small spherical blob of fluid is,

$$\mathbf{H} = \frac{1}{2}I\boldsymbol{\omega} \tag{2.26}$$

where I is its moment of inertia. Now consider a blob which is instantaneously spherical. Then $\mathbf{H}$ will change as a result of the tangential surface stresses alone. The pressure has no influence at the instant at which the blob is spherical since the pressure forces point radially inward. Therefore, at one particular instant in time,

$$\frac{D\mathbf{H}}{Dt} = (\text{viscous torque on spherical element}).$$

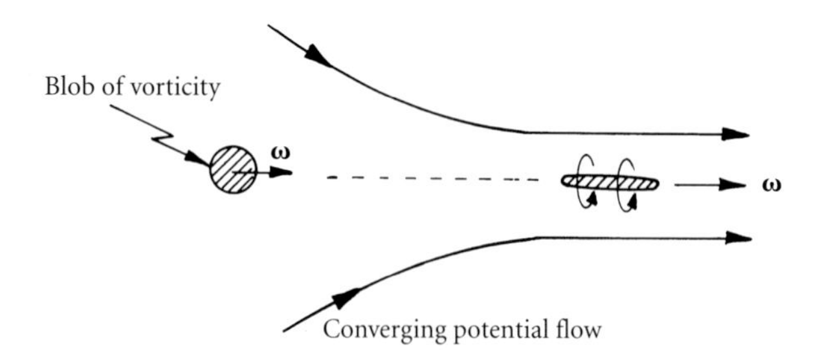

Figure 2.6 Stretching a fluid element can intensify the vorticity.

Since the convective derivative obeys the usual rules of differentiation this yields,

$$I\frac{D\boldsymbol{\omega}}{Dt} = -\boldsymbol{\omega}\frac{DI}{Dt} + 2 \times (\text{viscous torque on spherical element}). \tag{2.27a}$$

If viscosity is negligible we have, as a special case,

$$\frac{D(I\boldsymbol{\omega})}{Dt} = \mathbf{0}. \tag{2.27b}$$

Now (2.27a) holds only at one particular instant and for a blob that is instantaneously spherical. Nevertheless, at any instant and at any location we are always free to define such a fluid element. Thus (2.27a) really holds at every point and at all times, although the material which instantaneously constitutes the sphere will change with time and from place to place. There are three immediate consequences of (2.27a, b). First, since pressure is absent from (2.27a) we would expect $\boldsymbol{\omega}$ to evolve independently of p. Second, if $\boldsymbol{\omega}$ is initially zero, and the flow is inviscid, then $\boldsymbol{\omega}$ should remain zero in each fluid particle. This is the basis of potential flow theory in which we set $\boldsymbol{\omega} = \mathbf{0}$ in the upstream fluid. Third, if I decreases in a fluid element (and the viscous torque is small) then (2.27b) suggests that the vorticity of that element should increase. For instance, if a blob of vorticity is embedded in an otherwise potential flow field consisting of converging streamlines, as shown in Figure 2.6, then the moment of inertia of the element about an axis parallel to $\boldsymbol{\omega}$ will decrease, and so $\boldsymbol{\omega}$ will rise to conserve $\mathbf{H}$. Thus we can intensify vorticity by stretching fluid elements. This is referred to as *vortex stretching*.

We shall confirm all three of these propositions shortly. We close this section by noting that, since $\nabla \cdot (\nabla \times \boldsymbol{\omega}) = 0$, the vorticity field, like the velocity field, is solenoidal. Consequently, we can invoke the idea of vortex tubes and vortex lines which are the analogues of streamtubes and streamlines. Vortex tubes are simply bundles of vortex lines. Two familiar examples are a tornado and a vortex ring (smoke ring) as shown in Figure 2.7. Note that the flux of vorticity along a vortex tube, Φ, is constant along the tube since there is no

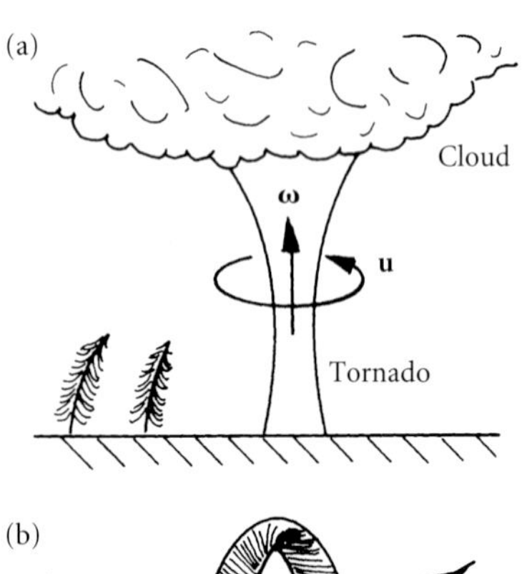

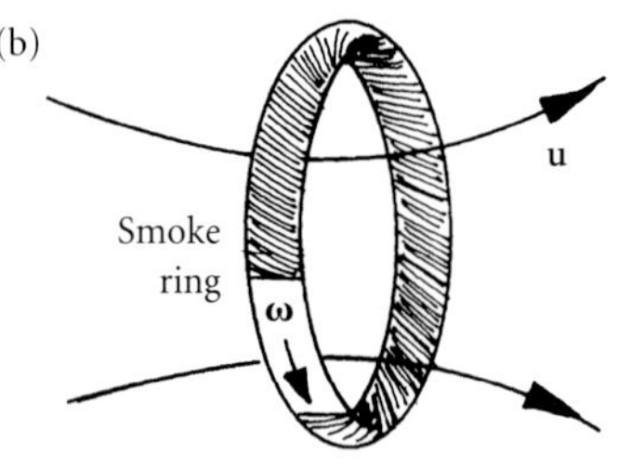

Figure 2.7 Examples of vortex tubes: (a) tornado and (b) smoke ring.

leakage of vorticity out the sides of the tube. Moreover, the flux Φ is related to the line integral of $\mathbf{u}$ around the tube by Stokes' theorem:

$$\Phi = \int \boldsymbol{\omega} \cdot d\mathbf{A} = \oint \mathbf{u} \cdot d\mathbf{l}.$$

Finally, we note that, in infinite domains, we may invert the relationship $\boldsymbol{\omega} = \nabla \times \mathbf{u}$ using the Biot–Savart law. This gives

$$\mathbf{u}(\mathbf{x}) = \frac{1}{4\pi} \int \frac{\boldsymbol{\omega}(\mathbf{x}') \times \mathbf{r}}{r^3} d\mathbf{x}', \qquad \mathbf{r} = \mathbf{x} - \mathbf{x}'. \tag{2.28}$$

Of course, there is an analogy to electromagnetism here, where the current density $\mathbf{J}$ is related to the magnetic field, $\mathbf{B}$, by $\nabla \times \mathbf{B} = \mu_0 \mathbf{J}$. We have an analogy in which $\mathbf{u} \leftrightarrow \mathbf{B}$ and $\boldsymbol{\omega} \leftrightarrow \mu_0 \mathbf{J}$. (Here μ_0 is the permeability of free space.) Thus, just as a current loop induces a poloidal magnetic field, so a vortex ring induces a poloidal velocity field (Figure 2.7).

Example 2.2 Consider an isolated blob of vorticity and any spherical volume, V, which encloses the vorticity. Show that the total angular momentum in V is,

$$\mathbf{H} = \int \mathbf{x} \times \mathbf{u} dV = \int_V (\boldsymbol{\omega} \cdot \mathbf{x}) \mathbf{x} dV = -\frac{1}{2} \int_V (\mathbf{x}^2 \boldsymbol{\omega}) dV$$
$$= \frac{1}{3} \int_V (\mathbf{x} \times (\mathbf{x} \times \boldsymbol{\omega})) dV.$$

2.3.2 *The vorticity equation*

Let us now derive the governing equation for $\boldsymbol{\omega}$. We start by rewriting (2.11) in the form,

$$\frac{\partial \mathbf{u}}{\partial t} = \mathbf{u} \times \boldsymbol{\omega} - \nabla C + \nu \nabla^2 \mathbf{u}, \qquad C = \frac{p}{\rho} + \frac{u^2}{2} \tag{2.29}$$

where C is Bernoulli's function. This follows from the identity

$$\nabla(u^2/2) = (\mathbf{u} \cdot \nabla)\mathbf{u} + \mathbf{u} \times \boldsymbol{\omega}.$$

We now take the curl of (2.29) which yields an evolution equation for $\boldsymbol{\omega}$

$$\frac{\partial \boldsymbol{\omega}}{\partial t} = \nabla \times [\mathbf{u} \times \boldsymbol{\omega}] + \nu \nabla^2 \boldsymbol{\omega}. \tag{2.30}$$

Since,

$$\nabla \times (\mathbf{u} \times \boldsymbol{\omega}) = (\boldsymbol{\omega} \cdot \nabla)\mathbf{u} - (\mathbf{u} \cdot \nabla)\boldsymbol{\omega}$$

(a)

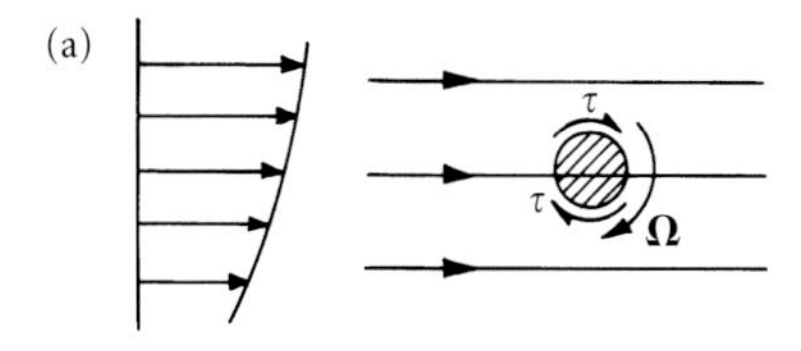

(b)

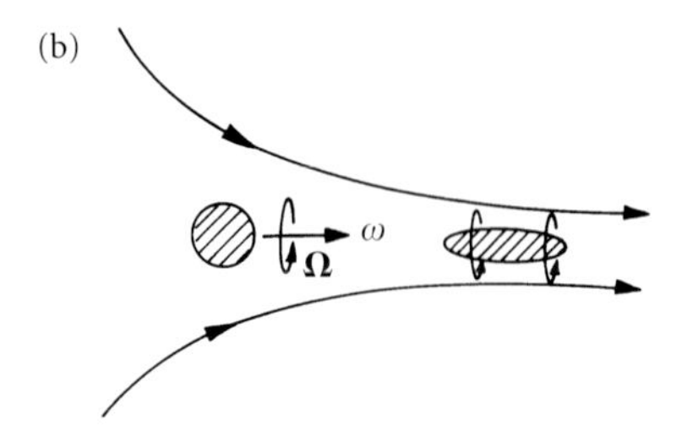

Figure 2.8 Vorticity can change because: (a) viscous forces spin up (or slow down) a fluid element or (b) because the moment of inertia of that element is changed.

this is often rewritten in the alternative form

$$\frac{D\boldsymbol{\omega}}{Dt} = (\boldsymbol{\omega} \cdot \nabla)\mathbf{u} + \nu\nabla^2\boldsymbol{\omega}. \tag{2.31}$$

We might compare this with our angular momentum equation (2.27),

$$I\frac{D\boldsymbol{\omega}}{Dt} = -\boldsymbol{\omega}\frac{DI}{Dt} + 2 \times (\text{viscous torque on spherical element}).$$

It would seem that the terms on the right of (2.31) represent: (i) the change in moment of inertia of a fluid element due to stretching of that element; and (ii) the viscous torque on the element. In short, the vorticity of a fluid blob may change because the blob is stretched, causing a change in moment of inertia, or else because the blob is spun up or slowed down by the viscous stresses (Figure 2.8).

It is instructive to consider the case of two-dimensional motion: $\mathbf{u}(x, y) = (u_x, u_y, 0)$, $\boldsymbol{\omega} = (0, 0, \omega)$. Here the first term on the right of (2.31) disappears. Consequently, there is no vortex stretching in planar flows and the vorticity in a fluid element will change because of viscous forces alone.[2] We have,

$$\frac{D\omega}{Dt} = \nu\nabla^2\omega. \tag{2.32}$$

Compare this with the governing equation for temperature, T, in a fluid

$$\frac{DT}{Dt} = \alpha\nabla^2 T \tag{2.33}$$

where α is the thermal diffusivity. Equations of this type are referred to as *advection–diffusion equations*. Evidently, in planar flows, vorticity is swept around by the flow and diffuses just like heat.

In order to gain some sense of what equations like (2.32) and (2.33) represent, consider the case of a wire of diameter d which sits in a cross flow u and is pulsed with electric current (Figure 2.9). Each time the wire is pulsed a packet of hot fluid is formed which is then swept downstream. The temperature field is governed by

$$\frac{DT}{Dt} = \frac{\partial T}{\partial t} + u\frac{\partial T}{\partial x} = \alpha\nabla^2 T. \tag{2.34}$$

If u is very small then heat soaks through the material by conduction alone, as if the fluid were a solid:

$$\frac{\partial T}{\partial t} \approx \alpha\nabla^2 T.$$

[2] This is because the stretching of fluid elements is confined to the x–y plane, whereas the vorticity points in the z-direction.

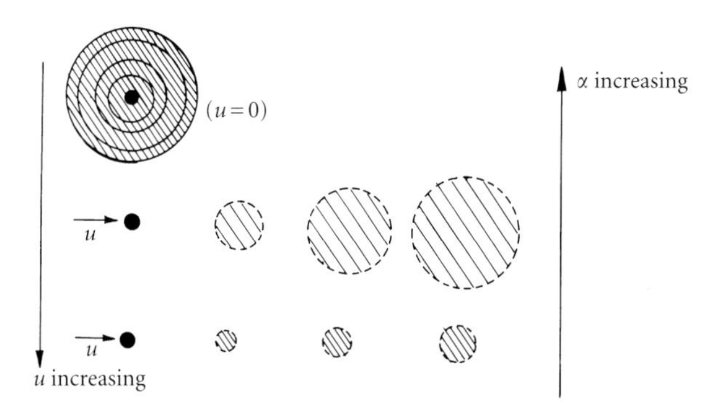

Figure 2.9 Advection and diffusion of heat from a pulsed wire.

The wire is then surrounded by concentric isotherms. Conversely, if α is very small, so there is very little diffusion of heat, then

$$\frac{DT}{Dt} \approx 0.$$

In this case each lump of fluid conserves its heat, and hence its temperature, as it is swept downstream. Thus we obtain a range of flows depending on the relative values of α and u (Figure 2.9). In fact, it is the Peclet number, $\mathrm{Pe} = ud/\alpha$, which determines the behaviour.

Of course, heat is neither destroyed nor created in the interior of the fluid. Thus $\int TdV$ is conserved for each of the dotted volumes shown in Figure 2.9. This is readily confirmed by integrating (2.33) over such a volume and invoking Gauss' theorem,

$$\frac{D}{Dt}\int_V TdV = \int_V \frac{DT}{Dt}dV = \alpha \oint_S (\nabla T)\cdot d\mathbf{S} = 0.$$

(We note in passing that, when dealing with a material volume—a volume always composed of the same particles—the operation of $D(\cdot)/Dt$ and $\int$ commute, that is, $D(T\delta V)/Dt = (DT/Dt)\delta V$.)

Equation (2.32) tells us that vorticity in a two-dimensional flow is advected and diffused just like heat, and that the analogue of the Peclet number is $\mathrm{Re} = ul/\nu$. The implication is that vorticity, like heat, cannot be created or destroyed within the interior of a two-dimensional flow. It can spread, by diffusion, and it can be moved from place to place by advection, but $\int \omega dV$ is conserved for all localized blobs of vorticity. A simple illustration of this, which is analogous to the blobs of heat above, are the vortices in the Karman street behind a cylinder. The vortices are advected by the velocity field and spread by diffusion, but the total vorticity within each eddy remains the same.

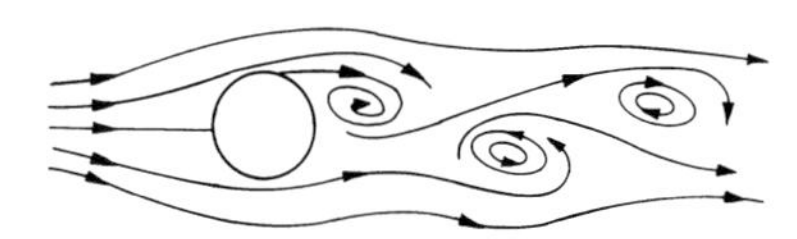

Figure 2.10 Karman street behind a cylinder.

Since vorticity cannot be created in the interior of the flow one might ask where the vorticity in Figure 2.10 came from. After all, the fluid particles upstream of the cylinder clearly have no angular momentum (vorticity) yet those downstream do. Again the analogy to heat is useful. The hot blobs in Figure 2.9 gained their heat from the

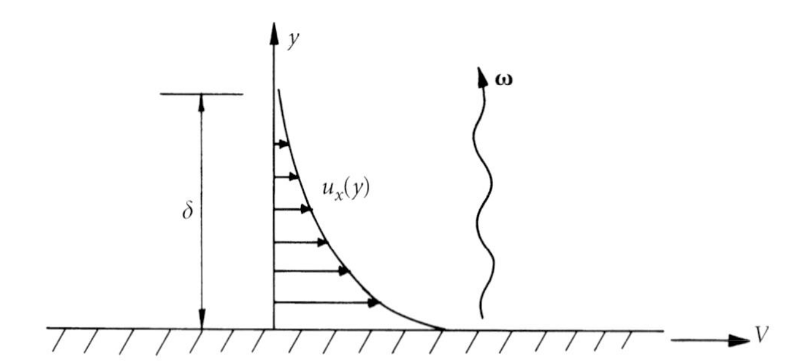

Figure 2.11 Motion adjacent to an impulsively started plate.

surface of the wire. Similarly, the vorticity (angular momentum) in the Karman street originated at the surface of cylinder. In fact, boundary layers are filled with intense vorticity which has diffused out from the adjacent surface. This gives us a new way of thinking about boundary layers: they are diffusion layers for the vorticity generated on a surface.

Let us push this idea a little further. We contend that boundaries are sources of vorticity, and that vorticity oozes out of them by diffusion, just like heat. The simplest example of this, which may be confirmed by exact analysis, is the case of a flat plate which sits in a semi-infinite fluid and is suddenly set into motion with a speed V (Figure 2.11).

The velocity field $\mathbf{u} = (u_x(y), 0, 0)$ is associated with a vorticity $\omega = -\partial u_x / \partial y$ and this, in turn, is governed by the diffusion equation,

$$\frac{\partial \omega}{\partial t} = \nu \frac{\partial^2 \omega}{\partial y^2}.$$

The situation is analogous to the diffusion of heat from an infinite plate whose surface temperature is suddenly raised from $T = 0$ (the temperature of the ambient fluid) to $T = T_0$. Here we have,

$$\frac{\partial T}{\partial t} = \alpha \frac{\partial^2 T}{\partial y^2}; \quad T = T_0 \text{ at } y = 0.$$

This sort of diffusion equation can be solved by looking for self-similar solutions of the type $T = T_0 f(y/\delta)$, $\delta = (2\alpha t)^{1/2}$. That is, heat diffuses out of the plate and the thickness of the heated region grows with time according to $\delta \sim (2\alpha t)^{1/2}$. The quantity δ is called the diffusion length and it turns out that f is an error function, but the details are unimportant. Our experience with the thermal problem suggests that we look for a solution of our vorticity equation of the form, $\omega = (V/\delta) f(y/\delta)$, $\delta = (2\nu t)^{1/2}$. Substituting this guess into (2.32) yields, after a little algebra,

$$\omega = (2/\pi)^{1/2} (V/\delta) \exp\lfloor -y^2/(4\nu t) \rfloor.$$

Thus vorticity is created at the surface of the plate by the shear stress acting on that surface. This vorticity then diffuses into the fluid in exactly the same way that heat diffuses away from a heated surface. It diffuses a distance $\delta \sim (2\nu t)^{1/2}$ in a time t.

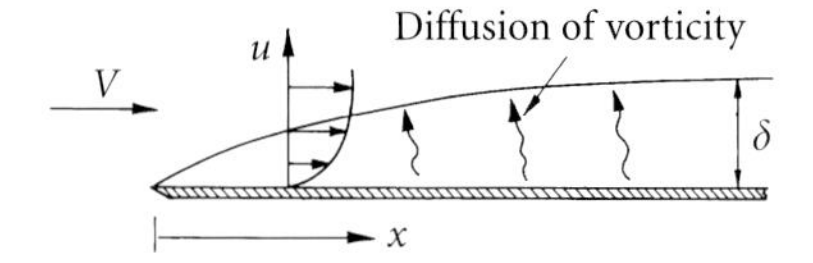

Figure 2.12 A boundary layer may be regarded as a diffusion layer for vorticity.

Consider now the laminar boundary layer shown in Figure 2.12. Here the plate is at rest and the fluid moves over the plate. We know that the vorticity in the boundary layer is intense, while that outside is weak. This follows simply from the definition $\boldsymbol{\omega} = \nabla \times \mathbf{u}$ and the fact that velocity gradients are large in the boundary layer. We interpret this as follows. Vorticity is generated at the surface of the plate, just like the previous example. This diffuses out from the plate at a rate $\delta \sim (2\nu t)^{1/2}$. Meanwhile, material particles are being swept downstream at a speed $\sim V$. A particle at a distance y from the plate will first feel the influence of the plate (by gaining some vorticity) after a time $t \sim y^2/\nu$, by which time it has moved a distance $x \sim Vt$ from the leading edge. So we expect the thickness of the diffusion layer to grow as $\delta \sim (\nu x/V)^{1/2}$. Of course, this is indeed the thickness of a laminar boundary layer on a plate.

Thus we see that boundary layers are prolific generators of vorticity and in fact this is the source of the vorticity in most turbulent flows. The wind gusting down a street is full of vorticity because boundary layers are generated on the sides of the buildings. These boundary layers are full of vorticity and when they separate at the downwind side of the building this vorticity is swept into the street (Figure 1.12).

Let us now return to three dimensions. Here the analogy to heat is lost since the governing equation is

$$\frac{D\boldsymbol{\omega}}{Dt} = (\boldsymbol{\omega} \cdot \nabla)\mathbf{u} + \nu\nabla^2\boldsymbol{\omega}. \tag{2.35}$$

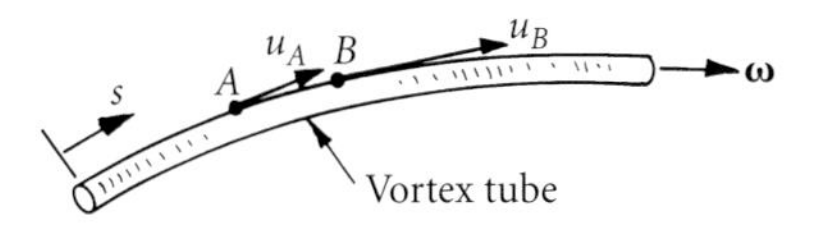

Figure 2.13 Stretching of a vortex tube.

We have a new term, $(\boldsymbol{\omega} \cdot \nabla)\mathbf{u}$, to contend with. Now we have already suggested that this represents the intensification (or diminution) of vorticity through the stretching (compression) of material elements. This is readily confirmed as follows. Consider a thin tube of vorticity, as shown in Figure 2.13. Let $u_{/\!/}$ be the component of velocity parallel to the vortex tube and s be a coordinate measured along the tube. Then

$$|\boldsymbol{\omega}|\frac{du_{/\!/}}{ds} = (\boldsymbol{\omega} \cdot \nabla)u_{/\!/}.$$

Now the tube is being stretched if $u_{/\!/}$ at B is greater than that at A; that is, if $du_{/\!/}/ds > 0$. Thus $(\boldsymbol{\omega} \cdot \nabla)u_{/\!/}$ is positive if the vortex tube is being stretched and from (2.35) we see that $\boldsymbol{\omega}$ will increase in such cases. This is simply the intensification of vorticity (angular velocity) through the conservation of angular momentum.

If we let the cross-section of the tube shrink to almost zero we get something approaching a vortex line. The same argument can then be repeated for any vortex line in the flow. So, if vortex lines are stretched, their vorticity is intensified.

The process of intensification of vorticity by stretching is often written in terms of the *enstrophy equation*. Enstrophy, $\boldsymbol{\omega}^2/2$, is governed by the equation

$$\frac{D}{Dt}\left(\frac{\boldsymbol{\omega}^2}{2}\right) = \omega_i\omega_j S_{ij} - \nu(\nabla\times\boldsymbol{\omega})^2 + \nu\nabla\cdot[\boldsymbol{\omega}\times(\nabla\times\boldsymbol{\omega})]. \tag{2.36}$$

This comes from taking the scalar product of (2.35) with $\boldsymbol{\omega}$. The divergence on the right integrates to zero for a localized distribution of vorticity and is often unimportant. The other two terms on the right correspond to the generation (or reduction) of enstrophy via vortex line stretching (or compression) and the destruction of enstrophy by viscous forces. Thus we see that enstrophy, just like mechanical energy, is destroyed by friction. We shall use (2.36) repeatedly in our discussion of turbulence.

2.3.3 *Kelvin's theorem*

No fluid is inviscid. Nevertheless, there are a few (but not many) circumstances where viscous effects can be neglected. In such cases a powerful theorem, called Kelvin's theorem, applies. Its most important consequence is that vortex lines, the analogue of streamlines, are convected by the flow as if frozen into the fluid.

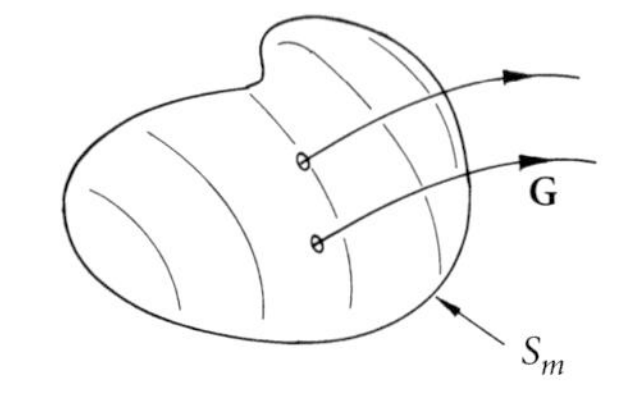

Figure 2.14 Defining sketch for equation (2.37).

To prove Kelvin's theorem we first need a simple kinematic result. Suppose S_m is a material surface (a surface always composed of the same fluid particles), **G** is any solenoidal vector field which lies in the fluid, and **u** is the velocity field of the fluid. We wish to determine the rate of change of the flux of **G** through S_m as it moves with the fluid (Figure 2.14).

It turns out that this is given by

$$\frac{d}{dt}\int_{S_m}\mathbf{G}\cdot d\mathbf{S} = \int_{S_m}\left[\frac{\partial\mathbf{G}}{\partial t} - \nabla\times(\mathbf{u}\times\mathbf{G})\right].d\mathbf{S} \tag{2.37}$$

We will not give a formal proof of (2.37) since this is readily found elsewhere. However, we note that the idea behind (2.37) is the following. The flux of **G** through S_m changes for two reasons. First, even if S_m were fixed in space, there is a change in flux whenever **G** is time-dependent. This is the first term on the right of (2.37). Second, since the boundary of S_m moves with velocity **u**, it may expand at points to include additional flux, or perhaps contract at other points to exclude flux. Suppose that the bounding curve for S_m, say C_m, is composed of line elements $d\mathbf{l}$. Then it happens that, in a time δt, the surface adjacent to the line element $d\mathbf{l}$ increases by an amount $d\mathbf{S} = (\mathbf{u}\times d\mathbf{l})\delta t$ and so the increase in flux due to movement of the boundary C_m is

$$\delta\int_{S_m}\mathbf{G}\cdot d\mathbf{S} = \oint_{C_m}\mathbf{G}\cdot(\mathbf{u}\times d\mathbf{l})\delta t = -\oint_{C_m}(\mathbf{u}\times\mathbf{G}).d\mathbf{l}\delta t.$$

The line integral may be converted into a surface integral using Stokes' theorem and (2.37) follows.

Now for an inviscid fluid we have,

$$\frac{\partial \boldsymbol{\omega}}{\partial t} = \nabla \times (\mathbf{u} \times \boldsymbol{\omega}).$$

Combining this with (2.37) we see that, for any material surface, S_m,

$$\frac{d}{dt}\int_{S_m} \boldsymbol{\omega} \cdot d\mathbf{S} = 0. \tag{2.38a}$$

To put it more simply, this states that the flux of vorticity through any material surface remains constant as the surface moves. If C_m is the bounding curve for S_m this may be rewritten in the form

$$\Gamma = \oint_{C_m} \mathbf{u} \cdot d\mathbf{l} = \text{constant}. \tag{2.38b}$$

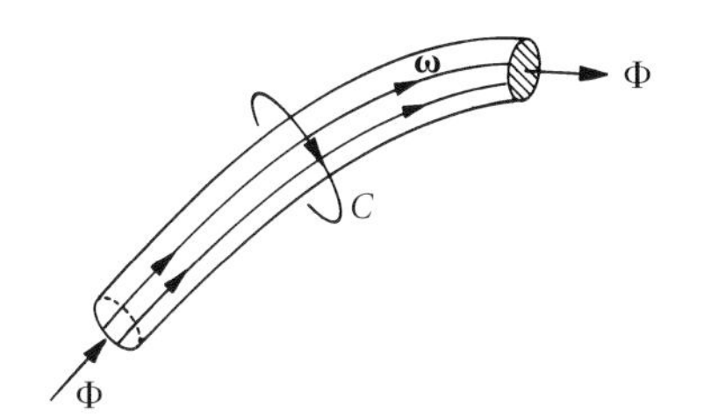

Figure 2.15 A vortex tube.

This is Kelvin's theorem. The quantity Γ is called the circulation. At first sight Kelvin's theorem may seem a little abstract but it has important physical consequences. To illustrate this it is instructive to apply (2.38b) to an isolated vortex tube: that is, a tube composed of an aggregate of vortex lines (Figure 2.15). Since $\boldsymbol{\omega}$ is solenoidal,

$$\oint \boldsymbol{\omega} \cdot d\mathbf{S} = 0$$

and it follows that the flux of vorticity, $\Phi = \int \boldsymbol{\omega} \cdot d\mathbf{S}$, is constant along the length of the vortex tube. From Stokes' theorem we have

$$\Phi = \oint_C \mathbf{u} \cdot d\mathbf{l} = \Gamma$$

where C is any curve which encircles the tube. It follows that Γ has the same value irrespective of the path C, provided, of course, that C encircles the vortex tube.

Now suppose that C is a material curve, C_m, which moves with the flow. Moreover, suppose that C_m at some initial instant encircles the vortex tube as shown in Figure 2.15. From Kelvin's theorem we know that Γ is conserved as the flow evolves. It follows that the flux through C_m is conserved by the flow and the implication is that C_m must always encircle the vortex tube. Since C_m moves with the fluid this suggests, but does not prove, that the tube itself moves with the fluid, as if frozen into the medium. This, in turn, suggests that every vortex line in an inviscid fluid moves with the fluid, since we could let the tube have a vanishingly small cross-section. We have arrived at one of the central results of vortex dynamics:

> Vortex lines are frozen into a perfectly inviscid fluid in the sense that they move with the fluid.

Of course, in two-dimensional motion we have already seen this at work since, when $\nu = 0$, $D\omega/Dt = 0$ in a planar flow.

A formal proof of the 'frozen-in' property of vortex lines actually requires a little more work. It normally proceeds along the following lines. Consider a short line $d\mathbf{l}$ drawn in the fluid at some instant and suppose $d\mathbf{l}$ subsequently moves with the fluid, like a dye line. Then the rate of change of $d\mathbf{l}$ is $\mathbf{u}(\mathbf{x} + d\mathbf{l}) - \mathbf{u}(\mathbf{x})$ where $\mathbf{x}$ marks the start of the line. It follows that

$$\frac{D}{Dt}(d\mathbf{l}) = \mathbf{u}(\mathbf{x} + d\mathbf{l}) - \mathbf{u}(\mathbf{x}) = (d\mathbf{l} \cdot \nabla)\mathbf{u}. \tag{2.39}$$

Compare this with the inviscid equation for vorticity

$$\frac{D\boldsymbol{\omega}}{Dt} = (\boldsymbol{\omega} \cdot \nabla)\mathbf{u}. \tag{2.40}$$

Evidently, $\boldsymbol{\omega}$ and $d\mathbf{l}$ obey the same equation. Now suppose that, at $t = 0$, we draw $d\mathbf{l}$ so that $\boldsymbol{\omega} = \lambda d\mathbf{l}$ for some λ. Then from (2.39) and (2.40) we have $D\lambda/Dt = 0$ at $t = 0$ and so $\boldsymbol{\omega} = \lambda d\mathbf{l}$ for all subsequent times. That is to say, $\boldsymbol{\omega}$ and $d\mathbf{l}$ evolve in identical ways under the influence of $\mathbf{u}$ and so the vortex lines are frozen into the fluid.

So we can reach the 'frozen-in' property of vortex lines either directly or else via Kelvin's theorem. Of course, it does not matter which route we take, it is the result itself which is important. It is a crucial result because it allows us to visualize the evolution of a high-Re flow. We simply need to track the movement of the vortex lines as the flow evolves. Equation (2.28) then gives $\mathbf{u}$ at each instant.

2.3.4 *Tracking vorticity distributions*

Much of nineteenth-century fluid mechanics focused on so-called potential flow theory. This is exemplified by Figure 2.16(a) which shows an aerofoil moving through still air. There is a boundary layer, which is filled with vorticity, and an external flow. In a frame of reference moving with the aerofoil the flow well upstream of the wing is uniform and hence free of vorticity. Since the vorticity generated on the surface of the foil is confined to the boundary layer (and subsequent wake), almost the entire external flow is irrotational. The problem of computing the external motion is now reduced to solving the two kinematic equations: $\nabla \cdot \mathbf{u} = 0$ and $\nabla \times \mathbf{u} = \mathbf{0}$. This is potential flow theory and it is really a branch of kinematics rather than dynamics. However, potential flows are extremely rare in nature, being largely confined to external flow over a streamlined body (in which upstream conditions are irrotational) and to certain types of water waves. In practice, virtually all real flows are laden with vorticity. This vorticity is generated in boundary layers and then released

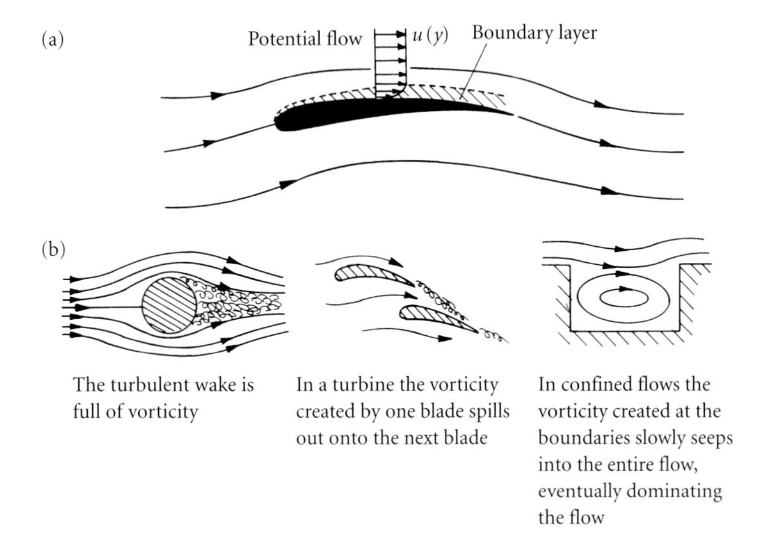

Figure 2.16 (a) Potential flow theory typified classical aerodynamics. (b) Virtually all real flows are laden with vorticity.

into the bulk flow as a result of boundary layer separation (as in the case of flow over a cylinder) or else in the form of a turbulent wake. Some simple examples are shown in Figure 2.16(b). Note that internal flows are *never* potential flows. (You can prove this using Stokes' theorem.)

So there are two types of flow: potential flow and vortical flow. The former is easy to compute but infrequent in nature, while the latter is commonplace but more difficult to quantify. The art of understanding vortical flows is to track the vorticity as it spills out from the boundary layer into the bulk flow.

Let us now summarize what we know about tracking the vorticity distribution in an evolving flow field. When Re is large the vortex lines are frozen into the fluid. Thus, for example, two interlinked vortex tubes in an inviscid fluid preserve their relative topology, remaining interlinked for all time (Figure 2.17).

However, we also know that vorticity can diffuse when Re is finite. Thus, in practice, the two vortex tubes shown in Figure 2.17 will sooner or later change their relative topology as excessive straining at one or more points induces significant diffusion which, in turn, causes the tubes to sever and reconnect.

The main point, however, is that we can track the vorticity from one moment to the next. It can spread by material movement or else by diffusion. In either case this is a localized process. This is not the case with linear momentum, $\mathbf{u}$. As discussed in Section 2.2 of this chapter, a localized distribution of momentum can be instantaneously redistributed throughout all space by the pressure force, $-\nabla p$. Thus we can talk about compact regions of vorticity evolving in space and time, while it makes no sense to talk of a compact region of $\mathbf{u}$. So when we allude to turbulent eddies of a finite size evolving in some

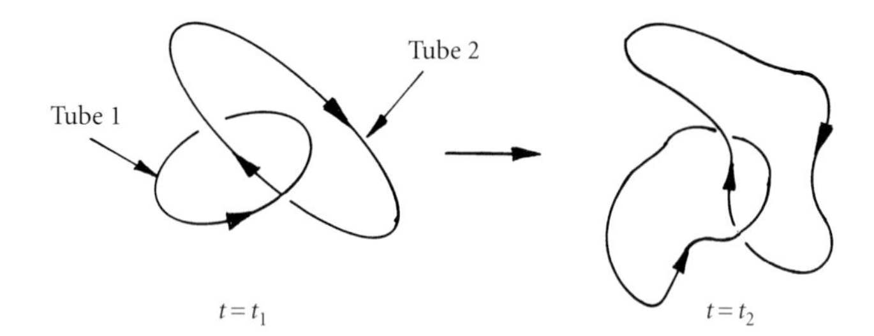

Figure 2.17 Two interlinked vortex lines in an inviscid fluid preserve their topology as they are swept around by the flow.

coherent fashion, we really mean blobs of vorticity (and their associated motion) evolving in a flow field.

Of course, we cannot divorce the velocity field from the vorticity since $\boldsymbol{\omega} = \nabla \times \mathbf{u}$. In a sense, the vorticity field advects itself. Consider the two interlinked vortex tubes shown in Figure 2.17. At any instance each induces a velocity field in accordance with the Biot–Savart law (2.28). This velocity field then advects the tubes as if they were frozen into the fluid. A short time later we have a new vorticity distribution and hence, from (2.28), a new velocity field. We then advect the vortex tubes a little more, based on the new velocity field. This gives yet another vorticity distribution and, by inference, yet another velocity field. In this type of flow the art of tracking the motion is one of tracking the development of the vorticity field.

Of course, in general not all of the velocity field need originate from the vorticity. We can add any potential flow, $\mathbf{u} = \nabla\phi$, to (2.28) and not change the vorticity distribution. For example, the tornado shown in Figure 2.7 might sit in an irrotational cross-wind. There are then two sources of motion: the cross-wind and the swirl associated with the vortex tube (tornado). Formally, we may decompose any incompressible flow into two components. One arises from the vorticity and is defined by $\nabla \cdot \mathbf{u}_\omega = 0$, $\nabla \times \mathbf{u}_\omega = \boldsymbol{\omega}$, or else by the Biot–Savart law (2.28). The other is a potential flow, $\mathbf{u}_p = \nabla\phi$, $\nabla^2\phi = 0$. The total velocity field is the sum of the two: $\mathbf{u} = \mathbf{u}_\omega + \mathbf{u}_p$. Such a decomposition is known as a Helmholtz decomposition.

2.4 A definition of turbulence

When you are a Bear of Very Little Brain, and you Think of Things, you find sometimes that a Thing which seemed very Thingish inside you is quite different when it gets out into the open and has others looking at it.
(A.A. Milne, House at Pooh Corner)

There has been a longstanding tradition in turbulence of studiously avoiding any formal definition of what we mean by a 'turbulent eddy', or for that matter 'turbulence'. It is almost as if we fear that, as soon as we try to define an eddy, the entire concept will melt away, proving

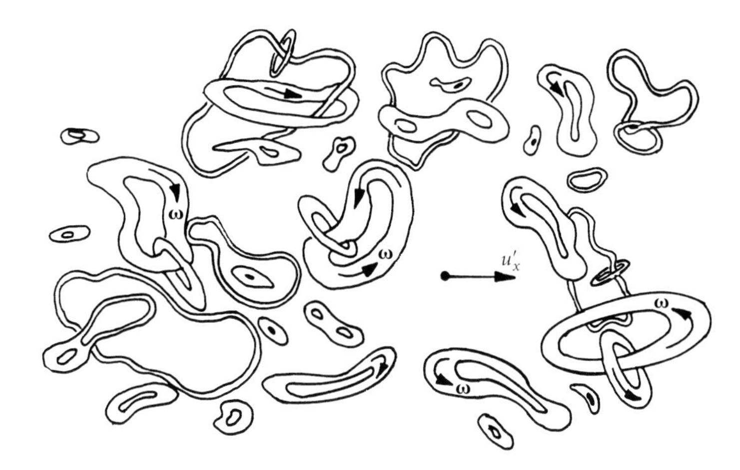

Figure 2.18 We may think of turbulence as a tangle of interacting vortex tubes.

to be entirely illusory, just like Pooh Bear's 'Thing'. Well, we have already stated what we interpret an eddy to be (i.e. a blob of vorticity and its associated velocity field) and so perhaps we should now take a stab at a definition of turbulence.

The idea that a turbulent eddy is a blob of vorticity gives us a means of picturing the evolution of a field of turbulence. The turbulence comprises of a sea of eddies (lumps of vorticity). These vortices are stretched and twisted by the velocity field, which is itself dictated by the instantaneous vorticity distribution through (2.28). Thus the vortices evolve and interact via their induced velocity field, with diffusion being restricted to regions where large gradients in vorticity develop. So we might picture turbulence as a seething tangle of vortex tubes, evolving under the influence of their self-induced velocity field (Figure 2.18). This suggests that, following the suggestion of Corrsin (1961), we might define turbulence as follows:

> Incompressible hydrodynamic turbulence is a spatially complex distribution of vorticity which advects itself in a chaotic manner in accordance with (2.31). The vorticity field is random in both space and time, and exhibits a wide and continuous distribution of length and time scales.

Note that (unlike Corrsin) we use the term chaotic, rather than random, to describe the self-advection of a turbulent vorticity field. We do so because we wish to emphasize that turbulence is extremely sensitive to its initial conditions and it is for this reason that no two realizations of a turbulent experiment are ever exactly the same. This extreme sensitivity to initial conditions is the hallmark of mathematical chaos. Note, however, that chaotic advection does not, in itself, guarantee turbulence. A small number of point vortices (four or more) can advect themselves in a chaotic manner in the x–y plane, yet this is not turbulence. Such a vorticity distribution is excluded from our definition because it is neither spatially complex nor does it exhibit a wide range of length scales. Note also that chaotic particle paths are not necessarily a sign of turbulence.[3] Simple, non-random Eulerian velocity fields (laminar flows) can cause fluid particles to follow complex trajectories which have certain chaotic properties. Such flows are also excluded from our definition as we require the vorticity field itself to be chaotic, not just the particle trajectories.

This concludes our brief introduction to fluid mechanics. We now return to the difficult task of quantifying turbulence. However, the key message you should carry away from this chapter is the following. Whenever we refer to a 'turbulent eddy', we really mean a blob, tube, or sheet of vorticity and the associated flow. Such a blob (eddy) evolves in the velocity field induced by itself and by all the other vortical structures. It remains coherent (localized) for a certain time because vorticity can spread only by material movement or else by diffusion. This is not true of linear momentum, $\mathbf{u}$, since the pressure force can instantaneously redistribute linear momentum throughout all space. In short, a turbulent flow is a complex tangle of vortex tubes, sheets, and blobs evolving in accordance with (2.30).

The other thing worth remembering is that blobs or sheets of vorticity are usually unstable and soon develop into spatially complex structures. In this sense the smoke ring and tornado shown in Figure 2.7 are atypical. Consider Figure 1.7: as the turbulent smoke rises it twists and turns forming highly convoluted shapes. If you imagine that the smoke marks the vorticity field you have a useful cartoon for turbulent flow.

Exercises

2.1 Flow down a cylindrical pipe has a component of swirl. The velocity field is given in (r, θ, z) coordinates by $(0,\ u_\theta(r),\ u_z(r))$. Calculate the vorticity and sketch the vortex lines and tubes.

2.2 Consider an axisymmetric flow consisting of swirl, $(0,\ u_\theta,\ 0)$, and poloidal motion, $(u_r, 0, u_z)$. Show that the instantaneous velocity field is completely determined by the angular momentum, $\Gamma = ru_\theta$, and the azimuthal vorticity, ω_θ. Now show that,

$$\frac{D\Gamma}{Dt} = 0, \qquad \frac{D(\omega_\theta/r)}{Dt} = \frac{\partial}{\partial z}\left[\frac{\Gamma^2}{r^4}\right].$$

Explain the origin of the source term in the evolution equation for ω_θ.

2.3 A thin plate is aligned with a uniform flow u_0 and a laminar boundary layer develops on the plate. The thickness of the boundary layer is of the order of $\delta \sim (\nu x/u_0)^{1/2}$ and the velocity profile may be approximated by the self-similar form $u/u_0 = \sin(\pi y/2\delta)$. (The coordinate x is measured in the

[3] See, for example, Ottino (1989).

streamwise direction from the leading edge of the plate.) Calculate the vorticity distribution and sketch it.

2.4 A vortex ring, for example, a smoke ring, is often modelled as a thin, circular vortex tube. Sketch the velocity distribution associated with a vortex ring. [Hint: recall that there is an exact analogy between $\mathbf{u}$ and $\boldsymbol{\omega}$ on the one hand and electrical currents and magnetic fields on the other. The electrical current plays the role of vorticity and the magnetic field plays the role of velocity.] Explain how the vortex ring propagates through a quiescent fluid. An unusual vortex motion may be observed in rowing. At the places where the oar breaks the surface of the water just previous to being lifted, a pair of small dimples (depressions) appear on the surface. Once the oar is lifted from the water this pair of dimples propagate along the surface. They are the end points of a vortex arc (half a vortex ring). Explain what is happening. Try this experiment using the blade of a knife or a spoon in place of the oar.

2.5 Potential flows are governed by the two kinematic equations $\nabla \cdot \mathbf{u} = 0$ and $\nabla \times \mathbf{u} = \mathbf{0}$. Where does dynamics enter such flows?

2.6 Estimate the rate of spreading of the two-dimensional blobs of heat shown in Figure 2.9. Now do the same calculation for the vortices in a Karman street at the rear of a cylinder.

2.7 The helicity of a region of inviscid fluid is defined as

$$H = \int_V \mathbf{u} \cdot \boldsymbol{\omega} dV.$$

Show that, if the vortex lines are closed in V, then H is an invariant of the motion. [Hint: first obtain an expression for the rate of change of $\mathbf{u} \cdot \boldsymbol{\omega}$, i.e., $D(\mathbf{u} \cdot \boldsymbol{\omega})/Dt = \mathbf{u} \cdot (D\boldsymbol{\omega}/Dt) + \boldsymbol{\omega} \cdot (D\mathbf{u}/Dt)$.]

2.8 Consider a region of fluid in which sit two interlinked, thin vortex tubes. Evaluate the net helicity for this region and show that, if the tubes are linked just once, as in Figure 2.17, then

$$H = \pm \Phi_1 \Phi_2$$

where Φ_1 and Φ_2 are the fluxes of vorticity in each tube. Show that, for this simple configuration, the conservation of H in an inviscid flow is a direct consequence of the 'frozen-in' behaviour of vortex lines.

Suggested reading

Acheson, D.J. (1990) *Elementary Fluid Dynamics*. Clarendon Press. (Consult chapter 5 for a nice overview of vortex dynamics.)

Batchelor, G.K. (1967) *An Introduction to Fluid Dynamics*. Cambridge University Press. (Chapter 3 introduces the Navier–Stokes equation and chapters 5 and 7 give a comprehensive overview of vortex dynamics.)

Ottino, J.M. (1989) *The Kinematics of Mixing*. Cambridge University Press. (This book describes how chaotic particle trajectories can appear in relatively simple flows.)

CHAPTER 3

The origins and nature of turbulence

> The next great era of awakening of human intellect may well produce a method of understanding the qualitative content of equations. Today we cannot. Today we cannot see that the water flow equations contain such things as the barber pole structure of turbulence that one sees between rotating cylinders. Today we cannot see whether Schrödinger's equation contains frogs, musical composers, or morality—or whether it does not. We cannot say whether something beyond it like God is needed, or not. And so we can all hold strong opinions either way.
>
> R.P. Feynman (1964)

It is an understatement to say that there is much we do not understand about turbulence. So, as with religion, we can all hold strong opinions! Of course the ultimate arbiter in such situations is the experimental evidence, and so experiments play a special role in turbulence theory.[1] They teach the theoretician to be humble and point the way to refining and clarifying our ideas. In this chapter we discuss one particular type of experiment at length: that is, grid turbulence in a wind tunnel. In many ways this is the archetypal example of turbulence and it has been extensively studied. It therefore provides a convenient vehicle for airing such issues as: 'does turbulence remember its initial conditions?' and 'is turbulence deterministic in a statistical sense?'.

We start, however, by revisiting the ideas of transition to turbulence and of chaos in a fluid. We have seen that, for all but the smallest of velocities, solutions of Navier–Stokes equations are usually chaotic. This is quite different from the behaviour of other, more familiar, partial differential equations, such as the diffusion equation or wave equation. Studying solutions of the diffusion equation is about as exciting and unpredictable as watching paint dry, and it is as well for devotees of Bach and Mozart that solutions of the wave equation *are*

[1] The dangers of the theoretician ignoring the experimental evidence is nicely summed up by Bertrand Russell who observed that "Aristotle maintained that women had fewer teeth than men: although he was twice married, it never occurred to him to verify this statement by examining his wives' mouths."

well behaved. It would seem, therefore, that there is something special about the Navier–Stokes equation and we claimed in Chapter 1 that it is the non-linear term, $(\mathbf{u} \cdot \nabla)\mathbf{u}$, which lies at the root of fluid chaos. The link between non-linearity, chaos and turbulence is our first topic of study.

3.1 The nature of chaos

Does the flap of a butterfly's wings in Brazil set off a tornado in Texas?
(E. Lorenz)

Our discussion of chaos is brief and limited to three main themes. First we wish to show that even a very simple evolution equation can lead to chaotic behaviour. We take as our example the much studied and rather elegant *logistic equation*. This equation contains both linear and non-linear terms. We shall see that the chaotic behaviour of solutions of the logistic equation is a direct result of non-linearity. When the non-linear term is relatively weak the solutions are well behaved. However, as the relative magnitude of the non-linear term is increased the solutions become increasingly complex, passing through a sequence of bifurcations (sudden changes), each bifurcation leading to a more complex state. Eventually the solutions become so intricate and complex that they are, to all intents and purposes, unpredictable. In short, the solutions are chaotic.

Our second theme is the idea that this 'transition to chaos' is not particular to the logistic equation. Rather, it is a general property of many non-linear systems. This leads us to Landau's theory of the transition to turbulence. Here a laminar flow is predicted to pass through a sequence of bifurcations, leading to increasingly complex states as Re is increased. In particular, Landau envisaged an infinite sequence of bifurcations, in which the jump in Re, $[\Delta \text{Re}]_n$, required to move from the nth bifurcation to the next becomes progressively smaller as n increases. In this way an infinite number of bifurcations (to ever more complex states) can occur as $\text{Re} \rightarrow \infty$, or possibly even within a finite range of Re. This picture of the transition to turbulence is now thought to be too simplistic. However, it does at least capture the spirit of the emergence of turbulence in certain geometries.

Our third and final topic has more of a thermodynamic flavour and harks back to an old dilemma in classical statistical mechanics. The issue relates to the so-called *arrow of time*. We have suggested that the chaotic mixing induced by a turbulent flow is due to the non-linearity of the Navier–Stokes equation. If this is true we might expect solutions of the invicid equation of motion (the Euler equation) to exhibit the same property, that is, that of continually mixing any 'frozen in' marker, such as die or vorticity. However, the Euler equation is time reversible, and so a computer simulation of an Euler flow played backwards would also be evolving according to the Euler equation. We might, for example, imagine a computer simulation of a two-dimensional Euler flow in which we start with all of the fluid on the left marked red and that on the right marked blue. As the simulation proceeds the colours begin to mix. After a while we stop the simulation and then play it backwards, like a movie played in reverse. This time the colours separate as the fluid returns to its original state. But this 'reverse flow' is also governed by the Euler equation. Thus the (alleged) tendency of a Euler flow to create ever greater mixing seems inconsistent with the underlying structure of the Euler equation. So how can an evolution equation which apparently does not distinguish between increasing or decreasing time lead, in practice, to a unidirectional increase in mixing? This is our third topic of discussion.

The literature on chaos theory, and its relationship to fluid turbulence, is vast. We can barely scratch the surface of the subject here. However, interested readers will find a more detailed discussion of many of these issues in Drazin (1992). Let us start, then, with the link between non-linearity and chaos.

3.1.1 *From non-linearity to chaos*

We discussed transition to turbulence in Chapter 1. For example, we noted that, in the experiments of Taylor and Bénard, the flow appears to pass through a number of states of increasing complexity as ν is decreased. Let us use the symbol R to represent the appropriate dimensionless control parameter in these types of experiments. In Taylor's experiments we might take $R = (\mathrm{Ta})^{\frac{1}{2}}$, while in Reynolds' experiment, or flow over a cylinder, we have $R = \mathrm{Re}$. In either case $R \sim \nu^{-1}$. In the Taylor or Bénard experiment it seems that the base configuration becomes unstable and bifurcates (changes) to a relatively simple flow at a certain critical value of R. This new flow itself becomes unstable at a slightly higher value of R and a more complex motion is established. In general, it seems that as R is increased we pass through a sequence of flows of ever increasing complexity until a fully turbulent regime is established. In Chapter 1 we emphasized that this transition to chaos is the result of the non-linearity of the Navier–Stokes equation.

Of course, chaotic behaviour is not unique to fluid mechanics. Chaos is seen in many, much simpler, mechanical, biological, and chemical systems. It is necessary only that the system is non-linear. Perhaps the simplest example of a non-linear equation which leads to chaos is the *logistic equation*, and we now digress for a moment to discuss its behaviour. The simple algebraic formula

$$x_{n+1} = F(x_n) = ax_n(1 - x_n); \quad 1 < a \leq 4 \tag{3.1}$$

was introduced in 1845 by Verhulst to model the growth of the population of a biological species. Here x_n (the normalized population of the nth generation) may take any value from 0 to 1. Note that difference equations of this type have many similarities to ordinary differential equations (ODEs) of the form $\dot{x} = G(x)$. Indeed, the numerical solution of $\dot{x} = G(x)$ by finite differences leads to a difference equation of the form $x_{n+1} = F(x_n)$, with the integer n now playing the role of time. The analogue of a steady solution of an ODE is a solution of the form $X = F(X)$, that is, $X_{n+1} = X_n$. Such solutions are called *fixed points* of the difference equation. In the case of the logistic equation (3.1) the fixed points are $X = 0$ and $X = (a - 1)/a$.

Now, just as a steady solution of an ODE can be unstable, so fixed points may be unstable. We say a fixed point is linearly unstable if $x_0 = X - \delta x$ leads to a sequence $x_0, x_1, \ldots, x_n$ which diverges from X for an infinitesimal perturbation δx. It is readily confirmed that the null point, $X = 0$, is unstable for all $a > 1$, while $X = (a - 1)/a$ is linearly stable for $1 < a \leq 3$ but unstable for $a > 3$ (see Exercise 3.2).

For $a > 3$ something interesting happens. Just as we lose stability of the fixed point $X = (a - 1)/a$, a new, periodic solution appears. This has the form $X_2 = F(X_1)$ and $X_1 = F(X_2)$ so the variable x_n flips back and forth between X_1 and X_2. It is left as an exercise for the reader to confirm that (see Exercise 3.1),

$$X_1, X_2 = \left[a + 1 \pm \left((a + 1)(a - 3)\right)^{1/2}\right]/2a.$$

Such a periodic solution is called a *two-cycle* of F and we talk of the fixed-point solution bifurcating to a two-cycle through a *flip bifurcation*. It turns out that the two-cycle is linearly stable for $3 < a \leq 1 + \sqrt{6}$. The flip bifurcation is shown in Figure 3.1(a).

Of course, linear (small perturbation) theory tells us nothing about the fate of iterates starting from arbitrary values of x_0. Nevertheless, it can be shown that for x_0 in the range $0 < x_0 < 1$ the iterates converge to the fixed point $X = (a - 1)/a$, provided that $1 < a \leq 3$, and to the two-cycle if $3 < a \leq 1 + \sqrt{6}$. We talk of the fixed point having a *domain of attraction* of $x_0 = 0 \rightarrow 1$ for $1 < a \leq 3$ and the two-cycle having a similar domain of attraction for $3 < a \leq 1 + \sqrt{6}$ (Figure 3.1(a)).

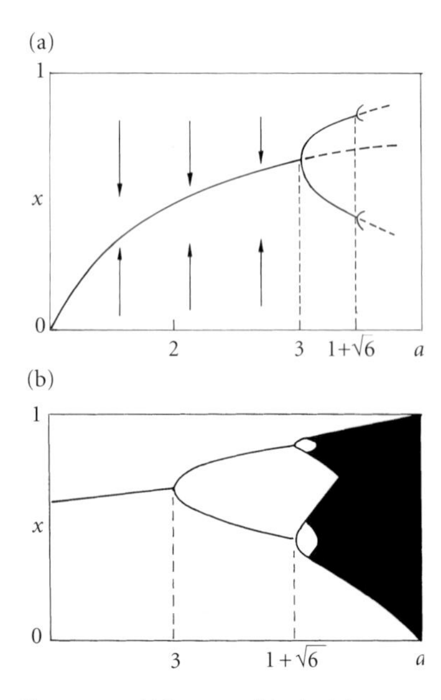

Figure 3.1 (a) Iterates of the logistic equation. Only the first bifurcation is shown. Stable fixed points and two-cycles are denoted by a continuous curve and unstable fixed points and two-cycles by a dotted curve. (b) Iterates of the logistic equation. Only the first two bifurcations are shown.

This is the situation for $a < 1 + \sqrt{6} = 3.449$. At $a = 3.449$ there is another bifurcation to a more complicated periodic state, called a *four-cycle*, in which x_n flips around between four branches of the 'solution curve'. (This type of transition is called *period doubling*.) Yet another bifurcation occurs at $a = 3.544$ to an even more complicated periodic solution (an eight-cycle). In fact, it turns out that there is an *infinite* sequence of period doubling bifurcations to ever more complex states and that these all occur for $3 < a < 3.5700$.[2] So what happens for values of a larger than this? Well, for $a > 3.5700$ the solutions become aperiodic and the sequence x_n becomes chaotic in the sense that it may be regarded as a sample of a random variable (Figure 3.1(b)).

Although chaos is the rule for $a > 3.5700$ there are islands of tranquillity within the sea of chaos. In particular, for $3.5700 < a < 4$ there are narrow windows (small intervals of a) in which we recover periodic behaviour. Actually, there is an infinity of such windows, although most of them are extremely narrow.

Some hint as to the behaviour of the logistic equation for $a > 3.5700$ may be obtained by considering the special case of $a = 4$ (von Neumann 1951). Rather remarkably this has an exact solution. Consider $x_n = \sin^2(\pi\theta_n)$ where θ_n is restricted to the range $0 \leq \theta_n < 1$ and θ_{n+1} is defined as the *fractional part* of $2\theta_n$. Then we have

$$x_{n+1} = \sin^2(\pi\theta_{n+1}) = \sin^2(2\pi\theta_n) = 4\sin^2(\pi\theta_n)\cos^2(\pi\theta_n)$$

which is a solution of

$$x_{n+1} = 4x_n(1 - x_n)$$

as required. Thus the solution for $a = 4$ is simply $x_n = \sin^2(2^n\pi\theta_0)$. To study the properties of this solution we note that, since $0 \leq \theta_0 < 1$, we can write θ_0 as a binary number,

$$\theta_0 = b_1/2 + b_2/2^2 + b_3/2^3 + \cdots$$

where $b_i = 0$ or 1. (This kind of expansion is possible since $1/2 + 1/4 + 1/8 + \cdots = 1$ and so, setting certain of the b_i's to zero gives a number between 0 and 1.) Evidently,

$$\begin{aligned}
\theta_1 &= b_2/2 + b_3/2^2 + b_4/2^3 + \cdots \\
\theta_2 &= b_3/2 + b_4/2^2 + b_5/2^3 + \cdots \\
\theta_3 &= b_4/2 + b_5/2^2 + b_6/2^3 + \cdots
\end{aligned}$$

This property of θ_n is known as the *Bernoulli shift* and it can be used to show that, if θ_0 is irrational, then the sequence θ_n is aperiodic. If θ_0 is rational, on the other hand, the sequence θ_n is periodic, that is, $\theta_{n+p} = \theta_n$ for some p, or else terminates at $\theta = 0$ after a finite number of steps. For example, $\theta_0 = \frac{1}{3}$ leads to a two cycle $\frac{1}{3}, \frac{2}{3}, \frac{1}{3}, \ldots$ while

[2] This infinite sequence of period doubling bifurcations is not restricted to solutions of the logistic equation. Precisely the same behaviour is seen in many other non-linear systems, and is known as a Feigenbaum sequence. It is remarkable that an infinite sequence of bifurcations occur within a finite range of the control parameter.

$\theta_0 = \frac{1}{4}$ leads to $\frac{1}{4}, \frac{1}{2}, 0, 0, \ldots$. Of course, all periodic sequences are unstable because, for any given rational value of θ_0, we can find an irrational value of θ_0, which is arbitrarily close to it. Thus a periodic solution can be changed into an aperiodic one through an infinitesimal change in the initial conditions. (These unstable p-cycles are the continuation of the stable p-cycles which arise for $3 < a < 4$.) Moreover, it may be shown that, if we let n get large enough, each aperiodic sequence θ_n eventually gets arbitrarily close to any given point in the range $0 \rightarrow 1$. Since any numerical experiment is subject to rounding error, and any physical experiment to imperfections, it is the more common aperiodic solutions, and not the unstable periodic ones, which are more important in practice.

Now consider two infinite sequences θ_n and θ_n^* which are generated from irrational initial conditions. Let $\gamma_n = \theta_n - \theta_n^*$ be the difference between the sequences and suppose that $\gamma_0 = \varepsilon$ where ε is small. Then it may be shown that, in general, γ_n grows exponentially with n. In short, two initially close sequences diverge exponentially fast. This extreme sensitivity to initial conditions is the hallmark of chaos, and it is not restricted to the case $a = 4$ or indeed to the logistic equation.[3]

Now we have seen that, if left for long enough, θ_n and hence x_n visits (or gets close to) every point in the range $0 \rightarrow 1$. In fact, it may be shown that, provided θ_0 is irrational, θ_n may be regarded as the sample of a random variable which visits all points in the range with equal likelihood. From this we conclude that x_n itself may be regarded as a sample of a random variable. Moreover, its probability density function (p.d.f.) can be shown to be

$$f(x) = \left[\pi^2 x(1-x)\right]^{-1/2}.$$

(A p.d.f. like $f(x)$ is defined by the fact that the relative likelihood of finding x in the range $x \rightarrow x + dx$ is $f(x)dx$.) Thus x_n is, on average, more likely to be found near the edges of range $0 \rightarrow 1$ than in the middle.

In summary then, when $a = 4$, the sequence x_n is, in principle, fixed by the simple deterministic equation

$$x_{n+1} = 4x_n(1 - x_n).$$

In practice, however, x_n appears to jump around in a chaotic fashion as if the sequence x_n were a sample of a random variable. This type of

[3] Typically, in non-linear difference equations a small perturbation in x_0 of size ε ensures that the trajectories of the original and the perturbed solutions diverge at a rate $\varepsilon \exp(\lambda n)$ where the constant λ is called the Liapounov exponent. In the case of the logistic equation (for $a = 4$) this sensitivity may be illustrated by considering the binary expansion for θ_1, θ_2, etc. given above. Suppose that we perturb θ_0 by changing the coefficient b_{10} from 1 to 0, or vise versa. In general this represents only a minute change to θ_0. However, after only 10 iterations this small change to θ_0 has become a first-order change to θ_{10}. In general, two irrational initial conditions θ_0 and $\theta_0 + \varepsilon$ diverge at a rate $\varepsilon.2^n$.

chaos is therefore called *deterministic chaos*. Moreover, the sequence x_n is extremely sensitive to initial conditions in that two arbitrarily close initial states will lead to exponentially diverging sequences. Thus we cannot, in practice, accurately track a given sequence for very long since a minute amount of rounding error eventually swamps the attempted calculation. The *statistical properties* of x_n, on the other hand, appear to be rather simple.

So the simple logistic equation embodies such complex phenomena as bifurcations, period doubling, the Feigenbaum sequence, and deterministic chaos. All of this from such a benign looking equation and all of this because it is non-linear! Since such a simple difference equation gives rise to so rich a behaviour it is little wonder that the Navier–Stokes equation embodies such diverse phenomena as tornadoes and turbulence.

One of the qualitative features of the chaotic regime in Figure 3.1(b), which ties in with our experience of turbulence, is the extreme sensitivity to initial conditions. Infinitesimal changes in x_0 lead to very diverse trajectories for x_n, and this is also true of turbulence (Figure 1.8). There are other qualitative similarities, such as the contrast between the complexity of individual trajectories and the simplicity of the statistical behaviour of the system. Finally we recall that, in the experiments of Taylor and Bénard the flow passes through a sequence of ever more complex states as ν^{-1} is increased, until eventually chaos is reached. Moreover, period doubling is sometimes observed in such a transition. This is all rather reminiscent of the logistic equation. So can we make the leap from population growth to the transition to turbulence? We shall pursue this idea a little further in the next section.

3.1.2 More on bifurcations

Although the turbulent motion has been extensively discussed in literature from different points of view, the very essence of this phenomenon is still lacking sufficient clearness. To the author's opinion, the problem may appear in a new light if the process of initiation of turbulence is examined thoroughly. (Landau 1944)

The idea that bifurcation theory might be relevant to hydrodynamic stability was probably first suggested by Hopf in 1942. However, perhaps it was Landau who (in 1944) initiated the debate on the role of bifurcations in the early stages of transition to turbulence. We start with Landau's theory.

Suppose that R_c is the critical value of our control parameter R at which infinitesimal disturbances first grow in a hydrodynamic experiment. For example, R_c might correspond to the critical value of

$(\mathrm{Ta})^{1/2}$ at which Taylor cells first appear. Let $A(t)$ and σ be the amplitude and growth rate of the unstable normal mode which sets in at $R = R_c$. According to linear (small perturbation) theory

$$A(t) = A_0 \exp[(\sigma + j\omega)\, t].$$

Now let us keep $R - R_c$ small so that all other normal modes are stable. We have $\sigma < 0$ where $R < R_c$ and $\sigma > 0$ for $R > R_c$. Usually one finds

$$\sigma = c^2(R - R_c) + 0[(R - R_c)^2], \quad |R - R_c| \ll R_c \tag{3.2}$$

for some constant, c. Now as the unstable mode grows it soon becomes large enough to distort the mean flow. A small perturbation analysis is then no longer valid and Landau suggested that the magnitude of A (averaged over many cycles) is governed by

$$\frac{d|A|^2}{dt} = 2\sigma|A|^2 - \alpha|A|^4 - \beta|A|^6 + \cdots \tag{3.3}$$

Provided $|A|$ is not too large we can neglect the higher-order terms and we arrive at the *Landau equation*:

$$\frac{d|A|^2}{dt} = 2\sigma|A|^2 - \alpha|A|^4. \tag{3.4}$$

The coefficient α is called *Landau's constant* and Landau suggested that $\alpha > 0$ for external flows while $\alpha < 0$ for pipe flow. When $\alpha = 0$ we recover the linear result and so $\alpha|A|^4$ represents the non-linear self-interactions of the disturbance which can either accentuate or moderate the growth rate. Interestingly, there is some resemblance between equation (3.4) and the logistic equation.

The nice thing about Landau's equation is that it has an exact solution:

$$\frac{|A|^2}{A_0^2} = \frac{e^{2\sigma t}}{1 + \lambda(e^{2\sigma t} - 1)}, \qquad \lambda = \frac{\alpha A_0^2}{2\sigma}. \tag{3.5}$$

The form of this solution depends on the sign of α. Consider first the case of $\alpha > 0$. Then we find that, as $t \to \infty$, $|A| \to 0$ when $R < R_c$ and $|A| \to (2\sigma/\alpha)^{1/2} = A_\infty$ for $R > R_c$. This is called a *supercritical bifurcation* and it is illustrated in Figure 3.2(a). Note that, since, $\sigma \sim (R - R_c)$, we have $A_\infty \sim (R - R_c)^{1/2}$. Thus, even though the flow is linearly unstable for $R > R_c$, it soon settles down to a new laminar motion and the difference between the original flow and the new flow grows with R according to $(R - R_c)^{1/2}$. This behaviour is reminiscent of the emergence of Taylor cells between rotating, concentric cylinders.

When $\alpha < 0$ the situation is quite different (Figure 3.2(b)). For $R > R_c$ the disturbance grows very rapidly and $|A| \to \infty$ within a finite

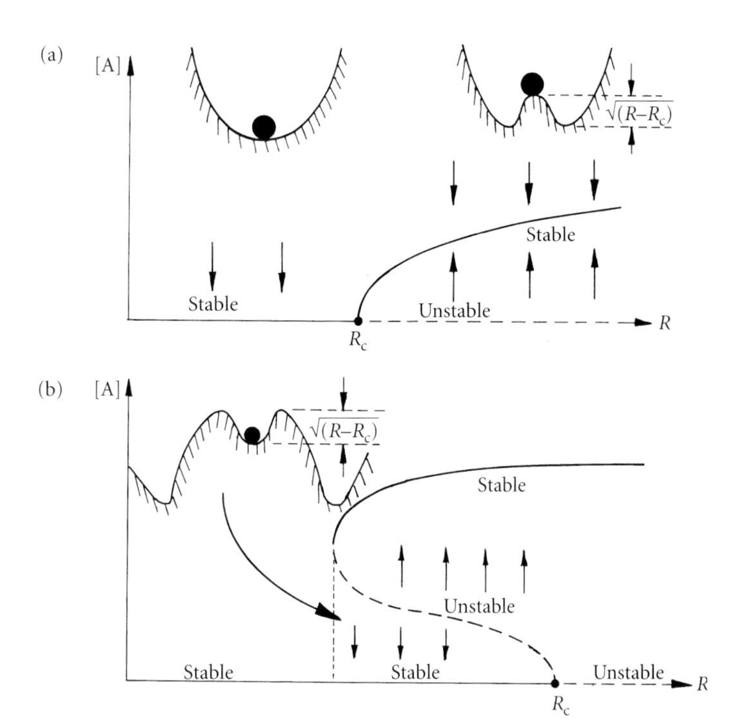

Figure 3.2 Bifurcations and their mechanical analogues. (a) Supercritical bifurcation. (b) Subcritical.

time, $t^\star = (2\sigma)^{-1} \ln[1 + |\lambda|^{-1}]$. Of course, long before t reaches $t^\star$ the Landau equation would cease to be valid and higher-order terms need to be considered, such as $-\beta|A|^6$. These higher-order effects can lead to a restabilization of the disturbance at larger amplitude, as indicated by the solid line in Figure 3.2(b). For $R < R_c$ we have $|A| \to 0$ as $t \to \infty$ provided that $A_0 < (2\sigma/\alpha)^{1/2}$, that is, the initial disturbance is small. However, if the system is given a big enough push, $A_0 > (2\sigma/\alpha)^{1/2}$, $|A| \to \infty$ within a finite time. (Of course, long before this occurs the higher-order terms cut in, moderating the growth.) This is called a *subcritical bifurcation* (Figure 3.2(b)). In summary, then, when $\alpha < 0$ the regime $R > R_c$ is both linearly and non-linearly unstable, while $R < R_c$ is linearly stable but non-linearly unstable if the initial disturbance is large enough. This sort of behaviour is reminiscent of the transition to turbulence in a pipe.

Landau went on to speculate as to what happens when $R - R_c$ ceases to be small. For the case of a supercritical bifurcation he suggested that, as $R - R_c$ increases, the flow resulting from the first bifurcation will itself become unstable and yet another, more complex, flow will emerge. He envisaged that, as R rises, more and more complex (but still laminar) flows will appear through a succession of bifurcations. He also suggested that the difference in the value of R between successive bifurcations will decrease rapidly so very soon the flow becomes complex and confused. The implication is that turbulence is the result of an infinite sequence of bifurcations. The idea is an appealing one, particularly in view of the behaviour of the logistic equation and it is remarkable that Landau suggested this some 30 years before the development of chaos theory. However, we shall see shortly that this is too simplistic a view of transition to turbulence.

Two years before Landau's conjectures Hopf had investigated the bifurcation of ordinary differential equations. What is now called a Hopf bifurcation is, in a sense, a subset of Landau's supercritical bifurcation. The main feature of such a bifurcation is that the state to which the system bifurcates is oscillatory, rather than steady. For example, one might anticipate that the sudden appearance of unsteady Taylor vortices at a certain value of Ta represents a Hopf bifurcation. The periodic state which emerges from a Hopf bifurcation is called a *limit cycle*.

Landau's 1944 paper is remarkable in the sense that it marked the first (pre-emptive) step in answering Feynman's call for a means of developing a qualitative understanding of the influence of non-linearity. However, it turns out that Landau's picture is a little simplistic and that life is much more complicated. For example, while Landau foresaw the emergence of increasingly random behaviour through a sequence of bifurcations, as exhibited by the logistic equation for $3 < a < 3.57$, he did not foresee the sudden transition to chaotic behaviour which occurs at $a = 3.57$. Rather, he pictured turbulence as the superposition of a very large number of 'modes' of varying frequencies and *random* phases, each new frequency being introduced though a bifurcation. Moreover, the infinite sequence of bifurcations exhibited by the logistic equation turns out to be only one of many *routes to chaos*. In fact, it now seems likely that there are several routes to chaos and to turbulence. One scenario is that three or four Hopf-like bifurcations lead directly to chaotic motion which, if analysed using the Fourier transform, exhibits a continuous spectrum of frequencies. Another route involves intermittent bursts of chaos in an otherwise ordered state.

3.1.3 *The arrow of time*

If someone points out to you that your pet theory of the universe is in disagreement with Maxwell's equations—then so much for Maxwell's equations. If it is found to be contradicted by observation—well, these experimentalists do bungle things sometimes. But if your theory is found to be against the second law of thermodynamics I can give you no hope; there is nothing for it but to collapse in deepest humiliation. (Eddington 1928)

We have pointed the finger at the inertial term, $(\mathbf{u} \cdot \nabla)\mathbf{u}$, as the source of chaos in the Navier–Stokes equation. It is this chaos, induced by non-linearity, which gives turbulence its ability to mix rapidly any contaminant, say dye. It is obvious, you might say, that if we drop some red dye into a turbulent stream, it will progressively mix. First it will be teased out into a tangle of spaghetti like streaks by the eddying motion, and then, when the streaks are thin enough for diffusion to be effective, the mass of red spaghetti will dissolve into a pink cloud. This turbulent mixing arises from chaotic advection, and this chaos, we claim, comes from non-linearity. Certainly, it is difficult to conceive of a situation where we start out with a tangled mess of dye strands and the turbulence acts spontaneously to 'unmix' the dye, concentrating it into a small blob. It seems as if chaotic advection, induced by non-linearity, gives turbulence an 'arrow of time'.

There is, however, a problem with this argument. Since non-linearity is responsible for chaos we might expect a turbulent flow governed by the *Euler equation*

$$\frac{\partial \mathbf{u}}{\partial t} + (\mathbf{u} \cdot \nabla)\mathbf{u} = -\nabla(p/\rho)$$

to exhibit the same arrow of time, progressively mixing fluid and any passive contaminant which marks it. Suppose, therefore, that we conduct a thought experiment. We take an extremely accurate computer and provide it with a computer code capable of integrating the Euler equation forward in time. We give it an initial condition of a turbulent velocity field confined to a box, and start the computation. In order to picture the mixing induced by the turbulence we tag all the fluid particles on the left of the box with a blue marker and all the fluid to the right with a red marker. The computation starts and, of course, we see that the two colours begin to mix due to the chaotic motion. After a while we stop the numerical experiment and reverse time, so that t becomes $-t$ and $\mathbf{u} = d\mathbf{x}/dt$ becomes $-\mathbf{u}$. We now restart the computation, only this time we are moving backwards in time. If our computer were infinitely accurate, subject to no rounding error, then the computation would resemble a film played backward. The fluid would appear to unmix, progressively moving back to a state in which the left of the box is all blue and the right all red. So what is the problem? Let us examine the 'backward' Euler equation in which t becomes $-t$ and $\mathbf{u}$ becomes $-\mathbf{u}$. It is

$$\frac{\partial \mathbf{u}}{\partial t} + (\mathbf{u} \cdot \nabla)\mathbf{u} = -\nabla(p/\rho).$$

The Euler equation is unchanged under a reversal of time. So now we do have a problem. When we stopped our calculation and reversed $\mathbf{u}$, we simply created a new initial condition. As we then marched backward in time the dye in the fluid separated rather than mixed, yet we are integrating exactly the same equation as before, which we believed had the property of creating greater disorder by progressive mixing. We are led to the conclusion that chaotic motion in an inviscid fluid need not lead to mixing. Yet, in practice, turbulence is always observed to lead to greater disorder!

There are two points we have overlooked which, when taken into account, help resolve our dilemma. The first is that all fluids have a finite viscosity and so all of our intuition relates to turbulence in a viscous fluid. The second is that we must distinguish between the *statistical likelihood* and *absolute certainty* of increased mixing.

Let us tackle the second of these issues first. A similar dilemma arises in classical statistical mechanics. At a fundamental level the classical laws of physics remain unchanged under a reversal of time, just like the Euler equation (which is, after all, just Newton's second law). Yet the macroscopic laws thermodynamics possess an arrow of time: entropy (disorder) increases in an isolated system. The thermodynamic analogue of our computer simulation is the following. Imagine a box filled with two gases, say helium and oxygen. At $t = 0$ all the helium is to the left and all the oxygen to the right. We then observe what happens as the gases are allowed to mix. (We burst the membrane separating the gases.) As time progresses they become more and more mixed and entropy rises in accord with the second law of thermodynamics. Now we play God. We suddenly stop the experiment and reverse t as well as the velocity of every molecule. The gases will then proceed to separate, returning to their unmixed state. However, the equations governing the reverse evolution of the gases are identical to the 'real' equations which dictated the natural (forward) evolution. That is, the fundamental equations of classical physics do not change as t becomes $-t$. Yet entropy (disorder) is seen to decrease as we march forward in 'backward time'. Have we violated the second law of thermodynamics? The consensus is that we have not. The second law tells us only about probabilities. It tells us that, statistically, disorder is extremely likely to increase. However, we are allowed to conceive of exceptional initial conditions which will lead to an increase in order, rather than disorder, and suddenly stopping all the molecules in a cloud of gas and reversing their velocities provides just such an exceptional initial condition.

Essentially the same logic applies to our computer simulation of the Euler equation. By suddenly stopping the computation and reversing the velocity of every fluid particle, we have created an exceptional initial condition: an initial condition which leads to the creation of order, rather than disorder, under the influence of turbulence. In practice, however, our thought experiment would be difficult to realize. When we stopped the computer and reversed time, rounding error would come into play. When we tried to march backward in time we would increasingly diverge from the true 'reverse trajectory'. It is probable that the initial conditions created by the imperfect reversal will be of the more usual, rather than exceptional, type and lead to yet more disorder as we tried to march backwards in time.

This idea that the *statistical behaviour* of a system can have an arrow of time, despite the reversibility of its governing equation, is important in turbulence closure modelling. Even though the Euler equation is time-reversible, *statistical* closure models of turbulence should possess an arrow of time, even in the absence of viscosity. Some of the early models, such as the so-called *quasi-normal scheme*, failed to take this into account and, as we shall see in Chapter 8, this led to problems.

Let us now return to our original dilemma: the contrast between the reversibility of the Euler equation and our intuition that turbulence should always leads to greater disorder. There is a second point to note. All real fluids have a finite viscosity and the Navier–Stokes equation (on which our intuition is based), unlike the Euler equation, is not reversible in time. Now it is true that the chaos in a turbulent flow is driven by the non-linear inertial forces, and that the viscous forces are very small. However, as we shall see, no matter how small we make the viscosity, the viscous stresses still play a crucial role, breaking and reconnecting the vortex lines at the small scales in a manner which cannot be realized in an ideal fluid.

In summary, then, the fact that the Euler equation is time reversible is not incompatible with the notion that turbulence leads to increased mixing and that the primary driving force for this is chaos resulting from the non-linearity of the equations. On the one hand, we have seen that time-reversibility of the inviscid equations is not inconsistent with an irreversible trend to greater disorder in the statistical properties of these equations. On the other hand, the small but finite viscosity possessed by all real fluids ensure that the real equations of motion are ultimately irreversible.

This more or less concludes our survey of mathematical chaos. We end, however, with a note of warning. Most studies of chaos theory have focused on model equations with only a few degrees of freedom. It is still unclear how these models relate to the onset of turbulence, which involves a near-infinite number of degrees of freedom, and indeed to turbulence itself. For example, chaos does not necessarily imply turbulence: a dripping tap can be made to drip chaotically. Enthusiasts of chaos theory have, in the past, suggested that the advent of chaos theory has resolved the 'problem of turbulence', whatever that is. This claim is strongly rejected by most hard core turbulence researchers. Chaos theory has taught us to be more comfortable with the idea that a simple non-linear equation (the Navier–Stokes equation) can lead to chaotic solutions, and it is highly suggestive as to potential transition routes to fluid chaos, but so far it has told us little about fully developed turbulence. The gap between chaos theory and turbulence is still large, and is likely to remain so for a considerable period of time.

3.2 Some elementary properties of freely evolving turbulence

We put our faith in the tendency for dynamical systems with a large number of degrees of freedom, and with coupling between those degrees of freedom, to approach a statistical state which is independent (partially, if not wholly) of the initial conditions. (G.K. Batchelor 1953)

Let us now turn to wind-tunnel turbulence, which is usually generated by passing a quiescent air stream through a grid or mesh of bars (Figure 3.3(a)). In some ways this represents turbulence in its simplest, purest form. That is to say, once the turbulence has been generated, there is virtually no interaction between the mean flow, which is more or less uniform, and the turbulence itself. The only function of the mean flow is to carry the turbulence through the tunnel. Indeed, the behaviour of grid turbulence is virtually identical to that observed in large tank of water which has been vigorously stirred and then left to itself. Such turbulence is called *freely evolving* turbulence. In a sense it is atypical since, in more complex flows, such as jets, wakes, and boundary layers, there is a continual interaction between the mean flow and the turbulence, with the former supplying energy to the

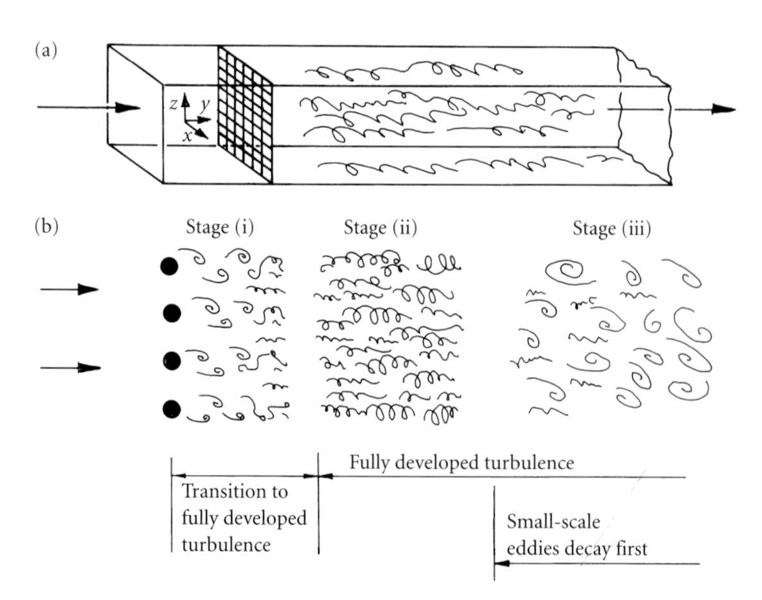

Figure 3.3 (a) The generation of grid turbulence. (b) Different stages of development for grid turbulence. The figure is schematic only and greatly exaggerates the rate of development of the turbulence in the axial direction.

latter. We shall tackle this more complex problem in the next chapter. However, here we restrict ourselves to freely evolving turbulence as it allows us to probe the structure of turbulence in its simplest environment.

A great deal of experimental data have been acquired over the years on the structure of grid turbulence. These data provide a critical test for any 'theory of turbulence'. If the theory is at odds with the data it must be rejected. We shall use grid turbulence as a vehicle for discussing a number of issues such as the extent to which turbulence is, or is not, deterministic. We shall also revisit Batchelor's assertion that the precise details of the initial conditions are unimportant as far as the ultimate state of the turbulence is concerned so that, perhaps, turbulence possesses some universal features. In particular we ask: 'how much does the turbulence remember?' and 'are some of its statistical features universal?'. The contents of this section are divided up as follows:

1. *Various stages of development*. Here we describe how fully developed (mature) turbulence evolves from an initial array of vortices. We emphasize that fully developed turbulence has a wide range of length scales, from the integral scale, l, to Kolmogorov's microscale, η. The energy is concentrated in the large eddies (at the integral scale) while enstrophy, and hence dissipation, is largely confined to the small eddies of size $\sim\eta$.
2. *The rate of destruction of energy in fully developed turbulence*. Here we revisit the idea of the energy cascade, first introduced in Chapter 1, Section 1.6. We discuss the rate of energy decay and touch on the paradoxical result that this decay is independent of the value of ν (provided Re is large). It is emphasized that viscosity plays only a passive role in the energy cascade, mopping up whatever energy is passed down to the small scales as a consequence of eddy break-up.
3. *How much does the turbulence remember?* It is often stated that turbulence has a short memory, rapidly forgetting where it has come from. In this section we explain why this need not be true. A cloud of turbulence can retain certain information throughout its evolution. For example, it might remember how much linear or angular momentum it had.
4. *The need for a statistical approach and different methods of taking averages*. In Chapter 1 we emphasized that, although $\mathbf{u}(\mathbf{x}, t)$ is chaotic and unpredictable, the statistical properties of $\mathbf{u}$ are perfectly reproducible in any experiment. This suggests that any theory of turbulence should be a statistical one and the first step in such a theory is to introduce a method of taking averages. There are several types of averages employed in turbulence theory and these are discussed in this section.

5. *Velocity correlations, structure functions, and the energy spectrum.* Here we introduce the various statistical parameters which are used to characterize the instantaneous state of a field of turbulence. In particular, we introduce the *velocity correlation function* which is the workhorse of turbulence theory. This tells us about the extent to which the velocity components at any one location are correlated to those at a second location. The Fourier transform of the velocity correlation function leads to something called the *energy spectrum* which is a particularly useful means of distinguishing between the energy held in eddies of different sizes.
6. *Is the asymptotic state universal? Kolmogorov's theory.* Here we discuss the question of whether or not mature turbulence possesses certain universal features. This leads to Kolmogorov's theory (probably the most celebrated theory in turbulence) which predicts how kinetic energy is distributed across the different eddy sizes.
7. *The probability distribution of the velocity field.* We end with a discussion of the probability distribution of $\mathbf{u}(\mathbf{x}, t)$. Here we examine the extent to which the probability distribution can be treated as Gaussian, an assumption which underpins certain heuristic models of turbulence. In fact, we shall see that turbulence is anything but Gaussian. This has profound ramifications for certain 'closure models' of turbulence.

There is a great deal of important information packed into these seven subsections, much of it empirical but some of it deductive. Although the discussion is framed in the context of grid turbulence, many of the concepts are quite general, and so it is well worth mastering these ideas before moving on to the more mathematical chapters which follow.

3.2.1 *Various stages of development*

Remember that we defined turbulence as a spatially complex distribution of vorticity which advects itself in a chaotic manner in accordance with the vortex evolution equation (2.31). The velocity field, which is the quantity we measure, plays the role of an auxiliary field, being determined at any instant by the vorticity distribution in accordance with the Biot–Savart law. So, if we are to create turbulence, we first need to generate vorticity. In the absence of body forces this can come only from a solid surface, and in a wind tunnel this is achieved most readily by using a grid.

Consider the flow shown in Figure 3.3(a). A grid is inserted into the path of a relatively quiescent stream with the intention of generating a field of turbulence downstream of the grid. The flow undergoes a sequence of transitions as indicated in Figure 3.3(b).

Table 3.1 Notation for this chapter

Total velocity field	$\mathbf{u} = \mathbf{V} + \mathbf{u}'$
Mean flow in wind tunnel	$\mathbf{V}$
Turbulent fluctuations	$\mathbf{u}'$
Size of large eddies (integral scale)	l
Typical fluctuating velocity of large eddies	u
Size of smallest eddies (Kolmogorov scale)	η
Typical fluctuating velocity of smallest eddies	v

Initially the motion consists of a discrete set of vortices which are shed from the bars of the grid (stage (i)). Although the vortices which peel off the bars may be initially laminar, they rapidly develop turbulent cores and then interact and mingle until some distance downstream we find a field of *fully developed turbulence* (stage (ii)). This is turbulence which contains the full range of scales from the integral scale (the size of the large eddies) down to the Kolmogorov microscale (the size of the smallest, most dissipative eddies).[4] This state is sometimes referred to as the *asymptotic state* of the turbulence.

We now have what is known as *freely evolving* or *freely decaying* turbulence. It so happens, for reasons which we shall discuss in Chapter 6, that the smallest eddies decay fastest in freely decaying turbulence. Thus, after a while, we find that the turbulence is dominated by large, slowly rotating eddies (stage (iii)). Eventually, the turbulence becomes so depleted that Re, based on the large eddy size, approaches unity. We then enter the so-called *final period of decay* (stage (iv)). Inertia is now relatively unimportant and we have, in effect, a spatially complex laminar flow (this last phase is not shown in Figure 3.3(b)).

Let us suppose that our wind tunnel is large so that we may obtain a reasonably high value of $\text{Re} = ul/\nu$, say 10^4. (Here l is a measure of the large eddies and u is a typical value of the turbulent velocity fluctuations.) We now insert a velocity probe into the tunnel and arrange for it to move with the mean flow, $\mathbf{V}$, following a 'cloud' of turbulence as it passes through the test section. The probe records the velocity at a point in the turbulent cloud from which we calculate $\frac{1}{2}(\mathbf{u}')^2(t)$. (Here, $\mathbf{u}' = \mathbf{u} - \mathbf{V}$, $\mathbf{u}$ being the absolute velocity of the fluid, and $\mathbf{u}'$ the turbulent fluctuations. See Table 3.1.) We now repeat the experiment many times under nominally identical conditions. Each time $\mathbf{u}'(t)$ is different because there are always infinitesimal differences in the way the experiment is carried out and it is in the nature of turbulence to amplify these differences. (This is reminiscent of the

[4] The idea of the Kolmogorov scale and the integral scale is introduced in Chapter 1, Section 1.6. You may wish to consult this before reading this section.

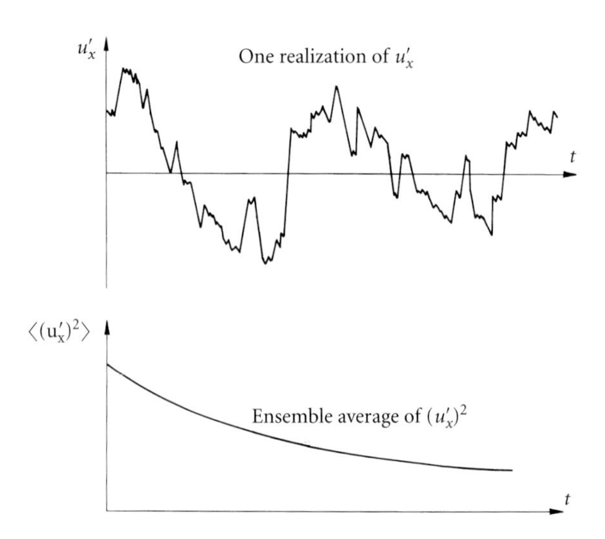

Figure 3.4 $u'_x(t)$, and hence $(u'_x)^2(t)$, is a highly irregular function of time in any one realization. However, the ensemble average is a smooth function of time.

behaviour of the logistic equation.) To overcome this problem we average $\frac{1}{2}(\mathbf{u}')^2(t)$ over the many realizations of the experiment to give a so-called *ensemble average*, denoted by $\frac{1}{2}\langle(\mathbf{u}')^2\rangle(t)$. This tells how rapidly, on average, the fluid loses its kinetic energy as a result of viscous dissipation as it passes through the test section.

It turns out that, while $\mathbf{u}'$ is apparently random in any one realization, the ensemble average, $\langle(\mathbf{u}')^2\rangle$, is a smooth function of time as illustrated in Figure 3.4. That is, although is $\mathbf{u}'$ chaotic, the average rate of loss of energy is a smooth, repeatable function of time. This is yet another manifestation of the fact that, although a turbulent velocity field appears to be quite random, and is different from one realization to the next, the statistical properties of a turbulent flow are perfectly reproducible.

Note that there appears to be many 'frequencies' contributing to u'_x in Figure 3.4. This is because the velocity at any one point is the result of a multitude of eddies (lumps of vorticity) in the vicinity of that point, each contributing to $\mathbf{u}$ via the Biot–Savart law. These vortices have a wide range of spatial scales and turn-over times, and so contribute different characteristic frequencies to $u'_x(t)$.

We now go further and divide up $\frac{1}{2}\langle(\mathbf{u}')^2\rangle$ according to the instantaneous size of the eddies surrounding the probe. That is, we calculate the relative contribution to $\frac{1}{2}\langle(\mathbf{u}')^2\rangle$ which comes from each of the various eddy sizes at each instant. The manner in which this can be done (via Fourier analysis) is discussed briefly in Section 3.2.5 of this chapter and in more detail in Chapters 6 and 8, but for the moment the details are unimportant. You need only know that it *can* be done.

When we plot kinetic energy against eddy size, r, we find something like that shown in Figure 3.5. Actually it is conventional to plot

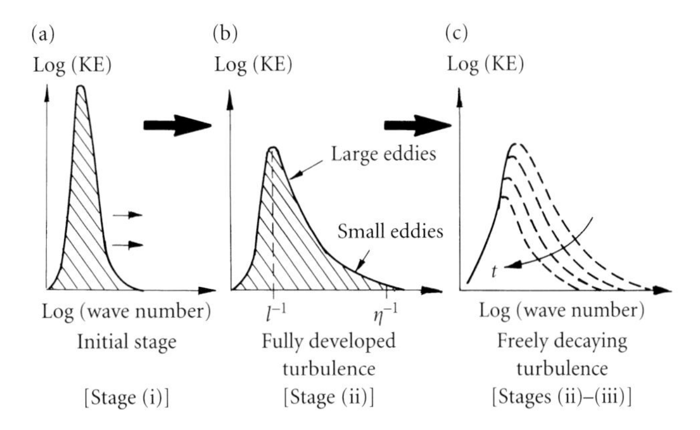

Figure 3.5 The variation of energy with eddy size at different times in the decay of grid turbulence.

energy versus wave number, $k \sim \pi / r$ rather than r itself, and, just for the moment, we shall conform to this convention. Initially much of the energy is centred around a wavelength of ~(bar size), corresponding more or less to the size of the coherent turbulent eddies shed from the bars of the grid. The non-linear term in the Navier–Stokes equation then starts to redistribute this energy over a broader range of eddy sizes. Eventually we reach stage (ii): fully developed turbulence. We now have energy distributed over a wide range of vortical structures (eddies) from the integral scale l (the size of the large eddies) down to the Kolmogorov microscale η (the size of the smallest, dissipative eddies). However, the bulk of the energy is contained in large (so-called *energy-containing*) eddies. If u is a typical fluctuating velocity of the large eddies, then $\langle (\mathbf{u}')^2 \rangle \sim u^2$.

Once the full range of length scales have developed, from l down to η, we enter a process referred to as the decay of *freely evolving turbulence* (Figure 3.5(c)). The total energy of the turbulence now starts to decline due to viscous dissipation, with the smallest eddies decaying fastest.

Now in Chapter 1, Section 1.6, we made two claims about the dissipation of energy in fully developed turbulence. First, we asserted that virtually all of the energy is dissipated in the small eddies. Second, we suggested that the small scales have a size $\eta \sim (ul/\nu)^{-3/4}\, l$ and a typical velocity $\mathrm{v} \sim (ul/\nu)^{-1/4}\, u$. One way to confirm the first of these claims is to plot the manner in which the square of the vorticity, $\langle \boldsymbol{\omega}^2 \rangle$, is distributed across the different eddy sizes (remember that $\nu \langle \boldsymbol{\omega}^2 \rangle$ is a measure of dissipation—equation (2.20)). If we do this, then we get something like Figure 3.6. The enstrophy, $\langle \frac{1}{2} \boldsymbol{\omega}^2 \rangle$, is indeed concentrated near eddies of size η, and these are very small indeed.

To fix thoughts, suppose that $u = 5\,\mathrm{m/s}$, $l = 2\,\mathrm{cm}$, and $\nu = 10^{-5}\,\mathrm{m^2/s}$. Then the small, dissipative eddies have a size $\eta \sim 0.02\,\mathrm{mm}$, a velocity $\mathrm{v} \sim 0.5\,\mathrm{m/s}$ and a characteristic timescale (called the turn-over-time) of $\eta/\mathrm{v} \sim 0.04 \times 10^{-3}\,\mathrm{s}$. Compare this with the turn-over-time of the

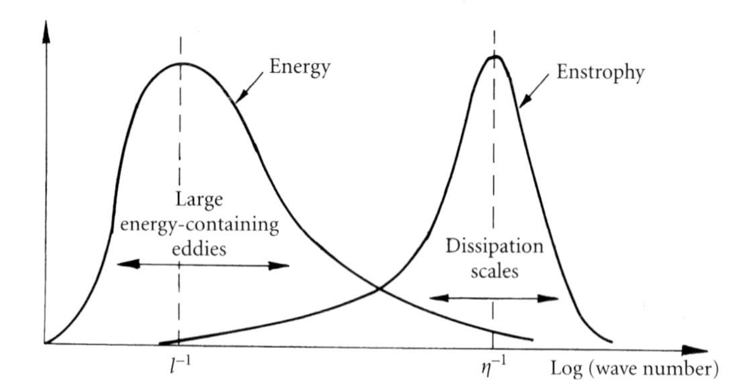

Figure 3.6 The distribution of energy and enstrophy in fully developed turbulence.

larger eddies: $l/u \sim 4 \times 10^{-3}$ s. Evidently, the small eddies have a very small turn-over-time, and since $|\boldsymbol{\omega}|$ is a measure of the rate of rotation of an eddy, this explains why the vorticity is concentrated in the small scales.

So, loosely speaking, we may say that the bulk of the energy, and the bulk of the enstrophy, are held in two mutually exclusive groups of eddies: the energy is held in the large eddies and the enstrophy in the small eddies. Actually, a more accurate statement would be to say that the vorticity which underpins the large eddies (via the Biot–Savart law) is weak and dispersed, and makes little contribution to the net enstrophy. Conversely, the small eddies are composed of intense patches of vorticity, and so they dominate the enstrophy. However, they make little contribution to the net kinetic energy because they are small and randomly orientated. This general picture is true for all high-Re turbulent flows. For example, in a typical pipe flow most of the turbulent energy is held in eddies of size comparable with the pipe radius, yet the majority of the enstrophy is held in eddies which are only a fraction of a percent of the pipe diameter.

3.2.2 *The rate of destruction of energy in fully developed turbulence*

We might note that the initial process of adjustment of the flow, from stage (i) to stage (ii), is almost completely non-dissipative. Energy is merely redistributed from one eddy size to another as the eddies evolve (break-up?). The net energy loss is small because the Kolmogorov scale is excited only once stage (ii) is reached.

In this section we are interested in the next phase: the decay of fully developed turbulence. This is shown schematically in Figure 3.5(c). There is now significant dissipation of energy and it turns out that the smallest eddies decay fastest. This is because their turn-over-time, which also turns out to be their break-up time, is much smaller than that of the large eddies, $\eta/\text{v} \ll l/u$.

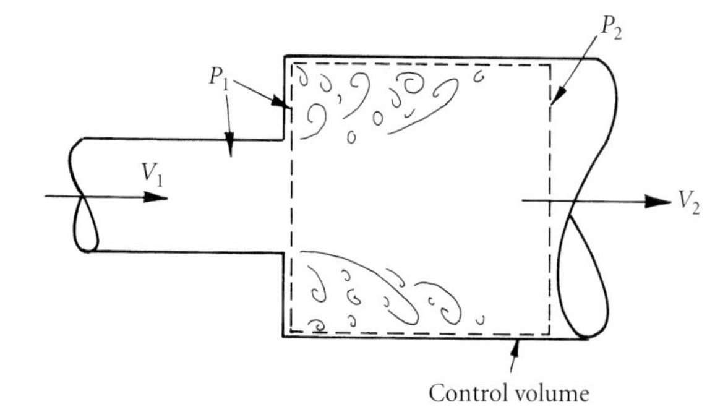

Figure 3.7 Energy loss due to a sudden expansion in a pipe.

Now we have already noted that $\langle(\mathbf{u}')^2\rangle \sim u^2$ and so du^2/dt is a measure of the rate of destruction of kinetic energy. It is an empirical observation that

$$\frac{du^2}{dt} = -\frac{Au^3}{l}, \quad A \sim 1. \tag{3.6}$$

Here A is more or less constant and has a value of the order of unity. (The exact value of A depends on the precise details of how we chose to define u and l.) Of course, we met this result in Chapter 1. It may be interpreted as

$$\frac{du^2}{dt} \sim -\frac{u^2}{l/u} \sim -\frac{(\text{energy of large eddies})}{(\text{turn-over-time of large eddies})}. \tag{3.7}$$

Evidently, the time scale of the decay of energy is l/u, which is the characteristic time of the energy-containing eddies. So the turbulence loses most of its energy in a few turn-over-times. This is because l/u is the timescale for the evolution of the large, energy-containing eddies and the process of decay is, in effect, the process of the destruction of the large eddies. Note, however, that the decay time is very long by comparison with the characteristic evolution time of the small eddies, η/v.

At first sight it seems paradoxical that the rate of dissipation of energy is independent of ν.[5] After all, it is the viscous stresses which cause the dissipation. However, hydraulic engineers have been aware of this kind of behaviour for over a century. Consider, for example, a sudden expansion in a pipe as shown in Figure 3.7. A great deal of

[5] Actually, it has been suggested that the rate of destruction of energy in fully developed turbulence might have a weak, logarithmic dependence on Reynolds number (Re) which is hard to detect experimentally (e.g. see, Hunt et al. 2001, and references therein). There is no fundamental reason why this should not be the case, although we might note that recent direct numerical simulations by Kaneda et al. (2003) suggest that, provided $\text{Re} > 10^4$, the dissipation is indeed independent of Reynolds number. Consequently, in this book we shall ignore the possibility of a logarithmic dependence of dissipation on Re.

turbulence is generated in the shoulder of the expansion and this leads to an abrupt loss in mechanical energy. The momentum equation, coupled to a few judicious assumptions about the nature of the flow, allows us to predict the loss of mechanical energy (per unit mass). It is

$$(\text{loss of energy}) = \frac{1}{2}(V_1 - V_2)^2$$

where V_1 and V_2 are the mean velocities in the pipe at sections 1 and 2. This is the famous 'Borda head loss' equation. (Try deriving this for yourself—see Exercise 3.6.) Now the important thing about the expression above is that, just like equation (3.6), the rate of loss of energy seems to be independent of ν.

Some instructors like to tease undergraduates about this. How can the dissipation be independent of ν? It is always reassuring to see how quickly those who have already met the boundary layer see the point. Recall that the key feature of a laminar boundary layer is that, no matter how small ν might be, the viscous shear stresses are always comparable with the inertial forces. This has to be so since it is the viscous stresses which pull the velocity down to zero at the surface, allowing the no-slip condition to be met. Thus, if we make ν smaller and smaller, nature simply retaliates by making the boundary layer thinner and thinner, and it does it in such a way that $\nu\partial u_x/\partial y$ remains an order one quantity: $\nu(\partial u_x/\partial y) \sim (\mathbf{u}\cdot\nabla)\mathbf{u}$. Thus, in a laminar boundary layer, $\text{Re} = u\delta/\nu$ is always of the order of unity (Figure 3.8(a)). The same sort of thing is happening in the pipe expansion and in grid turbulence. As we let ν become progressively smaller we might expect that the dissipation becomes less, but it does not. Instead we find that finer and finer structures appear in the fluid, and these fine-scale structures have a thickness which is just sufficient to ensure that the rate of viscous dissipation per unit mass remains finite and equal to $\sim u^3/l$. In short, the vorticity distribution becomes increasingly singular (and intermittent) as $\text{Re} \to \infty$, consisting of extremely thin sheets and tubes of intense vorticity.

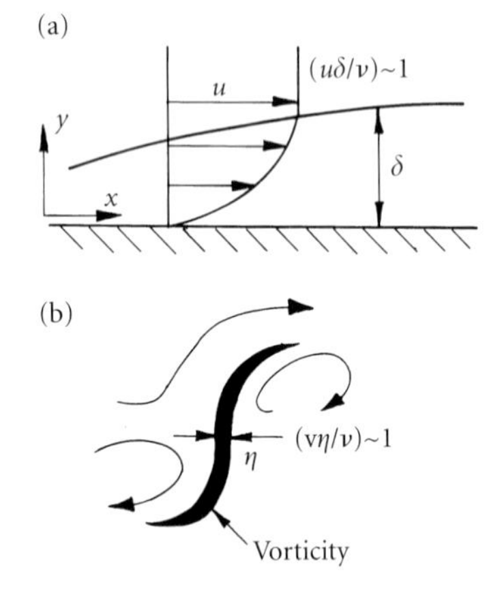

Figure 3.8 (a) A boundary layer and (b) an analogous small-scale structure in a turbulent flow. In both cases Re is of order unity.

We might also interpret (3.6) in terms of Richardson's energy cascade, which was introduced in Chapter 1, Section 1.6. Recall that most of the energy is held in large-scale eddies, while the dissipation is confined to the very small eddies (of size η). The question is how the energy transfers from the large scales to the small. Richardson suggested that this is a multistage process, involving a hierarchy of eddies from l down to η. The idea is that large eddies break up into smaller ones, which in turn produce even smaller structures, and so on. (Figure 3.9(a)). This process is driven by inertia since the viscous forces are ineffective except at the smallest scales. So the rate, Π, at which energy is passed down the cascade is controlled by the

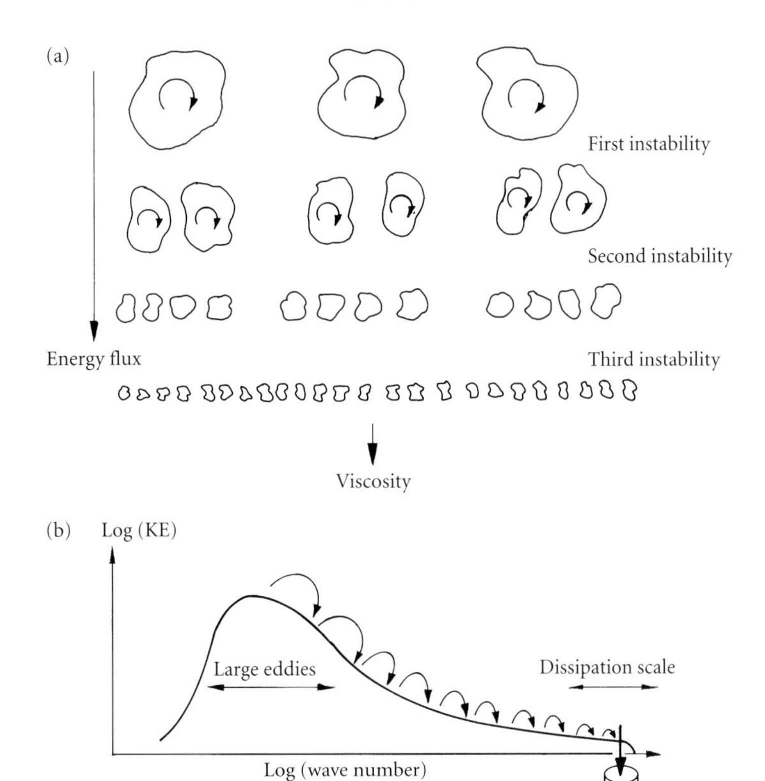

Figure 3.9 (a) A schematic representation of the energy cascade (after Frisch). (b) The energy cascade in terms of energy versus wave number.

'break-up' of the large eddies:[6]

$$\Pi \sim \frac{u^2}{l/u} \sim \frac{u^3}{l}.$$

Since the rate of destruction of energy at the small scales, ε, must be equal to Π in statistically steady turbulence, and is still approximately equal to Π in decaying turbulence, we have

$$\varepsilon \approx \Pi \sim u^3/l. \tag{3.8}$$

In summary then, there are two things that we know about the small scales. First, they have a characteristic size, η, and velocity, v, such that $(\mathrm{v}\eta/\nu) \sim 1$. (Smaller eddies would be rubbed out by viscosity while larger ones would not feel the viscous stresses.) Second, the dissipation at the small scales is

$$\varepsilon = 2\nu S_{ij} S_{ij} \sim \nu \mathrm{v}^2/\eta^2$$

[6] The term 'break-up' is being used rather loosely here. We describe the mechanism by which energy is passed to smaller scales in Chapter 5. For the present purposes we might think of an eddy as a blob or filament of vorticity, and imagine that the chaotic velocity field associated with all of the other eddies tends to tease out the vorticity (eddy) into finer and finer filaments. This process represents a transfer of energy from large to small scales.

and this is of the order of u^3/l. If we combine the expressions $(\mathrm{v}\eta/\nu) \sim 1$ and $(\nu \mathrm{v}^2/\eta^2) \sim u^3/l$ we arrive at the Kolmogorov estimates of Chapter 1,

$$\eta \sim \mathrm{Re}^{-3/4}\, l \sim (\nu^3/\varepsilon)^{1/4} \tag{3.9}$$

$$\mathrm{v} \sim \mathrm{Re}^{-1/4}\, u \sim (\nu\varepsilon)^{1/4}. \tag{3.10}$$

Note that the analogy to a boundary layer raises the question as to the shape of these small-scale structures. The term 'eddy' conjures up a picture of a spherically shaped structure, but perhaps they are really vortex sheets or tubes (Figure 3.8(b)). We shall return to this issue in Chapter 5. Note also that viscosity plays a rather passive role in the so-called *energy cascade*. Large structures in a turbulent flow 'break-up' into smaller ones which in turn pass their energy onto even smaller ones and so on. This entire process is driven by inertia and viscosity plays a role only when the eddy size reaches the dissipation scale η. In short, viscosity provides a dustbin for energy at the end of the cascade but does not (cannot) influence the cascade itself (Figure 3.9(b)).

3.2.3 *How much does the turbulence remember?*

Let us now return to our wind-tunnel experiment. Suppose that we repeat the experiment many times using different types of grid. We find that, provided the velocity of the large scales, u, and the integral scale of the turbulence, l, remain the same, then the statistical properties of the fully developed turbulence do not change by much. It seems that, as claimed by Batchelor, the precise details of the initial conditions are unimportant as far as the asymptotic (mature) turbulence is concerned. Most of the information associated with the initial conditions is lost in the process of creating the turbulence.

This is not the complete picture, however. It turns out that there *are* certain things which the turbulence remembers. That is, certain information contained in the initial conditions is retained throughout the evolution of the flow, despite all of the complex non-linear interactions. This robust information is associated with the dynamical invariants of the flow. That is, information is retained by the turbulence as a direct result of the laws of conservation of linear and angular momentum.

Consider the portion of a grid shown in Figure 3.10. The fluid upstream possesses no angular momentum (no vorticity) while that immediately downstream clearly does. The angular momentum (vorticity) has been generated in the boundary layers on the various bars in the grid and this is then swept downstream in the form of

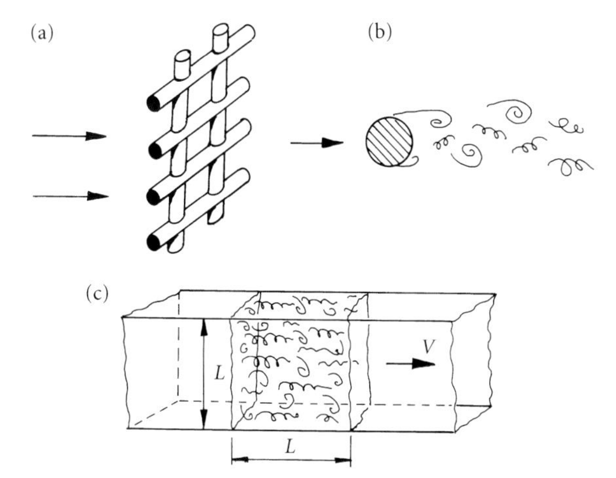

Figure 3.10 (a) Part of a grid. (b) Vortices shed from one of the bars in a grid introduce angular momentum into the flow. (c) A cloud of turbulence passes down the tunnel. In general the cloud possesses a finite amount of angular momentum.

turbulent vortices. In fact, we might suspect that the grid injects angular momentum into the flow because, if loosely suspended, it will shake and judder under the influence of the hydrodynamic forces exerted on it by the flow. By implication, the grid must exert torques and forces on the fluid and this results in the angular momentum of the turbulence.

Now suppose that we latch onto a particular cloud of turbulence as it passes down the tunnel. That is, we change to a frame of reference moving with the mean flow and observe the properties of a fixed volume of fluid (a control volume) in our new frame. This volume might, for example, be a cubic control volume of size L, as illustrated in Figure 3.10(c). From the discussion above we might expect our moving cloud (control volume) to have non-zero angular momentum, $\mathbf{H} = \int \mathbf{x} \times \mathbf{u}' dV$, $\mathbf{H}$ being the sum of the angular momentum of all of the vortices within the cloud. Moreover, this angular momentum will be conserved by the cloud except to the extent that the lateral boundaries exert forces on the turbulence, or there is a flux of angular momentum out of the open faces of the moving control volume. Both of these are surface effects, however, and in the limit that $L \gg l$, L being the cloud size, we might anticipate that they make only a small change to the net angular momentum of the cloud.

It might be thought that, because of the random way in which turbulent vortices are orientated, the net angular momentum $\mathbf{H} = \int (\mathbf{x} \times \mathbf{u}') dV$ of a large cloud of turbulence will be virtually zero. That is, when $L \gg l$, there are a very large number of randomly orientated eddies and so the contribution to $\mathbf{H}$ from individual vortices should tend to cancel when added together. This is true in the sense that $\mathbf{H}/V$, the angular momentum per unit volume, tends to zero as $L/l \to \infty$. One manifestation of this is the fact that the time-averaged of the stream-wise torque, $T_{//}$, exerted by the grid on the flow is zero,

$$\overline{T}_{/\!/} = \lim_{\tau\to\infty} \frac{1}{\tau}\int_0^{\tau} T_{/\!/} dt = 0.$$

However, for a *finite* volume, V, there will always be an incomplete cancellation of angular momentum and it is possible to show that, for large V, the residual angular momentum is at least of order $V^{1/2}$, that is, $|\int \mathbf{x}\times\mathbf{u}' dV| \sim V^{1/2}$, or greater. Moreover, as we shall see, this residual angular momentum is sufficiently strong as to exert some influence on the evolution of the turbulence.

In order to understand why $|\mathbf{H}| \sim V^{1/2}$ (or greater) we need to borrow a theorem from probability theory, called the *central limit theorem*. We state this here in full since we shall find many subsequent occasions in which the theorem is needed.

The central limit theorem

This theorem says the following: Suppose that $X_1, X_2, \ldots, X_N$ are independent random variables which have the same probability distribution, whose density function is $f(x)$. (Remember that $f(x_0)dx$ gives the relative number of times that x acquires a value in the range $x_0 \to x_0 + dx$, with $\int_{-\infty}^{\infty} f(x)dx = 1$.) We suppose that the p.d.f. has zero mean and a variance of σ^2, that is,

$$\int_{-\infty}^{\infty} xf dx = 0, \qquad \sigma^2 = \int_{-\infty}^{\infty} x^2 f dx.$$

We now form the new random variable $Y_N = X_1 + X_2 + \cdots + X_N$. Then the central limit theorem says the following:

(1) the probability density function for Y_N has zero mean and variance $N\sigma^2$;
(2) the p.d.f. for Y_N is asymptotically normal (Gaussian) in the limit $N \to \infty$.

Let us now return to our cloud of turbulence and see if we can use the central limit theorem to show that $|\mathbf{H}| \sim V^{1/2}$. Consider the following thought experiment. Suppose that we construct an initial condition for a cloud of turbulence as follows. We take the velocity field to be composed of a large set of discrete vortices (N in total) randomly located in space and randomly orientated. The nth vortex possesses a certain amount of angular momentum, say $\mathbf{h}_n$, and let us suppose that the ith component of this, $(h_i)_n$, is a random variable chosen from the probability distribution $f(x)$. Then $(h_i)_n$ has zero mean and a variance of σ^2. Now consider the ith component of the total angular momentum of the cloud, $H_i = (h_i)_1 + (h_i)_2 + \cdots + (h_i)_N$.

From the central limit theorem H_i is a random variable of zero mean and a variance of $\sigma^2_{H_i} = N\sigma^2$.

We now do the same thing for a sequence of turbulent clouds, that is, we perform many realizations of our thought experiment. Each time we arrange for the density of vortices in these clouds to be the same, so that volume of the cloud is proportional to N, $V \propto N$. Now consider the square of the angular momentum in a typical cloud, $\mathbf{H}^2$. Unlike H_i, H_i^2 has a non-zero mean. In fact, by definition the ensemble average $\langle H_i^2 \rangle$ is equal to the variance of H_i. So $\mathbf{H}^2$ has a mean (or expected) value of

$$\langle \mathbf{H}^2 \rangle = 3N\sigma^2 \propto V.$$

In this simple thought experiment, therefore, we expect the mean $\mathbf{H}^2$ to be non-zero with an expected value proportional to the volume of the cloud. This, in turn, tells us that the expected value of $|\mathbf{H}|$ is of order $V^{1/2}$, as anticipated above. Now, as the cloud subsequently evolves, $\mathbf{H}^2$ can change only as a result of the forces exerted on the fluid at the boundary of the cloud, or else as a result of the flux of angular momentum in or out of the cloud. However, if the domain is very large (relative to the integral scale of the turbulence) then we might anticipate that these surface effects will be small. If this is so (and it is not obvious that it is!), then $\mathbf{H}^2$ is a dynamical invariant of the fluid (for each realization of the turbulence) and so $\langle \mathbf{H}^2 \rangle / V$ is a statistical invariant of turbulence generated in this way.

Let us now return to our wind tunnel. If (and it is a big 'if') we imagine that the initial conditions for the turbulence (somewhat downstream of the grid) consist of a random set of vortices which have been shed from the grid, then we have the same situation as in our thought experiment. If we track a large volume of the turbulence as it moves downstream we might anticipate that $\langle \mathbf{H}^2 \rangle / V$ is an invariant of that volume. So there *is* the possibility that the turbulence remembers some of the information embedded in its initial conditions, a phenomenon which used to be called the *permanence of the big eddies*.

Invariants of the form $\langle \mathbf{H}^2 \rangle / V$ are called *Loitsyansky invariants* and there is a similar invariant, $\langle \mathbf{L}^2 \rangle / V$, which is based on the linear momentum of the turbulence, $\mathbf{L} = \int \mathbf{u}' dV$. This is usually called the *Saffman invariant*, though a historically more accurate name would be the *Saffman–Birkhoff invariant*. (Here $\mathbf{u}'$ is the velocity in a frame of reference moving with the mean flow.) It should be said, however, that there is considerable controversy over the existence or otherwise of these invariants. $\langle \mathbf{H}^2 \rangle / V$ may be finite or divergent (as $V \to \infty$) depending on how we envisage the turbulence to be created. Moreover, even if $\langle \mathbf{L}^2 \rangle / V$ or $\langle \mathbf{H}^2 \rangle / V$ are finite, there is the possibility that they vary due to surface effects over the volume V. In fact, the entire subject of integral invariants is a can of worms—but it is an important can of worms.

We take up the story again in Chapter 6 where we shall see that there are two schools of thought. Some argue that $\langle \mathbf{L}^2 \rangle / V$ is zero in grid turbulence, because of the way in which the turbulence is generated, but that $\langle \mathbf{H}^2 \rangle / V$ is non-zero and very nearly, but not exactly, a constant. (It is not strictly conserved due to surface effects.) Others argue that $\langle \mathbf{L}^2 \rangle / V$ is non-zero and strictly conserved, and that in such cases $\langle \mathbf{H}^2 \rangle / V$ diverges as $V \to \infty$. The near invariance of $\langle \mathbf{H}^2 \rangle / V$, or exact invariance of $\langle \mathbf{L}^2 \rangle / V$, turns out to impose a powerful constraint on the evolution of grid turbulence. In particular, we shall see that our two possibilities lead to $\langle \mathbf{H}^2 \rangle / V \sim u^2 l^5 \approx \textit{constant}$, or else $\langle \mathbf{L}^2 \rangle / V \sim u^2 l^3 = \textit{constant}$. Opinion is sharply divided over which of these two options is correct. Computer simulations of turbulence show that, depending on the initial conditions, both may be realized, but the wind-tunnel experiments themselves are not sufficiently clear-cut to distinguish unambiguously between the two possibilities. We shall return to this difficult issue in Chapter 6. In the meantime, the important point to note is that there is a possibility that grid turbulence *does* remember aspects of its initiation and this memory is closely related to the principle of conservation of momentum.

In the rest of this chapter we shall leave aside the possibility of Loitsyansky or Saffman-type invariants and take the position, espoused by G.K. Batchelor, that the turbulence has a short memory and that fully developed turbulence approaches a statistical state which is independent of the precise form of the initial conditions. This is a flawed view. However, it does provide a convenient starting point.

Example 3.1 Kolmogorov's decay law
Show that the energy equation (3.6), combined with the constraint imposed by the conservation of angular momentum, $u^2 l^5 = \text{constant}$, yields $u^2 \sim t^{-10/7}$. (This is known as Kolmogorov's decay law and is observed in some computer simulations of turbulence—see Chapter 6.)

3.2.4 The need for a statistical approach and different methods of taking averages

We have already emphasized the need for a statistical approach to turbulence. Despite the fact that the Navier–Stokes equation is perfectly deterministic, the turbulent velocity field, $\mathbf{u}(t)$, appears to be quite random. Consider, for example, measurements of one of the transverse components of velocity, $u_\perp(t)$, made at a particular location in our wind tunnel. Suppose we measure $u_\perp(t)$ for, say 100 s, wait a

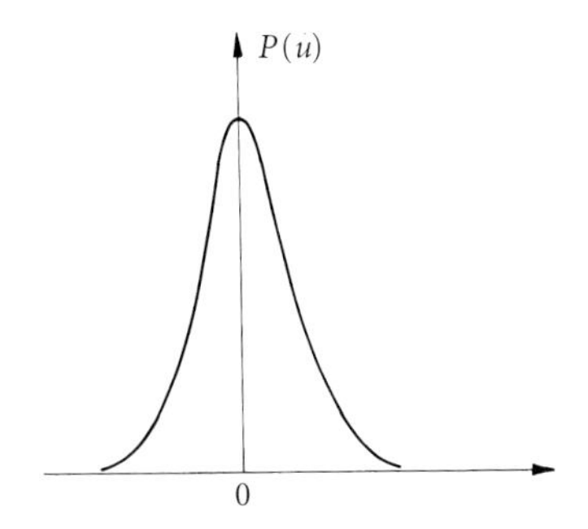

Figure 3.11 The p.d.f. for $u_\perp(t)$ in grid turbulence.

while, and then measure the velocity for a further 100 s. Although the trace of $u_\perp(t)$ will look quite different in the two readings, the *statistical properties* of the two signals will be very similar. For example, suppose we plot the p.d.f. of $u_\perp(t)$, that is, the relative number of times that $u_\perp$ attains a particular value. Provided the two traces of $u_\perp(t)$ represent large enough samples, the p.d.f. for the two readings will look virtually identical (Figure 3.11).

In Section 3.2.7 of this chapter we shall see that the probability distribution shown above is very nearly Gaussian. For the present purposes, though, this is unimportant. The essential point is that, although the detailed properties of $u_\perp(t)$ seem to be highly disorganized and unpredictable, its statistical properties are reproducible. The implication, of course, is that any theory of turbulence must be a statistical theory. The basic object of such a theory—the quantity which we wish to predict—is the *velocity correlation tensor*. This plays the same role in turbulence theory as velocity does in laminar flow. We shall discuss the velocity correlation tensor, and its various relatives, in the next section. First, however, we need to introduce a few more ideas from statistics. In particular, we need to formalize what is meant by an average and what we mean when we say that two quantities are statistically correlated.

Suppose we insert three small velocity measuring devices just downstream of one of the bars in our grid, at locations A, B, and C. The trace of $u_\perp(t)$ at A, B, and C might look something like that shown in Figure 3.12. What is happening at A looks a little like what is happening at B, though somewhat out of phase. This is because of the periodic vortex shedding. Even though the signals are quite different in detail, there is, on average, some similarity. In particular, if we multiply u_A and u_B and then time-average the product, we will get a non-zero value, $\overline{u_A u_B} \neq 0$. (As usual, an overbar represents a time-average.) Now consider point C, well downstream of the bar. Events at C are not strongly related to those at either A or B and so the trace of u_C looks quite different. If we form the product $u_A u_C$ or $u_B u_C$, and take a time-average, we find something close to zero, $\overline{u_A u_C} \sim 0$. We say that the points A and B are statistically correlated, while A and C are uncorrelated (or only weakly correlated). Note that statistical correlations are established via an averaging procedure—in this case a time-average.

There are other types of averaging procedures. Time-averages are convenient because they have a clear physical interpretation and are simple to visualize. However, they are meaningful only if the turbulence is statistically steady. There are two other types of averages in common use: the *ensemble average* and the *volume average*. Perhaps it is worth taking a moment to highlight the similarities and differences between these various forms of averaging.

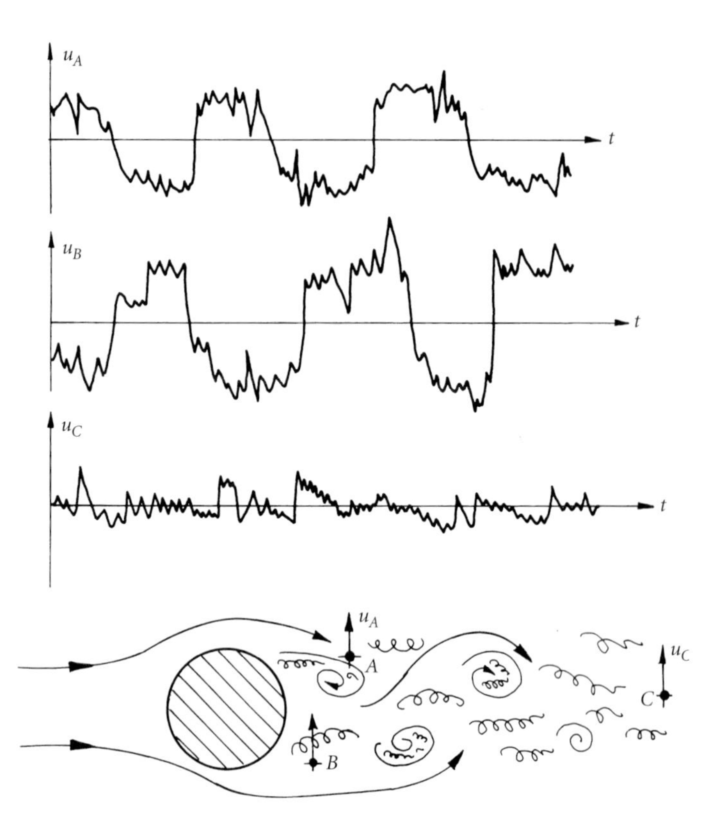

Figure 3.12 Traces of velocity against time. Points A and B are statically correlated while A and C are not.

Suppose we have a large tank of water which we vigorously stir (in some carefully prescribed manner). The water is then left to itself and we wish to study the statistics of the decay process. We might, for example, have a velocity probe which measures the velocity of the turbulence, $\mathbf{u}'$, at a particular location in the tank, A. A natural question to ask is, 'how does the average kinetic energy of the turbulence decay with time?'. The problem is, of course, that we do not have a clear definition of the word 'average' in this context. The turbulence is not statistically steady and so we cannot interpret 'average' to mean a time-average. One way out of this mess is to revisit the idea of ensemble averaging, which we introduced in Section 3.2.1 of this chapter.

Suppose we repeat the experiment 10,000 times. Each realization of the experiment is, as far as we are concerned, identical. However, as we have seen, minute variations in the initial conditions will produce quite different traces of $\mathbf{u}'(t)$ in each experiment. We now tabulate $(\mathbf{u}')^2$ against time for each realization and then, for each value of t, we find the average value of $(\mathbf{u}')^2$. This average, denoted $\langle(\mathbf{u}')^2\rangle$, is called an ensemble average. It is function of time but, unlike $(\mathbf{u}')^2(t)$, it is a smooth function of time (Figure 3.13). It is also

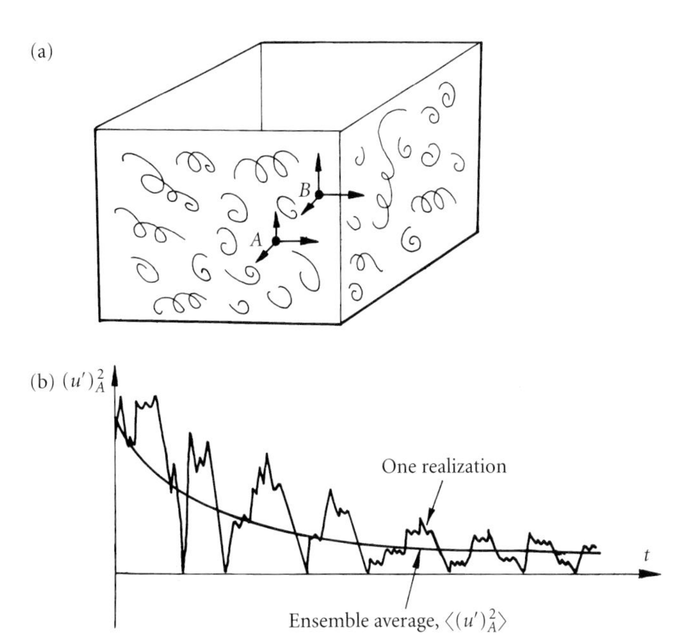

Figure 3.13 Decaying turbulence in a tank.

reproducible. If we come back the next day and do another 10,000 experiments, create a new ensemble average, $\langle (\mathbf{u}')^2 \rangle$, and compare it with the first one, we will find that they are identical. (This assumes, of course, that 10,000 realizations are sufficient to achieve statistical convergence.)

We can also use ensemble averaging to detect statistical correlations between different points in the flow. Suppose that in our 10,000 realizations we measure u'_x at both A and B. We wish to know if these points are statistically correlated. (We suspect that if the points are close they will be correlated, because an eddy can simultaneously span both points, ensuring that events at the two locations are related. If the points are remote, on the other hand, then the correlation should be weak.) One way to measure the correlation is to form the product $(u'_x)_A (u'_x)_B$ for each realization. Since $(u'_x)_A$ and $(u'_x)_B$ are both random signals we will find that $(u'_x)_A (u'_x)_B$ is also a random function of time. However if, for each instant t, we form an ensemble average over all of the experiments, $\langle (u'_x)_A (u'_x)_B \rangle$, we will obtain a reproducible quantity which, like $\langle (\mathbf{u}')^2 \rangle$, is a smooth function of time. If what is happening at A is very similar to what is happening at B, then $\langle (u'_x)_A (u'_x)_B \rangle$ will be non-zero. If A and B are very distant then $\langle (u'_x)_A (u'_x)_B \rangle \sim 0$.

A third form of averaging is the volume (or spatial) average. Suppose that, once again, we wish to determine how some average measure of $(\mathbf{u}')^2$ decays with time in our water tank. However, we are short of time and wish to perform only one experiment. We could, in principle, measure $(\mathbf{u}')^2$ at many locations in the tank (at

10,000 points). At any one instant we now average over all of the different measurements of $(\mathbf{u}')^2$ in the tank. The result, $(\mathbf{u}')^2_{AVE}$, will be a smooth, reproducible function of time, rather like $\langle(\mathbf{u}')^2\rangle$.

So we have three methods of averaging: the time-average, $\overline{(\sim)}$, the ensemble average, $\langle(\sim)\rangle$, and the volume average. It may be shown that, provided $\mathbf{u}'(\mathbf{x}, t)$ satisfies some rather mild conditions, which are usually met in practice, then:

(1) time-averages equal ensemble averages in steady-on-average flows, $\overline{(\sim)} = \langle(\sim)\rangle$;
(2) volume averages equal ensemble averages in homogeneous turbulence (turbulence in which the statistical properties do not depend on position).

3.2.5 Velocity correlations, structure functions, and the energy spectrum

The discussion so far has been more than a little qualitative. At some point we have to start developing quantitative arguments and this, in turn, means that we must introduce useful measures of the state of a turbulent flow. Consequently, in this section we introduce some statistical quantities which help quantify the state of a cloud of turbulence. There are three interrelated quantities commonly used for this purpose:

- The velocity correlation function.
- The second-order structure function.
- The energy spectrum.

The workhorse of turbulence theory is the velocity correlation function,

$$Q_{ij} = \langle u_i'(\mathbf{x})u_j'(\mathbf{x} + \mathbf{r})\rangle \tag{3.11}$$

In general Q_{ij} is a function of $\mathbf{x}$, $\mathbf{r}$, and t. If Q_{ij} does not depend on time we say the turbulence is statistically steady, and if Q_{ij} does not depend on $\mathbf{x}$ we say that the turbulence is statistically homogeneous, or just homogeneous. Note that Q_{ij} is defined in terms of the turbulent component of $\mathbf{u}$, that is, $\mathbf{u}'$. For example, in our wind tunnel we have to subtract out the mean velocity in the tunnel, $\mathbf{V}$, before we calculate Q_{ij}. Alternatively, we could measure $\mathbf{u}$ in a frame of reference moving with velocity $\mathbf{V}$.

So what does Q_{ij} represent? Consider $Q_{xx}(r\hat{\mathbf{e}}_x) = \langle u_x'(\mathbf{x})u_x'(\mathbf{x} + r\hat{\mathbf{e}}_x)\rangle$. This tells us whether or not u_x' at one point, A, is correlated to u_x' at an adjacent point, B (Figure 3.14). If the velocity fluctuations at A and B are statistically independent then $Q_{xx} = 0$. This might be the case if r is very much greater than the typical eddy size. On the other hand, as $r \to 0$ we have $Q_{xx} \to \langle(u_x')^2\rangle$. In general, Q_{ij} tells us about the degree to which,

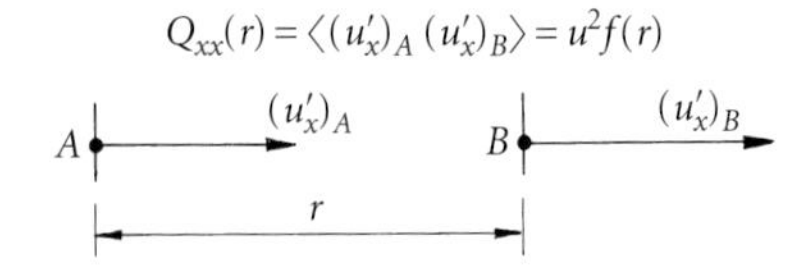

Figure 3.14 Definition of Q_{xx} and $f(r)$.

and the manner in which, the velocity components at different points are correlated to each other.

Now sometimes it turns out that a cloud of turbulence has statistical properties which are independent of direction. That is, all ensemble averages have reflectional symmetry and are invariant under rotations of the frame of reference.[7] This is referred to as isotropic turbulence. Turbulence in a wind tunnel is approximately homogeneous and isotropic.[8] The idealization of homogeneous, isotropic turbulence is such a useful concept that we shall devote the rest of this section to it. We start by introducing some notation. Let us define u through the equation

$$u^2 = \langle u_x^2 \rangle = \langle u_y^2 \rangle = \langle u_z^2 \rangle \tag{3.12}$$

This is consistent with our earlier definition of u as being a typical velocity of the large eddies. (Note that we have dropped the primes on $\mathbf{u}$ in (3.12). We shall do this in the remainder of this section on the assumption that, through a judicious choice of the frame of reference, the mean flow is zero.) Two typical components of Q_{ij} are

$$Q_{xx}(r\hat{\mathbf{e}}_x) = u^2 f(r) \tag{3.13}$$

$$Q_{yy}(r\hat{\mathbf{e}}_x) = u^2 g(r). \tag{3.14}$$

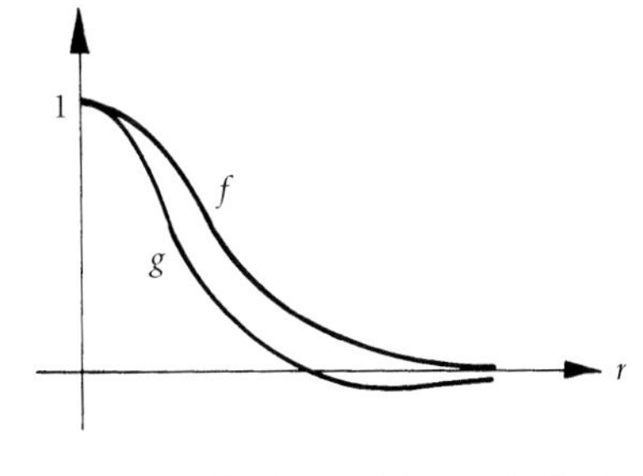

Figure 3.15 The shape of the longitudinal and lateral velocity correlation functions.

The functions f and g are called the longitudinal and lateral velocity correlation functions (or coefficients). They are dimensionless, satisfy $f(0) = g(0) = 1$, and have the shape shown in Figure 3.15. It turns out that f and g are not independent but are related by the continuity equation (2.3). In fact, we shall see in Chapter 6 that $2rg = (r^2 f)'$.

The integral scale, l, of the turbulence is often defined as

$$l = \int_0^\infty f(r)dr. \tag{3.15}$$

This provides a convenient measure of the extent of the region over which velocities are appreciably correlated, that is, the size of the large eddies.

We now come to an important point. It is evident from Figure 3.12 that there is a wide range of frequencies embedded in a typical turbulent signal. This reflects the fact that turbulence is, in effect, a hierarchy of tangled vortex tubes and ribbons of varying sizes,

[7] A less restrictive form of isotropy is where the turbulence has rotational symmetry but not reflectional symmetry. Such turbulence can possess helicity. However, we shall stay with the definition above.

[8] It cannot be perfectly homogeneous or isotropic because it decays along the length of the tunnel. Moreover, the grid itself introduces anisotropy, and if no particular effort is made to avoid this then typically $u_{/\!/}$ is some 10% greater than the perpendicular fluctuations. However, we shall gloss over these imperfections here.

advecting each other in a chaotic fashion. In mature turbulence the vortex blobs (eddies) vary in size from the minute to the large, and these all have different characteristic timescales, that is, frequencies. So the rapid fluctuations evident in Figure 3.12 correspond to small eddies in the vicinity of the velocity probe, while the low-frequency components are a reflection of larger, slowly rotating structures.

It is natural to try and build up a picture of what this population of eddies looks like. For example, we might ask: what is the size distribution of the eddies? Unfortunately, if we extract only mean quantities, such as $\langle (\mathbf{u}')^2 \rangle$, from a velocity signal then all of the information relating to the eddy size distribution is lost. So how can we reconstruct a picture of the eddy population, such as its size distribution, from a turbulent signal of the type shown in Figure 3.12? It turns out that this is a non-trivial task, made all the more difficult by the fact that we have carefully avoided giving a precise definition of an eddy. (Is it a vortex tube, a vortex ribbon, or a localized blob of vorticity?) Nevertheless, there are ways, all-be-it imperfect, of extracting some information about the distribution of eddy sizes from velocity measurements.

The first thing to note is that Q_{ij} does not, in of itself, tell us how the kinetic energy is distributed across the different eddy sizes. Rather, we must introduce two additional quantities which, in their own ways, attempt to fulfil this requirement. These are the energy spectrum and the structure function. Both are closely related to Q_{ij}.

We start with the *second-order longitudinal structure function*. It is defined in terms of the longitudinal *velocity increment*, $\Delta \mathrm{v} = u_x(\mathbf{x} + r\hat{\mathbf{e}}_x) - u_x(\mathbf{x})$, as follows:

$$\left\langle [\Delta \mathrm{v}]^2 \right\rangle = \left\langle [u_x(\mathbf{x} + r\hat{\mathbf{e}}_x) - u_x(\mathbf{x})]^2 \right\rangle. \tag{3.16}$$

More generally, the structure function of order p is $\langle [\Delta \mathrm{v}(r)]^p \rangle$. It seems plausible that only eddies of size $\sim r$, or less, can make a significant contribution to $\Delta \mathrm{v}$, and so $\langle [\Delta \mathrm{v}]^2 \rangle$ is often taken as an indication of the energy per unit mass contained in the eddies of size r or less. Of course $\langle [\Delta \mathrm{v}(r)]^2 \rangle$ and $f(r)$ are related by

$$\left\langle [\Delta \mathrm{v}(r)]^2 \right\rangle = 2u^2(1 - f) \tag{3.17}$$

and so for large r,

$$\left\langle [\Delta \mathrm{v}]^2 \right\rangle \to \frac{4}{3} \left\langle \frac{1}{2} \mathbf{u}^2 \right\rangle.$$

We might anticipate, therefore, that

$$\left\langle [\Delta \mathrm{v}(r)]^2 \right\rangle \sim \frac{4}{3} \left[\text{all energy in eddies of size } r \text{ or less} \right].$$

Actually this turns out to be a little too simplistic. Eddies larger than r do make a contribution to Δv and this contribution is of the order of $r \times$ (velocity gradient of eddy), or $r \times |\boldsymbol{\omega}|$. Thus a better interpretation of $\langle[\Delta v]^2\rangle$ would be,

$$\left\langle[\Delta v(r)]^2\right\rangle \sim \frac{4}{3}\,[\text{all energy in eddies of size } r \text{ or less}] + r^2\,[\text{all enstrophy in eddies of size } r \text{ or greater}]. \quad (3.18)$$

An alternative convention, however, is to work with wave number rather than eddy size, and to use the Fourier transform to identify structures of different sizes. In particular, a function, called the *energy spectrum*, is introduced via the transform pair:[9]

$$E(k) = \frac{2}{\pi}\int_0^\infty R(r)kr\sin(kr)dr \quad (3.19a)$$

$$R(r) = \int_0^\infty E(k)\frac{\sin(kr)}{kr}dk \quad (3.19b)$$

where $R(r) = \frac{1}{2}\langle\mathbf{u}(\mathbf{x})\cdot\mathbf{u}(\mathbf{x}+\mathbf{r})\rangle = u^2(g+f/2)$. It is possible to show (see Chapter 6) that $E(k)$ has three properties:

(1) $E(k) \geq 0$;
(2) for a random array of simple eddies of fixed size r, $E(k)$ peaks around $k \sim \pi/r$;
(3) from (3.19b), in the limit $r \rightarrow 0$, we have

$$\frac{1}{2}\langle\mathbf{u}^2\rangle = \int_0^\infty E(k)dk. \quad (3.20a)$$

In view of these properties it is customary to *interpret* $E(k)dk$ as the contribution to $\frac{1}{2}\langle\mathbf{u}^2\rangle$ from all eddies with wave numbers in the range $k \rightarrow k+dk$, where $k \sim \pi/r$. This provides a convenient measure of how energy is distributed across the various eddy sizes. It should be emphasized, however, that this is a flawed view of $E(k)$. Consider property (2) above. It is true that eddies of size r contribute primarily to $E(k)$ in the range $k \sim \pi/r$. However, they also contribute to $E(k)$ for all other values of k. In fact, we shall see in Chapter 6 that a random array of simple eddies of fixed size r has an energy spectrum which grows as k^4 for small k, peaks around π/r, and then declines exponentially (Figure 3.16). The key point is that eddies of given size contribute to $E(k)$ across the full range of wave numbers.

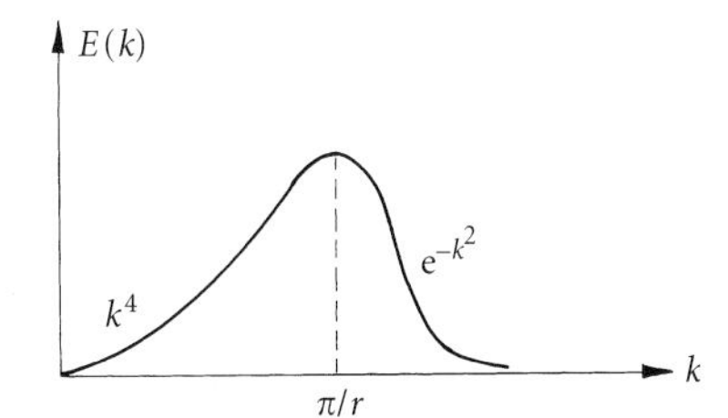

Figure 3.16 The shape of the energy spectrum for a random array of simple eddies of fixed size r.

[9] This definition of $E(k)$ is adequate for isotropic turbulence. However, if the turbulence lacks isotropy a more general definition is required (see Chapter 8).

Nevertheless, the conventional interpretation of $E(k)$, that it represents the energy in eddies of size π/k, is convenient and works well for many purposes. Consequently we shall adopt this as a kind of shorthand.[10] (It goes badly wrong, however, if we consider or $k < l^{-1}$ or $k > \eta^{-1}$.)

In Chapter 8 we shall see that $E(k)$ has one further property

$$\frac{1}{2}\langle \boldsymbol{\omega}^2 \rangle = \int_0^\infty k^2 E dk \tag{3.20b}$$

and so $k^2E(k)dk$ is interpreted as the contribution to the enstrophy, $\frac{1}{2}\langle \boldsymbol{\omega}^2 \rangle$, from the range of wave numbers $k \to k + dk$.

It seems that both $\langle [\Delta \mathrm{v}(r)]^2 \rangle$ and $E(k)$ make some claim to distinguish between scales, the energy spectrum through (3.20a, b) and $\langle [\Delta \mathrm{v}(r)]^2 \rangle$ via (3.18). It is natural, therefore, to try and find the relationship between these two functions. In principle this is straightforward. Equation (3.19b) relates $R(r)$ to an integral of $E(k)$, and $R(r)$ can, in turn, be related to f and hence to $\langle [\Delta \mathrm{v}]^2 \rangle$. The details are spelt out in the example at the end of this section where we find that,

$$\langle [\Delta \mathrm{v}]^2 \rangle = \frac{4}{3}\int_0^\infty E(k)H(kr)dk, \quad H(x) = 1 + 3x^{-2}\cos x - 3x^{-3}\sin x.$$

This does not look too promising, until we spot that a good approximation to $H(x)$ is,

$$H(x) \approx \begin{cases} (x/\pi)^2, & \text{for } x < \pi \\ 1, & \text{for } x > \pi. \end{cases}$$

It follows that

$$\langle [\Delta \mathrm{v}(r)]^2 \rangle \approx \frac{4}{3}\int_{\pi/r}^\infty E(k)dk + \frac{4r^2}{3\pi^2}\int_0^{\pi/r} k^2 E(k)dk \tag{3.21a}$$

In view of our approximate interpretation of $E(k)$ we might rewrite this as,

$$\langle [\Delta \mathrm{v}(r)]^2 \rangle \approx \frac{4}{3}\left[\text{all energy in eddies of size } r \text{ or less}\right] + \left(\frac{4}{3\pi^2}\right) r^2 \left[\text{enstrophy in eddies of size } r \text{ or greater}\right] \tag{3.21b}$$

[10] An alternative, but equivalent, definition of $E(k)$ is given in Chapter 8. Suppose that $\hat{\mathbf{u}}(\mathbf{k})$ represents the three-dimensional Fourier transform of $\mathbf{u}(\mathbf{x})$. If the turbulence is isotropic it may be shown that $E(k)\delta(\mathbf{k}-\mathbf{k}') = 2\pi k^2 \langle \hat{\mathbf{u}}^*(\mathbf{k}) \cdot \hat{\mathbf{u}}(\mathbf{k}') \rangle$ where $\mathbf{k}$ and $\mathbf{k}'$ are distinct wave numbers, * represents a complex conjugate, $k = |\mathbf{k}|$, and δ is the three-dimensional Dirac delta function. So $E(k)$ is a measure of the energy contained in the $\mathbf{k}$th mode of $\hat{\mathbf{u}}(\mathbf{k})$. Now when we take the Fourier transform of a random signal the rapid fluctuations tend to be associated with the large wave numbers and the slower fluctuations with the small wave numbers. Thus $E(k)$ is associated with small eddies if k is large, and large eddies if k is small.

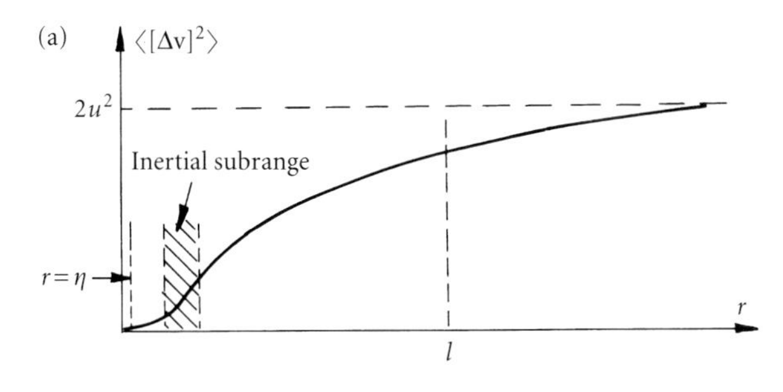

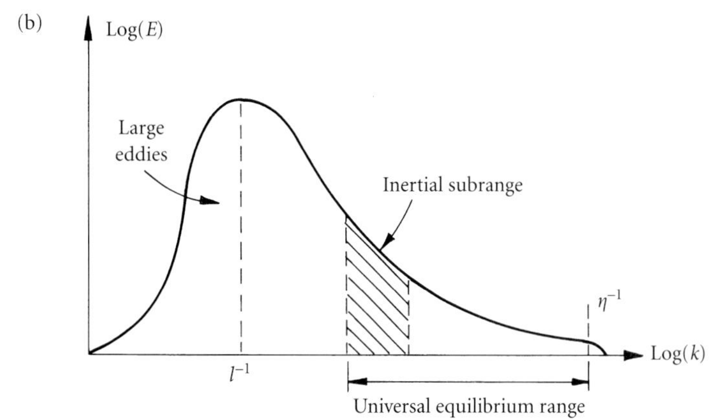

Figure 3.17 (a) The general shape of $\langle[\Delta \mathrm{v}]^2\rangle$. (b) The general shape of $E(k)$.

This has a reassuring similarity to our heuristic estimate (3.18). Of course (3.21b) cannot be exactly true because it is quite artificial to categorize the influence of eddies according to whether they are smaller or greater than r. The cut off is too sharp. Nevertheless, (3.21a) captures the main features of the relationship between $E(k)$ and $\langle[\Delta \mathrm{v}(r)]^2\rangle$. The general shapes of $\langle[\Delta \mathrm{v}]^2\rangle$ and $E(k)$ are shown in Figures 3.17 (a) and (b).

Earlier we saw that vorticity tends to be concentrated in the smallest eddies and is rather weak in the large scales (consult Figure 3.6). Thus, provided we are not too close to the dissipation scales, we might simplify (3.21b) to

$$\left\langle[\Delta \mathrm{v}(r)]^2\right\rangle \sim \frac{4}{3}\left[\text{all energy in eddies of size } r \text{ or less}\right]$$

and indeed this approximation is commonly used. However, this is often dangerous since the second term in (3.21a) can make a significant contribution to $(\Delta \mathrm{v})^2$. For example, it is readily confirmed from (3.21a) that, for small r, we have,

$$\left\langle(\Delta \mathrm{v})^2\right\rangle \approx \frac{2}{3\pi^2}\langle\boldsymbol{\omega}^2\rangle r^2 + \cdots = \frac{1.013}{15}\langle\boldsymbol{\omega}^2\rangle r^2 + \cdots .$$

Actually it turns out that the exact relationship is (consult Chapter 6), $\left\langle(\Delta \mathrm{v})^2\right\rangle = \frac{1}{15}\langle\boldsymbol{\omega}^2\rangle r^2 + \cdots$, the 1.3% error having crept in because of the approximation to $H(kr)$ made in deriving (3.21a). It seems that, for small r, the second-order structure function is controlled by the enstrophy, rather than energy. This highlights the danger of interpreting $\langle[\Delta v]^2\rangle$ purely in terms of energy, particularly when we are close to the dissipation scales.

In summary, then, we have two functions, $E(k)$ and $\langle[\Delta v]^2\rangle$, which claim to give an impression (all-be-it a rather imperfect one) of the spread of energy across the different eddy sizes. It is now clear how graphs of the type shown in Figures (3.5), (3.6), and (3.9b) can be constructed. They are simply plots of $E(k)$. It must be constantly borne in mind, however, that this is a naïve interpretation of $E(k)$. Suppose, for example, that we have a velocity field which is composed of compact structures (eddies) of size l and less. When we take the transform we find that $E(k)$ is a continuous function and in particular it has a significant component in the range $k = 0 \rightarrow \pi/l$. However, there are no eddies in that range. In short, the shape of $E(k)$ for small k is controlled by eddies of size much smaller than k^{-1}, as indicated in Figure 3.16. It is a common mistake to talk of the energy of eddies which do not exist!

Finally, we note that the whole notion of dividing up the energy and attributing the various parts to different eddies is a little artificial. Suppose, for example, that we have three eddies (small, medium, and large) with individual velocity fields $\mathbf{u}_1$, $\mathbf{u}_2$, and $\mathbf{u}_3$. We now suppose that all three simultaneously exist in the same region of space, so the total energy is

$$KE = \frac{1}{2}\int \left[\mathbf{u}_1^2 + \mathbf{u}_2^2 + \mathbf{u}_3^2 + 2\left(\mathbf{u}_1 \cdot \mathbf{u}_2 + \mathbf{u}_2 \cdot \mathbf{u}_3 + \mathbf{u}_3 \cdot \mathbf{u}_1\right)\right] dV.$$

Do we associate the cross terms like $\mathbf{u}_1 \cdot \mathbf{u}_2$ with the small or medium-sized eddies?[11] With these words of caution we now go on to consider some of the more general properties of $\langle[\Delta v]^2\rangle$ found in grid turbulence.

Example 3.2 Use the relationship to $2rg = (r^2 f)'$ show that $2r^2 R = u^2 (r^3 f)'$. Now use (3.19b) to confirm that

$$\langle[\Delta v]^2\rangle = \frac{4}{3}\int_0^\infty E(k)H(kr)dk, \quad H(x) = 1 + 3x^{-2}\cos x - 3x^{-3}\sin x.$$

3.2.6 *Is the asymptotic state universal? Kolmogorov's theory*

We now consider the structure of freely decaying, fully developed turbulence of the type produced by a grid. We adopt the position that

[11] This problem is resolved to some extent if the eddies are of very different sizes since the cross terms are then relatively small. (See Chapter 5.)

the details of the initial conditions are unimportant and ignore, for the moment, the existence of integral invariants such as Saffman's invariant and Loitsyansky's invariant. Our main aim is to show that arguments of a rather general kind can be used to make specific predictions about how energy is distributed across the different eddy sizes. Given the complexity of turbulence, it is rather surprising that this is possible, yet it seems that it is!

If we draw up a list of all of the things which might influence the instantaneous shape of $\langle [\Delta \mathrm{v}]^2 \rangle$ we might come up with

$$\left\langle [\Delta \mathrm{v}]^2 \right\rangle = \hat{F}(u, r, t, \nu, \mathrm{BC}, l)$$

where BC stands for boundary conditions and l is the integral scale. However, in grid turbulence we may take the turbulence to be approximately homogeneous and isotropic and so the boundary conditions may be dropped from this list:

$$\left\langle [\Delta \mathrm{v}]^2 \right\rangle = \hat{F}(u, \nu, l, r, t). \tag{3.22}$$

Note that u and l are themselves functions of time and so t makes three appearances in (3.22). In dimensionless terms we have

$$\left\langle [\Delta \mathrm{v}]^2 \right\rangle = u^2 \hat{F}\left(\frac{ut}{l}, \frac{r}{l}, \mathrm{Re}\right). \tag{3.23}$$

where $\mathrm{Re} = ul/\nu$ is the usual Reynolds number. As it stands, this expression is so general that it is of little use, so we must apply some physical reasoning in order to simplify (3.23). We have suggested that the behaviour of the largest eddies is independent of ν since (when Re is large) the shear stresses are very weak at the large scales. It follows that, for large r,

$$\left\langle [\Delta \mathrm{v}]^2 \right\rangle = u^2 \hat{F}\left(\frac{r}{l}, \frac{ut}{l}\right) \tag{3.24a}$$

In fact, the measurements suggest that we can go further and that, to a reasonable level of approximation, we can drop the parameter ut/l. This yields a self-similar expression in which t appears only implicitly in terms of u and l:

$$\left\langle [\Delta \mathrm{v}]^2 \right\rangle = u^2 \hat{F}\left(\frac{r}{l}\right) \qquad \text{(large eddies only)} \tag{3.24b}$$

Now consider eddies which are substantially smaller than l. They have a complex heritage, having come from the break-up of larger eddies, which in turn came from yet larger structures, and so on. Kolmogorov suggested that these eddies are aware of the large scales only to the extent that they feed kinetic energy down the energy cascade at a rate

$\Pi \sim \varepsilon \sim u^3/l$. Thus, for these smaller eddies, we should replace (3.22) by,

$$\langle [\Delta v]^2 \rangle = F(\varepsilon, \nu, t, r), \qquad r \ll l.$$

Next we note that the decay time of the turbulence is extremely long by comparison with the turn-over time of the small eddies. As far as they are concerned the turbulence is virtually in a state of statistical equilibrium, that is, it is almost steady-on-average. Thus we may anticipate that t is not a relevant parameter (except to the extent that ε varies with t) and we might speculate that

$$\langle [\Delta v]^2 \rangle = F(\varepsilon, \nu, r), \qquad r \ll l. \tag{3.25}$$

This is considerably simpler than (3.22) and was first suggested by Kolmogorov in 1941. Equation (3.25) should not be passed over lightly: in many ways it is a remarkable statement. Consider Plate 8(b), which shows small-scale eddies in a computer simulation. According to (3.25) the dynamics of this complex tangle of vortex tubes is controlled exclusively by ε and ν. It took the genius of Kolmogorov to see that such a simplification was possible. If we now introduce the Kolmogorov microscales of length and velocity, (3.9) and (3.10), we can rewrite (3.25) in the dimensionless form

$$\langle [\Delta v]^2 \rangle = v^2 F\left(\frac{r}{\eta}\right), \qquad r \ll l. \tag{3.26}$$

Here F should be some universal function valid for all forms of isotropic turbulence. However, many forms of turbulence, such as jets and wakes, are strongly anisotropic. Would we expect (3.26) to hold in such flows? Kolmogorov suggested that it should since the large-scale anisotropy in, say, a jet is not strongly felt by the small scales. That is, anisotropy is normally imposed at the large scales and its consequences become progressively weaker as we move down the energy cascade. So Kolmogorov suggested that (3.26) should hold for all turbulence—jets, wakes, boundary layers, etc.—and indeed there is considerable evidence that this is so. For this reason that part of the spectrum for which (3.26) applies is referred to as the *universal equilibrium range*. This is indicated in Figure 3.17(b).

Now suppose that Re is very large. It is possible, then, that there exists a range of eddies which satisfy (3.26) (i.e. they are in statistical equilibrium and 'feel' the large scales only to the extent that they determine ε), yet they are still large enough for the shear stresses to have no influence on their motion. In such a case (3.26) must take a special form since ν is no longer a relevant parameter. The only possibility is

$$\langle [\Delta v]^2 \rangle = \beta \varepsilon^{2/3} r^{2/3}, \qquad \eta \ll r \ll l \tag{3.27}$$

Table 3.2 Approximate form of the structure function for grid turbulence at high Re in the absence of a Loitsyansky or Saffman invariant

Regime	Range of r	Form of $\langle[\Delta v]^2\rangle$
Energy-containing eddies	$r \sim l$	$\langle[\Delta v]^2\rangle = u^2 F(r/l)$
Inertial subrange	$\eta \ll r \ll l$	$\langle[\Delta v]^2\rangle = \beta\varepsilon^{2/3} r^{2/3}$
Universal equilibrium range	$r \ll l$	$\langle[\Delta v]^2\rangle = v^2 F(r/\eta)$

where β (Kolmogorov's constant) is a universal constant thought to have a value of $\sim$2. This intermediate range is called the *inertial subrange* and it exists only for $\eta \ll r \ll l$. Equation (3.27) is known as *Kolmogorov's two-thirds law* and it is one of the most celebrated results in turbulence theory. Its validity, or otherwise, rests entirely on the hypothesis that the intermediate eddies are controlled exclusively by ε.

Of course, this entire discussion has been more than a little heuristic. What, for example, do we mean by an 'eddy', by 'eddy break-up', or by 'feeling the large scales'? The most that we can do is tentatively suggest these relationships and then study the wind-tunnel data to see if they do, or do not, hold true. Actually it turns out that it is difficult to get Re high enough in a typical wind tunnel to achieve a substantial inertial subrange. Nevertheless, this has been done on several occasions and the results are intriguing. It turns out that the data in the universal equilibrium range can indeed be collapsed using (3.26), and even more surprisingly, (3.27) is an excellent fit to data in the inertial subrange! So despite the rather vague nature of the physical arguments, the end results appear to be valid. These relationships are summarized in Table 3.2 and the various regimes are indicated in Figure 3.17(a) and (b).

However, this is not the end of the story. We shall see later that these arguments are too simplistic in at least two respects. First, we have ignored the existence of Saffman- or Loitsyansky-type invariants and it turns out that when these exist they dominate the behaviour of the large eddies. (They do not, however, call into question Kolmogorov's arguments.) Second, it turns out that the dissipation of energy at the small scales is very patchy in space. This raises the question of whether or not the ε which appears in Kolmogorov's arguments should be a global average for the flow as a whole or some local average of ε over a scale less than l. It turns out that this necessitates a modification to Kolmogorov's description of the small scales. We shall return to these issues in Chapters 5 and 6 where we take a more critical look at Kolmogorov's theory.

3.2.7 *The probability distribution of the velocity field*

Statistics show that of those who contract the habit of eating, very few will survive. (W.W. Irwin)

We close this chapter with a brief discussion of the probability distribution of $\mathbf{u}(\mathbf{x}, t)$ and its derivatives. (Again, we omit the prime on $\mathbf{u}$ on the understanding that we are talking only of the turbulent component of velocity.) One of the main points we wish to emphasize is that the probability distribution is not Gaussian. This is important because several 'models' of turbulence assume that aspects of the turbulence *are* Gaussian. Also, we wish to show how measurements of the probability distribution of $\mathbf{u}$ can provide hints as to the spatial structure of freely evolving turbulence.

Let us start by recalling some elementary statistics. The probability distribution of some random variable, X, is usually represented by a p.d.f. which is defined as follows. The probability that X lies in the range $a \to b$, which we write as $P(a < X < b)$, is related to the probability density function by

$$P(a < X < b) = \int_a^b f(x)dx.$$

Thus $f(x)dx$ represents the relative likelihood (sometimes called relative frequency) that X lies in the range $x \to x + dx$. Evidently f has the property

$$\int_{-\infty}^{\infty} f(x)dx = 1$$

since the sum of the relative likelihoods must come to 1. The mean of a distribution (sometimes called the expectation of X) is given by

$$\mu = \int_{-\infty}^{\infty} xf(x)dx = \langle X \rangle$$

while the variance, σ^2, is defined as

$$\sigma^2 = \int_{-\infty}^{\infty} (x - \mu)^2 f(x)dx.$$

We are primarily concerned with distributions with zero mean, in which case

$$\sigma^2 = \int_{-\infty}^{\infty} x^2 f(x)dx = \langle X^2 \rangle, \qquad \langle X \rangle = 0.$$

Of course, σ, is the standard deviation of the distribution. The *skewness factor* for a distribution of zero mean is defined in terms of the third moment of f:

$$S = \int_{-\infty}^{\infty} x^3 f(x) dx / \sigma^3 = \langle X^3 \rangle / \langle X^2 \rangle^{3/2}$$

while the *flatness factor* (or *kurtosis*) is a normalized version of the fourth moment

$$\delta = \int_{-\infty}^{\infty} x^4 f(x) dx / \sigma^4 = \langle X^4 \rangle / \langle X^2 \rangle^2.$$

A very common distribution is the *Gaussian* or *normal distribution*, which has the form

$$f(x) = \frac{1}{\sigma\sqrt{2\pi}} \exp\left[-(x-\mu)^2 / (2\sigma^2) \right].$$

When f has zero mean this simplifies to

$$f(x) = \frac{1}{\sigma\sqrt{2\pi}} \exp\left[-x^2 / (2\sigma^2) \right].$$

This is symmetric about the origin and so has zero skewness. It has a flatness factor of $\delta = 3$. The Gaussian distribution is important since the central limit theorem (see Section 3.2.3 of this chapter) tells us that a random variable, which is itself the sum of many other independent random variables, is approximately Gaussian.

Example 3.3 Intermittency and the flatness factor
Confirm that a Gaussian distribution has a flatness factor of 3. Now consider a distribution $g(x)$ which comprises a δ-function, of area $1-\gamma$, at the origin surrounded by a Gaussian-like distribution of area γ. That is:

$$g(x) = \frac{\gamma}{\sigma_* \sqrt{2\pi}} \exp\left[-\frac{x^2}{(2\sigma_*^2)} \right], \qquad x \neq 0.$$

This represents a signal which spends $(1-\gamma)\%$ of its time dormant, and occasionally bursts into life (Figure 3.18). Show that the variance of $g(x)$ is $\sigma^2 = \gamma \sigma_\star{}^2$ and that the flatness factor is $3/\gamma$, that is, *larger* than a Gaussian. Thus we have

$$g(x) = \frac{\gamma^{3/2}}{\sigma\sqrt{2\pi}} \exp\left[-\frac{\gamma x^2}{(2\sigma^2)} \right], \qquad x \neq 0.$$

If we compare this with a Gaussian distribution of the same variance we see that, for large x, $g(x)$ is greater than the equivalent Gaussian distribution. This is typical of a so-called intermittent signal, that is, a signal which is quiescent for much of the time, and occasionally burst

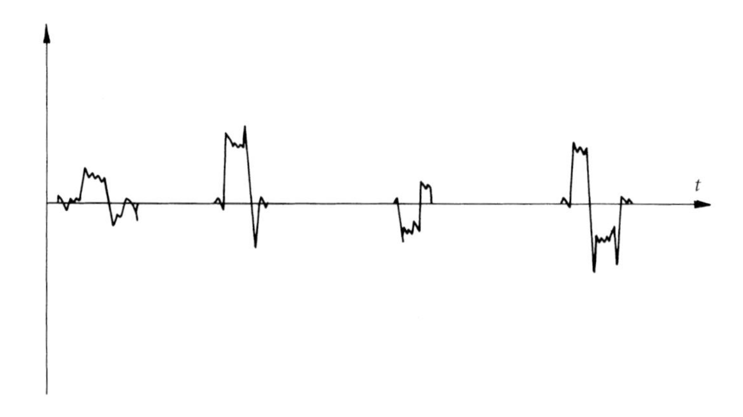

Figure 3.18 An example of an intermittent signal.

into life. Their p.d.fs have a high central peak and broad skirts, so that both near-zero and unexpectedly high values are more common than in a Gaussian signal. Intermittent signals can be recognized by their large flatness factor. □

With this brief introduction to statistics let us now return to turbulence. We start by looking at the probability distribution of **u** measured at a single point. (Later we shall look at the probability distribution of differences in **u** between two adjacent points.) Suppose we make many measurements of one component of **u**, say u_x, at a particular point in a wind tunnel. (Remember, we take **u** to represent the turbulent velocity so we are assuming the mean flow in the tunnel is subtracted out of the measurements.) We now plot the relative number of times that u_x attains a particular value, that is, we plot the density function $f(x)$ for u_x. The end result looks something like that shown in Figure 3.11.

It turns out that the probability density shown in Figure 3.11 can be fitted closely by a Gaussian p.d.f. It is symmetric and has a flatness factor, $\langle u_x^4 \rangle / \langle u_x^2 \rangle^2$ in the range 2.9–3.0. (Recall that the flatness factor for a normal distribution is 3.) One interpretation of this normal distribution for u_x is that the velocity at any one point is the consequence of a large number of randomly orientated vortical structures (i.e. blobs of vorticity) in its neighbourhood, the relationship between u_x and the surrounding vorticity being fixed by the Biot–Savart law. If these vortices are randomly distributed, and there are many of them, the central limit theorem (see Section 3.2.3 of this chapter) says that u_x should be Gaussian, which is exactly what is observed.

The story is very different, however, if we examine the probability distribution for gradients in **u**, or else the joint probability distribution for **u** at two points. Here we find that the p.d.fs are definitely not Gaussian and indeed this non-normal behaviour is essential to the dynamics of turbulence. Consider, for example, the fourth-order

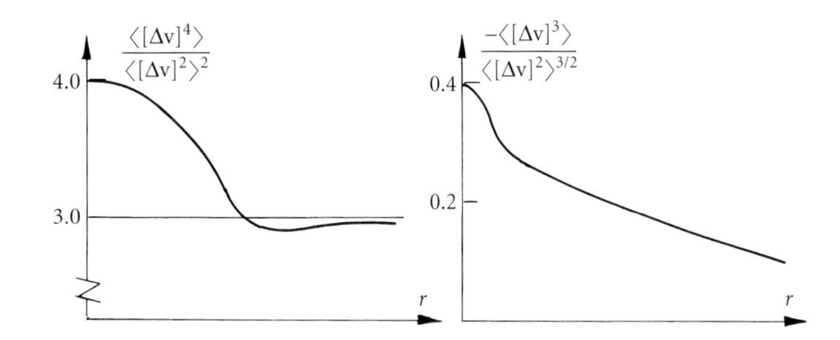

Figure 3.19 Flatness and skewness factors for $(u_x)_A - (u_x)_B$ in grid turbulence (schematic only).

structure function $\langle[\Delta v]^4\rangle$ normalized by $\langle[\Delta v]^2\rangle^2$. Of course, this is the flatness factor, δ, for $(u_x)_A - (u_x)_B$ where A and B are adjacent points, $\mathbf{x} + r\hat{\mathbf{e}}_x$ and $\mathbf{x}$. $\delta(r)$ has the shape shown in Figure 3.19. It approaches the Gaussian value of 3.0 only for large r, and is greater than 3 for small r. The fact that $\delta(r) = \langle[\Delta v]^4\rangle / \langle[\Delta v]^2\rangle^2$ approaches 3 as the separation, r, tends to infinity simply indicates that remote points are approximately statistically independent. That is, if $(u_x)_A$ and $(u_x)_B$ are statistically independent then it may be shown, by expanding $\langle[\Delta v]^4\rangle$, that[12]

$$\delta(r \to \infty) = \frac{\langle[\Delta v]^4\rangle}{\langle[\Delta v]^2\rangle^2} \to \frac{3}{2} + \frac{1}{2}\frac{\langle u_x^4\rangle}{\langle u_x^2\rangle^2}$$

and we have already noted that the flatness factor for u_x is ~ 3.0. For $r \to 0$ we find that δ is a function of the $\mathrm{Re} = ul/\nu$. For the modest values of Re found in a windtunnel, $\delta(0) \sim 4$. For higher values of Re, however, we find $\delta(0) \sim 4 \to 40$. The rule seems to be that the higher the value of Re, the larger the flatness factor.

The fact that δ is large, relative to a normal distribution with the same variance, tells us that the p.d.f. for Δv has a high central peak and relatively broad skirts, so that both near zero and unexpectedly high values of $|\Delta v|$ are common. This is consistent with a signal which is dormant much of the time and occasionally bursts into life. (See Example 3.3.)

The situation is similar for the third-order structure function. The skewness factor for u_x is $\langle u_x^3\rangle / \langle u_x^2\rangle^{3/2}$ and this is very close to zero since the probability density function for u_x is more or less symmetric. The skewness factor for $(u_x)_A - (u_x)_B$ is evidently $S(r) = \langle[\Delta v]^3\rangle / \langle[\Delta v]^2\rangle^{3/2}$. In contradistinction to $\langle u_x^3\rangle / \langle u_x^2\rangle^{3/2}$, this is not zero. It has a value of around -0.4 for $r \to 0$ and decays slowly with r. The precise shape of $\langle[\Delta v]^3\rangle / \langle[\Delta v]^2\rangle^{3/2}$ depends (slightly) on Re but a typical distribution is shown in Figure 3.19. Usually one finds $S(0) \sim -0.4 \pm 0.1$ for Re up to 10^6, with higher values of Re tending to favour slightly higher values of $|S|$.

[12] In performing this expansion it is necessary to recall that, if a and b are statistically independent, then $<ab> = <a><b>$.

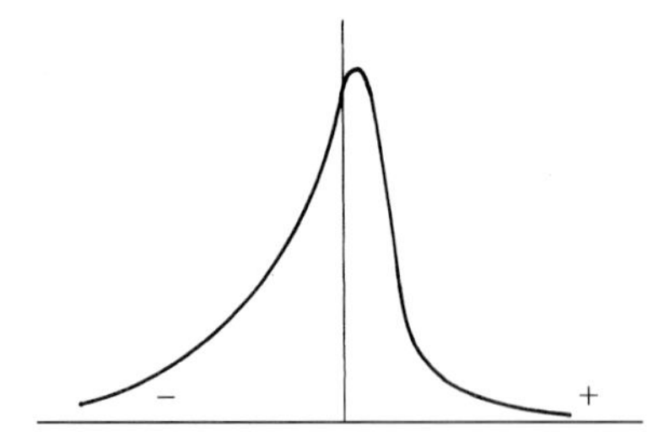

Figure 3.20 Schematic of the p.d.f. for $\partial u_x/\partial x$. (The departures from a Gaussian distribution are exaggerated.)

For small r the probability distributions for $(u_x)_A - (u_x)_B$ and $\partial u_x/\partial x$ become identical. For example, if $r = \delta x$ is small,

$$S(r \to 0) = S_0 = \frac{\langle [\Delta v]^3 \rangle}{\langle [\Delta v]^2 \rangle^{3/2}} = \frac{\langle (\partial u_x/\partial x)^3 \rangle}{\langle (\partial u_x/\partial x)^2 \rangle^{3/2}}. \tag{3.28}$$

It follows that $\partial u_x/\partial x$ is non-Gaussian, with a skewness factor of ~ -0.4 and a flatness factor of anything from 4 to 40. The p.d.f. for $\partial u_x/\partial x$ is shown schematically in Figure 3.20. Small positive values of $\partial u_x/\partial x$ are more likely than small negative ones. However, large negative values of $\partial u_x/\partial x$ are more likely than large positive ones, and this dominates the skewness, giving a negative value of S.

The fact that the flatness factor for $\partial u_x/\partial x$ is not Gaussian, growing to extremely high values as Re increases, tells us something important about the spatial structure of turbulence. Large values of δ correspond to a highly intermittent signal in which $\partial u_x/\partial x$ is nearly zero much of the time but periodically bursts into life. In short, the spatial distribution of the velocity gradients (i.e. the vorticity) is rather spotty, with the enstrophy being localized into small, intense patches of vorticity, the patches themselves being sparsely distributed throughout space. Moreover, it seems that the higher the value of Re, the more patchy the vorticity distribution becomes. This is the first hint that the vorticity in a turbulent flow is highly intermittent; an observation which turns out to have profound consequences for Kolmogorov's theories.

Perhaps we should not be surprised by the observed intermittency in vorticity. Think about the evolution of the vorticity field in grid turbulence. All of the vorticity originates from the boundary layers on the grid bars. This spills out into the flow in the form of turbulent Karman vortices. These vortices then interact, the velocity field induced by any one vortex advecting all the other vortices. The vortices start to intermingle and turbulence ensues. The resulting velocity field, which is chaotic, is simply a manifestation of the evolving vorticity field. That is, the instantaneous distribution of $\mathbf{u}$ can be attributed, via the Biot–Savart law, to the instantaneous distribution of $\boldsymbol{\omega}$. So, as the lumps of vorticity shed by the grid are swept downstream, they start to twist, turn, and stretch as a result of their self-induced velocity field. Moreover, this occurs at large Re, so that the vortex lines are more or less frozen into the fluid. A casual glance at a turbulent cloud of smoke, say that of a cigarette, is enough to convince most observers that a chaotic, turbulent velocity field continually teases out any 'frozen-in' marker, and this is also true of vorticity. As this mixing proceeds, sheets and tubes of vorticity are continually teased out into finer and finer structures, like a chef preparing filou pastry. This process continues until the tubes or sheets are thin enough for diffusion to set in, that is, we have reached the Kolmogorov microscales. So we might imagine that much of the vorticity in fully developed turbulence is teased out into extremely fine filaments and indeed this is consistent with the measured flatness factors for $(\Delta v)^2$. Moreover we would expect the measurements of intermittency, and hence of δ, to become more pronounced as Re is increased, since the cut-off scale, η, shrinks as Re grows. Again, this is consistent with the measurements.

Let us summarize our findings. The probability distribution of u_x at a single point is approximately normal, whereas the probability distribution of the difference in velocity between two neighbouring points is decidedly non-Gaussian, although departures from a normal distribution become less and less marked as r becomes large. This is hardly surprising. The value of u_x at a particular point is determined largely by chance, but the *difference* between u_x at two adjacent points is determined to a large extent by local dynamical considerations (the Navier–Stokes equations) and not by chance. For example, if the two points lie in the same eddy then the relationship between $(u_x)_A - (u_x)_B$ is determined by the dynamical behaviour of that eddy. Moreover, the fact that S is non-zero and negative is not a coincidence. We shall see in Chapter 5 that, in the inertial subrange,

$$S = \frac{\langle [\Delta v]^3 \rangle}{\langle [\Delta v]^2 \rangle^{3/2}} = -\frac{4}{5}\beta^{-3/2} \tag{3.29}$$

where β is Kolmogorov's constant. Given that $\beta \sim 2$, this predicts that $S \sim -0.3$, in line with the measurements. Thus Δv is necessarily *non-Gaussian*. We take up this story again in Chapter 5.

Exercises

3.1 Confirm that the two-cycle shown in Figure 3.1 is given by

$$x = \left[a + 1 \pm \{(a+1)(a-3)\}^{1/2}\right]/2a$$

3.2 Confirm that the fixed point $X = (a-1)/a$ of the logistic equation (3.1) is stable for $a < 3$ and unstable for $a > 3$. [Hint: look at what happens when $x_0 = X + \delta x$, $\delta x \ll X$.]

3.3 Confirm that the two-cycle shown in Figure 3.1 is stable for $3 < a < 1 + \sqrt{6}$. [Hint: consider the stability of the so-called second-generation map $x_{n+2} = F(F(x_n))$.]

3.4 Consider the system of differential equations,

$$\frac{dx}{dt} = -y + (a - x^2 - y^2)x$$

$$\frac{dy}{dt} = x + (a - x^2 - y^2)y.$$

It has a steady solution $\mathbf{x} = (x, y) = (0,0)$. Show that the steady solution is linearly stable if $a < 0$ and unstable if $a > 0$. Confirm that $r^2 = |\mathbf{x}|^2$ satisfies the Landau equation and find its solution. Now show that $\mathbf{x}$ undergoes a supercritical Hopf bifurcation at $a = 0$.

3.5 When we add the next term in the expansion of $|A|^2$ to the Landau equation we obtain

$$\frac{d|A|^2}{dt} = 2\sigma|A|^2 - \alpha|A|^4 - \beta|A|^6.$$

On the assumption that $\alpha < 0$ and $\beta > 0$ find the upper stability curve shown in Figure 3.2(b).

3.6 Use the integral form of the momentum equation (2.14) to show that the pressure rise in a sudden pipe expansion is

$$\Delta p = p_2 - p_1 = \rho V_2(V_1 - V_2)$$

where V_1 and V_2 are the mean velocities upstream and downstream of the expansion. [Hint: use the control volume shown in Figure 3.7.] You may ignore the shear stress acting on the boundary and take the time-averaged pressure on the shoulder of the expansion to be uniform and equal to the upstream pressure p_1. Now use the time-averaged energy equation (2.16) to show that, for the control volume in Figure 3.7,

$$\dot{m}\left[\left(p_2/\rho + V_2^2/2\right) - \left(p_1/\rho + V_1^2/2\right)\right] = -\rho \int \varepsilon dV$$

where $\dot{m}$ is the mass flow rate, $\rho A_1 V_1 = \rho A_2 V_2$, and ε is the dissipation per unit mass, $\varepsilon = 2\nu S_{ij}^2$. As before you may ignore the shear stresses acting on the boundary. Also, you may assume that the kinetic energy of the turbulence well upstream and downstream of the expansion is small by comparison with the kinetic energy of the mean flow, $V_2^2/2$ and $V_1^2/2$. Now confirm that the average dissipation per unit mass in the control volume is

$$\bar{\varepsilon} = \frac{1}{2}(V_1 - V_2)^2 V_2/L$$

where L is the length of the control volume. This is the rate at which energy is extracted from the mean flow and passed onto the turbulence. Note that it is independent of ν.

3.7 Use Kolmogorov-like arguments to find the form of $\langle(\Delta v)^3\rangle$ in the inertial subrange and hence show that the skewness factor S should be constant throughout the inertial subrange.

Suggested reading

Batchelor, G.K. (1953) *The Theory of Homogeneous Turbulence*. Cambridge University Press. (Despite its age this is still one of the definitive accounts of homogeneous turbulence. Kolmogorov's theory of a universal equilibrium range is carefully discussed in chapter 7, and the probability distribution of the velocity field is dealt with in chapter 9.)

Drazin, P.G. (1992) *Nonlinear Systems*. Cambridge University Press. (Bifurcation theory is introduced in chapter 1 and the logistic equation is dealt with at some length in chapter 3.)

Drazin, P.G. and Reid, W.H. (1981) *Hydrodynamic Stability*. Cambridge University Press. (The Landau equation is discussed in chapter 7.)

Frisch, U. (1995) *Turbulence*. Cambridge University Press. (The two-thirds law and the energy dissipation law are introduced in chapter 5.)

Hunt, J.C.R. et al. (2001) *J. Fluid Mech.*, **436**. (This review paper contains many useful references.)

Kaneda, Y. et al. (2003) *Phys. Fluids*, **15**(2).

Lesieur, M. (1990) *Turbulence in Fluids*. Kluwer Academic Publishers. (Consult chapter 3 for a discussion of transition to turbulence and chapter 6 for the phenomenological theory of Kolmogorov.)

Townsend, A.A. (1975) *The Structure of Turbulent Shear Flow*, (2nd edition). Cambridge University Press. (Chapter 1 gives a useful discussion of the physical significance of the energy spectrum and structure functions.)

Tritton, D.J. (1988) *Physical Fluid Dynamics*. Clarendon Press. (Transition to chaos is discussed in chapter 24.)

von Neumann, J. (1951) *J. Res. Nat. Bur. Stand.*, **12**.

CHAPTER 4

Turbulent shear flows and simple closure models

> Finally we should not altogether neglect the possibility that there is no such thing as 'turbulence'. That is to say, it is not meaningful to talk of the properties of a turbulent flow independently of the physical situation in which it arises. In searching for a theory of turbulence, perhaps we are looking for a chimera.... Perhaps there is no 'real turbulence problem', but a large number of turbulent flows and our problem is the self-imposed and possibly impossible task of fitting many phenomena into the Procrustean bed of a universal turbulence theory. Individual flows should then be treated on their merits and it should not necessarily be assumed that ideas valid for one flow situation will transfer to others. The turbulence problem may then be no more than one of cataloguing. The evidence is against such an extreme point of view as many universal features seem to exist, but nevertheless cataloguing and classifying may be a more useful approach than we care to admit.
>
> P.G. Saffman (1977)

We now turn to two subjects of great practical importance: (i) elementary 'models' of turbulence; and (ii) shear flows. By shear flows we mean flows in which the mean velocity is predominantly one-dimensional in nature, such as wakes, boundary layers, submerged jets, and pipe flows (Figure 4.1). Traditionally, many of the early studies of turbulence focused on such flows and there were some notable successes, such as the celebrated log-law of the wall. Moreover, many of the early 'models' of turbulence, which are still used today, were developed in the context of turbulent shear flows.[1] It is natural, therefore, to group these topics together.

Although there have been many successes in our attempt to understand shear flows, it would be misleading to suggest that there exists anything like a coherent theory. Rather, we have a hierarchy of

[1] Closure models used in the context of shear flows are called 'single-point closures' since they work with statistical quantities defined at a single point in space. Later, when we discuss homogeneous turbulence, we shall meet so-called 'two-point closures', which work with statistical quantities defined at two distinct points. Two-point closure models are generally too complex to be applied to inhomogeneous flows, although some small advances have been made in this direction.

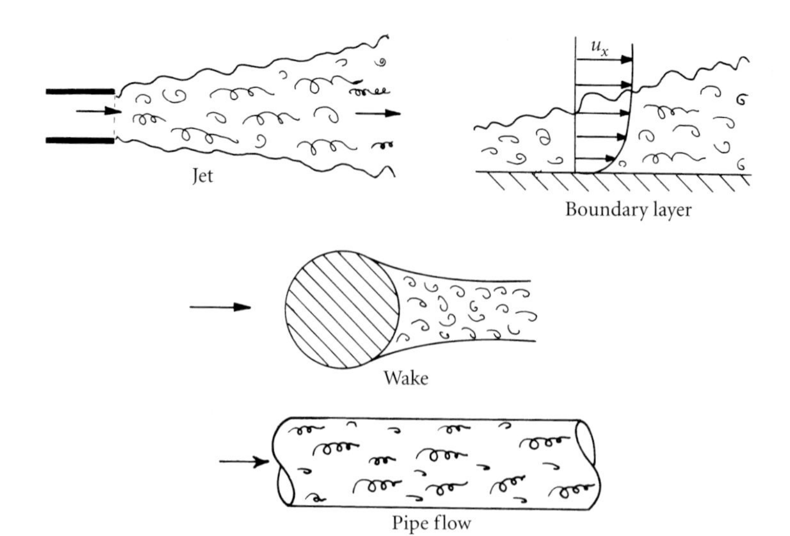

Figure 4.1 Different types of shear flows.

models of varying complexity, most of which are semi-empirical in nature. Perhaps, in view of Saffman's diagnosis, this should not come as a surprise. While certain unifying features exist, each class of flow seems to exhibit its own idiosyncratic behaviour. This is not good news for anyone who places their faith in the existence of universal turbulence closure models.

The flows shown above have a number of obvious common features. Perhaps the most striking of these is the fact that, in the external flows, there is a sharp interface between the turbulent and non-turbulent motion. Recall that we define turbulence as a spatially complex vorticity field which advects itself in a chaotic manner. In the external shear flows shown above this vorticity is generated on a solid surface, such as the inside surface of the nozzle, or the surface of the cylinder. It then spills out into the bulk flow where it is swept downstream. The sharp interface between the turbulent and non-turbulent fluid simply reflects the fact that, at high-Re, the vorticity is virtually frozen into the fluid, rather like a non-diffusive dye. The sort of questions we might pose are:

- how rapidly does the turbulence (vorticity) spread as it is swept downstream?
- how intense is the turbulence at different locations?
- what is the spatial distribution of the mean velocity?

By and large, we shall restrict ourselves to statistically steady flows, so that ensemble averages, $\langle(\sim)\rangle$, are equivalent to time-averages, $\overline{(\sim)}$. Since the latter are easier to understand we shall use time-averages throughout the chapter. As before, we take $\bar{\mathbf{u}}$ and $\mathbf{u}'$ to represent the mean and fluctuating components of motion.

4.1 The exchange of energy between the mean flow and the turbulence

Observe the motion of the surface of water, which resembles the behaviour of hair, which has two motions, of which one depends on the weight of the strands, the other on the line of its revolving; thus water makes revolving eddies, one part of which depends upon the impetus of the principle current, and the other depends on the incident and reflected motions. (Leonardo da Vinci 1513)

(Did Leonardo da Vinci foresee Reynolds' idea of dividing a turbulent flow into two components: a mean velocity and the turbulent fluctuations?)

One of the earliest concepts in turbulence is the idea of an *eddy viscosity*. In brief, this simply asserts that, as far as the mean flow is concerned, the net effect of the turbulence is to augment the laminar viscosity and replace it with a larger 'eddy viscosity'. This concept has had a grand tradition, starting with Saint-Venant and Boussinesq in the mid-nineteenth century, and culminating in Prandtl's 1925 *mixing-length* theory. These ideas may seem rather crude by modern standards, but they have had an enormous impact on the way in which engineers estimate the gross properties of turbulent flows. Many 'engineering models' of turbulence are eddy-viscosity models. These models are simple to use and often they work surprisingly well. On the other hand, sometimes they go badly wrong! Evidently, the discerning engineer must acquire some appreciation of the limitations of the eddy-viscosity concept.

In this section we review the early theories of Boussinesq and Prandtl and then give a brief introduction to their modern descendants, such as the so-called k–ε model. There is a significant change of emphasis between this section and the previous chapter. Up until now we have been concerned with the fundamental nature of turbulence itself. Any mean flow present simply acted as a mechanism for initiating the turbulence; thereafter it played little or no role. In shear flows, however, there is a complex and continual interaction between the turbulence and the mean flow; the mean flow generates, maintains, and redistributes the turbulence, while the turbulence acts back on the mean flow, shaping the mean velocity distribution. (Think of pipe flow; the mean flow continually generates turbulence, while the turbulence itself shapes the mean velocity profile.) In eddy-viscosity models the emphasis is on the second of these processes. We are interested in the turbulence only to the extent that it influences the mean flow, and the objective of an eddy-viscosity model is to try and parameterize this influence. Of course, this inevitably leads us back to the reverse problem, that of the generation of turbulence by the mean flow, since any eddy-viscosity model requires some knowledge of the local intensity of the turbulence.

As we shall see, the mean flow and the turbulence interact via a quantity called the *Reynolds stress*. This stress arises as a result of the turbulence and acts on the mean flow, shaping its evolution. It is also responsible for the maintenance of the turbulent fluctuations as it channels energy out of the mean flow and into the turbulence. We start, therefore, by introducing the idea of the Reynolds stress and some of its consequences.

4.1.1 *Reynolds stresses and the closure problem of turbulence*

Consider the Navier–Stokes equation (2.6) applied to a steady on-average flow

$$\rho\frac{\partial u_i}{\partial t} + \rho(\bar{\mathbf{u}}\cdot\nabla)u_i = -\frac{\partial p}{\partial x_i} + \frac{\partial \tau_{ij}}{\partial x_j} \tag{4.1}$$

where

$$\tau_{ij} = 2\rho\nu S_{ij} = \rho\nu\left[\frac{\partial u_i}{\partial x_j} + \frac{\partial u_j}{\partial x_i}\right]. \tag{4.2}$$

Here τ_{ij} represents the stresses (tangential and normal) associated with viscosity. We now time-average the Navier–Stokes equation. This yields[2]

$$\rho\left[(\bar{\mathbf{u}}\cdot\nabla)\bar{u}_i + \overline{(\mathbf{u}'\cdot\nabla)u_i'}\right] = -\frac{\partial \bar{p}}{\partial x_i} + \frac{\partial \bar{\tau}_{ij}}{\partial x_j}$$

which may be rearranged to give

$$\rho(\bar{\mathbf{u}}\cdot\nabla)\bar{u}_i = -\frac{\partial \bar{p}}{\partial x_i} + \frac{\partial}{\partial x_j}\left[\bar{\tau}_{ij} - \rho\overline{u_i'u_j'}\right]. \tag{4.3}$$

This gives us an equation for the mean quantities $\bar{\mathbf{u}}$ and $\bar{p}$, just as we would have expected. However, the quadratic (non-linear) term in (4.1) has given rise to a contribution to (4.3) involving turbulent quantities, $\rho\overline{u_i'u_j'}$. This couples the mean flow to the turbulence. It is as if the turbulent fluctuations have given rise to additional stresses, $\tau_{ij}^R = -\rho\overline{u_i'u_j'}$. These are the all-important *Reynolds stresses* and we can rewrite (4.3) as

$$\rho(\bar{\mathbf{u}}\cdot\nabla)\bar{u}_i = -\frac{\partial \bar{p}}{\partial x_i} + \frac{\partial}{\partial x_j}\left[\bar{\tau}_{ij} + \tau_{ij}^R\right]. \tag{4.4}$$

We might note in passing that we can also time-average the continuity equation, and this yields

$$\nabla\cdot\bar{\mathbf{u}} = 0, \quad \nabla\cdot\mathbf{u}' = 0 \tag{4.5a, b}$$

so that both $\bar{\mathbf{u}}$ and $\mathbf{u}'$ are solenoidal.

[2] Contributions of the form $(\bar{\mathbf{u}}\cdot\nabla)\mathbf{u}'$ and $(\mathbf{u}'\cdot\nabla)\bar{\mathbf{u}}$ vanish when time-averaged.

We can understand the physical origin of the Reynolds stresses as follows. Let us rewrite (4.1) in integral form:

$$\frac{d}{dt}\int_V (\rho u_i)dV = -\oint_S (\rho u_i)\mathbf{u}\cdot d\mathbf{S} + \oint_S \tau_{ij}dS_j - \oint_S p dS_i.$$

In words this states that the rate of change of momentum in the fixed volume V is equal to the sum of:

(1) minus the rate at which momentum flows out through the surface S;
(2) the net force arising from pressure and viscous forces acting on S.

If we now time-average this expression, noting that $\mathbf{u} = \bar{\mathbf{u}} + \mathbf{u}'$, we find,

$$\frac{d}{dt}\int_V (\rho \bar{u}_i)dV = -\oint_S (\rho \bar{u}_i)\bar{\mathbf{u}}\cdot d\mathbf{S} + \oint_S \left(\bar{\tau}_{ij} - \rho\overline{u_i'u_j'}\right)dS_j - \oint_S \bar{p} dS_i. \tag{4.6}$$

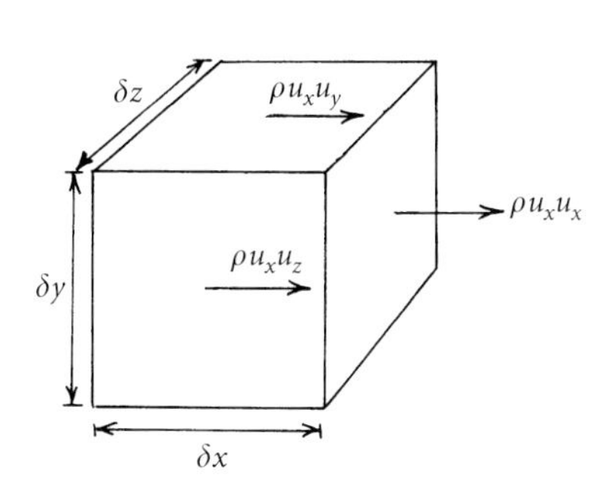

Figure 4.2 Momentum fluxes out of a cube.

We have new terms, $-\rho\overline{u_i'u_j'}$, which appear to act like stresses but really represent the flux of momentum in or out of V caused by the turbulent fluctuations. For example, if V is a small cube, $\delta x\, \delta y\, \delta z$, and we focus on the x-component of momentum, then the momentum fluxes through the sides of the cube which contribute to the rate of change of $(\rho u_x)dV$ are of the form (Figure 4.2),

$$\rho\left(\bar{u}_x + u_x'\right)\left(\bar{u}_x + u_x'\right)\delta y\,\delta z, \quad \text{(through the } \delta y\,\delta z \text{ faces)}$$
$$\rho\left(\bar{u}_x + u_x'\right)\left(\bar{u}_y + u_y'\right)\delta z\,\delta x, \quad \text{(through the } \delta z\,\delta x \text{ faces)}$$
$$\rho\left(\bar{u}_x + u_x'\right)\left(\bar{u}_z + u_z'\right)\delta x\,\delta y, \quad \text{(through the } \delta x\,\delta y \text{ faces)}$$

and when these are time-averaged the turbulence gives rise to the momentum fluxes,

$$\rho\left(\overline{u_x'u_x'}\right)\delta y\,\delta z, \quad \rho\left(\overline{u_x'u_y'}\right)\delta z\,\delta x, \quad \rho\left(\overline{u_x'u_z'}\right)\delta x\,\delta y.$$

These are the Reynolds stresses which appear in the time-averaged equation of motion. The important point to note, however, is that τ_{ij}^R is not really a true stress in the conventional sense of the word. Rather, it represents the mean momentum fluxes induced by the turbulence. It just so happens that, as far as the mean flow is concerned, we can capture the effects of these fluxes by pretending that τ_{ij}^R is a stress.

Returning now to (4.4), it is clear that if we are to predict the behaviour of the mean flow we need to know something about the

Reynolds stresses. Let us see if we can get an equation for τ_{ij}^R. If we subtract (4.3) from (4.1) then we obtain an equation of the form,

$$\frac{\partial u_i'}{\partial t} + \bar{u}_k \frac{\partial u_i'}{\partial x_k} + u_k' \frac{\partial \bar{u}_i}{\partial x_k} + u_k' \frac{\partial u_i'}{\partial x_k} = (\sim). \tag{4.7a}$$

We have a similar equation for u_j':

$$\frac{\partial u_j'}{\partial t} + \bar{u}_k \frac{\partial u_j'}{\partial x_k} + u_k' \frac{\partial \bar{u}_j}{\partial x_k} + u_k' \frac{\partial u_j'}{\partial x_k} = (\sim). \tag{4.7b}$$

We now multiply (4.7a) by u_j' and (4.7b) by u_i', add the resulting equations and time-average the result. After a little algebra we arrive at an important equation which we will have reason to revisit time and again,

$$\begin{aligned}\frac{\bar{D}}{Dt}\left[\rho\overline{u_i'u_j'}\right] = \bar{\mathbf{u}}\cdot\nabla\left(\rho\overline{u_i'u_j'}\right) = \tau_{ik}^R\frac{\partial \bar{u}_j}{\partial x_k} + \tau_{jk}^R\frac{\partial \bar{u}_i}{\partial x_k} + \frac{\partial}{\partial x_k}\left[-\rho\overline{u_i'u_j'u_k'}\right] \\ - \frac{\partial}{\partial x_i}\left[\overline{p'u_j'}\right] - \frac{\partial}{\partial x_j}\left[\overline{p'u_i'}\right] + 2\overline{p'S_{ij}'} \\ + \nu\nabla^2\left[\rho\overline{u_i'u_j'}\right] - 2\nu\rho\left[\overline{\frac{\partial u_i'}{\partial x_k}\frac{\partial u_j'}{\partial x_k}}\right].\end{aligned} \tag{4.8}$$

This is a bit complicated, but for the moment the main point to note is that our equation for τ_{ij}^R involves new quantities of the form $\overline{u_i'u_j'u_k'}$. We might then search for an equation for the triple correlations and indeed this can be readily found. Unfortunately it is of the form

$$\frac{\bar{D}}{Dt}\left(\overline{u_i'u_j'u_k'}\right) = \bar{\mathbf{u}}\cdot\nabla\left(\overline{u_i'u_j'u_k'}\right) = \frac{\partial}{\partial x_m}\left[-\overline{u_i'u_j'u_k'u_m'}\right] + (\sim).$$

So now we have yet another set of unknowns: $\overline{u_i'u_j'u_k'u_m'}$. Of course, the governing equation for these involves fifth-order correlations, and so it goes on. The key point is that we *always have more unknowns than equations*. This is the *closure problem of turbulence* introduced in Chapter 1.

So we have paid a heavy price in moving to a statistical description of turbulence. We started out with a perfectly deterministic equation—the Navier–Stokes equation—and ended up with an under-determined system. There is an irony here. If we take a non-statistical approach then we have a governing equation which is deterministic, yet the variable it predicts, **u**, behaves in a chaotic fashion. On the other hand, if we take a statistical approach, then the quantities we are interested in, $\overline{u_i'u_j'}$ etc., are non-random and perfectly reproducible in any experiment, yet we cannot find a closed set of equations which describes them!

The closure problem has profound ramifications. It is impossible to develop a predictive, statistical model of turbulence simply by manipulating the equations of motion. To close the system we need to introduce some additional information, and this information is necessarily *ad hoc* in nature. For almost a century, engineers have plugged this gap using the eddy-viscosity hypothesis and, indeed, this still forms the backbone many engineering models of turbulence.

Example 4.1 Derive equation (4.8) from first principles.

Example 4.2 Show that, by setting $i = j$ in (4.8), we obtain the energy equation

$$\bar{\mathbf{u}} \cdot \nabla\left[\tfrac{1}{2}\rho\overline{(\mathbf{u}')^2}\right] = \tau_{ik}^R \bar{S}_{ik} - \rho\nu\overline{\left[\frac{\partial u_i'}{\partial x_j}\right]^2} + \nabla \cdot \left[-\overline{p'\mathbf{u}'} + \nu\nabla\left[\tfrac{1}{2}\rho\overline{(\mathbf{u}')^2}\right] - \tfrac{1}{2}\rho\overline{(\mathbf{u}')^2\mathbf{u}'}\right].$$

Example 4.3 Confirm that the viscous terms in (4.8) can also be rewritten as

$$\frac{\partial}{\partial x_k}\left[\overline{u_i'\tau_{jk}'} + \overline{u_j'\tau_{ik}'}\right] - \overline{\tau_{ik}'\frac{\partial u_j'}{\partial x_k}} - \overline{\tau_{jk}'\frac{\partial u_i'}{\partial x_k}}$$

and that, when $i = j$, they reduce to,

$$2\left[\frac{\partial}{\partial x_k}\left[\overline{u_i'\tau_{ik}'}\right] - 2\rho\nu\overline{S_{ik}'S_{ik}'}\right].$$

Hence show that the energy equation of Example 4.2 can be rewritten as,

$$\frac{\bar{D}}{Dt}\left[\tfrac{1}{2}\rho\overline{(\mathbf{u}')^2}\right] = \bar{\mathbf{u}} \cdot \nabla\left[\tfrac{1}{2}\rho\overline{(\mathbf{u}')^2}\right] = \tau_{ik}^R\bar{S}_{ik} - 2\rho\nu\overline{S_{ik}'S_{ik}'} + \nabla \cdot \left[-\overline{p'\mathbf{u}'} + \overline{u_i'\tau_{ik}'} - \tfrac{1}{2}\rho\overline{(\mathbf{u}')^2\mathbf{u}'}\right].$$

4.1.2 *The eddy-viscosity theories of Boussinesq and Prandtl*

The first attempt to develop a 'turbulence model' probably dates back to Boussinesq's work in the 1870s. He proposed a shear-stress strain-rate relationship for time-averaged flows of a one-dimensional nature of the form,

$$\bar{\tau}_{xy} + \tau_{xy}^R = \rho(\nu + \nu_t)\frac{\partial \bar{u}_x}{\partial y} \tag{4.9}$$

where ν_t is an eddy viscosity. The general idea behind (4.9) is that the effect of the turbulent mixing of momentum, as characterized by τ^R_{xy}, is analogous to the molecular transport of momentum, which leads to the laminar stress τ_{xy}. Thus we might imagine that the role of turbulence is to bump up the effective viscosity from ν to $\nu + \nu_t$, where ν_t is, presumably, much greater than ν.

The concept of an eddy viscosity is now commonly used for flows of arbitrary complexity and the three-dimensional generalization of (4.9) is,

$$\tau^R_{ij} = -\rho\overline{u'_i u'_j} = \rho\nu_t\left[\frac{\partial\bar{u}_i}{\partial x_j} + \frac{\partial\bar{u}_j}{\partial x_i}\right] - \frac{\rho}{3}\overline{u'_k u'_k}\delta_{ij}. \tag{4.10}$$

The additional term on the right is necessary to ensure that the sum of the normal stresses adds up to $-\rho\overline{(u'_i)^2}$. We shall refer to (4.10) as Boussinesq's equation.

Of course, the question now is: what is ν_t? Evidently it is a property of the turbulence and not of the fluid. Prandtl was the first to propose a means of estimating ν_t, known as the *mixing-length* model. He was struck by the success of the kinetic theory of gases in predicting the macroscopic property of viscosity. In fact this theory predicts,

$$\nu = \tfrac{1}{3}lV \tag{4.11a}$$

where l is the mean free-path length of the molecules and V their rms speed. Prandtl noted that there was an analogy between Newton's law of viscosity and the Reynolds stress. In a laminar flow, layers of fluid which slide over each other experience a mutual drag (per unit area), τ_{xy}, because molecules bounce around between the layers exchanging momentum as they do so. This is illustrated in Figure 4.3(a). A molecule in the slow moving layer, A, may move up to B, slowing down the faster moving fluid. Conversely a molecule moving from C to D will tend to speed up the slower fluid. When these molecular processes are averaged out we obtain the macroscopic equation

$$\tau_{xy} = \frac{\rho lV}{3}\frac{\partial\bar{u}_x}{\partial y}.$$

Prandtl noted that a similar thing happens in a one-dimensional turbulent flow, only instead of thinking about thermally agitated molecules bouncing around, we must think of lumps of fluid being thrown around and jostled by the turbulence. Thus we can re-interpret Figure 4.3(a) as balls of fluid being thrown from one layer to another, carrying their momentum with them. The momentum exchange resulting from this process leads, when averaged over time, to the Reynolds stress $\tau^R_{xy} = -\rho\overline{u'_x u'_y}$. This analogy between molecular and

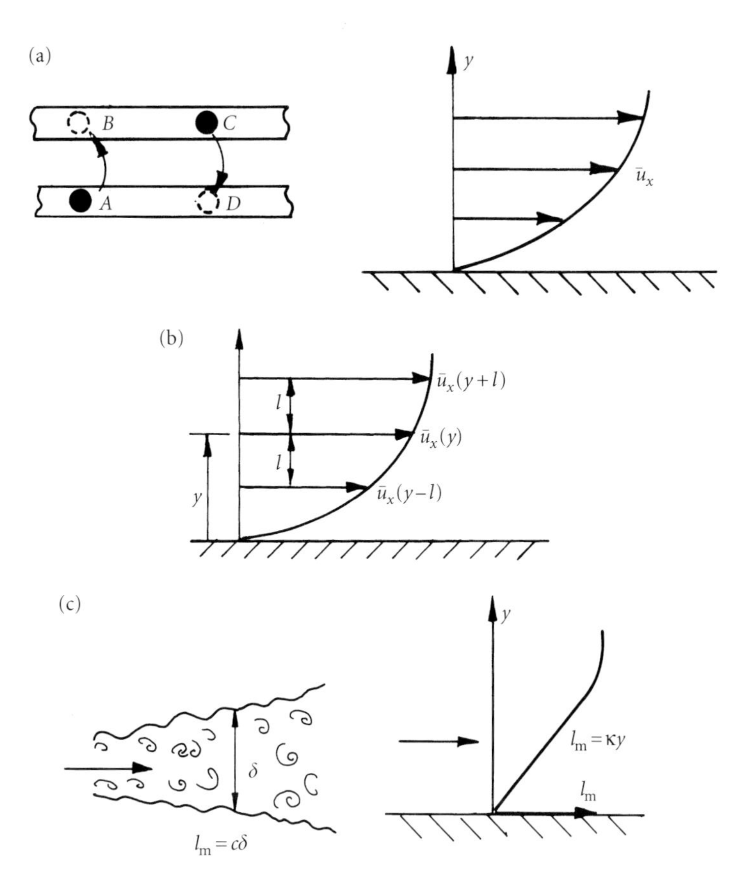

Figure 4.3 (a) Time-averaged flow corresponding to (4.9). (b) Prandtl's mixing-length theory. (c) The choice of mixing length in different geometries.

macroscopic processes led Prandtl to suggest that the macroscopic equivalent of (4.11a) is

$$\nu_t = l_m V_T \tag{4.11b}$$

where l_m is called the mixing length and V_T is a suitable measure of $|\mathbf{u}'|$. (This is consistent with the notion that the more energetic the turbulence, the greater the momentum exchange, and hence the greater is ν_t.) Of course (4.11b) is really just a dimensional necessity and merely transfers the problem from one of determining ν_t to one of determining l_m and V_T. Prandtl's *mixing-length theory* estimates these quantities as follows.

Consider a mean flow $\bar{u}_x(y)$ as shown in Figure 4.3(b). The fluid at y, which has mean velocity $\bar{u}_x(y)$ will, on average, have come from levels $y \pm l$ where l is some measure of the large eddies in the flow. Suppose that a fluid lump is thrown from $y + l$ to y, and it retains its linear momentum in the process, then on average it arrives at y with a velocity of $\bar{u}_x(y + l)$. So if l is small by comparison with $\partial\bar{u}_x/\partial_y$ then the spread of velocities at y will be

$$\bar{u}_x \pm l\frac{\partial \bar{u}_x}{\partial y}$$

and by implication

$$|u'_x(y)| \sim l|\partial \bar{u}_x/\partial y|$$

Next we note that, when $\partial \bar{u}_x/\partial_y > 0$, there is likely to be a negative correlation between u'_x and u'_y at y. That is, a positive u'_x is consistent with fluid coming from $y+l$, which requires a negative u'_y. Conversely, u'_x and u'_y are likely to have a positive correlation if $\partial \bar{u}_x/\partial_y < 0$. Moreover, we expect $|u'_x| \sim |u'_y|$ and so, whatever the sign of $\partial \bar{u}_x/\partial y$, we have the estimate

$$\overline{u'_x u'_y} \sim \pm \overline{(u'_x)^2} \sim -l^2 \left|\frac{\partial \bar{u}_x}{\partial y}\right| \frac{\partial \bar{u}_x}{\partial y}. \tag{4.12}$$

We now absorb the unknown constant in (4.12) into the definition of l and we find that, for this simple one-dimensional shear flow,

$$\tau^R_{xy} = -\rho \overline{u'_x u'_y} = \rho l_m^2 \left|\frac{\partial \bar{u}_x}{\partial y}\right| \frac{\partial \bar{u}_x}{\partial y}. \tag{4.13a}$$

If we compare this with the Boussinesq equation

$$\tau^R_{xy} = \rho \nu_t \frac{\partial \bar{u}_x}{\partial y}$$

we have,

$$\nu_t = l_m^2 \left|\frac{\partial \bar{u}_x}{\partial y}\right|. \tag{4.13b}$$

This is Prandtl's mixing-length model. If we can determine l_m, say by experiment, then we can find τ^R_{xy}. It has to be said, however, that this is a deeply flawed argument. The use of a Taylor expansion is not justified in estimating $\bar{u}_x(y+l) - \bar{u}_x(y)$ since l, the size of the large eddies, is usually comparable with the mean gradient in $\bar{u}_x$. Moreover, there is no justification for assuming that the lumps of fluid retain their linear momentum as they move between layers, suddenly giving up their momentum on arrival at a new layer.

Nevertheless, it seems that Prandtl's mixing-length model works reasonably well (at least better than might be expected) for simple one-dimensional shear flows, provided, of course, that l_m is chosen appropriately. For free shear layers it is found that l_m is reasonably uniform and of the order of $l_m = c\delta$, where δ is the local thickness of the layer. The constant c depends on the type of shear layer (mixing layer, wake, jet, etc.). In boundary layers one finds that, near the wall, $l_m = \kappa y$, where $\kappa \approx 0.4$ is known as Karman's constant (Figure 4.3(c)). This is often interpreted as the eddy size being proportional to the distance from the wall.

One defect of (4.13), however, is that it predicts that $\nu_t = 0$ on the centre-line of a jet or wake, something which is unlikely to be true in practice. However, this deficiency can be corrected for by adopting a slightly different expression for ν_t, as discussed in Section 4.3.1.

We shall return to the idea of a mixing length in Section 4.1.4 of this chapter, where we shall show that, for one-dimensional shear flows, the main results of mixing length really just follow from the laws of vortex dynamics. In the mean time we shall examine the process by which the mean flow and the turbulence exchange energy.

4.1.3 *The transfer of energy from the mean flow to the turbulence*

So far we have focused on the influence of τ_{ij}^R on the mean flow. We now show that the Reynolds stresses also act as a conduit for transferring energy from the mean flow to the turbulence, maintaining the turbulence in the face of viscous dissipation. Our first port of call, therefore, is to examine the rate of working of τ_{ij}^R.

In Chapter 2 Section 2.1.4 we saw that the rate of working of the viscous stresses in a laminar flow is equal to $\partial(\tau_{ij}u_i)/\partial x_j$. We found it convenient to split this into two parts,

$$\frac{\partial}{\partial x_j}\left[\tau_{ij}u_i\right] = u_i\frac{\partial \tau_{ij}}{\partial x_j} + \tau_{ij}\frac{\partial u_i}{\partial x_j} = u_i f_i + \tau_{ij}S_{ij} \tag{4.14a}$$

where f_i is the net viscous force per unit volume acting on the fluid. We interpret this equation as follows. The two terms on the right both represent rates of increase of energy per unit volume of the fluid. The first term, $u_i f_i$, represents the rate of increase of mechanical energy of the fluid, that is, the rate of working of f_i. The second term gives the rate of increase of internal energy. Together, they account for the total rate of working of τ_{ij}. Similarly, in a turbulent flow, the rate of working of τ_{ij}^R on the mean flow is

$$\frac{\partial}{\partial x_j}\left[\tau_{ij}^R\bar{u}_i\right] = \bar{u}_i\frac{\partial \tau_{ij}^R}{\partial x_j} + \tau_{ij}^R\bar{S}_{ij}. \tag{4.14b}$$

However, there is an important difference between (4.14a and b). The Reynolds stress is entirely fictitious, arising from our averaging procedure, and so it cannot create or destroy mechanical energy. Thus $\tau_{ij}^R\bar{S}_{ij}$ cannot represent the rate of change of internal energy in the fluid. Rather, we shall see that it represents the rate of generation of turbulent kinetic energy.

Now suppose we integrate (4.14b) over a closed volume, V. The term on the left is a divergence and integrates to zero since τ_{ij}^R vanishes on the boundary. In a global sense, therefore, the two terms on the right of (4.14b) must balance:

$$\int_V \left(-\bar{u}_i \frac{\partial \tau_{ij}^R}{\partial x_j}\right) dV = \int_V \tau_{ij}^R \bar{S}_{ij} dV.$$

But $\partial(\tau_{ij}^R)/\partial x_j$ is the net force per unit volume acting on the mean flow by virtue of the Reynolds stresses, and so $\bar{u}_i \partial(\tau_{ij}^R)/\partial x_j$ is the rate of working of this force. We would expect, therefore, that $-\bar{u}_i \partial(\tau_{ij}^R)/\partial x_j$ represents the rate of loss of mechanical energy from the mean flow as a result of the action of the Reynolds stresses, that is, as a result of the turbulence. (We shall see shortly that this is so.) Since τ_{ij}^R cannot create or destroy mechanical energy, we might anticipate that this mean-flow energy will reappear as kinetic energy in the turbulence, corresponding to the term $\tau_{ij}^R \bar{S}_{ij}$ in (4.14b):

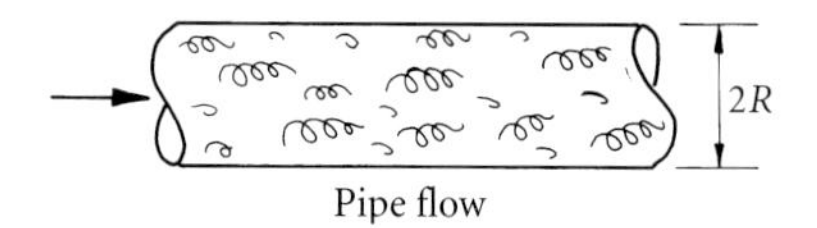

Figure 4.4 Turbulent flow in a pipe.

$$\int_V \left(-\bar{u}_i \frac{\partial \tau_{ij}^R}{\partial x_j}\right) dV = \int_V \tau_{ij}^R \bar{S}_{ij} dV$$

(rate of loss of KE from the mean flow) = (rate of gain of KE by the turbulence).

We shall now show that this is indeed the case.

To focus thoughts, let us consider a particularly simple flow, say a steady-on-average pipe flow. (Figure 4.4). Then (4.3) tells us that,

$$\rho(\bar{\mathbf{u}} \cdot \nabla)\bar{u}_i = -\frac{\partial \bar{p}}{\partial x_i} + \frac{\partial \tau_{ij}^R}{\partial x_j} + (\text{viscous forces}). \tag{4.15}$$

The Reynolds stress gives rise to a net force acting on the mean flow, $f_i = \partial(\tau_{ij}^R)/\partial x_j$. It turns out that the rate of working of this force, $f_i \bar{u}_i$, is negative in a pipe and so the mean flow loses energy to the agent which supplies the force, that is, to the turbulence. We talk of energy being transferred from the mean axial flow to the turbulence. This is why the turbulence in a pipe does not die away.

Of course this is all a little artificial. We have only one flow and one fluid. What we really mean is that we can divide $\frac{1}{2}\overline{\mathbf{u}^2}$ into two parts, $\frac{1}{2}(\bar{\mathbf{u}})^2$ and $\frac{1}{2}\overline{(\mathbf{u}')^2}$, and that there is a net transfer of energy from $\frac{1}{2}\bar{\mathbf{u}}^2$ to $\frac{1}{2}\overline{(\mathbf{u}')^2}$ when $f_i \bar{u}_i$ is negative. Now we know that the turbulence in the pipe does not die away (when the viscous forces are small) and so, by inference, $f_i \bar{u}_i$ *must* be negative. You might ask why this should be so. How does the turbulence extract energy from the mean flow? What is really happening is that turbulent eddies are continually being created and intensified through a distortion of the mean flow. This takes the form of warping or wrinkling of the vortex lines (they get punched out of shape by the turbulence) accompanied by a continual stretching and intensification of the turbulent vortices. The end result is a net transfer of energy from $\bar{\mathbf{u}}$ to $\mathbf{u}'$.

Now we can use (4.14b) to rewrite $f_i \bar{u}_i$ as

$$-f_i \bar{u}_i = \tau_{ij}^R \bar{S}_{ij} - \frac{\partial}{\partial x_j}\left[\tau_{ij}^R \bar{u}_i\right]. \tag{4.16}$$

The second term on the right is a divergence and, as we have seen, it integrates to zero in a closed domain. The first term, $\tau_{ij}^R \bar{S}_{ij}$, turns out to be the *local* rate of acquisition of kinetic energy by the turbulence as a result of τ_{ij}^R (see Example 4.4). This is called the *deformation work* (or the *turbulent energy generation*) and it represents the tendency for the mean shear to stretch and intensify the turbulent vorticity, leading to an increase in turbulent energy.

Equation (4.16) tells us that $-f_i \bar{u}_i$ and $\tau_{ij}^R \bar{S}_{ij}$ always balance in a global sense, so that any kinetic energy removed from the mean flow must end up as kinetic energy in the turbulence. However, they need not balance locally because the divergence term on the right of (4.16), which represents a flux of energy, can be non-zero. Thus, the energy removed from the mean motion by τ_{ij}^R at one point need not turn up as turbulent kinetic energy at exactly the same location. The reason for this is subtle and is related to the cross term $\rho \bar{\mathbf{u}} \cdot \mathbf{u}'$, as illustrated in Example 4.4.

Example 4.4 Let us write the equation of motion for the steady-on-average pipe flow as,

$$\frac{\partial}{\partial t}(\rho \mathbf{u}) + \mathbf{u} \cdot \nabla(\rho \mathbf{u}) = \sum \mathbf{F}$$

where $\sum \mathbf{F}$ represents the sum of the pressure and viscous forces. From this we have the kinetic energy equation

$$\frac{\partial}{\partial t}(\rho \mathbf{u}^2/2) + \mathbf{u} \cdot \nabla(\rho \mathbf{u}^2/2) = \sum \mathbf{F} \cdot \mathbf{u}.$$

Show that, when time-averaged, this yields

$$\bar{\mathbf{u}} \cdot \nabla\left(\rho \bar{\mathbf{u}}^2/2 + \rho \overline{(\mathbf{u}')^2}/2\right) + \overline{\mathbf{u}' \cdot \nabla\left(\rho (\mathbf{u}')^2/2\right)} = \frac{\partial}{\partial x_j}\left[\bar{u}_i \tau_{ij}^R\right] + \sum \overline{\mathbf{F} \cdot \mathbf{u}}.$$

This tells us that the transport of total kinetic energy by the mean velocity, plus the transport of turbulent kinetic energy by the turbulence, is equal to the rate of working of the Reynolds stresses plus the mean rate of working of the pressure and viscous forces. As anticipated above, $\partial(\bar{u}_i \tau_{ij}^R)/\partial x_j$ acts as a source of both mean flow and turbulent kinetic energy. Now show that, by first time-averaging the equation of motion, we obtain

$$\bar{\mathbf{u}} \cdot \nabla(\rho \bar{\mathbf{u}}^2/2) = \bar{u}_i \frac{\partial \tau_{ij}^R}{\partial x_j} + \sum \bar{\mathbf{F}} \cdot \bar{\mathbf{u}}.$$

Subtracting the two energy equations yields

$$\bar{\mathbf{u}} \cdot \nabla\left(\rho\overline{(\mathbf{u}')^2}/2\right) + \overline{\mathbf{u}' \cdot \nabla\left(\rho(\mathbf{u}')^2/2\right)} = \tau_{ij}^R\bar{S}_{ij} + \sum\overline{\mathbf{F}\cdot\mathbf{u}} - \sum\bar{\mathbf{F}}\cdot\bar{\mathbf{u}}.$$

Thus the two contributions to $\partial(\bar{u}_i\tau_{ij}^R)/\partial x_j$, namely $\bar{u}_i\partial\tau_{ij}^R/\partial x_j$ and $\tau_{ij}^R\bar{S}_{ij}$, act as sources of kinetic energy for the mean flow and turbulence respectively. Note that, although these two terms balance globally, they need not balance locally, the difference being $\partial(\bar{u}_i\tau_{ij}^R)/\partial x_j$. Show that this difference arises from the turbulent transport of the cross term $\rho\bar{\mathbf{u}}\cdot\mathbf{u}'$. □

An important corollary to our interpretation of $\tau_{ij}^R\bar{S}_{ij}$ is that, in the absence of body forces such as buoyancy, we need a finite rate of strain in the mean flow to keep the turbulence alive. The need for a finite strain, and the mechanism of energy transfer from the mean flow to the turbulence, can be pictured as follows. We might visualize the turbulent vorticity as a tangle of vortex tubes, rather like a seething mass of spaghetti. In the presence of a mean shear these tubes will be systematically elongated along the direction of maximum positive strain. As they are stretched their kinetic energy rises and this represents an exchange of energy from the mean flow to the turbulence (Figure 4.5). (Actually this is a little simplistic as some of the vorticity resides in sheets rather than tubes, but it does get the general idea across. We shall refine this picture in Chapter 5.)

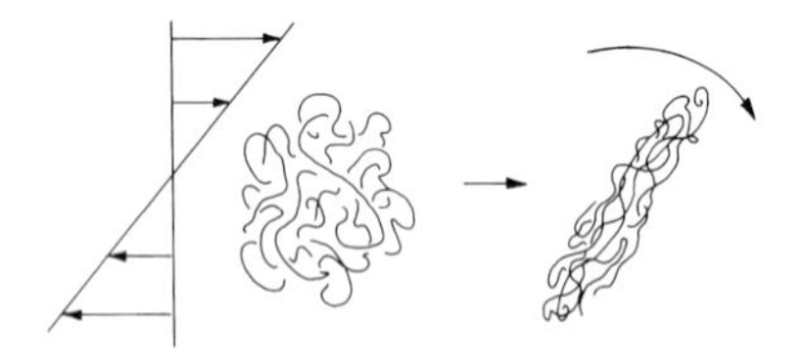

Figure 4.5 A mean shear teases out the vortex tubes in the turbulence. The kinetic energy of the turbulence rises as the tubes are stretched.

Let us take this idea of a transfer of energy a little further. We can derive an equation for the kinetic energy of the mean flow from (4.15). Multiplying this by $\bar{u}_i$ and rearranging terms yields,

$$\frac{\bar{D}}{Dt}\left(\frac{1}{2}\rho\bar{\mathbf{u}}^2\right) = \bar{\mathbf{u}}\cdot\nabla\left(\frac{1}{2}\rho\bar{\mathbf{u}}^2\right) = \frac{\partial}{\partial x_k}\left[-\bar{u}_k\bar{p} + \bar{u}_i\left(\bar{\tau}_{ik} + \tau_{ik}^R\right)\right] - \tau_{ik}^R\bar{S}_{ik} - 2\rho\nu\bar{S}_{ik}\bar{S}_{ik} \tag{4.17}$$

(rate of change of KE) = (flux of KE) − (loss of KE to turbulence) − (dissipation)

where $\bar{\tau}_{ij}$ is the mean value of the viscous stresses,

$$\bar{\tau}_{ij} = 2\rho\nu\bar{S}_{ij}.$$

We recognize the divergence on the right as representing the rate of working of the pressure, viscous and Reynolds stresses on the boundary of the domain of interest. For flow in a closed domain this term integrates to zero. The last two terms on the right represent: (i) the rate of transfer of energy to the turbulence; and (ii) the rate of destruction of mean energy by the mean viscous forces. The last term on the right is almost always negligible (except very close to boundaries).

We now turn to the turbulence itself. We can get an equation for the kinetic energy of the turbulence from (4.8). Setting $i = j$ in this expression yields[3]

$$\bar{\mathbf{u}} \cdot \nabla\left[\tfrac{1}{2}\rho\overline{(\mathbf{u}')^2}\right] = \frac{\partial}{\partial x_k}\left[-\overline{p'u'_k} + \overline{u'_i\tau'_{ik}} - \tfrac{1}{2}\rho\overline{u'_iu'_iu'_k}\right] + \tau^R_{ik}\bar{S}_{ik} - 2\rho\nu\overline{S'_{ij}S'_{ij}} \tag{4.18}$$

(material rate of change of KE) = (flux or transport of KE)
+ (generation of KE)
− (dissipation).

Note that $\tau^R_{ik}\bar{S}_{ik}$, which transfers energy from the mean flow to the turbulence, appears in both (4.17) and (4.18), but with an opposite sign.

Now $\tau^R_{ik}\bar{S}_{ik}$ represents the rate at which energy enters the turbulence and passes down the energy cascade. We represent this quantity by ρG. Also, we recognize the last term in (4.18) as the rate of dissipation of turbulent energy by the fluctuating viscous stresses. We denote this by $\rho\varepsilon$. So we have

$$\text{Generation:} \quad G = -\overline{u'_iu'_j}\,\bar{S}_{ij} \tag{4.19}$$

$$\text{Dissipation:} \quad \varepsilon = 2\nu\overline{S'_{ij}S'_{ij}}. \tag{4.20}$$

It is convenient to introduce one more label;

$$\text{Transport:} \quad \rho \mathrm{T}_i = \tfrac{1}{2}\rho\overline{u'_iu'_ju'_j} + \overline{p'u'_i} - 2\rho\nu\overline{u'_jS'_{ij}}$$

So our turbulence kinetic energy equation becomes,

$$\bar{\mathbf{u}} \cdot \nabla\left[\tfrac{1}{2}\overline{(\mathbf{u}')^2}\right] = -\nabla \cdot [\mathbf{T}] + G - \varepsilon. \tag{4.21}$$

When the turbulence is statistically homogeneous (which it never is near a wall) the divergences of all statistical quantities vanish and (4.21) reduces to

$$G = \varepsilon$$

which states that the local rate of generation of turbulence energy is equal to the rate of viscous dissipation. In Chapter 3 we introduced the additional quantity Π to represent the flux of energy down the turbulent energy cascade, from the large to the small eddies. If there is a continual depletion or accumulation of energy at some particular size range within the cascade then Π may vary in magnitude as we pass

[3] See Example 4.3.

down the cascade. However, in steady, homogeneous turbulence Π is the same at all points in the cascade and we have,

$$G = \Pi = \varepsilon$$

(KE into cascade) = (flux down cascade) = (dissipation at small scales).

Moreover, in Chapter 3 we saw that the large eddies in a turbulent flow tend to break up on a timescale of their turn-over time and so $\Pi \sim u^3/l$ where $u^2 \sim \overline{(\mathbf{u}')^2}$ and l is the integral scale (the size of the large eddies). So for steady, homogeneous turbulence we have

$$G = \Pi = \varepsilon \sim u^3/l$$
(homogeneous, steady).

However, shear flows are rarely homogeneous, and even if they are, there is no guarantee that they are steady. For example, the one-dimensional shear flow $\mathbf{u} = \bar{u}_x(y)\hat{\mathbf{e}}_x$, $\partial\bar{u}_x/\partial y = S =$ constant, tends to evolve towards an unsteady state in which the production of energy, $\tau_{ij}^R\bar{S}_{ij}$, exceeds the dissipation, $\rho\varepsilon$. In fact, wind-tunnel data suggests that $G/\varepsilon \sim 1.7$ where $G = \tau_{xy}^R S/\rho$ (Champagne, Harris and Corrsin 1970; Tavoularis and Corrsin 1981). (See Section 4.4 or else Table 4.1.) Nevertheless, even when the turbulence is unsteady or even inhomogeneous, G, Π, and ε often tend to be of the same order of magnitude, so that frequently

$$G \sim \Pi \sim \varepsilon \sim u^3/l$$
(inhomogeneous, unsteady).

The main exception to this rule is freely decaying turbulence in which $G = 0$ but nevertheless $\varepsilon \sim \Pi \sim u^3/l$, as discussed in Chapter 3.

Table 4.1 Wind-tunnel data for the asymptotic state of one-dimensional homogeneous turbulence. Note that u is defined through $u^2 = \frac{1}{3}\overline{(\mathbf{u}')^2}$

$\tau_{xy}^R/\rho u^2$	~0.42
Su^2/ε	~4.2
G/ε	~1.7
$\varepsilon/(u^3/l)$	~1.1

4.1.4 *A glimpse at the k–ε model*

We now turn our attention to 'engineering models of turbulence' which, by and large, are eddy-viscosity models. We start by re-examining Prandtl's mixing-length theory and then move on to one of the most popular models currently in use: the k–ε model. Recall that Prandtl's mixing-length model applied to a simple, one-dimensional flow predicts,

$$\tau_{xy}^R = \rho l_m^2 \left|\frac{\partial\bar{u}_x}{\partial y}\right| \frac{\partial\bar{u}_x}{\partial y} = \rho\nu_t \frac{\partial\bar{u}_x}{\partial y}. \tag{4.22}$$

We noted that there are several unjustified steps in the derivation of (4.22), but, nevertheless, it seems to work reasonably well provided its use is restricted to simple shear layers in which there is only one characteristic length scale. So why does it work at all? The answer is vortex dynamics. Sometimes for convenience we pretend that there

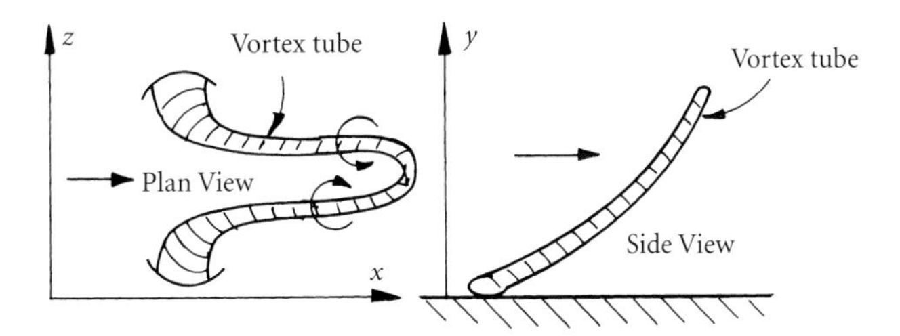

Figure 4.6 Hairpin vortices generated in a boundary layer.

are two flows—a mean flow and a turbulent flow—but of course this is fiction. There is only one vorticity field, and the mean and turbulent flows are simply different manifestations of this field. The vorticity associated with the mean flow points out of the x–y plane, $\bar{\omega} = (0, 0, \bar{\omega}_z)$, and that associated with the turbulence is random, but at any instant the real vorticity field is the sum of the two. So typically the vorticity of the large eddies and that of the mean flow are of the same order of magnitude, since they are just different manifestations of the same vortex lines. There is a simple (perhaps too simple) cartoon which gets this idea over. We might picture the z-directed vortex lines being continuously teased out into three-dimensional shapes by turbulent fluctuations. The eddies created by this process have a vorticity which is of the same order of magnitude as the vorticity of the mean flow (Figure 4.6). In any event, if u is a typical measure of $|\mathbf{u}'|$ and l the size of the large, energy-containing, eddies, then

$$u/l \sim \bar{\omega}_z \sim \frac{\partial \bar{u}_x}{\partial y}. \tag{4.23}$$

Now in turbulent shear flows u'_x and u'_y are strongly correlated for the reasons discussed in Section 4.1.2 of this chapter. Moreover, u'_x and u'_y are of the same order of magnitude. It follows that

$$\left|\overline{u'_x u'_y}\right| \sim u^2 \sim l^2 \left(\frac{\partial \bar{u}_x}{\partial y}\right)^2 \tag{4.24}$$

and Prandtl's mixing-length theory follows. The key point, though, is that Prandtl's mixing-length approach is unlikely to yield useful results in anything other than a simple shear flow.

Since mixing length is really restricted to one-dimensional shear flows, we need some more robust model for flows of greater complexity. Traditionally, engineers have fallen back on the Boussinesq–Prandtl equations

$$\tau^R_{ij} = 2\rho\nu_t \bar{S}_{ij} - (\rho/3)\overline{(u'_k u'_k)}\delta_{ij} \tag{4.25}$$

$$\nu_t = lV_T. \tag{4.26}$$

The idea behind (4.25) is that momentum is exchanged through eddying motion in a manner analogous to the microscopic transport of momentum by molecular action. Unfortunately this is a flawed concept because turbulent eddies are distributed entities which continually interact, whereas molecules are discrete and collide only intermittently. Moreover, the mean-free path length of the molecules is small by comparison with the macroscopic dimensions of the flow. This is not true of the meandering of turbulent eddies. Indeed, the large, energy containing eddies often have a size comparable with the characteristic scale of the mean flow.

Given that the molecular and turbulent transports of momentum are qualitatively different, we need an alternative means of justifying the eddy-viscosity hypothesis. One argument is to suggest that (4.25) merely defines ν_t, while (4.26) is simply a dimensional necessity. (We have not yet specified V_T.) However this argument is also inadequate. In fact there are three major shortcomings of the eddy-viscosity hypothesis. First, we have chosen to relate τ_{ij}^R and $\bar{S}_{ij}$ via a simple scalar, ν_t, rather than through a tensor. Thus, for example, the relationship between τ_{xy}^R and $\bar{S}_{xy}$ is the same as between τ_{yz}^R and $\bar{S}_{yz}$ and so on. This suggests that eddy-viscosity models will not work well when the turbulence is strongly anisotropic, as would be the case where stratification or rotation is important. Second, when $\bar{S}_{ij} = 0$ (4.25) predicts that the turbulence is isotropic, with $\langle (u_x')^2 \rangle = \langle (u_y')^2 \rangle = \langle (u_z')^2 \rangle$. Yet we know from studies of grid turbulence that anisotropy can persist for long periods of time, with or without a mean shear (see Section 4.6.1). Third (4.25) assumes that τ_{ij}^R is determined by the *local* strain rate, and not by the history of the straining of the turbulence. It is easy to see why this is not, in general, a valid assumption. The point is that the magnitude of the Reynolds stresses depends on the shapes and intensity of the local eddies (blobs of vorticity), and this, in turn, depends on the straining of the eddies prior to arriving at the point of interest. In short, we are not at liberty to assume that the turbulent eddies have relaxed to some sort of statistical equilibrium, governed by local conditions alone. All in all it would seem that there are a host of reasons for not believing in (4.25), many of which will come back to haunt us. Nevertheless, with these limitations in mind, let us proceed.

The question now is: what is ν_t? It seems natural to take V_T as $k^{1/2}$ where $k = \frac{1}{2}\langle \mathbf{u}'^2 \rangle$.[4] The idea is that the more energetic the turbulence the greater the momentum exchange and hence the larger the value of ν_t. Also, on physical grounds, we would expect l to be of the order of the integral scale since only the large eddies contribution to the momentum exchange. So we have,

$$\nu_t \sim k^{1/2} l. \tag{4.27}$$

[4] Unfortunately, it has become conventional to use k for both wave number and turbulent kinetic energy. This rarely leads to confusion, however.

The problem now is that we need to be able to estimate k and l at each point in the flow. The so-called *k–ε model* tackles this as follows. Recall that, in most forms of turbulence, $\varepsilon \sim u^3/l$. It follows that, if we believe (4.27), then

$$\nu_t \sim k^2/\varepsilon.$$

In the k–ε model this is usually written as

$$\nu_t = c_\mu k^2/\varepsilon \tag{4.28}$$

where the coefficient c_μ is given a value of $\sim$0.09. (This value is chosen to conform to the observed relationship between shear stress and velocity gradient in a simple boundary layer. See Section 4.2.3.) The model then provides empirical transport equations for both k and ε. The k equation is based on (4.18), which generalizes to

$$\frac{\partial k}{\partial t} + \bar{\mathbf{u}} \cdot \nabla[k] = -\nabla \cdot [\mathbf{T}] + (\tau_{ij}^R/\rho)\bar{S}_{ij} - \varepsilon \tag{4.29}$$

$$\rho T_i = \tfrac{1}{2}\rho \overline{u_i' u_j' u_j'} + \overline{p' u_i'} - 2\rho\nu \overline{u_j' S_{ij}'}$$

for unsteady flows. (We assume here that the timescale for changes in statistically averaged quantities is slow by comparison with the timescale for turbulent fluctuations, so that an averaging procedure based on time may be retained even in unsteady flows.) The problem, of course, is what to do with the unknowns, $\overline{p' u_i'}$ and $\overline{u_i' u_j' u_j'}$. (The viscous contribution to $\mathbf{T}$ is usually small.) In the k–ε model it is assumed that fluctuations in pressure induced by the turbulent eddies act to spread the turbulent kinetic energy from regions of strong turbulence to those of low intensity turbulence, and that this redistribution of energy is a diffusive process. The same assumption is made about the triple correlations and the vector $\mathbf{T}$ is written as,

$$\mathbf{T} = -\nu_t \nabla k.$$

This is a rather sweeping assumption, but it does have the pragmatic advantage of turning the k equation into a simple advection–diffusion equation with a source term, G, and a sink, ε. The net result is,

$$\frac{\partial k}{\partial t} + \bar{\mathbf{u}} \cdot \nabla k = \nabla \cdot (\nu_t \nabla k) + (\tau_{ij}^R/\rho)\bar{S}_{ij} - \varepsilon. \tag{4.30}$$

This equation at least guarantees that k is a well-behaved parameter. The ε equation, on the other hand, is almost pure invention. It contains three coefficients which are nominally arbitrary and these have been set to capture certain well-documented flows. In effect, the k–ε model is a highly sophisticated exercise in interpolating between data sets.

We shall return to this model in Section 4.6 of this chapter where the ε equation is described in some detail and the limitations of the k–ε model are aired. Perhaps it is sufficient for the moment to note that the k–ε model has been fairly successful, much more than one might have anticipated. In fact it has become the standard working model of turbulence currently used in engineering. It is flawed but simple, yields a reasonable estimate of the mean flow for a range of geometries, yet often goes badly wrong. Perhaps it is the combination of simplicity and familiarity which has made it just about the most popular model for the pragmatic engineer.

With this brief introduction to eddy-viscosity models we now examine the different types of shear flows which are important in engineering. We start with wall-bounded flows.

4.2 Wall-bounded shear flows and the log-law of the wall

The presence of a boundary has a profound influence on a turbulent shear flow. In part, this is because the velocity fluctuations must fall to zero near the wall. It is natural to divide up the subject of wall-bounded flows into internal flows (pipes, ducts, etc.) and external flows (boundary layers). We start with internal flows.

4.2.1 *Turbulent flow in a channel and the log-law of the wall*

Consider a fully developed, one-dimensional mean flow, $\bar{\mathbf{u}} = (\bar{u}_x(y), 0, 0)$ between smooth, parallel plates, as shown in Figure 4.7. The x and y components of (4.4) yield:

$$\rho \frac{\partial}{\partial y}\left[\nu \frac{\partial \bar{u}_x}{\partial y} - \overline{u'_x u'_y}\right] = \frac{\partial \bar{p}}{\partial x} \tag{4.31}$$

$$\rho \frac{\partial}{\partial y}\left[-\overline{u'_y u'_y}\right] = \frac{\partial \bar{p}}{\partial y} \tag{4.32}$$

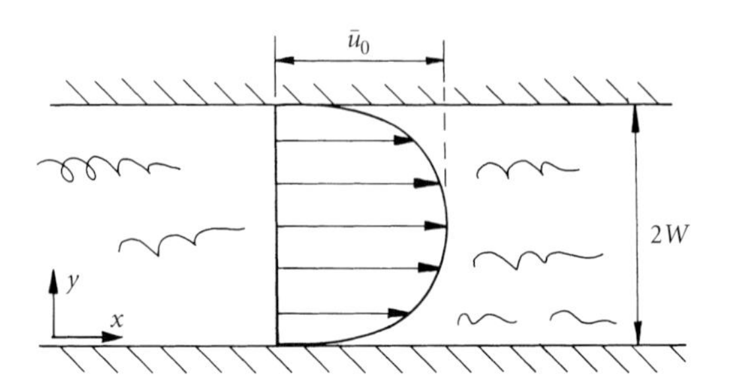

Figure 4.7 Flow between parallel plates.

(We have assumed here that, since the flow is fully developed, all statistical properties, except $\bar{p}$, are independent of x.)

Let us introduce the notation,

$$\bar{p}_w = \bar{p} + \rho\overline{u'_y u'_y}. \tag{4.33}$$

Then (4.32) tells us that $\bar{p}_w$ is a function of x alone and so, since $\mathbf{u}' = 0$ on $y = 0$,

$$\bar{p}_w = \bar{p}_w(x) = \bar{p}(y = 0).$$

Evidently $\bar{p}_w$ is the wall pressure. Noting that $\rho\overline{u'_y u'_y}$ is independent of x we rewrite (4.31) as

$$\rho\frac{\partial}{\partial y}\left[\nu\frac{\partial \bar{u}_x}{\partial y} - \overline{u'_x u'_y}\right] = \frac{d\bar{p}_w}{dx}. \tag{4.34}$$

Since the left-hand side is independent of x and the right is independent of y, this equation must be of the form

$$\frac{d}{dy}\left[\nu\frac{d\bar{u}_x}{dy} - \overline{u'_y u'_x}\right] = -K \tag{4.35}$$

where ρK is, of course, the magnitude of the pressure gradient in the pipe: a positive constant.

On integrating (4.35) we find that the total shear stress, $\bar{\tau}_{xy} + \tau^R_{xy}$, varies linearly with y, with the constant of integration being fixed by the fact that the flow is symmetric about $y = W$:

$$\tau = \bar{\tau}_{xy} + \tau^R_{xy} = \rho K(W - y) = \tau_w\left(1 - \frac{y}{W}\right). \tag{4.36}$$

It is conventional to introduce the notation

$$V_*^2 = \tau_w/\rho = KW \tag{4.37}$$

where τ_w is the wall shear stress and V_* is known as the *friction velocity*. Then (4.36) becomes

$$\tau/\rho = \nu\frac{d\bar{u}_x}{dy} - \overline{u'_x u'_y} = V_*^2 - Ky. \tag{4.38}$$

We seem to have reached an impasse since, in order to solve (4.38), we need to know the distribution of τ^R_{xy}. At this point there are two ways forward. We might invoke some closure model, such as mixing length, but this will leave us uncertain as to the validity of the results. An alternative approach is to deploy the rather general tools of dimensional analysis and asymptotic matching. Rather remarkably, this second strategy yields definite results, as we now show.

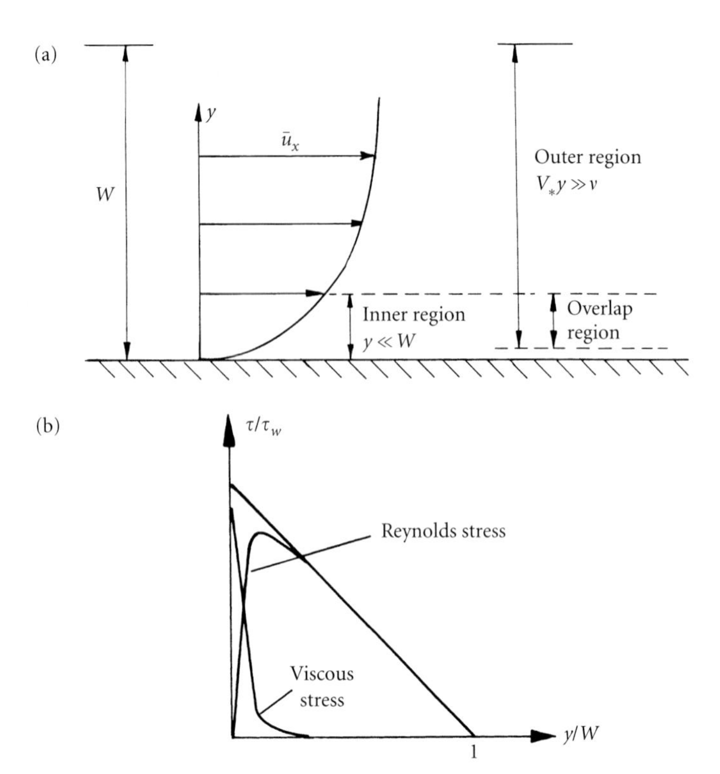

Figure 4.8 (a) Different regions in a turbulent duct flow (b) Variation of Reynolds stress and viscous stress with y.

Let us divide the flow into a number of regions as indicated by Figure 4.8(a). Close to the wall, $y \ll W$, the variation in shear stress given by (4.38) is negligible and we may assume that τ is constant and equal to τ_w. We may model the flow by

$$\tau/\rho = \nu \frac{d\bar{u}_x}{dy} - \overline{u'_x u'_y} = V_*^2, \quad y/W \ll 1 \tag{4.39}$$

We shall call this region the *inner layer* or *inner region*. We retain the viscous term in (4.39) because, adjacent to the wall, $\mathbf{u}'$ falls to zero and so the entire shear stress is laminar. The inner layer is characterized by rapid variations in $\bar{\tau}_{xy}$ and τ_{xy}^R. Although the sum of the two stresses is constant, we move rapidly from a situation in which τ is purely viscous at $y = 0$ to $\tau \approx \tau_{xy}^R$ a short distance from the wall (Figure 4.8(b)). This suggests that we introduce a second region, the *outer layer*, in which the laminar stress is negligible,

$$\tau/\rho = -\overline{u'_x u'_y} = V_*^2 - Ky, \quad V_* y/\nu \gg 1. \tag{4.40}$$

Note that we have interpreted 'far from the wall' in terms of the normalized distance $V_* y/\nu$. This seems plausible since, given the variables available, there are only two ways of normalizing y, $\eta = y/W$ and $y^+ = V_* y/\nu$. Only the second of these allows us to make the dimensionless y large. Still, we should check retrospectively that $V_* y/\nu \gg 1$ does indeed ensure a negligible viscous stress.

Now in the near-wall region the only parameters on which $\bar{u}_x$ can depend are V_*, y, and ν. The width W is not relevant because the eddies centred a distance y from the wall are not generally larger than y, since $\mathbf{u}' = \mathbf{0}$ at the wall. So near the wall the important eddies are typically very small and it seems plausible that the turbulence there does not know nor care about the presence of another boundary at a distance $2W$ away.[5]

In the outer region, on the other hand, we would not expect ν to be a relevant parameter because the viscous stresses are negligible. We would expect velocity gradients to scale with W since the largest eddies, which are most effective at transporting momentum, are of the order of W. So we might anticipate that departures from the centre-line velocity, $\Delta\bar{u}_x = \bar{u}_0 - \bar{u}_x$, will be independent of ν but a function of W. *If* all of this is true then we have,

$$\begin{aligned} &\text{Inner region:} \quad \bar{u}_x = \bar{u}_x(y, \nu, V_*), && (y/W) \ll 1 \\ &\text{Outer region:} \quad \bar{u}_0 - \bar{u}_x = \Delta\bar{u}_x(y, W, V_*), && (V_* y/\nu) \gg 1. \end{aligned}$$

In dimensionless form these become

$$\bar{u}_x/V_* = f(y^+), \quad \eta \ll 1 \tag{4.41}$$

$$\Delta\bar{u}_x/V_* = g(\eta), \quad y^+ \gg 1 \tag{4.42}$$

where $\eta = y/W$ and $y^+ = V_* y/\nu$. The first of these, (4.41), is known as the *law of the wall*, while the second, (4.42), is called the *velocity defect law*. Let us now suppose that $\text{Re} = W V_*/\nu \gg 1$ so that there exists an overlap region (which is sometimes called the *inertial sublayer*) in which y is small when normalized by W, but large when normalized by ν/V_*. Then this region has the property that τ is approximately constant (since $\eta \ll 1$) and the laminar stress is negligible (since $y^+ \gg 1$). As both (4.41) and (4.42) apply we have

$$y\frac{\partial \bar{u}_x}{\partial y} = V_* y^+ f'(y^+) = -V_* \eta g'(\eta). \tag{4.43}$$

Now y^+ and η are independent variables (we can change y^+ but not η by varying ν or V_*, and η but not y^+ by varying W). It follows that

$$y^+ f'(y^+) = -\eta g'(\eta) = \text{constant} = 1/\kappa.$$

[5] Actually, this is strictly not true as information is transmitted throughout the flow by pressure. Thus, in principle, eddies near a wall know about all of the other eddies in the flow. In fact, there is an influence of the large, central eddies, and hence W, on the near-wall region. However, it turns out that this influence is restricted to the distribution of turbulent kinetic energy and does not effect the Reynolds stress or the mean velocity profile. This issue is discussed later in Section 4.2.2.

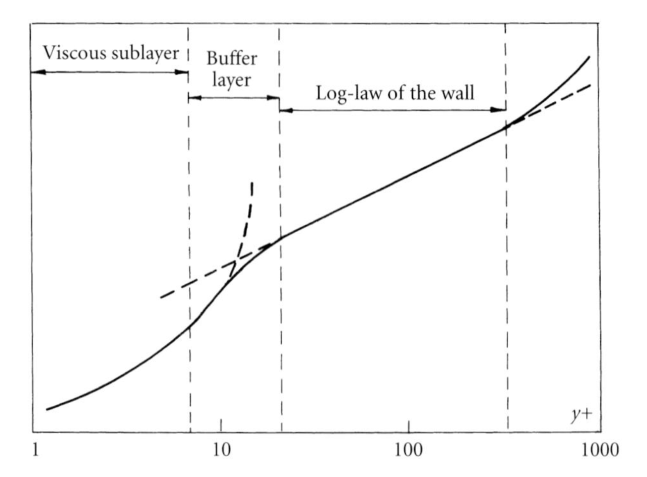

Figure 4.9 A plot of $\bar{u}_x/V_*$ versus y^+ showing the log-law of the wall.

On integration we find

$$\frac{\bar{u}_x}{V_*} = \frac{1}{\kappa}\ln y^+ + A \tag{4.44}$$

$$\frac{\bar{u}_0 - \bar{u}_x}{V_*} = -\frac{1}{\kappa}\ln \eta + B. \tag{4.45}$$

This is the famous *log-law of the wall*, and the constant κ is called Karman's constant. There is some scatter in the experimental estimates of κ, with most of the data lying in the range 0.38 to 0.43. Many researchers take $\kappa = 0.41$. However, because of the uncertainty in the second decimal place we shall simply adopt $\kappa = 0.4$.

The log-law of the wall is a remarkable result because we have used only very general arguments to derive it.[6] It is deeply satisfying to discover that it is an excellent fit to the experimental data. In particular, with the choice of $A \approx 5.5$ and $B \approx 1.0$ we find that (4.44) and (4.45) are a good fit in the range $y^+ > 40$ and $\eta < 0.2$. In the region $y^+ < 5$ it turns out that the flow is partially (but not completely) laminar and we have $\bar{u}_x \approx V_*^2 y/\nu$. This region is referred to as the *viscous sublayer* and the adjoining region, $5 < y^+ < 40$, is called the *buffer layer*. This information is summarized in Figure 4.9 and in Table 4.2.

There are other ways of deriving the log-law of the wall. For example, Prandtl's mixing-length theory, with $l = \kappa y$, leads to (4.44), as discussed in Exercise 4.5. (The conventional rationale for taking $l \propto y$ is that the average eddy size grows as we move away from the wall.)

[6] It is symptomatic of the field of turbulence that even the log-law of the wall, often held up as a land mark result, has its detractors. There are those who would prefer a power law and who question the need for universality. In particular, it has been suggested that (4.44) be replaced by $\bar{u}_x/V_* = a(y^+)^n$ where a and n are functions of Re. Nevertheless, the experimental data is well represented by (4.44), which has the advantage over the power law of being universal. More details of this controversy may be found in Buschmann and Gad-el-Hak (2003) and in Section 4.2.2.

Table 4.2 Different regions in channel flow and their associated velocity distributions

Outer region, $y^+ \gg 1$	Velocity defect law	$(\bar{u}_0 - \bar{u}_x)/V_* = g(y/W)$
Overlap region, $y^+ \gg 1$, $\eta \ll 1$	Log-law of the wall	$\bar{u}_x/V_* = (1/\kappa)\ln(V_* y/\nu) + A$
Inner region, $\eta \ll 1$	Law of the wall	$\bar{u}_x/V_* = f(V_* y/\nu)$
Viscous sublayer, $y^+ < 5$		$\bar{u}_x = V_*^2 y/\nu$

However, the derivation given above is the most satisfactory, as it makes the least number of assumptions. We note in passing that, in the outer region, it is conventional to write the defect law as

$$\frac{\Delta\bar{u}_x}{V_*} = \frac{\bar{u}_0 - \bar{u}_x}{V_*} = -\frac{1}{\kappa}\ln\eta + B - \Pi(\eta) \tag{4.46}$$

where $\Pi(\eta)$, the difference between the defect law and the log-law, is called the wake function. There have been several empirical suggestions for Π and the interested reader might consult Tennekes and Lumley (1972). We might also note that (4.44) and (4.45) may be combined to give

$$\frac{\bar{u}_0}{V_*} = \frac{1}{\kappa}\ln\left(\frac{WV_*}{\nu}\right) + A + B$$

which relates the centre-line velocity to V_*, and hence to K, the pressure gradient. This expression is an excellent fit to the experimental data for $\mathrm{Re} > 3000$.

4.2.2 *Inactive motion—a problem for the log-law?*

All in all, we seem to have a fairly complete picture of the Reynolds stress and mean velocity distributions, a picture which is well supported by the experiments. However, the scaling arguments which led us to the log-law have an Achilles' heel. It rests on the assumption that the turbulence near the wall is independent of W. A moment's thought is sufficient to confirm that this cannot be true. Remember that we have defined turbulence to be a vorticity field which advects itself in a chaotic manner. In a channel flow or a boundary layer this is vorticity which has been stripped off the rigid surfaces and thrown into the interior of the flow. This vorticity is (almost) frozen into the fluid and advects itself in a chaotic fashion in accordance with the vorticity transport equation. The turbulent velocity field is, in many ways, an auxiliary field, dictated by the instantaneous distribution of vorticity in accordance with the Biot–Savart law. It follows that an eddy (vortex blob) in the core of the flow induces a velocity field which pervades *all* of the fluid, including the near-wall region.

However, the core eddies are aware of, and depend on, the channel width $2W$. It follows that there *are* velocity fluctuations near the wall which depend on W, and it would seem that this is enough to discredit our derivation of the log-law.

So why does the log-law work so well in practice? Well, it turns out that the near-wall velocity fluctuations associated with the remote core eddies contribute little to τ^R_{xy}, and hence do not greatly influence the mean velocity profile in the vicinity of the wall. We can show that this is so as follows. First we note that

$$\frac{\partial}{\partial y}\left[-\overline{u'_x u'_y}\right] = \overline{u'_y \omega'_z} - \overline{u'_z \omega'_y} + \frac{1}{2}\frac{\partial}{\partial x}\left[\overline{(u'_x)^2} - \overline{(u'_y)^2} - \overline{(u'_z)^2}\right],$$

a relationship which can be confirmed by expanding the vorticity components. Since there is no x-dependence of the statistical variables this simplifies to

$$\frac{\partial}{\partial y}\left[\frac{\tau^R_{xy}}{\rho}\right] = \left[\overline{\mathbf{u}' \times \boldsymbol{\omega}'}\right]_x.$$

Evidently, the near-wall Reynolds stress depends only on the near-wall vorticity fluctuations and on those velocity perturbations which are strongly correlated to this vorticity. Let us now divide the near-wall turbulent velocity field into two parts: those fluctuations which are caused by the small-scale eddies near the wall, and those which originate from the remote, core eddies. The former are predominantly rotational, while the latter are largely irrotational (except to the extent that some of the core vortices will extend to the wall).

$$\underset{\text{(near-wall velocity)}}{\mathbf{u}'} = \underset{\text{(from near-wall eddies)}}{\mathbf{u}'_{\text{rot}}} + \underset{\text{(from remote, core eddies)}}{\mathbf{u}'_{\text{irrot}}}.$$

So in the near-wall region we have,

$$\frac{\partial}{\partial y}\left[\frac{\tau^R_{xy}}{\rho}\right] = \left[\overline{\mathbf{u}'_{\text{rot}} \times \boldsymbol{\omega}'}\right]_x + \left[\overline{\mathbf{u}'_{\text{irrot}} \times \boldsymbol{\omega}'}\right]_x.$$

However, the vorticity fluctuations near the wall are mostly small in scale, whereas the near-wall motion induced by the remote, core eddies consists of a large, planar, sweeping motion parallel to the surface. This irrotational motion operates over length and timescales much greater than those of the near-wall vorticity fluctuations. Thus we might expect $\mathbf{u}'_{\text{rot}}$ and $\boldsymbol{\omega}'$ to be reasonably well correlated but $\mathbf{u}'_{\text{irrot}}$ and $\boldsymbol{\omega}'$ to be only weakly correlated. Indeed, we might anticipate that, as far as the near-wall eddies are concerned, the slow sweeping motion induced by the core vortices looks a bit like a random fluctuation in the mean flow. If this is all true we would expect the large-scale irrotational fluctuations to make little or no contribution to the near-wall Reynolds stress,

$$\frac{\partial}{\partial y}\left[\frac{\tau_{xy}^R}{\rho}\right] \approx \left[\overline{\mathbf{u}'_{\text{rot}} \times \boldsymbol{\omega}'}\right]_x,$$

and this would account for the success of the log-law. For this reason $\mathbf{u}'_{\text{irrot}}$ is sometimes referred to as an *inactive motion* (Townsend 1976). Note, however, that $\mathbf{u}'_{\text{irrot}}$ will influence the kinetic energy distribution near the surface, so that the variation of $\overline{(u'_x)^2}$, $\overline{(u'_z)^2}$, and k in the log-layer should, in principle, depend on the channel width, W.

Townsend's idea that, as far as the near-wall dynamics is concerned, the slow sweeping action of the core eddies looks like a random modulation of the mean flow, has some interesting repercussions. In particular, it casts doubt over the assumed universality of Karman's constant, κ. The idea is the following. We know that the timescale of the core eddies is much greater than the turn-over time of the near-wall vortices. It follows that the inactive motion (the sweeping effect of the core eddies) leaves the flow near the wall in a state of quasi equilibrium. On averaging over the fast timescale of the wall eddies we have,

$$\frac{\partial(\bar{\mathbf{u}} + \mathbf{u}')}{\partial y} = \frac{\boldsymbol{\tau}/\rho}{(|\boldsymbol{\tau}|/\rho)^{1/2}\kappa y}.$$

Here $\mathbf{u}'$ represents the (unsteady) inactive motion and $\boldsymbol{\tau}$ must be interpreted as an unsteady wall stress, incorporating both the shear associated with the mean flow, $\tau_0\hat{\mathbf{e}}_x$, and that caused by the inactive motion, $\boldsymbol{\tau}'$. The streamwise component of our equation is, of course,

$$\frac{\partial(\bar{u}_x + u'_x)}{\partial y} = \frac{\tau_0 + \tau'_x}{\tau_0^{1/2}|\tau_0\hat{\mathbf{e}}_x + \boldsymbol{\tau}'|^{1/2}}\frac{V_0}{\kappa y}$$

where V_0 is the shear velocity associated with $\tau_0, V_0 = \sqrt{\tau_0/\rho}$. We now expand $|\boldsymbol{\tau}|^{-1/2}$ as a power series in $|\boldsymbol{\tau}'|/\tau_0$ to yield,

$$\frac{\partial(\bar{u}_x + u'_x)}{\partial y} = \left[1 + \frac{\tau'_x}{2\tau_0} - \frac{(\tau'_x)^2}{8\tau_0^2} - \frac{(\tau'_z)^2}{4\tau_0^2} + O(\tau'^3)\right]\frac{V_0}{\kappa y}.$$

Next we consider the relationship between and $\mathbf{u}'$ and $\boldsymbol{\tau}'$. Starting from $\boldsymbol{\tau} \sim |\mathbf{u}|\mathbf{u}$ and again expanding in a power series it is readily confirmed that

$$\tau'_x = \tau_0\left[2\frac{u'_x}{\bar{u}_x} + \frac{u'^2_x}{\bar{u}_x^2} + \frac{u'^2_z}{2\bar{u}_x^2} + O(u'^3)\right], \qquad \tau'_z = \tau_0\left[\frac{u'_z}{\bar{u}_x} + \frac{u'_x u'_z}{\bar{u}_x^2} + O(u'^3)\right]$$

and on substituting for $\boldsymbol{\tau}'$ in terms of $\mathbf{u}'$ our log-law simplifies to

$$\frac{\partial(\bar{u}_x + u'_x)}{\partial y} = \left[1 + \frac{u'_x}{\bar{u}_x} + O(u'^3)\right]\frac{V_0}{\kappa y}.$$

Finally we average over the long timescale of the inactive motion to obtain

$$\partial\bar{u}_x/\partial y = V_0/\kappa y$$

At first sight this seems unremarkable (i.e. consistent with the log-law), until we notice that the long time-average of our equation for τ'_x yields

$$\bar{\tau}_x = \tau_0 + \overline{\tau'_x} = \tau_0\left[1 + \frac{\overline{u'^2_x}}{\bar{u}^2_x} + \frac{\overline{u'^2_z}}{2\bar{u}^2_x} + O(u'^3)\right]$$

In short, the time-averaged wall stress is *not* τ_0, but rather $\tau_0 + \overline{\tau'_x}$. Consequently, the observed value of V_*, as measured by, say, the pressure gradient, differs from V_0,

$$V_* = \gamma V_0, \qquad \gamma = 1 + \frac{\overline{u'^2_x}}{2\bar{u}^2_x} + \frac{\overline{u'^2_z}}{4\bar{u}^2_x} + O(u'^3),$$

and so our long time-average of the log-law of the wall becomes

$$\frac{\partial\bar{u}_x}{\partial y} = \frac{V_*}{(\gamma\kappa)y} = \frac{V_*}{\kappa_{\text{eff}}y}.$$

The suggestion is that in any experiment the measured value of the Karman constant, κ_{eff}, will be greater than the supposed universal value, κ, by a factor of γ. Moreover, this amplification will depend on the value of Re. In practice, however, this effect is small (2% or less) since the magnitude of the inactive motion is always much less than that of the mean flow.

The observation that there may be more than one length scale of importance in the near-wall region has encouraged some researchers to look for an alternative to the log-law of the wall, particularly in pipe flow. One common suggestion is the power law

$$\bar{u}_x/V_* = a(y^+)^n,$$

where a and n are usually taken to be functions of Re. One proposal is $a = a_1\ln(\text{Re}) + a_2$, $n = a_3/\ln(\text{Re})$, the parameters a_1, a_2, and a_3 being assumed to be universal when Re is based on the mean velocity. Certainly a judicious choice of a_1, a_2, and a_3 provides a good fit to the experimental data, as good as that obtained using the log-law. At first sight this looks like a return to the pre-log-law empiricism of hydraulic engineering. However, theoretical arguments have been put forward in support of a power law and it is currently receiving considerable attention. (See Buschmann and Gad-el-Hak 2003 for a recent review.) Of course its weakness is the Re dependence of the coefficients a and n. Occam's razor might lead us to favour the log-law.

4.2.3 *Turbulence profiles in channel flow*

Many measurements of τ_{xy}^R, $k = \frac{1}{2}\overline{(\mathbf{u}')^2}$, and $S = \partial \bar{u}_x / \partial y$ have been made in channels. The value of τ_{xy}^R, as well as the individual contributions to $2k$, $\overline{(u_x')^2}$, $\overline{(u_y')^2}$, and $\overline{(u_z')^2}$, rise steadily through the viscous sublayer and the lower part of the buffer region, say $y^+ < 15$. In the upper part of the buffer region $\overline{(u_x')^2}$ drops somewhat, though τ_{xy}^R and the other two contributions to k continue to rise (Figure 4.10). By the time we reach the log-law region, $y^+ > 40$, the Reynolds stress and the rms turbulent velocity components have more or less (but not quite) settled down to the asymptotic values shown below.

$$\overline{u_x'^2}/k \approx 1.1, \quad \overline{u_y'^2}/k \approx 0.3, \quad \overline{u_z'^2}/k \approx 0.6, \quad \tau_{xy}^R/\rho k \approx 0.28$$

Actually, because of the large core eddies (i.e. the inactive motion), the ratio of k to τ_{xy}^R is not quite constant in the log region, but rather falls slowly as we move away from the wall. Townsend (1976) predicted this decay and, by making some plausible assumptions about the distribution of eddies in the core flow he was able to suggest

$$\rho k / \tau_{xy}^R = c_1 + c_2 \ln(W/y)$$

for the variation of k in the log region. Here c_1 and c_2 are constants which depend on the shape and distribution of the core eddies. (Townsends model is based on the idea that a typical eddy of diameter d is centred a distance $d/2$ from the wall, and so eddies of all size are in contact with the wall—the so-called *attached eddy hypothesis*.)

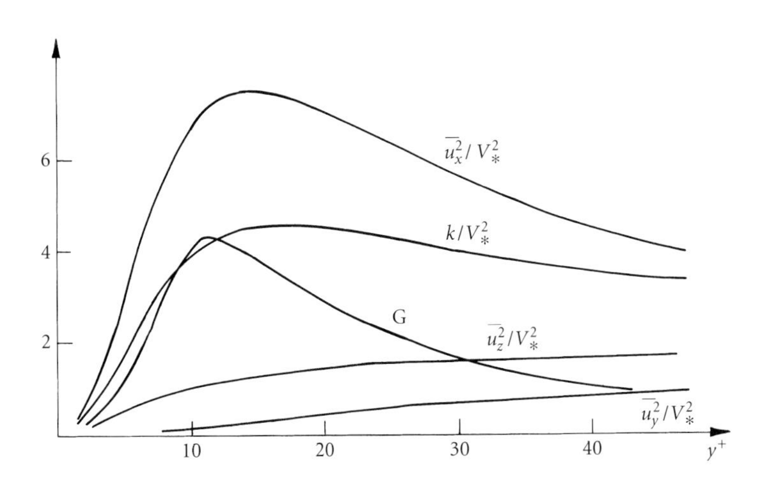

Figure 4.10 Plots of $\overline{(u_x')^2}$, $\overline{(u_y')^2}$, $\overline{(u_z')^2}$, and k all normalized by V_*^2.

Let us now return to the measurements of near-wall turbulence. The approximate distribution of turbulent energy generation, $G = (\tau^R_{xy}/\rho)\partial\bar{u}_x/\partial y$, is also shown in Figure 4.10. We know that G falls to zero at the wall and takes the value $G = V_*^3/\kappa y$ in the log layer. So we would expect G to be a maximum in the buffer region and this is exactly what is found. It rises steadily from $G = 0$ at the wall to a peak value at $y^+ \sim 12$ and then declines throughout the upper part of the buffer region and throughout the log layer. A similar behaviour is exhibited by k and by the dimensionless parameters G/ε and Sk/ε. They all peak in the lower part of the buffer region. Evidently, the buffer region $(5 < y^+ < 40)$ is the seat of the most violent turbulent activity. Most data suggests that G/ε and Sk/ε settle down to more or less constant values for $y^+ > 40$, $y/W < 0.2$. The approximate values of G/ε and Sk/ε in the log-law region are (Pope 2000),

$$G/\varepsilon \approx 0.91, \qquad Sk/\varepsilon \approx 3.2.$$

At first sight it may seem odd that G and ε do not balance. However (4.21) applied to channel flow requires $G - \varepsilon = \partial T_y/\partial y$ where $\mathbf{T}$ is the diffusive flux of kinetic energy arising from viscous effects, pressure fluctuations and triple correlations. Evidently there is some slight cross-stream diffusion of turbulent energy.

Note that the k–ε constant, c_μ, is readily evaluated for this flow. It is defined by (4.28) to be,

$$c_\mu = \nu_t[\varepsilon/k^2] = [\tau^R_{xy}/\rho S][\varepsilon/k^2]$$

Given that $\tau^R_{xy}/\rho k \approx 0.28$ and $Sk / \varepsilon \approx 3.2$, we obtain $c_\mu \approx 0.09$, which is precisely the value adopted in the k–ε model.

The peak in G/ε occurs around $10 < y^+ < 15$ and takes a value of between 1.5 and 2.0. Evidently, very near the wall the production of kinetic energy significantly out-ways the local dissipation. Thus there is a strong cross-stream diffusion of energy, both towards the wall and towards the core flow. For more details of the distributions of G, ε, and τ_{xy}, consult Townsend (1976).

4.2.4 *The log-law for a rough wall*

So far we have said nothing about the influence of wall roughness on the velocity profile. If the rms roughness height, $\hat{k}$, is large enough (greater than the viscous sublayer) then $\hat{k}$ becomes an important new parameter and the velocity profile in the inner region must be of the form,

$$\bar{u}_x/V_* = f(y/\hat{k}, V_* y/\nu), \quad y/W \ll 1.$$

For large values of $V_*\hat{k}/\nu$ viscous effects become negligible by comparison with the turbulence generated by the roughness elements and ν ceases to be a relevant parameter (except very close to the wall). The expression for $\bar{u}_x/V_*$ then simplifies to $\bar{u}_x/V_* = f(y/\hat{k})$. If we now repeat the arguments which lead up to (4.44) we find a modified form of the law of the wall:

$$\bar{u}_x/V_* = \frac{1}{\kappa}\ln(y/\hat{k}) + \text{constant} = \frac{1}{\kappa}\ln(y/y_0)$$
$$\text{(rough surface)}$$

where y_0 is defined in terms of $\hat{k}$ and the unknown additive constant. For sandgrain-type roughness we find (4.44) remains valid, and roughness may be ignored, for $V_*\hat{k}/\nu < 4$, while the fully rough expression above is legitimate for $V_*\hat{k}/\nu > 60$, with the additive constant equal to 8.5. (Interpolation formula exist for $4 < V_*\hat{k}/\nu < 60$.) Note that an additive constant of 8.5 gives $y_0 = \hat{k}/30$, which is much less than $\hat{k}$.

4.2.5 *The structure of a turbulent boundary layer*

The essence of the arguments above is that there is a region near the wall where the flow does not know or care about the gross details of the outer flow. Its properties are universal and depend only on V_*, y, and ν (or V_* and $\hat{k}$ if the wall is rough). We would expect, therefore, to find a log-law and a viscous sublayer adjacent to any smooth, solid surface in a shear layer. Moreover, the values of κ and A should be universal. The only requirements are that $\text{Re} \gg 1$ and that variations of $\bar{u}_x$, $\overline{u'_x u'_y}$, etc in the streamwise direction are small. This is exactly what is observed. For example, in a smooth pipe of radius R we find,

$$\begin{aligned} &\text{Inner region:} && \bar{u}_x = V_* f(y^+); && \text{for } y \ll R \\ &\text{Outer region:} && \bar{u}_0 - \bar{u}_x = V_* g(y/R); && \text{for } y^+ \gg 1 \\ &\text{Overlap region:} && \bar{u}_x = V_*\left[\frac{1}{\kappa}\ln y^+ + A\right]; && \text{for } y^+ \gg 1,\ y \ll R. \end{aligned}$$

Similarly, in a turbulent boundary layer on a flat plate we have

$$\begin{aligned} &\text{Inner region:} && \bar{u}_x = V_* f(y^+); && y \ll \delta \\ &\text{Outer region:} && \Delta\bar{u}_x = \bar{u}_\infty - \bar{u}_x = V_* g(y/\delta); && y^+ \gg 1 \\ &\text{Overlap region:} && \bar{u}_x = V_*\left[\frac{1}{\kappa}\ln y^+ + A\right]; && y^+ \gg 1,\ y \ll \delta. \end{aligned}$$

Here δ is the local thickness of the boundary layer, $\bar{u}_\infty$ is the free-stream velocity as shown in Figure 4.11, and the log-law is found to apply for $y^+ > 40$ and $y/\delta < 0.2$. Although the form of $\bar{u}_x$ in the inner region is universal, the details of the flow in the outer region

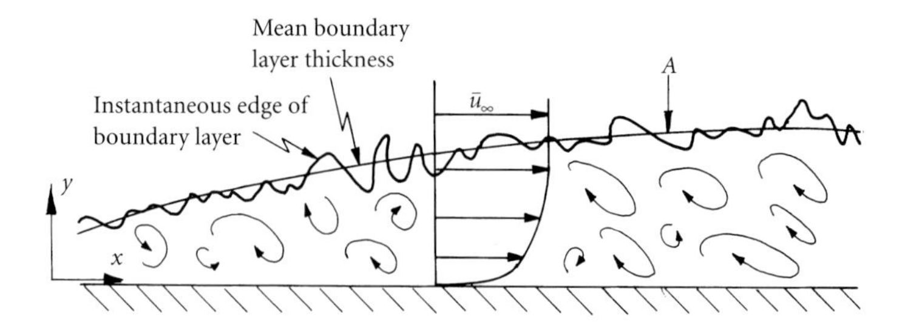

Figure 4.11 Turbulent boundary layer.

depend on the global characteristics of the flow, and especially on the free-stream pressure gradient.

It is important to note that the viscous sublayer is not completely quiescent. It turns out that it is subjected to frequent turbulent *bursts*, in which fluid is ejected from the wall, carrying its intense vorticity with it. Indeed, these bursts are thought to be one of the mechanisms by which high levels of vorticity are maintained in the boundary layer. That is to say, all of the vorticity in a boundary layer (both the mean and the turbulent vorticity) must originate from the rigid surface. This vorticity may spread upward by diffusion or else by advection. In a quiescent viscous sublayer the dominant mechanism is diffusion, whereby vorticity oozes out of the surface and into the adjacent fluid. However, this is a slow process, and when a turbulent burst occurs the local value of Re becomes large, allowing vorticity to be transported by the more efficient mechanism of material advection. Thus, each turbulent burst propels near-wall vorticity out into the core of the boundary layer. Perhaps it is useful to think of two parallel processes occurring near the wall. On the one hand we have vorticity continually diffusing out from the surface and building up large reserves of enstrophy adjacent to the wall. On the other hand these reserves are occasionally plundered by random turbulent burst which fling the vorticity out into the main boundary layer.

One of the remarkable features of a turbulent boundary layer, indicated in Figure 4.11, is the highly convoluted shape of the instantaneous edge of the layer. A velocity probe placed at position A, say, will experience intermittent bursts of turbulence. Perhaps we should explain what the outer edge of the boundary layer (called the viscous superlayer) shown in Figure 4.11 really represents. Below the convoluted edge we have vorticity, above it we have none. So by the phrase 'turbulent boundary layer' we mean: 'that part of the flow into which the vorticity originally generated at the surface has now spread'. Now, when Re is large, vorticity is virtually frozen into the fluid and moves with the fluid. Thus we see that the convoluted outer surface simply represents the material advection of vorticity by large-scale eddies as they tumble and roll along the boundary layer.

You might ask: why is turbulence restricted to that part of the flow field in which $\boldsymbol{\omega}$ is non-zero? This is an interesting question. In fact, if we make velocity or pressure measurements in the irrotational region (outside the boundary layer) then we do indeed detect random fluctuations in $\mathbf{u}$ and p. So, in a sense, this region is also turbulent. Actually, we know this must be true because (2.23) tells us that a fluctuation in velocity at any one point sends out pressure waves (which travel infinitely fast in an incompressible fluid) and these pressure waves induce irrotational motion. Thus, as an eddy rolls along the boundary layer, it induces pressure fluctuations (which fall off as y^{-3}) and hence pressure forces and velocity fluctuations in the fluid outside the boundary layer.[7] However, we do not choose to call the fluctuating, irrotational motion turbulent. Rather, we think of it as a passive response to the nearby turbulent vorticity field. This is a little arbitrary, but it reflects the fact that there can be no intensification of velocity fluctuations by vortex stretching, and hence no energy cascade, in an irrotational flow. Moreover, it is readily confirmed that the y-derivative of the Reynolds stress τ_{xy} is zero in the irrotational fluid and so the point at which $\bar{u}_x$ reaches the free-stream value of $\bar{u}_\infty$ is effectively the same as the time-averaged edge of the vortical region.

4.2.6 *Coherent structures*

Another striking feature of boundary layers is the existence of *coherent structures*. This is a rather imprecise term, but it is usually used to describe vortical structures which are robust in the sense that they retain their identity for many eddy-turn-over times and which appear again and again in more or less the same form. An example of such a structure is seen in Plate 5(b) which shows structures in a boundary layer visualized with the aid of smoke and a sheet of laser light. It is clear that mushroom-like eddies are a common feature of such a flow. Note that the plate shows activity in a single plane only, and so the precise interpretation of these mushroom-like structures has been a source of controversy. Most researchers claim that they represent a slice through a so-called *hairpin vortex*. These are vortex loops which span the boundary layer, arching up from the surface (see below). Others argue that they are vortex rings, localized in the outer part of the boundary layer.

In any event the most famous of the coherent structures is the so-called hairpin vortex (Figure 4.6). These are long, arch-like vortices of

[7] The fact that pressure fluctuations due to an eddy fall off as y^{-3} is established in Chapter 6, Section 6.3.4.

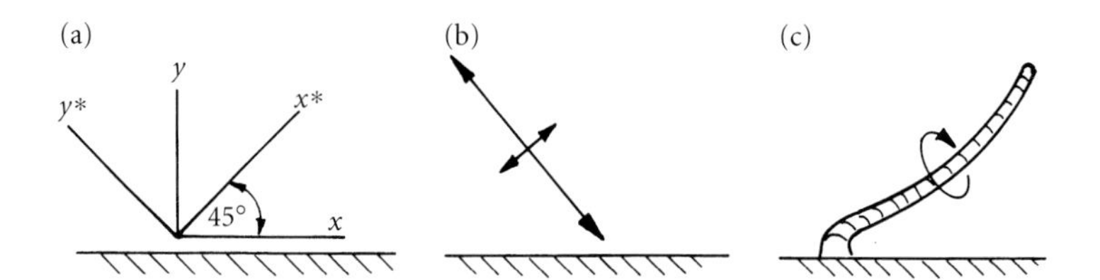

Figure 4.12 (a) Coordinate system x^*, y^*. (b) Velocity fluctuations needed for a large Reynolds stress. (c) Side view of a hairpin vortex.

modest diameter. The tubes have a maximum length of the order of δ and their diameter may be as small as $\sim 5\nu/V_*$. Flow visualization studies suggest that turbulent boundary layers are liberally populated with hairpin vortices of a variety of sizes, many of them orientated at about 45° to the mean flow (Head and Bandyopadhyay 1981). It is likely that they are far from passive and indeed they may contribute significantly to the Reynolds stress and hence to the production of turbulent energy.[8]

Recall that energy passes from the mean flow to the turbulence at a rate,

$$\tau_{xy}^R \bar{S}_{xy} \sim -\rho \overline{u'_x u'_y} \frac{\partial \bar{u}_x}{\partial y}.$$

Now consider a coordinate system (x^*, y^*), which is inclined at 45° to (x, y). Our Reynolds stress can be expressed in terms of these new coordinates as

$$\tau_{xy}^R = \frac{\rho}{2} \left\{ \overline{(u'_{y^*})^2} - \overline{(u'_{x^*})^2} \right\} \tag{4.47}$$

So, a high Reynolds stress is associated with large fluctuations in the y^* direction and weak fluctuations in the x^* direction (Figure 4.12(b)). However, this is precisely what a hairpin vortex achieves (Figure 4.12(c)) and so they are prime candidates for generating positive τ_{xy}^R and, by implication, positive $\tau_{xy}^R \bar{S}_{xy}$.[9]

This high rate of generation of turbulent energy may be explained as follows. The hairpin vortices are ideally orientated to be stretched by mean flow. This is because they are aligned with the principal strain rate, that is, the direction of maximum stretching. As they are stretched by the mean flow so the kinetic energy associated with the tubes is intensified, which represents an exchange of energy from the mean flow to the turbulence.

It is not difficult to see how hairpin vortices develop in the first place. The mean flow $\bar{u}_x$ is associated with a vorticity field

[8] Perry and Chong (1982) and Perry, Henbert, and Chong, (1986) have shown that many of the observed statistical features of a boundary layer may be reproduced by imagining that a boundary layer is composed of a hierarchy of hairpin vortices of different scales.

[9] Alternatively we may recall that $\frac{\partial}{\partial y}\left[\tau_{xy}^R\right] = \rho\left[\overline{\mathbf{u}' \times \boldsymbol{\omega}'}\right]_x$ so that a vertical velocity fluctuation combined with a cross-stream vorticity fluctuation gives rise to a gradient in Reynolds stress. This is exactly the situation at the tip of a rising hairpin vortex.

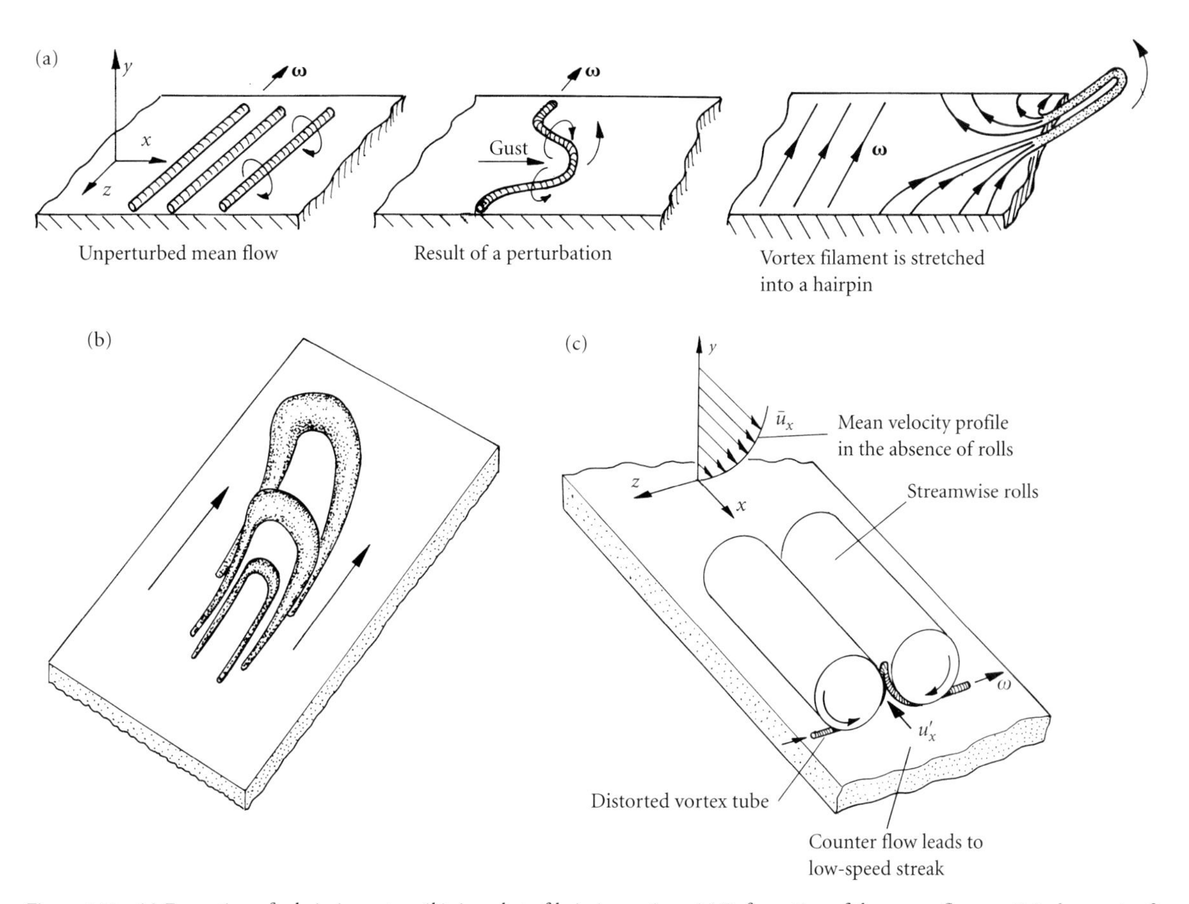

Figure 4.13 (a) Formation of a hairpin vortex. (b) A packet of hairpin vortices. (c) Deformation of the mean- flow vorticity by a pair of vortex rolls.

$\bar{\boldsymbol{\omega}} = (0, 0, \bar{\omega}_z)$, as shown in Figure 4.13(a). We might envisage this vorticity field as composed of a multitude of vortex filaments, or tubes. Now recall that, when Re is large, vortex tubes are (almost) frozen into the fluid. Any streamwise fluctuation in velocity (a gust) will therefore sweep out an axial component of vorticity and we have the beginning of a hairpin vortex.[10] It is readily confirmed that the curvature of a hairpin vortex induces a velocity field rather like one half of a vortex ring and that this tends to advect the tip of the hairpin upwards into the prevailing flow. As soon as the loop starts to rotate the tip of the loop finds itself in a region of high mean velocity relative to its base. The vortex loop then gets stretched out by the mean flow, intensifying the initial perturbation and promoting yet more rotation. Of course, the tendency for the hairpin to rotate, with its tip moving

[10] At the beginning of this process the turbulence is acting on the mean-flow vorticity by stretching its vortex lines. Later, the hairpin is regarded as part of the turbulence and the situation is reversed, with the mean flow doing work on the turbulence. There is no clear division between the two stages and this highlights the slightly artificial nature of pretending that there are two flows instead of one.

upwards, is countered, to some degree, by the mean shear which tends to rotate the vortex in the opposite direction. In order to maintain a quasi-steady orientation of 45° the two processes must roughly match and this, in turn, will favour a particular strength of vortex tube.

It seems probable that these hairpin vortices are initiated in the lower regions of the boundary layer, if for no other reason than the fact that the mean vorticity, which provides the nutrient in which these vortices grow, is most intense there. It is also likely that these vortices are eventually destroyed through their interaction with other boundary layer structures, or perhaps through cross-diffusion of vorticity between opposite legs of a single hairpin. In summary then, one possibility for the life cycle of a hairpin vortex is as follows:

(1) axial gust close to the wall + mean cross-stream vorticity → small, horizontal vortex loop;
(2) self-advection of small vortex loop → rotation of loop;
(3) inclined loop + mean shear → stretching and intensification of vortex loop;
(4) intensification of loop → accelerated rotation → more stretching;
(5) eventual destruction of loop through interaction with other eddies or else through the cross-diffusion of vorticity between adjacent legs.

It should be emphasized, however, that this is a highly idealized picture. Hairpins rarely appear as symmetric structures. Almost invariably one leg is much more pronounced than the other, and indeed often only one leg is apparent. Moreover, other explanations have been offered for the observed structures. One alternative scenario is that a large eddy in the upper part of the boundary layer is swept down towards the wall where it collides with the buffer region. The mean vortex lines in the wall layer are then bent out of shape and we rejoin the above sequence at point (2). In this picture, then, the formation of hairpin vortices is triggered by events which begin far from the wall. This is a sort of 'top–down' picture, as distinct from the 'bottom–up' view expressed earlier.

There are many other explanations for the observed structures. Some work better at low Re, others at high Re. Some capture the behaviour of rough surfaces, others seem to work best for smooth surfaces. All are cartoons. The unifying theme, however, is the power of the mean shear to stretch out the vortex tubes across the boundary layer to form inclined, elongated structures. Usually these structures are asymmetric, exhibiting one short and one long leg. Sometimes they appear in groups, with each hairpin initiating another in its wake (Figure 4.13(b)). In fact, hairpins seem to manifest themselves in a bewildering variety of forms. While many authors argue about the precise details of their origin and shape, all agree that hairpins of one form or another are present in a boundary layer.

Very close to the wall a different, though possibly related, type of structure is thought to exist. In particular, pairs of streamwise vortex tubes (or rolls), of opposite polarity, are thought by some to be the dominant structure for $y^+ < 50$ (Figure 4.13(c)).

The rotation in the tubes is such that fluid near the wall is swept horizontally towards the gap between the tubes and then pumped up and away from the wall. At any one time there may be many such pairs of rolls and so a marker introduced into the boundary layer near the wall, say hydrogen bubbles if the fluid is water, will tend to form long streamwise streaks, and this is precisely what is observed for $y^+ < 20$.[11]

There is an interaction between these rolls and the mean spanwise vorticity, in which the spanwise vortex lines are deflected upward in the gap between the rolls. This induces a velocity perturbation, u'_x, which is anti-parallel to the mean flow and so the streamwise velocity in a streak is below that of the ambient fluid. (See Example 4.5.) Hence they are known as *low-speed streaks*. These streaks are eventually ejected from the wall region once they get caught up in the updraft between the rolls. The ejection process is often followed by a so-called burst, in which there is a sudden loss of stability in the rising fluid and a more erratic motion ensues. Indeed, these near-wall bursts are thought by some to be one of the primary mechanisms of turbulent energy generation in a boundary layer. In some cartoons the streamwise rolls are associated with the lower regions of hairpin vortices,[12] those parts of the vortex which lie close to the wall having become highly elongated due to the strong local shear. Moreover, there is some evidence that the hairpins themselves appear in packets,[13] one following another (see Figure 4.13(b)), so that the low-speed streak which lies below such a packet may appear considerably longer than any one hairpin. (The low-speed streaks are typically 10^3 wall units long, $\sim 10^3 \, \nu / V_*$, while individual streamwise rolls might have a length of, say, $\sim 200 \, \nu / V_*$.) In this model, then, the streamwise streaks are an inevitable consequence of the rolls and the rolls themselves are the foot-points, or near-wall remnants, of one or more hairpin vortices. There are other cartoons, however.

In one alternative model the rolls are not attached to any vortical structure in the outer part of the boundary layer, but rather owe their

[11] Alternative explanations for the observed streaks have been offered. See Robinson (1991).

[12] In this picture the rolls are the foot-points of a hairpin vortex in which the lower portion of the hairpin has been subject to extensive streamwise straining, resulting in strong streamwise vorticity (see Figure 4.13(b)). This stretching also drives the two legs of the hairpin together, producing a structure reminiscent of that shown in Figure 4.13(c).

[13] Such packets arise as follows. Consider that region of a hairpin leg where the vortex begins to lift off the wall, developing a slight angle of inclination to the surface. The induced velocity at this point has a streamwise component which, in turn, bends the mean-flow vortex lines out of shape, that is, step (1) above. This initiates a new 'baby hairpin' in the wake of the parent vortex.

existence to a sort of near-wall cycle in which both the streaks and rolls are dynamically interacting structures. For example, some researchers believe in a regenerative cycle in which rolls produce streaks, the streaks become unstable, and the non-linear instability generates new rolls. Clearly there is still much debate over how exactly to interpret the experimental evidence, but see Jimenez (2002), Panton (2001), or Holme et al. (1996), for more details.

Example 4.5 Deformation of the mean-flow vorticity by a pair of vortex rolls

Consider the deformation of the mean-flow boundary-layer vorticity, $\overline{\omega}_z = -\partial \bar{u}_x / \partial y$, by a pair of vortex rolls, as shown in Figure 4.13(c). The mean-flow vortex lines are bent upward in the gap between the rolls, creating a vertical component of vorticity as shown. Use the Biot–Savart law to show that the perturbation in velocity midway between the rolls caused by this vortex-line deformation is anti-parallel to the mean flow, thus creating a low-speed streak.

Example 4.6 Winding up the mean-flow vorticity with a single streamwise vortex

Consider a homogeneous, z-directed shear flow, $\mathbf{u} = Sy\hat{\mathbf{e}}_z$. We are interested in how this flow is distorted by the introduction of a streamwise line vortex. Clearly the mean-flow vorticity, $\boldsymbol{\omega} = S\hat{\mathbf{e}}_x$, will be wound up by the line vortex as shown in Figure 4.14. This will result in oscillations of the axial velocity $u_z(x, y)$, possibly leading to an instability of the mean flow. In order to model this we consider an initial-value problem in which a line vortex is introduced at $t = 0$. In order to keep the analysis simple we assume no z-dependence in the flow at $t = 0$, which ensures a z-independent solution for all t. Confirm that the flow may written as the sum of two parts:

$$\text{axial flow:} \quad \mathbf{u}(x, y, t) = u_z(x, y, t)\hat{\mathbf{e}}_z, \qquad \boldsymbol{\omega}(x, y, t) = \omega_x\hat{\mathbf{e}}_x + \omega_y\hat{\mathbf{e}}_y$$

$$\text{line vortex:} \quad \mathbf{u}(x, y, t) = u_x\hat{\mathbf{e}}_x + u_y\hat{\mathbf{e}}_y, \qquad \boldsymbol{\omega}(x, y, t) = \omega_z(x, y, t)\hat{\mathbf{e}}_z$$

and that the governing equation for each part is,

$$\frac{\partial u_z}{\partial t} + \mathbf{u}_\perp \cdot \nabla u_z = \nu \nabla^2 u_z$$

$$\frac{\partial \omega_z}{\partial t} + \mathbf{u}_\perp \cdot \nabla \omega_z = \nu \nabla^2 \omega_z$$

where $\mathbf{u}_\perp = u_x\hat{\mathbf{e}}_x + u_y\hat{\mathbf{e}}_y$. Note that the line vortex is decoupled from the shear flow. In Section 5.3.3 of Chapter 5 we shall see that a line vortex freely decaying under the influence of viscosity takes the form

$$\omega_z = \frac{\Gamma_0}{\pi\delta^2}\exp(-r^2/\delta^2)$$

$$u_\theta = \frac{\Gamma_0}{2\pi r}[1 - \exp(-r^2/\delta^2)], \qquad u_r = 0$$

$$\delta^2 = \delta_0^2 + 4\nu t$$

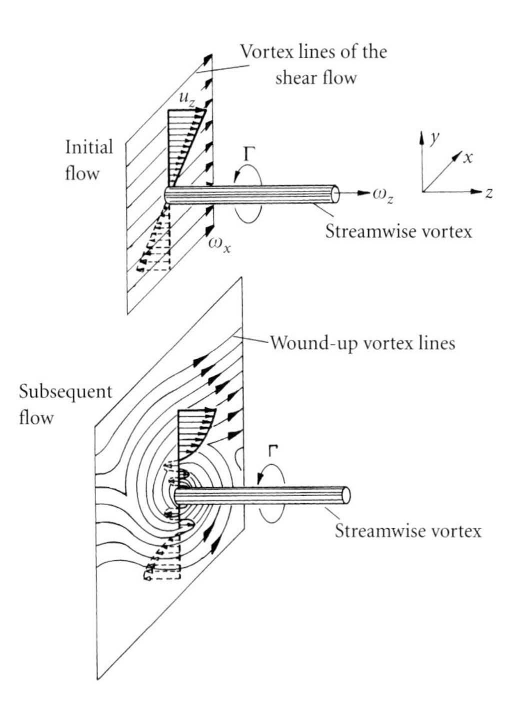

Figure 4.14 Winding up of the mean-flow vorticity by a streamwise vortex. (After F.S. Sherman 1990.)

where we have used cylindrical polar coordinates centred on the axis of the vortex. The constant Γ_0 represents the strength of the vortex and δ is its characteristic radius. For simplicity we shall take $\delta_0 = 0$.

The problem now is to determine the evolution of the axial flow, u_z. Confirm that the solution for $\nu = 0$ is

$$u_z = Sr\sin(\theta - \tau), \qquad \tau = \Gamma_0 t/2\pi r^2.$$

Sketch the variation of u_z and ω_θ with r and confirm that, as we approach the origin, they oscillate with ever increasing frequency. Now show that the solution for non-zero viscosity takes the form

$$\frac{u_z}{Sr} = A(\tau)\sin(\theta) + B(\tau)\cos(\theta)$$

and find the governing equations for A and B. Confirm that an asymptotic solution for small viscosity is

$$\frac{u_z}{Sr} = \exp\left(-\frac{8\pi\tau^3}{3\Gamma_0/\nu}\right)\sin(\theta - \tau).$$

4.2.7 *Spectra and structure functions near the wall*

So far we have discussed only one-point statistical properties of a boundary layer, that is, statistical properties measured at only one point in space. In Chapter 3 we emphasized that two-point statistical properties, such as the structure function

$$\left\langle \left[\Delta u'_x\right]^2 \right\rangle = \left\langle \left[u'_x(\mathbf{x} + r\hat{\mathbf{e}}_\mathbf{x}) - u'_x(\mathbf{x})\right]^2 \right\rangle$$

are required if we wish to determine the manner in which energy is distributed across the different eddy sizes in some region of the flow. We close our brief overview of boundary layers with a few comments about $\langle [\Delta u_x']^2 \rangle (r)$ which, as you will recall, provides a measure of the energy held in eddies of size r. (In this context both x and r are measured in the streamwise direction.)

When Re is large, it is observed that, in the outer part of the boundary layer, the Kolmogorov two-thirds law, $\langle [\Delta u_x']^2 \rangle \sim \varepsilon^{2/3} r^{2/3}$, holds true for $\eta \ll r \ll \delta$ where η now represents the Kolmogorov microscale. This is, of course, to be expected, since we anticipate that Kolmogorov's two-thirds law will hold whenever the inertial subrange may be treated as approximately homogeneous and isotropic (see Chapter 3). As we move closer to the wall, however, the eddies become increasingly inhomogeneous and anisotropic since the vertical fluctuations tend to zero much faster than the horizontal ones. This promotes a departure from the two-thirds law. In particular, we find that the $r^{2/3}$ inertial subrange becomes narrower and a new regime appears wedged between the $r^{2/3}$ range and the large scales (Perry, Henbest, and Chong 1986). It may be argued that this new regime is characterised by a constant value of $\langle [\Delta u_x']^2 \rangle$, which is of the order of $\sim V_*^2$. Since $\langle [\Delta u_x']^2 \rangle (r)$ is an estimate of the energy held in eddies of size r, these results tend to suggest that, near the wall, there exists a range of eddy sizes in which the kinetic energy density is more or less constant and equal to V_*^2. (Remember that $\overline{\mathbf{u}'^2}$, which is dominated by the large eddies, scales as V_*^2 in the log-law region.) Thus, as we approach the wall we find that $\langle [\Delta u_x']^2 \rangle \sim V_*^2$ to the right of the inertial subrange (large r), while the estimate $\langle [\Delta u_x']^2 \rangle \sim \varepsilon^{2/3} r^{2/3}$ continues to hold in the low-r end of the inertial subrange. This is shown schematically in Figure 4.15.

We may show that the cross-over between $\langle [\Delta u_x']^2 \rangle \sim V_*^2$ and $\langle [\Delta u_x']^2 \rangle \sim \varepsilon^{2/3} r^{2/3}$ occurs at $r \sim y$ as follows. Since $\varepsilon \sim G = V_*^3/\kappa y$ the form of $\langle [\Delta u_x']^2 \rangle (r)$ in the two regimes may be written as

$$\left\langle [\Delta u_x']^2 \right\rangle \sim V_*^2, \qquad \left\langle [\Delta u_x']^2 \right\rangle \sim V_*^2 (r/y)^{2/3}, \quad y \ll \delta.$$

Evidently there is a transition from one to the other at $r \sim y$. So we may summarize our results in the form

$$\left\langle [\Delta u_x']^2 \right\rangle \sim V_*^2, \qquad y < r < \delta, \quad (y \ll \delta)$$

$$\left\langle [\Delta u_x']^2 \right\rangle \sim V_*^2 (r/y)^{\frac{2}{3}}, \quad \eta \ll r < y, \quad (y \ll \delta)$$

Actually, data taken in the atmospheric boundary layer suggests that the $\langle [\Delta u_x']^2 \rangle \sim V_*^2$ law may be extended up to $r \sim 5\delta$ at high Re.

Usually these results are expressed in terms of the one-dimensional Fourier transform of u_x'. If $E_x(k_x)$ is the one-dimensional power

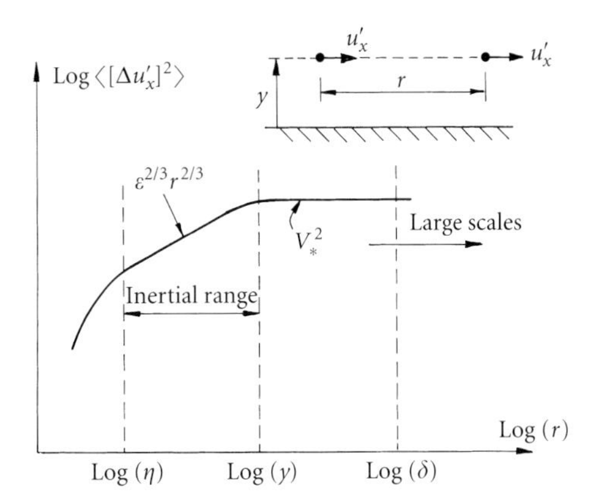

Figure 4.15 The possible shape of the structure function $\langle[\Delta u'_x]^2\rangle$ near the wall.

spectrum of u'_x, then the equivalent results in Fourier space are:

$$E_x(k_x) \sim V_*^2 k_x^{-1}, \qquad E_x(k_x) \sim \varepsilon^{2/3} k_x^{-5/3}$$

for the low-to-intermediate k and intermediate-to-high k ends of the spectrum respectively. This follows from the fact that

$$k_x E_x(k_x) \sim \left\langle \left[\Delta u'_x\right]^2 \right\rangle (r)$$

where $r \sim k_x^{-1}$ (see Chapter 8). Note that this k^{-1} behaviour is not exhibited by the one-dimensional spectrum of u_y. Note also that a large Re ($\text{Re} > 10^6$) is needed if the k^{-1} region is to be observed. Indeed, even at $\text{Re} = 10^6$ the extent of the k^{-1} region is so slight that there are those who remain unconvinced as to its existence.

This concludes our all too brief overview of wall-bounded shear flows. The subject is an important one and there exists a vast literature on it. Some suggestions for further reading are given in the references at the end of the chapter.

4.3 Free shear flows

We now turn to shear flows which are remote from boundaries: so-called free shear flows. This includes turbulent jets and wakes. In the interests of simplicity we start with two-dimensional jets and wakes.

4.3.1 *Planar jets and wakes*

Examples of planar jets and wakes are shown in Figure 4.16. The mean flow is characterized by $\bar{u}_x \gg \bar{u}_y$ and $\partial/\partial x \ll \partial/\partial y$. The turbulence, on the other hand, is characterized by the sudden transition from a turbulent vorticity field to an irrotational external motion. As in a boundary layer, the interface between the two regions is highly convoluted (see Plate 6), and so a probe placed at points A or B in Figure 4.16 will see intermittent bursts of turbulence. Curiously, though, the nature of this interface appears to change at around $\text{Re} \sim 10^4$ (Dimotakis 2000).

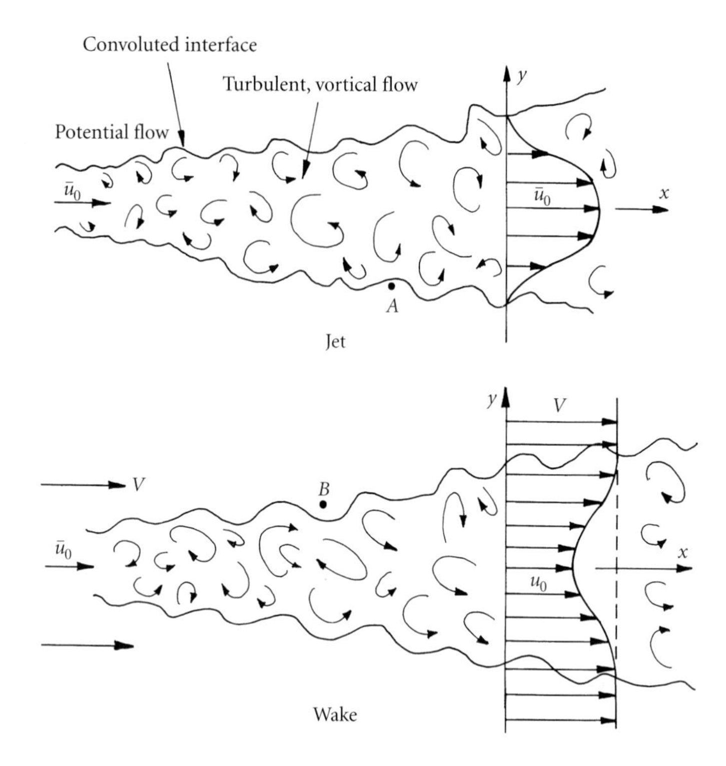

Figure 4.16 Planar jet and wake. See also Plate 6.

Below Re $\sim 10^3$ there is relatively little fine-scale turbulence and the interface between the turbulent and irrotational fluid undulates on the large scale only. Above Re $\sim 10^4$ the turbulence appears to be more fully developed, with a pronounced fine-scale structure. The turbulent/non-turbulent interface is now more intricate, exhibiting both large and small-scale wrinkles. As yet, there is no satisfactory explanation for this intriguing transition.

Let us now set out the governing equations for a jet and wake. We start by noting that there are three simplifying features of these flows:

(1) axial gradients in the Reynolds stresses, $\partial\tau_{ij}^R/\partial x$, are much smaller than transverse gradients;
(2) the laminar stresses are negligible;
(3) the transverse component of the mean inertia term, $(\bar{\mathbf{u}}\cdot\nabla)\bar{u}_y$, is of order $\bar{u}^2$/(radius of curvature of mean streamlines) and so is very small.

The axial and transverse equations of motion then simplify to

$$\rho(\bar{\mathbf{u}}\cdot\nabla)\bar{u}_x = \frac{\partial}{\partial y}\left[\tau_{xy}^R\right] - \frac{\partial\bar{p}}{\partial x} \tag{4.48}$$

$$0 = \frac{\partial}{\partial y}\left[\tau_{yy}^R\right] - \frac{\partial\bar{p}}{\partial y}. \tag{4.49}$$

The second of these tells us that

$$\bar{p} + \rho\overline{\left(u_y'\right)^2} = \bar{p}_\infty(x)$$

where $\bar{p}_\infty(x)$ is the pressure far from the jet or wake. However $\bar{p}_\infty$ is a constant if the external flow outside the wake is uniform (which we assume it is) and it is certainly constant for a jet. It follows that, in either case, $\bar{p}$ is a function of x only to the extent that $\overline{(u_y')^2}$ depends on x. Since longitudinal gradients in the Reynolds stresses may be neglected (4.48) simplifies to,

$$\rho(\bar{\mathbf{u}} \cdot \nabla)\bar{u}_x = \frac{\partial \tau_{xy}^R}{\partial y} \tag{4.50}$$

to which we might add

$$\nabla \cdot \bar{\mathbf{u}} = 0. \tag{4.51}$$

Combining these we obtain the simplified momentum equation,

$$\frac{\partial}{\partial x}\left[\rho\bar{u}_x^2\right] + \frac{\partial}{\partial y}\left[\rho\bar{u}_y\bar{u}_x\right] = \frac{\partial \tau_{xy}^R}{\partial y}. \tag{4.52}$$

In the case of the wake it is more convenient to rewrite this as

$$\frac{\partial}{\partial x}\left[\rho\bar{u}_x(V - \bar{u}_x)\right] + \frac{\partial}{\partial y}\left[\rho\bar{u}_y(V - \bar{u}_x)\right] = -\frac{\partial \tau_{xy}^R}{\partial y} \tag{4.53}$$

where V is the external velocity and the quantity $V - \bar{u}_x$ is known as the velocity deficit. Now $\bar{u}_x$ (in a jet) and $(V - \bar{u}_x)$ (in a wake) both tend to zero for large $|y|$, as does τ_{xy}^R. It follows that, if we integrate (4.52) and (4.53) from $y = -\infty$ to $y = +\infty$, we find,

$$M = \int_{-\infty}^{\infty} \rho\bar{u}_x^2 dy = \text{constant} \quad (\text{jet}) \tag{4.54}$$

$$D = \int_{-\infty}^{\infty} \rho\bar{u}_x(V - \bar{u}_x)dy = \text{constant} \quad (\text{wake}). \tag{4.55}$$

The first of these tells us that the momentum flux in a jet is conserved, that is, independent of x, while the second says that the *momentum deficit* in a wake is constant. Note that, although momentum, or momentum deficit, is conserved in a jet or a wake, the mass flux, $\dot{m}$, need not be. Indeed, a turbulent jet, like its laminar counterpart, drags ambient fluid into it, increasing its mass flux. This process is known as entrainment. In a laminar jet, entrainment is caused by viscous drag, while in the turbulent case it is a result of the convoluted outer boundary of the jet which continually engulfs external, irrotational

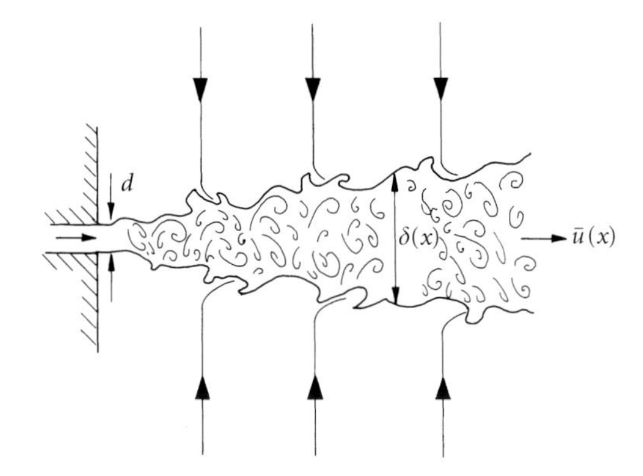

Figure 4.17 Entrainment into a planar jet.

fluid. Thus, far from the jet, there is a small but finite flow towards the jet which feeds its increasing girth (Figure 4.17).

We now consider jets and wakes separately. We start with the former. Traditionally, there have been two distinct approaches to modelling turbulent jets. As it turns out, both yield more or less the same result. One method of attack rests on the Boussinesq–Prandtl idea of an eddy viscosity; the other rests on characterizing the rate of entrainment into the jet. We start with the entrainment argument.

Let d be the initial width of the jet. It is an empirical observation that, after a distance of $\sim 30d$, the jet has forgotten the precise details of its initiation and so its local structure is controlled simply by the local jet speed, say its centre-line velocity, and the jet width, $\delta(x)$. Thus we may write $\bar{u}_x(x,y) = f(y, \bar{u}_0(x), \delta(x))$. It is then a dimensional necessity that the jet adopts the self-similar structure

$$\frac{\bar{u}_x}{\bar{u}_0(x)} = f(y/\delta(x)) = f(\eta) \tag{4.56}$$

where $\bar{u}_0$ is the centre-line velocity. It is this simple yet crucial observation about the self-similarity of $\bar{u}_x$ which allows us to make some progress. Let us pursue some of its consequences. Two immediate results are that, for large enough x,

$$M = \int_{-\infty}^{\infty} \rho \bar{u}_x^2 dy = \rho \bar{u}_0^2 \delta \int_{-\infty}^{\infty} f^2 d\eta = \text{constant}$$

$$\dot{m} = \int_{-\infty}^{\infty} \rho \bar{u}_x dy = \rho \bar{u}_0 \delta \int_{-\infty}^{\infty} f d\eta \propto \rho \bar{u}_0 \delta.$$

Now let us suppose that the rate of entrainment of mass at any one point is proportional to the local fluctuations in velocity, $\mathbf{u}'$. These, in turn, are proportional to the mean (local) velocity in the jet, and so on dimensional grounds we have

$$\frac{d}{dx}(\dot{m}) = \alpha \rho \bar{u}_0 \int_0^{\infty} f d\eta$$

where α is an example of an *entrainment coefficient*. (The integral on the right is included for algebraic convenience.) It follows that, for large enough distances downstream,

$$\bar{u}_0^2\delta = \text{constant}, \qquad \frac{d}{dx}(\bar{u}_0\delta) = \frac{1}{2}\alpha\bar{u}_0$$

If we take α to be constant we can integrate these equations to yield

$$\frac{\delta}{\delta_0} = 1 + \frac{\alpha x}{\delta_0} \tag{4.57a}$$

$$\frac{\bar{u}_0}{V_0} = \left[1 + \frac{\alpha x}{\delta_0}\right]^{-1/2} \tag{4.57b}$$

where the origin for x is taken at the start of the self-similar regime and δ_0 and V_0 are the values of δ and $\bar{u}_0$ at $x=0$. Actually it turns out that (4.57a) in the form

$$\frac{d\delta}{dx} = \alpha = \text{constant}$$

is an excellent fit to the experimental data with $\alpha \approx 0.42$, giving a semi-angle of $\sim 12^\circ$. This provides some support to the idea of using a constant entrainment coefficient. (It is also common to quote a semi-angle based on a wedge defined by the point where $\bar{u}_x/\bar{u}_0 = 0.5$. This angle is around 6°.)

Let us now see where the Boussinesq–Prandtl eddy-viscosity hypothesis leads for a planar jet. For simplicity we take ν_t to be independent of y and determined by the local values of $\bar{u}_0(x)$ and $\delta(x)$. On dimensional grounds we have $\nu_t \sim \delta(x)\bar{u}_0(x)$ and so we write $\nu_t = b\delta(x)\bar{u}_0(x)$ for some constant b. (This estimate of ν_t was first suggested by Prandtl in 1942. It is appropriate only to free shear flows and has the advantage over (4.13b) that it is simpler to use and does not give the unphysical result that $\nu_t = 0$ on the jet centre-line.) Our governing equation (4.50) now becomes

$$\bar{u}_x\frac{\partial\bar{u}_x}{\partial x} + \bar{u}_y\frac{\partial\bar{u}_x}{\partial y} = b\delta(x)\bar{u}_0(x)\frac{\partial^2\bar{u}_x}{\partial y^2}.$$

This admits a self-similar solution of the form (4.56) in which,

$$F'^2 + FF'' + \tfrac{1}{2}\lambda^2 F''' = 0, \qquad f = F'(\eta)$$

$$\frac{\delta}{\delta_0} = 1 + \frac{4bx}{\lambda^2\delta_0}$$

$$\frac{\bar{u}_0}{V_0} = \left[1 + \frac{4bx}{\lambda^2\delta_0}\right]^{-1/2}$$

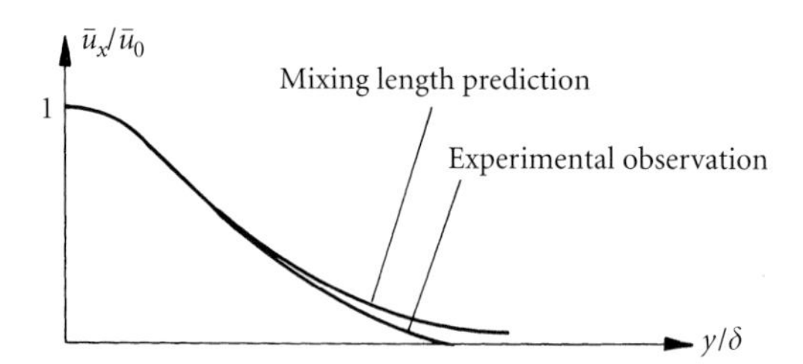

Figure 4.18 Comparison of theory and experiment for a plane jet.

where λ is an (as yet) undetermined coefficient. Evidently we have the same power law behaviour as before ($\delta \sim x$, $\bar{u}_0 \sim x^{-1/2}$) with the eddy-viscosity coefficient b related to the entrainment coefficient by

$$\alpha = \frac{d\delta}{dx} = 4b/\lambda^2$$

The equation for F is familiar from laminar jet theory and may be integrated to give (see Example 4.7),

$$\bar{u}_x/\bar{u}_0 = f(y/\delta) = \text{sech}^2(y/\lambda\delta)$$

So far, we have not given a precise definition of δ, and it is this which fixes the value of λ. Of course, any definition is somewhat arbitrary since the time-averaged velocity profile declines exponentially with y. Let us adopt the simple definition that the jet velocity has dropped to 10% of $\bar{u}_0$ when $y = \pm\delta/2$. This yields $\lambda = 0.275$ and, given that $\alpha \approx 0.42$, we find $b \approx 8.0 \times 10^{-3}$. The eddy viscosity is then given by

$$\nu_t \approx 0.0080\ \delta\bar{u}_0$$

So how do our mixing-length estimates fare against the experimental data? The answer is: surprisingly well! If we exclude the initial part of the jet, where it is not fully developed, then the estimates $\delta \sim x$, $\bar{u}_0 \sim x^{-1/2}$ and $f \sim \text{sech}^2(y/\lambda\delta)$ are all excellent fits to the data, as illustrated in Figure 4.18. Moreover the estimate $\nu_t \sim 8 \times 10^{-3}\delta\bar{u}_0$ is close to the observed value.

Example 4.7 Show that the plane jet equation

$$F'^2 + FF'' + \tfrac{1}{2}\lambda^2 F''' = 0, \qquad f = F'(\eta)$$

may be integrated twice to yield

$$\lambda^2(F' - 1) + F^2 = 0$$

and hence confirm that the jet profile is

$$\bar{u}_x/\bar{u}_0 = f(y/\delta) = \text{sech}^2(y/\lambda\delta).$$

Consider now the plane wake shown in Figure 4.16. It is usually found that, well downstream of the object which initiated the wake, the velocity deficit, $\bar{u}_d = V - \bar{u}_x$, is much smaller than V and so a good approximation to (4.50) is

$$\rho V \frac{\partial \bar{u}_d}{\partial x} = -\frac{\partial \tau_{xy}^R}{\partial y}.$$

Let $\bar{u}_0(x) = \bar{u}_d(y = 0)$ and suppose $\nu_t = b\delta(x)\bar{u}_0(x)$, as before. Then our simplified momentum equation yields

$$V \frac{\partial \bar{u}_d}{\partial x} = b\delta(x)\bar{u}_0(x)\frac{\partial^2 \bar{u}_d}{\partial y^2}.$$

This has a self-similar solution in which $\delta \sim x^{1/2}$, $\bar{u}_0 \sim x^{-1/2}$, and $\bar{u}_d$ satisfies

$$\bar{u}_d = \bar{u}_0(x) \exp\left[-y^2/(\lambda\delta^2)\right]$$

Note that the combination of $\delta \sim x^{1/2}$ and $\bar{u}_0 \sim x^{-1/2}$ satisfies (4.55) in the form

$$\int_{-\infty}^{\infty} \rho V(V - \bar{u}_x)dy = \text{constant}.$$

Measurements of planar wakes suggest that, well downstream of the obstacle which created the wake, δ and $\bar{u}_0$ do indeed scale as $x^{1/2}$ and $x^{-1/2}$. Moreover, the velocity profile is self-preserving in the sense that $\bar{u}_d = \bar{u}_0 f(y/\delta)$, and the particular form of f suggested by mixing length is a good fit to the data. As with the jet, λ can be pinned down through a suitable definition of δ. For example, we might define δ through the requirement $\bar{u}_d(\delta/2) = 0.05\bar{u}_0$, which fixes $\lambda \sim 0.083$.

4.3.2 *The round jet*

We now turn to the case of an axisymmetric jet. As with the planar jet this is characterised by $\bar{u}_z \gg \bar{u}_r$, and $\partial/\partial z \ll \partial/\partial r$. (We shall use polar coordinates, (r, θ, z) in this section.) Of course, $\bar{u}_z$ falls off with axial distance while δ, the time-averaged diameter of the jet, increases as the jet spreads. However, it is observed that, after ~30 diameters downstream of the source, the time-averaged velocity profile depends only on radial position, r, the local jet width, δ, and the local centre-line velocity. It is then a dimensional necessity that $\bar{u}_z(r, z)$ adopts the self-similar form

$$\frac{\bar{u}_z(r, z)}{\bar{u}_0(z)} = f(r/\delta(z)), \qquad \bar{u}_0 = \bar{u}_z(0, z).$$

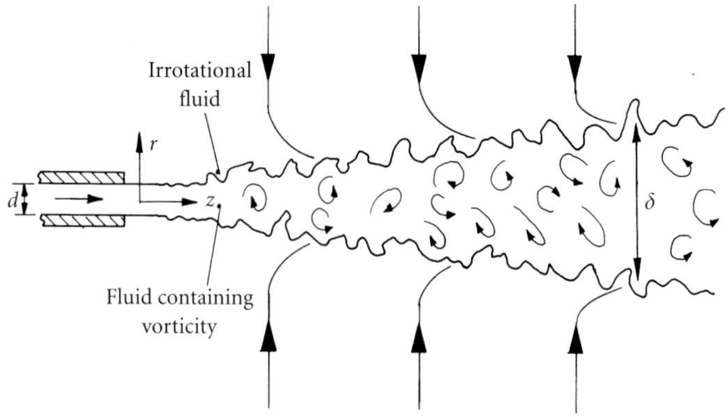

Figure 4.19 A round jet. The jet consists of vorticity which has been stripped off the inside of the nozzle and is then swept downstream. The convoluted outer edge of the jet represents the interface between fluid filled with vorticity and the external, irrotational fluid.

The instantaneous interface between the turbulent jet and its surroundings is highly convoluted, as we would expect, since the turbulence is a manifestation of the vorticity which is extruded from the inside of the nozzle and then swept downstream (Figure 4.19). When Re is large this vorticity is virtually frozen into the fluid and so the convoluted outer edge of the jet, which marks the outer limits of the vorticity, is an inevitable consequence of the eddying motion within the jet. As with a planar jet, there is entrainment of the ambient fluid as the convoluted outer edge engulfs irrotational fluid. Thus the mass flux of the jet increases with z.

The governing equations for a round jet may be simplified for three reasons:

(1) axial gradients in the Reynolds stresses are much weaker than radial gradients;
(2) laminar stresses are negligible;
(3) radial components of the mean inertial force are negligible.

The time-averaged Navier–Stokes equation then yields (see Appendix I)

$$\rho \bar{\mathbf{u}} \cdot \nabla \bar{u}_z = -\frac{\partial \bar{p}}{\partial z} + \frac{1}{r}\frac{\partial}{\partial r}\left[r\tau^R_{rz}\right]$$

$$0 = -\frac{\partial \bar{p}}{\partial r} + \frac{1}{r}\frac{\partial}{\partial r}\left[r\tau^R_{rr}\right] - \frac{\tau^R_{\theta\theta}}{r}.$$

The second of these may be integrated to give

$$\bar{p} = \tau^R_{rr} - \int_r^\infty \left[\frac{\tau^R_{rr} - \tau^R_{\theta\theta}}{r}\right] dr$$

where we have taken the pressure at $r \to \infty$ to be zero. Since axial gradients in the Reynolds stresses are much smaller than radial gradients, this tells us that we may take $\partial\bar{p}/\partial z = 0$ in the axial equation of motion. The end result is:

$$\rho\bar{\mathbf{u}}\cdot\nabla\bar{u}_z = \frac{1}{r}\frac{\partial}{\partial r}[r\rho\bar{u}_r\bar{u}_z] + \frac{\partial}{\partial z}\left[\rho\bar{u}_z^2\right] = \frac{1}{r}\frac{\partial}{\partial r}\left[r\tau_{rz}^R\right]$$

from which we see that the momentum flux is conserved:

$$M = \int_0^\infty \left[\rho\bar{u}_z^2\right]2\pi r dr = \text{constant}.$$

If we now invoke the self-similar approximation $\bar{u}_z = \bar{u}_0 f(r/\delta)$ we can express the mass and momentum fluxes as

$$\dot{m} = \rho\bar{u}_0\delta^2\int_0^\infty 2\pi\eta f(\eta)d\eta$$
$$M = \rho\bar{u}_0^2\delta^2\int_0^\infty 2\pi\eta f^2(\eta)d\eta = \text{constant}$$

where $\eta = r/\delta$. As with the planar jet, we may estimate the rate of change of $\bar{u}_0$ and δ using an entrainment argument or else an eddy-viscosity approximation. Let us start with the entrainment approach. It seems plausible that the rate of entrainment of mass per unit length is proportional to the perimeter of the jet and the local intensity of the turbulent fluctuations. These fluctuations are, in turn, proportional to the local value of $\bar{u}_0$. Thus we might write

$$\frac{d\dot{m}}{dz} = \alpha\rho\bar{u}_0\delta\int_0^\infty 2\pi\eta f(\eta)d\eta$$

where α is an entrainment coefficient and the integral on the right has been added for convenience. We now have the two equations,

$$\frac{d}{dz}\left[\bar{u}_0\delta^2\right] = \alpha\bar{u}_0\delta$$
$$\bar{u}_0^2\delta^2 = \text{constant}$$

which integrate to give

$$\frac{\delta}{\delta_0} = 1 + \frac{\alpha z^*}{\delta_0}$$

$$\frac{\bar{u}_0}{V_0} = \left[1 + \frac{\alpha z^*}{\delta_0}\right]^{-1}$$

where z^* is measured from the start of the self-similar portion of the jet and δ_0 and V_0 are the values of δ and $\bar{u}_0$ at $z^* = 0$. Note that, as for a planar jet, we have

$$\frac{d\delta}{dz} = \alpha = \text{constant}.$$

It turns out that a linear growth of δ is exactly what is observed, and if we define δ via the condition $\bar{u}_x/\bar{u}_0 = 0.1$ when $r = \delta/2$, then $\alpha \approx 0.43$.

The eddy-viscosity approach is similar to that for the planar jet. We take $\nu_t = b\delta(z)\bar{u}_0(z)$ and then look for a self-similar solution of

$$\bar{\mathbf{u}} \cdot \nabla \bar{u}_z = (b\delta\bar{u}_0)\frac{1}{r}\frac{\partial}{\partial r}\left[r\frac{\partial \bar{u}_z}{\partial r}\right].$$

It is readily confirmed that setting $\bar{u}_z = \bar{u}_0 f(r/\delta)$ leads to

$$\eta f'' + f' + (\alpha/b)\left[\eta f^2 + f'\int_0^\eta \eta f d\eta\right] = 0$$

$$\frac{d\delta}{dz} = \alpha = \text{constant}$$

$$\bar{u}_0^2\delta^2 = \text{constant}$$

where $\eta = r/\delta$. Evidently the eddy-viscosity approach leads to the same laws for δ and $\bar{u}_0$ as the entrainment coefficient method. The governing equation for f may be integrated to give

$$f = \frac{1}{[1 + a\eta^2]^2}, \qquad a = \alpha/(8b).$$

We now recall that δ is defined so that $\bar{u}_z/\bar{u}_0 = 0.1$ when $r = \delta/2$. This requires $f = 0.1$ when $\eta = \frac{1}{2}$, which in turn gives $a = 8.65$. The relationship between α and b is then $b = \alpha/69.2$ and since α is observed to have a value of around 0.43 we have $b \approx 6.2 \times 10^{-3}$.

The form of f predicted by the eddy-viscosity method gives a good fit to the experimental data and so, as for the planar jet, we see that the eddy-viscosity method works well. The comparison of prediction with experiment for a round jet is very similar to that shown in Figure 4.18 for a plane jet. In both cases the mixing length tends to slightly overestimate the velocity at the edge of the jet. This is because the assumption that ν_t is independent of r leads to an overestimate of ν_t near $r = \delta/2$ and this, in turn, leads to an underestimate in the cross-stream gradient in mean velocity.

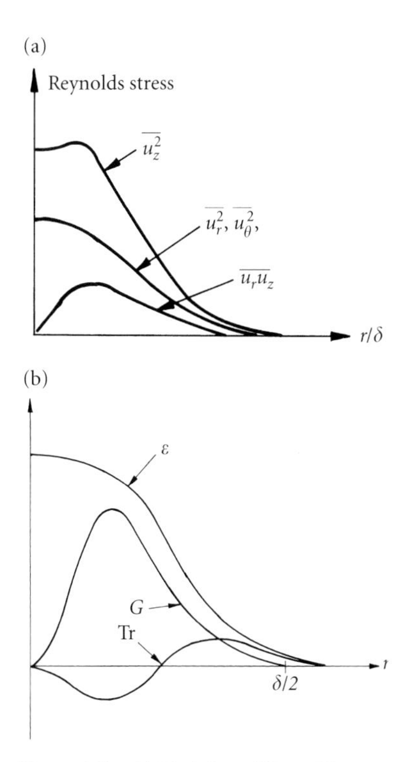

Figure 4.20 (a) Variation of Reynolds stresses with radius for a round jet. (b) Variation of kinetic energy generation, G, transport, Tr, and dissipation, ε, for a round jet as a function of radius.

The observed variation of the different Reynolds stresses with r/δ is shown in Figure 4.20(a). Notice that, although $\overline{u_r'^2} \approx \overline{u_\theta'^2}$, the large scales in the jet are far from isotropic, with $\overline{u_z'^2} \sim 2\overline{u_r'^2}$ in the core of the jet.

Figure 4.20(b) shows schematically the various contributions to the turbulent kinetic energy equation (4.21). As might be expected, the dissipation is fairly uniform over the centre of the jet, falling off as we move to edge, $r = \delta/2$. The generation of turbulent energy, on the other hand, is small near the centre-line since the mean rate of strain is weak at $r = 0$. It peaks around $r = 0.3(\delta/2)$, where $G/\varepsilon \approx 0.8$, and then falls off as we move towards the edge of the jet. The transport of turbulent kinetic energy by pressure fluctuations and the triple correlations, $\text{Tr} = -\nabla \cdot (\mathbf{T})$, is negative in the core of the jet and positive near the boundary. This represents a radial flux of energy from the region of most intense turbulence to that of weaker turbulence, which is consistent with the modelling of $\nabla \cdot (\mathbf{T})$ as a diffusive process in the k–ε model.

4.4 Homogeneous shear flow

The simplest shear flow which can be (approximately) realized in the laboratory is that of homogeneous shear. That is, we imagine that we are well-removed from any boundary, that the mean velocity is $\bar{\mathbf{u}} = (\bar{u}_x(y), 0, 0)$, where $\bar{u}_x(y) = Sy$, and that the spatial gradients in the turbulent quantities are negligible. Such an idealized situation is of little direct practical interest, but is highly instructive from a theoretical point of view. In particular, it illustrates in a simple way the interaction between turbulence and a mean shear, and highlights the role played by pressure forces in redistributing energy between the different components of motion. It also forms a useful test case against which to compare engineering 'models' of turbulence.

4.4.1 *The governing equations*

Suppose that we have a steady mean flow, $\bar{u}_x = Sy$, and turbulence whose statistics are homogeneous (independent of position) and possibly unsteady. Despite the fact that the turbulence statistics may evolve in time we retain time averages as a way of evaluating mean quantities, on the assumption that the rate of change of statistical quantities is slow by comparison with the timescale of the turbulent fluctuations. The turbulence has reflectional symmetry about the x–y plane and this ensures that $\tau^R_{xz} = \tau^R_{yz} = 0$. The quantities of interest, then, are τ_{xy}^R, $\overline{(u_x')^2}$, $\overline{(u_y')^2}$ and $\overline{(u_z')^2}$. For homogeneous turbulence the time-dependent generalization of (4.8) yields evolution equations for each of these:

$$\frac{\partial}{\partial t}\left[\rho\overline{(u_x')^2}\right] = 2\overline{p'S_{xx}'} - 2\rho\nu\overline{\frac{\partial u_x'}{\partial x_k}\frac{\partial u_x'}{\partial x_k}} + 2\tau_{xy}^R S$$

$$\frac{\partial}{\partial t}\left[\rho\overline{\left(u_y'\right)^2}\right] = 2\overline{p'S_{yy}'} - 2\rho\nu\overline{\frac{\partial u_y'}{\partial x_k}\frac{\partial u_y'}{\partial x_k}}$$

$$\frac{\partial}{\partial t}\left[\rho\overline{(u_z')^2}\right] = 2\overline{p'S_{zz}'} - 2\rho\nu\overline{\frac{\partial u_z'}{\partial x_k}\frac{\partial u_z'}{\partial x_k}}$$

$$\frac{\partial}{\partial t}\left[\tau_{xy}^R\right] = -2\overline{p'S_{xy}'} + 2\rho\nu\overline{\frac{\partial u_x'}{\partial x_k}\frac{\partial u_y'}{\partial x_k}} + \rho\overline{\left(u_y'\right)^2}S.$$

Now the turbulent motions which contribute most to the viscous terms above are the small-scale eddies. These are approximately isotropic and so we may replace the viscous tensor by its homogeneous, isotropic equivalent:

$$2\nu\overline{\frac{\partial u_i'}{\partial x_k}\frac{\partial u_j'}{\partial x_k}} = \frac{2}{3}\varepsilon\delta_{ij}.$$

Our governing equations now simplify to

$$\frac{\partial}{\partial t}\left[\frac{1}{2}\rho\overline{(u_x')^2}\right] = \overline{p'S_{xx}'} - \frac{1}{3}\rho\varepsilon + \tau_{xy}^R S \tag{4.58a}$$

$$\frac{\partial}{\partial t}\left[\frac{1}{2}\rho\overline{\left(u_y'\right)^2}\right] = \overline{p'S_{yy}'} - \frac{1}{3}\rho\varepsilon \tag{4.58b}$$

$$\frac{\partial}{\partial t}\left[\frac{1}{2}\rho\overline{(u_z')^2}\right] = \overline{p'S_{zz}'} - \frac{1}{3}\rho\varepsilon \tag{4.58c}$$

$$\frac{\partial}{\partial t}\left[\tau_{xy}^R\right] = -2\overline{p'S_{xy}'} + \rho\overline{\left(u_y'\right)^2}S. \tag{4.58d}$$

The first three of these combine to give the familiar energy equation

$$\frac{dk}{dt} = (\tau_{xy}^R S/\rho) - \varepsilon = G - \varepsilon. \tag{4.59}$$

(Remember that $S_{ii} = 0$ because of the continuity equation.) Evidently the turbulence is maintained by the rate of working of the Reynolds stress, G, and destroyed in the usual way by the small-scale eddies. If $G = \varepsilon$ we have steady turbulence, whereas an imbalance between G and ε will lead to the growth or decay of the turbulence.

It is clear from (4.58a–c) that τ_{xy}^R generates only $\overline{(u_x')^2}$, yet dissipation is present in all three equations and observations suggest that all three components of $\mathbf{u}'$ are of a similar magnitude. (Typically, after a while, we find that the various components settle down to $\overline{(u_x')^2} \sim k$, $\overline{(u_y')^2} \sim 0.4k$, and $\overline{(u_z')^2} \sim 0.6k$.) Evidently the role of the pressure-rate-of-strain correlation is to redistribute energy from u_x' to u_y' and u_z'. This is thought to be typical of the role of pressure fluctuations: they tend to scramble the turbulence, continually pushing it towards an isotropic state. However, they can neither create nor destroy turbulent energy and this is why they are absent from (4.59).

Notice that the mean shear S acts directly as a source for τ_{xy}^R in (4.58d). Thus, if the turbulence were initially isotropic, so that τ_{xy}^R starts out as zero, the Reynolds stress will not stay zero for long. That is,

$$\frac{\partial}{\partial t}\left[\tau_{xy}^R\right] = \rho\overline{(u_y')^2}S + [\text{pressure term}].$$

The growth in τ_{xy}^R caused by the mean shear tends to be offset, to some degree, by the 'scrambling' effect of the pressure term. Nevertheless, τ_{xy}^R invariably ends up as positive, leading to a net transfer of energy from the mean flow to the turbulence via the generation term, G.

We may understand the way in which S and τ_{xy}^R promotes turbulent energy as follows. The shear flow $\bar{u}_x = Sy$ may be divided into an irrotational plane strain plus one component of rotation:

$$\bar{\mathbf{u}} = \tfrac{1}{2}(Sy, Sx, 0) + \tfrac{1}{2}(Sy, \ -Sx, 0)$$

We recognize the second of these terms as having uniform vorticity, $\bar{\boldsymbol{\omega}} = (0,\ 0,\ -S)$, and representing rigid-body rotation. The first represents an irrotational straining motion whose principal axes of strain (the directions of maximum and minimum rate of straining) are inclined at 45° to the x and y axes (Figure 4.21). Evidently vortex lines tend to be teased out in the direction of maximum strain while undergoing some rotation. The process of stretching the vortex lines tends to intensify their kinetic energy and it is this which maintains the turbulence in the face of viscous dissipation (Figure 4.21). The vortices which result from this process tend to be aligned with the maximum principal strain and, as shown by (4.47) and Figure 4.12 they are ideally suited to generated a high value of τ_{xy}^R. So the production of a positive Reynolds stress and the generation of energy are both part of the same process.

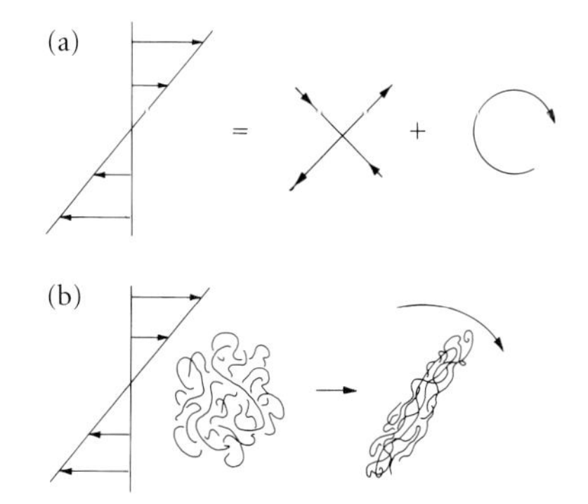

Figure 4.21 (a) A shear flow can be decomposed into an irrotational straining motion plus rigid-body rotation. (b) The influence of the irrotational straining on a tangle of vortex tubes is shown.

It is instructive to explore a little further this idea of turbulence generation by vortex stretching. Writing $\boldsymbol{\omega} = \bar{\boldsymbol{\omega}} + \boldsymbol{\omega}'$ the vorticity equation becomes

$$\frac{D\boldsymbol{\omega}'}{Dt} = \bar{\boldsymbol{\omega}} \cdot \nabla\mathbf{u}' + \boldsymbol{\omega}' \cdot \nabla\bar{\mathbf{u}} + \boldsymbol{\omega}' \cdot \nabla\mathbf{u}' + \nu\nabla^2\boldsymbol{\omega}'$$

and on substituting for $\bar{\mathbf{u}}$ and $\bar{\boldsymbol{\omega}}$ this yields

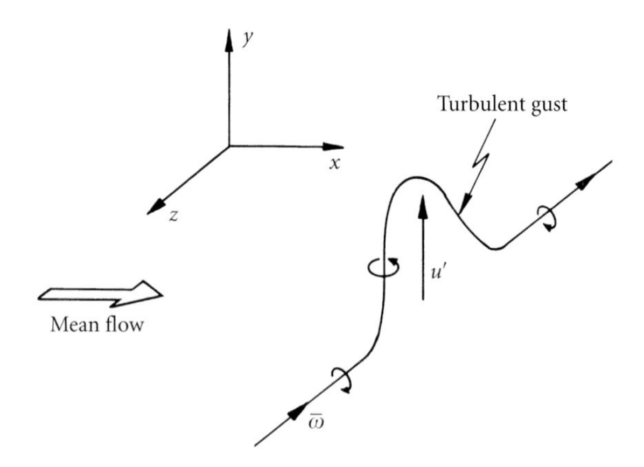

Figure 4.22 The creation of turbulent vorticity through the action of turbulent gusts acting on the mean vortex lines.

$$\frac{D\boldsymbol{\omega}'}{Dt} = -S\frac{\partial \mathbf{u}'}{\partial z} + S\omega'_y\hat{\mathbf{e}}_x + \boldsymbol{\omega}' \cdot \nabla \mathbf{u}' + \nu\nabla^2\boldsymbol{\omega}'.$$
$$(1) \qquad (2) \qquad (3) \qquad (4)$$

Let us consider the various contributions to the right-hand side of this equation.

(1) The first contribution, $-S\partial\mathbf{u}'/\partial z$, has its roots in the term $\nabla \times (\mathbf{u}' \times \bar{\boldsymbol{\omega}})$ and so represents the advection of the mean vortex lines by the turbulent velocity field. It may be pictured as shown above (Figure 4.22). A turbulent 'gust' distorts the mean vortex lines, producing some turbulent vorticity. For example, a vertical gust, u'_y, induces vertical vorticity, ω'_y. So $-S\partial\mathbf{u}'/\partial z$ acts to convert mean vorticity into turbulent vorticity.

(2) The second contribution, $S\omega'_y\hat{\mathbf{e}}_x$, derives from the term $\nabla \times (\bar{\mathbf{u}} \times \boldsymbol{\omega}')$ and so represents the action of the mean velocity on the turbulent vorticity. As discussed above, this process is one by which the mean flow intensifies the turbulent vorticity by vortex stretching (Figure 4.21). The reason why it appears as a source of streamwise vorticity can be understood as follows. The mean velocity, $\bar{u}_x = Sy$, tilts the vertical vorticity ω'_y, thus acting as a source of ω'_x.

(3) The term $\boldsymbol{\omega}' \cdot \nabla\mathbf{u}'$ represents the chaotic advection of the turbulent vorticity by $\mathbf{u}'$, that is, turbulence acting on itself. It is this process which feeds the energy cascade, passing energy down to small scales by vortex stretching. Since vorticity is intensified by the cascade, $\boldsymbol{\omega}' \cdot \nabla\mathbf{u}'$ is dominated by the small-scale vorticity.

(4) Of course $\nu\nabla^2\boldsymbol{\omega}'$ represents the diffusion of vorticity. This is important at small scales where it leads to the destruction of enstrophy by the cross diffusion of patches of oppositely signed vorticity.

Let us now take the product of $\boldsymbol{\omega}'$ with our vorticity equation. This yields

$$\frac{D}{Dt}\left[\tfrac{1}{2}(\boldsymbol{\omega}')^2\right] = -S\nabla\cdot\left[u_z'\boldsymbol{\omega}'\right] + S\omega_x'\omega_y' + \omega_i'\omega_j'S_{ij}' + \nu\boldsymbol{\omega}'\cdot\nabla^2\boldsymbol{\omega}'$$
$$\qquad\qquad(1)\qquad\qquad(2)\qquad\qquad(3)\qquad\qquad(4)$$

where we have used the fact that $\boldsymbol{\omega}'\cdot\nabla u_z' = \boldsymbol{\omega}'\cdot\partial\mathbf{u}'/\partial z$ to simplify the first term on the right. Next we ensemble average this equation while noting that homogeneity requires the divergence of an average to be zero. The end result is

$$\frac{\partial}{\partial t}\left\langle\tfrac{1}{2}(\boldsymbol{\omega}')^2\right\rangle = S\left\langle\omega_x'\omega_y'\right\rangle + \left\langle\omega_i'\omega_j'S_{ij}'\right\rangle - \nu\left\langle(\nabla\times\boldsymbol{\omega}')^2\right\rangle.$$
$$\qquad(2)\qquad\qquad(3)\qquad\qquad(4)$$

Interestingly effect (1) does not appear in this averaged equation. It seems that $\nabla\times(\mathbf{u}'\times\bar{\boldsymbol{\omega}})$ can create turbulent vorticity by bending the mean vortex lines, but that this does not influence the mean enstrophy when homogeneity is imposed.

So we may interpret the evolution of the vortex field, and by implication the turbulent velocity field, as follows. Turbulent vorticity is continually amplified by the mean strain $\bar{S}_{ij}$. This represents a transfer of energy from the mean flow to the turbulence. The large-scale vorticity is then passed down the energy cascade by $\boldsymbol{\omega}'\cdot\nabla\mathbf{u}'$ (turbulence acting on turbulence) until it is destroyed at small scales.

4.4.2 *The asymptotic state*

Now suppose that at $t=0$ we start with isotropic turbulence. The asymmetry in the equations (4.58a–c) means that it will not stay isotropic for long and it is natural to ask if the turbulence will tend to a new, anisotropic state determined simply by S. (It is not obvious whether such a state should be steady or unsteady.) It turns out that the numerical and experimental studies suggest that the turbulence tends to a state in which the ratio of G to ε is constant, as is the ratio of τ_{xy}^R to ρk. Since (4.59) may be rewritten in the form

$$\frac{d}{d(St)}[\ln k] = \frac{\tau_{xy}^R}{\rho k}[1-\varepsilon/G]$$

it follows that the asymptotic state is of the form

$$k = k_0\exp[\lambda St]$$

where λ is the constant,

$$\lambda = \frac{\tau_{xy}^R}{\rho k}[1-\varepsilon/G]$$

Estimates of $\tau_{xy}^R/\rho k$ and G/ε vary, but typical values are given in Table 4.1 in Section 4.1.3 of this chapter. These are

$$\tau_{xy}^R/\rho k = 0.28, \qquad G/\varepsilon = 1.7$$

from which

$$k = k_0 \exp[0.12\ St]$$
$$\varepsilon = \varepsilon_0 \exp[0.12\ St].$$

Also, since $\varepsilon \sim u^3/l$, the integral scale will grow as

$$l = k^{3/2}/\varepsilon \sim l_0 \exp[0.06\ St].$$

We shall return to homogeneous shear flows in Section 4.6.1 of this chapter where we shall see that they provide a convenient test case for various 'models' of turbulence.

4.5 Heat transfer in wall-bounded shear flows—the log-law revisited

4.5.1 Turbulent heat transfer near a surface and the log-law for temperature

The influence of turbulence on heat transfer is an important practical problem and perhaps now is the right time to say something about it. Consider the situation shown in Figure 4.23(a). Heat will be carried from the hot wall to the cold one by turbulent diffusion. That is, heat is materially transported by the random eddying motion and this mixing process will have the effect of dispersing the heat, carrying hot fluid away from the lower wall and cold fluid away from the upper surface. In the core of the flow (i.e. away from the boundaries), the large eddies are the most energetic, and span the largest distance, and so they are the most important ones for transporting heat. The small-scale eddies in the core, which have a much faster turn-over time, but are rather weak, simply smooth things out at the small scales, performing a sort of micro-mixing. Very close to the walls, however, the turbulence is suppressed, and the burden of transporting the heat falls to a combination of molecular diffusion and turbulent advection by small eddies. Thus wall regions tend to be resistant to turbulent heat transfer. The central question which concerns us here is: can we predict the influence of a (statistically) prescribed field of turbulence on the rate of heat transfer near a wall? Let us start with the heat equation.

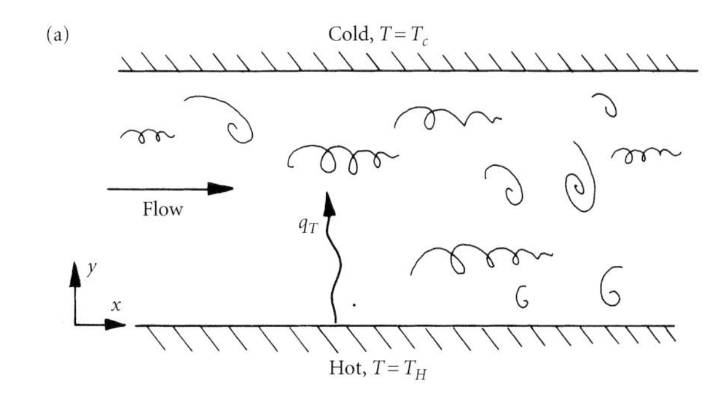

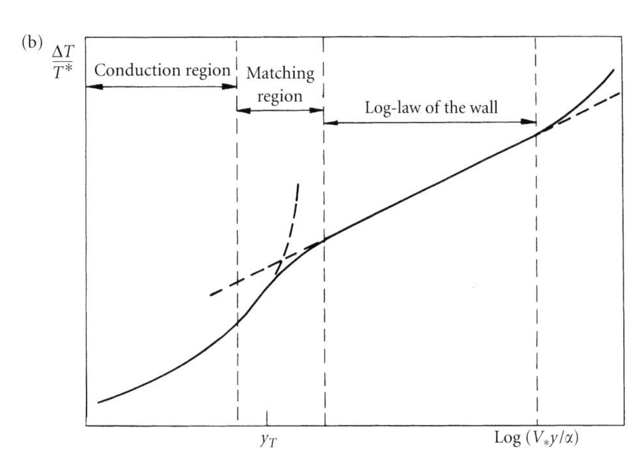

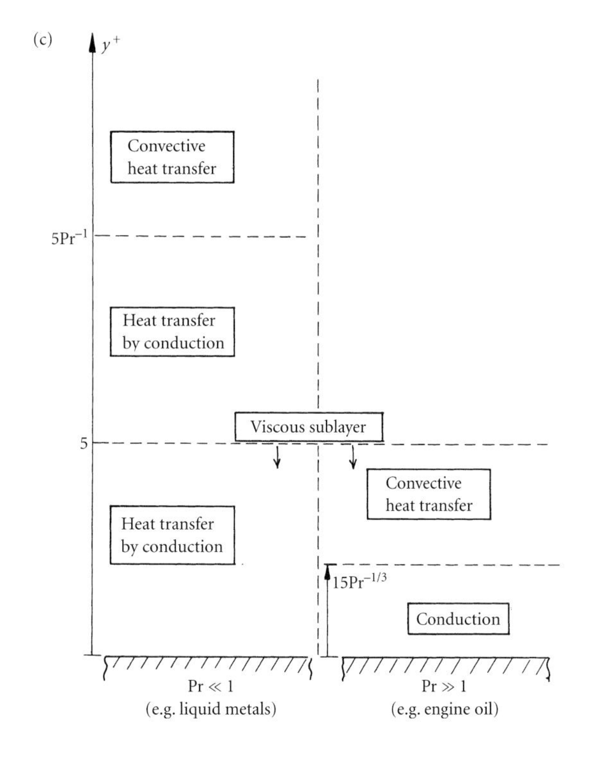

Figure 4.23 (a) Turbulent heat transfer. (b) Patching together the log-law and conduction–only regions at y_T determines A_T. (c) The structure of the near-wall region is determined by Pr. For small Pr (highly-conducting fluids) the viscous sublayer is smaller than the conduction-only region. For large Pr (poor conductors), on the other hand, the low levels of turbulence in the viscous sublayer dominate the heat transfer throughout the bulk of the viscous sublayer. The conduction-only region is now restricted to a very thin surface layer which dominates the near-wall thermal resistance.

4.5.1.1 Turbulent diffusion of heat and the gradient-diffusion approximation

The advection–diffusion equation for heat is

$$\frac{DT}{Dt} = \alpha \nabla^2 T, \qquad T = \overline{T} + T' \tag{4.60}$$

where $\overline{T}$ and T' are the mean and fluctuating temperatures. We can understand the origin of (4.60) more readily if we rewrite it as,

$$\frac{D}{Dt}\left(\rho c_{\mathrm{p}} T\right) = -\nabla \cdot (\mathbf{q}), \qquad \mathbf{q} = -k\nabla T \tag{4.61}$$

where $\mathbf{q}$ is the heat flux density associated with molecular diffusion, k is the thermal conductivity, and c_{p} the specific heat. Expression (4.61) comes from equating the rate of loss of thermal energy from a lump of fluid of fixed volume δV to the rate at which heat diffuses out of the lump as it moves around,

$$\frac{D}{Dt}(\rho c_{\mathrm{p}} T \delta V) = -\oint_{\delta S} \mathbf{q} \cdot d\mathbf{S} = -\int_{\delta V} \nabla \cdot \mathbf{q} dV \approx -\nabla \cdot \mathbf{q} \delta V.$$

Dividing through by δV brings us back to (4.61). (We ignore here the contribution to (4.61) which comes from the viscous generation of internal energy since it is usually negligible.) Now suppose $\bar{\mathbf{u}}$ and T are statistically steady. Then (4.60) yields

$$\bar{\mathbf{u}} \cdot \nabla\left(\rho c_{\mathrm{p}} \overline{T}\right) = -\nabla \cdot \left(-k\nabla\overline{T} + \rho c_{\mathrm{p}} \overline{T'\mathbf{u}'}\right). \tag{4.62}$$

We might rewrite this as

$$\bar{\mathbf{u}} \cdot \nabla\left(\rho c_{\mathrm{p}} \overline{T}\right) = -\nabla \cdot \left(\mathbf{q}_T\right) \tag{4.63}$$

$$\mathbf{q}_T = -k\nabla\overline{T} + \rho c_{\mathrm{p}} \overline{T'\mathbf{u}'} \tag{4.64}$$

where $\mathbf{q}_T$ is now the turbulent heat flux density which includes both molecular conduction and turbulent mixing. The turbulent term, $\overline{T'\mathbf{u}'}$, which has arisen from the averaging process, is clearly analogous to the Reynolds stresses which arise from averaging the momentum equation.

We are now faced with the problem of estimating $\overline{T'\mathbf{u}'}$. One approach, which has much in common with the Prandtl–Boussinesq approximation for τ_{ij}^R, is to write

$$\overline{T'\mathbf{u}'} = -\alpha_{\mathrm{t}} \nabla\overline{T} \tag{4.65}$$

so that (4.64) becomes

$$\bar{\mathbf{u}} \cdot \nabla \bar{T} = \nabla \cdot [(\alpha + \alpha_t) \nabla \bar{T}]. \tag{4.66}$$

Thus the net effect of turbulent mixing is, in this model, captured by bumping up the thermal diffusivity from α to $\alpha + \alpha_t$. An estimate of the form

$$\overline{T'\mathbf{u}'} = -\alpha_t \nabla \bar{T}$$

is known as a *gradient-diffusion* approximation and α_t is called the turbulent diffusivity. The idea here is that, on average, turbulent mixing will tend to eradicate gradients in mean temperature, just as molecular diffusion does, and the higher the gradient in $\bar{T}$, the more vigorous the heat transfer by mixing will be. (Actually, we could regard (4.65) as simply defining α_t.)

Let us now return to the simple case shown in Figure 4.23(a). If the flow is statistically independent of x and z then (4.63) and (4.64) tell us that

$$|\mathbf{q}_T| = q_{Ty} = \text{constant}$$

$$q_{Ty} = -k\frac{\partial \bar{T}}{\partial y} + \rho c_p \overline{T'u'_y}.$$

Molecular conduction is usually negligible except near the boundaries where the turbulence becomes somewhat muted. Thus, in the core of the flow, we have

$$\overline{T'u'_y} = \text{constant} = q_{Ty}/\rho c_p.$$

If we invoke the gradient-diffusion approximation this becomes

$$\alpha_t \frac{d\bar{T}}{dy} = \text{constant} = -q_{Ty}/\rho c_p$$

and the mean temperature profile $\bar{T}(y)$ can be calculated provided that α_t (which could be a function of y) is known. There are several ways forward at this point. One approach, called Reynolds analogy, relies on the fact that the same eddies which are responsible for the transport of momentum are responsible for the transport of heat. So we might make the approximation of $\alpha_t = \nu_t$, which yields

$$\frac{q_{Ty}}{c_p \tau^R_{xy}} = -\frac{\partial \bar{T}/\partial y}{\partial \bar{u}_x/\partial y}. \tag{4.67}$$

So if $\bar{u}_x(y)$ and $\tau^R_{xy}(y)$ are known, this allows us to determine the relationship between q_{Ty} and $\bar{T}(y)$.

An alternative approach is to invoke a mixing-length type argument and write $\alpha_t = u' l_m$ where $u'^2 = \overline{(u'_y)^2}$ and l_m is a mixing length. This amounts to the assertion that

$$\overline{T'u'_y} = -\left[\overline{(u'_y)^2}\right]^{1/2} l_m \frac{\partial \overline{T}}{\partial y}.$$

Actually this is nothing more than a definition of l_m, transferring the problem to one of determining the mixing length. Now presumably the magnitude of l_m at any one location is of the order of the mean size of the large eddies at that location. We might expect, therefore, that l_m will be a function of y, since the average eddy size gets smaller as we approach the boundaries. So, in line with simple mixing-length theory, we might anticipate that $l_m = \kappa y$ near the lower boundary, where κ is Karman's constant, and that l_m is approximately constant in the core of the flow.

These kind of naive mixing-length arguments tend to work well for simple shear flows, of the type shown in Figure 4.23(a), and also close to relatively flat boundaries. For more complex geometries, however, the entire gradient-diffusion approximation should be regarded with a certain amount of caution. Nevertheless, we have already noted that most of the thermal resistance to heat transfer tends to come from the near-wall regions, and so mixing-length arguments can indeed be useful. Near the lower wall in Figure 4.23(a) our mixing-length model yields the following relationship between q_{Ty} and $\partial \overline{T}/\partial y$,

$$q_{Ty} = -\rho c_p u' \kappa y \frac{\partial \overline{T}}{\partial y}.$$

So, if we know $u'(y)$ we can determine the distribution of the mean temperature, $\overline{T}(y)$, near the boundary.

Example 4.8 Consider the near-wall region in Figure 4.23(a) where $u' \sim V_*$, $l = \kappa y$, and $\bar{u}_x / V_* = \kappa^{-1} \ln y^+ + A$. Show that the mixing-length equation above leads to,

$$\frac{\Delta T}{T^*} = \frac{T_H - \overline{T}(y)}{T^*} = \frac{1}{\kappa_T} \ln\left[\frac{V_* y}{\alpha}\right] + A_T$$

where κ_T and A_T are dimensionless coefficients, T_H is the wall temperature at $y = 0$, and

$$T^* = \frac{q_T}{\rho c_p V_*}.$$

Now show that Reynolds' analogy (4.67), leads to precisely the same result, though with the restriction that $\kappa_T = \kappa$, the Karman constant.

4.5.1.2 The law of the wall for temperature

The appearance of a log-law of the wall for temperature in Example 4.8 is intriguing because of its similarity to the log-law for velocity. However, one might ask if this is a truly genuine result or just an artifact of the mixing-length approximation. Actually it turns out that the appearance of a log-law of the wall for temperature is not coincidence. In fact, following arguments analogous to those leading up to (4.44) we may show that, when $V_* y/\alpha$ is large, yet y/W is small, the expression

$$\frac{\Delta T}{T^*} = \frac{1}{\kappa_T} \ln\left[\frac{V_* y}{\alpha}\right] + A_T \tag{4.68a}$$

is a dimensional necessity. (W is the channel width.) This is an important equation as it constitutes one of the few rigorous results in turbulent heat transfer. The coefficient κ_T is a universal constant, with a value of ~ 0.48, while A_T is a function of the Prandtl number, $\mathrm{Pr} = \nu/\alpha$. (See Landau and Lifshitz 1959, for the first rigorous derivation of this law or, say, Bradshaw and Huang 1995, for a more recent account.) The fact that κ_T has a value close to the Karman constant lends support to Reynolds analogy.

This log-law must be matched to the near-wall region where heat is transported by conduction alone. Here we have $\Delta T = q_{\mathrm{T}} y/\rho c_{\mathrm{p}} \alpha$, which may be rewritten as

$$\Delta T/T^* = V_* y/\alpha.$$

It is the process of patching together these two expressions which determines A_T and fixes its Pr dependence (Figure 4.23(b)). That is,

$$A_T = y_T - \kappa_T^{-1} \ln y_T$$

where y_T is the value of $V_* y/\alpha$ at which the linear and log-laws intersect.

Let us see if we can determine $A_T(\mathrm{Pr})$ through this crude matching process. It is found that the thickness of the conduction dominated region, that is, y_T, is sensitive to the value of Pr. For Pr of the order of unity, y_T is more or less set by the thickness of the viscous sublayer, $y^+ \sim 5$, which gives $y_T \sim 5$. This estimate of y_T is also found to hold for $\mathrm{Pr} \ll 1$ (highly conducting fluids). So, in a low-Pr fluid, the near-wall conduction zone is thicker than the viscous sublayer, the ratio of the two thicknesses being Pr^{-1}. In such cases diffusion remains the dominant heat transfer mechanism up to a distance of the order of $y^+ \sim 5\,\mathrm{Pr}^{-1}$ from the wall. (For liquid metals, where Pr is extremely small, this can extend right into the core of the flow, so that turbulence has little impact on the bulk heat transfer.)

For large values of Pr, on the other hand, the transport of heat by conduction is very inefficient and the low levels of turbulence found in the viscous sublayer dominate the heat transfer throughout much of this region. In such cases the conduction-only layer is a small fraction of the viscous sublayer, and estimating this thickness involves the delicate issue of characterizing the intensity of the intermittent turbulence very close to the wall. Some authors suggest $y^+ \sim 15\,\mathrm{Pr}^{-1/3}$ as an upper bound on the conduction-only region, although Townsend (1976), offers the alternative estimate of $y^+ \sim 10\,\mathrm{Pr}^{-1/4}$. We shall stay with the minus-one-thirds estimate since, as we shall see, it is consistent with the analysis of Kader (1981). Note that this thin insulating layer can have a disproportionate influence on the net heat flux from the surface, since it dominates the near-wall thermal resistance. In summary, then, we have

$$y_T \sim 5, \quad (\mathrm{Pr} \leq 1); \qquad y_T \sim 15\,\mathrm{Pr}^{2/3}, \quad (\mathrm{Pr} \gg 1)$$

as illustrated in Figure 4.23(c). This allows us to estimate A_T using the expression $A_T = y_T - \kappa_T^{-1} \ln y_T$. In the limits of large and small Pr this yields:

$$A_T \sim 1.6, \quad (\mathrm{Pr} \leq 1); \qquad A_T \sim 15\,\mathrm{Pr}^{2/3}, \quad (\mathrm{Pr} \gg 1).$$

In practice the experimental data is reasonably well approximated by the simple curve fit,

$$A_T \approx \frac{5}{3}\left(3\,\mathrm{Pr}^{1/3} - 1\right)^2, \quad 10^{-2} < \mathrm{Pr} < 10^4$$

(Kader 1981), which is consistent with our estimates above. Our log-law can now be written as

$$\frac{\Delta T}{T^*} = \frac{1}{\kappa_T} \ln\left[\frac{V_* y}{\alpha}\right] + A_T, \qquad A_T \approx \frac{5}{3}\left(3\,\mathrm{Pr}^{1/3} - 1\right)^2. \tag{4.68b}$$

There are three particular cases of interest. For most simple gases at room temperature we have $\mathrm{Pr} \sim 0.7$ (this is true of He, H_2, O_2, N_2, and CO_2). In such cases

$$\frac{\Delta T}{T^*} = \frac{1}{\kappa_T} \ln\left[\frac{V_* y}{\alpha}\right] + 4.6, \quad (\mathrm{Pr} = 0.7)$$

which is very close to the velocity profile

$$\frac{\bar{u}_x}{V_*} = \frac{1}{\kappa} \ln\left[\frac{V_* y}{\nu}\right] + A, \quad A \approx 5.5$$

For poorly conducting fluids, on the other hand, such as engine oil ($\mathrm{Pr} \sim 10^4$) or ethylene glycol ($\mathrm{Pr} \sim 200$), we have

$$\frac{\Delta T}{T^*} = \frac{1}{\kappa_T}\ln\left[\frac{V_* y}{\alpha}\right] + 15\,\mathrm{Pr}^{2/3}, \qquad (\mathrm{Pr} \gg 1)$$

The inclusion of the $\mathrm{Pr}^{2/3}$ term means that the temperature drop here is much larger than that of a $\mathrm{Pr} \sim 1$ fluid. This reflects the fact that, when $\mathrm{Pr} \gg 1$, there is a highly insulating region adjacent to the surface where conduction is the only mechanism of transporting heat. Finally, for highly conducting fluids (liquid metals), we have,

$$\frac{\Delta T}{T^*} = \frac{1}{\kappa_T}\ln\left[\frac{V_* y}{\alpha}\right] + \frac{5}{3}, \qquad (\mathrm{Pr} \ll 1)$$

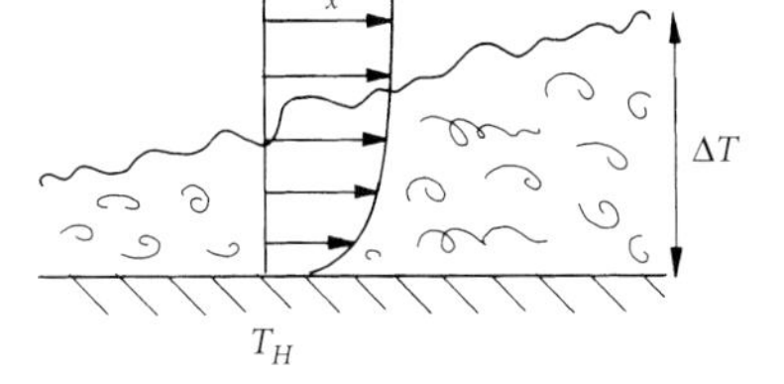

Figure 4.24 Heat transfer in a boundary layer.

Note that this expression is independent of ν, as it must be, since virtually all of the transport of heat is by conduction, even in the fully turbulent region.

4.5.1.3 Heat transfer across a boundary layer

Let us now turn our attention to boundary layers (Figure 4.24), rather than the channel flow of Figure 4.23(a).

In texts on heat transfer it is conventional to characterize the heat transfer across a turbulent boundary layer in terms of the parameters

$$\mathrm{St} = \frac{q_T}{\rho c_p u_\infty \Delta T} \qquad \text{(Stanton number)}$$

$$c_f = \frac{\tau_w}{(1/2)\rho u_\infty^2} = \frac{2V_*^2}{u_\infty^2} \qquad \text{(Friction coefficient).}$$

Here u_∞ is the velocity external to the boundary layer and ΔT is now the net temperature drop across the layer. If it is assumed that the log-law can be applied throughout the boundary layer, which is a reasonable approximation for flat-plate flow without a pressure gradient, then our log-law can be rewritten in terms of St and c_f. After a little algebra we find,

$$\mathrm{St} = \frac{c_f/2}{(\kappa/\kappa_T) + (c_f/2)^{1/2}\left[(1/\kappa_T)\ln(\mathrm{Pr}) + (5/3)(3\mathrm{Pr}^{1/3} - 1)^2 - (\kappa/\kappa_T)A\right]}.$$

(This is left as an exercise to the reader.) The expression involving Pr is somewhat messy and, since part of it has its origins in a curve fit, it makes sense to replace it by a simpler function of Pr. For the range of values of Pr encountered in practice (other than in liquid metals) a reasonable approximation is

$$\mathrm{St} = \frac{c_f/2}{\kappa/\kappa_T + (c_f/2)^{1/2}\left[13\,\mathrm{Pr}^{2/3} - 12\right]}, \qquad \mathrm{Pr} > 0.7.$$

A number of empirical expressions very similar to this are used in engineering for heat transfer calculations in flat-plate boundary layers and in pipes (Holman 1986).

4.5.2 *The effect of stratification on the log-law—the atmospheric boundary layer*

So far we have ignored the influence of T on $\mathbf{u}$, via the buoyancy force, and focused simply on the transport of heat by some prescribed set of velocity fluctuations. Let us now consider the reverse problem. Suppose that we have a turbulent shear flow over a rough, hot surface and that there is a prescribed flux of heat, q_T, away from that surface (Figure 4.25(a)). We are interested in how the buoyancy force associated with the hot fluid perturbs the mean and turbulent velocity fields. We adopt the so-called Boussinesq approximation for

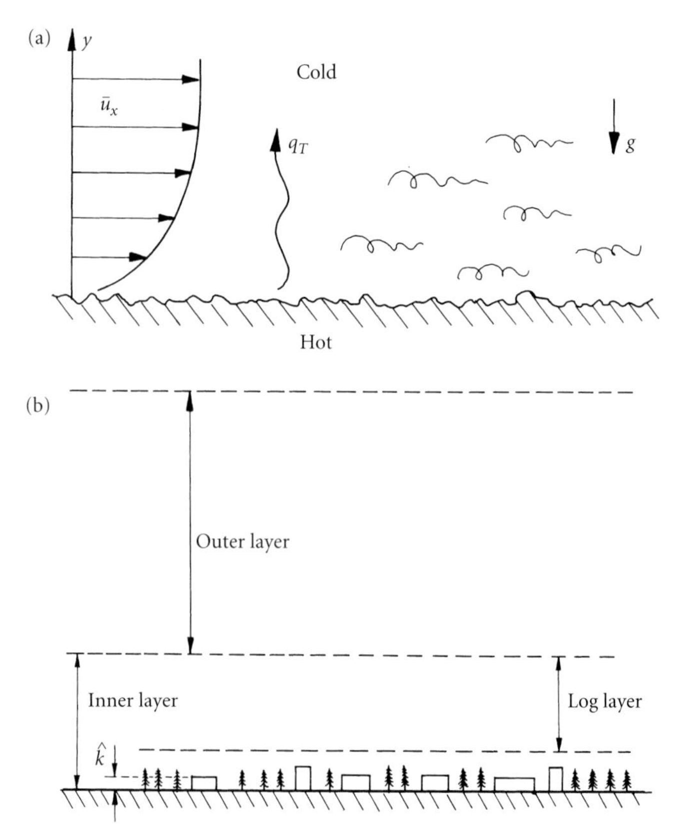

Figure 4.25 (a) Turbulence generated by both shear and buoyancy, (b) The atmospheric boundary layer. This is schematic only. In practice the inner region is only a small proportion of the over-all boundary layer and the log region begins at an altitude which is many multiples of the roughness height $\hat{k}$.

the buoyancy force in which changes in density are sufficiently small for ρ to be treated as a constant, except to the extent that there is a buoyancy force, $\delta\rho\mathbf{g}$, per unit volume. We may rewrite this as $-\rho\beta(T - T_0)\mathbf{g}$ where β is the so-called expansion coefficient, $\beta = -(d\rho/dT)\rho^{-1}$, and T_0 is a reference temperature representative of the ambient temperature.

4.5.2.1 The atmospheric boundary layer

The flow shown in Figure 4.25(a) is, perhaps, most relevant to the atmospheric boundary layer (ABL) and we now shift our attention to such flows. However, before embarking on a detailed analysis, perhaps we should say something about this boundary layer. (See Garrett 1992, for more details.) The ABL constitutes that part of the atmosphere where the direct effects of ground friction and surface heating (or cooling) are felt. It differs from more conventional boundary layers in at least two respects: both buoyancy and Coriolis forces are important. The ABL, which may be between 0.5 km and 5 km deep, is conventionally divided into two regions. The upper 90% of the layer is called the outer or Ekman layer. Here the Coriolis and pressure forces are dominant and the details of the ground cover (be it a corn field or a forest) is unimportant. Conversely, in the lower 10% of the boundary layer, called the inner layer, the Coriolis force is relatively unimportant but the flow is sensitive to the level of surface friction. In the inner layer the mean horizontal shear stress may be treated as more or less constant (independent of y) and so, in the absence of buoyancy forces, the mean velocity satisfies the conventional log-law (see Section 4.2.4),

$$\frac{\bar{u}_x}{V_*} = \frac{1}{\kappa}\ln\left[\frac{y}{y_0}\right]$$

Here y_0 is the surface roughness parameter which characterizes the height, shape, and packing density of the surface irregularities. (For sandgrain-type roughness $y_0 \approx \hat{k}/30$, where $\hat{k}$ is the rms roughness height.) Of course, this universal law does not apply immediately adjacent to the surface $(y \sim \hat{k})$ where the details of the flow depend crucially on the precise geometry of the ground cover. Measurements the mean velocity profile suggest that typical values of y_0 are as follows:

$$\begin{aligned}
\text{sand or soil;}\quad & y_0 = 0.001\,\text{m} \to 0.005\,\text{m}\\
\text{grass;}\quad & y_0 = 0.002\,\text{m} \to 0.02\,\text{m}\\
\text{various crops;}\quad & y_0 = 0.02\,\text{m} \to 0.1\,\text{m}\\
\text{woodland and bush;}\quad & y_0 = 0.2\,\text{m} \to 0.5\,\text{m}\\
\text{suburbs;}\quad & y_0 \sim 0.5\,\text{m}\\
\text{temperate forest;}\quad & y_0 = 0.5\,\text{m} \to 0.9\,\text{m}
\end{aligned}$$

By comparison, the viscous sublayer on an imaginary smooth ground would be only a fraction of a millimetre thick!

The turbulence intensity in the atmospheric boundary layer depends crucially on the nature of the buoyancy forces. These may be stabilizing (hot air over cold air), destabilizing (cold air over hot air), or neutral (negligible buoyancy forces). Neutral conditions tend to correspond to windy conditions and complete cloud cover. Unstable conditions are common during the day when the ground heats up by solar radiation, while stable conditions are more common at night as the ground cools via long wavelength radiation. One of the central questions in ABL theory is: can we parameterize the influence of the buoyancy forces on the mean and turbulent velocity fields? It turns out that this is most readily investigated through a consideration of the rate of working of these forces. For simplicity, we shall restrict ourselves to the inner layer where the Coriolis force may be neglected.

For a perfect gas it may be shown that and $\beta = T_0^{-1}$ so, in the atmosphere, the buoyancy force per unit volume is $-\rho[(T-T_0)/T_0]\mathbf{g}$.[14] Let us write $\theta = T - T_0$. Then the rate of working of the buoyancy force in a one-dimensional shear flow of the type shown in Figure (4.25(a)) is

$$\left(\overline{\theta u_y'}\right)\rho g/T_0 = q_T g/c_p T_0.$$

Equation (4.18) may be modified to incorporate this energy source. For a one-dimensional shear flow this equation yields

$$0 = -\frac{\partial}{\partial y}\left[\overline{p'u_y'} - \overline{u_i'\tau_{iy}'} + \tfrac{1}{2}\rho\overline{u_i'u_i'u_y'}\right] + \tau_{xy}^R\frac{\partial \bar{u}_x}{\partial y} + \left(\overline{\theta u_y'}\right)\rho g/T_0 - 2\rho\nu\overline{(S_{ij}')^2}$$

$$0 = (\text{redistribution of KE}) + (\text{shear production of KE}) + (\text{buoyant production of KE}) - (\text{dissipation of KE}).$$

The first term on the right merely redistributes turbulent kinetic energy without creating or destroying energy. Evidently we have two net sources of turbulent energy, $\tau_{xy}^R \partial\bar{u}_x/\partial y$ and $(\overline{\theta u_y'})\rho g/T_0$, the sum of which is ultimately matched by the small-scale viscous dissipation, $2\rho\nu\overline{(S_{ij}')^2}$. This simple energy equation forms the basis of most phenomenological models of buoyant shear flows.

[14] This estimate of β comes from the ideal gas law on the assumption that fluctuations in density are due exclusively to fluctuations in temperature, and that changes in ρ due to pressure fluctuations may be neglected. One consequence of this approximation is that unstable stratification (heavy fluid overlying light fluid) correspond to $dT/dy < 0$. In fact, a more careful thermodynamic analysis reveals that instability actually corresponds to $dT/dy < -G_a$, where $G_a = (\gamma-1)g/\gamma R$ is called the *adiabatic temperature gradient*. (Here γ is the usual ratio of specific heats and R the gas constant.) However, in the atmosphere G_a is of the order of 1°C/100 m, which is small, and so our approximation is reasonable.

4.5.2.2 The case of small wind shear: Prandtl's theory

The case where the shear is very small deserves, perhaps, particular attention. Here the generation of turbulence by buoyancy is more or less matched by the viscous dissipation, and so

$$q_T g / c_p T_0 \sim \rho\varepsilon.$$

In addition, the mixing-length estimate of q_T yields

$$\frac{q_T}{\rho c_p} \sim -(u'l)\frac{\partial \bar{T}}{\partial y} \sim \theta' u'$$

where θ' is some suitable measure of the fluctuations in $\theta = T - T_0$, and l is the mixing length. Now suppose that the buoyancy force does not significantly alter the conventional energy cascade. Then the dissipation per unit mass can be estimated as $\varepsilon \sim (u')^3/l$ and we may combine the estimates above to yield,

$$\frac{q_T}{\rho c_p} \sim \theta' u' \sim u'l\left|\frac{\partial \bar{T}}{\partial y}\right| \sim \frac{T_0}{g}\frac{u'^3}{l}.$$

Near the surface $y = 0$ we might expect the size of the eddies to grow as $l \sim \kappa y$ and so our estimates lead to the often quoted expressions (attributed to Prandtl)

$$u' \sim \left(\frac{q_T}{\rho c_p}\right)^{1/3}\left(\frac{T_0}{\kappa g}\right)^{-1/3} y^{1/3} \tag{4.69}$$

$$\theta' \sim \left(\frac{q_T}{\rho c_p}\right)^{2/3}\left(\frac{T_0}{\kappa g}\right)^{1/3} y^{-1/3} \tag{4.70}$$

$$\left|\frac{d\bar{T}}{dy}\right| \sim \left(\frac{q_T}{\rho c_p}\right)^{2/3}\left(\frac{T_0}{\kappa g}\right)^{1/3} y^{-4/3}. \tag{4.71}$$

In practice, $|d\bar{T}/dy|$ is found to decrease a little faster than $y^{-4/3}$, perhaps as $y^{-1.5}$, thus raising doubts about the assumption that $l \sim \kappa y$, which is strictly true only in simple, neutrally buoyant shear flows.

Actually, there is some difference of opinion in the literature as to the accuracy of (4.69) → (4.71). Monin and Yaglom (1975), provide considerable evidence in support of these expressions, while Townsend (1976), suggests that $|d\bar{T}/dy| \sim y^{-2}$ rather than $y^{-4/3}$. Wyngaard (1992), tends to come out in favour of Monin and Yaglom, while the data reproduced in Garrett (1992), suggests. $|d\bar{T}/dy| \sim y^{-1.5}$. The variation of the observed exponent in y^n may be explained in part by the fact that atmospheric measurements in the absence of a wind are hard to obtain. Moreover, there are theoretical reasons for suspecting that, in the absence of any shear, there is a y^{-2} layer nestling between the $y^{-4/3}$ region and the boundary (Kraichnan 1962).

4.5.2.3 The case of finite wind shear: the Monin–Obukhov theory

Let us now return to the more general case where both shear and buoyancy are important. The ratio of the rate of production of energy by these two forces is given by the so-called *flux Richardson number*, R_f:

$$R_f = -\frac{g}{T_0}\frac{\overline{\theta u'_y}}{\left(-\overline{u'_x u'_y}\right)\partial\bar{u}_x/\partial y} = -\frac{g}{\rho c_p T_0}\frac{q_T}{\left(-\overline{u'_x u'_y}\right)\partial\bar{u}_x/\partial y}. \quad (4.72)$$

Note that it is conventional to define R_f so that it is negative for an upward transfer of heat and positive for downward transfer. When R_f is large and negative the primary source of turbulence is the buoyancy force, while a small, negative R_f implies that buoyancy is negligible. The third case, of $0 < R_f < 0(1)$, corresponds to a stable stratification of T, and thus to a partial suppression of turbulence. Now recall that we are restricting attention to the near surface region of the ABL, which may be, say, ~100 m deep. Here τ_{xy} is constant and equal to ρV_*^2. If the velocity profile is logarithmic, $\partial\bar{u}_x/\partial y = V_*/(\kappa y)$, as it would be in the absence of buoyancy, then (4.72) reduces to,

$$R_f = \frac{y}{L}, \qquad L = -\frac{(T_0/\kappa g)V_*^3}{(q_T/\rho c_p)}. \quad (4.73)$$

The quantity L, called the *Monin–Obukhov length*, plays a central role in atmospheric flows (Monin and Yaglom 1975). Typically $|L|$ is a few tens of metres. For heights much less than $|L|$ buoyancy is relatively unimportant as far as the generation of turbulence is concerned, while for $y \gg |L|$, the generation of turbulence by shear is negligible and buoyancy dominates. One of the nice things about L is that it is effectively determined by just two parameters, V_* and q_T. As with R_f, a negative value of L indicates unstable stratification (enhanced turbulence) and a positive value of L signifies stable stratification (suppression of turbulence). When L is large and positive an interesting situation arises where the suppression of turbulence is only partial at low altitudes (because y/L is small) but more complete at higher altitudes, where y/L is large. In such cases the turbulent mixing is most intense near the ground.

Monin and Yaglom (1975) have suggested that, in general, near-surface (i.e. constant shear stress) atmospheric flows may be characterized by the expression

$$\frac{\partial\bar{u}_x}{\partial y} = \frac{V_*}{\kappa y}\phi(y/L)$$

where ϕ is the correction to the log-law introduced by buoyancy. This has turned out to be a profitable strategy in as much as the use of the normalized variable y/L has proven to be an effective way of compacting the experimental data for $(\partial\bar{u}_x/\partial y)(\kappa y/V_*)$ in atmospheric flows. Let us see if we can estimate ϕ.

When $y \ll |L|$, so that the buoyancy force only mildly perturbs the boundary layer, it is found that buoyancy effects can be taken into account by modifying the usual form of the log-law with a linear correction term. The modified version, valid for positive or negative L, is

$$\frac{\bar{u}_x}{V_*} = \frac{1}{\kappa}\left[\ln\left(\frac{y}{y_0}\right) + \gamma\frac{y}{L}\right], \quad y \ll |L|$$

where y_0 is the surface roughness parameter and γ is a constant. We can also write this as

$$\frac{\partial\bar{u}_x}{\partial y} = \frac{V_*}{\kappa y}\left[1 + \gamma\frac{y}{L}\right], \quad y \ll |L|$$

which effectively fixes ϕ for small $|y/L|$. Estimates of γ vary considerably, but typically γ lies in the range 3 to 8 with data corresponding to negative L tending to give lower values of γ (Monin and Yaglom 1975). Thus $\bar{u}_x$ changes less rapidly with height in unstable conditions ($L < 0$), due to increased vertical mixing, and more rapidly in a stable environment ($L > 0$).

Let us now consider the case of unstable stratification in which $y \gg |L|$. In such a situation the generation of energy is dominated by the buoyancy force, the mean shear having negligible effect on the energy budget. So we may return to (4.69), which gives us the estimates,

$$u' \sim V_*(y/|L|)^{1/3}, \quad l \sim \kappa y.$$

If we believe in the phenomenology of an eddy viscosity this suggests,

$$\nu_t \sim u'l \sim V_*(y/|L|)^{1/3}\kappa y$$

where ν_t is defined via the relationship

$$\tau_{xy}^R/\rho = V_*^2 = \nu_t\frac{\partial\bar{u}_x}{\partial y}.$$

This yields an expression for the gradient in mean velocity,

$$\frac{\partial\bar{u}_x}{\partial y} \sim \frac{V_*}{\kappa y}\left(\frac{y}{|L|}\right)^{-1/3}, \quad y \gg |L|$$

which might be compared with the equivalent expression for small, $y/|L|$,

$$\frac{\partial\bar{u}_x}{\partial y} = \frac{V_*}{\kappa y}\left[1 + \gamma\frac{y}{L}\right], \quad y \ll |L|$$

Evidently both of these estimates of the mean velocity gradient conform to the Monin and Yaglom equation

$$\frac{\partial \bar{u}_x}{\partial y} = \frac{V_*}{\kappa y}\phi(y/L).$$

This is as far as heuristic physical arguments will get us. We must now revert to empiricism to pin down the form of ϕ for intermediate values of y/L. Numerous semi-empirical expressions for ϕ have been suggested over the years. A typical prescription is

$$\phi(\zeta) \approx 1 + \gamma\zeta, \qquad 0 < \zeta < 0.2$$
$$\phi(\zeta) \approx (1 - \beta\zeta)^{-1/3}, \quad \zeta < 0$$

where estimates of γ lie in the range 4 to 8, while those of β vary from 5 to 14. Of course, this could be interpreted as a crude interpolation between our various theoretical estimates. (Note that an exponent of $-1/4$ is often used instead of $-1/3$ in the expression for $\phi(\zeta)$ for $\zeta < 0$.)

4.6 More on one-point closure models

We now return to the subject of 'engineering models' of turbulence: that is, closure schemes which seek to predict the Reynolds stresses and thus allow the mean flow to be calculated. These are referred to as *one-point closure models* as they focus attention on τ_{ij}^R, which relates to just one point in space, rather than Q_{ij}, which relates two points in space. (We shall discuss two-point closure models in Chapters 6 and 8.) Two of the most popular closure schemes are the k–ε model (and its relatives) and the so-called Reynolds stress model. We start with the k–ε model.

4.6.1 *A second look at the k–ε model*

We introduced this scheme in Section 4.1.4 of this chapter. Like Prandtl's mixing length, it is an eddy-viscosity model, and assumes that τ_{ij}^R is related to the mean strain rate $\bar{S}_{ij}$ by

$$\tau_{ij}^R = 2\rho\nu_t\bar{S}_{ij} - (\rho/3)\left(\overline{u_k' u_k'}\right)\delta_{ij}. \tag{4.74}$$

As noted earlier, three weaknesses of the eddy-viscosity hypothesis are:

1. τ_{ij}^R and $\bar{S}_{ij}$ are related by a scalar, ν_t, and not a tensor, which is unlikely to be valid in strongly anisotropic turbulence.

2. If $\overline{S}_{ij} = 0$, or else the mean flow is a one-dimensional shear flow, then (4.74) predicts $\overline{(u'_x)^2} = \overline{(u'_y)^2} = \overline{(u'_z)^2}$, which is not, in general, found to be the case.
3. There is an implicit assumption that τ_{ij}^R is controlled by the *local* rate of strain in the mean flow, and not by the *history* of the straining of the turbulence. This will lead to erroneous results in those cases where the turbulence is subject to rapid straining by the mean flow, so that the shape of the eddies, and hence the magnitude of the Reynolds stresses, depend upon the immediate history of the turbulence.

Let us now explore these three potential flaws in more detail. An example of the first difficulty arises in stratified flows where the buoyancy force tends to suppress vertical fluctuations in velocity. Consider, for example, a jet or wake evolving in a stratified medium. The turbulence in the jet soon becomes anisotropic and there is no reason to suppose that the relationship between the Reynolds stresses and the mean shear is the same in the horizontal plane as it is in the vertical plane. Similar problems arise when strong rotation or intense magnetic fields are present. Both of these can induce severe anisotropy, with the large eddies being elongated in the direction of the rotation axis or in the direction of the magnetic field.

As an example of the second difficulty above, consider the turbulence created by a grid in a wind tunnel (Figure 4.26). Here the mean flow is uniform and so $\overline{S}_{ij} = 0$. Nevertheless, turbulence produced by a biplane grid tends to be anisotropic just downstream of the grid, with the cross-stream fluctuations being $\sim$10% smaller than the streamwise fluctuations, $u_\perp / u_\parallel \sim 0.9$. This anisoptropy is very persistent. Indeed, if we track the turbulence for quite some distance downstream of the grid, so that its energy has decayed by a factor of 20 or 30, we find that there is still an observable difference between $u_\perp$ and $u_\parallel$ of around 5% (Townsend 1976). Yet all eddy-viscosity models would have us believe that the turbulence is isotropic, since $\overline{S}_{ij}$ is zero. Of course, you might say: 'who cares about a 10% error, there are more important issues at stake', and you would be right. However, it is possible to create even stronger anisotropy in a wind tunnel and, despite the absence of shear, it displays the same stubborn persistence seen in ordinary grid turbulence. One such experiment is described by Townsend (1976) in which grid turbulence is made to pass through a sequence of fine wire gauzes. The resulting flow has an anisotropy factor of $u_\perp / u_\parallel \sim 1.3$. In the experiment the flow was tracked downstream to the point where its kinetic energy had fallen by a factor of 3. There was no discernable decrease in $u_\perp / u_\parallel$. Now it is probably true that, if left to itself for long enough, a cloud of turbulence will become increasingly isotropic due to the

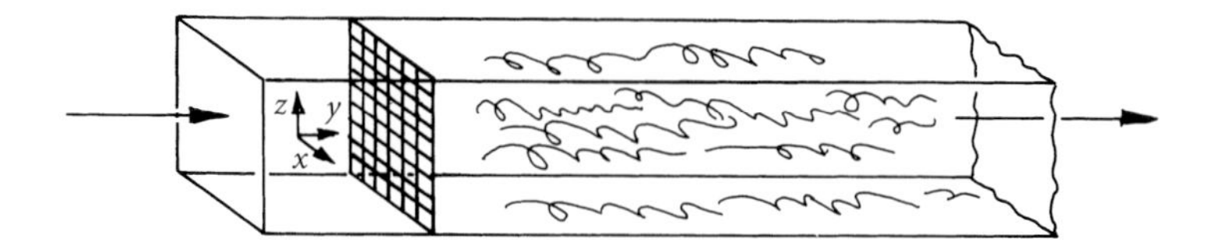

Figure 4.26 Turbulence induced by a grid.

random mixing of vorticity. However, these simple wind-tunnel experiments tell us that this can be a surprisingly slow process. So the assumption that, when $\bar{S}_{ij} = 0$, the turbulence is isotropic is not, in general, realistic. A similar problem arises in the one-dimension shear flow, $\bar{\mathbf{u}} = (\bar{u}_x(y), 0, 0)$ where, once again, the eddy-viscosity hypothesis gives $\overline{(u'_x)^2} = \overline{(u'_y)^2} = \overline{(u'_z)^2}$, yet experiments suggest that these three contributions to the kinetic energy are unequal. (See Section 4.1 of this chapter.)

Let us now turn to the third weakness of the eddy-viscosity hypothesis: it is assumed that τ^R_{ij} is completely determined by the *local* strain-rate in the flow, and not by the *history* of straining. Is this reasonable? Consider a simple shear flow, $\bar{\mathbf{u}} = (\bar{u}_x(y), 0, 0), \bar{u}_x(y) = Sy$. As discussed in Section 4.1 of this chapter, this may be divided up into an irrotational straining motion plus one component of vorticity:

$$\begin{aligned} \bar{\mathbf{u}} &= \tfrac{1}{2}(Sy, Sx, 0) + \tfrac{1}{2}(Sy, -Sx, 0) \\ \bar{\boldsymbol{\omega}} &= 0 \qquad\qquad\qquad - S\hat{\mathbf{e}}_z \\ \bar{S}_{xy} &= \tfrac{1}{2}S \qquad\qquad\quad + 0 \end{aligned}$$

The principal axes of strain (the directions of maximum rate of straining) are inclined at 45° to the x–y axes, as shown in Figure 4.21(a). Now imagine that a cloud of turbulence (i.e. a tangle of vortex tubes) is introduced into the mean flow at $t = 0$. We suppose that the turbulence is initially weak in the sense that $|\boldsymbol{\omega}'| < S$. It seems plausible that the turbulent vortex lines will, on average, tend to be stretched along the axis of maximum positive stain, as shown in Figure 4.21(b). This stretching process transfers energy from the mean flow to the turbulence since the elongation of a vortex tube increases its kinetic energy. Moreover, the inclined vortex tubes are particularly effective at generating the Reynolds stresses τ^R_{xy}. To see why, consider a coordinate system (x^*, y^*) inclined at 45° to the (x, y) axes, with x^* parallel to the maximum rate of straining. In terms of these coordinates we have

$$\tau^R_{xy} = \frac{\rho}{2}\left[\overline{(u'_{y^*})^2} - \overline{(u'_{x^*})^2}\right].$$

(See equation (4.47).) So vortical structures which lead to $|u'_{y^*}| \gg |u'_{x^*}|$ produce a significant contribution to τ^R_{xy}, and this is exactly what a vortex tube inclined at 45° to the mean flow achieves.

Of course, this discussion is more than a little heuristic. In particular, we have ignored the rotational contribution to $\bar{\mathbf{u}}$, which will tend to rotate the vortex lines in a clockwise direction. Still, it seems plausible that the magnitude of τ_{xy}^R will depend on the instantaneous structure of the tangle of vortex tubes and that this, in turn, will generally depend on the history of the straining of the turbulence.

In the eddy-viscosity hypothesis the turbulence is allowed no memory. The magnitude of τ_{xy}^R and hence the statistical structure of the vortex tubes, is assumed to depend only on local gradients in the mean flow. It is as if we have assumed that the turbulence at any one point has relaxed towards some sort of statistical equilibrium which is governed by the local conditions alone. In Figure 4.21, for example, this equilibrium is a balance between turbulence induced mixing of the vortex tubes, the teasing out of the mean-flow vorticity by the turbulence, the tendency of $\bar{S}_{xy}$ to stretch the turbulent vortex tubes, and the tendency of $\bar{\boldsymbol{\omega}}$ to rotate them.

The notion of a statistical equilibrium governed by local conditions would be plausible if the timescale for the turbulent fluctuations, k/ε, were rapid by comparison with the mean-flow timescale. However, it is the large eddies which contribute most to τ_{ij}^R, and these usually have a turn-over time comparable with the inverse of the mean vorticity. So we are not, in general, at liberty to ignore the memory of the turbulence. This limitation of the eddy-viscosity hypothesis is likely to be particularly damaging when the mean flow subjects the turbulence to rapid, irrotational straining, that is, $|\bar{S}_{ij}|k/\varepsilon \gg 1$.

All in all, it would seem that the eddy-viscosity hypothesis must be regarded with caution. Of course, the three shortcomings listed above are all interrelated. For example, the third shortfall amounts to an assumption of no memory, and this lack of memory also characterizes the second limitation since, in the absence of mean shear, the turbulence will head towards an isotropic state. Perhaps the main point to emphasis, however, is that any model based on (4.74) must be subject to suspicion in circumstances where: (i) the turbulence is strongly anisotropic; or (ii) the turbulence is subject to rapid irrotational straining.

So far, the k–ε model does not seem to be fairing too well, and we have not yet detailed all of the *ad hoc* modelling which goes into predicting the eddy viscosity at each point in the flow. It is all the more surprising, therefore, that the k–ε model has turned out to be moderately successful. With only five empirical constants at its disposal, constants which are fixed by certain standard flows, the k–ε model seems to provide passable estimates of the Reynolds stresses under a wide range of conditions. Given that it is particularly easy to implement, perhaps it is not so surprising that it has become the standard closure model for routine engineering calculations. So let us now set aside the limitations of the eddy-viscosity hypothesis and detail the main features of the k–ε model, the primary function of which is to provide an estimate of the eddy viscosity at each point in the flow.

Since we are assuming that it is the local conditions which determine τ_{ij}^R, we would expect ν_t to be determined by the characteristics of the large eddies at the point in question. We are also excluding severe anisotropy, and so these eddies are presumably characterized by a single velocity scale, V_T, and a single timescale τ, say the eddy turnover time. So we have $\nu_t = f(V_T, \tau)$ and the only dimensionally consistent possibility is,

$$\nu_t \sim V_T^2 \tau.$$

It is conventional to take the turbulence kinetic energy to represent V_T^2 and so our estimate can be rewritten as

$$\nu_t \sim k\tau, \qquad k = \tfrac{1}{2}\overline{(\mathbf{u}')^2}. \tag{4.75}$$

The implication is that the more energetic the turbulence, the higher the momentum exchange by turbulent fluctuations. Let us now introduce three alternative ways of writing (4.75). The vorticity of the large scale eddies, which we will denote ω, is of the order τ^{-1} and so we could write,

$$\nu_t \sim k/\omega. \tag{4.76a}$$

Alternatively, we could note that $\tau \sim l/V_T$ where l is a measure of the size of the large eddies. This leads to

$$\nu_t \sim k^{1/2} l. \tag{4.76b}$$

Finally we recall that the rate of destruction of turbulent energy, ε, by small-scale viscous forces is of the order of $k^{3/2}/l$, and so we could also propose

$$\nu_t \sim k^2/\varepsilon. \tag{4.76c}$$

All three estimates (4.76a–c) are essentially saying the same thing and it is, perhaps, a little arbitrary which we adopt. Partly for historical reasons the engineering community has, by and large, plumped for (4.76c). So in the k–ε model we write

$$\nu_t = c_\mu k^2/\varepsilon \tag{4.77}$$

where c_μ is a constant ($c_\mu \sim 0.09$), and then specify semi-empirical transport equations for k and ε. Relatives of the k–ε model work with k

and l or else k and ω. Conceptually there is a little difference between these different schemes, and so we shall stay with the k–ε model as it is the most popular. As discussed in Section 4.1.4 of this chapter the transport equation for k is based on the exact energy equation (4.29),

$$\frac{\partial k}{\partial t} + \bar{\mathbf{u}} \cdot \nabla(k) = -\nabla \cdot [\mathbf{T}] + \left(\tau_{ij}^R / \rho\right) \bar{S}_{ij} - \varepsilon \tag{4.78}$$

where $\mathbf{T}$ involves unknown correlations such as $\overline{\mathbf{u}'p'}$ and $\overline{\mathbf{u}'u_i'u_i'}$. The key modelling step here is to assume that the influence of the triple correlations and pressure–velocity correlations is to spread the turbulence energy in a diffusive manner from regions of high intensity fluctuations to regions of low intensity. Specifically, the k–ε model proposes

$$\mathbf{T} = -\alpha_t \nabla k$$

where α_t is some unknown diffusivity, normally taken equal to ν_t. The transport equation for k then becomes a simple advection–diffusion equation containing a source term, $\tau_{ij}^R \bar{S}_{ij}$, and a sink term, ε. It is

$$\frac{\partial k}{\partial t} + \bar{\mathbf{u}} \cdot \nabla(k) = \nabla \cdot [\nu_t \nabla k] + \left(\tau_{ij}^R / \rho\right) \bar{S}_{ij} - \varepsilon. \tag{4.79a}$$

Sometimes this is written in the more general form

$$\frac{\partial k}{\partial t} + \bar{\mathbf{u}} \cdot \nabla(k) = \nabla \cdot [(\nu + \nu_t/\sigma_k) \nabla k] + \left(\tau_{ij}^R / \rho\right) \bar{S}_{ij} - \varepsilon \tag{4.79b}$$

where σ_k is a constant. This gives greater freedom in the choice of the diffusivity α_t. In practice, however, σ_k is nearly always set equal to unity. Note that, in homogeneous turbulence, the divergence on the right of (4.78) and (4.79) is equal to zero and in such cases our 'model equation' is exact. The ε equation, on the other hand, is pure invention. It is

$$\frac{\partial \varepsilon}{\partial t} + \bar{\mathbf{u}} \cdot \nabla(\varepsilon) = \nabla \cdot \left(\left(\nu + \frac{\nu_t}{\sigma_\varepsilon} \right) \nabla \varepsilon \right) + c_1 \frac{G\varepsilon}{k} - c_2 \frac{\varepsilon^2}{k} \tag{4.80}$$

where $G = \left(\tau_{ij}^R / \rho\right) \bar{S}_{ij}$ and σ_ε, c_1, and c_2 are tunable coefficients which have been set to capture a range of standard flows. The values of σ_ε, c_1, and c_2 are usually taken to be

$$\sigma_\varepsilon = 1.3, \quad c_1 = 1.44, \quad c_2 = 1.92.$$

At this point cynics might protest: 'This is pure empiricism! How can an entirely fictional equation, plucked out of thin air, and forced, through the judicious choice of some arbitrary coefficients, to reproduce one or two laboratory results, possibly hope to anticipate the evolution of a wide range of flows?' The extraordinary thing, however, is that, by and large, it works reasonably well, at least much better than it ought to. So perhaps there is more to (4.80) than meets the eye. Perhaps there is some underlying rationale for this equation. In retrospect, it turns out that there is.

One interpretation of the ε equation is given by Pope (2000). It goes something like this. The turn-over time of a large-scale eddy is $\sim l/u \sim k/\varepsilon$, since ε is of the order of u^3/l. Thus the characteristic vorticity of the large, energy containing eddies is, $\omega \sim \varepsilon/k$. Now consider a one-dimensional, homogenous shear flow in which $\bar{\mathbf{u}} = \bar{u}_x(y)\hat{\mathbf{e}}_x$, $\partial\bar{u}_x/\partial y = S$. The vorticity of the energy-containing eddies comes from a distortion of the mean-flow vortex lines (they get pushed out of shape by the turbulence) and so we would expect that, after a while, ω will settle down to some value of the order of S, say $\omega = S/\lambda$. (Table 4.1 in Section 4.1.3 suggests the value of λ is 6.3.) If the initial value of ω is different to S/λ then we might expect ω to relax towards S/λ on timescale of S^{-1}. A heuristic equation which captures this behaviour is,

$$\frac{d\omega}{dt} = a^2\left[(S/\lambda)^2 - \omega^2\right], \quad \omega = \varepsilon/k. \tag{4.81}$$

The k–ε model, on the other hand, would have us believe,

$$\frac{dk}{dt} = G - \varepsilon$$
$$\frac{d\varepsilon}{dt} = c_1\frac{G\varepsilon}{k} - c_2\frac{\varepsilon^2}{k}$$
$$G = \nu_t S^2 = c_\mu k^2 S^2/\varepsilon$$

for this simple, one-dimensional shear flow. However, these may be rearranged to give

$$\frac{d\omega}{dt} = (c_1 - 1)c_\mu S^2 - (c_2 - 1)\omega^2, \quad \omega = \varepsilon/k.$$

Of course, this is the same as our heuristic model equation (4.81), with $a^2 = (c_2 - 1)$ and $a^2/\lambda^2 = c_\mu(c_1 - 1)$. Thus, provided c_1 and c_2 are both greater than unity, the particular form of the ε equation used in the k–ε model ensures that the large-scale vorticity behaves in a plausible manner in homogeneous shear flow. Thus we might think of the ε equation as being a transport equation for large-scale vorticity, tending to push the large-scale vorticity towards that of the mean flow.

Let us now summarize what we know about the k–ε model. We have the Boussinesq equation

$$\tau_{ij}^{R}/\rho = 2\nu_t \bar{S}_{ij} - (2/3)k\delta_{ij} \tag{4.82}$$

combined with a Prandtl-like estimate of ν_t,

$$\nu_t = c_\mu k^2/\varepsilon \tag{4.83}$$

plus two empirical transport equations for k and ε,

$$\frac{\partial k}{\partial t} + \bar{\mathbf{u}} \cdot \nabla(k) = \nabla \cdot \left[\left(\nu + \frac{\nu_t}{\sigma_k} \right) \nabla k \right] + G - \varepsilon \tag{4.84}$$

$$\frac{\partial \varepsilon}{\partial t} + \bar{\mathbf{u}} \cdot \nabla(\varepsilon) = \nabla \cdot \left[\left(\nu + \frac{\nu_t}{\sigma_\varepsilon} \right) \nabla \varepsilon \right] + c_1 \frac{G\varepsilon}{k} - c_2 \frac{\varepsilon^2}{k} \tag{4.85}$$

where $G = \left(\tau_{ij}^{R}/\rho\right)\bar{S}_{ij}$. The tunable coefficients are usually assumed to have the values of

$$c_\mu = 0.09; \quad \sigma_\varepsilon = 1.3; \quad \sigma_k = 1; \quad c_1 = 1.44; \quad c_2 = 1.92.$$

So how does this model fair in practice? Well, there are certain well-documented flows against which we may compare the k–ε predictions. Consider first the free decay of homogeneous, isotropic turbulence in which there is no mean flow, $\bar{\mathbf{u}} = 0$. The k–ε model predicts

$$\frac{dk}{dt} = -\varepsilon$$
$$\frac{d\varepsilon}{dt} = -c_2 \frac{\varepsilon^2}{k}$$
$$\overline{(u_x')^2} = \overline{(u_y')^2} = \overline{(u_z')^2}$$

These may be integrated to give

$$k = k_0(1 + t/\tau)^{-n}, \qquad n = (c_2 - 1)^{-1} = 1.08$$

where τ is proportional to the initial turn-over time, $\tau = nk_0/\varepsilon_0$. In practice it is found that k decays considerably faster than this, as $k \sim t^{-1.4}$ (see Chapter 6), and so the model has not performed particularly well here. Another failing of the k–ε model is that, when $\bar{\mathbf{u}} = 0$, it does not distinguish between homogeneous, anisotropic turbulence and homogeous, isotropic turbulence: both are governed by the same model equations. In effect, the k–ε model assumes that freely decaying turbulence is always isotropic. Yet we know from studies of grid turbulence that anisotropy is easily generated (in fact it is hard to get rid of!) and that any anisotropy introduced by the grid is remarkably persistent.

Next we consider the log-law region of a boundary layer. Here we know that τ^R_{xy} is constant and equal to $\tau_w = \rho V_*^2$. The rate of generation of turbulent energy is therefore

$$G = \frac{\tau^R_{xy}}{\rho}\frac{\partial \bar{u}_x}{\partial y} = \frac{V_*^3}{\kappa y}.$$

We have also seen that, throughout the log-law region, Sk/ε, k/V_*^2, and G/ε are all more or less constant and equal to

$$\frac{G}{\varepsilon} \approx 0.91, \qquad \frac{Sk}{\varepsilon} \approx 3.2, \qquad \frac{k}{V_*^2} = \frac{\varepsilon}{G}\frac{Sk}{\varepsilon} \approx 3.52$$

where S is the shear $\partial \bar{u}_x/\partial y$. (See Section 4.2.1.) Given that k is constant, let us see what k–ε predicts for the log-law region. The k equation (4.84) simply reduces to

$$G = \varepsilon$$

requiring a local balance between the generation and dissipation of energy. The eddy-viscosity estimate (4.83) then becomes

$$\nu_t = \frac{V_*^2}{V_*/\kappa y} = c_\mu \frac{k^2}{V_*^3/\kappa y}$$

which, when simplified, yields

$$c_\mu = \left(k/V_*^2\right)^{-2}, \quad (G = \varepsilon).$$

Given the measured values of k/V_*^2, this effectively fixes c_μ at $\sim$0.09. Next we turn to the ε equation. Given that $\varepsilon = G = V_*^3/\kappa y$, and that c_μ is fixed by the expression above, (4.85) reduces to

$$\sigma_\varepsilon c_\mu^{1/2}(c_2 - c_1) = \kappa^2.$$

For specified values of c_2 and c_1, this effectively determines σ_ε. (The model values of $\sigma_\varepsilon = 1.3$ and $c_2 - c_1 = 0.48$ correspond to $\kappa = 0.43$.) Thus we begin to see how the various coefficients in the k–ε model can be chosen to reproduce certain standard flows, and in fact the k–ε model does not do too badly in the log-law region because it has been tuned to do so.

Let us finally consider a one-dimensional, homogeneous shear flow, $\bar{u}_x(y) = Sy$. Here the k–ε model predicts that the flow evolves towards a self-similar solution in which,

$$S/\omega = kS/\varepsilon = \left[c_\mu(c_1 - 1)/(c_2 - 1)\right]^{-1/2} = 4.8$$

$$\tau^R_{xy}/(\rho k) = \left[c_\mu(c_2 - 1)/(c_1 - 1)\right]^{1/2} = 0.43$$

$$G/\varepsilon = (c_2 - 1)/(c_1 - 1) = 2.1.$$

(Readers can verify this for themselves. A good starting point is the relaxation equation (4.81) for ε/k.) We might compare this with the wind-tunnel data given in Table 4.1 and discussed in Section 4.4 of this chapter:

$$kS/\varepsilon \sim 6.3$$
$$\tau_{xy}^R/(\rho k) \sim 0.28$$
$$G/\varepsilon \sim 1.7.$$

Evidently, the k–ε model somewhat overestimates the ratio of production to dissipation, but at least it admits a self-similar solution of the form: $G/\varepsilon = \text{constant}$, $kS/\varepsilon = \text{constant}$. Moreover the discrepancies between the actual and predicted values of kS/ε and G/ε are more or less acceptable for many engineering purposes. These results are fairly typical. In general, the standard k–ε model performs well in simple shear flows (the various coefficients having been chosen to encourage this), but can go badly wrong in more complex configurations such as: stagnation-point flows; flows with a rapid mean rate of strain; boundary layers with a strong adverse pressure gradient or large curvature; and highly anisotropic turbulence (flows with buoyancy or strong swirl).

While there are various ways of patching up the standard model on a case by case basis, this is not entirely satisfactory. Problems also arise very close to boundaries, where we have to match the k and ε equations to the viscous sublayer, which turns out to be a matter of some delicacy.

The k–ε model has proved extremely popular in the engineering community. In part, this is because it is simple to use and provides reliable results for simple shear flows. Moreover, those situations where it does not fair so well (e.g. flows with strong rotation or turbulence which is suddenly subjected to a large strain field) are now reasonably well documented. So it is an imperfect tool whose deficiencies are well known. There are many variants of the k–ε model, such as the k–ω formulation in which we write $\omega = \varepsilon/k$ and transport equations are provided for k and ω. Each formulation has its own strengths and weaknesses and these are nicely summed up in Durbin and Petterson-Reif (2001).

However, one should not lose sight of the fact that the k–ε model and its variants are ultimately highly sophisticated exercises in interpolating between data sets. Perhaps we should leave the last word to the physicist George Gamow, who, in a moment of whimsy, noted, 'With five free parameters a theorist could fit the profile of an elephant' (1990).

4.6.2 *The Reynolds stress model*

We now give a brief discussion of the so-called Reynolds stress model of turbulence. This is, perhaps, the most sophisticated (or perhaps we should say, complex) of the heuristic models used by engineers to estimate the influence of turbulence on the mean flow. It is simple enough to be applied to flows of some geometric complexity, yet offers the possibility of remedying some of the worst excesses of the mixing-length or k–ε models. Like the k–ε model, it is referred to as a *one-point closure model*, as it involves statistical quantities evaluated at only one point in space. (We will meet so-called *two-point closure models* in Chapter 6. These models make a more serious attempt to emulate the physics of the energy cascade, but are so complex that they are of limited value in an engineering context.)

The motivation behind the Reynolds-stress model is the inability of the k–ε model, or indeed any eddy-viscosity model, to cope with severe anisotropy in the turbulence, or to allow for a non-local relationship between τ_{ij}^R and $\bar{S}_{ij}$, that is, history effects. So it rejects the Boussinesq equation and works instead with a system of equations of the form,

$$\rho\frac{\partial \bar{u}_i}{\partial t} + \rho(\bar{\mathbf{u}}\cdot\nabla)\bar{u}_i = -\frac{\partial \bar{p}}{\partial x_i} + \frac{\partial}{\partial x_j}\left[\bar{\tau}_{ij} + \tau_{ij}^R\right] \tag{4.86}$$

$$\frac{\partial \tau_{ij}^R}{\partial t} + (\bar{\mathbf{u}}\cdot\nabla)\tau_{ij}^R = (\sim) \tag{4.87}$$

$$\frac{\partial \varepsilon}{\partial t} + (\bar{\mathbf{u}}\cdot\nabla)\varepsilon = (\sim). \tag{4.88}$$

Here the ε equation is very similar, though not identical, to that used in the k–ε model. (The only difference is that the diffusivity, ν_t/σ_ε, appearing in the divergence term on the right of (4.80) is replaced by an anisotripic diffusivity—see Hanjalic and Jakirlic 2002.) The transport equations for τ_{ij}^R, on the other hand, are based on (4.8). In principle, such an approach should work better than the k–ε model, since it is not burdened with the restrictions of the eddy-viscosity hypothesis. Nevertheless, there is still a great deal of *ad hoc* modelling involved, in which a variety of tunable coefficients are set to capture a number of standard flows. So, like the k–ε model, it is essentially an extremely sophisticated exercise in interpolating between data sets.

In moving from k–ε to Reynolds stress modeling the key new equation is, of course (4.87). Let us write it out in full. Rearranging the terms in (4.8) we have,

$$\frac{\partial \tau_{ij}^R}{\partial t} + (\bar{\mathbf{u}}\cdot\nabla)\tau_{ij}^R = -2\overline{p'S'_{ij}} - \left[\tau_{ik}^R\frac{\partial \bar{u}_j}{\partial x_k} + \tau_{jk}^R\frac{\partial \bar{u}_i}{\partial x_k}\right] + \rho\varepsilon_{ij} + \frac{\partial}{\partial x_k}\left[H_{ijk}\right] \tag{4.89}$$

where

$$\varepsilon_{ij} = 2\nu \overline{\frac{\partial u_i'}{\partial x_k} \frac{\partial u_j'}{\partial x_k}}$$

and

$$H_{ijk} = \rho \overline{u_i' u_j' u_k'} + \nu \frac{\partial \tau_{ij}^R}{\partial x_k} + \delta_{ik} \overline{p' u_j'} + \delta_{jk} \overline{p' u_i'}.$$

Since the small scales in a turbulent flow are approximately isotropic (except close to boundaries), and it is these which contribute most to ε_{ij}, we may approximate this tensor by its homogeneous, isotropic counterpart,

$$\varepsilon_{ij} = \tfrac{1}{3} \varepsilon_{kk} \delta_{ij} = \tfrac{2}{3} \varepsilon \delta_{ij} \qquad \text{(except near surfaces).}$$

Moreover, the viscous term in H_{ijk} is usually negligible by comparison with the other terms, and so our transport equation for τ_{ij}^R simplifies to

$$\frac{\partial \tau_{ij}^R}{\partial t} + (\bar{\mathbf{u}} \cdot \nabla) \tau_{ij}^R = -2\overline{p' S_{ij}'} - \left[\tau_{ik}^R \frac{\partial \bar{u}_j}{\partial x_k} + \tau_{jk}^R \frac{\partial \bar{u}_i}{\partial x_k} \right] + \frac{2}{3} \rho \varepsilon \delta_{ij} + \nabla \cdot \left[\mathbf{H}_{ij} \right] \tag{4.90}$$

where

$$H_{ijk} = \rho \overline{u_i' u_j' u_k'} + \overline{p' \left(\delta_{ik} u_j' + \delta_{jk} u_i' \right)}. \tag{4.91}$$

Evidently, we have two new tensors which require some sort of closure approximation: H_{ijk} and $\overline{p' S_{ij}'}$. If these can be modelled in terms of τ_{ij}^R, ε, and $\bar{\mathbf{u}}$ then we have a closed system of equations. In the simplest of Reynolds stress models the pressure-rate-of-strain terms are estimated as

$$2\overline{p' S_{ij}'} = -\rho c_R \frac{\varepsilon}{k} \left[\overline{u_i u_j} - \tfrac{2}{3} \delta_{ij} k \right] + [\text{terms involving the mean strain rate}] \tag{4.92}$$

where c_R is a constant. (Typically $c_R = 1.8$.) On the other hand, H_{ijk} is usually modeled so that the divergence in (4.90) acts like a diffusion term for τ_{ij}^R, with an anisotropic diffusivity. (Again, see Hanjalic and Jakirlic 2002.) This is reminiscent of the modelling of $\mathbf{T}$ in the kinetic energy equation (4.78). For flows in which the turbulence is homogeneous the divergence of H_{ijk} vanishes and nature of the closure assumption for H_{ijk} is unimportant. Consequently, it is closure approximation (4.92) which has attracted most debate. Let us rewrite (4.92) in the form

$$\overline{p' S_{ij}'} = -\rho c_R \varepsilon b_{ij} + \left[\text{terms involving } \bar{S}_{ij} \right] \tag{4.93}$$

where

$$b_{ij} = \frac{\overline{u_i' u_j'}}{2k} - \frac{1}{3}\delta_{ij}. \tag{4.94}$$

Two questions now arise: (i) what lies behind (4.93)? and; (ii) to what extent is (4.93) generally applicable? Let us start with the second of these. The central problem with (4.93) is that $p'(\mathbf{x}_0)$ depends upon $\bar{\mathbf{u}}$ and $\mathbf{u}'$ at all points in the flow, and not just on $\mathbf{u}$ at location $\mathbf{x}_0$. That is, from (2.23),

$$p(\mathbf{x}) = \frac{\rho}{4\pi}\int \frac{\left[\nabla \cdot (\mathbf{u} \cdot \nabla \mathbf{u})\right]''}{|\mathbf{x} - \mathbf{x}''|} d\mathbf{x}'' \tag{4.95}$$

so eddies at all locations contribute to p' at position $\mathbf{x}_0$. In the Reynolds stress model, however, we try to estimate $\overline{p'S_{ij}'}$ at $\mathbf{x} = \mathbf{x}_0$ purely in terms of events at $\mathbf{x}_0$, a strategy which is philosophically unsound from the outset.

Still, let us accept this limitation and try to understand where the closure approximation (4.93) comes from. The first thing to note is that b_{ij} and $\overline{p'S_{ij}'}$ are both zero in isotropic turbulence (see Chapter 6) and so in some sense they represent departures from isotropy. Moreover, continuity requires that $\overline{p'S_{ii}'} = 0$ and so this term makes no contribution to the kinetic energy equation (which is obtained from (4.89) by setting $i = j$). Rather, we may think of $\overline{p'S_{ij}'}$ as redistributing energy between the different components $\overline{(u_x')^2}$, $\overline{(u_y')^2}$, and $\overline{(u_z')^2}$, and it seems plausible that the random pressure fluctuations associated with $\overline{p'S_{ij}'}$ will tend to push the turbulence towards an isotropic state. This is certainly the case in homogeneous shear flow, as we saw in Section 4.4.2.

To fix thoughts, let us consider freely decaying, homogeneous turbulence in which $\bar{\mathbf{u}} = 0$. In such a flow we may use (4.90) to derive an evolution equation for b_{ij}. It is

$$\frac{db_{ij}}{dt} = \frac{\varepsilon}{k}\left[b_{ij} + \left(\overline{p'S_{ij}'}/\rho\varepsilon\right)\right]. \tag{4.96}$$

Notice that, in the absence of the pressure-rate-of-strain term, b_{ij} would increase exponentially on a timescale of ε/k; something which is not observed. So, as anticipated above, the role of $\overline{p'S_{ij}'}$ is to encourage isotropy. If we now combine (4.96) with the *ad hoc* estimate $\overline{p'S_{ij}'} = -\rho\varepsilon c_R b_{ij}$, we find

$$\frac{db_{ij}}{dt} = -\frac{\varepsilon}{k}\left(c_R - 1\right)b_{ij}, \qquad c_R \sim 1.8 \tag{4.97}$$

So the closure hypothesis (4.93) applied to freely decaying turbulence predicts that b_{ij}, which is a mark of anisotropy, is eventually destroyed as a result of random pressure fluctuations. For a while this was thought plausible and so the estimate $\overline{p'S'_{ij}} = -\rho c_R \varepsilon b_{ij}$ has become the starting point for modelling the pressure-rate-of-strain correlation in Reynolds stress models. The decline of b_{ij} in freely decaying turbulence is sometimes referred to as the *return to isotropy*, and the coefficient c_R is known as *Rotta's coefficient*. However, wind-tunnel experiments involving grid turbulence show that any anisotropy present in the initial condition is stubbornly persistent, and need not die away in a time of order l/u. (See Section 4.6.1.) This was nicely summed up by Bradshaw (1971) who notes: 'It is curious that vortex stretching produces isotropy with increasing wavenumber much more effectively than with increasing time.' All of this suggests we should be cautious of (4.93) as an estimate of $\overline{p'S'_{ij}}$, even in the simple case of homogeneous turbulence with no mean shear.

Now it is clear from (4.95) that p', and hence $\overline{p'S'_{ij}}$, will depend not only on the turbulence, $\mathbf{u}'$, but also on the mean flow. This is why a second term appears in (4.92) and (4.93). The full closure estimate of $\overline{p'S'_{ij}}$ must therefore involve the mean strain. One approach is to separate (4.95) into those parts which involve the mean flow and those which do not. Provided we are well removed from boundaries, so that (4.95) can be applied without surface corrections, we find

$$p'(\mathbf{x}) = \frac{\rho}{4\pi}\int \frac{\partial^2}{\partial x''_i \partial x''_j}\left(u'_i u'_j - \overline{u'_i u'_j}\right)'' \frac{d\mathbf{x}''}{|\mathbf{x}-\mathbf{x}''|} + \frac{\rho}{4\pi}\int \frac{\partial^2}{\partial x''_i \partial x''_j}\left(2\bar{u}_i u'_j\right)'' \frac{d\mathbf{x}''}{|\mathbf{x}-\mathbf{x}''|}$$

from which

$$2\overline{p'S'_{ij}}(\mathbf{x}) = \frac{\rho}{2\pi}\int \overline{\frac{\partial^2 (u'_n u'_m)''}{\partial x''_n \partial x''_m} S'_{ij}(\mathbf{x})} \frac{d\mathbf{x}''}{|\mathbf{x}-\mathbf{x}''|} + \frac{\rho}{\pi}\int \overline{\frac{\partial^2 (\bar{u}_n u'_m)''}{\partial x''_n \partial x''_m} S'_{ij}(\mathbf{x})} \frac{d\mathbf{x}''}{|\mathbf{x}-\mathbf{x}''|}$$

The two contributions to $\overline{p'S'_{ij}}$ above are known as the 'slow' and 'rapid' terms, respectively. It is common to assume that $\bar{\mathbf{u}}$ varies slowly in space (compared to the turbulent eddies) so that this simplifies to

$$2\overline{p'S'_{ij}}(\mathbf{x}) = \underbrace{\frac{\rho}{2\pi}\int \overline{\frac{\partial^2 (u'_n u'_m)''}{\partial x''_n \partial x''_m} S'_{ij}(\mathbf{x})} \frac{d\mathbf{x}''}{|\mathbf{x}-\mathbf{x}''|}}_{\text{(slow term)}} + \underbrace{\frac{\rho}{\pi}\frac{\partial \bar{u}_n}{\partial x_m}\int \overline{\frac{\partial (u'_m)''}{\partial x''_n} S'_{ij}(\mathbf{x})} \frac{d\mathbf{x}''}{|\mathbf{x}-\mathbf{x}''|}}_{\text{(rapid term)}}$$

We have already seen how the slow term is modelled. The question now is what should be done about the rapid term. A common model is:

$$2\overline{p'S'_{ij}} = -\rho c_R \frac{\varepsilon}{k}\left[\overline{u_i u_j} - \frac{2}{3}\delta_{ij}k\right] - \rho\hat{c}_R\left[P_{ij} - \frac{2}{3}G\delta_{ij}\right] \tag{4.98}$$

(slow term) (rapid term)

where $\hat{c}_R$ is yet another coefficient (typically given a value of 0.6) and

$$\rho P_{ij} = \tau^R_{ik}\frac{\partial \bar{u}_j}{\partial x_k} + \tau^R_{jk}\frac{\partial \bar{u}_i}{\partial x_k}, \qquad G = \frac{1}{2}P_{ii}.$$

You might ask why the 'rapid' term is taken to be proportional to $P_{ij} - \frac{1}{3}P_{ll}\delta_{ij}$, but that is another story and we will not pursue it. We note only that it comes, in part, from assuming that the integral

$$\int \overline{\frac{\partial (u'_m)''}{\partial x''_n} S'_{ij}(\mathbf{x})} \frac{d\mathbf{x}''}{|\mathbf{x} - \mathbf{x}''|}$$

can be approximated by a linear function of the Reynolds stress tensor evaluated at $\mathbf{x}$, so that

$$\int \overline{\frac{\partial (u'_m)''}{\partial x''_n} S'_{ij}(\mathbf{x})} \frac{d\mathbf{x}''}{|\mathbf{x} - \mathbf{x}''|} \sim \sum \overline{u_p u_q}.$$

Certain additional assumptions are then required to reach (4.98).

Let us now summarize what we might call the basic Reynolds stress model. We have the Reynolds averaged equation of motion

$$\rho\frac{\partial \bar{u}_i}{\partial t} + \rho(\bar{\mathbf{u}}\cdot\nabla)\bar{u}_i = -\frac{\partial \bar{p}}{\partial x_i} + \frac{\partial}{\partial x_j}\left[\bar{\tau}_{ij} + \tau^R_{ij}\right].$$

To this we must add the Reynolds stress transport equation

$$\frac{\partial \tau^R_{ij}}{\partial t} + (\bar{\mathbf{u}}\cdot\nabla)\tau^R_{ij} = -2\overline{p'S'_{ij}} - \left[\tau^R_{ik}\frac{\partial \bar{u}_j}{\partial x_k} + \tau^R_{jk}\frac{\partial \bar{u}_i}{\partial x_k}\right] + \rho\varepsilon_{ij} + \frac{\partial}{\partial x_k}\left[H_{ijk}\right]$$

where the unknown terms on the right are modelled as

$$\varepsilon_{ij} = \frac{2}{3}\varepsilon\delta_{ij},$$

$$H_{ijk} = 0.22\alpha_{kl}\frac{\partial \tau^R_{ij}}{\partial x_l}, \qquad \alpha_{ij} = \overline{u_i u_j}k/\varepsilon,$$

$$2\overline{p'S'_{ij}} = -\rho c_R \frac{\varepsilon}{k}\left[\overline{u_i u_j} - \tfrac{2}{3}\delta_{ij}k\right] - \rho\hat{c}_R\left[P_{ij} - \frac{2}{3}G\delta_{ij}\right].$$

Finally we need a dissipation equation, which is taken to be similar to (4.85) but with ν_t/σ_ε replaced by the anisotropic diffusivity $0.15\alpha_{ij}$:

$$\frac{\partial \varepsilon}{\partial t} + \bar{\mathbf{u}} \cdot \nabla(\varepsilon) = \frac{\partial}{\partial x_i}\left[0.15\alpha_{ij}\frac{\partial \varepsilon}{\partial x_j}\right] + c_1\frac{G\varepsilon}{k} - c_2\frac{\varepsilon^2}{k}.$$

The single most important, and controversial, step in the model is the way in which the pressure-rate-of-strain correlation is handled. It turns out that expression (4.98) works well for free shear layers, but has been found wanting in more complex flows, particularly near surfaces where major discrepancies arise. Of course, there have been many attempts to improve on (4.98). For example, it is now common to add wall correction terms which attempt to model the effects of the anisotropy found near impermeable boundaries. Also, some researchers advocate replacing the 'slow' term, $\overline{p'S'_{ij}} = -\rho c_R \varepsilon b_{ij}$, which constitutes a linear relationship between $\overline{p'S'_{ij}}$ and b_{ij}, by a non-linear expression of the form

$$\overline{p'S'_{ij}} = -\rho c_R \varepsilon \left(c'_R b_{ij} + c''_R \left(b_{ik}b_{jk} - \tfrac{1}{3}b_{mn}b_{mn}\delta_{ij}\right)\right).$$

(Here c'_R and c''_R are yet more coefficients.) Analogous non-linear expressions for the 'rapid' term have also been proposed, but these are extremely complex. All in all there appears to be a bewildering variety of Reynolds stress models, but a comprehensive review of the various schemes may be found in Hanjalic and Jakirlic (2002).

Despite the difficulties inherent in modelling the pressure-rate-of-strain tensor, there are some who think that the Reynolds stress formulation will provide the basis for the next generation of engineering models of turbulence. Certainly, it has notched up some notable successes in recent years. Like the k–ε model, it seems to fair better than cold logic might suggest. However, it has a rival, called large eddy simulation (LES).

4.6.3 *Large eddy simulation: a rival for one-point closures?*

In LES we abandon traditional turbulence modelling and resign ourselves to the task of computing both the mean flow and the evolution of all of the large-scale eddies. In effect, we integrate the Navier–Stokes equation forward in time, resolving all turbulent structures (eddies) down to a certain scale. The unresolved eddies, which are mostly responsible for dissipation, are parameterized using some heuristic *sub-grid model* whose function is simply to mop up all of the kinetic energy which cascades down from the large scales. The advantage of LES is that we can throw out most of the *ad hoc* modelling associated with one-point closure models. Its disadvantages are that it requires extremely large computation times (compared to one-point closures), and that it is ineffective near boundaries where the small eddies are dynamically important. It seems, therefore, that Reynolds stress modelling has the edge on LES when computer resources are limited or else boundaries play an important role in the flow.

Despite these limitations LES has long been used in meteorology, and is beginning to make a real impact on the engineering community, finding applications in, for example, modelling the dispersion of pollutants in the urban environment. We shall air some of the technical issues in Chapter 7, particularly the problem of handling the small scales using a sub-grid model. Here we merely give a hint as to what LES has to offer. We shall take, as an example, flow over a cube of height H, as shown in Figure 4.27.

(a)

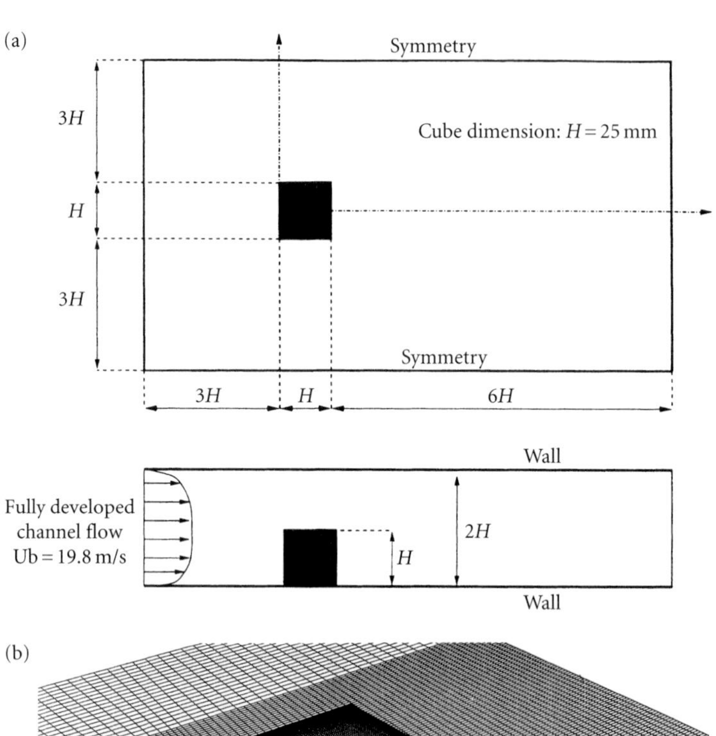

(b)

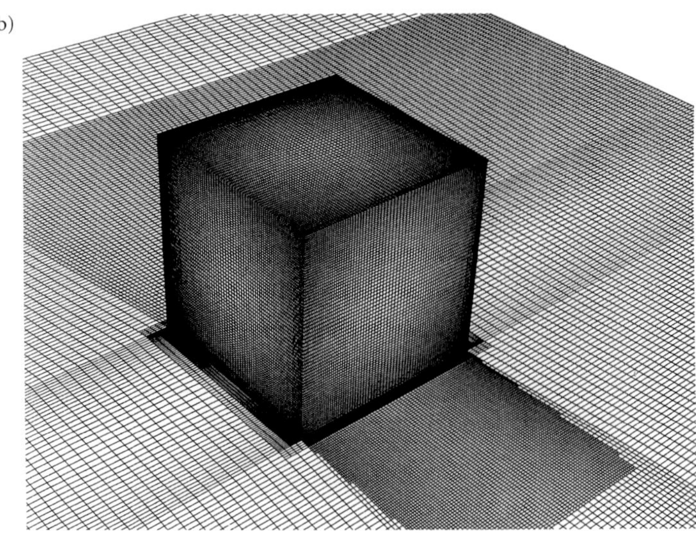

Figure 4.27 Geometry used in one LES study. (a) Computational domain, (b) finite volume mesh. (Courtesy of F. Boysan and D. Cokljat, Fluent Europe Ltd.)

This unsteady, turbulent flow has become a popular test case for LES (Rodi 2002). The particular simulation shown below has been computed using the code FLUENT, employing two different sub-grid models: the much used *Smagorinsky model* and another model which happens to be called WALE. (The details of the sub-grid models are not important for the present discussion. However, the reader may wish to consult Chapter 7 for a description of the Smagorinsky model.)

The time-averaged results obtained by averaging over many eddy-turnover times are shown in Figures 4.28–4.30 along with experimental measurements. Figure 4.28 shows the mean flow obtained using the Smagorinsky sub-grid model. The computed reattachment points upstream and downstream of the cube are $1.18H$ and $1.78H$, respectively, which might be compared with the experimental measurements of $1.04H$ and $1.61H$.

The time-averaged velocity profiles at $x = H$, $z = H$ and $x = 1.5H$, $z = 0$ are shown in Figure 4.29, while those at $x = 2.5H$, $z = 0$ and $x = 4H$, $z = 0$ are shown in Figure 4.30. By and large there is good agreement with the experimental data, although the non-zero centre-plane values of u_z suggest that the simulation may not have been quite long enough to obtain statistical convergence. Also, the comparison is only with the mean flow, and not with the turbulence itself.

Of course, one successful computation hardly validates an entire methodology! Nevertheless, there are several interesting features of this calculation. First, a conventional finite-difference method was used (second-order central differences). Second, although the geometry is relatively simple, it does contain much of the physics relevant to problems of practical interest, such as flow over a group of buildings. For example, there are rapid changes in the turbulence level in the streamwise direction, a stretching of the mean-flow vorticity around the block, and interactions between distinct shear layers (Figure 4.31). These are all features which are characteristic of complex, separated flows around obstacles.

All of this tentatively suggests that LES has the potential to make predictions of engineering interest and at a cost which may be acceptable to some sectors of the engineering community. We shall take up the story again in Chapter 7 where we discuss some of the problems of LES as well as some of its successes.

This concludes our brief discussion of turbulent shear flows. There is a truly vast literature on shear flows and one-point closure models, and despite 70 years of intense study there are still many unanswered questions and much controversy. Interested readers are urged to consult the references given at the end of the chapter, which at least provide a starting point for further study.

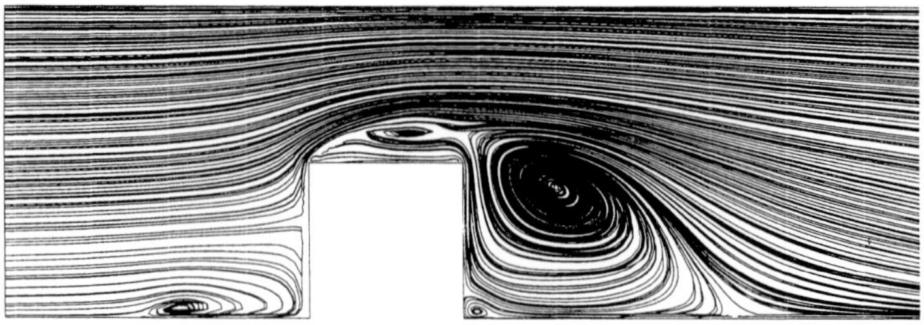

Figure 4.28 Time-averaged flow pattern calculated by LES using the Smagorinsky sub-grid model. (Courtesy of F. Boysan and D. Cokljat, Fluent Europe Ltd.)

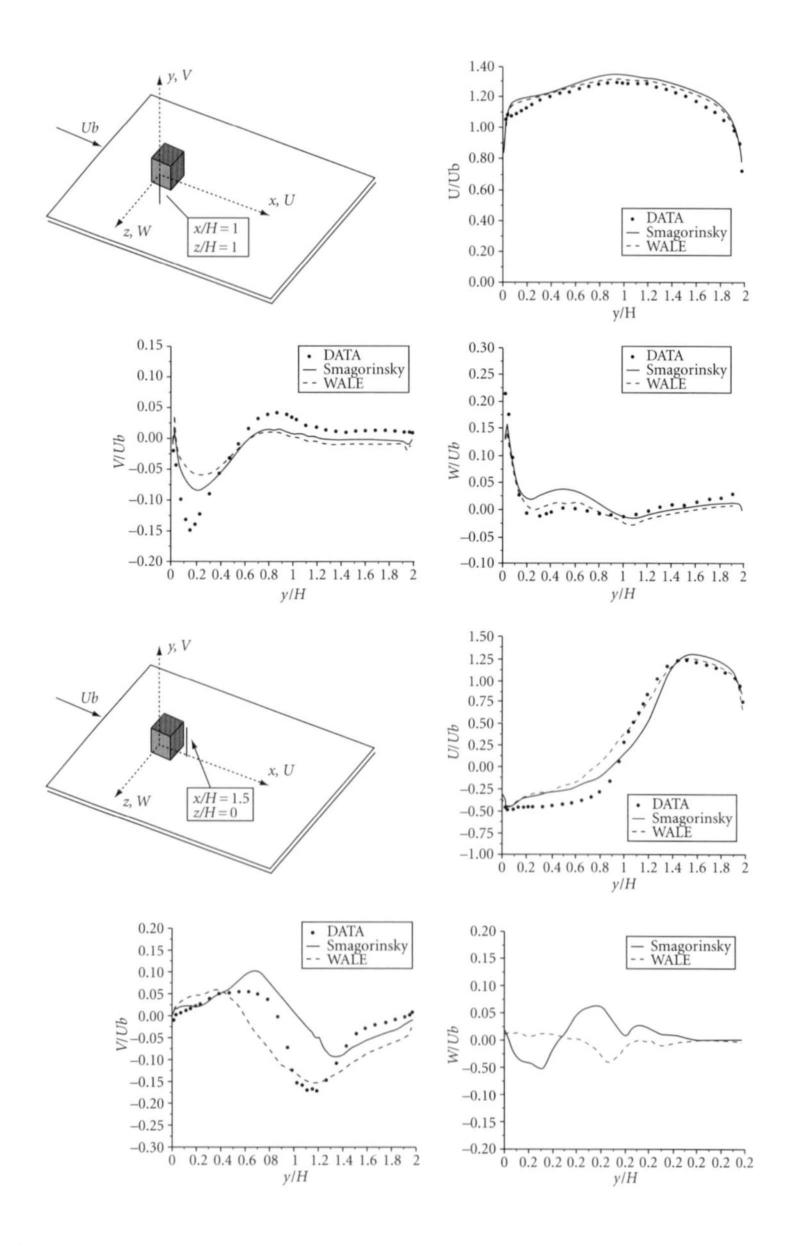

Figure 4.29 Results of the LES. Calculated and measured time-averaged velocity profiles. (Courtesy of F. Boysan and D. Cokljat, Fluent Europe Ltd.)

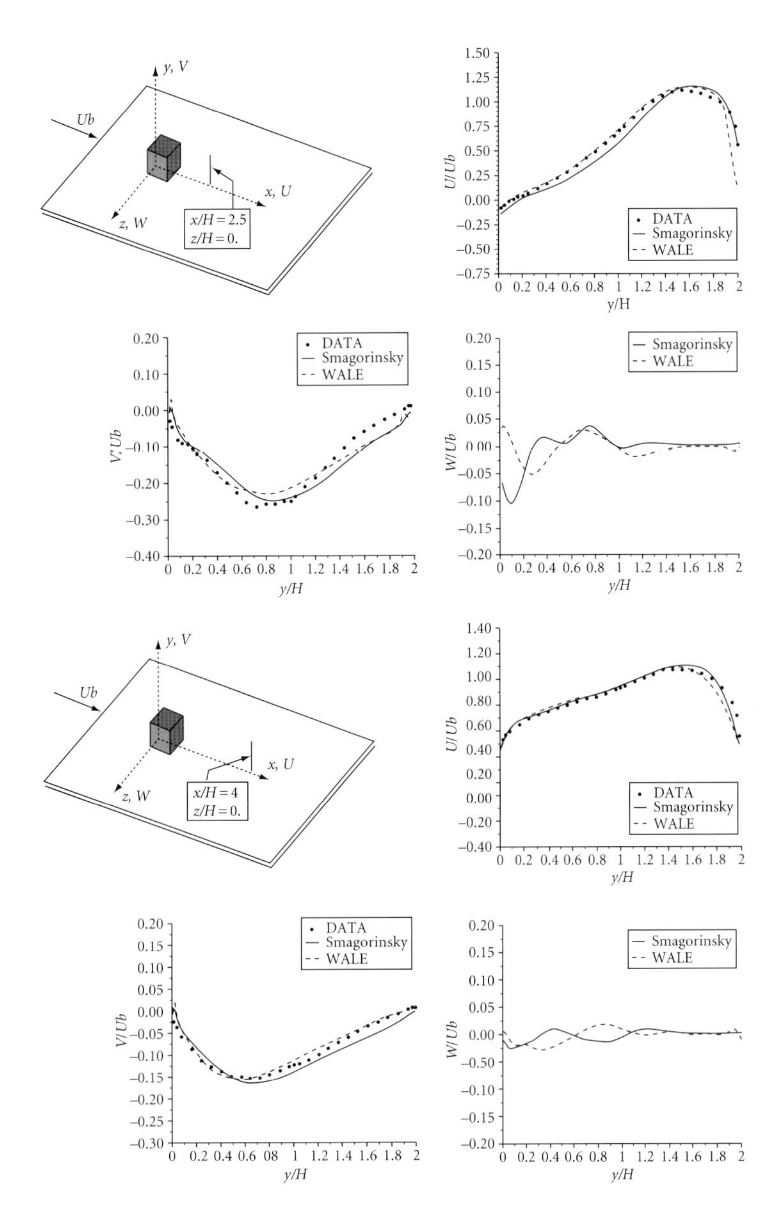

Figure 4.30 Results of the LES. Calculated and measured time-averaged velocity profiles. (Courtesy of F. Boysan and D. Cokljat, Fluent Europe Ltd.)

Exercises

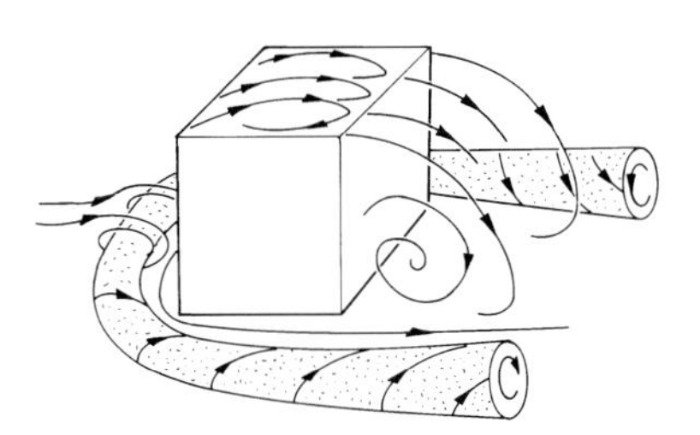

Figure 4.31 Schematic of flow around a cube.

4.1 Consider an unsteady simulation of flow around a building using a Reynolds-stress model. Suppose that the mesh is sufficiently fine to capture much of the vortex shedding, and that only the small-scale turbulence is modeled by the Reynolds-stress closure. Now consider an LES of the same flow. Is their any fundamental difference between the two approaches to simulation?

4.2 Show that the transport equation for correlations of order n involves correlations of order $(n + 1)$.

4.3 Consider a round jet generated by a nozzle of diameter d. The initial speed of the jet is $\bar{u}_0$. The fluid into which the jet issues is not stationary, but rather moves in the same direction as the jet, at a speed of $V < \bar{u}_0$. Show how to adapt the analysis of Section 4.3.2 to describe the evolution of the jet.

4.4 (a) Derive the transport equation for the triple correlations $\overline{u_i' u_j' u_k'}$. (b) A more accurate scheme than the Reynolds stress model might be to model the various unknown terms in the transport equation for $\overline{u_i' u_j' u_k'}$. Do you think this is a practical scheme for flows of interest to the engineer?

4.5 Consider a one-dimensional shear layer $\bar{u}_x(y)$ adjacent to a wall. Show that, if $\partial \bar{p}/\partial x$ can be neglected, then $\bar{\tau}_{xy}(y) + \tau_{xy}^R(y) = \text{constant}$. Now use mixing length, with $l_m = \kappa y$, to find $\bar{u}_x(y)$. You may assume that $\bar{\tau}_{xy}$ is negligible except in the immediate vicinity of the wall. Confirm that $\bar{u}_x/V_* = \kappa^{-1} \ln y + \text{const.}$, where $V_* = \left(\tau_{xy}^R/\rho\right)^{1/2}$.

4.6 Estimate the value of $V_* \hat{k}/\nu$ at which roughness first becomes important in the log-law for flow over a rough surface. ($\hat{k}$ is the rms roughness height.)

4.7 Show that, if an incipient hairpin vortex is created by an axial 'gust', as shown in Figure 4.13(a), then the tip of the vortex will rise by self-induction.

Suggested reading

Bradshaw, P. (1971) *An Introduction to Turbulence and its Measurement*. Pergamon Press. (A short but readable introduction to turbulence.)

Bradshaw, P. and Huang, G.P. (1995) *Proc. R. Soc. Lond.* A, **451**, 165–188 (Discusses the log-law of the wall in some depth.)

Buschmann, M.H. and Gad-el-Hak, M. (2003) *AIAA Jn.* **41**(4), 565–572. (Discusses the merits of power law and log-law models of wall turbulence.)

Champagne, F.H., Harris, V.G., and Corrsin, S. (1970) *J. Fluid Mech.*, **41**(1), 81. (An early paper on homogeneous shear flow.)

Dimotakis, P.E. (2000) J. Fluid Mech., **409**, 69–98.

Durbin P.A. and Pettersson Reif, B.A. (2001) *Statistical Theory and Modeling for Turbulent Flows*. Wiley. (Chapters 6–8 give an extensive discussion of single-point closures.)

Garrett, J.R. (1992) *The Atmospheric Boundary Layer*. Cambridge University Press. (A recent monograph on the atmospheric boundary layer.)

Hanjalic, K. and Jakirlic, S. (2002) In: *Closure Strategies for Turbulent and Transitional Flows* (eds. B. Launder and N. Sandham). Cambridge University Press. (Discusses the Reynolds stress closure model.)

Head, M. R. and Bandyopadhyay, P. (1981) *J. Fluid Mech.*, **107**. (Discusses the role of hairpin vortices in boundary layers.)

Hinze, J.O. (1975) *Turbulence*. McGraw-Hill. (Homogeneous shear-flow is discussed in Chapter 4. Not for the faint-hearted though.)

Holman, J.P. (1986) *Heat Transfer*, McGraw Hill. (A simple introduction to heat transfer.)

Holmes, P., Lumley, J.L., and Berkooz, G. (1996) *Turbulence, Coherent Structures, Dynamical Systems and Symmetry*. Cambridge University Press. (This contains a discussion of coherent structures in boundary layers.)

Hunt, J.C.R. et al. (2001) *J. Fluid Mech.*, **436**, 353. (A review article which contains a host of useful references.)

Jimenez, J. (2002) In: *Tubes, Sheets and Singularities in Fluid Mechanics* (eds. K. Bajer and H.K. Moffatt). Kluwer Academic Publishers, pp. 229–240.

Kader, B.A. (1981) *Int. J. Heat Mass Transfer*, **24**(9), 1541–1544. (Discusses heat transfer in boundary layers.)

Kraichnan, R.H. (1962) *Phys. Fluids*, **5**, 1374.

Landau, L.D. and Lifshitz, E.M. (1959) *Fluid Mechanics*. Pergamon Press. (Chapter 5 discusses the log-law for temperature.)

Libby, P.A. (1996) *Introduction to Turbulence*. Taylor and Francis. (Chapter 10 contains an extensive discussion of engineering models of turbulence.)

Mathieu, J. and Scott, J. (2000) *An Introduction to Turbulent Flow*. Cambridge University Press. (Shear flows are discussed in Chapter 5.)

Monin A.S. and Yaglom, A.M. (1975) *Statistical Fluid Mechanics I*. MIT Press. (Chapter 4 discusses turbulence in a heated medium.)

Panchev, S. (1971) *Random Functions and Turbulence*. Pergamon Press. (Chapter 9 discusses the atmospheric boundary layer.)

Panton, R.L. (2001) *Prog. Aerospace Sci.*, **37**, 341–383. (A review of coherent structures in boundary layers.)

Perry, A.E. and Chong, M.S. (1982) *J. Fluid Mech.*, **119**, 173–217. (On the role of hairpin vortices in boudary layers.)

Perry, A.E., Henbest, S., and Chong, M.S. (1986) *J. Fluid Mech.*, **165**, 163–199.

Pope, S.B. (2000) *Turbulent Flows*. Cambridge University Press. (One-point closure models are discussed in Chapters 10 and 11, wall-bounded flows in Chapter 7, and free shear flows in Chapter 5.)

Robinson, S.K. (1991) *Ann. Rev. Fluid Mech.*, **23**, 601–639. (A comprehensive review of coherent structures in boundary layers.)

Rodi, W. (2002) In: *Closure Strategies for Turbulent and Transitional Flows* (eds. B. Launder and N. Sandham). Cambridge University Press. (Discusses large eddy simulation.)

Tavoularis, S. and Corrsin, S. (1981) *J. Fluid Mech.*, **104**, 311–347. (Contains much useful data on homogeneous shear flow.)

Tennekes, H. and Lumley, J.L. (1972) *A First Course in Turbulence*. MIT Press. (Consult Chapter 2 for a discussion of mixing length, Chapter 3 for a discussion of the exchange of energy between the mean flow and the turbulence, and Chapter 4 for shear flows.)

Townsend, A.A. (1976) *The Structure of Turbulent Shear Flows*, 2nd edition. Cambridge University Press. (This contains a wealth of information on shear flows.)

Wyngaard, J.C. (1992) *Ann. Rev. Fluid Mech.*, **24**, 205–233. (A review of atmospheric turbulence.)

CHAPTER 5

The phenomenology of Taylor, Richardson, and Kolmogorov

> The general pattern of turbulent motion can be described (according to Taylor and Richardson) in the following way. The mean flow is accompanied by turbulent fluctuations imposed on it and having different scales, beginning with maximal scales of the order of the 'external scale' of turbulence l (the 'mixing length') to the smallest scales of the order of the distance η at which the effect of viscosity becomes appreciable (the 'internal scale' of turbulence)... Most large-scale fluctuations receive energy from the mean flow and transfer it to fluctuations of smaller scales. Thus there appears to be a flux of energy transferred continuously from fluctuations of larger scales to those of smaller scales. Dissipation of energy, that is, transformation of energy into heat, occurs mainly in fluctuations of scale η. The amount of energy ε dissipated in unit time per unit volume is the basic characteristic of turbulent motion for all scales.
>
> Kolmogorov (1942)

We have already discussed Richardson's idea of an energy cascade as well as Kolmogorov's theory of the small scales. Richardson's hypothesis says that, in a turbulent flow, energy is continually passed down from the large-scale structures to the small scales, where it is destroyed by viscous stresses. Moreover, this is a multi-stage process involving a hierarchy of vortex sizes. Kolmogorov's theory, on the other hand, asserts that the statistical properties of the small scales depend only on ν and on the rate at which energy is passed down the energy cascade. In addition, it states that the small scales are statistically isotropic and have a structure which is statistically universal, that is, the same for jets, wakes, boundary layers, and so on.

These statements cannot be formally 'proven' in any deductive way. The best that we can do is examine whether or not they are plausible, check that they are self-consistent, and then see how they hold up against the experimental data. That is the primary purpose of this chapter. En route, we shall discuss the Richardson–Taylor theory of turbulent dispersion and also vortex stretching in turbulent flows, since it is vortex stretching which underlies Richardson's cascade. We shall see that the existence of vortex stretching implies that the turbulent velocity field *cannot* have a Gaussian probability distribution.

This is important because certain 'theories of turbulence' assume near Gaussian behaviour.

Perhaps it is worthwhile giving an overview of this chapter. The layout is as follows:

5.1 Richardson revisited
- 5.1.1 Time and length scales in a turbulent flow
- 5.1.2 The energy cascade pictured as vortex stretching
- 5.1.3 The dynamic properties of turbulent eddies

5.2 Kolmogorov revisited
- 5.2.1 Assumptions and weaknesses of Kolmogorov's theory
- 5.2.2 The extension of the theory to passive scalar fluctuations

5.3 The stretching of vortices and material lines
- 5.3.1 Enstrophy production by vortex stretching
- 5.3.2 Are the vortices tubes, sheets or blobs?
- 5.3.3 Examples of stretched vortex tubes and sheets
- 5.3.4 Can finite-time singularities develop in the vorticity field?
- 5.3.5 The stretching of material lines

5.4 Turbulent diffusion
- 5.4.1 The turbulent diffusion of a single particle (Taylor diffusion)
- 5.4.2 The relative diffusion of two particles (Richardson's law)
- 5.4.3 The influence of mean shear on turbulent dispersion

5.5 Turbulence is never Gaussian
- 5.5.1 The experimental evidence
- 5.5.2 The consequences

We start the chapter by re-examining Richardson's idea of an energy cascade in which, it is claimed, energy is passed from large to small scales by a repeated sequence of discrete steps. But why should energy always pass from large to small scales, and why must this involve a distributed hierarchy of eddy sizes? Is it not possible for energy to travel upscale? Or perhaps we can transfer energy directly from the large eddies to the Kolmogorov microscale in one single action, thus bypassing the cascade. We shall show that Richardson's energy cascade is a direct consequence of vortex stretching, and that, by and large, it is in the nature of chaotic vortex dynamics to pass energy from large to small scale. So, Richardson's picture is more or less correct. However, there *are* exceptions. For example, it is not difficult to conceive of circumstances in which energy travels from small to large scale, thus violating the conventional picture of the cascade. Two-dimensional turbulence provides one such example. (This is discussed in Chapter 10.)

Next, in Section 5.2, we revisit Kolmogorov's theory of the small scales. This is undoubtedly one of the greatest success stories in turbulence, but is it correct? Certainly its foundations are, at best, a little tenuous. We discuss Kolmogorov's theory in some detail, re-examining the assumptions which underpin the analysis. We shall see that these assumptions are not always valid and that, in certain circumstances, Kolmogorov's theory breaks down. We shall also discuss how Kolmogorov's ideas can be extended to describe the fluctuations of a passive scalar, say smoke or dye in a turbulent flow.

In Section 5.3, we return to the idea of vortex stretching. This crucial non-linear process drives the energy cascade and lies at the heart of turbulence theory. We start by deriving an equation for the rate of generation of enstrophy, $\frac{1}{2}\langle \boldsymbol{\omega}^2 \rangle$. In the process we shall see that enstrophy is generated by the stretching of vortex lines and destroyed at the small scales by the cross-diffusion of oppositely signed vorticity. Moreover, the rate of generation of enstrophy is directly proportional to the skewness of the velocity difference, Δv. This is important because it tells us that the statistics of the velocity fluctuations are *inherently non-Gaussian*, a fact which will come back to haunt us time and again when we discuss two-point closure models of turbulence.

Since stretched vortices are the 'sinews of turbulence', it is natural to ask what shape these vortices have? Do they take the form of tubes, sheets, or ribbons? As we shall see, both vortex tubes and vortex sheets are prime candidates, and there has been much debate as to which is the dominant structure. The current view is that both are important, though they play somewhat different roles in the energy cascade. Examples of stretched tubes and sheets are discussed at length in Section 5.3.3.

Of course, in a typical turbulent flow, the Reynolds number (Re) is high at all but the smallest of scales. Since vortex lines are frozen into the fluid at high Re, the tendency for vortex lines to be stretched by chaotic motion is closely related to the observation that material lines are, on average, extended in a field of turbulence. This underlies the ability of turbulence to mix any frozen-in marker, be it vortex lines or dye lines. This is achieved through a repeated process of *stretch* and *fold*, whereby material lines and surfaces are simultaneously stretched and wrinkled to form highly convoluted shapes. We close Section 5.3 with a discussion of material line stretching.

We pick up the theme of mixing again in Section 5.4 where we discuss turbulent diffusion: that is, the ability of turbulence to accelerate the mixing of a contaminant. The turbulent diffusion of a contaminant through the jostling of fluid particles is of great practical importance. It is often divided into two classes: the diffusion of a single particle and the relative diffusion of two particles. In the first case, we are interested in how far, on average, a marked lump of fluid migrates from its point of release under the influence of random turbulent fluctuations. This is somewhat analogous to the well-known 'random walk', and is referred to as Taylor diffusion. It is relevant to those situations where a contaminant is released continuously from a single source (say a chimney) into a turbulent flow, and provides an estimate of how fast the contaminant spreads. The second case in which we are interested, the relative dispersion of two adjacent particles, was first analysed by Richardson in 1926. Accordingly, the equation which describes the average rate of separation of two particles is known as Richardson's law. The early concepts which underlie this law are closely related to the later theories of Kolmogorov and indeed they helped pave the way for Kolmogorov's famous two-thirds law.

We close, in Section 5.5, by returning to a consideration of the probability distribution of velocity differences. We emphasis that turbulence is fundamentally a non-Gaussian process. Without a finite skewness there is no vortex stretching and no energy cascade. We shall see how this has bedevilled attempts to develop closure models based on near-Gaussian behaviour.

The theme of this chapter, then, is that of cascades, vortex stretching, and mixing. The emphasis is on physical ideas, rather than mathematical models. (We shall introduce the mathematics in Chapter 6.) Let us start at the beginning, with Richardson.

5.1 Richardson revisited

5.1.1 *Time and length-scales in turbulence*

A French five minutes is ten minutes shorter than a Spanish five minutes, but slightly longer than an English five minutes which is usually ten minutes.
(*Guy Bellamy*)

It is an empirical observation that turbulence contains a wide range of time and length scales. In a high wind, for example, the velocity field in a street might exhibit fluctuations over scales from 1 m to 0.1 mm. It is also an empirical observation that the vorticity in a turbulent flow is concentrated at the smallest scales. Since $\nu\langle\boldsymbol{\omega}^2\rangle$ is a measure of the rate of dissipation of mechanical energy, this implies that dissipation is associated primarily with the smallest structures. In our high wind, for example, much of the energy will be dissipated in structures of size 1 mm or less.

Now the turbulence usually receives its energy from the mean flow. In a shear flow, for example, the rate of generation of turbulent energy is,

$$\rho G = \tau_{ij}^R \bar{S}_{ij} \tag{5.1}$$

where $\tau_{ij}^R = -\rho\langle u_i' u_j'\rangle$ and $\bar{S}_{ij}$ is the strain-rate of the mean flow $\bar{S}_{ij} = \frac{1}{2}(\partial\bar{u}_i/\partial x_j + \partial\bar{u}_j/\partial x_i)$ (see Chapter 4). Physically, this corresponds to turbulent vortices being stretched by the mean shear, increasing their energy. The eddies which are primarily responsible for this energy transfer are the largest in the flow, and these have a size dictated by the nature of their birth. Often the large turbulent vortices arise through a

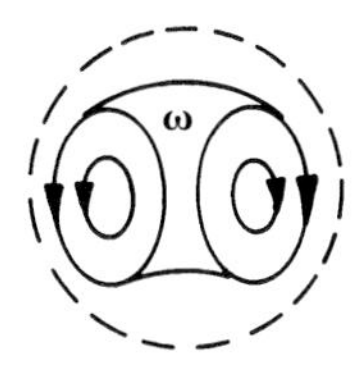

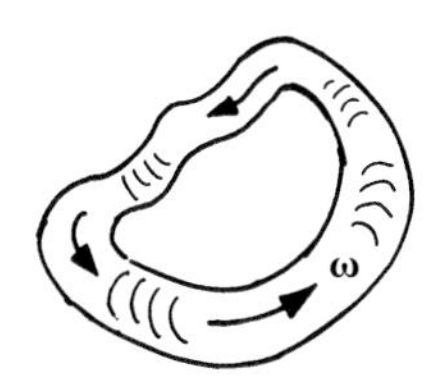

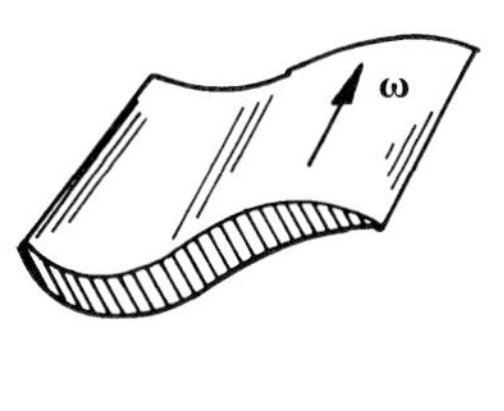

Figure 5.1 Eddies are 'blobs' of vorticity. They may be spherical, tubular or sheet-like in shape, or even more complex.

distortion or instability of the mean flow vortex lines. Their size then corresponds to a length scale characteristic of the mean flow, for example, the length associated with gradients in the mean velocity field.

Therefore, we have mechanical energy transferred to the turbulence at a large scale, and extracted at a much smaller one. The question, of course, is how does the energy get from the large-scale to the small-scale structures. Richardson attempted to bridge this gap by invoking the idea of an *energy cascade*. He suggested that the large structures pass their energy onto somewhat smaller ones which, in turn, pass energy onto even smaller vortices and so on. We talk of a *cascade* of energy from large scale down to small. The essential claim of Richardson is that this cascade is a multistage process, involving a hierarchy of vortices of varying size. It is conventional to talk of these different sized structures as *eddies*, which conjures up a picture of spherical-like objects of different diameters. However, this is a little misleading. The structures may be sheet-like or tubular in shape (Figure 5.1). It is also customary to talk of the energy cascade in terms of eddies continually 'breaking-up' into smaller ones as a consequence of 'instabilities'. Again, this is a little misleading and is just a kind of shorthand. By *break-up* we really just mean that energy is being transferred from one scale to the next through a distortion of the eddy shape. Also, the use of the word *instability* is possibly a little inappropriate, since an 'eddy' does not represent a steady base state. The word is intended to imply that large structures can evolve into smaller ones via familiar mechanisms, some of which we might encounter in stability theory. (We shall return to this shortly.)

Richardson also suggested that, at high Re, viscosity plays no part in the energy cascade, except at the smallest scales. Let u be a typical large-scale velocity and l the characteristic size of the large eddies (the integral scale). Inevitably we have $ul/\nu \gg 1$ and so viscous effects are quite ineffective at the large scales, and indeed at scales somewhat smaller than l. So Richardson envisaged an inviscid cascade of energy down to smaller and smaller scales, the cascade being driven by inertial forces alone. The cascade is halted, however, when the structures become so small that Re based on the small-scale eddy-size is of the order of unity. That is, the very smallest eddies are dissipated by viscous forces and for viscosity to be significant we need

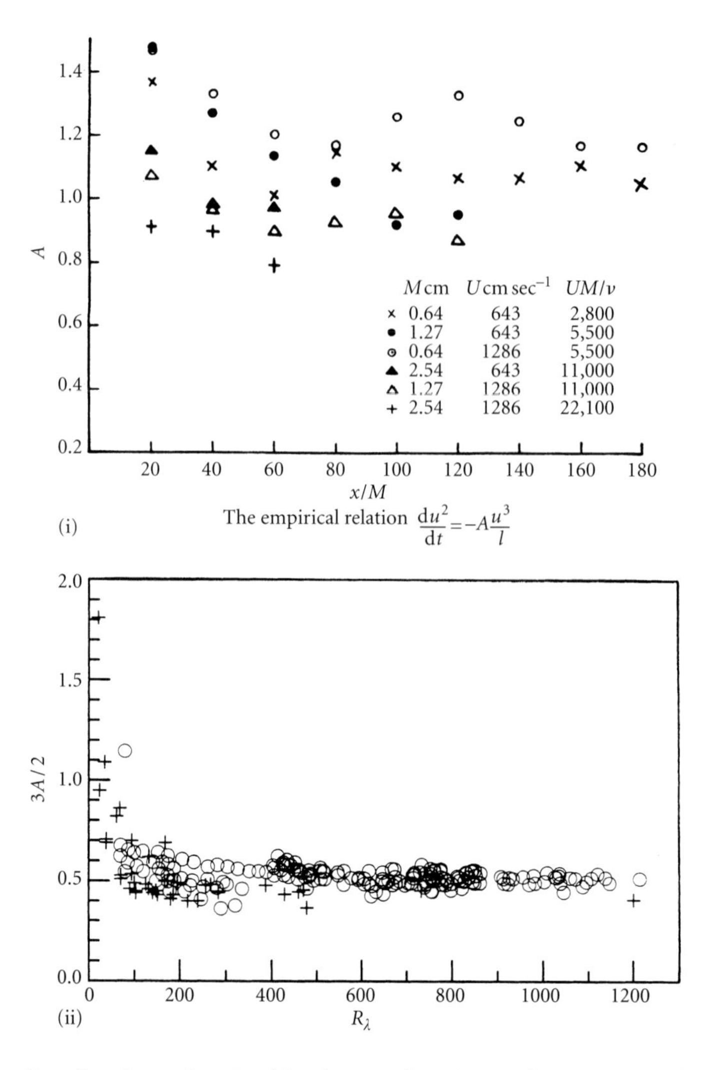

Figure 5.2 The rate of loss of energy in grid turbulence from (i) Batchelor (1953) and (ii) Pearson et al. (2004). In (ii) the circles are experiments, the crosses DNS, and R_λ the Reynolds number based on the Taylor scale. The parameter A is defined by the energy equation $du^2/dt = -Au^3/l$. For high R_λ, $A \sim 0.5$.

Re of order unity. In this picture, the viscous forces are passive in nature, mopping up whatever energy cascades down from above.

Now the large-scale eddies are observed to evolve (break up?) on a timescale of l/u, and so the rate at which energy passes down the cascade from above is[1]

$$\Pi \sim u^3/l. \tag{5.2}$$

Evidence for the validity of (5.2) has been accumulated in a wide range of flows. A typical example is shown in Figure 5.2. It relates to grid turbulence and is taken from (i) Batchelor (1953) and (ii) Pearson et al. (2004). Here l is defined in terms of the longitudinal correlation function $l = \int_0^\infty f\,dr$ (see equation 3.15) and $u^2 = \langle u_x^2 \rangle$. The data corresponds to a range of grid sizes and to a variety of Reynolds numbers.

[1] Recall that we use G to represent the rate of *generation* of turbulent energy by the mean flow, and Π to represent the *flux* of energy down the cascade.

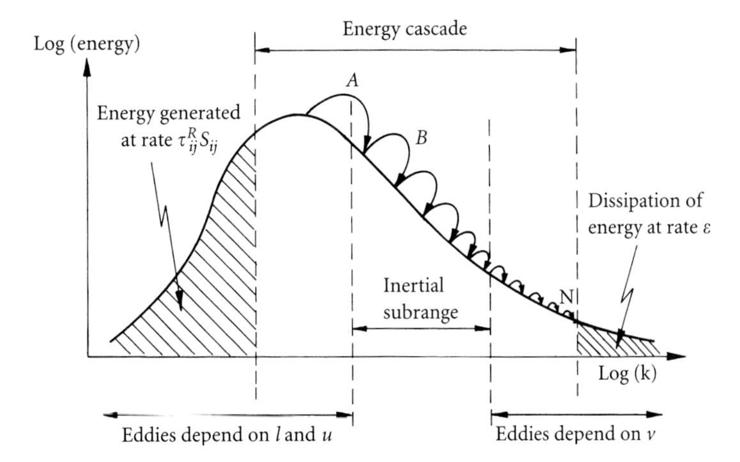

Figure 5.3 Schematic representation of the energy cascade.

Evidently, in the case of Batchelor's data we have $\Pi = \frac{3}{2}Au^3/l$, where A is approximately constant during the decay and of the order of unity, $A \sim 1.1 \pm 0.2$. Note, however, that the value of Re in these early experiments is modest. The more recent data of Pearson is at much higher Re and suggests that the asymptotic value of A is $A \sim 0.3$. This is consistent with Kaneda et al.'s (2003) findings.

Now consider the smallest scales. Suppose they have a characteristic velocity v and length scale η. Since the rate of dissipation of mechanical energy is $\nu\langle\boldsymbol{\omega}^2\rangle$ we have

$$\varepsilon \sim \nu \mathrm{v}^2/\eta^2. \tag{5.3}$$

In homogeneous, statistically steady turbulence the rate of extraction of energy from the mean flow, $\tau_{ij}^R \bar{S}_{ij}/\rho$, must equal the rate at which energy is passed down the energy cascade from the large scales, $\Pi_A \sim u^3/l$. This must also equal the rate of transfer of energy at all points in the cascade since we cannot lose or gain energy at any particular scale in a steady-on-average flow. In particular, if Π_A, $\Pi_B, \ldots, \Pi_N$ represents the energy flux at various stages of the cascade then we have $\Pi_A = \Pi_B = \cdots = \Pi_N \sim u^3/l$ (Figure 5.3). So the energy transfer even in the small eddies is controlled by the rate of break-up of the large eddies. Finally we note that the energy flux at the end of the cascade, Π_N, must equal the viscous dissipation rate, ε. In summary, then, for homogeneous, statistically steady turbulence,

$$G = \underbrace{\rho^{-1}\tau_{ij}^R \bar{S}_{ij}}_{\text{transfer of energy to the turbulence}} = \underbrace{\Pi_A = \Pi_B = \cdots = \Pi_N}_{\text{energy flux down the cascade}} = \underbrace{\varepsilon.}_{\text{dissipation at small scales}} \tag{5.4}$$

Combining (5.2)–(5.4) we have

$$\Pi \sim u^3/l \sim \varepsilon \sim \nu \mathrm{v}^2/\eta^2. \tag{5.5}$$

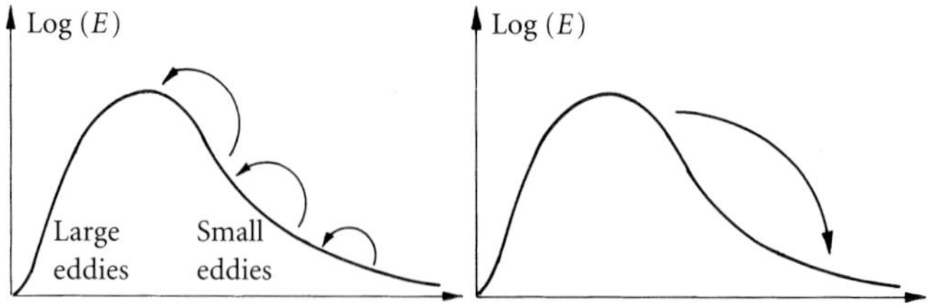

Log (E)

Large eddies

Small eddies

Log (wave number)

(i) Can energy move to larger scales through vortex merger?

Log (E)

Log (wave number)

(ii) Can energy transfer directly to samll scales, bypassing the cascade?

Figure 5.4 Processes which violate Richardson's cascade. Why do these not occur? (Actually, sometimes they do.)

We also know that

$$\mathrm{v}\eta/\nu \sim 1 \tag{5.6}$$

and so from (5.5) and (5.6) we can derive estimates for η and v:

$$\eta \sim l(ul/\nu)^{-3/4} \sim (\nu^3/\varepsilon)^{1/4} \tag{5.7}$$

$$\mathrm{v} \sim u(ul/\nu)^{-1/4} \sim (\nu\varepsilon)^{1/4}. \tag{5.8}$$

These are, of course, the Kolmogorov microscales introduced in Chapter 1. When the flow is neither homogeneous nor statistically steady then G and ε need not balance. However, it is usually found that Π and ε are of the same order of magnitude and so (5.5)–(5.8) still hold.

Now the Richardson picture is not entirely implausible, but it does raise at least two fundamental questions. First, is there some generic process which causes energy always to pass from large to small scale, and not from small to large? Second, why must the process be a multi-stage one? Perhaps there is a simple dynamical process, involving only two or three steps, by which structures of scale (η, v) can emerge directly from structures of scale (l,u), as shown in Figure 5.4. In fact, these apparent violations of Richardson's cascade can and do occur, though they are probably uncommon at very high Re.[2] Let us examine this in a little more detail.

5.1.2 *The energy cascade pictured as the stretching of turbulent eddies*

To focus thoughts we now outline some simple inviscid processes by which energy passes from large to small scale. We do not suggest that these are the mechanisms responsible for the cascade in a turbulent flow.

[2] Actually, both processes shown in Figure 5.4 can, and probably do, occur. The first is typical of two-dimensional turbulence where there is an inverse cascade of energy. The second is discussed at length by Tsinober (2001), who warns against the notion of a multi-stage cascade in real space. Two-dimensional turbulence, with its inverse transfer of energy, is discussed in Chapter 10.

They are meant only to illustrate that an inviscid transfer of energy to small scales is not so surprising and indeed entirely natural. Our three examples are: (i) the stretching of a vortex tube; (ii) the self-induced bursting of a vortex blob; and (iii) the roll-up of a vortex sheet.

As our first example, suppose that we have some weak, rather dispersed large-scale vorticity, $\boldsymbol{\omega}_2$. In some particular region of space it sets up a large-scale straining motion, $\mathbf{u}_2$, as illustrated in Figure 5.5. ($\boldsymbol{\omega}_2$ is not shown in this figure). Now consider a vortex tube, $\boldsymbol{\omega}_1$, with a scale somewhat smaller than $\boldsymbol{\omega}_2$, which finds itself in the strain field $(S_{ij})_2$. The tube will be stretched, and in the process its kinetic energy $\int \left(\mathbf{u}_1^2/2\right) dV$ will rise. Thus we may think of two velocity fields, $\mathbf{u}_1$ and $\mathbf{u}_2$, each related to their respective vorticity fields, $\boldsymbol{\omega}_1$ and $\boldsymbol{\omega}_2$, via the Biot-Savart law. The total kinetic energy does not reside exclusively with $\mathbf{u}_1$ and $\mathbf{u}_2$ acting individually, since there is a cross-term $\int \mathbf{u}_1 \cdot \mathbf{u}_2 \, dV$. Nevertheless, as $\int \left(\mathbf{u}_1^2/2\right) dV$ rises due to vortex stretching, the remaining energy $\int \left(\mathbf{u}_1 \cdot \mathbf{u}_2 + \mathbf{u}_2^2/2\right) dV$ must fall, and we may think of energy being transferred from large to small scale.

As a second example, suppose we have an isolated blob of vorticity (an eddy) sitting in an otherwise quiescent fluid (Figure 5.6). For simplicity, we take the initial velocity field to be axisymmetric and we assume that it remains so for at least some period of time after $t = 0$. At $t = 0$ we take

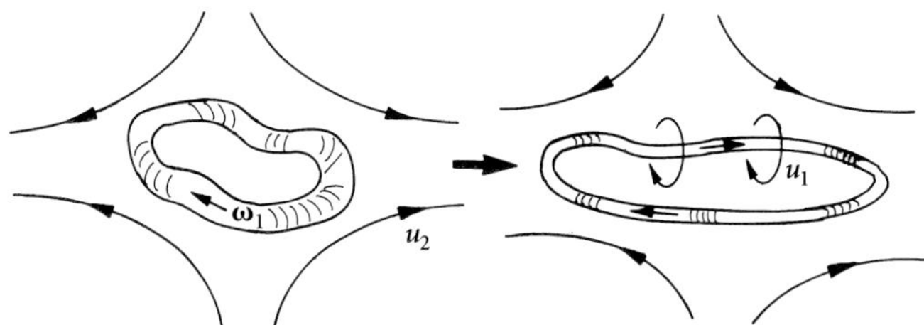

Figure 5.5 Stretching of a vortex tube.

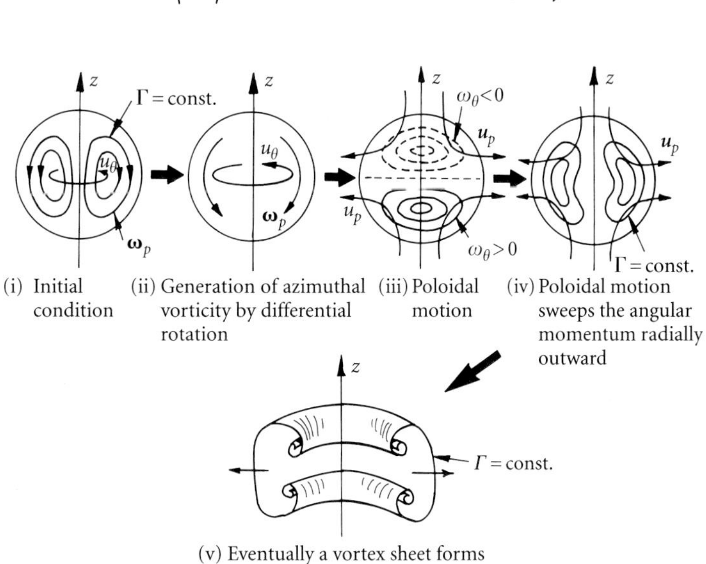

Figure 5.6 Various stages in the bursting of a vortex. The initial vortex blob eventually ends up as a thin vortex sheet. See also Plate 7.

$\mathbf{u} = (0, u_\theta, 0)$ in (r, θ, z) coordinates and so the vortex is essentially a compact distribution of angular momentum, $\Gamma = ru_\theta$, orientated parallel to z. The vortex does not stay compact for long, however. Rather, it 'bursts' radially outward through the action of the centrifugal force. In the process, it creates a thin vortex sheet (Figure 5.6(v)). The bursting of the vortex and the subsequent formation of the vortex sheet is a consequence of a secondary poloidal flow, $(u_r, 0, u_z)$, which, although absent in the initial condition, is induced by vortex stretching.

Perhaps it is worth digressing for a moment to explain this process. The azimuthal components of the inviscid momentum and vorticity equations are readily shown to be

$$\frac{D\Gamma}{Dt} = \left(\frac{\partial}{\partial t} + \mathbf{u}_p \cdot \nabla\right)\Gamma = 0 \tag{5.9}$$

$$\frac{D}{Dt}\left(\frac{\omega_\theta}{r}\right) = \frac{\partial}{\partial z}\left(\frac{\Gamma^2}{r^4}\right). \tag{5.10}$$

Here ω_θ is the vorticity associated with the poloidal velocity $\mathbf{u}_\mathrm{p} = (u_r, 0, u_z)$. Conversely, the azimuthal velocity, $(0, u_\theta, 0)$, is associated with the poloidal vorticity, $\boldsymbol{\omega}_\mathrm{p} = (\omega_r, 0, \omega_z)$. Indeed, it is readily confirmed that Γ is the Stokes streamfunction for $\boldsymbol{\omega}_\mathrm{p}$. At $t = 0$ both $\mathbf{u}_p$ and ω_θ are zero. However, it is evident from (5.10) that they will not stay zero for long. It seems that axial gradients in swirl act as a source of azimuthal vorticity. At first sight, this may appear a little mysterious until we realize that the term $\partial(\Gamma^2/r^4)/\partial z$ has its origins in $\nabla \times (\mathbf{u}_\theta \times \boldsymbol{\omega}_\mathrm{p})$. Thus ω_θ is produced through a process of self-induction in which differential rotation (axial gradients in u_θ) distorts the initial $\boldsymbol{\omega}_\mathrm{p}$-lines, spiralling out an azimuthal component of vorticity (Figure 5.6(ii)). It is clear that $\omega_\theta < 0$ in the upper half of the vortex and $\omega_\theta > 0$ in the lower half. Also, ω_θ is confined to regions where Γ is, or has been, present.

Now Γ is materially conserved (equation (5.9)) and so the poloidal flow associated with ω_θ sweeps the angular momentum radially outward as shown in Figure 5.6 (iv). In effect, the vortex starts to centrifuge itself outward. The skew-symmetric distribution of ω_θ is maintained as the vortex expands and it is readily confirmed that $\int_{z<0} (\omega_\theta/r)\, dV$ grows monotonically since, from (5.10), we have

$$\frac{d}{dt}\int_{z<0} (\omega_\theta/r)\, dV = 2\pi \int_0^\infty \left(\Gamma_0^2/r^3\right) dr > 0 \tag{5.11}$$

where $\Gamma_0(r)$ is $\Gamma(r,\ z = 0)$. Thus the vortex bursts radially outward as shown below. However, this is not the end of the story. As the Γ-contours are swept outward, they start to develop into a thin, axisymmetric sheet as shown in Figure 5.6(v). When viewed in the r–z plane this has the appearance of a mushroom-like structure, reminiscent of the head of a thermal plume (Plate 7). Actually, this similarity to buoyancy-driven flow is not coincidental. As noted in Chapter 1, there is a direct analogy between thermally driven and centrifugally driven flows. For the present purposes the important point to note is that Γ is the streamfunction for $\boldsymbol{\omega}_p$ and so this axisymmetric sheet is, in fact, a vortex sheet. The formation of the sheet is due to the continual straining at the edge of the vortex which progressively thins the sheet down. If the flow were to remain axisymmetric (which it does not) the eventual thickness of the sheet would be determined by diffusion. So, in this simple example we go from a blob of vorticity to a thin vortex sheet. Actually, this busting mechanism, which is, of course, a Rayleigh-like centrifugal instability, is quite common. An impulsively rotated rod, for example, builds up an annulus of swirling fluid which then disintegrates via this instability into a sequence of axisymmetric, sheet-like eddies of the type shown in Figure 5.6. (This is shown in Plate 7.)

As a third example, consider the flow shown in Figure 5.7. This is the famous Kelvin–Helmholtz instability in which a vortex sheet rolls up to form a sequence of vortex tubes. Of course, whether or not this is regarded as a transfer of energy to smaller scales depends on whether one regards the initial sheet as characterized by its thickness or its larger transverse dimension. In any event, one popular cartoon of homogeneous turbulence is the continual formation of vortex sheets, through the stretching action of the large-scale eddies, followed by a disintegration of the sheets (via the Kelvin–Helmholtz instability) into thin vortex tubes (called worms). In this cartoon the energy cascade is fuelled by the formation and subsequent roll-up of vortex sheets. The vortex tubes, on the other hand, are passive debris whose main role is to provide centres of intense dissipation.

There are four things to note about these simple examples, over and above the fact that energy accumulates in the small scales. First, the descriptions are in terms of the evolution of the vorticity field, not the velocity field. It is more meaningful to talk about the evolution of the $\boldsymbol{\omega}$-field, since vorticity can move from place to place only if it is materially advected, or if it diffuses. Linear momentum, on the other hand, can be instantaneously redistributed over all space by the pressure field (see Chapter 2). Second, these mechanisms can occur at all scales, from the large to the small. For example, the bursting vortex in Figure 5.6 gives rise to an axisymmetric vortex sheet. This sheet

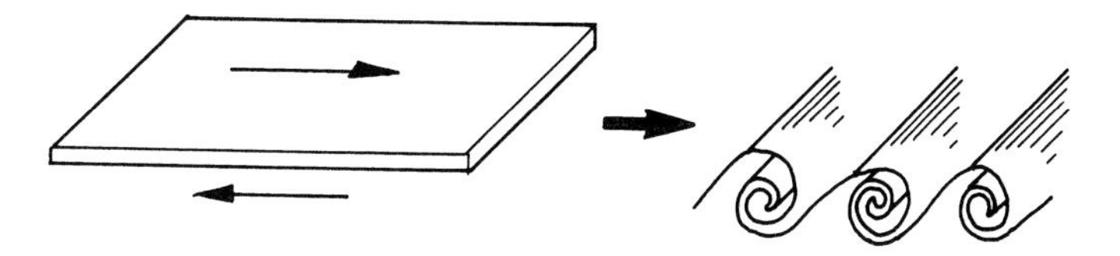

Figure 5.7 Roll-up of a vortex sheet.

might, in turn, become unstable by the roll-up mechanism of Figure 5.7 producing vortex tubes and, of course, these tubes can then become intensified by the vortex-stretching mechanism of Figure 5.5, and so on. So, one can envisage a complex sequence of events, by which vorticity is continually teased out into thinner and thinner films or tubes. Third, the transfer of enstrophy to smaller and smaller scales can arise either because eddies of two different scales interact (example shown in Figure 5.5) or else because a single structure evolves under a process of self-advection (examples shown in Figures 5.6 and 5.7). Fourth, there is little or no *helicity* in these flows. (The helicity is exactly zero for examples shown in Figures 5.6 and 5.7.) Recall that helicity, $h = \mathbf{u} \cdot \boldsymbol{\omega}$, is globally conserved in an inviscid flow:

$$\int \mathbf{u} \cdot \boldsymbol{\omega} \, dV = \text{constant}$$

(see Exercise 2.7 in Chapter 2). It is thought by some (but not all) researchers that those eddies which are most effective in passing energy down the cascade in a turbulent flow have relatively low values of h. The argument, which is a little tentative, rests on the fact that

$$\frac{\langle (\mathbf{u} \cdot \boldsymbol{\omega})^2 \rangle + \langle (\mathbf{u} \times \boldsymbol{\omega})^2 \rangle}{|\mathbf{u}^2||\boldsymbol{\omega}^2|} = 1$$

so that regions of relatively high helicity tend to correspond to regions of low $\mathbf{u} \times \boldsymbol{\omega}$ (for a given energy and enstrophy). However, $\mathbf{u} \times \boldsymbol{\omega}$ corresponds to the non-linear term in the Navier–Stokes equation which, in turn, drives the energy cascade. So perhaps regions of high helicity tend to exhibit a somewhat depleted energy cascade, with a relatively low value of ε.

Of course, the three examples above represent highly idealized processes. Turbulence is a lot more complicated than simple axisymmetric vortices or planar vortex sheets. Perhaps a more realistic cartoon of turbulence is to think of the vorticity field, part of which comprises vortex tubes, as like a seething tangle of spaghetti[3] constantly evolving under the influence of its self-induced velocity field (Plate 8). The constant stirring action of the velocity field teases out the vortex tubes (spaghetti) into finer and finer strands. However, this is still a highly simplistic view—a rather naïve cartoon.

If we accept that vortex stretching is the primary mechanism by which energy is passed down the cascade to the small scales (it underlies the bursting vortex and the spaghetti cartoon) then it is natural to ask why, in a random velocity field, the vortex tubes are not

[3] Connoisseurs would probably prefer a combination of spaghetti and lasagne since there is ample evidence that much of the vorticity resides in the form of sheets or ribbons rather than tubes! (see Sections 5.3.2 and 7.3.1).

compressed as fast as they are stretched. Well, in some sense they are (there is a lot of compression going on), but we can use a simple example to illustrate why stretching has the dominant influence on the kinetic energy transfer between scales.

Consider the steady two-dimensional shear flow $\bar{S}_{xx} = -\bar{S}_{yy} = \alpha =$ constant, all other components of the strain-rate tensor being zero. This corresponds to the steady, irrotational velocity field $\bar{\mathbf{u}} = (\alpha x, -\alpha y, 0)$. Now suppose that two vortex tubes are placed in this flow, aligned with the x and y axes (Figure 5.8). These are stretched or compressed in line with

$$\frac{D\boldsymbol{\omega}}{Dt} = \boldsymbol{\omega} \cdot \nabla \mathbf{u}.$$

Let us assume that $|\bar{\mathbf{u}}| \gg |\mathbf{u}'|$, where $\mathbf{u}'$ is the velocity associated with the vortex tubes. Then, from (2.36),

$$\frac{D}{Dt}\left(\frac{\omega^2}{2}\right) = \omega_i \omega_j \bar{S}_{ij} \tag{5.12}$$

and it is readily confirmed that the x-orientated vortex grows in intensity as

$$\int_{V_m} \omega_x^2 \, dV = \int_{V_m} (\omega_x)_0^2 \, dV \exp[2\alpha t]$$

where V_m is a material volume corresponding to a finite portion of the tube. The y-orientated vortex, on the other hand, is compressed and

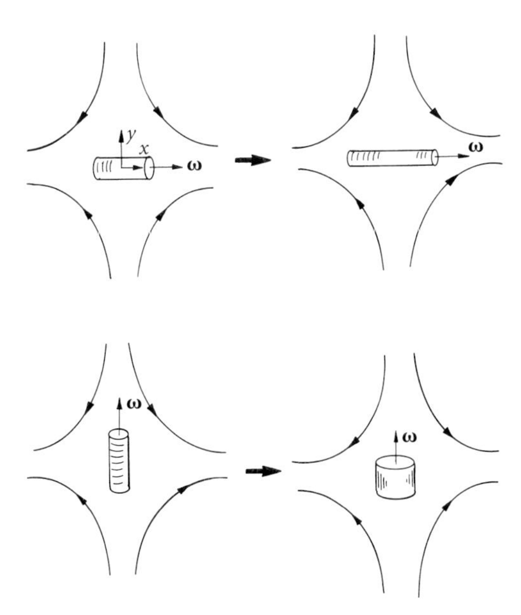

Figure 5.8 The effect of a strain field on two vortices of different orientations.

we find that ω_y^2 declines:

$$\int_{V_m} \omega_y^2 \, dV = \int_{V_m} (\omega_y)_0^2 \, dV \exp[-2\alpha t].$$

If the initial strengths of the two vortex tubes are the same then,

$$\int_{V_m} \boldsymbol{\omega}^2 \, dV = \int_{V_m} \boldsymbol{\omega}_0^2 \, dV \cosh[2\alpha t] \tag{5.13}$$

so the total enstrophy grows. To estimate the energy transfer we note that

$$\frac{\bar{D}\mathbf{u}'}{Dt} = -\mathbf{u}' \cdot \nabla\bar{\mathbf{u}} - \bar{\mathbf{u}} \cdot \nabla\bar{\mathbf{u}} - \nabla(\bar{p} + p')$$

However, the term $\bar{\mathbf{u}} \cdot \nabla\bar{\mathbf{u}}$ cancels with $\nabla\bar{p}$ and so we have

$$\frac{\bar{D}\mathbf{u}'}{Dt} = -\mathbf{u}' \cdot \nabla\bar{\mathbf{u}} - \nabla p'$$

from which

$$\frac{\bar{D}}{Dt}\left(\frac{(\mathbf{u}')^2}{2}\right) = -u_i' u_j' \bar{S}_{ij} + \nabla \cdot (\sim) = \alpha\left(\left(u_y'\right)^2 - \left(u_x'\right)^2\right) + \nabla \cdot (\sim). \tag{5.14}$$

The divergence on the right integrates to zero and so the kinetic energy of the vortices grows or falls according to whether $\alpha((u_y')^2 - (u_x')^2)$ is positive or negative. However, $(u_y')^2$ grows due to the stretching of the x-orientated vortex while $(u_x')^2$ falls due to compression of the y-orientated vortex (Figure 5.8). So the net change in energy, $\frac{1}{2}(\mathbf{u}')^2$, is positive. It seems that, despite the fact that there is both stretching and compression of vorticity, there is an overall gain in both energy and enstrophy. In short, stretching of vorticity has the dominant influence on the energy transfer.

In summary, then, this toy problem suggests that stretching outweighs compression in the long run. In fact, we could have reached the same conclusion from a consideration of Figure 5.5. It does not matter what the initial orientation of the vortex tube happens to be. It seems probable that sooner or later it will end up being stretched as shown, with the long axis of the vortex orientated with the strain field. This idea is generalized in the following example.

Example 5.1 Rapid distortion theory applied to an isolated eddy
Consider a small-scale eddy, such as that shown in Figure 5.5. It sits in the irrotational velocity field induced by an adjacent, much larger eddy. Let $\boldsymbol{\omega}^s$ be the vorticity field of the smaller eddy, $\mathbf{u}^s$ be the velocity field of the small eddy calculated from $\boldsymbol{\omega}^s$ using the Biot–Savart law, and $\mathbf{u}^L$ be the large-scale, irrotational flow. The total

velocity field is $\mathbf{u} = \mathbf{u}^L + \mathbf{u}^s$. We shall suppose that: (i) the fluid may be treated as inviscid; (ii) $\mathbf{u}^L$ may be considered as quasi steady; (iii) the gradients of $\mathbf{u}^L$ may be considered as uniform on the scale of the smaller eddy, $u_i^L = \left(u_i^L\right)_0 + \alpha_{ij}x_j$; and (iv) the smaller eddy is weak in the sense that $|\mathbf{u}^s| \ll |\mathbf{u}^L|$. Show that, in such a situation, $\boldsymbol{\omega}^s$ is governed by,

$$\frac{\bar{D}\boldsymbol{\omega}^s}{Dt} = \frac{\partial \boldsymbol{\omega}^s}{\partial t} + (\mathbf{u}^L \cdot \nabla)\boldsymbol{\omega}^s = \boldsymbol{\omega}^s \cdot \nabla \mathbf{u}^L$$

where the quadratic term in $\mathbf{u}^s$ has been neglected. Note that this equation is linear in $\boldsymbol{\omega}^s$. Moreover, $\boldsymbol{\omega}^s$ is 'frozen' into the large-scale irrotational flow and so is stretched and twisted by $\mathbf{u}^L$. This kind of linear problem is sometimes referred to as *rapid distortion theory* because it relates to a vortex being rapidly stained by a larger, adjacent eddy. Confirm that

$$\frac{\bar{D}^2}{Dt^2}\left(\frac{\boldsymbol{\omega}^2}{2}\right) = 2(\alpha_{ij}\omega_i)^2 > 0$$

and hence show that, sooner or later, the enstrophy of the small-scale eddy is bound to rise due to vortex stretching. Now suppose that x, y, and z, are aligned with the principal axis of strain of $\mathbf{u}^L$. In such a case, we have $\alpha_{xx} = a$, $\alpha_{yy} = b$, $\alpha_{zz} = c$, and $\alpha_{ij} = 0$ if $i \neq j$. Conservation of mass requires $a + b + c = 0$ and we shall order a, b, and c such that $a > b > c$. Show that ω_x^2 grows exponentially on a timescale of $2a$, while ω_z^2 declines exponentially on a timescale of $2c$. Hence confirm that, in the long run, there is an exponential growth in enstrophy. □

There is general agreement that vortex stretching underlies the energy cascade. The intensification of vorticity by straining can be quantified by the entrophy equation (see Section 2.3.2),

$$\frac{D}{Dt}\left(\frac{\boldsymbol{\omega}^2}{2}\right) = \omega_i\omega_j S_{ij} - \nu(\nabla \times \boldsymbol{\omega})^2 + \nabla \cdot [\nu\boldsymbol{\omega} \times (\nabla \times \boldsymbol{\omega})]. \quad (5.15)$$

Enstrophy, $\boldsymbol{\omega}^2/2$, is destroyed at the small scales by the viscous forces but is intensified at the larger scales by the strain field. Indeed, we shall see shortly (in Section 5.3.1) that $\langle \omega_i' \omega_j' S_{ij}' \rangle$ is positive in conventional turbulence.

We have already suggested a simple cartoon of turbulence in which the vorticity field is pictured as a tangle of spaghetti (Plate 8). If this is at all realistic then we would expect the vorticity field at the small scales to be highly intermittent (spotty). That is, as the vortex tubes and ribbons are teased out by their self-induced strain field they become thinner and thinner, concentrating the vorticity into small, localized regions of space. While it is uncertain how representative a cartoon this really is, it is certainly true that the small-scale vorticity is indeed highly intermittent. We shall return to this issue in Chapter 6.

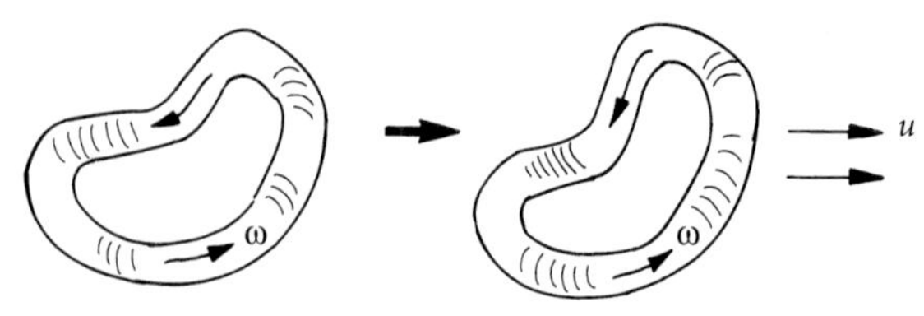

(i) Large-scale velocity simply moves the vortex tube around

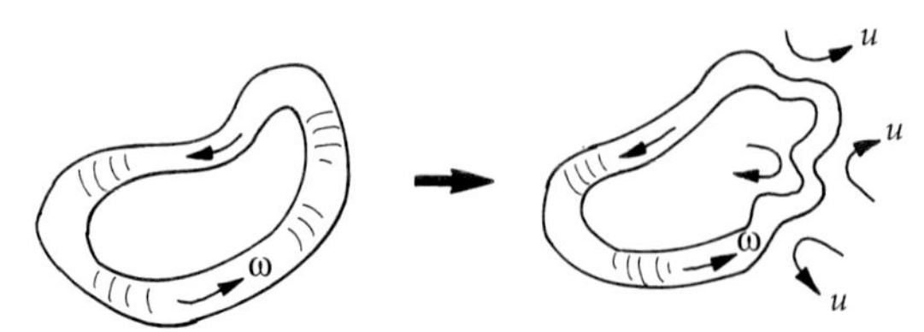

(ii) Small-scale velocity wrinkles surface of the vortex tube

Figure 5.9 Neither very large nor very small-scale velocity fields have much influence on an intermediate sized vortex.

We close this section by returning to the issue of whether or not energy is really transferred to the small scales via a multistage cascade. There is no universal agreement here, but most people believe that, by and large, the cascade model is a reasonable approximation. The usual argument goes something like this. Consider the vortex tube shown in Figure 5.9. Suppose that this is characteristic of some intermediate-sized vortex in the turbulent cascade. Very large structures have a velocity field which is almost uniform on the scale of our tube and so they simply advect it around in a passive manner (Figure 5.9(i)). Also, we know that the strain rate of the very small vortices, v/η, is much greater than that of the large vortices, $(v/\eta) \sim (ul/\nu)^{1/2}(u/l)$. We would expect, therefore, that the strain rate increases monotonically as we pass from large to small scales, and this is precisely what is observed. Thus the structures which most effectively strain the vortex tube shown in Figure 5.9 are those which are similar in size to the tube itself. (Very small structures will have negligible influence on the tube because they simply wrinkle the surface of the vortex (Figure 5.9(ii)).) So, if we believe that the energy cascade is driven by vortex stretching, then it seems likely that the most efficient transfer of energy to smaller scales occurs when vortices of similar sizes interact. We might, for example, picture an intermediate vortex falling prey to the strain field of a somewhat larger one, transferring energy to the smaller vortex, or else two like-sized vortices mutually interacting and straining each other, eventually giving rise to smaller-sized structures. In any event, it is thought that the strongest interactions tend to involve structures of similar sizes.

5.1.3 The dynamic properties of turbulent eddies (an exercise in vortex dynamics)

We close Section 5.1 by examining the dynamic properties of turbulent eddies (vortex blobs). The main point we wish to emphasis is that we can characterize much of the behaviour of turbulent eddies, and indeed a turbulent cloud, in terms of the linear and angular momentum of individual eddies. The key ideas are set out below in the form of a sequence of examples. Those readers who are well versed in vortex dynamics will be familiar with much of this material.

Example 5.2 The linear impulse (or linear momentum) of a turbulent eddy

Eddies (blobs of vorticity) move around partly because they get caught up in the irrotational velocity field of other, adjacent eddies, and partly because, even in the absence of other eddies, they can propel themselves through otherwise still fluid. This ability of an eddy to advect itself is related to the *linear impulse* (a measure of the linear momentum) of the eddy, as this example shows.

A standard result from elementary magnetostatics is the following. If a set of currents are confined to a spherical region of space, V, then the average magnetic field in V is proportional to the dipole moment of the current distribution. More precisely,

$$\int_V \mathbf{B}\,dV = \frac{1}{3}\mu_0 \int_V \mathbf{x} \times \mathbf{J}\,dV$$

where $\mathbf{B}$ is the magnetic field and $\mathbf{J}$ is the current density (Figure 5.10). (These fields are related by Ampere's law $\nabla \times \mathbf{B} = \mu_0 \mathbf{J}$, μ_0, being the permeability of free space.) If all of the current lies outside the sphere, on the other hand, we have

$$\int_V \mathbf{B}\,dV = \mathbf{B}_0 V$$

where $\mathbf{B}_0$ is the value of $\mathbf{B}$ at the centre of the sphere (see, for example, Jackson, 1998, Chapter 5).

Now consider an eddy (blob of vorticity) in a turbulent flow. Suppose that the eddy is spatially compact in the sense that its vorticity, $\boldsymbol{\omega}_e$, is negligible outside some spherical volume V_e. Let $\mathbf{u}_e$ be the velocity

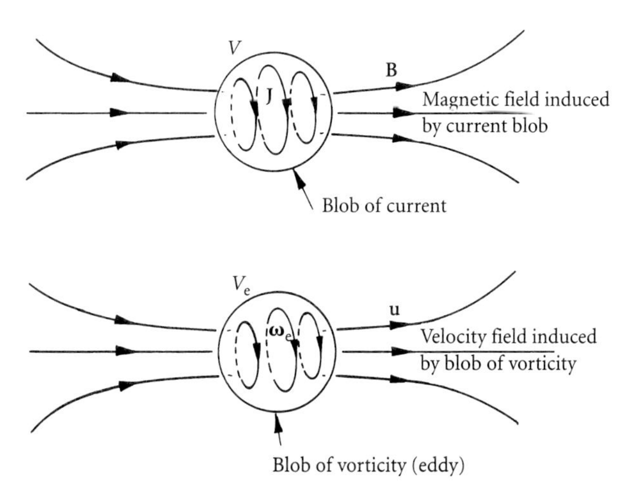

Figure 5.10 An isolated blob of current of volume V generates an average magnetic field within V of magnitude $\bar{\mathbf{B}} = (\mu_0/3V)\int \mathbf{x}\times \mathbf{J}dV$. The equivalent result in fluid mechanics is that an isolated blob of vorticity of volume V_e generates an average velocity within V_e of $\mathbf{v}_e = (1/3V_e)\int \mathbf{x} \times \boldsymbol{\omega}_e dV$.

field induced in the fluid by the presence of the eddy, $\nabla \times \mathbf{u}_e = \boldsymbol{\omega}_e$, and $\hat{\mathbf{u}}$ be the velocity field associated (by the Biot–Savart law) with all the other eddies (vortex blobs). The total velocity field is $\mathbf{u} = \mathbf{u}_e + \hat{\mathbf{u}}$. We define the spatially-averaged translational velocity of the eddy, $\mathbf{v}_e$, to be

$$\mathbf{v}_e = \frac{1}{V_e}\int_{V_e} \mathbf{u}\, dV$$

Use the magnetostatic result to show that $\mathbf{v}_e$ is composed of two terms, as follows:

$$\mathbf{v}_e = \frac{1}{V_e}\int_{V_e} \mathbf{u}\, dV = \hat{\mathbf{u}}_0 + \frac{1}{3V_e}\int_{V_e} \mathbf{x} \times \boldsymbol{\omega}_e\, dV$$

Here $\hat{\mathbf{u}}_0$ is the velocity at the centre of V_e induced by all the other turbulent eddies. Thus the eddy moves partly because it caught up in the irrotational velocity field of the other eddies, and partly through a process of self-advection. This second effect is related to the magnitude of

$$\mathbf{L}_e = \frac{1}{2}\int_{V_e} \mathbf{x} \times \boldsymbol{\omega}_e\, dV \quad \text{(linear impulse)}.$$

$\mathbf{L}_e$ is referred to as the *linear impulse* of the eddy. It is also, on occasions, called the linear momentum of the eddy since it may be shown that $\mathbf{L}_e$ is the net linear momentum introduced into the fluid by virtue of the presence of $\boldsymbol{\omega}_e$ (see Appendix II). In terms of $\mathbf{L}_e$ we have (Figure 5.11)

$$\mathbf{v}_e = \hat{\mathbf{u}}_0 + \frac{2\mathbf{L}_e}{3V_e}$$

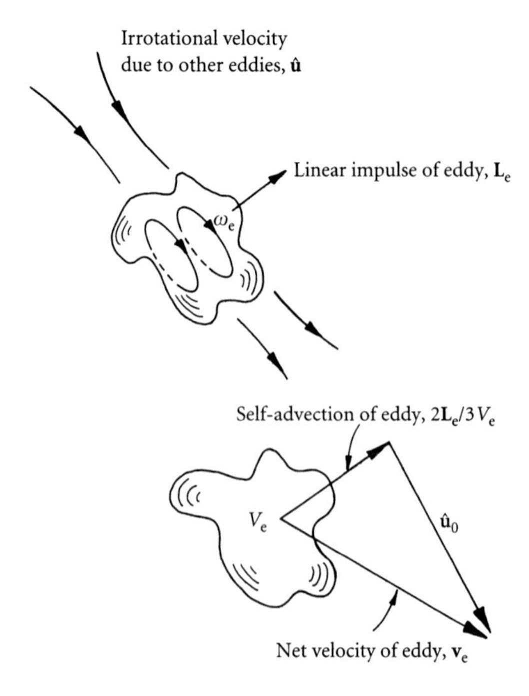

Figure 5.11 An eddy moves partly because it gets caught up in the irrotational velocity field of other eddies, $\hat{\mathbf{u}}_0$, and partly through a process of self-advection, $2\mathbf{L}_e/3V_e$. The total velocity of the eddy, averaged over its volume, is $\mathbf{v}_e = \hat{\mathbf{u}}_0 + 2\mathbf{L}_e/3V_e$.

Note that, because the integral of $\boldsymbol{\omega}$ over V_e is zero,

$$\int \omega_i \, dV = \int \nabla \cdot (\boldsymbol{\omega} x_i) \, dV = \oint x_i \boldsymbol{\omega} \cdot d\mathbf{S} = 0$$

the value of $\mathbf{L}_e$ is independent of the choice of origin for $\mathbf{x}$. We may summarize these statements as follows:

> The linear impulse of an eddy, $\mathbf{L}_e = \frac{1}{2} \int \mathbf{x} \times \boldsymbol{\omega}_e \, dV$, is a measure of the linear momentum introduced into the fluid by virtue of the presence of that eddy. The eddy moves around partly because it gets caught up in the irrotational velocity field, $\hat{\mathbf{u}}$, of the other eddies and partly as a result of its own linear impulse, $\mathbf{v}_e = \hat{\mathbf{u}}_0 + 2\mathbf{L}_e / 3V_e$.

Example 5.3 The linear impulse (or linear momentum) of a turbulent cloud
Use the vorticity evolution equation to derive an expression for the material rate of change of $\mathbf{x} \times \boldsymbol{\omega}$. Show that this equation is of the form

$$\frac{D(\mathbf{x} \times \boldsymbol{\omega})}{Dt} = \text{grad}(\sim) + \text{div}(\sim) + \text{curl}(\sim)$$

and deduce that, for a localized distribution of vorticity evolving in an infinite fluid (which is otherwise free from vorticity),

$$\mathbf{L} = \frac{1}{2} \int \mathbf{x} \times \boldsymbol{\omega} \, dV = \text{constant} \quad \text{(conservation of linear impulse).}$$

Now consider a cloud of turbulence composed of many eddies, $\boldsymbol{\omega} = \sum \boldsymbol{\omega}_e$ (Figure 5.12). From the above we may say that:

> The net linear impulse of a turbulent cloud evolving in an infinite fluid, $\mathbf{L} = \frac{1}{2} \int \mathbf{x} \times \boldsymbol{\omega} \, dV$, is the sum of the impulses of the individual eddies, $\mathbf{L} = \sum \mathbf{L}_e$, and is a dynamical invariant of the motion.

In Chapter 6, we shall see that the conservation of $\mathbf{L}$ leads to a statistical invariant (called the Saffman–Birkhoff integral) for freely

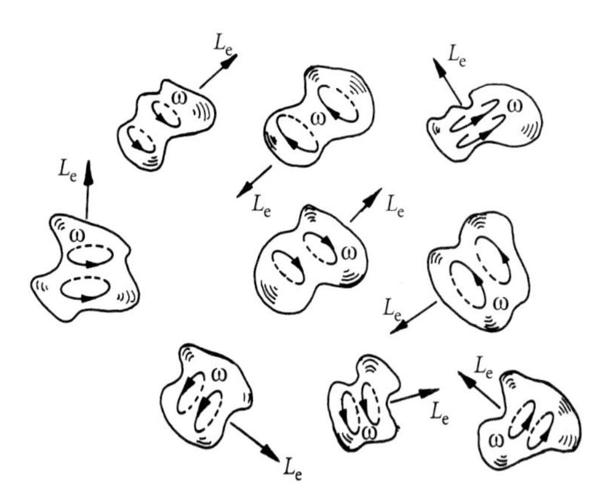

Figure 5.12 The total linear impulse (or linear momentum) of a cloud of turbulence, $\mathbf{L} = \Sigma \mathbf{L}_e = \frac{1}{2} \int \mathbf{x} \times \boldsymbol{\omega} dV$ is an invariant of the motion.

evolving turbulence. However, in some types of turbulence, $\mathbf{L}$ is approximately zero because negligible linear momentum is imparted to the turbulence during its formation. In such cases attention is transferred to the angular momentum of the eddies.

Example 5.4 The angular impulse (or angular momentum) of a turbulent eddy
Each turbulent eddy carries with it a certain amount of angular momentum, as this example shows. Suppose we have a turbulent flow consisting of a number of discrete, non-overlapping eddies (blobs of vorticity). Consider one particular eddy which is spatially compact in the sense that its vorticity, $\boldsymbol{\omega}_e$, is negligible outside some spherical control volume V_e. (V_e encloses the eddy in question, but no other eddy.) Use the vector identity

$$6(\mathbf{x} \times \mathbf{u}) = 2\mathbf{x} \times (\mathbf{x} \times \boldsymbol{\omega}) + 3\nabla \times (r^2\mathbf{u}) - \boldsymbol{\omega} \cdot \nabla(r^2\mathbf{x})$$

to show that the net angular momentum within V_e, measured about the centre of V_e, is given by,

$$\mathbf{H}_e = \int_{V_e} \mathbf{x} \times \mathbf{u}\, dV = \frac{1}{3}\int_{V_e} \mathbf{x} \times (\mathbf{x} \times \boldsymbol{\omega}_e)\, dV \quad \text{(angular momentum = angular impulse).}$$

Thus the net angular momentum, $\mathbf{H}_e$, of an eddy, defined in the sense above, is uniquely determined by its vorticity distribution, and independent of the vorticity outside V_e. (Remote eddies make no contribution to $\mathbf{H}_e$.) The integral on the right is called the *angular impulse* of the eddy. Note that, since the angular impulse is independent of the radius of V_e, so is the angular momentum. That is, different concentric spheres enclosing our eddy (but no other eddy) all give the same value of $\mathbf{H}_e$.

This result holds only if the origin for $\mathbf{x}$ lies at the centre of the control volume V_e. Suppose we shift the origin for $\mathbf{x}$ to some other location within V_e. Let $\mathbf{x}_e$ be the geometric centre of V_e in the new coordinate system, and $\mathbf{x}'$ be the position vector measured from $\mathbf{x}_e$, that is $\mathbf{x} = \mathbf{x}_e + \mathbf{x}'$. (Here $\mathbf{x}'$ plays the role formally adopted by $\mathbf{x}$.) Use the identity

$$2[\mathbf{x}' \times (\mathbf{x}_e \times \boldsymbol{\omega})]_i = [\mathbf{x}_e \times (\mathbf{x}' \times \boldsymbol{\omega})]_i + \nabla \cdot \left[\mathbf{x}' \times (\mathbf{x}_e \times \mathbf{x}')_i \boldsymbol{\omega}\right]$$

to show that,

$$\frac{1}{3}\int_{V_e} \mathbf{x} \times (\mathbf{x} \times \boldsymbol{\omega}_e)\, dV = \frac{1}{3}\int_{V_e} \mathbf{x}' \times (\mathbf{x}' \times \boldsymbol{\omega}_e)\, dV + \mathbf{x}_e \times \mathbf{L}_e.$$

Thus we must be careful in our choice of origin. However, in certain types of turbulence, $\mathbf{L}_e \cong 0$ in a typical eddy. In such cases the choice of origin for $\mathbf{x}$ is immaterial.

Example 5.5 The angular impulse of a turbulent cloud
Consider a cloud of turbulence composed of many discrete eddies, $\boldsymbol{\omega} = \sum \boldsymbol{\omega}_e$. The net angular momentum of the fluid within any spherical control volume V_c which encloses $\boldsymbol{\omega}$ is

$$\mathbf{H} = \frac{1}{3}\int_{V_c} \mathbf{x} \times (\mathbf{x} \times \boldsymbol{\omega})\, dV = \sum \frac{1}{3}\int_{V_e} \mathbf{x} \times (\mathbf{x} \times \boldsymbol{\omega}_e)\, dV$$

where $\mathbf{x}$ is measured from the centre of V_c. Find an expression for $D(\mathbf{x} \times (\mathbf{x} \times \boldsymbol{\omega}))/Dt$ and show that, for a turbulent cloud evolving in an infinite fluid (which is otherwise free from vorticity), $\mathbf{H}$ is a dynamical invariant of the motion:

$$\mathbf{H} = \frac{1}{3}\int_{V_c} \mathbf{x} \times (\mathbf{x} \times \boldsymbol{\omega})\, dV = \text{constant}$$

(conservation of angular impulse).

In certain types of turbulence $\mathbf{L}_e \cong 0$ in a typical eddy. In such cases the conservation of $\mathbf{H}$ leads to a statistical quantity which is almost, but not precisely, conserved (see Chapter 6).

Example 5.6 Angular momentum constraint in confined, freely decaying turbulence
Consider a cloud of turbulence confined to a closed spherical domain of radius R. Suppose that $R \gg l_0$ where l_0 is the integral scale of the turbulence at $t = 0$. The energy density of the turbulence declines on a time-scale of l/u in accordance with (5.2). The angular momentum, on the other hand, changes only as a result of the surface stresses and Landau and Lifshitz (1959) suggest that, since this is a surface effect, these will influence the bulk angular momentum on a much larger timescale, say τ_H. The implication is that the cascade-enhanced dissipation of energy is much faster than the rate of decay of global angular momentum. This suggests that, as long as $R \gg l$, we might treat the free decay of confined turbulence as a monotonic decay in energy subject to the conservation of angular momentum (Figure 5.13).

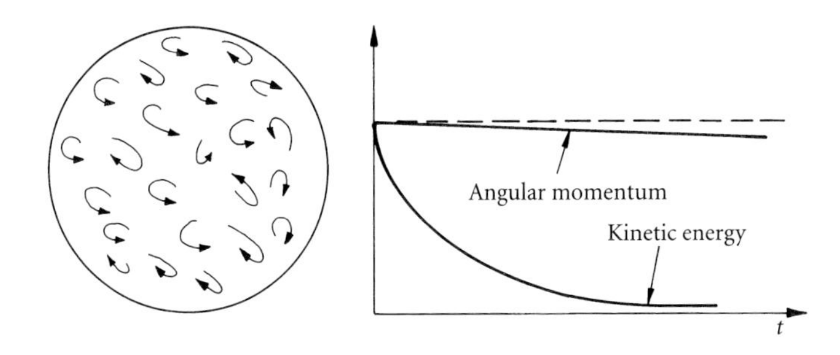

Figure 5.13 In a large confined domain the kinetic energy of a cloud of turbulence decreases much faster that the angular momentum.

Use the Schwartz inequality to show that the conservation of angular momentum imposes a lower bound on the global kinetic energy. Now show that the variational problem of minimizing kinetic energy subject to the constraint of conservation of angular momentum leads, when viscous stresses on the outer boundary are ignored, to a flow which consists of one large, axisymmetric vortex. (This suggests that, as long as t is large relative to l/u, but small relative to τ_H, the integral scale, l, rises. In practice, though, the idea that there are two timescales, l/u and τ_H, breaks down long before we reach the point where a single vortex is predicted to emerge.)

Example 5.7 The exchange of impulse (momentum) between eddies Consider a cloud of freely decaying turbulence composed of many discrete eddies, $\boldsymbol{\omega} = \sum \boldsymbol{\omega}_e$. The cloud evolves in an infinite domain which is otherwise free from vorticity. The global impulse, $\mathbf{L} = \frac{1}{2}\int \mathbf{x} \times \boldsymbol{\omega}\, dV$, and angular impulse, $\mathbf{H} = \frac{1}{3}\int \mathbf{x} \times (\mathbf{x} \times \boldsymbol{\omega})\, dV$, of the cloud are conserved during the decay (see Examples 5.3 and 5.5 above). However the linear and angular impulse of an individual eddy, $\mathbf{L}_e$ and $\mathbf{H}_e$, may change due to an exchange of momentum between eddies. We wish to determine the nature of this exchange. Consider a single eddy within the cloud which is characterized by an isolated region of vorticity, $\boldsymbol{\omega}_e$, confined to a volume V_e. Show that, if viscous diffusion is neglected,

$$\frac{D\mathbf{L}_e}{Dt} = \int_{V_e} \mathbf{u} \times \boldsymbol{\omega}\, dV, \quad \mathbf{L}_e = \frac{1}{2}\int_{V_e} \mathbf{x} \times \boldsymbol{\omega}\, dV.$$

This represents an exchange of momentum between the eddy V_e and the surrounding eddies. The velocity which appears in the integral above has two components, $\mathbf{u} = \mathbf{u}_e + \hat{\mathbf{u}}$, where $\mathbf{u}_e$ is associated (though the Biot–Savart law) with the vorticity of the eddy, $\boldsymbol{\omega}_e = \nabla \times \mathbf{u}_e$, and $\hat{\mathbf{u}}$ is the irrotational velocity field generated within V_e by the other, remote eddies. Show that $\mathbf{u}_e$ makes no net contribution to the integral, and so

$$\frac{D\mathbf{L}_e}{Dt} = \int_{V_e} \hat{\mathbf{u}} \times \boldsymbol{\omega}\, dV.$$

(Hint: rewrite $\mathbf{u}_e \times \boldsymbol{\omega}_e$ as $\nabla(u_e^2/2) - \mathbf{u}_e \cdot \nabla \mathbf{u}_e$ and convert the volume integral into a surface integral.) Since the volume integral of $\boldsymbol{\omega}_e$ over V_e is zero this can be rewritten as

$$\frac{D\mathbf{L}_e}{Dt} = \int_{V_e} (\hat{\mathbf{u}} - \hat{\mathbf{u}}_0) \times \boldsymbol{\omega}\, dV$$

where $\hat{\mathbf{u}}_0$ is the value of $\hat{\mathbf{u}}$ at the centre of V_e. Next we suppose that the irrotational strain field of the surrounding eddies may be

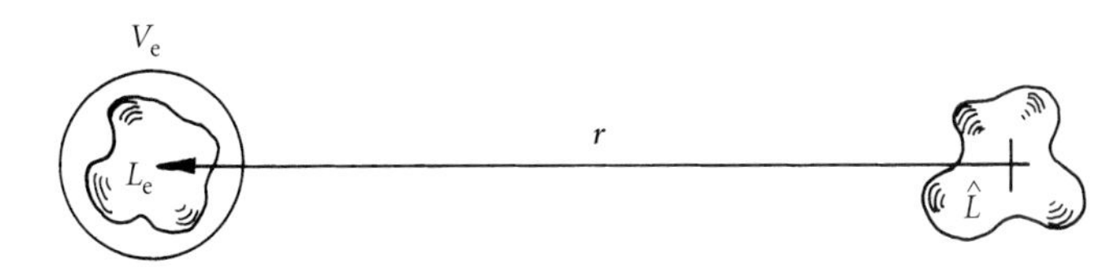

Figure 5.14 Eddies in a cloud of turbulence exchange linear momentum at a rate determine by their linear impulses and by their separation, r. If they both have finite linear impulse then they exchange momentum at a rate proportional to r^{-4}. If they both have zero initial impulse then the exchange is no greater than $0(r^{-6})$.

considered as uniform on the scale of V_e. Confirm that, in such cases, the equation above reduces to

$$\frac{D\mathbf{L}_e}{Dt} = -[(\mathbf{L}_e \cdot \nabla)\hat{\mathbf{u}}]_0.$$

We now consider the interaction of our eddy in V_e with just one of the surrounding eddies, as shown in Figure 5.14. If the two eddies are widely separated then we can use the far-field approximation to $\hat{\mathbf{u}}$:

$$\hat{\mathbf{u}} = -(4\pi)^{-1}(\hat{\mathbf{L}} \cdot \nabla)(\mathbf{r}/r^3) + \cdots$$

(see Appendix II). Here $\hat{\mathbf{L}}$ is the linear impulse of the distant eddy and $\mathbf{r}$ is the displacement between the two eddies, measured from the remote eddy to V_e. Combining these result yields

$$\frac{D\mathbf{L}_e}{Dt} = \frac{1}{4\pi}(\mathbf{L}_e \cdot \nabla)(\hat{\mathbf{L}} \cdot \nabla)(\mathbf{r}/r^3) + \cdots$$

Thus, to leading order in $1/r$, the two eddies exchange linear momentum at a rate proportional to $\hat{\mathbf{L}}$, $\mathbf{L}_e$, and r^{-4}. It turns out that, if both $\hat{\mathbf{L}}$ and $\mathbf{L}_e$ are initially zero then the exchange of linear momentum is much weaker, being at most of order r^{-6}. This equation extends in an obvious way to incorporate the simultaneous interaction of N eddies (vortex blobs) in a turbulent cloud. Thus we may conclude the following:

> The strength of the interaction of two remote eddies in a turbulent cloud, as measured by their exchange of linear momentum, depends crucially on whether or not the eddies posses a non-zero linear impulse. If their linear impulses are both finite, then the interaction is of order r^{-4}. If the linear impulses are both zero then the interaction is no greater than r^{-6}.

Finally we consider the angular impulse of our isolated eddy, $\mathbf{H}_e = \frac{1}{3}\int_{V_e} \mathbf{x} \times (\mathbf{x} \times \boldsymbol{\omega})\, dV$. Show that $\mathbf{H}_e$ evolves according to

$$\frac{D\mathbf{H}_e}{Dt} = \int_{V_e} \mathbf{x} \times (\mathbf{u} \times \boldsymbol{\omega})\, dV$$

and that, as for the linear impulse, the rotational velocity $\mathbf{u}_e$ makes no contribution to the integral on the right. It follows that

$$\frac{D\mathbf{H}_e}{Dt} = \int_{V_e} \mathbf{x} \times (\hat{\mathbf{u}} \times \boldsymbol{\omega})\, dV$$

where $\hat{\mathbf{u}}$ is the irrotational velocity of the remote eddies. This represents the exchange of angular momentum between the eddy in V_e and the surrounding eddies.

Example 5.8 Is the enstrophy and energy arising from different scales additive?
We often talk as if the enstrophy of a field of turbulence can be unambiguously distributed amongst the different scales. That is, we talk as if eddies of one scale make a certain contribution to $\frac{1}{2}\langle\boldsymbol{\omega}^2\rangle$, while those of another scale make a second contribution, and so on. However, this is not the case. Suppose, for example, that $\boldsymbol{\omega} = \sum \boldsymbol{\omega}_n$ where the $\boldsymbol{\omega}_n$'s represent the vorticity fields of different sized eddies. Then $\boldsymbol{\omega}^2 = \sum\sum \boldsymbol{\omega}_n \cdot \boldsymbol{\omega}_m$ which leads to cross terms of the form $\boldsymbol{\omega}_n \cdot \boldsymbol{\omega}_m$, $n \neq m$. These cross terms cannot be unambiguously associated with one particular eddy size. Luckily, though, in those cases where the eddies are of very different size, the cross terms are small, as we shall now show.

Consider a small and large eddy occupying a common space. Let their vorticity distributions be $\boldsymbol{\omega}^S$ and $\boldsymbol{\omega}^L$. The total enstrophy is then

$$\frac{1}{2}\int_{V_L} \boldsymbol{\omega}^2\, dV = \frac{1}{2}\int_{V_L} (\boldsymbol{\omega}^L)^2\, dV + \frac{1}{2}\int_{V_S} (\boldsymbol{\omega}^S)^2\, dV + \int_{V_S} \boldsymbol{\omega}^L \cdot \boldsymbol{\omega}^S\, dV$$

where V_L and V_S are the volumes occupied by the large and small eddies, respectively ($V_L \gg V_S$). Show that, because $\boldsymbol{\omega}^S$ is solenoidal and negligible outside V_S, the cross term is of the order of

$$\int_{V_S} \boldsymbol{\omega}^L \cdot \boldsymbol{\omega}^S\, dV = -\int_{V_S} \frac{\partial \omega_j^L}{\partial x_i} x_j \omega_i^S\, dV \sim \frac{l_S}{l_L}\omega^L \omega^S V_S$$

where l_S and l_L are the characteristic length-scales of the two eddies. This is small relative to the integral of $(\boldsymbol{\omega}^S)^2$ provided that $l_S \ll l_L$.

Now show that, in those cases where the turbulent eddies have negligible linear impulse, that is,

$$\mathbf{L}_e = \frac{1}{2}\int_{V_e} \mathbf{x} \times \boldsymbol{\omega}_e\, dV = 0$$

a similar result holds for the relative contributions to the kinetic energy. That is, the cross terms are relatively small for eddies of very different size. (You will need to make use of the results in Example 5.2.)

5.2 Kolmogorov revisited

5.2.1 *Dynamics of the small scales*

Let us now consider Kolmogorov's (1941) theory of a universal equilibrium range.[4] In many ways this is a remarkable theory because it makes a very specific prediction (the two-thirds law) which turns out to be quite robust. Such results are few and far between in turbulence. Our starting point is the structure function,

$$\left\langle [\Delta \mathrm{v}(r)]^2 \right\rangle = \left\langle \left[u'_x(\mathbf{x} + r\hat{\mathbf{e}}_x) - u'_x(\mathbf{x}) \right]^2 \right\rangle$$

In Chapter 3, we noted that $\langle [\Delta \mathrm{v}]^2 \rangle$ is of the order of all of the energy contained in eddies of size r or less. In isotropic turbulence, for example,

$$\left\langle [\Delta \mathrm{v}(r)]^2 \right\rangle \sim \frac{4}{3} \text{ [energy in eddies of size } r \text{ or less].}$$

Let us temporarily restrict ourselves to homogeneous, freely-decaying turbulence. (We shall remove this restriction shortly.) If we follow the received wisdom then, after a while, freely decaying turbulence will have largely forgotten the precise details of its initial conditions and we would expect the number of parameters which influence $\langle [\Delta \mathrm{v}]^2 \rangle$ to be rather limited. In fact it is often supposed that,

$$\left\langle [\Delta \mathrm{v}(r)]^2 \right\rangle = \hat{F}(u, l, r, t, \nu). \tag{5.16}$$

Here u is a typical velocity of the large eddies, say $u^2 = \left\langle (u'_x)^2 \right\rangle$, and l is the integral scale, perhaps defined in terms of the longitudinal correlation function $l = \int f dr$. Actually, (5.16) is incomplete. We shall see in Chapter 6 that there is at least one missing parameter in $\hat{F}$.

[4] The ideas described in this section were first published by Kolmogorov in Russian in 1941 in two short papers. (A convenient English translation of these two papers is listed in the references.) These papers were brought to the attention of western scientists by Batchelor (1947) who discovered an English translation of the 1941 papers in Cambridge in 1945. It is remarkable that English language editions of the Russian journal made it from the USSR to Britain during such turbulent times. (Moffatt, 2002, notes that Batchelor's lucky find may be due to the fact that bound copies of Soviet journals were used as ballast in supply ships making their way through Arctic waters to the west.) In any event, it was Batchelor who promoted and popularized Kolmogorov's work. However, as is often the case in science, similar ideas where developed independently by a number of other scientists around the same time. For example, Heisenberg, who, in 1945, was being held under military restraint just outside Cambridge, disclosed to G. I. Taylor that he and Von Weizsacker had developed a statistical theory of the small scales. It turns out that this theory had much in common with Kolmogorov's work. Of course, Heisenberg could not have known of Kolmogorov's work and indeed in the 1946 edition of Sommerfeld's Lectures on Theoretical Physics the new theories are attributed exclusively to the German scientists. The physical chemist L. Onsager also published similar ideas in 1945.

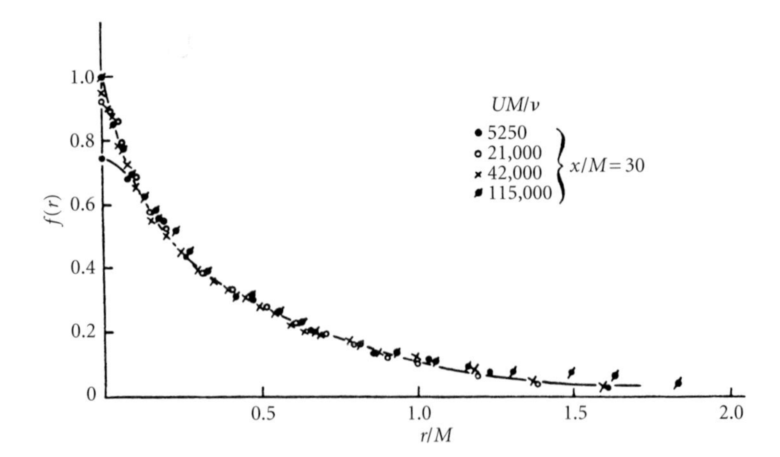

Figure 5.15 Measurements of the longitudinal correlation function taken at different times during the decay of grid turbulence (from Batchelor, 1953).

In particular, freely decaying turbulence possesses a statistical invariant associated with either the conservation of linear momentum or else the conservation of angular momentum. This influences the shape of $\hat{F}$ for large r, though it is probably unimportant as far as the small eddies are concerned. In any event, we shall proceed for the moment on the assumption that the list of parameters in $\hat{F}$ is complete. Let us consider first the form of the spectrum for the large eddies.

When $r \gg \eta$ (η being the Kolmogorov scale) the viscous forces are negligible and we may drop ν from the list in (5.16). In dimensionless form (5.16) becomes

$$\left\langle [\Delta \mathbf{v}(r)]^2 \right\rangle = u^2 F(r/l, ut/l)$$

where F, unlike $\hat{F}$, is dimensionless. In decaying grid turbulence it is usually found that this simplifies to

$$\left\langle [\Delta \mathbf{v}(r)]^2 \right\rangle = u^2 F(r/l) \tag{5.17}$$

where F is related to the longitudinal correlation function, f, by

$$F = 2(1 - f). \tag{5.18}$$

Expressions of the form (5.17), which are self-similar, are referred to as self-preserving spectra. Examples of self-preserving spectra, given by Batchelor (1953), are shown in Figure 5.15. (Actually this shows f rather than F.) These spectra represent grid turbulence measured at different times in the decay.

Let us now remove the restriction of homogeneity and turn to the small eddies, of size $r \ll l$. This is the regime which interested Kolmogorov. He claimed that these small vortices are statistically isotropic (this is known as *local isotropy*), in *statistical equilibrium*, and of *universal form*. Let us try to explain what these three terms mean and why they represent a plausible approximation.

If we accept the phenomenology of the energy cascade then there are two important things we can say about the eddies of size $r \ll l$. First, they have a complex heritage. They are the offspring of larger eddies, which came from yet larger parents, and so on. It seems plausible, therefore, that they do not retain any of the information which relates to their great-great-great-grandparents. Moreover, it is unlikely that they feel the instantaneous effect of the large-scale structures since these have a velocity field which is almost uniform on the scale of the small eddies, and this simply advects the small structures around in a passive manner (Figure 5.9).

Second, the characteristic timescale of these small structures is very fast by comparison with the large eddies. For example, at the Kolmogorov microscale we have $(\eta / \mathrm{v}) \sim (l/u)(ul/\nu)^{-1/2} \ll l/u$, and we would expect the turn-over time of eddies of intermediate size to decrease monotonically from large to small scale. In summary, then, the small scales do not feel the large scales directly and the large scales evolve very slowly by comparison with the small eddies.

Now in general the large scales are both anisotropic and statistically unsteady. However, the anisotropy arises from the mechanism which generates, or maintains, the turbulence. Since the scales of size $r \ll l$ do not feel the large eddies directly, and since the large eddies evolve very slowly (relative to the small), it seems probable that the small structures do not feel the large-scale anisotropy, nor do they feel the overall time-dependence of the flow except to the extent that the flux of energy down the energy cascade changes, $\Pi = \Pi(t)$. So, at any instant the small eddies are in approximate statistical equilibrium with the large scales and they are more or less isotropic. This is what Kolomogorov meant by local isotropy and statistical equilibrium. The regime $r \ll l$ is known as the *universal equilibrium range*.

An appealing (if somewhat simplified) cartoon which illustrates the tendency for the small scales to become statistically isotropic is given by Bradshaw (1971). Suppose we have a portion of a large-scale vortex tube which is being stretched by the mean strain. If it is aligned with the z-axis, say, then its kinetic energy, which grows due to the stretching, is primarily associated with u'_x and u'_y. These enhanced velocity components will now stretch smaller vortex tubes in the vicinity of the original large-scale tube. For example, x-directed gradients in the x component of $\mathbf{u}'$ will tend to stretch the smaller vortex lines along the x-axis, increasing u''_z and u''_y. (Here the double prime indicates the velocity field associated with the smaller vortex tubes.) Gradients in u'_y, on the other hand, tend to increase u''_z and u''_x. We now consider the influence of $\mathbf{u}''$ on yet smaller vortex tubes and so on. In this way we can build up a 'family tree' showing how the consequences of large-scale stretching progressively feed down to smaller and smaller scales. The picture which emerges is shown in

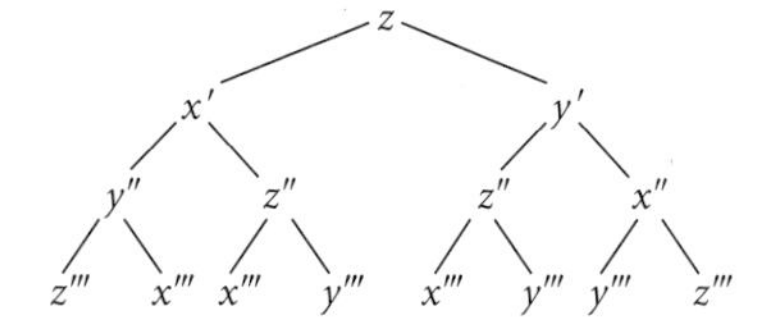

Figure 5.16 Three generations of vortex stretching resulting from the initial elongation of a z-directed vortex tube. By the time we have reached the third generation there is little hint of the large-scale anisotropy.

Figure 5.16. The important point to note is that the anisotropy of the large-scale vortex tube is rapidly lost as we pass down to smaller and smaller eddies.

Let us now return to (5.16). It would seem that t is relevant only to the extent that it influences $\Pi \sim \varepsilon \sim u^3/l$. This is also true of u and l and so (5.16) reduces to

$$\langle [\Delta \mathrm{v}]^2 \rangle = \hat{F}(\varepsilon, \nu, r) \quad (r \ll l). \tag{5.19}$$

This is a special case of Kolmogorov's *First Similarity Hypothesis*, which states that:

> When Re is large enough, and $r \ll l$, the statistical properties of $[\Delta \mathrm{v}](r)$ have a universal form which depends on only $\varepsilon = \langle 2\nu S_{ij} S_{ij} \rangle$, r and ν.

In dimensionless form we may rewrite (5.19) as

$$\langle [\Delta \mathrm{v}]^2 \rangle = \mathrm{v}^2 F(r/\eta) \tag{5.20}$$

where v and η are the Kolmogorov microscales, $\mathrm{v} = (\nu\varepsilon)^{1/4}$, $\eta = (\nu^3/\varepsilon)^{1/4}$. Since the large scales have only an indirect influence on the small eddies, and since the global geometry impacts only on the large scales, we might expect $F(r/\eta)$ to be a universal function, valid for all forms of the turbulence. (Hence the claim of universality in the first similarity hypothesis.) This is the basis of Kolmogorov's *universal equilibrium theory*.

When the experimental data corresponding to high-Re flows are examined, it turns out that (5.20) is a remarkably good way of compacting the data, and that F does indeed appear to be universal. For example, Figure 5.17, which is taken from Saddoughi and Veeravalli (1994), shows data taken from boundary layers, wakes, grids, ducts, pipes, jets, and even the oceans. The energy spectrum, normalized by the Kolmogorov microscales, is plotted instead of $\langle [\Delta \mathrm{v}]^2 \rangle / \mathrm{v}^2$. However, we shall see shortly that whenever $\langle [\Delta \mathrm{v}]^2 \rangle / \mathrm{v}^2$ is a universal function of r/η then $kE(k)/\mathrm{v}^2$ must be a universal function $k\eta$. Put another way, a test of (5.20) is to look to see if $E(k)/\mathrm{v}^2\eta = E(k)/(\varepsilon\nu^5)^{1/4}$ is a universal function of $k\eta$. For $r \ll l$ all of the data collapse to a single universal curve when k is normalized by η and $E(k)$ by $\mathrm{v}^2\eta$. (Different sets of data correspond to different values of Re and so peel off the universal curve at different values of $k\eta$.) So there is convincing support for (5.20). This is a great triumph for Kolmogorov's (1941) theory!

We must be cautious, however, of endorsing all aspects of Kolmogorov's theory on the basis of such data. For example, the existence of a universal equilibrium range does not, in itself, offer direct confirmation of Kolmogorov's *local isotropy* hypothesis. Indeed, there is now convincing evidence that, although local isotropy usually holds at very high k, the universal equilibrium range frequently begins before

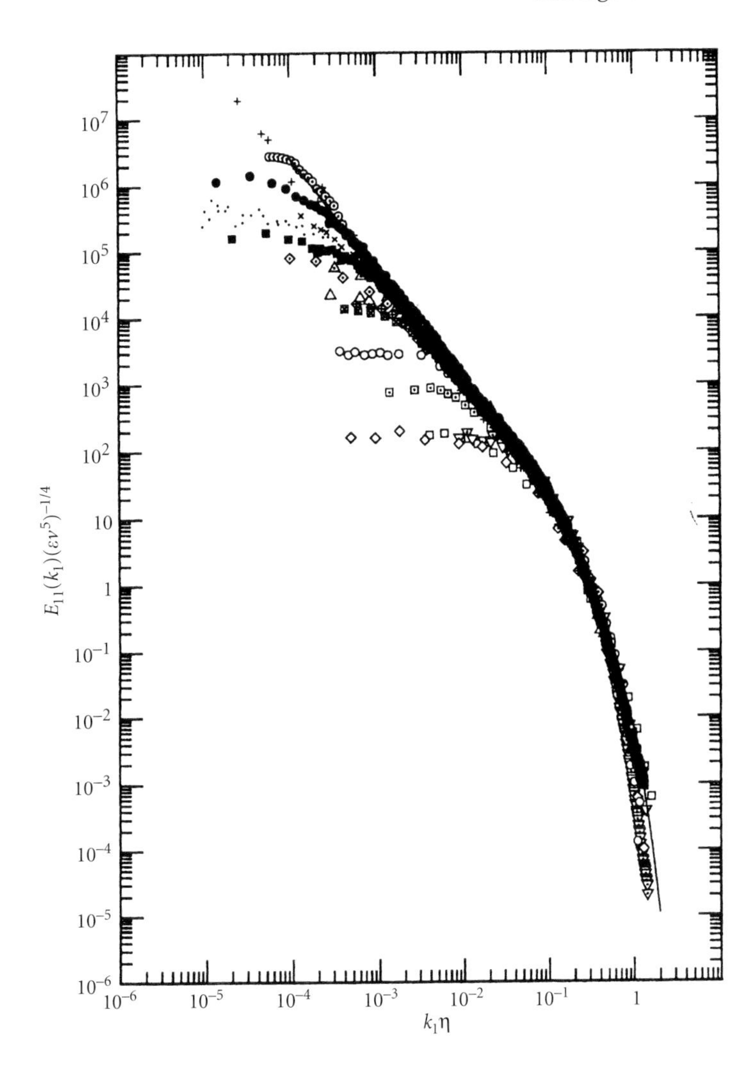

Figure 5.17 Energy spectrum versus wave number normalised by the Kolmogorov scales. The data is taken from Saddoughi and Veeravalli (1994) and incorporates measurements compiled from many experiments including measurements made in boundary layers, wakes, grids, ducts, pipes, jets, and the oceans. All the data corresponding to $kl \gg 1$ fits on a universal curve when E and k are normalised by the Kolmogorov scales. This gives direct support for eqn (5.20) and Kolmogorov's universal equilibrium theorem. (With permission of Cambridge University Press.)

local isotropy is fully achieved. In some extreme cases the influence of the large-scale anisotropy can be felt almost all the way down to the dissipation range (Ishihara, Yoshida and Kaneda, 2002).

So one of the cornerstones of Kolmogorov's (1941) theory appears to be a little tenuous. Another of Kolmogorov's claims, that of universality, has also come under attack. Indeed, even from a very early stage, Landau objected to Kolmogorov's claim that the structure of turbulence in the equilibrium range has a universal form. In the first edition of Landau and Lifshitz's *Fluid Mechanics* (English translation, 1959) there appears a footnote which was to have great ramifications for turbulence theory. It states:

It might be thought that a possibility exists in principle of obtaining a universal formula, applicable to any turbulent flow, which should give $\langle[\Delta v]^2\rangle$ for all distances r that are small compared with l. In fact, however, there can be no such formula, as we see from the following argument. The instantaneous

> value of $\langle[\Delta v]^2\rangle$ might in principle be expressed as a universal function of the energy dissipation ε at the instant considered. When we average these expressions, however, an important part will be played by the law of variation of ε over times of the order of the periods of the large eddies (of size $\sim l$), and this law is different for different flows. The result of the averaging therefore cannot be universal.

The physical insight crammed into these few words turns out to be staggering, and parts of the turbulence community are still unpicking some of its consequences. It is common to reinterpret Landau's objection in the spatial domain rather than the temporal domain. The difficulty which Landau foresaw is the following: in Kolmogorov's theory, it is not the globally averaged dissipation, $\langle\varepsilon\rangle$, which is important but rather the local dissipation averaged over a volume somewhat larger than r but much smaller than l. This local average of ε is itself a random function of position and time and in principle its manner of fluctuation can vary from one flow to the next. This is a subtle point which turns out to have important consequences.

In summary, then, the experimental data seems to lend strong support for Kolmogorov's first similarity hypothesis (at least in the restricted form of 5.20) but Kolmogorov's claim that the equilibrium range is both isotropic and universal (the same for all types of flow) may not be strictly correct. Still, encouraged by Figure 5.17, let us pursue Kolmogorov's theory a little further, temporarily glossing over Landau's objection.

We now consider a sub-domain of the universal equilibrium range. This sub-domain, called the *inertial sub-range*, satisfies $\eta \ll r \ll l$. In this range we would not expect ν to be a relevant parameter and this leads to Kolmogorov's *Second Similarity Hypothesis*, which states that:

> When Re is large, and in the range $\eta \ll r \ll l$, the statistical properties of $[\Delta v](r)$ have a universal form which is uniquely determined by r and $\varepsilon = \langle 2\nu S_{ij}S_{ij}\rangle$ alone.

The only possibility of eliminating ν from (5.20) is if $F(x) \sim x^{2/3}$. So, in the inertial sub-range we have,

$$\left\langle[\Delta v]^2\right\rangle = \beta\varepsilon^{2/3}r^{2/3} \quad (\eta \ll r \ll l) \tag{5.21}$$

where β is (according to this theory) a universal constant, found to have a value of ~ 2. This is known as Kolmogorov's *two-thirds law*. When formulated in terms of the energy spectrum, rather than $\langle[\Delta v]^2\rangle$, we have

$$E(k) = \alpha\varepsilon^{2/3}k^{-5/3}. \tag{5.22}$$

It turns out that, for $\mathrm{Re} \to \infty$, $\alpha \approx 0.76\beta$ (see Landau and Lifshitz, 1959). Equation (5.22) is known as Kolmogorov's *five-thirds law*. We

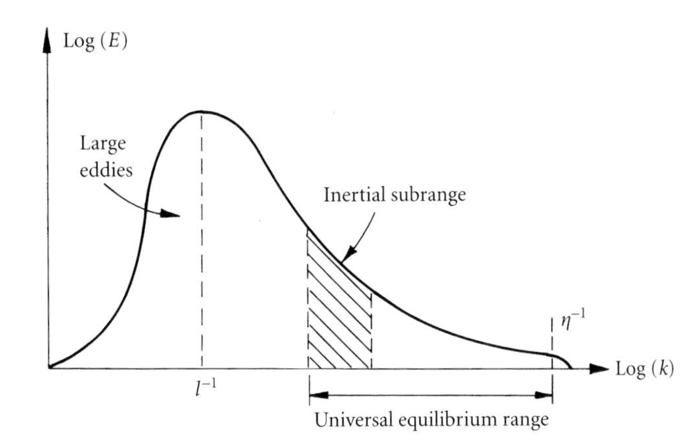

Figure 5.18 Different ranges in the energy spectrum.

can derive (5.22) either by dimensional arguments, as we did to obtain (5.21), or else by noting that

$$\left\langle [\Delta v(r)]^2 \right\rangle \sim \int_{\pi/r}^{\infty} E(k)\, dk$$

(see Section 2.5 of Chapter 3).

In summary, then, we may divide the spectrum of eddies up into three ranges (Figure 5.18), as shown below:

Name	Range	Form of $\langle[\Delta v]^2\rangle$
Energy containing eddies (freely decaying turbulence only)	$r \sim l$	$\langle[\Delta v]^2\rangle = u^2 F(r/l)$
Inertial sub-range (all types of turbulence)	$\eta \ll r \ll l$	$\langle[\Delta v]^2\rangle = \beta\varepsilon^{2/3} r^{2/3}$
Universal equilibrium range (all types of turbulence)	$r \ll l$	$\langle[\Delta v]^2\rangle = v^2 F(r/\eta)$

We shall return to this table in Chapters 6 and 8 where we see that it must be modified. In the meantime we note that very high values of Re are necessary in order to obtain an inertial sub-range. Recall that $\eta \sim (ul/\nu)^{-3/4} l$. If we are to obtain a range of r in which $\eta \ll r \ll l$, then we need $\text{Re}^{3/8} \gg 1$. This is difficult to achieve in a wind-tunnel, though it can and has been done.

There is an alternative means of deriving the two-thirds law which is often attributed to Obukhov. Obukhov's theory is usually framed in terms of $E(k)$, rather than $\langle[\Delta v(r)]^2\rangle$, but we shall stay in real space in our description. Suppose that eddies of size r have a typical velocity v_r. We have seen that the flux of energy down the cascade is constant (provided we have statistical equilibrium) and so the cascade of energy at each point in the universal equilibrium range, $\Pi(r)$, must be equal to ε. Also, we have seen that $\Pi(l) \sim u^3/l$ because the large eddies evolve (break up?) on a timescale of their turn-over time. Now suppose this is true of all eddies, that is eddies of size r evolve on a timescale of r/v_r. Then we have $\Pi(r) \sim \mathrm{v}_r^3/r$. So, in that part of the spectrum which is in statistical equilibrium, we have $\Pi(r) \sim \mathrm{v}_r^3/r \sim \varepsilon$, from which $\mathrm{v}_r^2 \sim \varepsilon^{2/3}r^{2/3}$. However, $\langle[\Delta\mathrm{v}]^2\rangle$ is of the order of all of the energy in eddies of size r or less, and the dominant contribution will come from eddies of size r since these are the most energetic. Thus $\mathrm{v}_r^2 \sim \langle[\Delta\mathrm{v}]^2\rangle$ and it follows that $\langle[\Delta\mathrm{v}]^2\rangle \sim \varepsilon^{2/3}r^{2/3}$. It seems that there is more than one way of explaining the $r^{2/3}$ variation of $\langle[\Delta\mathrm{v}]^2\rangle$! (Actually, Kolmogorov himself gave two quite distinct derivations of the two-thirds law—see Section 5.3.1.)

There is no doubt that, to within the limits of experimental accuracy, the $r^{2/3}$ law (or equivalently the $k^{-5/3}$ law) appears to be correct. Wind-tunnel data in support of this law is given, for example, by Townsend (1976) and Frisch (1995), and there is an extensive discussion of the two-thirds law and its experimental validation in Monin and Yaglom (1975). While some experiments show a slight deviation from $r^{2/3}$, this deviation is usually smaller than the scatter in the experimental data. Interestingly, there is anecdotal evidence that, before formulating his theory, Kolmogorov was aware of the 1935 measurements of Godecke, which strongly suggest an $r^{2/3}$ law. (Godecke's data is shown in Monin and Yaglom, 1975.) So, while Kolmogorov's universal equilibrium hypothesis is undoubtedly a work of inspiration, it seems that nature provided a few hints on the way.

We conclude this section with a note of warning. In Chapter 3, we noted that structure functions of order p can be defined as

$$\langle[\Delta\mathrm{v}]^p\rangle = \langle[u_x(\mathbf{x} + r\hat{\mathbf{e}}_x) - u_x(\mathbf{x})]^p\rangle. \tag{5.23}$$

Following Kolmogorov's second similarity hypothesis the form of $\langle[\Delta\mathrm{v}]^p\rangle$ in the inertial range must be of the form,

$$\langle[\Delta\mathrm{v}]^p\rangle = \beta_p(\varepsilon r)^{p/3} \quad (\eta \ll r \ll l). \tag{5.24}$$

No other combination would eliminate ν as a parameter. For $p = 2$ we recover the two-thirds law. For $p = 3$ we have $\langle[\Delta\mathrm{v}]^3\rangle = \beta_3\varepsilon r$. In fact it can be shown that, when the turbulence is *globally isotropic*, $\beta_3 = -4/5$ (see Chapter 6), and so

$$\langle[\Delta\mathrm{v}]^3\rangle = -\frac{4}{5}\varepsilon r. \tag{5.25}$$

This is known as Kolmogorov's four-fifths law. Like the two-thirds law, this is well supported by the experimental data. Moreover, it is encouraging that β_3 is universal, as suggested by Kolmogorov's theory. So far, so good. Unfortunately, things start to go wrong as p is increased above 3. The exponent n in the relationship $\langle[\Delta\mathrm{v}]^P\rangle \sim r^n$ starts to drop below $p/3$. The discrepancy is small at first but by the time p reaches 12 the exponent n has a value of ~ 2.8, rather than the expected 4. Clearly there is something incomplete about Kolmogorov's theory and this is related to Landau's cryptic (but in retrospect profound) objection.

The problem is the following. The dissipation $2\nu S_{ij}S_{ij}$ is extremely intermittent in space. There are regions of large dissipation and regions of small dissipation. In a particular region of size r (r being assumed much less than l) the average flux of energy to the small scales, $\Pi(r)$, should be equal to the spatially averaged dissipation in that region. It follows that the dynamics of the eddies of size r, which are the eddies responsible for the flux $\Pi(r)$, should be controlled by $2\nu S_{ij}S_{ij}$ averaged over a volume of size r, rather than the globally averaged dissipation $\varepsilon = \langle 2\nu S_{ij}S_{ij}\rangle$. Let us define

$$\varepsilon_{AV}(r, \mathbf{x}, t) = \frac{1}{V_r}\int_{V_r} (2\nu S_{ij}S_{ij})\, dV$$

where V_r is a spherical volume of radius r centred on $\mathbf{x}$. Then Kolmogorov's second similarity hypothesis might be amended to read:

When Re is large, and r lies in the range $\eta \ll r \ll l$, the statistical properties of $[\Delta v](r)/(r\varepsilon_{AV}(r))^{1/3}$ have a universal form, being the same for all types of flow and independent of ν.

This suggests that (5.24) should be replaced by,

$$\langle [\Delta v]^p(r)\rangle = \beta_p \left\langle \varepsilon_{AV}^{p/3}(r)\right\rangle r^{p/3}, \quad \eta \ll r \ll l$$

which is sometimes known as Kolmogorov's refined similarity hypothesis, (Kolmogorov, 1962). As in the original theory the β_p's are universal, that is, the same for all types of flow. Noting that $\langle \varepsilon_{AV}(r)\rangle = \varepsilon$, the globally averaged dissipation, we see that the four-fifth law is unchanged by this refined view of events. However, for $p \neq 3$, we have the possibility that $\langle [\Delta v]^p\rangle$ no longer scales as $r^{p/3}$. In order to determine the relationship between $\langle [\Delta v]^p\rangle$ and r we need to examine the statistics of $\varepsilon_{AV}(r)$ and estimate $\langle \varepsilon_{AV}^{p/3}(r)\rangle$ in terms of, say, r, l and $\varepsilon = \langle 2\nu S_{ij}S_{ij}\rangle$.

We shall return to this question in Chapter 6 where we shall see that, some 20 years after his original theory, Kolmogorov proposed a simple statistical model for $\varepsilon_{AV}(r)$ (called the log-normal model) which led to a correction to (5.24) of the form

$$\langle [\Delta v]^p\rangle = C_P(\varepsilon r)^{p/3}(l/r)^{\mu\, p(p-3)/18}$$

where μ is known as the *intermittency exponent*. C_p is taken to be a universal constant in the theory and is usually given a value in the range $0.2 < \mu < 0.3$. Actually, the precise form of Kolmogorov's (1962) correction is often criticized, although the essential idea, that the intermittency of the dissipation necessitates a correction to Kolmogorov's original hypotheses, is now generally accepted. Note that, for $p = 2$, the correction to the two-thirds law is small and possibly lies within the range of uncertainty associated with experimental error. Note also that in this modified version of his theory Kolmogorov did not retain universality, in the sense that the C_p's are not regarded as universal constants, that is, the same for jets, wakes, boundary layers, etc. This is consistent with Landau's view that the form of intermittency would change from one type of flow to another. In fact, in Chapter 6 we shall see that, in those cases where there is pronounced intermittency at the large scales, neither the β_p's of the original theory, nor the C_p's in the refined one, are universal.

Example 5.9 Filtering the velocity field to distinguish between small & large scales
Consider the trace of one component of velocity, say u_x, measured along a straight line in a field of turbulence, say $u_x(x)$ measured along $y = z = 0$. This signal will contain both small- and large-scale fluctuations corresponding to the presence of small and large eddies. It is desired to distinguish between the different scales in u_x. Consider the new function

$$u_x^L(x) = \int_{-\infty}^{\infty} u_x(x - r)G_1(r)dr$$

where $G_1(r)$ is a filter function defined by

$$G_1(r) = 1/L, \quad |r| < L/2$$

$$G_1(r) = 0, \quad |r| > L/2.$$

Evidently u_x^L is a smoothed out, or filtered, version of u_x in which fluctuations of scales much less than L are absent. That is, $u_x^L(x)$ is the average value of u_x in the neighbourhood of x, the average being performed over the length L (Figure 5.19). Other filter functions which perform similar tasks are

$$G_2(r) = \frac{\exp(-r^2/L^2)}{\pi^{1/2}L}$$

$$G_3(r) = \frac{\sin(\pi r/L)}{\pi r}$$

(All three functions, G_i, are even in r, have integrals equal to unity, and are small for $r \gg L$.) Use the convolution theorem for Fourier transforms to show that, in general, the Fourier transform of u_x^L is given by

$$\hat{u}_x^L(k) = 2\pi\hat{u}_x(k)\hat{G}(k)$$

where $\hat{u}_x(k)$ and $\hat{G}(k)$ are the Fourier transforms of $u_x(x)$ and $G(r)$. Now show that, for the particular case of filter function G_3, $\hat{u}_x^L(k)$ is simply

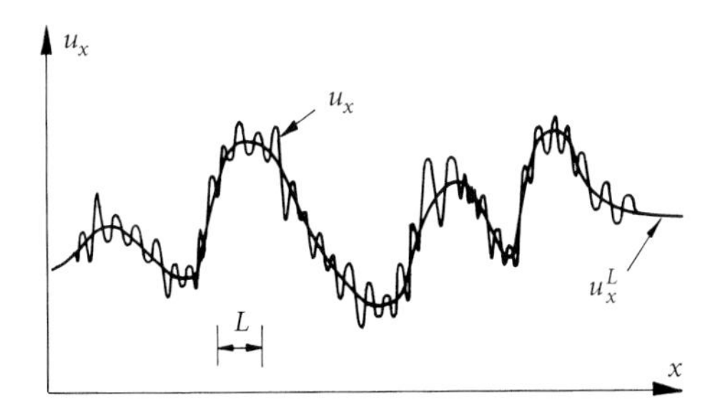

Figure 5.19 Filtering the velocity $u_x(x)$.

the transform of $u_x(x)$ in which all contributions to $\hat{u}_x(k)$ corresponding to $|k| > \pi/L$ are annihilated:

$$\hat{u}_x^{\mathrm{L}}(k) = \hat{u}_x(k), \quad |k| < \pi/L$$

$$\hat{u}_x^{\mathrm{L}}(k) = 0, \quad |k| > \pi/L.$$

It would seem that, while $\hat{u}_x^{\mathrm{L}}$ is dominated by structures of size L or greater, its transform is non-zero only for k less than π/L. We conclude that most of the information about the rapid fluctuations in $\hat{u}_x$ are held in the high-k part of $\hat{u}_x(k)$, while information about the slow fluctuations in $u_x(x)$ are held in the low-k part of $\hat{u}_x(k)$. Thus the Fourier transform can be used to distinguish between scales in $u_x(x)$. This idea is pursued in Chapter 8.

Example 5.10 Kinetic energy transfer from large to small-scale eddies
Consider homogeneous turbulence of integral scale l and microscale η. Let r be some length intermediate between l and η. We divide the velocity field into two parts, $\boldsymbol{\omega} = \boldsymbol{\omega}^{\mathrm{L}} + \boldsymbol{\omega}^{\mathrm{S}}$. Here $\boldsymbol{\omega}^{\mathrm{L}}$ is the contribution to $\boldsymbol{\omega}$ which arises from structures greater than r and $\boldsymbol{\omega}^{\mathrm{S}}$ is associated with eddies smaller than r. (This division is not unambiguous and implies the use of a filter of the type discussed in Example 5.9 above to distinguish between scales.) The Biot–Savart law now allows us to divide $\mathbf{u}$ into $\mathbf{u}^{\mathrm{L}}$ and $\mathbf{u}^{\mathrm{S}}$, where $\nabla \times \mathbf{u}^{\mathrm{L}} = \boldsymbol{\omega}^{\mathrm{L}}$, $\nabla \times \mathbf{u}^{\mathrm{S}} = \boldsymbol{\omega}^{\mathrm{S}}$ and both $\mathbf{u}^{\mathrm{L}}$ and $\mathbf{u}^{\mathrm{S}}$ are solenoidal. Show that the Navier–Stokes equation yields

$$\frac{\partial}{\partial t}\left\langle \tfrac{1}{2}\rho(\mathbf{u}^{\mathrm{L}})^2 \right\rangle + \left\langle \rho\mathbf{u}^{\mathrm{L}} \cdot \frac{\partial \mathbf{u}^{\mathrm{S}}}{\partial t} \right\rangle = \left\langle \tau^{\mathrm{L}}_{ij} S^{\mathrm{S}}_{ij} - \tau^{\mathrm{S}}_{ij} S^{\mathrm{L}}_{ij} \right\rangle + \nu(\sim)$$

$$\frac{\partial}{\partial t}\left\langle \tfrac{1}{2}\rho(\mathbf{u}^{\mathrm{S}})^2 \right\rangle + \left\langle \rho\mathbf{u}^{\mathrm{S}} \cdot \frac{\partial \mathbf{u}^{\mathrm{L}}}{\partial t} \right\rangle = \left\langle \tau^{\mathrm{S}}_{ij} S^{\mathrm{L}}_{ij} - \tau^{\mathrm{L}}_{ij} S^{\mathrm{S}}_{ij} \right\rangle + \nu(\sim)$$

where S_{ij} is the strain rate, $\tau^{\mathrm{L}}_{ij} = -\rho u^{\mathrm{L}}_i u^{\mathrm{L}}_j$ and $\tau^{\mathrm{S}}_{ij} = -\rho u^{\mathrm{S}}_i u^{\mathrm{S}}_j$. Through a careful choice of filter (say G_3 in Example 5.9 above) the cross terms on the left can be set to zero (Frisch, 1995) and we end up with

$$\frac{\partial}{\partial t}\left\langle \frac{1}{2}(\mathbf{u}^{\mathrm{L}})^2 \right\rangle = -\Pi_r - \nu\left\langle (\boldsymbol{\omega}^{\mathrm{L}})^2 \right\rangle$$

$$\frac{\partial}{\partial t}\left\langle \frac{1}{2}(\mathbf{u}^{\mathrm{S}})^2 \right\rangle = \Pi_r - \nu\left\langle (\boldsymbol{\omega}^{\mathrm{S}})^2 \right\rangle$$

$$\rho\Pi_r = \left\langle \tau^{\mathrm{S}}_{ij} S^{\mathrm{L}}_{ij} - \tau^{\mathrm{L}}_{ij} S^{\mathrm{S}}_{ij} \right\rangle.$$

The function $\Pi_r(r)$, the precise shape of which depends on our choice of filter, represents the flux of energy from large to small scales driven by inertia.

We expect $\left\langle \tau_{ij}^{S} S_{ij}^{L} \right\rangle$ to be non-zero and positive for the same reason that $\tau_{ij}^{R} \bar{S}_{ij}$ is positive in a shear flow. That is, the large-scale strain S_{ij}^{L} shapes the small eddies, stretching the small vortex filaments along the axis of maximum principal stain, increasing their energy. The idea that $\left\langle \tau_{ij}^{s} S_{ij}^{L} \right\rangle$ controls the energy transfer to smaller scales underlies the Smagorinsky sub-grid model in large-eddy simulations (see Section 7.1.2) and Heisenberg's closure hypothesis (see Section 8.2.2). Let us now introduce the function $V(r)$ defined by

$$\left\langle \frac{1}{2}(\mathbf{u}^{s})^{2} \right\rangle = \int_{0}^{r} V(s)ds.$$

Evidently $V(r)$ represents a sort of energy density, characterising the eddies of size r. In the inertial subrange, where viscous effects are unimportant and the turbulence is in quasi-equilibrium, we might expect that Π_r depends on the behaviour of eddies of size r, as characterized by $V(r)$ and r, but it should not depend on ν, nor be an explicit function of time. This suggests that $\Pi_r = \Pi_r(V, r)$ Show that, if this is true, then dimensional considerations demand $\Pi_r \sim V^{3/2} r^{1/2}$. Hence confirm that, for high-Re turbulence,

$$\left\langle \tfrac{1}{2}(\mathbf{u}^{s})^{2} \right\rangle \sim \varepsilon^{2/3} r^{2/3}$$

as required by Kolmogorov's two-thirds law.

5.2.2 *Turbulence induced fluctuations of a passive scalar*

We shall now show how Kolmogorov's ideas may be extended to describe the turbulent mixing of a contaminant.

There are many cases in which one is interested in the influence of a turbulent velocity field on the distribution of some scalar quantity; say the distribution of temperature, smoke or dye. If the scalar has no dynamic influence on the turbulence then we refer to it as a *passive scalar*.

Most passive scalars obey an advection–diffusion equation of the form,

$$\frac{\partial C}{\partial t} + (\mathbf{u} \cdot \nabla)C = \alpha \nabla^{2} C$$

where C is the scalar contaminant (temperature or dye concentration) and α is its diffusivity. When the Peclet number, $\text{Pe} = ul/\alpha$, is large, which it usually is in a turbulent flow, diffusion is negligible at the

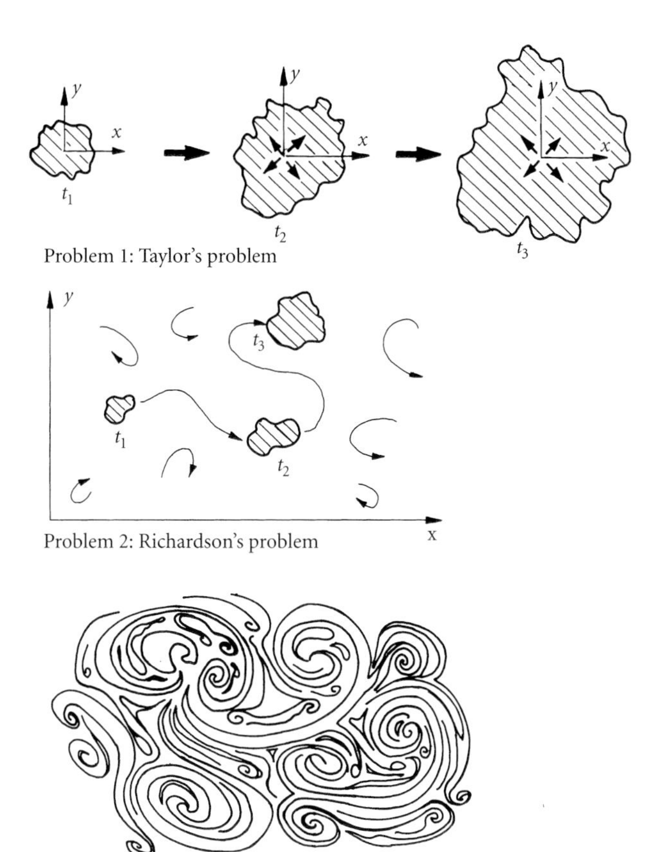

Figure 5.20 In Taylor's problem we consider the continuous release of a contaminant from a fixed point and estimate the average rate of growth of the contaminant cloud. In Richardson's problem a small puff of contaminant is released at $t = 0$ and we aim to determine the rate of growth of the puff as it bounces around in the turbulent flow. In Problem 3 the contaminant is unevenly dispersed throughout the medium.

scale of the large eddies. The scalar C then acts like a marker which tags the fluid particles. Diffusion is important only at scales which are characterized by $\text{Pe} \sim 1$. We shall take both $ul/\alpha \gg 1$ and $ul/\nu \gg 1$ throughout this section so that diffusive effects are restricted to the microscales of the turbulence. We shall also assume that there is no mean flow. Although this second restriction is rarely achieved in practice, it has the merit of greatly simplifying the analysis.

There are three interrelated problems which often arise in the study of passive scalars (Figure 5.20):

Problem 1 Taylor diffusion
Imagine that dye is released continuously into a turbulent flow from a single point source. As time progresses the cloud of dye will spread by turbulent mixing. A natural question to ask is; how large, on average, will the cloud be after a time t? Since $ul/\alpha \gg 1$ this is equivalent to asking how far, on average, will a fluid particle released from the source migrate in a time t. This problem is called the *Taylor problem of single particle diffusion*.

Problem 2 Richardson's law

Now suppose that, instead of continuously releasing dye, we create a single cloud, or puff, of dye at time $t = 0$. Let the initial size of the puff be smaller than the integral scale, l, but larger than the Kolmogorov scale. The centroid of the puff will move as described in Problem 1 as the blob of dye is swept around by the large-scale eddies. In addition, the size of the puff will increase with time as a result of the small-scale turbulence. In Problem 2 we wish to determine the average rate of spreading of the puff. In effect, we want to determine the average rate of separation of two adjacent fluid particles (which mark opposite sides of the puff) as a result of turbulent mixing. This is known as *Richardson's problem* or the problem of the *relative dispersion* of two particles.

Problem 3 Local turbulent fluctuations in scalar intensity

In our third problem we abandon the idea of a local release of contaminant and suppose that the passive scalar is distributed unevenly throughout the field of turbulence. For example, we might envisage a large water tank filled with numerous patches of dye. The tank is then subject to agitation and, of course, eventually the water and dye become well mixed. However, during the intermediate stages of mixing the dye concentration will be non-uniform. That is, although the course-grain picture may be one of nearly uniform dye across the tank, the fine-scale dye concentration will remain non-uniform until such time as small-scale mixing has eradicated all fluctuations in concentration. In such a case we might be interested in the spatial structure of the concentration field and the time required to achieve near perfect mixing.

The first two problems, which are essentially ones of particle tracking in a field of turbulence, will be discussed in Sections 5.4.1 and 5.4.2. Here we shall focus on the third class of problem, where it turns out that Kolmogorov-type arguments furnish a great deal of useful information. Our aim is to characterize the spatial fluctuations in C and to determine the time required for complete mixing.

Let us suppose that the distribution of C is statistically homogeneous and isotropic, with zero mean, $\langle C \rangle = 0$. (A mean of $\langle C \rangle = 0$ can be enforced by choosing an appropriate datum from which to measure C.) Then a convenient measure of the non-uniformity of the contaminant is provided by the variance of C, defined as $\langle C^2 \rangle$. We can get an expression for the rate of change of the variance from our advection–diffusion equation. Multiplying throughout by C yields,

$$\frac{\partial}{\partial t}\left[\frac{1}{2}C^2\right] + \nabla \cdot \left[\left(\frac{1}{2}C^2\right)\mathbf{u}\right] = \nabla \cdot [\alpha C \nabla C] - \alpha(\nabla C)^2.$$

When this is ensemble averaged those terms which take the form of divergences disappear by virtue of our assumption of homogeneity. The net result is

$$\frac{d}{dt}\left\langle\frac{1}{2}C^2\right\rangle = -\alpha\left\langle(\nabla C)^2\right\rangle.$$

Thus the fluctuations in contaminant are destroyed by diffusion at a rate proportional to

$$\varepsilon_c = \alpha\left\langle(\nabla C)^2\right\rangle.$$

Physically this process represents the cross-diffusion of the contaminant between positive and negative regions of C. One of the curious features of this equation is the absence of a convective term. Evidently, the convection of C cannot, by itself, reduce the variance. However, convection still plays a crucial role. Think of cream being stirred into coffee. The stirring disperses the cream and then teases it out into finer and finer filaments. When the filaments are so thin that diffusion can act the destruction term ε_c cuts in, eradicating the variance and producing near perfect mixing at the small scales. In short, convection is required in order to generate the large gradients in C required for the diffusive elimination of $\langle C^2\rangle$. Now suppose that η_c is the characteristic length-scale of the most rapid spatial fluctuations in C. This is the analogue of the Kolmogorov microscale for $\mathbf{u}'$, and represents the length-scale at which diffusion becomes important. In terms of η_c we have,

$$\varepsilon_c \sim \alpha\left[\frac{(\Delta C)_{\eta_c}}{\eta_c}\right]^2$$

where $(\Delta C)_{\eta_c}$ is the characteristic fluctuation in C over distances of order η_c.

We shall now show that we can attribute to ε_c and η_c a role somewhat analogous to those of ε and η in Richardson's cascade. We start by noting that, since C is materially conserved for scales greater than η_c, Richardson's cascade of eddies should be accompanied by a corresponding cascade of $\langle C^2\rangle$. That is, as a parent eddy fractures into smaller daughter eddies (i.e. a vortex blob gets teased out into several smaller vortex blobs) the contaminant presumably becomes more finely mixed, with the characteristic length-scale for C going from l_{parent} to l_{daughter}. Let us suppose that this picture is accurate. Then, just as there is a flux of kinetic energy down the cascade, so there is a flux, $\Pi_c = \varepsilon_c$, of the scalar variance. This cascade of $\langle C^2\rangle$ is halted when the characteristic length-scale for the fluctuations in C reaches η_c and diffusion sets in. However, just as in Richardson's cascade, we might expect that, for $r \gg \eta_c$, the details of the cascade are independent of the magnitude of the diffusivity.

Perhaps we should note here that not everyone believes in this picture of two analogous cascades, one for energy and one for the scalar variance. Sreenivasan (1991), for example, points out that the analogy is rather weak and questions whether or not there is a passive scalar cascade at all. Others are less critical. In any event, we shall see that the phenomenology of a passive scalar cascade yields certain predictions which are consistent with at least some of the experiments, and so we shall tentatively adopt this picture here.

Let us now determine the analogue of Kolmogorov's two-thirds law for passive scalars. Let l be the integral scale of the turbulence, l_c be the characteristic length of the large-scale variations in C, and $\langle[\Delta C]^2\rangle$ be the structure function

$$\left\langle[\Delta C]^2\right\rangle = \left\langle[C(\mathbf{x}+\mathbf{r}) - C(\mathbf{x})]^2\right\rangle.$$

In isotropic turbulence $\langle[\Delta C]^2\rangle$ is independent of orientation and so is a function of $r = |\mathbf{r}|$ only. Now consider an intermediate range of length-scales characterized by,

$$\eta_{\max} = \max[\eta, \eta_c] \ll r \ll \min[l, l_c] = l_{\min}.$$

This defines the so-called *inertial-convective subrange*. The name derives from the fact that the restriction $r \gg \eta$ guarantees the dominance of inertia over viscous forces, while the requirement that $r \gg \eta_c$ ensures that the convection of C is much greater than diffusion. Let us now rework Kolmogorov's ideas. Since $\max[\eta, \eta_c] \ll r$, we would expect that neither ν nor α will influence $\langle[\Delta C]^2\rangle$ in the inertial-convective subrange. On the other hand, the restriction $r \ll \min[l, l_c] = l_{\min}$ suggests that $\langle[\Delta C]^2\rangle$ depends on the large scales only to the extent that they determine the flux of energy and scalar variance from large to small scale, that is, they determine ε and ε_c. Thus, following a line of reasoning close to that of Kolmogorov we might expect that, in the inertial-convective subrange,

$$\left\langle[\Delta C]^2\right\rangle = f(\varepsilon, \varepsilon_c, r).$$

We must now revert to dimensional analysis. First we note that ε and r do not contain the units of C, which might be temperature say. However, the dimensions of ε_c scale as C^2, and so the function f must be linear in ε_c:

$$\left\langle[\Delta C]^2\right\rangle = \varepsilon_c f(\varepsilon, r).$$

Dimensional arguments are then sufficient to uniquely determine the form of f. It is readily confirmed that the only possibility is

$$\left\langle[\Delta C]^2\right\rangle \sim \varepsilon_c \varepsilon^{-1/3} r^{2/3}, \quad \eta_{\max} \ll r \ll l_{\min}$$

which fixes $\langle[\Delta C]^2\rangle$ to within a constant of proportionality. This result is the analogue of Kolmogorov's two-thirds law for passive scalars and was first suggested by Obukhov in 1949, and independently by Corrsin in 1951. It is somewhat reassuring to discover that its predictions are in close agreement with most numerical and physical experiments (Lesieur, 1990). More generally, the Kolmogorov–Obukhov–Corrsin argument leads to

$$\langle[\Delta C]^p\rangle \sim \varepsilon_c^{p/2}\varepsilon^{-p/6}r^{p/3}, \quad \eta_{\max} \ll l_{\min}$$

for any positive integer p. However, there is growing evidence that the Kolmogorov–Obukhov–Corrsin theory is imperfect, with discrepancies arising from the strong intermittency of the scalar concentration in the inertial-range, particularly for large p. Also, some experiments display a surprising lack of isotropy in the scalar concentration at the small scales, in contravention of Kolmogorov's theory of *local isotropy* (Sreenivasan, 1991, Warhaft, 2000). For example, Sreenivasan (1991) looked at the small-scale structure of a passive scalar in simple shear flows. He found that local isotropy is almost never achieved in shear flows at terrestrial values of Re. One illustration of this is the fact that the skewness of $\partial C/\partial x$, that is, $\langle(\partial C/\partial x)^3\rangle/\langle(\partial C/\partial x)^2\rangle^{3/2}$, which should be zero if local isotropy exists, is found to be of the order of unity, and not zero. Curiously, though, despite this lack of local isotropy, the two-thirds law seems to hold true at high Re. Evidently, great care must be exercised when extrapolating Kolmogorov's cascade ideas to the statistical distribution of a passive scalar.

One exact result, however, can be obtained directly form the governing equation for a passive scalar. It is (see the Appendix at the end of this chapter)

$$\langle\Delta u_{\|}[\Delta C]^2\rangle = -\tfrac{4}{3}\varepsilon_c r, \quad \eta_{\max} \ll r \ll l_{\min}$$

where $\Delta u_{\|}$ is the component of $\mathbf{u}(\mathbf{x}+\mathbf{r}) - \mathbf{u}(\mathbf{x})$ which is parallel to $\mathbf{r}$. This is the scalar analogue of Kolmogorov's four-fifths law.

Note that the two-thirds law allows us to estimate the magnitude of $\langle[\Delta C]^2\rangle$ for the scale $r \sim l_{\min}$. That is to say, although the two-thirds variation in r breaks down long before we reach $r \sim l_{\min}$, we might at least expect the magnitude of $\langle[\Delta C]^2\rangle$ at the top of the inertial-convective range to be of the same order as the magnitude of $\langle[\Delta C]^2\rangle$ at $r \sim l_{\min}$. *If* this is true then we have

$$\langle[\Delta C]^2\rangle_{l_{\min}} \sim \varepsilon_c\varepsilon^{-1/3}l_{\min}^{2/3}.$$

Similarly we might expect,

$$\langle[\Delta C]^2\rangle_{\eta_{\max}} \sim \varepsilon_c\varepsilon^{-1/3}\eta_{\max}^{2/3}.$$

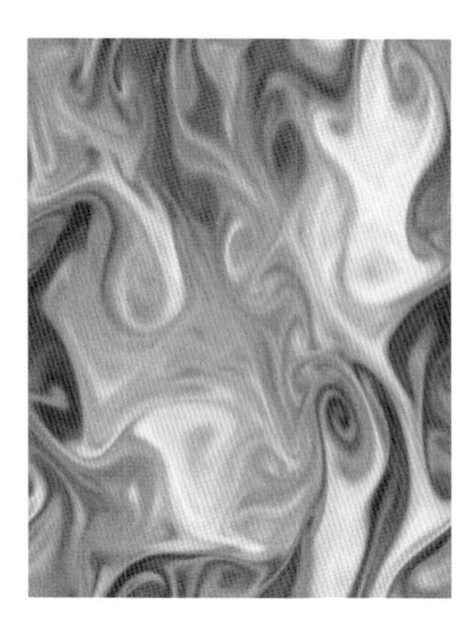

Figure 5.21 Computer simulation of a passive scalar in isotropic turbulence at a Schmidt number of 25. (Picture by G. Brethouwer and F. Nieuwstadt, University of Delft. Courtesy of efluids.com.)

There is now the issue of determining the contaminant microscale η_c. Here we must distinguish carefully between low and high Schmidt numbers, ν/α. We start with the latter. When the Schmidt number is greater than unity, $\nu > \alpha$, the diffusion of C is less effective than the diffusion of vorticity. We expect, therefore, that a fine-scale structure of C will develop in which $\eta_c < \eta$. In particular, it was suggested by Batchelor that very thin sheets or ribbons of contaminant will be teased out by eddies of size η and velocity v, that is, the Kolmogorov microscales. A similar process of sheet formation is considered in Section 5.3.3 where we show that the resulting sheet thickness is proportional to $(\text{diffusivity})^{1/2}$ and to the strain of the relevant eddies raised to the power of minus one half. Such sheets are evident in the computer simulation of isotropic turbulence shown in Figure 5.21. The figure shows the concentration distribution of a passive scalar at a Schmidt number of 25. The spirals indicate regions where the scalar sheets have become wrapped around turbulent vortices.

The thickness of the sheets in Figure 5.21 is of the order of $\alpha^{1/2}(\text{v}/\eta)^{-1/2}$, in line with the theoretical estimate above. Thus $\eta_c \sim \alpha^{1/2}(\text{v}/\eta)^{-1/2}$, or equivalently

$$\frac{(\text{v}/\eta)\eta_c^2}{\alpha} \sim 1 \quad (\nu > \alpha).$$

Since the Kolmogorov microscales are related by $\text{v}\eta/\nu \sim 1$, this yields

$$\eta_c \sim \left(\frac{\alpha}{\nu}\right)^{1/2} \eta \quad (\nu > \alpha)$$

which is consistent with the assertion that $\eta_c < \eta$. The range of scales between η_c and η is referred to as the *viscous-convective subrange*, for obvious reasons.

Now consider the case where the Schmidt number is less than unity, $\nu < \alpha$. Since the diffusion of C now is more effective than that of vorticity we have $\eta_c > \eta$. The range of scales between η and η_c represents the so-called *inertial-diffusive subrange*. The defining characteristic of η_c in such a flow is a little different to that of the high Schmidt number problem discussed above. Here we require that the Pe based on η_c is of the order of unity so that diffusion can compete with advection to smear out gradients in C. Thus we expect

$$\frac{\text{v}_c \eta_c}{\alpha} \sim 1 \quad (\nu < \alpha)$$

where v_c is the characteristic velocity fluctuation at the scale of η_c. From Obukhov's derivation of Kolmogorov's two-thirds law we have $\text{v}_r \sim (\varepsilon r)^{1/3}$ in the inertial subrange. This expression is also satisfied by the Kolmogorov microscales v and η: $\text{v} \sim (\varepsilon\eta)^{1/3}$. Thus, provided η_c

lies in the inertial or dissipative ranges, we might take $v_c \sim (\varepsilon\eta_c)^{1/3}$. It follows that,

$$\frac{(\varepsilon\eta_c)^{1/3}\eta_c}{\alpha} \sim 1 \quad (\nu < \alpha)$$

or equivalently,

$$\eta_c \sim \left(\frac{\alpha}{\nu}\right)^{3/4}\eta \quad (\nu < \alpha).$$

This information about how the microscales vary with Schmidt number is summarized below:

High Schmidt number	$\eta_c \sim \left(\frac{\alpha}{\nu}\right)^{1/2}\eta$ $(\eta_c < \eta)$	Inertial-convective range: $\eta \ll r \ll l$	Viscous-convective range: $\eta_c < r < \eta$
Small Schmidt number	$\eta_c \sim \left(\frac{\alpha}{\nu}\right)^{3/4}\eta$ $(\eta_c > \eta)$	Inertial-convective range: $\eta_c \ll r \ll l$	Inertial-diffusive range: $\eta < r < \eta_c$

There remains the question of how rapidly the scalar will mix. Let us assume that $l_c \leq l$. Then the two-thirds law yields the estimate

$$\langle C^2 \rangle \sim \varepsilon_c \varepsilon^{-1/3} l_c^{2/3}$$

which can be rearranged to give

$$\varepsilon_c \sim \langle C^2 \rangle \varepsilon^{1/3} l_c^{-2/3} \sim \langle C^2 \rangle u l^{-1/3} l_c^{-2/3}$$

since $\varepsilon \sim u^3/l$. Thus the scalar variance declines at a rate

$$\frac{d}{dt}\frac{1}{2}\langle C^2 \rangle = -\varepsilon_c \sim -\frac{u}{l^{1/3} l_c^{2/3}}\langle C^2 \rangle.$$

Perhaps the most important situation is when the same mechanism is used to create the turbulence and the scalar fluctuations, for example, a heated grid in a wind tunnel. In such a case $l_c \sim l$ and our equation simplifies to

$$\frac{d}{dt}\left\langle \frac{1}{2} C^2 \right\rangle = -\varepsilon_c \sim -\frac{u}{l}\langle C^2 \rangle.$$

Compare this with the equation for the destruction of energy in freely decaying turbulence,

$$\frac{d}{dt}\left\langle \frac{1}{2}\mathbf{u}^2 \right\rangle = -\varepsilon \sim -\frac{u}{l}\langle \mathbf{u}^2 \rangle$$

Evidently, for such cases, $\langle C^2 \rangle$ and $\mathbf{u}^2$ decay on the same timescale, which is the large-eddy turnover time. Let us now return to the more general situation where the two lengths l and l_c need not be equal.

Then our expressions above for the decay of $\langle C^2 \rangle$ and $\mathbf{u}^2$ can be combined for freely decaying turbulence to give,

$$\frac{d\ln\langle C^2\rangle}{d\ln(u^2)} = m\left(\frac{l}{l_c}\right)^{2/3}$$

where m is, presumably, a number of the order of unity.

This concludes our brief introduction to the mixing of passive scalars. We have left a great deal out, but those interested in filling in the gaps could do a lot worse than consult Tennekes and Lumley (1972), or Warhaft (2000) for a more recent review.

5.3 The intensification of vorticity and the stretching of material lines

5.3.1 Enstrophy production, the skewness factor, and scale invariance

In Section 5.1.2, we suggested that the energy cascade is maintained by vortex-line stretching which tends to pass energy down to smaller and smaller scales. That is, we imagine the vorticity field advecting itself in a chaotic manner, teasing out the vortex tubes and sheets into finer and finer structures (Plate 8). This process of *stretch and fold* produces, we claimed, a highly intermittent vorticity field. We now return to this issue. In particular we want to explore the intimate relationship between vortex stretching (i.e. enstrophy production) and the skewness factor. Our starting point is the vorticity equation

$$\frac{D\boldsymbol{\omega}}{Dt} = (\boldsymbol{\omega}\cdot\nabla)\mathbf{u} + \nu\nabla^2\boldsymbol{\omega} \tag{5.26}$$

from which we can obtain an equation for enstrophy,

$$\frac{D}{Dt}\left(\frac{\boldsymbol{\omega}^2}{2}\right) = \omega_i\omega_j S_{ij} - \nu(\nabla\times\boldsymbol{\omega})^2 + \nabla\cdot[\nu\boldsymbol{\omega}\times(\nabla\times\boldsymbol{\omega})]. \tag{5.27}$$

(Note that some authors define the enstrophy as $\boldsymbol{\omega}^2/2$ and others as $\boldsymbol{\omega}^2$.) For simplicity, let us consider the case of freely decaying turbulence (no mean velocity) which is statistically homogeneous. Then, on taking averages, the divergence on the left-hand side of (5.27) disappears since $\langle\sim\rangle$ commutes with $\nabla\cdot[\sim]$ and $\nabla\cdot[\langle\sim\rangle]=0$ in homogeneous turbulence. The same is true of the term $\mathbf{u}\cdot\nabla(\omega^2/2) = \nabla\cdot(\omega^2\mathbf{u}/2)$. So we are left with

$$\frac{\partial}{\partial t}\langle\boldsymbol{\omega}^2/2\rangle = \langle\omega_i\omega_j S_{ij}\rangle - \nu\langle(\nabla\times\boldsymbol{\omega})^2\rangle. \tag{5.28}$$

(Strictly, according to our convention, we should use a prime on $\boldsymbol{\omega}$ and S_{ij} to indicate that these are turbulent quantities. However, since there is no mean flow we may omit the prime without fear of ambiguity.) Equation (5.28) tells us that enstrophy can be created or destroyed by the strain field, and is destroyed by viscous forces. Now we have already seen that the flow as a whole evolves on the time-scale of the large eddies, l/u, with the small eddies continually adjusting to the local conditions of the large scales. Also, we have seen that, for fully developed turbulence, $\boldsymbol{\omega}$ and S_{ij} are concentrated at the Kolmogorov end of the spectrum (Figure 3.6). So we have, from (5.7) and (5.8)

$$\frac{\partial}{\partial t}\langle\boldsymbol{\omega}^2/2\rangle \sim \frac{u}{l}\left(\frac{\mathrm{v}}{\eta}\right)^2 \sim \frac{u}{l}\left(\frac{\varepsilon}{\nu}\right)$$

$$\langle\omega_i\omega_j S_{ij}\rangle \sim \frac{\mathrm{v}}{\eta}\left(\frac{\mathrm{v}}{\eta}\right)^2 \sim \frac{\mathrm{v}}{\eta}\left(\frac{\varepsilon}{\nu}\right)$$

$$\nu\langle(\nabla\times\boldsymbol{\omega})^2\rangle \sim \frac{\nu}{\eta^2}\left(\frac{\mathrm{v}}{\eta}\right)^2 \sim \frac{\mathrm{v}}{\eta}\left(\frac{\varepsilon}{\nu}\right).$$

When $\mathrm{Re} \gg 1$ we know that $u/l \ll \mathrm{v}/\eta$ and so the rate of change of $\boldsymbol{\omega}^2$ is relatively small by comparison with the other terms in (5.28). It follows that, for large Re, the two terms on the right-hand side must have similar magnitudes

$$\langle\omega_i\omega_j S_{ij}\rangle = \nu\langle(\nabla\times\boldsymbol{\omega})^2\rangle\left[1 + 0\left(\mathrm{Re}^{-1/2}\right)\right]. \tag{5.29}$$

This tells us two things.

(i) The stretching of vorticity outweighs compression of the vortex lines so that, as suggested in Section 5.1.2, the net effect of the strain field is to create enstrophy, that is, $\langle\omega_i\omega_j S_{ij}\rangle$ is positive.

(ii) There is an approximate balance between the production of enstrophy and viscous dissipation.

The implication is that vortex stretching transfers vorticity (and its associated energy) from large and intermediate scales down to the small, with dissipation concentrated at the small scales.

Now in Chapter 3, we introduced the skewness factor for the probability distribution of $\Delta\mathrm{v} = u_x(\mathbf{x} + r\hat{\mathbf{e}}_x) - u_x(\mathbf{x})$:

$$S = \frac{\langle[\Delta\mathrm{v}]^3\rangle}{\langle[\Delta\mathrm{v}]^2\rangle^{3/2}}. \tag{5.30}$$

In the limit of small r we have

$$S_0 = S(r \to 0) = \frac{\langle(\partial u_x/\partial x)^3\rangle}{\langle(\partial u_x\partial x)^2\rangle^{3/2}}. \tag{5.31}$$

We noted that for a Gaussian probability distribution S is zero, whereas in practice S_0 has a value of around -0.4 (in grid turbulence). It is no accident that S_0 is negative, since it may be shown that, for isotropic turbulence,

$$\langle \omega_i \omega_j S_{ij} \rangle = -\frac{7}{6\sqrt{15}} S_0 \langle \boldsymbol{\omega}^2 \rangle^{3/2} \tag{5.32}$$

(see Section 6.2.3 in Chapter 6). Evidently, a negative value of S_0 is needed in order to ensure that $\langle \omega_i \omega_j S_{ij} \rangle$ is positive, in line with (5.29). We shall return to the idea of skewness, and its physical interpretation, in Section 5.5.1 where we shall see that its partner, the flatness factor $\delta = \langle [\Delta v]^4 \rangle / \langle [\Delta v]^2 \rangle^2$, also tells us a great deal about the vorticity field. In particular, measurements of δ provide direct evidence that the vorticity field is highly intermittent, something we would expect in a cascade driven by vortex stretching and folding.

We conclude this section with an aside about the skewness distribution in the inertial subrange. Recall that, in the inertial range, $\langle [\Delta v]^2 \rangle = \beta \varepsilon^{2/3} r^{2/3}$ while $\langle [\Delta v]^3 \rangle = -\frac{4}{5}\varepsilon r$. It follows that

$$S = \frac{\langle [\Delta v]^3 \rangle}{\langle [\Delta v]^2 \rangle^{3/2}} = -\frac{4}{5}\beta^{-3/2} \quad (\eta \ll r \ll l) \tag{5.33}$$

which was introduced in Section 3.2.7. It is common to interpret the constancy of the skewness, S, across the inertial subrange as follows. The Euler equation exhibits a property known as *scale-invariance*. That is, suppose $\mathbf{u}(\mathbf{x}, t)$ represents one solution of the Euler equation. Then $\mathbf{u}^*(\mathbf{x}^*, t^*)$ is also a solution provided that,

$$\mathbf{u}^* = \lambda^n \mathbf{u}, \quad \mathbf{x}^* = \lambda \mathbf{x}, \quad t^* = \lambda^{1-n} t,$$

where λ is a scale factor and n a *scaling exponent*. Thus a solution at one scale has its counterparts at all other scales. This has led to some speculation that the statistical behaviour of $\mathbf{u}(\mathbf{x}, t)$ should be *scale invariant* in the inertial subrange, in the sense that

$$\langle [\Delta v]^p (\lambda r) \rangle = \lambda^{pn} \langle [\Delta v]^p (r) \rangle, \quad \eta \ll r \ll l.$$

What does this mean? Imagine that we performed a high Re computer simulation of turbulence, with Re large enough to yield many decades of the inertial subrange. (Such a simulation may not be practical at present, but that is not important for this discussion.) Suppose that we home in on some small region of the flow, say a cube of volume $(l/5)^3$, and plot the isovortical surfaces using some appropriate threshold. We might get a picture which looks something like that shown in Figure 5.22 or else Plate 8. We now zoom in and select a sequence of cubes of ever smaller size, say $(l/50)^3$, $(l/500)^3$, and $(l/5000)^3$. For each cube we plot the vorticity contours, taking care each time to rescale

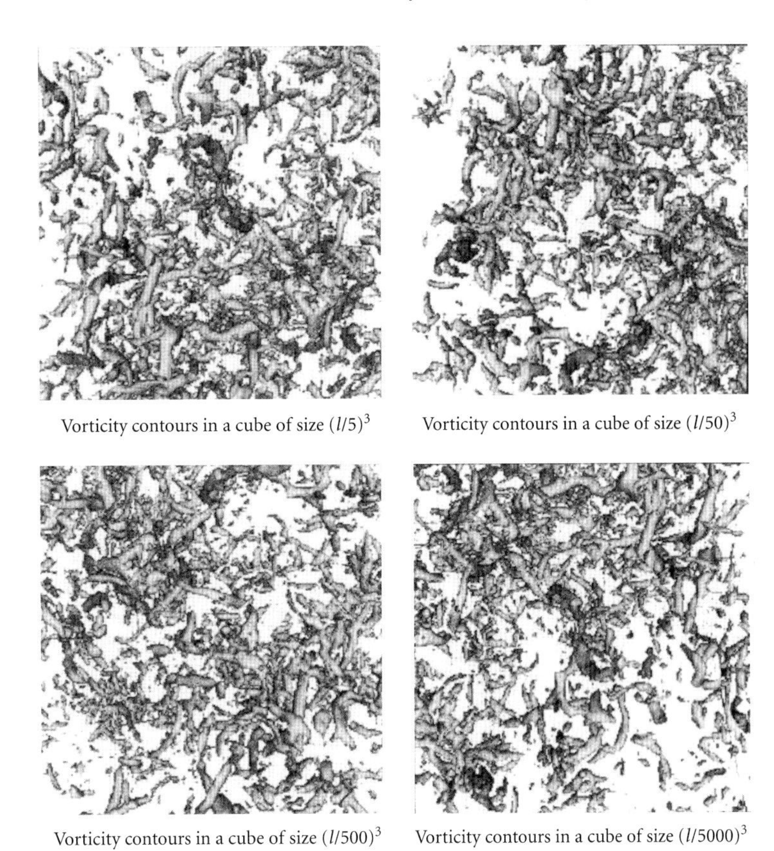

Vorticity contours in a cube of size $(l/5)^3$

Vorticity contours in a cube of size $(l/50)^3$

Vorticity contours in a cube of size $(l/500)^3$

Vorticity contours in a cube of size $(l/5000)^3$

Figure 5.22 Vorticity contours in a sequence of cubes of decreasing size taken from the same turbulent flow. If the turbulence is scale-invariant then the plots should all look qualitatively similar provided that the threshold for vorticity is reset each time in the appropriate manner.

the threshold for $|\boldsymbol{\omega}|$ by the appropriate factor, that is, 10^{1-n} for the $(l/50)$ cube, 100^{1-n} for the $(l/500)$ cube, and so on. Of course, for each cube the precise details of the vorticity field will look different. However, if the turbulence is scale-invariant, then the sequence of vorticity plots should look qualitatively similar, exhibiting the same statistical properties.

If scale invariance does hold then the scaling exponent, n, is set by the four-fifth's law,

$$\left\langle [\Delta v]^3(r) \right\rangle = -\tfrac{4}{5}\varepsilon r, \quad \eta \ll r \ll l.$$

That is

$$\left\langle [\Delta v]^3(\lambda r) \right\rangle = \lambda^{3n}\left\langle [\Delta v]^3(r) \right\rangle = \lambda^{3n}\left(-\tfrac{4}{5}\varepsilon r\right) = \lambda^{3n-1}\left(-\tfrac{4}{5}\varepsilon\lambda r\right) = -\tfrac{4}{5}\varepsilon\lambda r$$

from which $n = 1/3$. One immediate consequence of this scale invariance (if it is true) is that the skewness should be constant in the inertial subrange, as predicted by a combination of Kolmogorov's two-thirds and four-fifth's laws, and as observed in the experiments. (Try proving for yourself that scale invariance demands $S = \text{constant}$.) Thus the constancy of S, as represented by (5.33), is seen by some as providing evidence for scale invariance of the statistical properties of $\mathbf{u}$ in the inertial range.

Actually, Kolmogorov noticed that he could use these ideas to formulate an alternative derivation of his two-thirds law. Recall that the conventional procedure is to derive the two-thirds law from Kolmogorov's two similarity hypotheses, and then show that this demands that the skewness is constant. This is the line adopted in this book and indeed it was the argument which appeared in the first of Kolmogorov's 1941 papers. However, suppose instead that we start with the hypothesis that the skewness is constant in the inertial subrange (perhaps based on an argument of scale invariance), rather than with the two similarity hypotheses. One can then combine the constant skewness hypothesis with the four-fifth's law (which is exact provided the turbulence is homogeneous) to deduce the two-thirds law. In short, it is possible to turn the original argument on its head. This was the strategy proposed in Kolmogorov's second 1941 paper, though he made no direct appeal to scale invariance. It is, perhaps, a matter of taste as to which approach is to be preferred.

5.3.2 *Sheets or tubes?*

You asked, 'What is this transient pattern?'
If we tell the truth of it, it will be a long story;
It is a pattern that came up out of an ocean
And in a moment returned to the ocean's depth. (Omar Khayyam)

Equation (5.32) is an intriguing result. Since we need stretching, and not compression, for enstrophy generation we might have expected a positive value of $\langle \omega_i \omega_j S_{ij} \rangle$ to be associated with positive $\langle (\partial u_x / \partial x)^3 \rangle$. Evidently this is not the case. In fact, quite the reverse: we need $S_0 < 0$ for enstrophy generation! The reason for this is rather subtle. The first point to note is that a variable with zero mean and negative skewness is characterized by the fact that positive excursions from zero are long and shallow, while negative excursions are less frequent but deeper. Thus $\partial u_x / \partial x$ is positive much of the time but is subject to large negative fluctuations. The physical interpretation of these negative (compressive) excursions may be explained as follows. It turns out, though it is not obvious, that S_0 is proportional to, and takes the same sign as, $\langle abc \rangle$ where a, b, and c are the three principal rates of strain at any point and at any time in the flow (see Section 5.3.6). Now incompressibility requires $a + b + c = 0$, so if we order the rates of strain according to $a > b > c$ then a will be positive, c will be negative, and b will be positive if $|c| > |a|$ or negative if $|a| > |c|$. It follows that the intermediate principal strain rate, b, has the opposite sign to that of abc. The fact that S_0 is negative tells us that, on average, we encounter a situation where $c < 0$ and a, $b > 0$, that is, we have one large compressive strain rate accompanied by two weaker extensional rates. So material is stretched in the a–b plane and compressed in the direction

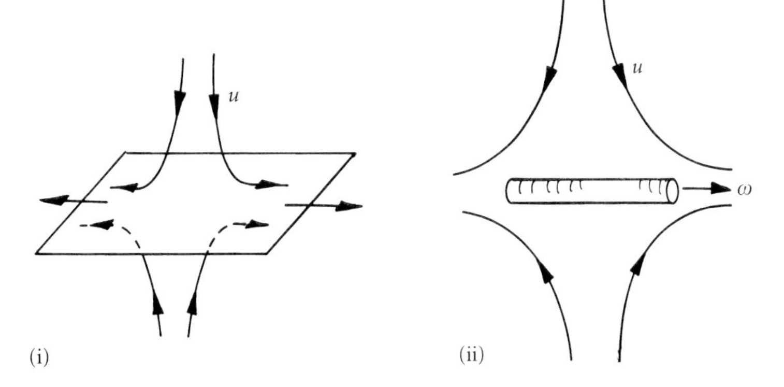

Figure 5.23 (i) A negative skewness requires one compressive strain rate and two extensional ones. This tends to lead to the formation of vortex sheets. (ii) A positive skewness requires one extensional strain rate and two compressive ones. This tends to lead to the formation of vortex tubes. In isotropic turbulence S_0 is negative, favouring the formation of sheets.

of 'c'. Such a situation, which we might call biaxial strain, tends to be associated with the formation of vortex sheets (Batchelor, 1953, Chapter 7, Section 4). That is, vorticity is compressed into sheets by the large compressive strain while the sheets themselves are stretched by the two weaker extensional strains. This has led to one particular cartoon of turbulence in which the strain field is continually teasing out vortex sheets and these sheets then roll up via a Kelvin-Helmholtz instability to produce vortex tubes. In fact, this classical view of the vorticity field is receiving renewed interest as a result of recent computer simulations which suggest that sheets are indeed formed ahead of the tubes, and that $\langle a, b, c\rangle \sim (3,1,-4)$ (see Section 7.3.1).

We note in passing that a positive value of S_0, and hence a negative value of b, would lead to a situation in which we have only one large extensional rate of strain accompanied by two weaker compressive strain rates (axial strain). This would tend to favour the formation of vortex tubes (Batchelor, 1953), with the large extensional strain stretching a tube along its axis and the weaker compressive strains compressing the sides of the tube. The two situations are shown in Figure 5.23.

So it would seem that the statistical bias towards a negative skewness favours the formation of sheets rather than tubes. However, the situation is not that simple. When a vortex tube is being stretched it has its own strain field and this adds to the external, imposed strain, tending to convert axial strain to biaxial strain (Moffatt et al. 1994). On the other hand, as noted above, recent computer simulations support the view that the dominant process is the formation of sheets, and that those tubes which are observed are simply the debris resulting from the (Kelvin–Helmholtz) disintegration of the sheets. So should we picture the vorticity field as a seething mass of spaghetti or lasagne? The relative importance of vortex sheets and vortex tubes is still a matter of debate and it is likely to remain that way until higher-Re computer simulations can be performed. In the meantime perhaps it is prudent to consider both sheets and tubes as potentially important.

Whatever the generic structure of turbulent eddies, there is no doubt that vortex stretching underpins the energy cascade. Both the vortex tube and vortex sheet shown in Figure 5.23 have their vorticity aligned with a positive rate of strain, so that the vortex lines are stretched. Of course, in isotropic turbulence there is always some compression of vortex lines as well as stretching. However, the stretching wins out in terms of enstrophy production.[5]

Example 5.11

Consider a small eddy (a localized blob of vorticity) which falls prey to the stain field of a larger adjacent eddy. The linear impulse of the smaller eddy is

$$\mathbf{L}_e = \frac{1}{2}\int_{V_e} \mathbf{x} \times \boldsymbol{\omega}_e \, dV$$

where $\boldsymbol{\omega}_e$ is the vorticity of the smaller eddy and V_e is the volume occupied by $\boldsymbol{\omega}_e$. Suppose that the strain field of the larger, remote eddy may be considered as quasi-steady and uniform on the scale of the small eddy, and let $\hat{\mathbf{u}}$ be the (locally) irrotational velocity field of the larger, remote eddy. Let us take x, y, and z to be aligned with the principal axes of strain of $\hat{\mathbf{u}}$ and a, b, and c to be the principal rates of strain. (Conservation of mass requires $a + b + c = 0$.) In such a case we have,

$$\frac{d}{dt}[L_{ex}, L_{ey}, L_{ez}] = -[aL_{ex}, bL_{ey}, cL_{ez}]$$

(see Example 5.7 in Section 5.1.3). It seems that, when we have biaxial strain ($c < 0$, and $a, b > 0$), L_{ez} grows but L_{ex} and L_{ey} decay. Show that this is compatible with the vorticity of the eddy being squashed into a pancake-like structure. On the other hand, when we have axial strain ($a > 0$, and $b, c < 0$), we obtain a growth of L_{ey} and L_{ez} and a decay of L_{ex}. Show that this is consistent with the vorticity being teased out into filaments. Finally, use the Schartz inequality to estimate the enstrophy change.

5.3.3 Examples of concentrated vortex sheets and tubes

There are a number of simple (but informative) mathematical cartoons which illustrate how vortex sheets and tubes can be intensified

[5] Actually, it turns out that, on average, the vorticity tends to be aligned with the intermediate principal rate of strain, b, which is positive most of the time. Nevertheless most of the vortex stretching is associated with the largest positive principal strain, a. That is, although the vorticity is infrequently aligned with a, on those occasions when it is, a great deal of stretching occurs (see Tsinober, 2001, or Section 5.3.6).

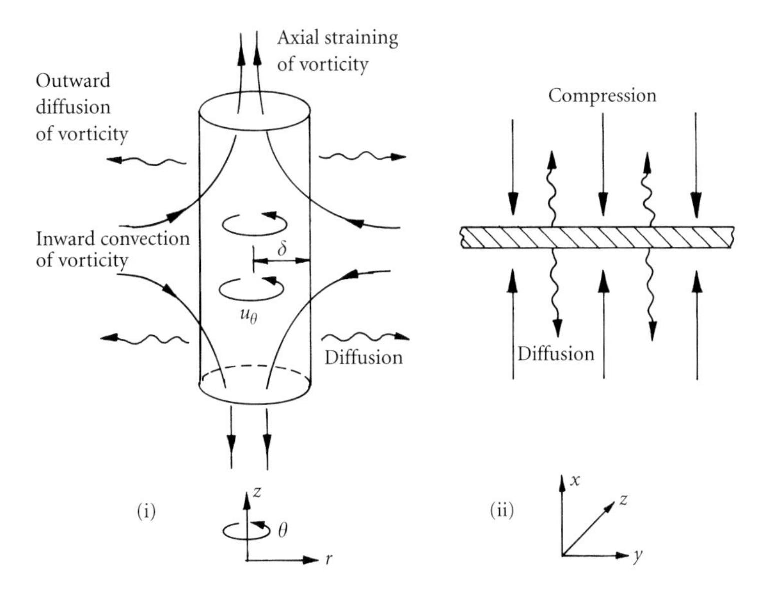

Figure 5.24 Straining of vorticity: (i) Burgers' vortex; (ii) formation of a vortex sheet.

by straining. The most famous one is, perhaps, Burgers' vortex. Suppose we have a vortex tube,

$$\boldsymbol{\omega} = \frac{\alpha\Gamma_0}{4\pi\nu}\exp[-r^2/\delta^2]\hat{\mathbf{e}}_z$$

sitting in an axisymmetric, irrotational strain field, $u_r = -\frac{1}{2}\alpha r,\ u_z = \alpha z$. Here Γ_0 is the strength of the vortex tube, $\Gamma_0 = \int_0^\infty \omega 2\pi r\,dr$, and $\delta = (4\nu/\alpha)^{1/2}$ is its characteristic width (Figure 5.24). It is readily confirmed that this represents an exact solution of the steady Navier–Stokes equation. The irrotational motion sweeps vorticity radially inward while simultaneously straining the vortex tube in the axial direction. These processes exactly counterbalance the tendency for vorticity to diffuse radially outward. In a turbulent flow the strain rate, α, might be associated with the large-scale eddies, $\alpha \sim u/l$, in which case

$$\delta \sim (ul/\nu)^{-1/2}l \sim \mathrm{Re}^{-1/2}l \sim \lambda.$$

The quantity $\lambda \sim \mathrm{Re}^{-1/2}l$, which lies somewhere between η and l, is called the *Taylor microscale*. On the other hand, the strain might be associated with somewhat smaller eddies, in which case α is larger and δ smaller. In fact, if we believe in the phenomenology of the energy cascade, we would expect a range of tube sizes. So, if anything resembling a Burgers' vortex is to be found in real turbulent flows then we might expect to see vortex tubes of diameters ranging from the Taylor microscale down to η. It is interesting that the direct numerical simulations of turbulence show exactly that (see Plate 8), although just how important these tubes are for the energy cascade is still a matter of debate. (We will return to this issue in Chapter 7.)

It is left to an exercise for the reader to confirm that, in the limit of small viscosity, the global viscous dissipation produced by Burgers' vortex is finite, independent of ν, and equal to $\alpha\Gamma_0^2/(8\pi)$ per unit length of the vortex (see Exercise 5). This is reminiscent of the dissipation in a turbulent flow where, according to Kolmogorov's theory, we may take ε and ν as independent parameters. Indeed it was precisely this feature of the vortex which Burgers emphasized in his original paper (Burgers, 1948).

Burgers' vortex may be generalized to unsteady flows in an obvious way. Let us assume that the irrotational strain field is $u_r = -\frac{1}{2}\alpha(t)r, u_z = \alpha(t)z$ (i.e. we have axial stain), and look for an unsteady, axisymmetric, tubular vortex of the form

$$\boldsymbol{\omega} = \frac{\Gamma_0}{\pi l^2}\exp[-r^2/l^2]\ \hat{\mathbf{e}}_z, \quad l = l(t).$$

Then it is readily confirmed that we have an exact solution of the unsteady Navier–Stokes equation provided that the vortex radius, $l(t)$, satisfies

$$\frac{dl^2}{dt} + \alpha(t)l^2 = 4\nu.$$

Several special cases immediately come to mind. If α is constant and $t \to \infty$ then we recover the steady solution given above, with $l^2 = 4\nu/\alpha$. Here the inward advection of vorticity exactly balances the outward diffusion. If $\alpha = 0$, on the other hand, we find $l^2 = l_0^2 + 4\nu t$ (see Exercise 3). In this case there is no inward advection of vorticity and the vortex core grows by diffusion. The general solution for arbitrary but constant α is

$$l^2 = l_0^2 \mathrm{e}^{-\alpha t} + (4\nu/\alpha)\left[1 - \mathrm{e}^{-\alpha t}\right].$$

If the initial vortex radius, l_0, exceeds the steady-state value of $[4\nu/\alpha]^{1/2}$ then the vortex core shrinks due to vortex stretching and approaches the steady-state value on a timescale of α^{-1}. On the other hand, if l_0 is less than $[4\nu/\alpha]^{1/2}$ then the vortex core grows by diffusion until the steady-state value is reached. Of course, it is the first of these two options which is of most relevance to turbulence.

When l greatly exceeds $[4\nu/\alpha]^{1/2}$ the viscous term can be neglected and the general solution for l, allowing for a time-dependent strain, is clearly,

$$l^2 = l_0^2 \exp\left[-\int_0^t \alpha(t)dt\right].$$

Thus the vortex core shrinks exponentially fast and it is readily confirmed that the rate of growth of kinetic energy per unit length of the vortex is $\alpha\Gamma_0^2/(8\pi)$.

As a second example, we consider a vortex sheet embedded in an externally imposed biaxial strain field. In particular, consider the vortex sheet

$$\boldsymbol{\omega} = \omega_0 \exp\left[-x^2/\delta^2\right]\hat{\mathbf{e}}_z, \quad \delta = (2\nu/\alpha)^{1/2}$$

sitting in the irrotational flow, $u_x = -\alpha x$, $u_z = \alpha z$. Again, when α is constant, it is readily confirmed that this represents a steady solution of the Navier–Stokes equation in which the tendency of vorticity to diffuse outward is countered by the inward advection and (compressive) straining of the vorticity field (Figure 5.24). Of course, this sheet is prone to the instabilities of the Kelvin–Helmholtz type and so it will not survive for long. Rather, it will rapidly disintegrate into a set of vortex tubes.

Example 5.12 Formation of a vortex sheet
Confirm that the example above of a vortex sheet sitting in a biaxial strain field generalises, in cases where $\delta(0) \neq (2\nu/\alpha)^{1/2}$, to

$$\boldsymbol{\omega} = (u_0/l)\exp[-x^2/l^2]\hat{\mathbf{e}}_z, \quad u_0 = \text{constant}, \quad l = l(t)$$

where l is governed by

$$\frac{dl^2}{dt} + 2\alpha l^2 = 4\nu.$$

Find the solution for $l(t)$ and show that, if $l(0)$ exceeds $(2\nu/\alpha)^{1/2}$, then the sheet is compressed until $l = (2\nu/\alpha)^{1/2}$, whereas $l(0)$ less than $(2\nu/\alpha)^{1/2}$ leads to a thickening of the sheet by diffusion. In either case, l tends to $(2\nu/\alpha)^{1/2}$ on a timescale of α^{-1}.

5.3.4 *Are there singularities in the vorticity field?*

It seems likely that, since any real fluid possesses a finite amount of viscosity, singularities cannot develop in the vorticity field. This is illustrated in Example 5.12 where the strained vortex sheet always tends to a thickness of $(2\nu/\alpha)^{1/2}$. If $l(0)$ happens to be less than $(2\nu/\alpha)^{1/2}$, then the sheet thickens by diffusion. The same behaviour is exhibited by Burgers' vortex. Thus it is probable that a small amount of viscosity is all that is needed to unlock any potential singularity in the vorticity field.

There remains the more academic question as to whether or not a singularity, defined, say, as a blow-up of the enstrophy, can develop in an inviscid fluid. This issue has been the subject of much discussion and some controversy. There is no doubt that, given enough time, singularities can begin to emerge in an inviscid fluid. Consider, once again, the example of a strained vortex sheet. It is readily confirmed that the solution for $\nu = 0$ is

$$l = l(0)\exp[-\alpha t]$$

$$\frac{D}{Dt}\frac{1}{2}\omega^2 = \alpha\omega^2$$

so that ω^2 grows exponentially. A more important (and controversial) question is whether or not inviscid singularities can emerge within a finite period of time: so-called finite-time singularities. There was once a widespread belief that this is indeed the case, though the hunt for such a singularity has proved to be more than a little problematic. Incidental (though not very convincing) evidence in favour of finite-time singularities in an inviscid fluid includes the following:

(i) Some models of Richardson's energy cascade suggest that the energy transfer time from l down to η is of the order l/u, which remains finite as ν tends to zero (Frisch, 1995). Moreover, η scales as $\nu^{3/4}$ while $\boldsymbol{\omega}^2$ scales as ν^{-1} as ν becomes small. This suggests that, for a vanishingly small viscosity, energy piles up at a distribution of near singular points in a finite time.

(ii) Certain turbulence closure models, such as the quasi-normal closure scheme, predict finite-time singularities for $\nu = 0$.

One motivation for establishing whether or not such singularities can form is the hypothesis that, if they do exist, then similar structures should appear in very high Re turbulence. This is, however, an assumption. In any event, heuristic arguments in favour of finite-time singularities have appeared in a number of texts. Typically, they proceed as follows. We start with the vorticity equation for a real (viscous) fluid and try and estimate the rate of generation of enstrophy, perhaps with the help of some plausible hypothesis. The next step, which is completely unjustified, and almost certainly wrong, is to assume that this rate of generation is the same in an inviscid fluid. Although the conclusions from such an analysis carry little weight, it is worth going through the details because they bring out a number of interesting points. To keep the analysis simple we shall restrict ourselves to isotropic turbulence.

Perhaps the two most common heuristic arguments are the so-called *quasi-normal* and *constant skewness* models. In either model the starting point is

$$\frac{D\boldsymbol{\omega}}{Dt} = (\boldsymbol{\omega}\cdot\nabla)\mathbf{u} + \nu\nabla^2\boldsymbol{\omega}$$

from which,

$$\frac{D}{Dt}\left(\frac{\boldsymbol{\omega}^2}{2}\right) = [\omega_i\omega_j S_{ij}] + \nu(\sim).$$

Here the last term on the right simply indicates a viscous contribution, the precise details of which are not important here. In the constant skewness model we use (5.32) to rewrite this as

$$\frac{d}{dt}\langle\boldsymbol{\omega}^2/2\rangle = \langle\omega_i\omega_j S_{ij}\rangle + \nu(\sim) = -\frac{7}{6\sqrt{15}}S_0\langle\boldsymbol{\omega}^2\rangle^{3/2} + \nu(\sim).$$

Now in fully-developed turbulence at high Re it is observed that the skewness, S_0, of the velocity derivative $\partial u_x / \partial x$ is more or less constant and equal to -0.4. *Suppose* that this were also true for an inviscid fluid. Then we have, for an inviscid flow,

$$\frac{d}{dt}\langle \boldsymbol{\omega}^2/2 \rangle = -\frac{7}{6\sqrt{15}} S_0 \langle \boldsymbol{\omega}^2 \rangle^{3/2}$$

where S_0 is now a negative constant. This integrates to give

$$\left\langle \tfrac{1}{2}\boldsymbol{\omega}^2 \right\rangle \sim (t_0 - t)^{-2}$$

where t_0 is a constant proportional $\left(\boldsymbol{\omega}_0^2\right)^{-1/2}$. This predicts a blow-up of the enstrophy at time t_0. Of course, the problem with this argument is that we have absolutely no right to assume that the skewness exhibits the same behaviour in both viscous and inviscid flows. Let us turn, therefore, to the quasi-normal model and see if it fairs any better. (We shall see that it does not.)

In the quasi-normal model it is assumed that, as far as calculating the fourth-order correlations is concerned, we may take the velocity field to have a Gaussian probability distribution. This is an approximation for which, at one time, it was thought that there was a certain amount of experimental support. This scheme is discussed in detail in Chapter 8, where its deficiencies are highlighted. For the present purposes, however, the important point to note is that the quasi-normal (QN) model leads to a second-order differential equation for $\langle \boldsymbol{\omega}^2 \rangle (t)$. So, before looking at the QN model in detail, let us see if we can get a second-order equation for the evolution of $\boldsymbol{\omega}^2$ in a viscous fluid using simple physical arguments. If we go back to

$$\frac{D}{Dt}\left(\frac{\boldsymbol{\omega}^2}{2}\right) = [\omega_i \omega_j S_{ij}] + \nu(\sim)$$

and differentiate once, we find

$$\frac{D^2}{Dt^2}\left(\frac{\boldsymbol{\omega}^2}{2}\right) = 2[\omega_i S_{ij}]^2 + \omega_i \omega_j \frac{D}{Dt} S_{ij} + \nu(\sim).$$

The problem now is to estimate the terms on the right. One useful observation in this respect is that, for fully-developed isotropic turbulence, it is observed that,

$$\left\langle \omega_i \omega_j \frac{D}{Dt} S_{ij} \right\rangle \approx -\frac{4}{3}\left\langle [\omega_i S_{ij}]^2 \right\rangle + \nu(\sim)$$

(see, for example, Tsinober, 2001). Thus we have

$$\frac{d^2}{dt^2}\left\langle \frac{\boldsymbol{\omega}^2}{2} \right\rangle \approx \frac{2}{3}\left\langle [\omega_i S_{ij}]^2 \right\rangle + \nu(\sim) \quad \text{(empirical)}.$$

Interestingly, it turns out that the QN model leads to an equation of similar form, though without the factor of 2/3 on the right-hand side:

$$\frac{d^2}{dt^2}\left\langle\frac{\boldsymbol{\omega}^2}{2}\right\rangle = \left\langle[\omega_i S_{ij}]^2\right\rangle + \nu(\sim) \quad \text{(quasi-normal model only).}$$

The question now is: can we estimate the size of $[\omega_i S_{ij}]^2$? One might be tempted to take it to be of the order of $\langle\boldsymbol{\omega}^2\rangle^2$ and indeed this is precisely what the QN closure scheme predicts. In fact, we shall see in Chapter 8 that the QN model suggests,

$$\frac{d^2}{dt^2}\left\langle\frac{1}{2}\boldsymbol{\omega}^2\right\rangle = \frac{2}{3}\left[\left\langle\frac{1}{2}\boldsymbol{\omega}^2\right\rangle\right]^2 + \nu(\sim) \quad \text{(quasi-normal model only).}$$

Actually this is consistent with the observation that the skewness of $\partial u_x/\partial x$ remains finite and constant in fully-developed, high-Re turbulence. That is, if we take the skewness S_0 to be constant then it is readily confirmed that the derivative of the enstrophy equation yields

$$\frac{d^2}{dt^2}\left\langle\frac{1}{2}\boldsymbol{\omega}^2\right\rangle = \frac{49}{45}S_0^2\left[\left\langle\frac{1}{2}\boldsymbol{\omega}^2\right\rangle\right]^2 + \nu(\sim) \quad \text{(constant skewness model).}$$

So we see that our quasi-normal equation is a form of constant skewness model in which the skewness is given a particular value.[6] Note that, for fully-developed, high-Re turbulence, eqn (5.29) tells us that the two terms on the right-hand side of this equation are very nearly equal and opposite. It follows that the left-hand side is close to zero, or to be more precise, is of the order of $(ul/\nu)^{-1}$ times one of the terms on the right.

So far we have taken ν to be small but finite. Let us now set the viscosity to zero while assuming that the form of the inertial terms in our model equations remain the same. Of course, this is a dangerous strategy as the experimental evidence which tentatively supports the QN model (or our assumption that S_0 is constant) comes from *real* (i.e. viscous) flows. Still, let us see where this leads. We have, from the QN model,

$$\frac{d^2}{dt^2}\left\langle\frac{1}{2}\boldsymbol{\omega}^2\right\rangle = \frac{2}{3}\left[\left\langle\frac{1}{2}\boldsymbol{\omega}^2\right\rangle\right]^2 \quad \text{(quasi-normal model only).}$$

One integration yields

$$\frac{d}{dt}\left\langle\frac{1}{2}\boldsymbol{\omega}^2\right\rangle = \frac{2}{3}\left[\left\langle\frac{1}{2}\boldsymbol{\omega}^2\right\rangle^3 - \left\langle\frac{1}{2}\boldsymbol{\omega}_0^2\right\rangle^3\right]^{1/2}$$

[6] By comparing these two equations it is readily confirmed that the skewness in the QN model is equal to -0.782. In practice, however, the skewness for high-Re turbulence is closer to -0.4, indicating that the QN model significantly overestimates the rate of generation of enstrophy, a failure of the model which was a source of concern for Lighthill as far back as 1957 (unpublished manuscript).

where, for convenience, the initial value of $d\langle \boldsymbol{\omega}^2 \rangle / dt$ has been set equal to zero.[7] On integrating once more we find,

$$\left\langle \tfrac{1}{2}\boldsymbol{\omega}^2 \right\rangle \rightarrow 9/(t_0 - t)^2$$

for $t \rightarrow t_0$, where t_0 is a timescale of the order of $\langle \boldsymbol{\omega}_0^2 \rangle^{-1/2}$. (This may be confirmed by neglecting $\langle \boldsymbol{\omega}_0^2 \rangle$ in the differential equation above during the second integration.) Thus the quasi-normal closure scheme predicts a finite-time blow-up of the enstrophy, just like the constant skewness model (of which it is a special case). However, we shall see later that there are good reasons for *not* believing in the quasi-normal scheme. (It leads to a number of erroneous predictions.) Perhaps a better approach is to go back to the approximate expression,

$$\frac{d^2}{dt^2}\left\langle \frac{\boldsymbol{\omega}^2}{2} \right\rangle \approx \frac{2}{3}\left\langle [\omega_i S_{ij}]^2 \right\rangle + \nu(\sim).$$

If we compare this with the QN equation we see that it is implicit in the QN model that the rate of strain is of the order of the local vorticity. It is this coupling between S_{ij} and $\boldsymbol{\omega}$ which lies behind the predicted finite-time singularity. However, the examples of strained vortex sheets and tubes given in the previous section suggest that an alternative estimate of the right-hand side might be

$$\frac{d^2}{dt^2}\left\langle \frac{1}{2}\boldsymbol{\omega}^2 \right\rangle \sim \alpha^2 \left\langle \frac{1}{2}\boldsymbol{\omega}^2 \right\rangle$$

where α is a strain rate which is characteristic of the eddies somewhat larger than the enstrophy containing eddies. (This is the picture which emerges from the simulations of Laval et al., 2001, and of Jimenez and Wray, 1998—see section 7.3.2.) If α is decoupled from $\boldsymbol{\omega}^2$ this yields a benign, exponential growth in enstrophy, rather than the finite-time singularity predicted by the QN model.

So it seems that various plausible, but heuristic, estimates of $\langle [\omega_i S_{ij}]^2 \rangle$ lead to very different predictions for the fate of the enstrophy. Everything hinges on the assumed magnitude of the strain field responsible for teasing out the small-scale vortex tubes. Clearly, heuristic models of the type discussed above are very unsatisfactory. In particular, they do not build in the sort of detailed dynamic information needed to distinguish reliably between an exponential or a finite-time singularity. The issue can be resolved only if more rigorous arguments are deployed, or else exact solutions can be found which exhibit singularities.

There have been many attempts to construct simple model problems which exhibit finite-time singularities. An early contender was

[7] Gaussian initial conditions, for example, would give $d\langle \boldsymbol{\omega}^2 \rangle / dt = 0$ at $t = 0$.

the self-centrifuging vortex blob shown in Figure 5.6. Consider the final sketch in Figure 5.6. Pumir and Siggia (1992) suggested that the poloidal vortex sheet, which is represented by lines of constant $\Gamma = ru_\theta$, is unstable. That is, a wrinkling of the sheet leads to enhanced axial gradients in Γ which, in turn, act as a source of azimuthal vorticity in accordance with (5.10). This then augments the local strain field and advects Γ (i.e. the sheet), leading to additional wrinkling and thinning of the vortex sheet. Pumir and Siggia (1992) suggested that this might lead to a runaway situation, giving rise to a singularity. However, it is now thought probable that such a singularity emerges exponentially in time, rather than the algebraic growth needed for a finite-time blow up of the enstrophy.

Other model problems have involved so-called Leray self-similar flows (see Example 5.7 at the end of this chapter) and/or entangled vortex tubes, but these too have proved problematic. The failure of these simple models has shifted the emphasis towards the numerical simulation of more complex flows. However such simulations are difficult to perform since they require the ability to track the emergence of a latent singularity. The results of these computations are, as yet, inclusive, but the interested reader may consult Bajer and Moffatt (2002) or Pelz (2001).

Example 5.13
Combine (5.28) and (5.32) to show that, for an inviscid fluid, the enstrophy evolves according to

$$\frac{d}{dt}\langle \boldsymbol{\omega}^2/2\rangle = \langle \omega_i \omega_j S_{ij}\rangle = -\frac{7}{6\sqrt{15}} S_0 \langle \boldsymbol{\omega}^2\rangle^{3/2}.$$

Integrate this equation on the assumption that S_0 is time dependant and negative. Show that, for large times, the magnitude of S_0 must decline faster than t^{-1} if a finite-time blow-up of enstrophy is to be avoided.

5.3.5 The stretching of material line elements

Now at high Re the vortex lines are frozen into the fluid, except at the Kolmogorov microscales where diffusion takes place. Since the vortex lines are continually stretched, increasing the enstrophy, we might anticipate that, on average, material line elements are also stretched in a turbulent flow. The experimental evidence suggest that this is indeed true, although a formal proof that this is so turns out to be difficult to formulate!

A hint as to how a proof might proceed is provided by the following simple example. Consider a small element of fluid which, at $t = 0$, is spherical, of radius δ. A short time later the sphere, which has been subjected to an infinitesimal amount of strain, is deformed into a triaxial ellipsoid whose principal axes align with the axes of principal strain. Let γ_1, γ_2, and γ_3 be the three principal strains, that is, the eigenvalues of the strain tensor. Then the length of the three principal axes of the ellipsoid are $w_1\delta = (1+\gamma_1)\delta$, $w_2\delta = (1+\gamma_2)\delta$, and $w_3\delta = (1+\gamma_3)\delta$, where w_i are scaling factors for the axes. Since volume is conserved during the deformation we have $w_1 w_2 w_3 = 1$. Moreover, an arithmetic mean is always greater than, or equal to, a geometric mean and so we may conclude that

$$\tfrac{1}{3}(w_1 + w_2 + w_3) \geq (w_1 w_2 w_3)^{1/3} = 1.$$

Clearly some fluid elements have contracted during the deformation while others have extended. What is most important, however, is the expected behaviour of a randomly orientated line element which spans the sphere at $t = 0$. If we take the average length of the three principal axes of our ellipse as indicative of the expected length, after straining, of our randomly orientated line element then it will, on average, grow in a turbulent flow. This is the key idea which underpins many attempts to prove that, on average, line elements grow rather than contract. However, there are a number of gaps to be filled in. For example, we need to justify equating the expected extension of a random line element to the arithmetic mean of the extensions of the principal axes of strain. Also, we would like to consider what happens to our line element over a finite period of time, implying that we must consider finite strains. All-in-all, there is a lot more to do.

A more formal proof proceeds along the following lines. Let $\mathbf{x}(\mathbf{a}, t)$ be a particle path in the turbulence with $\mathbf{x}(\mathbf{a}, 0) = \mathbf{a}$, that is, $\mathbf{a}$ is the initial position. Suppose that we release a sequence of particles from the same point, $\mathbf{a}$, in a statistically steady flow. Then $\mathbf{a}$ is non-random although the path of the nth particle, $\mathbf{x}_n(\mathbf{a}, t)$, is. Now consider a short line element $\delta\mathbf{a}$ located at $\mathbf{a}$ at $t = 0$. After a time t the two ends of the element will have moved in rather different ways (Figure 5.25) and so for each realization

$$\delta x_i(t) = \frac{\partial x_i}{\partial a_j}\delta a_j$$

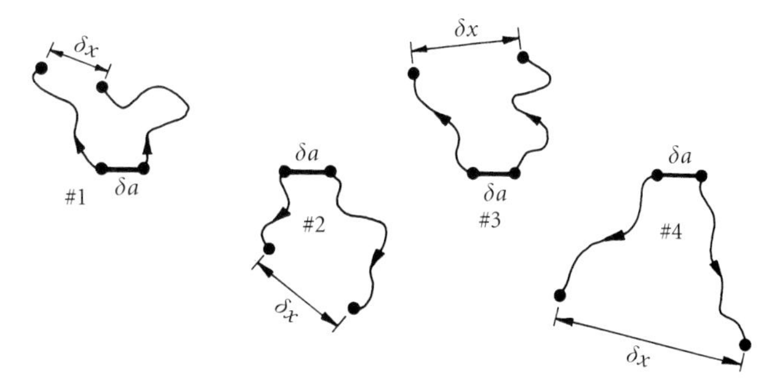

Figure 5.25 Stretching of line elements. The histories of finite, rather than infinitesimal, line elements are shown for clarity.

where $\delta \mathbf{x}$ is the separation of the two ends after time t. The tensor $\partial x_i / \partial a_j$, which is called the deformation tensor, has the volume-conserving property that $\det(\partial x_i / \partial a_j) = 1$ (see Sokolnikoff, 1946). We now repeat the experiment many times with identical values of $\delta \mathbf{a}$ at $t = 0$. Ensemble averaging the results yields,

$$\left\langle (\delta \mathbf{x})^2 \right\rangle = \left\langle \frac{\partial x_i}{\partial a_j} \frac{\partial x_i}{\partial a_k} \right\rangle \delta a_j \, \delta a_k. \tag{5.34}$$

Note that the terms δa_j and δa_k may be taken outside the averaging operator since $\delta \mathbf{a}$ is non-random (i.e. the same for all realizations).

We now introduce the symmetric matrix $A_{jk} = (\partial x_i / \partial a_j)(\partial x_i / \partial a_k)$. It has three real eigenvalues, λ_1, λ_2 and λ_3 and matrix theory tells us that

$$\lambda_1 + \lambda_2 + \lambda_3 = \text{Trace}\,(\mathbf{A})$$
$$\lambda_1 \cdot \lambda_2 \cdot \lambda_3 = \det\,(\mathbf{A}).$$

Moreover, since $\det(\partial x_i / \partial a_j) = 1$, we have $\det(\mathbf{A}) = 1$. It follows that, for each realization,

$$\tfrac{1}{3}(\lambda_1 + \lambda_2 + \lambda_3) \geq (\lambda_1 \lambda_2 \lambda_3)^{1/3} = 1. \tag{5.35}$$

Now suppose that the turbulence is isotropic. Then $\langle A_{jk} \rangle$, which is a statistical property of the flow, must also be isotropic and it follows that $\langle A_{jk} \rangle$ is diagonal:

$$\left\langle \frac{\partial x_i}{\partial a_j} \frac{\partial x_i}{\partial a_k} \right\rangle = \alpha(t)\delta_{jk}, \quad 3\alpha = \langle \text{Trace}(\mathbf{A}) \rangle = \langle (\lambda_1 + \lambda_2 + \lambda_3) \rangle.$$

However, for each realization the eigenvalues of A_{ij} satisfy (5.35) and so, after ensemble averaging, we find that $\alpha \geq 1$. Note that the equality sign is extremely unlikely to occur as it would require the equality sign in (5.35) for each realization which, in turn, requires the λ_i's are all equal to unity in each experiment. Equation (5.34) now reduces to

$$\left\langle (\delta \mathbf{x})^2 \right\rangle = \alpha(t)(\delta \mathbf{a})^2, \quad \alpha \geq 1. \tag{5.36}$$

This is a promising result as it indicates that, on average, the particles are further apart at time t than at $t = 0$. A slightly different argument yields the more general result $\langle |\delta \mathbf{x}|^P \rangle = \alpha(t) |\delta \mathbf{a}|^P$, $\alpha \geq 1$ (see Monin and Yaglom, 1975). It is tempting to conclude from this that $(\delta \mathbf{x})^2$ continually grows, but in fact (5.36) does not establish this fact. For example, $\langle (\delta \mathbf{x})^2 \rangle = (\delta \mathbf{a})^2 + [(\delta \mathbf{b}) \cos \omega t]^2$ satisfies (5.36) (see Figure 5.26(a)). In order to obtain a more useful result an approximation, called *finite memory*, is usually introduced. In effect this endows the turbulence with an arrow of time. We divide time up into lots of increments. In the first increment (5.36) applies and the line element grows. At the end of this period we assume that the velocity field is

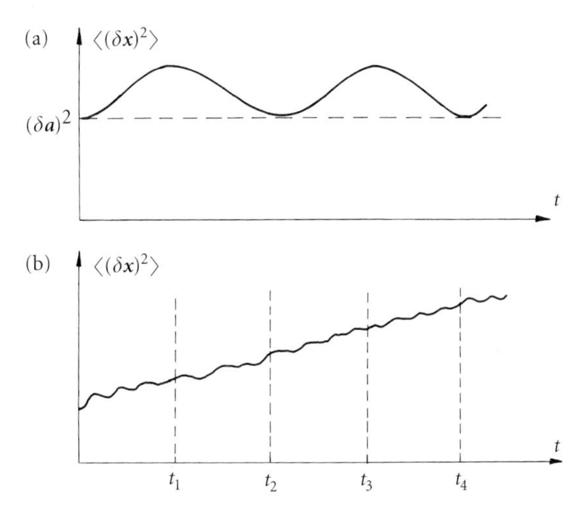

Figure 5.26 (a) Equation (5.36) does not necessarily imply a continual increase in the expected size of the line element. For example, $(\delta \mathbf{x})^2 = (\delta \mathbf{a})^2 + [(\delta \mathbf{b})\cos \omega t]^2$ satisfies (5.36). (b) The use of a 'finite memory' can be used to ensure a continual growth of the expected line element length. (After Moffatt, 1997, Lecture Notes on Turbulence, Ecole Polytechnique, Paris.)

statistically independent of the velocity at the start of the increment. We can now apply (5.36) once again and we obtain a second elongation of the element, and so on from one increment to the next (Figure 5.26(b)).

Although a *finite memory* is, perhaps, a little hard to justify theoretically, it is clear from visual observations that line elements in a turbulent flow are indeed continually stretched. In fact, it is found that, on average, material lines (or surfaces) increase exponentially in length (or area) with time (Monin and Yaglom, 1975). In the case of surfaces this involves a continual stretching, twisting and folding of the surface. It is generally agreed that this underlies the intensification of vorticity by stretching and, probably, the mechanism of transferring energy from scale to scale.[8] The stretching of material elements is, perhaps, most readily seen in the cloud of smoke from a cigar or cigarette (Figure 5.27).

[8] Actually, one has to be cautious in making a direct link between the stretching of line elements and the stretching of vortex lines. The argument above is purely kinematic and applies to any random velocity field. Indeed it would even apply to a velocity field with Gaussian statistics, and we know there is no mean vortex stretching in such a case since the skewness is then zero. The point is that vortex lines are not dynamically passive, like dye lines. Nor can they be specified independently of **u**. So vortex lines directly influence the flow and indeed in a sense they *are* the flow. When we talk about vortex lines being stretched we really have in mind a flow which is dynamically admissible (satisfies the Navier–Stokes equation) as well as kinematically admissible, since only then are the vortex lines partially frozen into the fluid. Such dynamically admissible flows are never Gaussian. However, since material-line stretching seems to be the norm for the broader class of kinematically admissible fields, it should also be the norm for the narrower class of dynamically admissible velocity fields, and so one should not be surprised that vortex-line stretching, like material line stretching, is seen in practice.

Figure 5.27 Sketch of cloud of smoke. Visual observations of such a cloud show material lines and surfaces being stretched by the turbulence.

5.3.6 *The interplay of the strain and vorticity fields*

We close our review of turbulent vortex dynamics with a brief discussion of the structure of the turbulent strain field and its relationship to the vorticity field. This is still a poorly understood topic, though it is of great importance if we are to understand how dissipation is distributed throughout space, or how to model the vortex stretching term in the enstrophy equation,

$$\frac{D}{Dt}\left(\frac{\boldsymbol{\omega}^2}{2}\right) = [\omega_i \omega_j S_{ij}] + \nu(\sim).$$

Our primary aim is to introduce the theories of Betchov (1956) and of Perry and Chong, though we have a certain amount of background material to cover first.

5.3.6.1 Enstrophy and dissipation are not coincident

Let us start with dissipation. In homogeneous turbulence the mean dissipation rate can be written in terms of the enstrophy:

$$\varepsilon = \langle 2\nu S_{ij} S_{ij} \rangle = \nu \langle \boldsymbol{\omega}^2 \rangle.$$

This works because the difference between $2\nu S_{ij}S_{ij}$ and $\nu\boldsymbol{\omega}^2$ takes the form of a divergence and the divergence of an average is zero in homogeneous turbulence, that is, $\nabla \cdot [\langle \sim \rangle] = 0$. It is common, therefore, when discussing homogeneous turbulence to associate enstrophy with dissipation. While this is permissible when dealing with global budgets, it can be misleading when it comes to identifying the instantaneous spatial distribution of the dissipation. In fact, at a given instant, the regions of high dissipation are not, in general, coincident with the regions of large enstrophy. That is, the unaveraged functions $2\nu S_{ij}S_{ij}$ and $\nu\boldsymbol{\omega}^2$ tend to look rather different. This is just one manifestation of the fact that the rate-of-strain tensor and vorticity field, though related, have somewhat different spatial structures.

Consider, by way of an example, the Burgers vortex of section 5.3.3. It is readily confirmed that, for large values of Γ_0/ν, the enstrophy and dissipation take the form,

$$\nu\boldsymbol{\omega}^2 = \nu \left[\frac{\alpha\Gamma_0}{4\pi\nu} \right]^2 F_\omega(x)$$

$$2\nu S_{ij}S_{ij} = \nu \left[\frac{\alpha\Gamma_0}{4\pi\nu} \right]^2 F_s(x).$$

Here Γ_0 is the circulation of the vortex, and $x = r^2/\delta^2$ where r is the radial coordinate and δ the characteristic radius of the vortex. The two dimensionless functions F_ω and F_s are

$$F_\omega = e^{-2x}$$

$$F_s = x^{-2}[1 - (1 + x)e^{-x}]^2 \cdot$$

The volume integrals of $2\nu S_{ij}S_{ij}$ and $\nu\boldsymbol{\omega}^2$ are equal so that, as expected, $2\nu S_{ij}S_{ij}$ and $\nu\boldsymbol{\omega}^2$ match globally. However, the enstrophy falls off exponentially with radius, whereas the dissipation initially grows as $\varepsilon \sim r^4$ and then falls as $\varepsilon \sim r^{-4}$ at large r. In fact, the spatial overlap between $2\nu S_{ij}S_{ij}$ and $\nu\boldsymbol{\omega}^2$ is slight, with the dissipation rate reaching a maximum at the point where the enstrophy has fallen to less than 3% of its peak value. Roughly speaking, we have a central core of enstrophy embedded in an annulus of dissipation (Figure 5.28).

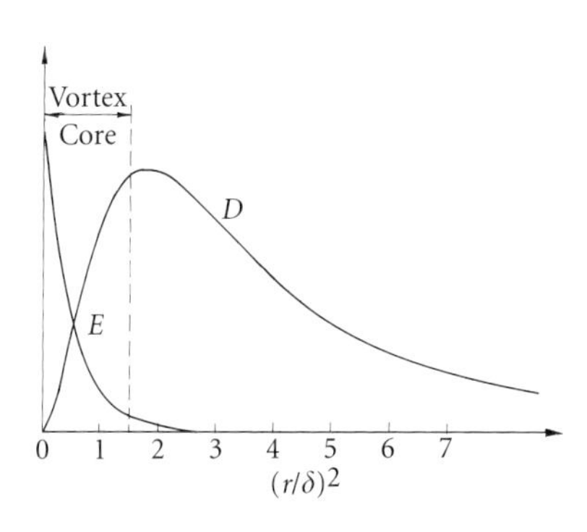

Figure 5.28 Enstrophy distribution (curve E) and dissipation (curve D) for a Burgers vortex. The vertical scale for D is a factor of ten greater than that for E.

Since a turbulent vorticity field is invariably intermittent, possibly dominated by well-separated vortex tubes (Plate 8), this raises a number of questions about the structure of the strain field, such as:

- Is the strain field equally intermittent and does this intermittency have a spatial structure which is similar to that of the vorticity?

- Do the regions of high enstrophy coincide, roughly, with those of peak strain?
- Is the vorticity aligned, on average, with the maximum rate of strain, as might be expected on the basis of a simple vortex stretching argument?

We shall air some of these issues here, although perhaps we should note from the outset that, to date, there are many more questions than answers. In order to keep the discussion brief we shall restrict the arguments to homogeneous turbulence in which the mean velocity is zero. Let us start with some kinematics.

5.3.6.2 Different strategies for finding the relationship between strain and vorticity

Formally, the instantaneous distribution of S_{ij} is uniquely determined by the spatial distribution of vorticity. That is, given $\boldsymbol{\omega}(\mathbf{x})$, the Biot–Savart law fixes $\mathbf{u}(\mathbf{x})$ as

$$\mathbf{u}(\mathbf{x}) = \frac{1}{4\pi}\int \frac{\boldsymbol{\omega}(\mathbf{x}') \times \mathbf{r}}{r^3} d\mathbf{x}', \quad \mathbf{r} = \mathbf{x} - \mathbf{x}'.$$

The rate-of-stain field S_{ij} follows. Unfortunately, this is of little direct help in our attempt to picture the structure of S_{ij}. The problem is that the Biot–Savart integral constitutes a non-local relationship between S_{ij} and $\boldsymbol{\omega}$ (Figure 5.29). That is, we need a detailed, global description of $\boldsymbol{\omega}$ in order to reconstruct the strain field at any one point. Nor is the local equation

$$2\frac{\partial S_{ij}}{\partial x_j} = -(\nabla \times \boldsymbol{\omega})_i$$

of much help, as it merely tells us that *gradients* in strain will tend to be coincident with gradients in vorticity. What we would really like to know is whether or not regions of large strain tend to coincide with those of intense vorticity.

Since the Biot–Savart law has, so far, yielded little, a somewhat different strategy has been adopted by most researchers. In fact there have been two, interrelated approaches, one theoretical and the other numerical. They are:

(i) to take advantage of statistical homogeneity to establish kinematic relationships between means of the form $\langle \omega_i \omega_j S_{ij} \rangle$ and $\langle S_{ij} S_{jk} S_{ki} \rangle$;

(ii) to examine data sets obtained by the direct numerical simulation of turbulence to construct a qualitative picture of the strain and vorticity fields. Here one of the problems is that the flow fields are very complex and so one needs a simple classification scheme to differentiate between, say, regions of large strain and strong vorticity.

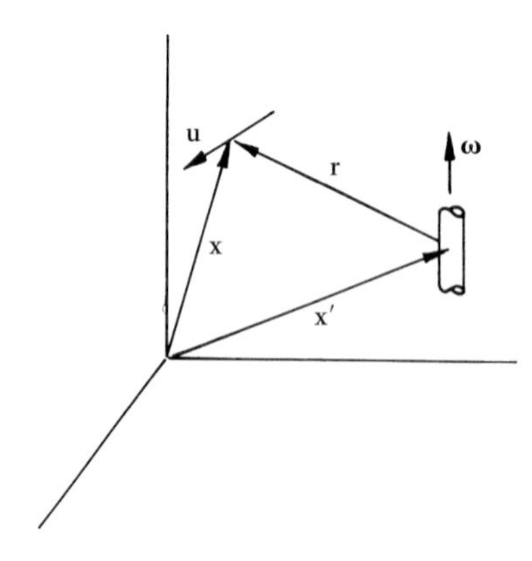

Figure 5.29 The Biot–Savart law. The relationship between vorticity and velocity is non-local in the sense that a blob of vorticity induces a velocity field at all points.

We discuss both of these approaches below. The first dates back to the classical work of Betchov (1956), while the second has bloomed in recent years as a result of the growth in the computer simulations of turbulence. The common theme is that both strategies focus attention on the velocity gradient (or rate-of-deformation) tensor, $\dot{A}_{ij} = \partial u_i / \partial x_j$. This tensor is important as it uniquely determines the local flow field as seen by an observer moving with a fluid element:

$$\delta u_i = u_i(\mathbf{x}_0 + \delta\mathbf{x}) - u_i(\mathbf{x}_0) = \left[\partial u_i / \partial x_j\right]_{\mathbf{x}_0} \delta x_j.$$

Betchov's idea was to investigate the implications of statistical homogeneity for products of the type $\langle \dot{A}_{ij}\dot{A}_{ji} \rangle$ and $\langle \dot{A}_{ij}\dot{A}_{jk}\dot{A}_{ki} \rangle$. This turns out to be an extremely fruitful line of enquiry, yielding a relationship between $\langle \omega_i \omega_j S_{ij} \rangle$ and $\langle S_{ij} S_{jk} S_{ki} \rangle$. By a coincidence, the same quantities, $\langle \dot{A}_{ij}\dot{A}_{ji} \rangle$ and $\langle \dot{A}_{ij}\dot{A}_{jk}\dot{A}_{ki} \rangle$, turn out to play a useful role in the examination of numerical data sets, providing a convenient means of classifying different regions in a flow according to their relative levels of strain, vorticity, vortex stretching, and so on.

5.3.6.3 Properties of the velocity gradient tensor

Perhaps we should take a moment to review some of the properties of the rate-of-deformation tensor. In particular, we need to know a little about the properties of its eigenvalues and associated invariants. In Chapter 2, we saw that the symmetric and anti-symmetric parts of $\dot{A}_{ij} = \partial u_i / \partial x_j$ are the rate-of-strain tensor, S_{ij}, and rate-of-rotation tensor, W_{ij}:

$$\begin{aligned}\dot{A}_{ij} &= \frac{\partial u_i}{\partial x_j} = \frac{1}{2}\left[\frac{\partial u_i}{\partial x_j} + \frac{\partial u_j}{\partial x_i}\right] + \frac{1}{2}\left[\frac{\partial u_i}{\partial x_j} - \frac{\partial u_j}{\partial x_i}\right] \\ &= S_{ij} + W_{ij} = S_{ij} - \tfrac{1}{2}\varepsilon_{ijk}\omega_k.\end{aligned}$$

Now at any point in the flow we can always find principal axes for S_{ij} and if we align our coordinates with these axes then S_{ij} becomes diagonal and $\dot{A}_{ij}$ takes the simple form

$$\dot{A}_{ij} = S_{ij} + W_{ij} = \begin{vmatrix} a & 0 & 0 \\ 0 & b & 0 \\ 0 & 0 & c \end{vmatrix} + \frac{1}{2}\begin{vmatrix} 0 & \omega_c & \omega_b \\ \omega_c & 0 & -\omega_a \\ -\omega_b & \omega_a & 0 \end{vmatrix}$$

were a, b, and c are the three principal rates of strain and ω_a, ω_b, and ω_c are the components of $\boldsymbol{\omega}$ along the principal axes. We shall order a, b, and c such that

$$a \geq b \geq c.$$

Also, continuity requires

$$\dot{A}_{ii} = S_{ii} = a + b + c = 0$$

from which we obtain

$$a^2 + b^2 + c^2 = -2(ab + bc + ca)$$
$$a^3 + b^3 + c^3 = 3(abc)$$
$$a^4 + b^4 + c^4 = \tfrac{1}{2}(a^2 + b^2 + c^2)^2.$$

We shall return to these relationships shortly. In the meantime we note that the local structure of the flow field, as seen by an observer moving with the fluid, is determined by:

(i) the three principal rates of strain, of which only two are independent;
(ii) the magnitude of the vorticity;
(iii) the orientation of $\boldsymbol{\omega}$ relative to the principal axes.

Thus a total of five parameters are required to classify the flow at a point; say a, b, $\boldsymbol{\omega}^2$, and the two angles required to fix the direction of $\boldsymbol{\omega}$ relative to the principal axes. This is a lot of information and it is natural to ask if some subset of these parameters, or perhaps some simple combination of them, is sufficient to yield useful information.

In recent years a number of researchers have focussed on just two quantities, denoted Q and R, which turn out to be simple combinations of our five independent parameters. Q and R are two of the so-called *invariants* of $\dot{A}_{ij}$. (The term invariant is used here in its kinematic, rather than dynamic, sense to mean a scalar quantity constructed from $\dot{A}_{ij}$ whose value is independent of the orientation of the coordinate system.) We shall see that Q and R play a central role in Betchov's theory, as well as in the classification of numerical data sets.

It is well known that there are three invariants of a second-order tensor which are directly related to its eigenvalues. Suppose that the eigenvalues of $\dot{A}_{ij}$ are denoted λ_i and that the characteristic equation for these eigenvalues is

$$\lambda^3 - P\lambda^2 + Q\lambda - R = 0.$$

Since the λ_i's are independent of the choice of coordinates, so are P, Q, and R. These are referred to as the first, second, and third invariants of $\dot{A}_{ij}$. The properties of the roots of a cubic equation, along with elementary eigenvalue theory, tell us that

$$P = (\lambda_1 + \lambda_2 + \lambda_3) = \mathrm{Trace}(\dot{A}_{ij}) = \dot{A}_{ii}$$
$$Q = (\lambda_1\lambda_2 + \lambda_2\lambda_3 + \lambda_3\lambda_1) = \tfrac{1}{2}P^2 - \tfrac{1}{2}\dot{A}_{ij}\dot{A}_{ji}$$
$$R = (\lambda_1\lambda_2\lambda_3) = \det(\dot{A}_{ij}).$$

Moreover, continuity requires

$$\lambda_1 + \lambda_2 + \lambda_3 = 0$$

so that the first invariant is zero, $P = 0$. Also, the Cayley–Hamilton theorem tells us that $\dot{A}_{ij}$ satisfies its own characteristic equation

$$\dot{\mathbf{A}}^3 + Q\dot{\mathbf{A}} - R\mathbf{I} = \mathbf{0}$$

the trace of which yields

$$R = \tfrac{1}{3}\dot{A}_{ij}\dot{A}_{jk}\dot{A}_{ki}.$$

In summary, then, the second and third invariants of the velocity gradient tensor are

$$Q = -\tfrac{1}{2}\dot{A}_{ij}\dot{A}_{ji} = (\lambda_1\lambda_2 + \lambda_2\lambda_3 + \lambda_3\lambda_1) = -\tfrac{1}{2}(\lambda_1^2 + \lambda_2^2 + \lambda_3^2)$$
$$R = \tfrac{1}{3}\dot{A}_{ij}\dot{A}_{jk}\dot{A}_{ki} = (\lambda_1\lambda_2\lambda_3) = \tfrac{1}{3}(\lambda_1^3 + \lambda_2^3 + \lambda_3^3).$$

Perhaps these are most usefully expressed in terms of S_{ij} and W_{ij}. Writing $\dot{A}_{ij} = S_{ij} + W_{ij}$ and expanding products of $\dot{A}_{ij}$ yields

$$Q = -\tfrac{1}{2}S_{ij}S_{ji} + \tfrac{1}{4}\boldsymbol{\omega}^2 = -\tfrac{1}{2}(a^2 + b^2 + c^2) + \tfrac{1}{4}\boldsymbol{\omega}^2$$
$$R = \tfrac{1}{3}(S_{ij}S_{jk}S_{ki} + \tfrac{3}{4}\omega_i\omega_j S_{ij}) = \tfrac{1}{3}(a^3 + b^3 + c^3) + \tfrac{1}{4}\omega_i\omega_j S_{ij}.$$

(Note that most authors define R with the opposite sign to the expression above, that is, $R = -\tfrac{1}{3}\dot{A}_{ij}\dot{A}_{jk}\dot{A}_{ki}$.)

By way of an illustration, consider the case where the vorticity is locally zero. Then the invariants Q and R reduce to those of the rate-of-strain tensor:

$$Q_s = -\tfrac{1}{2}S_{ij}S_{ji} = -\tfrac{1}{2}(a^2 + b^2 + c^2)$$
$$R_s = \tfrac{1}{3}S_{ij}S_{jk}S_{ki} = \tfrac{1}{3}(a^3 + b^3 + c^3) = abc.$$

Conversely, when there is no strain the invariants reduce to

$$Q_\omega = \tfrac{1}{4}\boldsymbol{\omega}^2, \quad R_\omega = 0.$$

In the more general case, where there is both strain and vorticity, Q gives a measure of the relative intensity of the two, with large negative Q indicating regions of strong strain and large positive Q marking regions of intense enstrophy. We shall see shortly that R can be used in a similar way to differentiate between regions of stretching and compression.

So the proposal (of some researchers) is to use these two invariants to try and classify each point in a flow field, and in particular to differentiate between regions of intense vorticity and regions of strong strain. It is important to note, however, that these are not the only invariants of the motion. Indeed, we know that there are five independent invariants, which we might take as

$$I_1 = S_{ij}S_{ji}, \quad I_2 = S_{ij}S_{jk}S_{ki}, \quad I_3 = \boldsymbol{\omega}^2,$$
$$I_4 = \omega_i\omega_j S_{ij}, \quad I_5 = \omega_i S_{ij}\omega_k S_{kj}$$

Q and R are simple combinations of $I_1 \rightarrow I_4$.

5.3.6.4 Betchov's theory

So far we have been thinking about Q and R as a means of classifying different regions in a turbulent flow. This is very much in the spirit of the numerical approach to investigating the strain and vorticity fields (approach (ii) above). However, they also play a central role in Betchov's (1956) theory, to which we now turn. (We shall return to the idea of using Q and R to classify flow fields shortly.) Betchov noticed that Q and R can both be written as divergences

$$Q = -\frac{1}{2}\frac{\partial}{\partial x_j}\left[u_i \frac{\partial u_j}{\partial x_i}\right]$$

$$R = \frac{1}{3}\frac{\partial}{\partial x_i}\left[\frac{\partial u_i}{\partial u_j}\frac{\partial u_j}{\partial u_k}u_k - \frac{1}{2}u_i\frac{\partial u_k}{\partial x_j}\frac{\partial u_j}{\partial x_k}\right]$$

and since the divergence of an average is zero in homogeneous turbulence, we have

$$\langle Q \rangle = \left\langle -\tfrac{1}{2}\dot{A}_{ij}\dot{A}_{ji}\right\rangle = 0$$

$$\langle R \rangle = \left\langle \tfrac{1}{3}\dot{A}_{ij}\dot{A}_{jk}\dot{A}_{ki}\right\rangle = 0.$$

It follows that both the mean enstrophy, $\langle \boldsymbol{\omega}^2 \rangle$, and the mean rate of generation of enstrophy, $\langle \omega_i \omega_i S_{ij} \rangle$, can be written in terms of the principal strains:

$$\langle \boldsymbol{\omega}^2 \rangle = 2\langle S_{ij}S_{ji}\rangle = 2\langle a^2 + b^2 + c^2 \rangle$$

$$\langle \omega_i \omega_j S_{ij} \rangle = -\tfrac{4}{3}\langle S_{ij}S_{jk}S_{ki}\rangle = -\tfrac{4}{3}\langle a^3 + b^3 + c^3 \rangle.$$

The first of these expressions is, of course, familiar in the context of the mean dissipation of kinetic energy. The second, however, is new. It is a remarkable result, first discovered by Townsend (1951) and then later rediscovered by Betchov in 1956, who expressed it in the form

$$\langle \omega_i \omega_j S_{ij} \rangle = -4\langle abc \rangle.$$

It is extraordinary that the mean rate of generation of enstrophy by vortex-line stretching can be expressed purely in terms of the strain field. This equation has a number of important consequences. We know that the enstrophy generation is positive, and so Betchov's equation tells us that $\langle abc \rangle < 0$. Yet abc takes the same sign as $-b$, partly as a result of the continuity equation, $a + b + c = 0$, and partly as a result of our ordering of the strains $a \geq b \geq c$. Thus, as Betchov noted, turbulence favours a situation in which $c > 0$ and a, $b > 0$, that is, we have one large compressive strain accompanied by two weaker extensional ones. The implications of this for vortex sheet formation are discussed in Section 5.3.2.

The question now arises, which principal strain is primarily responsible for the generation of enstrophy? In terms of principal coordinates we have

$$\langle \omega_i \omega_j S_{ij} \rangle = \langle a\omega_a^2 + b\omega_b^2 + c\omega_c^2 \rangle = -4\langle abc \rangle.$$

The evidence of the direct numerical simulations of turbulence (which relate to modest values of R_λ) is that:

(i) the ratio of the principal strains is $\langle a, b, c \rangle \sim (3,1,-4)$;

(ii) $\boldsymbol{\omega}$ is, on average, aligned with the intermediate principal strain, b (see Section 7.3.1). We might speculate, therefore, that $\langle b\omega_b^2 \rangle$ makes the largest contribution to $\langle \omega_i \omega_j S_{ij} \rangle$. However, the results of the computer simulations suggest that this is not the case. Rather, well over 50% of $\langle \omega_i \omega_j S_{ij} \rangle$ comes from the largest principal rate of strain, $\langle a\omega_a^2 \rangle$. The implication is that, although $\boldsymbol{\omega}$ is rarely aligned with a, on those occasions when it is, a great deal of stretching occurs.

Betchov established a number of other results relating to the principal strains. For example, he showed that, when the turbulence is isotropic

$$\left\langle \left(\frac{\partial u_x}{\partial x}\right)^2 \right\rangle = \frac{2}{15}\langle a^2 + b^2 + c^2 \rangle$$

$$\left\langle \left(\frac{\partial u_x}{\partial x}\right)^3 \right\rangle = \frac{8}{105}\langle a^3 + b^3 + c^3 \rangle = \frac{24}{105}\langle abc \rangle$$

$$\left\langle \left(\frac{\partial u_x}{\partial x}\right)^4 \right\rangle = \frac{8}{105}\langle a^4 + b^4 + c^4 \rangle = \frac{4}{105}\langle (a^2 + b^2 + c^2)^2 \rangle$$

from which we can construct expressions for the flatness and skewness factors for $\partial u_x / \partial x$:

$$S_0 = \frac{\langle (\partial u_x/\partial x)^3 \rangle}{\langle (\partial u_x/\partial x)^2 \rangle^{3/2}} = \frac{12\sqrt{15}}{7\sqrt{2}} \frac{\langle abc \rangle}{\langle a^2 + b^2 + c^2 \rangle^{3/2}}$$

$$= -\frac{6\sqrt{15}}{7} \frac{\langle \omega_i \omega_j S_{ij} \rangle}{\langle \boldsymbol{\omega}^2 \rangle^{3/2}}$$

$$\delta_0 = \frac{\langle (\partial u_x/\partial x)^4 \rangle}{\langle (\partial u_x/\partial x)^2 \rangle^2} = \frac{15}{7} \frac{\langle (a^2 + b^2 + c^2)^2 \rangle}{\langle a^2 + b^2 + c^2 \rangle^2}.$$

Note that this expression for the skewness is equivalent to (5.32). It is an important equation because it provides a simple link between a readily measured quantity, S_0, and the all-important vortex stretching term, that is,

$$\langle \omega_i \omega_j S_{ij} \rangle = -\frac{7}{6\sqrt{15}} S_0 \langle \boldsymbol{\omega}^2 \rangle^{3/2}.$$

We shall derive this equation via a different (though equivalent) route in Chapter 6.

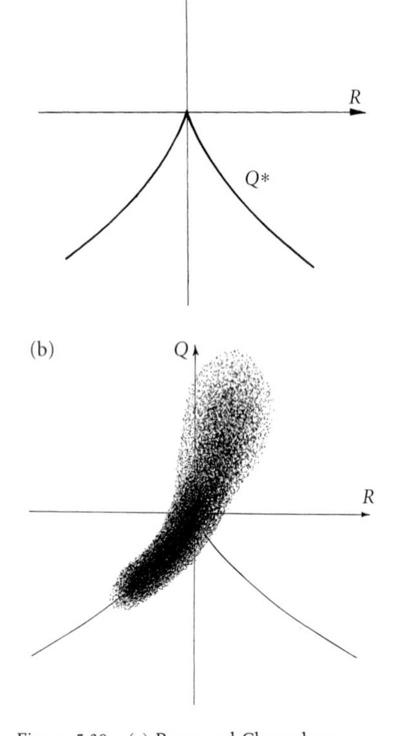

Figure 5.30 (a) Perry and Chong have suggested that the Q–R plane be subdivided into two regions corresponding to $Q < Q^*$ and $Q > Q^*$, with different behaviour in the two regions. (b) Typical Q–R scatter plot. In most flows Q and R take the same sign at most locations.

5.3.6.5 Numerical Simulations and the Q–R Classification

Let us now return to our invariants Q and R and their use in the classification of different regions in a turbulent flow. Although Q and R do not uniquely define the local structure of a flow (five quantities are needed to do this) some authors have adopted them as a means of providing a broad-brush system of classification. Recall that

$$Q = \tfrac{1}{4}\boldsymbol{\omega}^2 - \tfrac{1}{2}(a^2 + b^2 + c^2), \quad \langle Q \rangle = 0$$
$$R = \tfrac{1}{4}\omega_i\omega_j S_{ij} + abc, \quad \langle R \rangle = 0.$$

When Q is large and negative we have a lot of strain and very little vorticity. Conversely, a large positive Q indicates the dominance of vorticity. In the Burgers vortex, for example, Q is positive in the core of the vortex and negative outside the core where dissipation is dominant. Now suppose Q is large and positive. Then the strain is locally weak and $R \sim \frac{1}{4}\omega_i\omega_j S_{ij}$. In such cases a positive R implies vortex stretching while a negative R indicates a region of vortex compression. On the other hand if Q is large and negative, then $R \sim abc$. This time a positive R implies a region of axial strain ($a > 0$; b, $c < 0$), while a negative R indicates a region of bi-axial strain ($c < 0$; a, $b > 0$). The four regimes are summarized in Table 5.1.

Of course, the problem with this kind of argument is that we have not defined what we mean by 'large and positive' or 'large and negative'. Large by comparison with what? (Hence the question marks in the left-hand column in Table 5.1).

In a sequence of papers, Perry and Chong suggest that there is a sharp division between $Q < Q^*$ and $Q > Q^*$ where $Q^* = -3(R^2/4)^{1/3}$ (see, for example, Ooi et al., 1999, for a recent assessment of the status of this work). The rationale for this distinction is the observation that the eigenvalues of $\partial u_i/\partial x_j$ are all real for $Q < Q^*$, but contain a pair of complex conjugates for $Q > Q^*$. This has certain implications for the topological structure of the flow. So, according to these authors, the question marks in Table 5.1 should be replaced by Q^* (see Figure 5.30(a)).

Whether or not such a sharp division is justified is, perhaps, a matter of debate. In any event it is common to adopt Q and R as a sort of broad-brush form of classification. For example, instantaneous velocity fields are reproduced as *scatter plots* in the Q–R plane. (Each point in the flow is

Table 5.1 A crude classification of points in a turbulent flow in terms of Q and R

	$R < 0$	$R > 0$
$Q > ?$	Vortex compression $\omega_i\omega_j S_{ij} < 0$	Vortex stretching $\omega_i\omega_j S_{ij} > 0$
$Q < ?$	Bi-axial strain $abc < 0$	Axial strain $abc > 0$

assigned a point in the Q–R plane.) Such plots seem to have the same qualitative shape for a wide range of flows (both homogeneous and inhomogeneous flows), with the bulk of the flow field lying in just two quadrants; $Q, R < 0$ and $Q, R > 0$ (see Figure 5.30(b)) In other words, there is a strong positive correlation between Q and R, $\langle QR \rangle > 0$. Thus the two most common states are vortex stretching, $\omega_i \omega_j S_{ij} > 0$, and bi-axial strain, $abc < 0$, which is precisely what Betchov predicted in 1956.

Another common observation is that regions of intense dissipation are closely correlated to those where the value of $|S_{ij}S_{jk}S_{ki}| = 3|abc|$ is high, and this correlation is stronger than that between dissipation and vortex stretching, $\omega_i \omega_j S_{ij}$. This observation is, perhaps, not so surprising. Consider Figure 5.28, which shows the enstrophy and dissipation in a Burgers vortex. There is a core of high enstrophy embedded in an annulus of intense dissipation. The inner region is also the seat of strong vortex stretching while the outer annulus marks the region where $S_{ij}S_{jk}S_{ki}$ is large. So, in Burgers-like vortex tubes we would expect $S_{ij}S_{jk}S_{ki}$, but not $\omega_i \omega_j S_{ij}$, to be well correlated in space with the dissipation. Of course this does not in any way contradict the view that it is vortex stretching, $\omega_i \omega_j S_{ij}$, which lies behind the energy cascade.

It is possible to establish dynamic equations for Q and R. For example, the evolution equations for the enstrophy, rate-of-deformation tensor, and strain-rate are

$$\frac{D}{Dt}\frac{1}{2}\boldsymbol{\omega}^2 = \omega_i \omega_j S_{ij} + \nu(\sim),$$

$$\frac{D}{Dt}\dot{A}_{ij} = -P_{ij} - \dot{A}_{ik}\dot{A}_{kj} + \nu(\sim),$$

$$\frac{D}{Dt}\frac{1}{2}S_{ij}S_{ji} = -S_{ij}S_{jk}S_{ki} - \frac{1}{4}\omega_i \omega_j S_{ij} - S_{ij}P_{ij} + \nu(\sim)$$

where P_{ij} is the *pressure Hessian*, $\partial^2(p/\rho)/\partial x_i \partial x_j$. Combining the first and last of these we obtain

$$\frac{DQ}{Dt} = 3R + S_{ij}P_{ij} + \nu(\sim)$$

where $\nu(\sim)$ indicates a viscous term, the precise form of which does not concern us. A similar equation can be found for the rate of change of R. Ooi et al. (1999) have exploited such relationships to predict temporal variations in Q and R. Actually, it is not yet clear that much is to be gained by formulating dynamic equations for Q and R. It may be that their main value is purely diagnostic. Perhaps a more interesting result comes from combining our expression for $D\dot{A}_{ij}/Dt$ with the vorticity evolution equation to give

$$\frac{D^2\omega_i}{Dt^2} = -P_{ij}\omega_j + \nu(\sim).$$

Suppose that λ_1, λ_2, and λ_3 are the eigenvalues of P_{ij}, which are all real. Then, for timescales over which the λ_i may be treated as quasi-constant, negative eigenvalues will give rise to an exponential growth in vorticity. Since $\lambda_1 + \lambda_2 + \lambda_3 = P_{ii} = 2Q$, it would appear that regions of negative Q (strong strain) tend to favour such a growth. But perhaps this should not come as a surprise.

Example 5.14

Use the fact that the geometric mean of a^2, b^2, c^2 is less than the arithmetic mean, combined with the Schwarz inequality, to show that

$$\langle |abc| \rangle \leq \tfrac{1}{3\sqrt{3}} \left\langle (a^2 + b^2 + c^2)^{3/2} \right\rangle$$

$$\leq \tfrac{1}{3\sqrt{3}} \left\langle (a^2 + b^2 + c^2)^2 \right\rangle^{1/2} \langle (a^2 + b^2 + c^2) \rangle^{1/2}.$$

Now use Betchov's expressions for skewness and flatness to show that

$$|S_0| \leq \tfrac{2\sqrt{2}}{\sqrt{21}} \sqrt{\delta_0}.$$

Example 5.15

Consider a dilute suspension of very small particles in a moving fluid. Let $n(\mathbf{x}, t)$ be the number density of particles and $\mathbf{v}$ the velocity of a typical particle. If the particles are very small then they act almost as passive tracers in the flow, the small difference between $\mathbf{v}$ and the fluid velocity $\mathbf{u}$ giving rise to a Stokes drag force which is linear in $\mathbf{u} - \mathbf{v}$. The equation of motion for a particle is then

$$\frac{d\mathbf{v}}{dt} = \frac{\mathbf{u} - \mathbf{v}}{\tau}$$

where τ is the particle relaxation time (assumed small). Conservation of particle numbers also yields

$$\frac{\partial n}{\partial t} = -\nabla \cdot (n\mathbf{v}).$$

Show that, in the limit of $|\mathbf{u} - \mathbf{v}| \ll |\mathbf{u}|$, which corresponds to τ being much less than the typical timescale of the flow, we have

$$\frac{Dn}{Dt} = n\nabla \cdot (\mathbf{u} - \mathbf{v}) = n\tau\nabla \cdot (D\mathbf{u}/Dt).$$

Hence confirm that

$$\frac{D}{Dt} \ln n = \tau \frac{\partial u_i}{\partial x_j} \frac{\partial u_j}{\partial x_i} = -2\tau Q = \tau \left(S_{ij} S_{ij} - \tfrac{1}{2} \boldsymbol{\omega}^2 \right).$$

It would seem that particles accumulate in regions where Q is negative (large strain) at the expense of regions where Q is positive (large enstrophy). Now rewrite the equation as

$$\frac{Dn}{Dt} = -n\tau\nabla^2 (p/\rho).$$

Noting that ∇p is related to streamline curvature, explain this effect in terms of particles migrating away from the centre of curvature of curved streamlines.

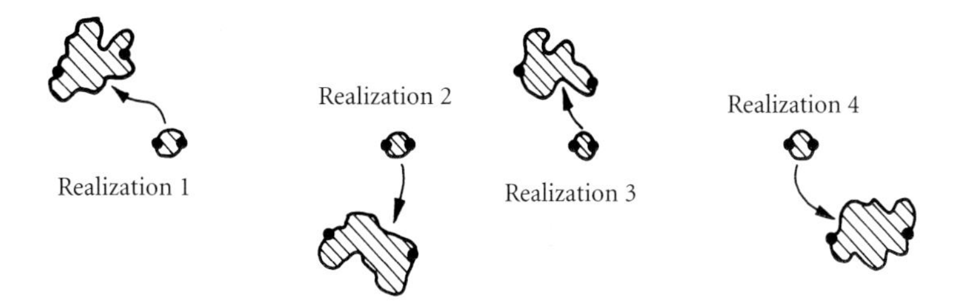

Figure 5.31 Spreading of a small patch of contaminant in several realizations.

5.4 Turbulent diffusion by continuous movements

The fact that, on average, two points in a turbulent flow tend to move apart underlies the phenomenon of turbulent dispersion. Consider, for example, Figure 5.25. Suppose that $\delta\mathbf{a}$ at $t=0$ represents the location and size of a spherical cloud of some chemical contaminant, say smoke or dye. That is, at $t=0$ the contaminant is restricted to a sphere of diameter $|\delta\mathbf{a}|$, centred on the midpoint of $\delta\mathbf{a}$. Let C represent the concentration distribution of the contaminant, which we take to be non-reacting and dynamically passive. It satisfies an advection–diffusion equation similar to that of heat

$$\frac{DC}{Dt} = \alpha\nabla^2 C$$

where α is the diffusivity for C. If we assume that ul/α is large, which is typical of nearly all passive scalars in a turbulent flow, then C is more or less materially conserved and so follows the fluid particles as they move around. However, we saw in Section 5.3.5 that, on average, the two ends of $\delta\mathbf{a}$ move apart as the flow evolves. This is true of any line element, $\delta\mathbf{a}$, which spans the spherical cloud at $t=0$. It follows, therefore, that the cloud of contaminant will spread (disperse) under the action of the turbulent fluctuations (Figure 5.31). This is an example of *turbulent diffusion*.[9]

We now give a fairly classical account of turbulent diffusion which dates back to Taylor (1921) and Richardson (1926). We have in mind the situation where there is a local release of a contaminant into a field of turbulence, say dye released into a mixing tank. For simplicity we shall take the Peclet number, ul/α, to be large, the mean velocity to be zero, and the turbulence to be isotropic and statistically steady, which is not at all typical but provides a convenient starting point. (We allow for a mean shear in Section 5.4.3 where we shall see that gradients in the mean velocity have a radical effect on dispersion.)

Since Pe is large, the problem of predicting the spread of the contaminant comes down to that of tracking the turbulent migration of individual fluid particles. When the mean velocity is zero it turns out

[9] The term turbulent dispersion is also used in this context. However, this phrase carries a more general meaning in that it may include the dispersive effects of a mean shear as well as that of the random eddies (see Section 5.4.3).

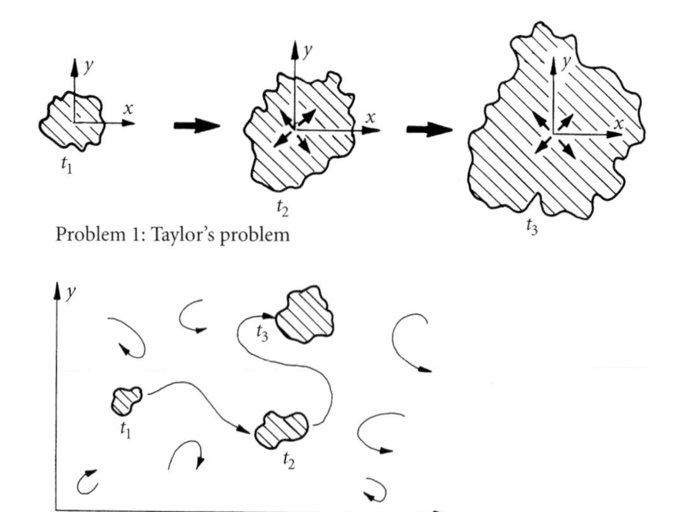

Figure 5.32 In Taylor's problem we consider the continuous release of a contaminant from a fixed point and estimate the average rate of growth of the contaminant cloud. In Richardson's problem a small puff of contaminant is released at $t = 0$ and we aim to determine the rate of growth of the puff as it bounces around in the turbulent flow.

that there are two rather different questions which need to be addressed. They are (Figure 5.32):

(i) *Taylor's problem* of single particle diffusion, which asks—'how far, on average, will a single fluid particle migrate from its point of release in a time t as a result of the turbulent eddying?' (Taylor, 1921);

(ii) *Richardson's problem* of relative dispersion, which asks—'how rapidly, on average, will a pair of particles separate from each other in a turbulent flow?' (Richardson, 1926).

The first of these problems is relevant to the dispersal of a pollutant which is *continuously* released at a single location. That is to say, a contaminant continuously released into a turbulent flow will spread like ink soaking through blotting paper, the rate of spreading of the contaminant being controlled by the rate of migration of fluid particles from the point of release.

The second problem, that of the relative dispersion of a pair of particles, is relevant to the *discrete* release of a small patch of contaminant. (By small we mean that the size of the patch is significantly smaller than the size of the large eddies, though larger than the Kolmogorov scale.) Let us consider what happens to such a patch. Its centroid moves around chaotically as the small patch becomes engulfed in, and transported around by, the large eddies. Simultaneously, the size of the patch will grow as a result of small-scale turbulent mixing. While Taylor's problem will tell us how the centroid of the patch moves around, Richardson problem is concerned with the rate of growth of the patch. That is, the rate of spreading of a small patch, or puff, of contaminant is determined by the relative rate of separation of two fluid particles which mark opposite sides of the patch at its time of release.

We have seen how difficult it is to quantify the relative movement of two particles in a turbulent flow and so we start with the simpler problem, that of single-particle dispersion. The method employed is Lagrangian, rather than Eulerian, and so the notation used is a little different to that of the rest of this chapter.

5.4.1 *Taylor diffusion of a single particle*

Before embarking on an analysis of turbulent diffusion, perhaps it is worthwhile reminding ourselves of a simpler, more familiar problem—that of the *random walk*. This is frequently discussed in terms of the meandering path followed by the much maligned 'drunken sailor'. The idea is the following. The sailor leaves the bar and takes a sequence of steps of length l. However, each successive step is in a random direction. The question is: after N steps of length l, how far, on average, will the sailor have progressed from the bar. Since the direction of each successive step is random we might think that the answer is zero. But it is not. With each step he is more likely to have stumbled farther from the bar. We can estimate the mean progress of the sailor as follows. Let $\mathbf{R}_N$ be his position after the Nth step is taken, and $\mathbf{L}_N$ the displacement vector of the Nth step. Then

$$\mathbf{R}_N^2 = \mathbf{R}_{N-1}^2 + 2\mathbf{R}_{N-1} \cdot \mathbf{L}_N + \mathbf{L}_N^2.$$

Now the random direction of $\mathbf{L}_N$ implies that $\langle \mathbf{L}_N \rangle = 0$. Moreover, the direction of the Nth step is assumed independent of the location $\mathbf{R}_{N-1}$, and so $\langle \mathbf{L}_N \cdot \mathbf{R}_{N-1} \rangle = 0$. (As the sailor is about to take his Nth step he does not care where he is.) It follows that, after a large number of steps

$$\langle R_N^2 \rangle = \langle R_{N-1}^2 \rangle + \langle L_N^2 \rangle = \langle R_{N-1}^2 \rangle + l^2$$

and by induction we have

$$\langle R_N^2 \rangle = Nl^2.$$

In short, the poor sailor progresses only slowly from the bar, the likely distance travelled being proportional to $N^{1/2}$, rather than N, as would be the case for a more sober man walking in a straight line.

It is natural to suppose that a particle in a random, turbulent flow will behave in a similar way. Surely, we might think, it will travel a distance of order $t^{1/2}$ in a time t. It turns out that this is substantially correct, though there are exceptions, as we now show.

Let $\mathbf{x} = \mathbf{X}(t)$ be the position of a fluid particle released from the origin at $t = 0$ (Figure 5.33). The Lagrangian velocity of the particle is $\mathbf{v} = d\mathbf{X}/dt$. In order to simplify matters we shall assume that the

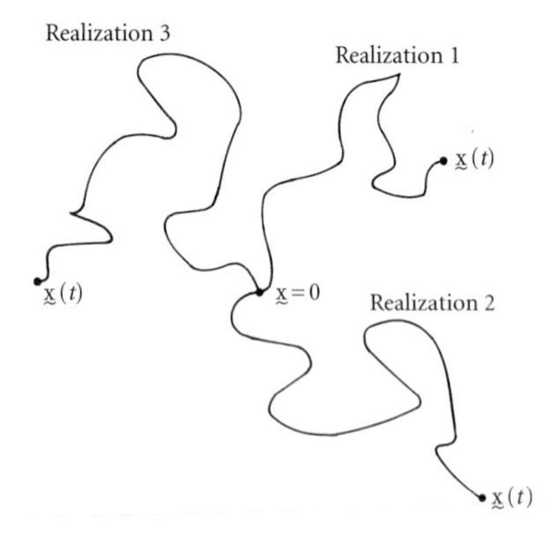

Figure 5.33 Dispersion of particles released from a point source into a turbulent environment. (We may think of this as the progress of three drunken sailors all leaving the same bar at the same time.)

turbulent flow into which the particle is released has zero mean velocity, is homogeneous, and is statistically steady.

The rate at which the particle distances itself from its point of release is given by

$$\frac{d}{dt}[\mathbf{X}^2] = 2\mathbf{X}\cdot\mathbf{v} = 2\mathbf{v}(t)\cdot\int_0^t \mathbf{v}(t')dt'$$

which we might rewrite as

$$\frac{d}{dt}[\mathbf{X}^2] = 2\int_0^t \mathbf{v}(t)\cdot\mathbf{v}(t-\tau)d\tau, \quad \tau = t - t'.$$

We now imagine releasing many such particles, one after the other. We track each particle and ensemble average the results. We find

$$\frac{d}{dt}\langle\mathbf{X}^2\rangle = 2\int_0^t \langle\mathbf{v}(t)\cdot\mathbf{v}(t-\tau)\rangle d\tau.$$

Let us now consider the integrand on the right, which we write as

$$Q^L_{ii}(\tau) = \langle\mathbf{v}(t)\cdot\mathbf{v}(t-\tau)\rangle.$$

This tells us about the degree to which $\mathbf{v}$ at time t is correlated to $\mathbf{v}$ at time t'. If $t = t'$ then $Q^L_{ii} = \langle\mathbf{v}^2\rangle$. If $\tau = t - t'$ is much greater than the characteristic timescale of the turbulence, say the large eddy turn-over time, then we expect $Q^L_{ii} \approx 0$. The quantity

$$Q^L_{ij}(\tau) = \langle v_i(t)v_j(t-\tau)\rangle$$

is called the Lagrangian velocity correlation tensor. It is similar to the velocity correlation tensor introduced in Chapter 3, $Q_{ij}(\mathbf{r})$, except that it involves the Lagrangian (rather than Eulerian) velocity and a separation in time rather than a separation in space. Note that Q^L_{ij} is a function only of τ, and not t, since we have assumed that the turbulence is statistically steady. It is conventional to write

$$\langle\mathbf{u}^2\rangle t_L = \int_0^\infty Q^L_{ii}(\tau)d\tau$$

which defines t_L, the *Lagrangian correlation time*. This timescale is, in effect, a measure of the time for which a fluid particle retains some memory of the circumstances of its release. Returning now to our expression for the rate of change of $\langle\mathbf{X}^2\rangle$ we have,

$$\frac{d}{dt}\langle\mathbf{X}^2\rangle = 2\int_0^t Q^L_{ii}(\tau)d\tau.$$

There are two cases of particular interest: small t and large t. When $t \ll t_L$ we may take $Q^L_{ii} = \langle\mathbf{u}^2\rangle$ and so

$$\frac{d}{dt}\langle\mathbf{X}^2\rangle \approx 2\langle\mathbf{u}^2\rangle t, \quad t \ll t_L$$

which yields

$$\langle \mathbf{X}^2 \rangle \approx \langle \mathbf{u}^2 \rangle t^2, \quad t \ll t_L.$$

This reflects the fact that, for small t, a particle simply moves with its initial velocity, so that $\mathbf{X} \approx \mathbf{v}(0)t$. (This is analogous to our drunken sailor taking his first stride.) For large time, on the other hand, we expect that

$$\frac{d}{dt}\langle \mathbf{X}^2 \rangle \approx 2\int_0^\infty Q_{ii}^L(\tau)d\tau, \quad t \gg t_L$$

which gives the estimate

$$\langle \mathbf{X}^2 \rangle \approx [2\langle \mathbf{u}^2 \rangle t_L]t, \quad t \gg t_L.$$

In this case the r.m.s. displacement is proportional to $t^{1/2}$, which is typical of Brownian motion, or of a random walk. Note that $t_L \sim l/u'$ and so $t \gg t_L$ corresponds to cases where $\langle \mathbf{X}^2 \rangle \gg l^2$, l being the integral scale.

So what does all of this mean for the dispersion of a contaminant? Suppose that a contaminant is continuously released from a fixed point into a turbulent flow. The radius of the contaminant cloud will grow, on average, at a rate $R \sim t$ for $t \ll t_L$ and $t^{1/2}$ for $t \gg t_L$, by which time the cloud diameter is much greater than the integral scale of the turbulence. Note that the $R \sim t^{1/2}$ behaviour is exactly what we would get from a mixing length approximation

$$\frac{\partial \overline{C}}{\partial t} = \alpha_t \nabla^2 \overline{C}, \quad \alpha_t = u' l_m$$

where $\overline{C}$ is the expected (ensemble averaged) concentration of the contaminant, α_t the turbulent diffusivity, and l_m a mixing length which is of the order of the size of the large eddies. (This is discussed in the context of the turbulent diffusion of heat in Section 4.5.1.)

5.4.2 *Richardson's law for the relative diffusion of two particles*

The situation becomes more complicated when we consider the *discrete*, rather than continuous, release of a contaminant. This is often described as *relative diffusion* and relates to the diffusion of a small, single *puff*, or cloud, which is released at $t = 0$. In effect, this brings us back to the relative dispersion of a pair of particles, as discussed in Section 5.3.5. That is, the average rate at which two particles (which span the cloud) separate gives us a measure of the rate of spreading of the cloud.

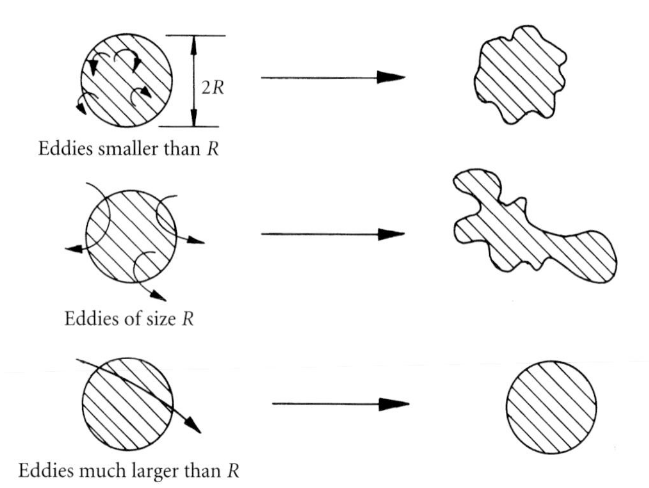

Figure 5.34 The effect of different eddy sizes on a cloud of contaminant.

It is important to understand that this problem is very different to the diffusion of a single particle. To understand why, consider two particles which are adjacent at $t = 0$. To some extent they tend to move as a pair as they are swept around by the turbulent eddies. However, as they bounce around they will also tend to separate and it is that relative separation, rather than the movement of the pair, which is of interest here. If the particles are initially very close we might anticipate that this relative separation is much slower that the absolute movement of the pair, and indeed this is the case. So a small cloud of contaminant will spread by turbulent diffusion, but at a rate somewhat slower than the average rate of movement of its centre of mass. The question, though, is how much slower?

Let $R(t)$ be the mean radius of the cloud at time t. We are interested in clouds for which $\eta \ll R \ll l$, where l is the integral scale. Evidently eddies of size R are most important for the growth of the puff, since smaller eddies merely ripple the surface of the cloud while larger ones tend to advect it without a change in shape. This is illustrated in Figure 5.34.

Now we have already suggested that eddies of size r tend to break up on a timescale of their turn-over time, and so $\varepsilon \sim v_r^3/r$ where ε is the rate of dissipation of turbulent energy per unit mass. It follows that $v_r \sim (\varepsilon r)^{1/3}$. Consider now a cloud which has grown sufficiently to have forgotten the conditions of its initial release, yet whose radius is still significantly smaller than the integral scale, l. Since only eddies of size R contribute to the rate of change of R, we might expect that the average rate of change of R should be independent of ν, u and l and depend only on v_R and t: $dR/dt = f(v_R, t)$. The only dimensionally consistent possibility is

$$\frac{dR}{dt} \sim v_R \sim (\varepsilon R)^{1/3}.$$

Note that we could have arrived at the same expression using Kolmogorov's theory of the small scales. That is, in the inertial subrange the longitudinal velocity increments $\langle[\Delta v]^p\rangle$ depend on ε and r only. For example, $\langle[\Delta v]^2\rangle = \beta\varepsilon^{2/3}r^{2/3}$ and $\langle[\Delta v]^3\rangle = -\frac{4}{5}\varepsilon r$. However, Δv represents the instantaneous rate of separation of two particles along the line connecting those particles. The implication is that dR/dt should be a function of ε and R only. Dimensional reasoning then tells us that dR/dt must have the form $dR/dt \sim (\varepsilon R)^{1/3}$, as given above. This expression is usually rewritten as

$$\frac{dR^2}{dt} \sim \varepsilon^{1/3}R^{4/3}, \quad \eta \ll R \ll l$$

which is known as *Richardson's four-thirds law.*[10] This law implies that the rate of growth of R is of the order of $u(R/l)^{1/3}$ where u is a measure of the large-scale velocity fluctuations.

Although Richardson was interested in the diffusion of a small cloud of contaminant, it is often more convenient to think in terms of the relative separation of a pair of marked fluid particles. The connection between the relative dispersion of two marked particles and the spreading of a small cloud of contaminant is, perhaps, intuitively clear. Nevertheless, it is useful to clarify the precise relationship between the two problems. The formal link was established by Brier in 1950 and a little later by Batchelor in 1952. (There is an extensive discussion of these papers in Monin and Yaglom, 1975.) The key result is the following. Let $\delta\mathbf{x}(t)$ be the instantaneous separation of two marked particles in the cloud. It turns out that the value of $\langle|\delta\mathbf{x}|^2\rangle$, averaged over all particle pairs in the cloud, is equal to twice the dispersion, σ, of the cloud about its center of mass:

$$\left\langle|\delta\mathbf{x}|^2\right\rangle_{AV} = 2\sigma = 2\frac{\int(\mathbf{x}-\mathbf{x}_c)^2\langle C\rangle d\mathbf{x}}{\int\langle C\rangle d\mathbf{x}}$$

where C is the concentration of the contaminant, $\mathbf{x}_c$ is the center of mass of the cloud, and the subscript AV indicates an average over all particle pairs in the cloud. Thus the value of $\left\langle|\delta\mathbf{x}|^2\right\rangle_{AV}^{1/2}$ provides a convenient measure of the effective diameter of the cloud. For example, if the concentration, $\langle C\rangle$, falls off in a Gaussian manner then the radius $\frac{1}{2}\left\langle|\delta\mathbf{x}|^2\right\rangle_{AV}^{1/2}$ marks the point where the concentration has fallen to $\sim$50% of the maximum.

[10] This law was originally determined in an empirical fashion by Richardson who released balloons into the atmosphere and then followed their trajectory. Although its roots are empirical, the four-thirds law has proved to be fairly robust. Hunt (1998) provides a whimsical account of how Richardson offered a demonstration of the four-thirds law for a visiting colleague by tossing parsnips into a nearby loch and monitoring their subsequent movement.

The four-thirds law, when applied to two-particle dispersion, becomes

$$\frac{d}{dt}\langle|\delta\mathbf{x}|^2\rangle \sim \varepsilon^{1/3}\langle|\delta\mathbf{x}|^2\rangle^{2/3}, \quad \eta \ll |\delta\mathbf{x}| \ll l$$

where $\delta\mathbf{x}(t)$ is the instantaneous separation of two particles which were released at $t=0$ from two fixed adjacent points. Remember that this equation is valid only if:

(i) $\eta \ll |\delta\mathbf{x}| \ll l$;
(ii) sufficient time has passed for the precise conditions of the release of the particles to have been forgotten, the 'memory time' being of the order of $\left(|\delta\mathbf{x}|_0^2/\varepsilon\right)^{1/3}$.

In statistically steady flows Richardson's law integrates to yield,

$$\langle|\delta\mathbf{x}|^2\rangle = g\varepsilon t^3$$

for some coefficient g, known as the *Richardson constant*. It might be thought that g has a universal value, valid for all types of turbulence, in the same way that Kolmogorov's constant was once thought to be universal. However, the experimental estimates of g quoted in the literature display a bewildering range of values, from 0.06 to $\sim$1 (see, for example, Monin and Yaglom, 1975). Closure models of relative dispersion also produce a wide range of values of g, from around 0.1 to $\sim$5 (see Sawford, 2001, for a recent review). Numerical simulations, on the other hand, seem to suggest $g \sim 0.4 \rightarrow 0.7$ (Boffetta and Sokolov, 2002; Ishihara and Kaneda, 2002), which is consistent with more recent experiments (Ott and Mann, 2000). Quite why there is so wide a range of estimates of g is still a matter of debate, though it is noticable that the value of Re in some of the experiments and numerical simulations is rather modest.

Interestingly there is some evidence to suggest that the manner in which particles separate is highly intermittent. That is to say, adjacent particles tend to stay close together for a long time and then separate quite suddenly (Jullien et al., 1999). If this is correct then one might suppose that two particles will separate through a sequence of sudden bursts, their spacing remaining more or less constant between bursts. Some have argued that the location of these bursts are the rapidly diverging streamlines found, say, downstream of two counter-rotating vortices, or at an irrotational stagnation point. Such regions are characterized by a relatively high pressure, suggesting that sudden particle separations may occur in regions where

$$\nabla^2(p/\rho) = \tfrac{1}{2}\boldsymbol{\omega}^2 - S_{ij}S_{ij} < 0$$

that is, regions of low enstrophy and high strain.

There is an alternative (though possibly related) cartoon which could explain the sporadic nature of two-particle separation. This relies on the observation that the vorticity in the inertial range is spatially intermittent, possibly consisting of randomly orientated vortex tubes (Plate 8). The acceleration field is similarly intermittent, being very intense in the thin annuli which envelope the vortex tubes, but somewhat weaker in the irrotational regions between the tubes. A single fluid particle will therefore follow a very irregular path, consisting of a sequence of almost straight-line trajectories (sometimes called ballistic trajectories) interspersed by sudden changes in velocity. These sporadic changes in trajectory arise when the particle approaches a vortex tube, falling prey to the locally high values of acceleration. (A crude analogy might be the motion of a ball in a pin-ball machine.) If we now consider a pair of marked fluid particles we might envisage sudden changes in relative separation when one, but not both, particles falls prey to the intense acceleration field of a vortex tube. Note that the vortex tubes which play the most important role here are those whose diameter are more or less matched to the separation of the pair of particles. That is to say, vortices much larger than the separation are unlikely to cause the pair to diverge, as they induce similar velocities in each particle. On the other hand the characteristic velocity of a vortex of size r is proportional to $r^{1/3}$, so that eddies much smaller than the particle separation are less effective than those matched to it.

Let us now return to our small turbulent puff of radius R and explore the implications of Richardson's law for the rate of growth of the cloud. For simplicity we shall restrict ourselves to statistically steady turbulence, so that the equation above implies $R^2 \sim \varepsilon t^3$. This expression can be rewritten as $R \sim (ult)^{1/2}(R/l)^{2/3}$, and since Richardson's law applies only for $R \ll l$ we conclude that $R \ll (ult)^{1/2}$. Now recall that the single-particle random walk argument of the previous section gives $\langle \mathbf{X}^2 \rangle \sim (ult)$ for large t. When such an argument is applied to the centroid of our small puff we see that the average distance over which the centroid will migrate in a time t is $(ult)^{1/2}$, which is much greater than Richardson's estimate of R. Of course, this is to be expected. A small puff migrates faster than it spreads.

Once the diameter of the puff exceeds the large-eddy size, $R > l$, Richardson's law fails because there are no eddies of size R. At this point particles on opposite sides of the cloud are more or less statistically independent of each other and since each moves as if in a state of Brownian motion the edges of the cloud will spread at a rate $R \sim t^{1/2}$. That is, the dispersion of the puff is essentially the same as that of a cloud resulting from a continuous release, and so the puff grows as $R \sim (ult)^{1/2}$.

When $R \gg l$ (and the turbulence is statistically steady and isotropic) the spread of a cloud of contaminant may be described, at least approximately, using an eddy diffusivity argument. That is

$$\frac{\partial \langle C \rangle}{\partial t} = \alpha_t \nabla^2 \langle C \rangle, \quad \alpha_t = u' l_m$$

where $\langle C \rangle$ is the ensemble averaged concentration field, α_t is the eddy diffusivity, and l_m, the mixing length, is of the order of the integral scale of the turbulence. For an initially spherical cloud this has the well-known solution

$$\langle C \rangle = \frac{I_0}{(4\pi \alpha_t t)^{3/2}} \exp\left[-\frac{r^2}{4\alpha_t t}\right], \quad I_0 = \int \langle C \rangle dV = \text{const.}$$

Note that the cloud radius grows as $R \sim (\alpha_t t)^{1/2}$, as expected.

5.4.3 *The influence of mean shear on turbulent dispersion*

So far we have assumed that the mean velocity is zero. When it is non-zero, and in particular when there exists a mean shear, the arguments of the previous two sections do not apply and a new phenomenon, known as shear dispersion, comes into play. Perhaps this is most readily illustrated by the following simple (if somewhat artificial) example. Suppose that we have a steady, homogeneous, turbulent shear flow in which $\bar{\mathbf{u}} = Sy\hat{\mathbf{e}}_x$. We are interested in the long-term spread of a cloud of contaminant which is released at $t = 0$ and whose mean radius is much larger than the integral scale l. Since $R \gg l$ and $t \gg l/u$, we might tentatively model the process using a turbulent eddy diffusivity, $\alpha_t \sim ul$. Strictly speaking we should use an anisotropic eddy diffusivity since, as we saw in Section 4.4.1, the turbulence in a shear flow is anisotropic. However, since this example is intended for illustrative purposes only we shall take the diffusivity to be a simple scalar. The governing equation for the expected (ensemble averaged) concentration is then

$$\frac{\partial \langle C \rangle}{\partial t} + Sy \frac{\partial \langle C \rangle}{\partial x} = \alpha_t \nabla^2 \langle C \rangle.$$

It is readily confirmed that, if the cloud is initially spherical, this has the solution

$$\langle C \rangle = \frac{I_0}{(4\pi \alpha_t t)^{3/2}} \frac{1}{(1+\lambda^2)^{1/2}} \exp\left[-\frac{(x-\sqrt{3}\lambda y)^2}{4\alpha_t t(1+\lambda^2)} - \frac{y^2}{4\alpha_t t} - \frac{z^2}{4\alpha_t t}\right],$$
$$\lambda = St/\sqrt{12}.$$

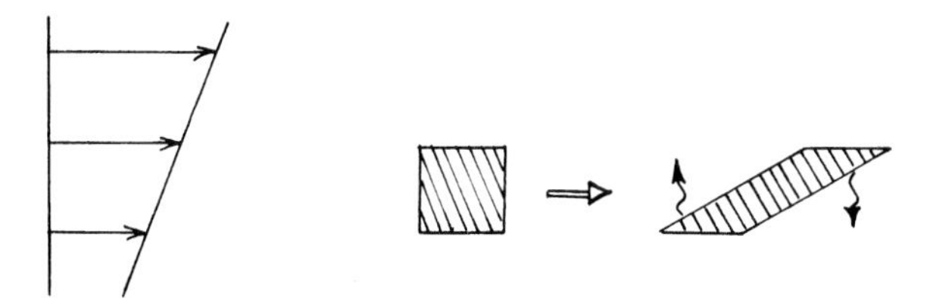

Figure 5.35 A patch of dye is sheared as it moves downstream. A combination of shearing plus cross-stream diffusion causes the patch to grow rapidly in the steamwise direction.

Note that, for small St, we recover the solution given above for no shear. For large St, on the other hand, we have

$$\langle C \rangle = \frac{I_0}{(4\pi\alpha_t t)^{3/2}} \frac{1}{\lambda} \exp\left[-3(\hat{x} - \hat{y})^2 - \hat{y}^2 - \hat{z}^2\right],$$

$$\hat{x} = \frac{x}{St\sqrt{\alpha_t t}}, \quad \hat{y} = \frac{y}{\sqrt{4\alpha_t t}}, \quad \hat{z} = \frac{z}{\sqrt{4\alpha_t t}}.$$

Evidently the cloud spreads along the y and z-axes in the usual way, growing at a rate $R_y \sim (4\alpha_t t)^{1/2}$, $R_z \sim (4\alpha_t t)^{1/2}$. However, it spreads along the x-axis at the much faster rate of $R_x \sim St\,(\alpha_t t)^{1/2}$. Thus an initially spherical cloud progressively deforms into an inclined ellipsoid. The mechanism for the enhanced dispersion in the axial direction is clear. Suppose some of the contaminant diffuses from y to $y + \delta y$ in a time t. Since the cross-stream diffusion proceeds in the usual way we have $\delta y = (\alpha_t t)^{1/2}$. However, the fluid at $y + \delta y$ moves at a different rate to that at y, the difference in axial velocity being $S\delta y$. Thus, as the contaminant diffuses upward it is also sheared by the mean flow. The difference in horizontal movement of material at y and $y + \delta y$ is, in a time t, $S\delta y t \sim St\,(\alpha_t t)^{1/2}$. It is this which fixes the length scale for x in this shear-enhanced diffusion process (Figure 5.35).

Although the toy problem above illustrates rather nicely the impact of a mean shear on turbulent dispersion, its specific conclusions should be applied with care. We have assumed an infinite fluid (no boundaries) and uniform shear. When we have both a mean shear and *solid boundaries* the behaviour can be quite different. Perhaps the simplest illustration of this is turbulent dispersion in pipe flow. This was investigated by Taylor in 1954, although the key ideas can be found in a slightly earlier paper, also by Taylor, on diffusion in laminar pipe flow (Taylor, 1953). Perhaps it is worth taking a moment to summarize Taylor's findings, if only to show how different things can be when boundaries are present.

Let us start with laminar flow in a circular pipe in which the velocity distribution can be written as $u = u_0[1 - r^2/R^2]$, u_0 being the centreline velocity. Suppose that we inject a cloud or slug of contaminant into the pipe. The slug will be swept downstream and also spread under the action of diffusion. Two natural questions are: how rapidly will the centroid of the slug be swept downstream?; and how rapidly will the slug disperse (spread) relative to its centre? Taylor was interested particularly in the case where the cross-stream diffusion of a

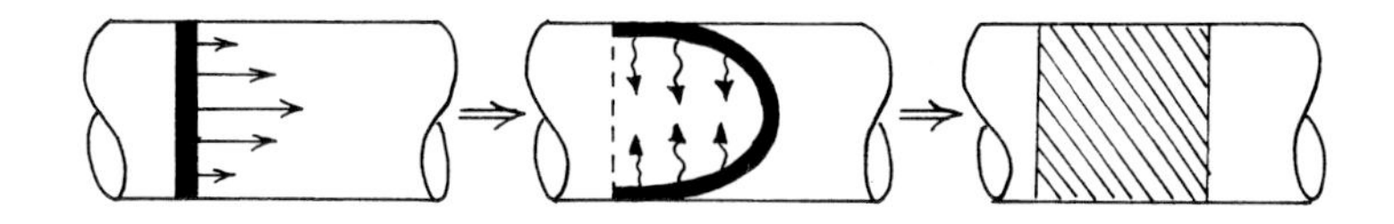

Figure 5.36 A sheet of dye is sheared into a parabola as it moves downstream. A combination of shearing plus cross-stream diffusion causes the sheet to grow into a slug of length $u\delta t$.

pollutant is much faster than the axial dispersion caused by radial gradients in the mean velocity. (This turns out to be the important case in turbulent pipe flow.) In some sense this implies that the pipe is narrow, or else the diffusivity is high. We can make this more precise as follows. The time taken for a pollutant to diffuse from the axis to the wall is of order $\tau \sim R^2/14\alpha$. (The factor of 14 comes from the radial diffusion equation, whose solution involves a Bessel function.) In the same time the cross-stream gradient in mean velocity will cause two adjacent packets of fluid, which initially lie in the same plane, but at different radii, to separate by an amount $\delta z \sim u_0\tau$. Or equivalently, points initially separated by an axial distance $\delta z \sim u_0\tau$ can find themselves in the same cross-section after a time τ. This differential movement in the axial direction causes radial variations in C of the order of

$$\delta C \sim (\partial C/\partial z)\delta z \sim (\partial C/\partial z)u_0\tau.$$

Thus, provided that

$$u_0 R^2/\alpha L \ll 1,$$

diffusion will homogenize things in the radial direction much faster than radial gradients in C are induced by the differential axial velocity. (Here L is the characteristic axial length-scale of the contaminant cloud.) Taylor showed that, when this condition is satisfied, the centroid of the slug moves downstream with a speed equal to the mean speed in the pipe, $\bar{u} = u_0/2$. Essentially this is because of the rapid homogenization in the radial direction, as can be seen from the following heuristic argument. Suppose that the peak in C, say $\hat{C}$, at some instant takes the form of a plane surface normal to the pipe axis (Figure 5.36). Then the various points on this plane will move at different rates corresponding to different radial positions and the contour of $C = \hat{C}$ will start to bow out to form a parabola. However, significant radial gradients in C cannot develop because rapid cross-stream diffusion eradicates them as soon as they start to form. Thus the would-be parabolic contour of $C = \hat{C}$, which in the absence of diffusion would be $u_0\delta t$ deep, diffuses to form a short cylinder of length $u_0\delta t$, with the centre of the cylinder located at a point $\bar{u}\delta t$ down-stream of the starting point. The suggestion is that the peak in C moves with speed $\bar{u}$. (Actually this argument is too simplistic, and the reader is urged to consult Taylor's original paper for a rigorous derivation.)

Taylor also showed that the slug of pollutant spreads in a diffusive manner relative to its centre with an effective axial diffusivity of

$$k_z = \bar{u}^2 R^2/48\alpha.$$

At first sight the prediction that the centre of the slug moves at a speed of $\bar{u} = u_0/2$ seems quite unremarkable. However, a moments thought confirms that this is, in fact, an extraordinary result. Consider a packet of fluid which lies on the centreline of the pipe and upstream of the slug of pollutant. It moves at twice the mean flow speed and so approaches the slug, passes through it, and then re-emerges at the far end. As it passes into the slug it absorbs the pollutant, which is not so surprising. What is surprising, though, is that the packet eventually re-emerges pollution free! The explanation for this curious behaviour is the following. As a blob of fluid on the centreline of the pipe begins to emerge from the downstream end of the pollutant cloud, it finds itself surrounded by fluid depleted in the contaminant. It then starts to lose its pollutant to the surrounding fluid by cross-stream diffusion. Since, in a narrow pipe, this is a very rapid process, the fluid does not have to travel far before it has given up all of its contaminant. So, this bizarre behaviour does have an explanation after all. Still, this phenomenon is sufficiently counter-intuitive that Taylor felt compelled to verify his predictions by experiment. In his 1953 paper Taylor reports, with a note of quiet satisfaction, that experiments did indeed confirm his analysis.

Many of these ideas carry over to turbulent pipe flow. For example, as in the laminar case, the centre of a slug of pollutant moves at the mean flow speed. Moreover, the slug grows in length (relative to its centre) in a diffusive manner with an effective axial diffusivity of

$$k_z \sim 10RV_*$$

where V_* is the shear velocity. We can rationalize this estimate of k_z using an eddy diffusivity argument. The laminar result suggests $k_z = \bar{u}^2R^2/48\alpha_t \sim \bar{u}^2R/48V_*$, since the eddy diffusivity will be of the order of the eddy viscosity which, in turn, is of order RV_*. Now $\bar{u}/V_*$ is a weak function of Re, but for the range of Reynolds numbers encountered in practice it is around 24–26. Substituting for $\bar{u}/V_*$ we arrive at $k_z \sim 13RV_*$. A more careful analysis changes the factor of 13 to one of 10.

This concludes our brief discussion of turbulent diffusion. We have barely scratched the surface of what is a difficult but intriguing area of ongoing research.

5.5 Why turbulence is never Gaussian

We close this chapter by revisiting the question of the nature of the probability distribution of **u**. The main point which we wish to emphasize is that the dynamics of turbulence is such that the statistical distribution of **u** is anything but Gaussian. This is important because certain popular closure schemes have been introduced on the assumption that aspects of the turbulent velocity field may be treated as Gaussian. In their purest form these closure schemes rapidly run into trouble, and a variety of *ad hoc* fixes are required to avoid the most obvious breaches of the laws of nature. So our main message here is that great caution must be employed if you adopt such a closure scheme.

5.5.1 *The experimental evidence and its interpretation*

Let us start by reminding you of some of the elementary ideas of statistics. Suppose we have a random variable u which is determined by the result of some experiment. For example, u may be the lifetime of a light-bulb selected at random from a large collection of bulbs. Let $P(u)$ be the *probability density function* for u. So, if we choose N light bulbs at random, the likely number of times a light-bulb fails in the time-interval $u_0 \to u_0 + du$ is $NP(u_0)du$. In short, $P(u_0)du$ is, on average, the fraction of time the random variable u spends in the range $u_0 \to u_0 + du$. By definition

$$\int_{-\infty}^{\infty} P(u)du = 1$$

since the sum of all the probabilities must add up to one. Now the ensemble average of some function $f(u)$ is,

$$N^{-1}\sum_{1}^{N} [f(u_i) \times (\text{number of times } u_i \text{ is realized})]$$

where N is the total number of realizations. Since $P(u_i)du$ is the relative number of times u_i is likely to appear we can rewrite this as

$$\langle f \rangle = \int_{-\infty}^{\infty} f(u)P(u)du.$$

For example, the relative number of times a particular value of u is likely to occur is $P(u)du$, so the mean of u is

$$\langle u \rangle = \int_{-\infty}^{\infty} uP(u)du.$$

Let us now restrict ourselves to random variables which have zero mean, $\langle u \rangle = 0$, since this is the case of most interest in turbulence theory. Then we may introduce three quantities which characterise the probability density function $P(u)$. These are,

$$\langle u^2 \rangle = \int_{-\infty}^{\infty} u^2P(u)du = \sigma^2 \quad \text{(variance of } u\text{)}$$

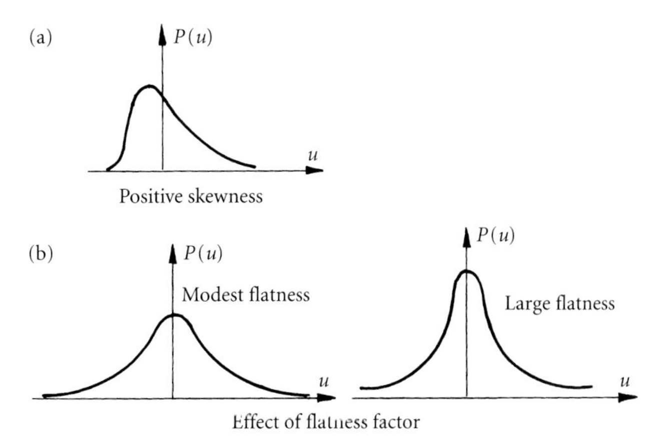

Figure 5.37 Different probability density functions: (a) a positive skewness factor; (b) large and small flatness factors.

$$\langle u^3 \rangle = \int_{-\infty}^{\infty} u^3 P(u) du = S\sigma^3 \quad \text{(skewness of } u\text{)}$$

$$\langle u^4 \rangle = \int_{-\infty}^{\infty} u^4 P(u) du = \delta\sigma^4 \quad \text{(flatness of } u\text{).}$$

The parameters S and δ are the skewness and flatness factors, respectively. They are, of course, dimensionless. The variance, σ^2, is a measure of the spread of $P(u)$ about zero. If σ^2 is small, then u spends most of its time close to zero. If σ^2 is large, on the other hand, then u may take large excursions about zero. Of course, σ is the standard deviation, or r.m.s value, of u.

S is a measure of lop-sidedness of $P(u)$ (Figure 5.37(a)). If $P(u)$ is symmetric about $u = 0$, so that positive and negative probabilities of u are symmetrically distributed, then $S = 0$. The sign of S tells us whether u is more likely to cling to $u = 0$ for positive or negative excursions about $u = 0$. If u clings to $u = 0$ more often for negative excursions than positive ones, then S is positive.

For a given value of σ^2, the flatness factor tells us how far, and for how much time, u departs from $u = 0$ (Figure 5.37(b)). Probability density functions with narrow peaks and broad skirts have relatively large values of δ, and a signal in which δ is large is usually very intermittent, exhibiting periods of quiescence interspersed by large transient excursions away from $u = 0$ (see the example in Section 3.2.7).

In Section 3.2.7, we introduced the idea of the probability distribution of u'_x and of $\Delta v = u'_x(\mathbf{x} + r\hat{\mathbf{e}}_x) - u'_x(\mathbf{x})$. We saw the probability density function for u'_x in grid turbulence is very nearly Gaussian. We interpreted this as follows. The velocity at any one point $\mathbf{x}$ is the consequence of a large number of randomly oriented vortical structures in the vicinity of $\mathbf{x}$, the relationship between $\mathbf{u}$ and the surrounding vorticity being fixed by the Biot–Savart law (Figure 5.38). If the number of relevant vortices is large, and they are

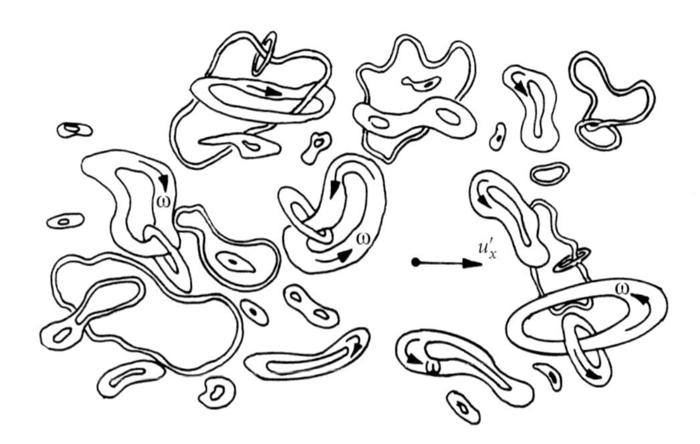

Figure 5.38 The probability density of $u_x'(x_0)$ is approximately Gaussian because it is the result of a large number of randomly orientated vortices in the neighbourhood of x_0.

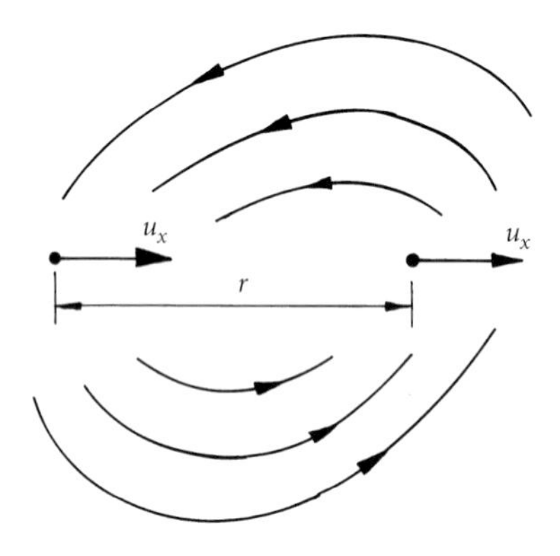

Figure 5.39 The probability density function for Δv is not determined by chance but rather by the local dynamics of the flow.

indeed randomly distributed, then the central limit theorem tells us that **u** should have a Gaussian distribution (see Section 3.2.3 for a brief discussion of the central limit theorem).

The probability distribution of Δv, on the other hand, is not at all Gaussian. This is not surprising since *differences in velocity* at any instant are not dictated by pure chance. As eddies sweep through the region surrounding $\mathbf{x}$ and $\mathbf{x} + r\hat{e}_x$ the velocities at the two points are strongly correlated and so the probability distribution of Δv is largely influenced by the dynamics of the turbulence (Figure 5.39). Figure 5.40 shows the flatness and skewness factors for Δv in grid turbulence. For $r \leq l$ they depart significantly from the Gaussian values of $\delta(r) = 3$, $S(r) = 0$.

This departure from a Gaussian distribution is not coincidental. Consider turbulence which is homogeneous and isotropic. Combining the enstrophy equation (5.28) with (5.32) we have

$$\frac{\partial}{\partial t}\langle \boldsymbol{\omega}^2/2 \rangle = -\frac{7}{6\sqrt{15}} S_0 \langle \boldsymbol{\omega}^2 \rangle^{3/2} - \nu \langle (\nabla \times \boldsymbol{\omega})^2 \rangle \tag{5.37}$$

where $S_0 = S(0)$. If the skewness of Δv were zero, as in a Gaussian distribution, $\langle \boldsymbol{\omega}^2 \rangle$ would simply monotonically decline due to viscous dissipation. There would be no vortex stretching and no energy cascade. So a non-zero skewness is an essential part of the dynamics of turbulence.[11]

Consider now the flatness factor of Δv, which exceeds the Gaussian value of 3. A large flatness factor implies that the probability density function has a narrow, higher peak and broader skirts than a Gaussian distribution of the same variance. This implies that very small and very large values of Δv are both more likely than a Gaussian distribution

[11] Betchov (1956) has shown that, in isotropic turbulence, the incompressibility of the fluid imposes a constraint on the value of S_0: $|S_0| \leq [4\delta(0)/21]^{1/2}$. Thus the skewness of Δv sets the rate of generation of enstrophy, but within limits imposed by the flatness factor.

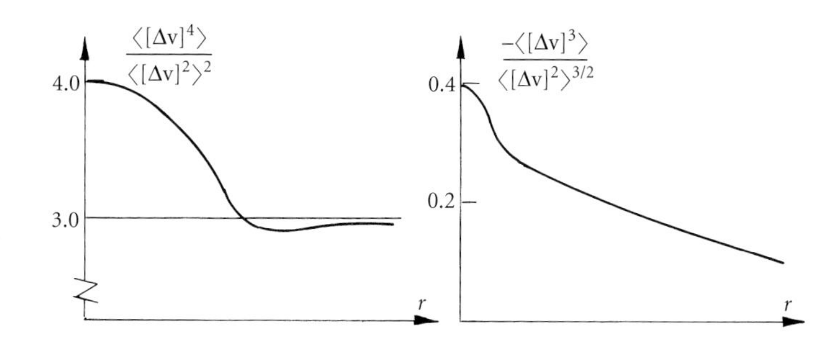

Figure 5.40 Skewness and flatness factors for Δv in grid turbulence.

would suggest. Now for small r we have $\Delta v / r \sim \partial u_x / \partial x$ and so

$$\delta(0) = \delta_0 = \left\langle (\partial u_x / \partial x)^4 \right\rangle \Big/ \left\langle (\partial u_x / \partial x)^2 \right\rangle^2.$$

We have seen that, in grid turbulence at $\text{Re} \sim 500$, $\delta_0 \sim 4$ (Figure 5.40). Measurements at higher Re, say in the atmospheric boundary layer, suggest[12] $\delta_0 \sim 3 + \frac{1}{2}(ul/\nu)^{0.25}$ for $10^3 < \text{Re} < 10^7$. So the flatness factor increases with Re. Now suppose we measure the flatness factors of $\partial^n u_x / \partial x^n$, rather than $\partial u_x / \partial x$. Let

$$(\hat{\delta})_n = \left\langle (\partial^n u_x / \partial x^n)^4 \right\rangle \Big/ \left\langle (\partial^n u_x / \partial x^n)^2 \right\rangle^2.$$

There is considerable scatter in the results, but to a first approximation the wind-tunnel data indicates that, for grid turbulence at modest Re, say $\text{Re} \sim 10^3$,

$$(\hat{\delta})_n = (n + 3) \pm 20\% \tag{5.38}$$

so we have $(\hat{\delta})_0 \sim 3$, $(\hat{\delta})_1 \sim 4$, $(\hat{\delta})_2 \sim 5$ and so on (Batchelor, 1953). In summary, then, the higher the derivative the less Gaussian the probability density function looks and the higher the Reynolds number, the greater the departure from a normal distribution. Now, as we have already noted, very high flatness factors are associated with a signal which is highly intermittent, spending large amounts of time near zero, suddenly bursting into life, and then returning to zero once again. So it would seem that gradients in velocity (which are most pronounced at the Kolmogorov scales) are spatially intermittent. The implication is that there is a significant fraction of space in which $\boldsymbol{\omega}$ is very weak, and then areas of concentrated vorticity. In short, the vorticity field is very spotty at the small scales.

This observation is consistent with the 'spaghetti cartoon' of turbulence in which the vorticity tubes are pictured as a seething tangle of spaghetti (Plate 8). The velocity field induced by the vortex tubes (via the Biot–Savart law) continually stirs the vortex filaments (spaghetti), teasing them out into finer and finer structures. Such a process would result in the small scale vorticity being highly intermittent, concentrated in fine-scale vortex filaments. Moreover, the degree of intermittency should rise as Re increases. All of this is consistent with the observations.

[12] See Sreenivasan and Antonia (1997).

However, one must be cautious in drawing too many conclusions from the observed intermittency. There are other, equally plausible, mechanisms which would explain spatial intermittency, such as the formation and subsequent roll-up of large-scale vortex sheets, resulting in small-scale vortex tubes. Indeed, as we have seen, another popular cartoon of turbulence asserts that much of the vorticity resides in thin sheets. These sheets are continually rolling up through a Kelvin–Helmholtz-like mechanism to produce fine-scale tubes. In this cartoon most of the energy transfer to small scales is associated with the formation and roll-up of the sheets, and the fine-scale tubes, called *worms*, represent passive debris. (The evidence for this picture is laid out in Section 7.3.1.)

In any event, whatever be the source of the intermittency, it is clear from the experimental evidence that the probability distribution of Δv is anything but Gaussian. Indeed, (5.37) tells us that Gaussian statistics are incompatible with the Kolmogorov picture of turbulence, with its energy cascade induced by vortex stretching. Curiously, though, an assumption of near Gaussian statistics forms the basis of certain closure schemes.

5.5.2 *A glimpse at closure schemes which assume near-Gaussian statistics*

In Chapter 4 we saw that taking successive averages of the Navier–Stokes equation leads to a hierarchy of equations of the form,

$$\frac{\partial}{\partial t}\langle uu\rangle = \frac{\partial}{\partial x}\langle uuu\rangle + \cdots$$

$$\frac{\partial}{\partial t}\langle uuu\rangle = \frac{\partial}{\partial x}\langle uuuu\rangle + \cdots$$

$$\frac{\partial}{\partial t}\langle uuuu\rangle = \frac{\partial}{\partial x}\langle uuuuu\rangle + \cdots$$

where $\langle uu\rangle$ is a symbolic representation of the second-order correlations, $\langle uuu\rangle$ represents the third-order correlations, and so on. Truncation of this series at any point always leaves more unknowns than equations, that is, we are up against the closure problem of turbulence. This impasse has been circumvented by Millionshtchikov's suggestion in 1941 that we treat the fourth-order velocity correlations as if the probability distribution of $\mathbf{u}$ is Gaussian. This is discussed in more detail in Chapter 6, but the main point to note is that the assumption that $\mathbf{u}$ is Gaussian allows us to write the fourth-order correlations in terms of products of second-order correlations:

$$\langle uuuu\rangle = \sum \langle uu\rangle\langle uu\rangle.$$

At a stroke we have solved the closure problem! Of course, the assumption of Gaussian statistics cannot be applied to the third-order correlations as this would lead to a complete absence of vortex stretching, and hence to the absence of an energy cascade. Thus the proposal of Millionshtchikov was to apply the Gaussian assumption to the fourth-order correlations only—the so-called *quasi-normal hypothesis*.[13] Our dynamical equations now become,

$$\frac{\partial}{\partial t}\langle uu\rangle = \frac{\partial}{\partial x}\langle uuu\rangle + \cdots$$
$$\frac{\partial}{\partial t}\langle uuu\rangle = \frac{\partial}{\partial x}\sum\langle uu\rangle\langle uu\rangle + \cdots$$

which may, in principle, be solved. In practice the resulting scheme, though closed, is still very complex and progress can be made only for idealized cases such as homogeneous turbulence. Nevertheless, the quasi-normal approximation has formed the basis of a number of two-point closure schemes. We shall discuss these schemes in Chapters 6 and 8, but perhaps it is worth noting now that, in their purest form, they rapidly run into trouble, producing unphysical results. It is only through the judicious use of a variety of *ad hoc* modifications that they can be made to work. In this respect they are not unlike the one-point closure models discussed in Chapter 4. Despite the obvious shortcomings of these closure schemes they are widely used in the physics and mathematics communities and yield detailed information about the structure of the turbulence: information which cannot be extracted from one-point closures.

5.6 Closure

This concludes our brief overview of the phenomenological models of Richardson, Taylor and Kolmogorov. These theories are quite unlike the usual laws of physics, in that they attempt to make very general statements about a highly complex, non-linear system on the basis of rather hand-waving arguments.

We have seen that there is some support for Richardson's energy cascade, with its many stages of energy transfer. On the other hand, we cannot completely rule out 'non-local' interactions, in which energy is passed directly from large to small scale, bypassing the cascade.[14] We have also seen that Kolmogorov's two-thirds law is well supported by

[13] This idea is given weight by measurements made in grid turbulence where we find that, for *well-separated points*, $\langle uuuu\rangle$ is indeed very nearly Gausssian. See, for example, Van Atta and Yeh (1970). Crucially, however, Gaussian behaviour is not observed for adjacent points.

[14] As a warning against placing too much reliance on Richardson's multistage cascade, Betchov suggested replacing Richardson's famous poem (the parody of Swift's sonnet) by: *Big whirls lack smaller whirls, to feed on their velocity. They crash to form the finest curls, permitted by viscosity* (see Section 1.6 for Richardson's poem).

the experimental data, which is rather remarkable in view of the shaky foundations on which it is based. Finally we have argued that the mechanism by which energy is passed down the cascade is vortex stretching and that this implies that the probability distribution of $\Delta \mathrm{v}$ cannot be Gaussian. The reason why vorticity is stretched is that, when ν is small, the vortex-lines are frozen into the fluid and of course material lines are, on average, extended in a turbulent flow. You will recall that, although it is physically obvious that material lines and surfaces are continually stretched in a turbulent flow, we had considerable difficulty formulating a rigorous proof that this is so!

The fact that a rigorous proof of the continual stretching of a material element is so hard to construct tells us something important about turbulence. Even the simplest and intuitively obvious hypotheses are extremely difficult to quantify and justify in a rigorous manner. In this respect the study of turbulence is somewhat different from many other branches of applied mathematics or theoretical physics. We have a wealth of experimental data, some plausible intuitive ideas, and very little in the way of rigorous results! This makes it all the more miraculous that Kolmogorov could predict something as concrete as $\langle[\Delta \mathrm{v}]^2\rangle \sim r^{2/3}$, $\langle[\Delta \mathrm{v}]^3\rangle \sim r$. In the next chapter we shall see what the great might of applied mathematics offers turbulence. We introduce correlation tensors of great complexity, massage the Navier–Stokes equation into a sequence of evolution equations for these complex quantities, and manipulate them as best we can. And still we obtain little in the way of rigorous results. At every turn the overwhelming power of nature puts paid to the attempts of applied mathematics to rationalize the dynamics of turbulence. Those few rigorous (or nearly rigorous) results we have, stem directly from simple physical laws. For example, conservation of the Saffman–Birkhoff invariant in freely evolving turbulence (see Section 3.2.3) is simply a manifestation of the law of conservation of linear momentum.

The closure problem of turbulence has proven to be a formidable barrier to developing a rigorous, predictive model of turbulence and indeed most of the 'problems of turbulence' remain unsolved. It is unlikely that this situation will change in the foreseeable future and perhaps we just have to get used to living in this twilight world of heuristic cartoons, supported by experimental data, and bolstered by the occasional theoretical finding.

Exercises

5.1 Consider the bursting vortex of Figure 5.6. Confirm that

$$\frac{d}{dt}\int_{z<0} (\omega_\theta / r)dV = 2\pi \int_0^\infty (\Gamma_0^2 / r^3)dr > 0$$

where $\Gamma_0 = \Gamma(z = 0)$. Now estimate the rate at which the vortex propagates radially outward.

5.2 Suppose that we approximate the probability distribution of $\partial^n u_x / \partial x^n$ by a combination of a δ-function at the origin, occupying an area $(1 - \gamma)$, and a normal distribution (of area γ) for all other values of $\partial^n u_x / \partial x^n$. This represents an intermittent random variable which spends $(1 - \gamma)$ per cent of its time at zero but the rest of the time is distributed in a Gaussian manner. Show that the flatness factor for $\partial^n u_x / \partial x^n$ is then $3/\gamma$. Given that, when $\mathrm{Re} \sim 10^3$, $(\hat{\delta})_n \sim (n + 3)$, estimate the intermittency factor γ for arbitrary n.

5.3 Show that, if the external strain field is removed from Burgers vortex, the vortex spreads and decays according to

$$\omega = (\Gamma_0 / 4\pi \nu t) \exp[-r^2/(4\nu t)].$$

5.4 Consider the vortex sheet shown in Figure 5.24. If the strain field which generated the sheet is associated with the large scales, $S_{ij} \sim u/l$, estimate the thickness of the sheet. Now consider the stability of the sheet and estimate the diameter of the vortex tubes which emerge from a Kelvin–Helmholtz instability.

5.5 Calculate the distribution of enstrophy and dissipation in a steady Burger's vortex and show that, in the limit of vanishing viscosity, the net dissipation per unit length of the vortex is independent of ν and equal to $\alpha\Gamma_0^2/(8\pi)$. Confirm that this is equal to the net inward flux of energy from infinity.

When ν is finite there is a difference between the volume integrals of $2\nu S_{ij} S_{ij}$ and $\nu \boldsymbol{\omega}^2$. If the domain of integration is a cylinder of large radius, concentric with the vortex, show that the difference in these integrals is equal to $(3\nu\alpha^2)Vol$. Show also that this difference arises from the work done by the normal viscous stresses acting on the surface of the cylindrical volume.

5.6 Consider the steady, viscous vortex sheet of Section 5.3.3. Calculate the distribution of enstrophy and dissipation and show that $2\nu S_{ij} S_{ij} = \nu \boldsymbol{\omega}^2 + 4\nu\alpha^2$. Explain why there is a difference in the volume integrals of dissipation and $\nu \boldsymbol{\omega}^2$.

5.7 Consider scaled velocity and coordinates, $\hat{\mathbf{u}}$ and $\hat{\mathbf{x}}$, defined by $\mathbf{u}(\mathbf{x}, t) = \hat{\mathbf{u}}(\hat{\mathbf{x}})/[2f(t_0 - t)]^{1/2}$ and $\hat{\mathbf{x}} = \mathbf{x}/[2f(t_0 - t)]^{1/2}$, where f and t_0 are positive constants. Show that these so-called Larey solutions, which explode at $t = t_0$, satisfy the Euler equation provided that $(\hat{\mathbf{u}} + f\hat{\mathbf{x}}) \times \hat{\boldsymbol{\omega}} = \hat{\nabla}\Pi$, $\hat{\boldsymbol{\omega}} = \hat{\nabla} \times \hat{\mathbf{u}}$, for some Π.

Appendix: The statistical equations for a passive scalar in isotropic turbulence: Yaglom's four-thirds Law and Corrsin's integral.

This appendix is a footnote to Section 5.2.2. Consider a passive scalar whose concentration C is governed by the advection–diffusion equation

$$\frac{\partial C}{\partial t} + \mathbf{u} \cdot \nabla C = \alpha \nabla^2 C. \tag{A.1}$$

Let us suppose that $\mathbf{u}$ and C have zero mean and that their distributions are statistically homogeneous and isotropic. In Section 5.2.2 we saw that the variance of C, $\langle C^2 \rangle$, is governed by

$$\frac{d}{dt}\left\langle \frac{1}{2}C^2 \right\rangle = -\alpha\left\langle (\nabla C)^2 \right\rangle = -\varepsilon_C.$$

Our aim is to determine the governing equation for the two-point correlations $\langle C(\mathbf{x})C(\mathbf{x}+\mathbf{r})\rangle = \langle CC' \rangle$. From (A.1) we have

$$\frac{\partial C}{\partial t} = -\nabla \cdot [\mathbf{u}C] + \alpha \nabla^2 C$$
$$\frac{\partial C'}{\partial t} = -\nabla' \cdot [\mathbf{u}'C'] + \alpha \nabla'^2 C'$$

where a prime indicates a quantity evaluated at $\mathbf{x}' = \mathbf{x} + \mathbf{r}$. On multiplying the first by C', and the second by C, adding the two and averaging, we obtain

$$\frac{\partial}{\partial t}\langle CC' \rangle = -\left\langle C'\frac{\partial}{\partial x_i}(u_i C) + C\frac{\partial}{\partial x_i'}(u_i' C')\right\rangle + \alpha\langle C'\nabla_x^2 C + C\nabla_{x'}^2 C'\rangle.$$

Next we note that the operations of averaging and differentiation commute, C' is independent of $\mathbf{x}$ and C independent of $\mathbf{x}'$, and $\partial/\partial x_i$ and $\partial/\partial x_i'$ operating on averages may be replaced by $-\partial/\partial r_i$ and $\partial/\partial r_i$, respectively. Our evolution equations for $\langle CC' \rangle$ now simplifies to

$$\frac{\partial}{\partial t}\langle C'C \rangle = -\frac{\partial}{\partial r_i}\langle (u_i' - u_i)C'C \rangle + 2\alpha \nabla_r^2 \langle C'C \rangle.$$

From (6.30) the isotropic tensor $\langle (u_i' - u_i)C'C \rangle$ can be written in the form,

$$\langle (u_i' - u_i)C'C \rangle = A(r) r_i$$

where $A(r)$ might be defined by, say,

$$rA(r) = \langle \Delta u_{\|} C'C \rangle.$$

Here $\Delta u_{\|}$ is the component of $(\mathbf{u}' - \mathbf{u})$ parallel to $\mathbf{r}$. It follows that

$$\frac{\partial}{\partial r_i}\langle (u_i' - u_i)C'C \rangle = rA'(r) + 3A = \frac{1}{r^2}\frac{d}{dr}(r^3 A)$$

from which we conclude that our evolution equation for $\langle CC' \rangle$ can be rewritten as

$$\frac{\partial}{\partial t}\langle CC' \rangle = -\frac{1}{r^2}\frac{\partial}{\partial r}\left[r^2 \langle \Delta u_{\|} CC' \rangle\right] + 2\alpha \frac{1}{r^2}\frac{\partial}{\partial r} r^2 \frac{\partial}{\partial r}\langle CC' \rangle$$

or equivalently

$$\frac{\partial}{\partial t}[r^2\langle CC'\rangle] = -\frac{\partial}{\partial r}\left[r^2\langle \Delta u_{\|} CC'\rangle\right] + 2\alpha\frac{\partial}{\partial r}r^2\frac{\partial}{\partial r}\langle CC'\rangle. \tag{A.2}$$

As always we have hit the closure problem in that, to determine the evolution of $\langle CC'\rangle$, we need to know about $\langle \Delta u_{\|} CC'\rangle$. Nevertheless we can extract useful information from (A.2). Our first step is to note that, if γ is any scalar quantity then, as shown in Chapter 6, continuity demands

$$\langle \mathbf{u}\gamma'\rangle = 0$$

It follows that

$$\langle \Delta u_{\|}(\Delta C)^2\rangle = -2\langle \Delta u_{\|} CC'\rangle, \quad \Delta C = C' - C$$

and (A.2) becomes

$$\frac{\partial}{\partial t}\langle CC'\rangle = \frac{1}{r^2}\frac{\partial}{\partial r}r^2\left[\frac{1}{2}\langle \Delta u_{\|}(\Delta C)^2\rangle + 2\alpha\frac{\partial}{\partial r}\langle CC'\rangle\right]. \tag{A.3}$$

When working in the inertial-convective subrange it is convenient to use the structure function $\langle(\Delta C)^2\rangle$ rather than $\langle CC'\rangle$, the two being related by

$$\langle(\Delta C)^2\rangle = 2\langle C^2\rangle - 2\langle CC'\rangle.$$

In the universal equilibrium range, where $r \ll l$, the left-hand side of (A.3) simplifies to

$$\frac{\partial}{\partial t}\langle CC'\rangle = \frac{\partial}{\partial t}\langle C^2\rangle - \frac{1}{2}\frac{\partial}{\partial t}\langle(\Delta C)^2\rangle \approx \frac{\partial}{\partial t}\langle C^2\rangle = -2\varepsilon_c$$

since the time derivative of $\langle(\Delta C)^2\rangle$ is much less than that of $\langle C^2\rangle$. Thus, in the equilibrium range (A.3) integrates to yield

$$\langle \Delta u_{\|}(\Delta C)^2\rangle - 2\alpha\frac{\partial}{\partial r}\langle(\Delta C)^2\rangle = -\frac{4}{3}\varepsilon_c r, \quad (r \ll l). \tag{A.4}$$

In the inertial-convective subrange, where diffusion is negligible, this simplifies to

$$\langle \Delta u_{\|}(\Delta C)^2\rangle = -\frac{4}{3}\varepsilon_c r, \quad (\eta \ll r \ll l) \tag{A.5}$$

which is the passive scalar equivalent of Kolmogorov's four-fifths law. Equation (A.5) is known as *Yaglom's four-thirds law*. On the other hand, for $r \to 0$ the convective term is negligible and (A.4) integrates to give

$$\langle(\Delta C)^2\rangle = \frac{\varepsilon_c}{3\alpha}r^2 + \cdots \tag{A.6}$$

Let us now return to (A.2) and integrate the equation from $r = 0$ to $r \to \infty$. Assuming that $\langle CC' \rangle$ decays faster than r^{-3} at large r we have,

$$\frac{d}{dt}\int_0^\infty r^2 \langle CC' \rangle dr = -\left[r^2 \langle \Delta u_{\|} CC' \rangle\right]_\infty.$$

If, and it is a significant *if*, the convective correlation $\langle \Delta u_{\|} CC' \rangle$ decays faster than r^{-2} for large r, then this yields the integral invariant

$$K = \int \langle CC' \rangle d\mathbf{r} = \text{constant}. \tag{A.7}$$

This is known as *Corrsin's integral*. We can rewrite K as,

$$K = \frac{1}{V}\left[\int C\, dV\right]^2$$

where V is some large volume. The physical interpretation of (A.7) is therefore one of the global conservation of the contaminant in a large volume. The requirement that $\langle \Delta u_{/\!/} CC' \rangle_\infty$ decays faster than r^{-2} is equivalent to specifying that the turbulent advection of C through the bounding surface of V is negligible.

Suggested reading

Bajer, K. and Moffatt, H.K. (2002) *Tubes, Sheets, and Singularities in Fluid Dynamics*. Kluwer Academic Publishers.

(This contains several articles on finite-time singularities.)

Batchelor, G.K. (1947) *Proc. Camb. Phil. Soc.*, **43**, 533–559.

Batchelor, G.K. (1953) *The Theory of Homogeneous Turbulence*. Cambridge University Press.

(A measured description of Kolmogorov's theory is given in Chapter VI, and a nice discussion of the probability distribution of **u** in Chapter VIII.)

Betchov, R. (1956) *J. Fluid Mech.* **1**(5), 497.

Boffetta, G. and Sokolov, I.M. (2002) *Phys. Rev. Lett.*, **88**(9).

Bradshaw, P. (1971) *An Introduction to Turbulence and its Measurement*. Pergamon Press.

(A short, introductory text.)

Burgers, J.M. (1948) A mathematical model illustrating the theory of turbulence. *Adv. Appl. Mech.*, **1**, 171–199.

Frisch, U. (1995) *Turbulence*. Cambridge University Press.

(Virtually the entire text is devoted to Kolmogorov's theories, some of which are reinterpreted in a novel way. We have taken a more conventional, pedestrian route.)

Hunt, J.C.R. (1998) *Ann. Rev. Fluid Mech.*, **30**, 8–35.

Hunt, J.C.R., et al. (2001) *J. Fluid Mech.*, **436**, 353.

Ishihara, T. and Kaneda, Y. (2002) *Phys. Fluids*, **14**(11), L69.

Ishihara, T., Yoshida, K., and Kaneda, Y. (2002) *Phys. Rev. Lett.*, **88**(15).

Jackson, J.D. (1998) *Classical Electrodynamics*. Wiley, 3rd edition.

(A superb reference text on electrodynamics.)

Jimenez, J. and Wray, A.A. (1998) *J. Fluid Mech.*, **373**.
Jullien, M.C. et al. (1999) *Phys. Rev. Lett.*, **82**.
Kaneda, Y. et al. (2003) *Phys. Fluids*, **15**(2), L21.
Kolmogorov, A.N. (1962) *J. Fluid Mech.* **13**, 82–85.
Kolmogorov, A.N. (1991) *Proc. Roy. Soc.* A, **434**, 9–13, 15–17. (These are English translations of Kolmogorov's 1941 papers on the small scales.)
Landau, L.D. and Lifshitz, E.M. (1959) *Fluid Mechanics*. Pergamon Press, 1st edition.
(Chapter 3, on turbulence, provides one of the most beautiful and compact introductions to turbulence.)
Laval, J.P. et al. (2001) *Phys Fluids*, **13**(7), 1995–2012.
Lesieur, M. (1990) *Turbulence in Fluids*. Kluwer Academic Publishers.
(Chapter 6 discusses Kolmogorov's theory and Obokhov's law.)
Monin, A.S. and Yaglom, A.M. (1975) *Statistical Fluid Mechanics II*. MIT Press.
(This is the bible of homogeneous turbulence. Chapter 8 covers locally isotropic turbulence.)
Moffatt, H.K. (2002) *Ann. Rev. Fluid Mech.*, **34**, 19–35.
Moffatt, H.K., Kida, S., and Ohkitani, K., (1994) *J. Fluid Mech.*, **259**, 241–264.
Ooi, A. et al. (1999) *J Fluid Mech.*, **381**, 141–174.
Ott, S. and Mann, J. (2000) *J. Fluid Mech.*, **422**, 207.
Pearson, B.R. et al. (2004) In: *Reynolds Number Scaling in Turbulent Flows* (ed. A.J. Smits). Kluwer Academic Publishers.
Pelz, R.B. (2001) In: *An Introduction to the Geometry and Topology of Fluid Flows* (ed. R.L. Ricca). Nato Science Series.
Pumir, A. and Siggia E.D. (1992) *Phys. Fluids*, A**4**, 1472–1491.
Richardson, L.F. (1926) *Proc. Roy. Soc. London* A, **110**, 709–737.
Saddoughi, S.G. and Veeravalli S.V., *J. Fluid Mech.*, **268**, 1994, 333–372.
Sawford, B. (2001) *Ann. Rev. Fluid Mech.*, **33**, 289.
Sokolnikoff, I.S. (1946) *Mathematical Theory of Elasticity*. McGraw Hill.
(Section 1.11 discusses the influence of finite deformations on a continuous medium.)
Sreenivasan, K.R. (1991) *Proc. Roy. Soc.* A, **434**, 165–182.
Sreenivasan, K.R., and Antonia, R.A. (1997) *Annual Review of Fluid Mechanics*, **29**, 1997, pp 435–472.
Taylor, G.I. (1921) *Proc. London Maths Soc.*, **20**, 196–211.
Taylor, G.I. (1953) *Proc. Roy. Soc* A, **CCXIX**, 186–203.
Taylor, G.I. (1954) *Proc. Roy. Soc* A, **CCXXIII**, 446–468.
Tennekes, H. and Lumley, J.L. (1972) *A first course in turbulence*. MIT Press.
(Chapters 6 and 8 give excellent introductions to a statistical description of turbulence and the energy cascade.)
Townsend, A.A. (1951) *Proc. Roy. Soc. A*, **208**, 534.
Townsend, A.A. (1976) *The Structure of Turbulent Shear Flow*. Cambridge University Press, 2nd edition.
(Chapters 1 and 3 provide a simple but informative introduction to turbulence.)
Tsinober, A. (2001) *An Informal Introduction to Turbulence*. Kluwer.
(Contains useful data on the structure of isotropic turbulence.)
Van Atta, C.W. and Yeh, T.T. (1970) *J. Fluid Mech.*, **41**(1), 169–178.
Warhaft, Z. (2000) *Ann. Rev. Fluid Mech.*, **32**, 203.

PART II

Freely decaying, homogeneous turbulence

The intense and enduring difficulty of the problem of homogeneous turbulence is associated with the fact that all linearizable features have been stripped away, and the naked nonlinearity of the problem is all that remains.

H. K. Moffatt (2002)

In the next three chapters we turn to what is probably the most difficult problem in turbulence: that of predicting the evolution of freely decaying, homogeneous turbulence. By 'freely decaying' we mean turbulence free from a mean shear or any body force which might maintain and shape the turbulence. The archetypal example of this is grid turbulence. It may seem paradoxical that turbulence, which is free from all of the complexities of a mean shear, or of body forces, is the most difficult to predict, but this does seem to be the case. The point is this. When turbulence is maintained by, say, a mean shear, the forcing mechanism tends to promote large-scale eddies of a particular type or shape, such as hairpin vortices in a shear layer. Moreover, we can gain at least a little insight as to the nature of these eddies by experiment, or perhaps through some form of linear analysis, such as rapid distortion theory. Given that the large scales in the turbulence have some degree of structure, and that experiment or linear analysis can provide at least a hint as to what that structure might be, one can begin to make estimates of useful quantities, such as Reynolds stresses or turbulent energy levels. And so the process of assembling the jigsaw puzzle begins.

The problem with freely decaying turbulence is that there is no mechanism to shape or organize the large eddies, and all we are left with is a shapeless tangle of turbulent vortex tubes interacting in a chaotic, non-linear fashion. In the words of Moffatt, we have stripped away everything but the 'naked nonlinearity of the problem'. So, while it is true that freely decaying turbulence is conceptually the simplest form of turbulence, it has also proven to be the most resistant to attack, essentially because it isolates the most formidable aspect of the problem: the non-linear interaction of turbulent eddies.

The conceptual simplicity of freely evolving, homogeneous turbulence means that the governing statistical equations are relatively easy to write down. (They are much simpler than, say, the equivalent shear flow equations.) On the other hand, we lack any intuitive relationship to the subject, and so it is by no means clear how to attack the problem from a physical point of view. In such circumstances it seems natural to take a rather formal, mathematical approach, in the hope that the mathematics might flush out results which our physical intuition could not foresee. So studies of homogeneous turbulence tend to be characterized by a high level of mathematical sophistication. Gone are the simple intuitive hypotheses such as Prandtl's mixing length or Taylor's entrainment coefficient for a jet. In their place comes a more rigorous dissection of the governing statistical equations. So in many ways the chapters which follow are not just an introduction to homogeneous turbulence, they are also an introduction to the formal mathematics of statistical fluid mechanics.

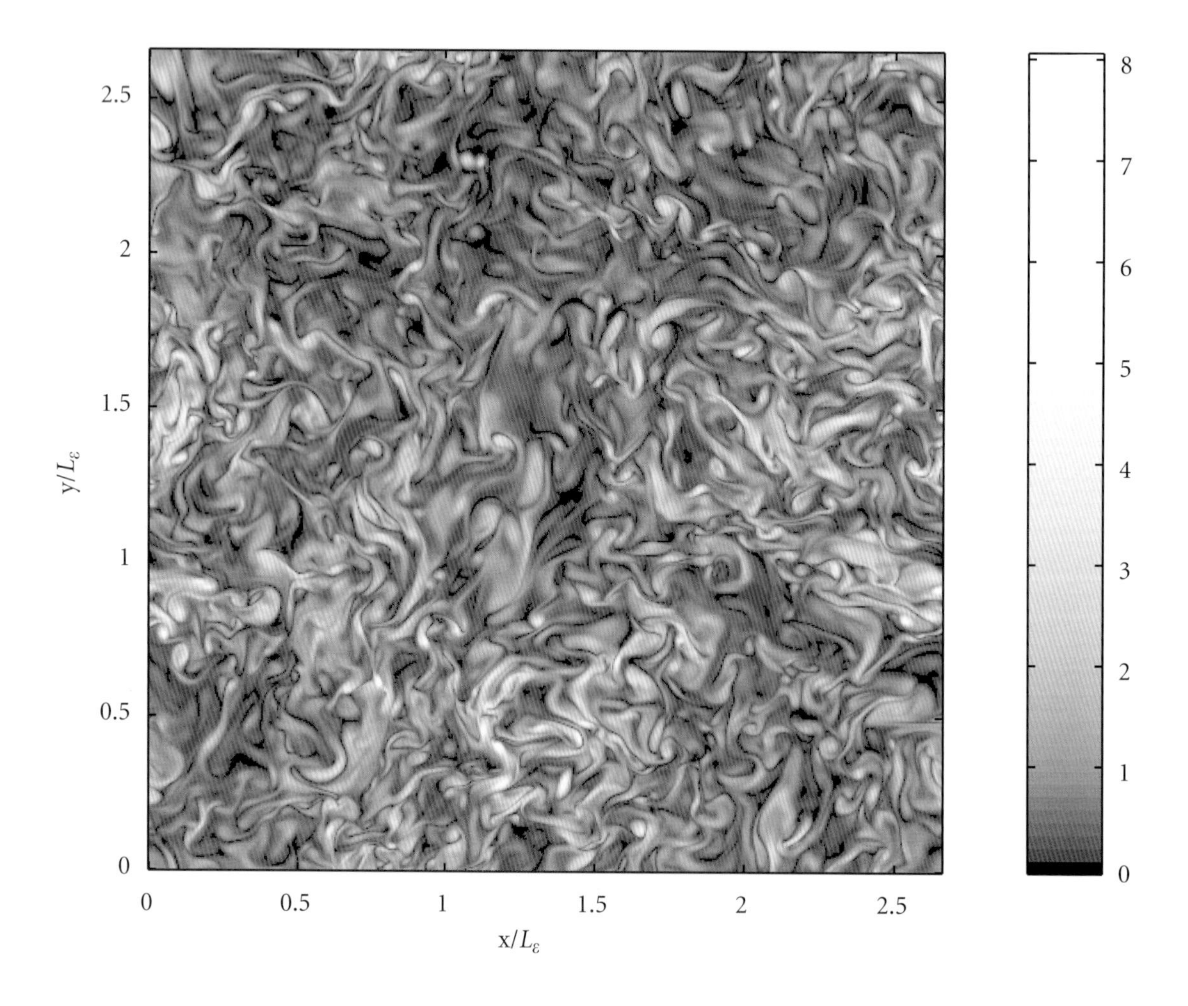

CHAPTER 6

Isotropic turbulence (In real space)

Moriaty: How are you at Mathematics?
Seegoon: I speak it like a native.

Spike Milligan—The Goon Show

So far we have studiously avoided mathematics of any great complexity. We have relied instead on simple physical arguments, often of a rather hand-waving nature. Of course, such an approach is rather limited and in the final analysis, deeply unsatisfactory. At some point we simply have to bite the bullet and lay out the equations, laws and proofs in a formal, rigorous fashion. For the newcomer to turbulence this can seem daunting. The formal language of turbulence involves tensors of great complexity, and the equations are equally off-putting. They have none of the elegance of, say, Maxwell's equations of electromagnetism. Rather, they are complex, non-linear, partial differential equations involving tensors of more than one order. One of the aims of this chapter is to begin the process of putting in place those equations. In order to make the process as painless as possible we shall make two simplifications: one physical and the other mathematical. First, we shall restrict ourselves to isotropic turbulence. This allows us to use symmetry arguments, which greatly simplify the governing equations. Second, we shall avoid, as far as it is possible to do so, the Fourier transform, which many believe provides the natural language of homogeneous turbulence.

6.1 Introduction: exploring isotropic turbulence in real space

Surfaces of constant vorticity in a numerical simulation of turbulence. The vorticity advects itself in a chaotic manner by its self-induced velocity field. The vorticity field is so complex that it is natural to adopt a statistical approach to homogeneous turbulence. (Courtsey of J. Jimenez)

Before embarking on a blow-by-blow description of the statistical theory of homogeneous turbulence, perhaps it is worth taking a step back and asking what we might expect of such models and what price is paid by remaining in real space, rather than moving into Fourier space. Let us start by addressing the question: what use are statistical models?

6.1.1 *Deterministic cartoons versus statistical phenomenology*

Perhaps the simplest problem in turbulence is the following. Suppose that we create a large cloud of turbulence, whose radius, R, is very much greater than the size of a typical large-scale eddy, l. This turbulent cloud is, in effect, a seething tangle of vorticity, which advects itself in a chaotic manner in accordance with the vortex evolution equation. Let us choose R so that the spherical surface $|\mathbf{r}| = R$ encloses virtually all of the vorticity. (It cannot enclose *all* of $\boldsymbol{\omega}$ since, in the presence of viscosity, vorticity will diffuse causing $\boldsymbol{\omega}$ to be finite but exponentially small in the far field.) The question at hand is: what can we say about the evolution of this turbulent cloud? If we restrict ourselves to rigorous statements the answer is somewhat depressing: we can say very little! In fact, there are probably only three useful dynamical statements that can be made. They are (Figure 6.1):

(1) The linear momentum of the cloud, as measured by the *linear impulse*, $\mathbf{L} = \frac{1}{2}\int \mathbf{x} \times \boldsymbol{\omega}\, dV$, is an invariant of the motion (see Section 5.1.3 or else Appendix II);
(2) The angular momentum of the cloud, as measured by the *angular impulse*, $\mathbf{H} = \frac{1}{3}\int \mathbf{x} \times (\mathbf{x} \times \boldsymbol{\omega})\, dV$, is an invariant of the motion (see Section 5.1.3 or else Appendix II);
(3) The total kinetic energy of the motion declines according to

$$\frac{d}{dt}\int \frac{1}{2}\mathbf{u}^2\, dV = -\nu \int \boldsymbol{\omega}^2\, dV.$$

To this we might add two plausible hypotheses, both of which follow from the observation that, at high Reynolds number (Re), vorticity tends to be teased out into finer and finer filaments by the chaotic velocity field. These are:

(1) energy will, on average, pass down to smaller and smaller scales;
(2) much of the vorticity will end up with a fine-scale structure and this small-scale vorticity will almost certainly be spatially intermittent (spotty).

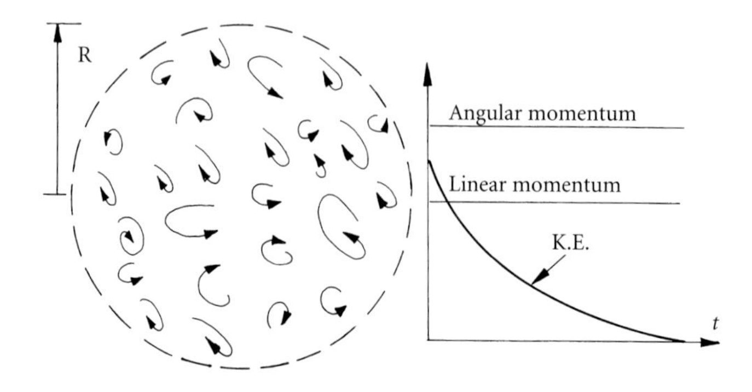

Figure 6.1 There are probably only three rigorous dynamical statements, which can be made about a cloud of freely decaying turbulence: (i) it conserves its linear momentum; (ii) it conserves its angular momentum; and (iii) its kinetic energy falls monotonically.

Unfortunately, these statements are, by themselves, of little help in determining the detailed behaviour of the turbulent cloud. If we are to make progress, additional hypotheses must be introduced, perhaps based on the evidence of physical or numerical experiments. Traditionally, two distinct approaches have been adopted, which we might label *statistical phenomenology* and *deterministic cartoons*. Most of the literature on homogeneous turbulence is of the former category, of which Kolmogorov's theory of the small scales is the most obvious success. In this approach we work only with statistical quantities, such as the energy spectrum, $E(k)$, or structure function, $\langle[\Delta v]^2\rangle$, and seek to make only statistical statements about the flow. The *ad hoc* hypotheses, which are introduced to circumvent the closure problem, are also of a statistical nature. Although Kolmogorov's theory represents a shining example of what can be achieved by this approach, there are few other success stories. One of the major drawbacks of the statistical approach is that, often, one feels that the equations of fluid mechanics play only a marginal role, somewhat subservient to the statistical hypotheses and their associated formalisms. This seems to be particularly true of certain two-point closure models and of the various attempts which have been made to characterize the small-scale intermittency of the vorticity field. Indeed, this uneasy situation led Pullin and Saffman (1998) to the somewhat provocative conclusion that such phenomenological statistical models *'have negligible dynamical and kinematic content and hardly any connection with fluid dynamics'*! We do not take quite such an extreme point of view. In any event most of the research in homogeneous turbulence has centred around these statistical models and so, in this chapter, we shall spend some time establishing the mathematical landscape in which such models are postulated and deployed.

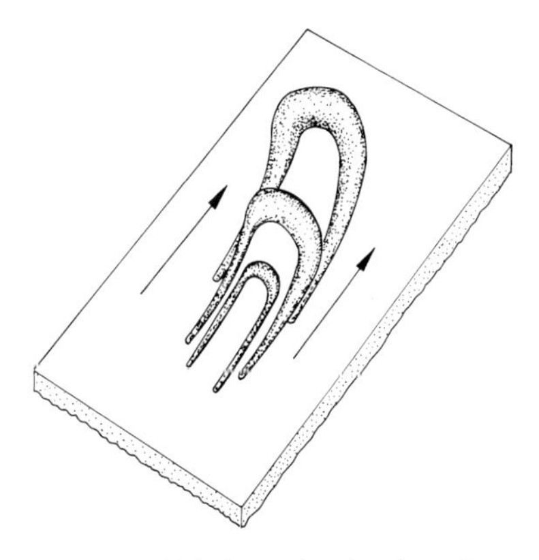

Figure 6.2 Hairpin vortices in a boundary layer.

The alternative strategy, of *deterministic cartoons*, is almost as old as the statistical approach, but is less well developed. Here one starts by making an hypothesis as to the generic shape of a typical eddy. In a boundary layer, for example, we might suppose that the large-scale eddies are dominated by hairpin structures (Figure 6.2). The small scales in isotropic turbulence, on the other hand, might be characterized as Burgers-like vortex tubes (Plate 8). Having identified the key vortical structure, all-be-it in the form of an idealized cartoon, one then examines the detailed dynamical behaviour of such vortices. (At this point we re-enter the world of classical fluid mechanics.) Finally, by making a few judicious assumptions about the birth, death, and distribution of these model eddies, one can reconstruct statistical information about the turbulence as a whole. This approach was championed by Townsend and Perry in the context of shear flows (see references in Chapter 4), and has been deployed with some success by Lundgren (1982) and Pullin and Saffman (1998) in homogeneous turbulence. It is a strategy that is likely to become increasingly popular (and successful) as computer simulations of turbulence provide more and more information as to the form of the dominant vortical structures in different types of flow.

Both strategies (statistical phenomenology and deterministic cartoons) have their strengths and weaknesses. Nevertheless, one might speculate that, in the decades to come, deterministic cartoons will play an increasingly important role, if only because they allow us to tap into our highly developed intuition as to the behaviour of individual vortices. We do not have the same intuitive relationship to the statistical theories, which in any event are plagued by the curse of the closure problem.

Perhaps one surprise for the newcomer to turbulence is the fact that neither of the above approaches make much use of the few things we can say for certain about a cloud of turbulence! In particular, neither strategy exploits the fact that the cloud conserves its linear momentum, **L**, and angular momentum, **H**. At first sight this is astonishing. After all, these conservation laws impose powerful constraints on the way in which the cloud develops. We would expect such conservation principles to be particularly relevant to the behaviour of the large-scale vortices, as these contribute most to the net linear and angular momentum of the cloud.

Actually, it turns out that there *is* a statistical theory of the large scales based on these conservation principles, though it is somewhat controversial and little discussed. Broadly speaking there are two theories and two camps. There are those who believe that the eddies, which make up a field of homogeneous turbulence, possess a significant impulse (linear momentum) in the sense that the linear impulse of a typical eddy, $\mathbf{L}_e = \frac{1}{2}\int_{V_e} \mathbf{x} \times \boldsymbol{\omega}_e dV$, is of order $\sim |\boldsymbol{\omega}_e| V_e l_e$. (Here the subscript e indicates an individual eddy, $|\boldsymbol{\omega}_e|$ is a characteristic vorticity of the eddy, and V_e and l_e are the volume and size of that eddy. In the interests of brevity, we shall generally omit the ρ when discussing linear or angular momentum.) In such cases the global linear momentum of the cloud, **L**, will be non-zero, despite the random orientation of the vortices. That is, for a finite sized cloud there will always be some imperfect cancellation of linear momentum when we form the sum $\sum \mathbf{L}_e$ (Figure 6.3). Saffman (1967) (see also Birkhoff 1954) has shown that turbulence of this type possesses the statistical invariant $\langle \mathbf{L}^2 \rangle / V_R = \int \langle \mathbf{u} \cdot \mathbf{u}' \rangle \, d\mathbf{r}$ and that, as a consequence, its kinetic energy decays as $u^2 \sim t^{-6/5}$. (Here **u** and **u**′ are the velocities measured at two points separated by the displacement vector **r**, and V_R is the volume of the turbulent cloud.) It so happens that the energy spectrum in such cases grows as $E \sim k^2$ for small k.

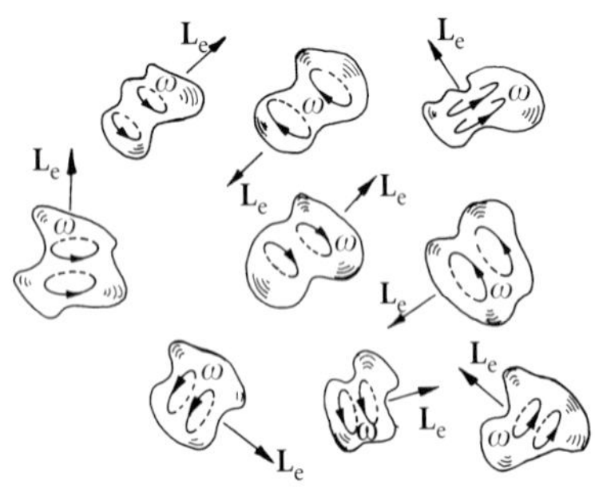

Figure 6.3 When turbulence is composed of a sea of eddies of finite linear impulse, $\mathbf{L}_e = \frac{1}{2}\int_{V_e} \mathbf{x} \times \boldsymbol{\omega}_e \, dV$, the turbulence possesses the statistical invariant $\langle \mathbf{L}^2 \rangle / V_R = \int \langle \mathbf{u} \cdot \mathbf{u}' \rangle \, d\mathbf{r}$.

There are others who believe that typical eddies have negligible linear impulse (linear momentum) in the sense that,

$$\mathbf{L}_e = \frac{1}{2}\int_{V_e} \mathbf{x} \times \boldsymbol{\omega}_e \, dV \ll |\boldsymbol{\omega}_e| V_e l_e.$$

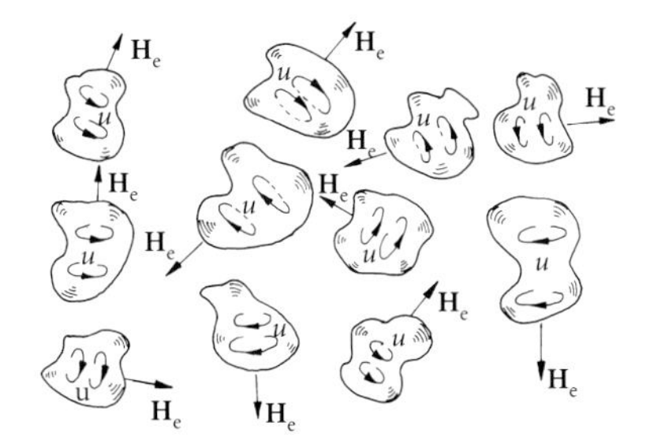

Figure 6.4 When the eddies have negligible linear impulse but finite angular impulse (angular momentum), $\mathbf{H}_e = \frac{1}{3}\int \mathbf{x} \times (\mathbf{x} \times \boldsymbol{\omega}_e)\, dV$, the statistical quantity $\langle \mathbf{H}^2 \rangle / V_R \approx -\int \mathbf{r}^2 \langle \mathbf{u} \cdot \mathbf{u}' \rangle\, d\mathbf{r}$ is almost conserved. Here $\mathbf{H}$ is the net angular momentum of the turbulent cloud and V_R is its volume.

(This is not to say that the eddies are stationary! Remember that eddies translate partly because of their own linear impulse and partly because they get caught up in the irrotational velocity field of adjacent eddies. See example 5.2 of Section 5.1.3.) In such cases it can be shown that $\int \langle \mathbf{u} \cdot \mathbf{u}' \rangle\, d\mathbf{r} = 0$, but that $\int \mathbf{r}^2 \langle \mathbf{u} \cdot \mathbf{u}' \rangle\, d\mathbf{r}$ is finite, negative and (almost) constant (Figure 6.4). As we shall see, the kinetic energy now decays as $u^2 \sim t^{-10/7}$ and E grows as $E \sim k^4$ at small k. In fact, this is the classical view, which predates Saffman's (1967) landmark paper. The integral $-\int \mathbf{r}^2 \langle \mathbf{u} \cdot \mathbf{u}' \rangle\, d\mathbf{r}$ is an approximate measure of the angular momentum (squared) of the turbulent cloud,

$$\langle \mathbf{H}^2 \rangle / V_R \approx -\int \mathbf{r}^2 \langle \mathbf{u} \cdot \mathbf{u}' \rangle\, d\mathbf{r}.$$

So which camp is correct? The short answer is both. That is, both $E \sim k^2$ and $E \sim k^4$ spectra are observed in the numerical simulations of turbulence. It is the initial conditions that dictates which form of turbulence emerges. If the initial linear momentum is large enough, then a Saffman spectrum is guaranteed. If it is small, an $E \sim k^4$ spectrum is obtained. The main controversy centres on the delicate issue of which form of turbulence is likely to be encountered in a real flow, such as grid turbulence. Here there is little agreement. A second controversy relates to the degree to which $-\int \mathbf{r}^2 \langle \mathbf{u} \cdot \mathbf{u}' \rangle\, d\mathbf{r}$ is really conserved is cases where $\mathbf{L}_e \approx 0$. (There are technical difficulties with the proof that the integral is equal to $\langle \mathbf{H}^2 \rangle / V_R$ and is thus conserved, though in practice any time dependence of the integral is slight.) Both controversies are discussed in Section 6.3.

So, after half a century of concerted effort, where do we stand in our understanding of homogeneous turbulence? We have Kolmogorov's approximate theory of the small scales, various competing (and controversial) theories of the large scales, a vast amount of carefully compiled experimental data, and a great many questions. It seems likely that phenomenological theories of the statistical type have largely run their course, while deterministic cartoons of turbulence are still plagued by our inability to agree as to what the dominant vortical structures look like. It is possible that this impasse may be broken by the numerical simulations of individual turbulent flows. Certainly this is an outcome which would have pleased Von Neumann who urged a concerted numerical attack on the problem as far back as 1949. In any event, this is a question for the future.

Given the open-ended nature of the problem, it is inevitable that our account of isotropic turbulence will seem incomplete: in effect, a story without an ending. In any event, we shall take a fairly conventional approach to the subject, placing much emphasis in the well-developed (if incomplete) statistical theories. Less attention is given to deterministic cartoons as these have yet to make their mark. We break with convention in only one respect: the discussion is almost exclusively developed in real space, rather than in Fourier space. That is to say, we work with $\mathbf{u}$ rather than its Fourier transform.

The attraction of real space lies in the fact that we can develop the theory in terms of familiar physical quantities, such as velocity correlations. This seems better suited to the beginner. Certainly one of the most off-putting aspects of the conventional Fourier-space approach lies in the somewhat obscure physical meaning of the central objects of that theory, that is, the Fourier transform of the velocity correlations. However, there is a price to pay for remaining in real space, as we shall now see.

6.1.2 *The strengths and weaknesses of Fourier space*

Many (possibly most) monographs on the statistical theory of homogeneous turbulence develop their arguments using the language of Fourier space. There are, perhaps, two reasons for this; one is relatively unimportant and the other is of considerable significance. The less important reason is that certain mathematical manipulations become less tedious in Fourier space. The more substantial argument in its favour is that Fourier space provides us with the energy spectrum $E(k)$.

The idea is the following. We would like to find a method of resolving the velocity field into a number of components which reflect the hierarchy of eddy sizes present in the turbulence. Ideally, these components would make additive contributions to $\frac{1}{2}\langle \mathbf{u}^2 \rangle$ in the sense that, if $\mathbf{u} = \Sigma_n \mathbf{u}_n$, then $\frac{1}{2}\langle \mathbf{u}^2 \rangle = \Sigma_n \frac{1}{2}\langle \mathbf{u}_n^2 \rangle$. In other words, we would like the 'modes' into which $\mathbf{u}$ is decomposed to be mutually orthogonal in the sense that $\langle \mathbf{u}_n \cdot \mathbf{u}_m \rangle = 0$ if $n \neq m$. (It may be useful to think of ensemble averages as volume averages here.) Fourier analysis seems well suited to this purpose. Indeed, we shall see in Chapter 8 that if $\hat{\mathbf{u}}(\mathbf{k})$ is the three-dimensional transform of $\mathbf{u}(\mathbf{r})$ then the energy spectrum, $E(k)$, is given by,

$$E(k)\delta(\mathbf{k} - \mathbf{k}') = 2\pi k^2 \langle \hat{\mathbf{u}}^\dagger(\mathbf{k}) \cdot \hat{\mathbf{u}}(\mathbf{k}') \rangle, \quad k = |\mathbf{k}|.$$

Here $\mathbf{k}$ is the three-dimensional wavevector, $\mathbf{k}$ and $\mathbf{k}'$ are distinct wavevectors, δ is the three-dimensional delta function, and $\dagger$ represents a complex conjugate. Thus the energy held in the $\mathbf{k}$th mode of $\hat{\mathbf{u}}(\mathbf{k})$ is proportional to $E(k)$. Moreover, the product of the modes of different wavevectors make no contribution to $E(k)$.

We *choose to interpret* $E(k)$ as the distribution of energy across the different eddy sizes because: (i) it is non-negative; (ii) it has the property that,

$$\frac{1}{2}\langle \mathbf{u}^2 \rangle = \int_0^\infty E(k)\, dk \tag{6.1}$$

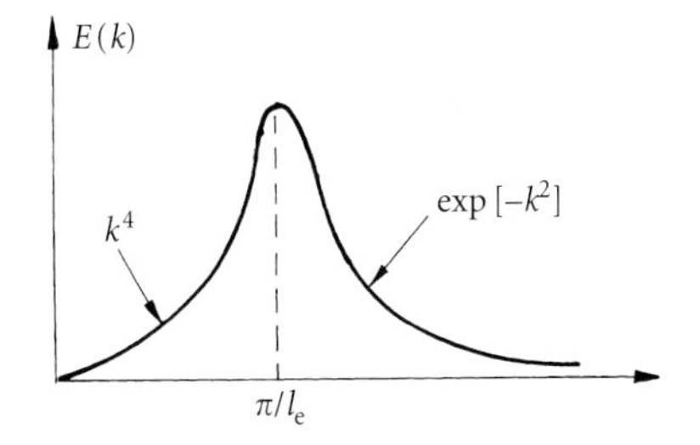

Figure 6.5 A random distribution of eddies of size l_e has an energy spectrum peaked around $k \sim \pi/l_e$. (See Section 6.4.1.)

and; (iii) a collection of randomly distributed eddies of fixed size l_e has an energy spectrum peaked around $k \sim l_e^{-1}$ (see Section 6.4.1 and Figure 6.5). So, if we know $E(k)$ at any instant then, perhaps, we can say something about how energy is distributed amongst the various eddy sizes.[1] (Of course, this assumes that you are willing to accept the notion of an energy cascade with its hierarchy of eddy sizes.)

Now it turns out that we can massage the Navier–Stokes equation into the form,

$$\frac{\partial E}{\partial t} = (\text{Non-linear terms}) + (\text{Viscous terms}) \tag{6.2}$$

Those who believe in cascades like this, because they can focus on the non-linear term and interpret it as a mechanism for redistributing energy in Fourier space (assuming that our interpretation of E is correct). It turns out that, by and large, the non-linear terms in (6.2) have the effect of passing energy from large to small scale (small k to large k), and this fits nicely with Richardson's picture of turbulence (Figure 6.6).

In a formal sense, of course, we have gained little in writing down (6.2). The closure problem of turbulence (see Section 4.1) tells us that we need another equation to determine the non-linear terms in (6.2), and this involves yet more unknowns. It is not possible to write down a closed set of rigorous statistical equations which describe turbulence. Nevertheless, the general form of (6.2) fits nicely with the phenomenology of Richardson's cascade and so many researchers use it as a starting point for developing *closure models*. That is, they introduce *ad*

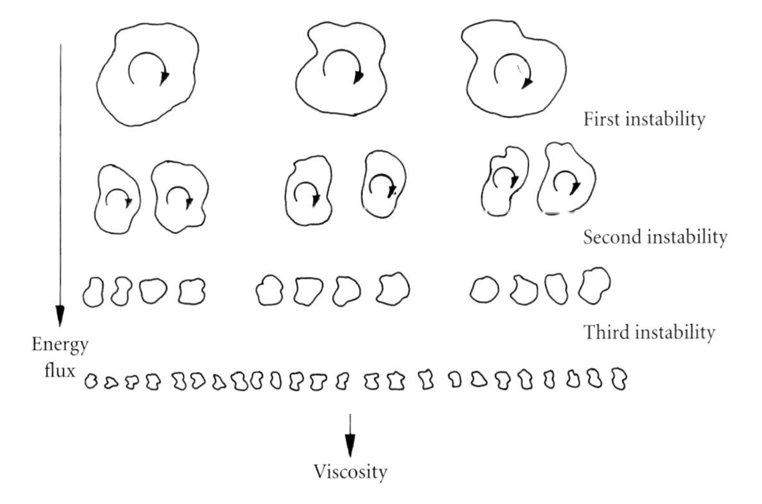

Figure 6.6 Schematic representation of Richardson's energy cascade. (After Frisch 1995.)

[1] Actually, we shall see in Section 6.4 that this interpretation is not strictly true. In particular, it fails for $k \ll l^{-1}$ and $k \gg \eta^{-1}$ where there are no eddies yet E is finite.

hoc assumptions about the behaviour of the non-linear terms (i.e. the way in which energy is passed down the cascade) in order to supplement (6.2) and close the system.

For the novice, however, Fourier space is less attractive. It is bad enough that we have to work with tensors of great complexity, governed by unfamiliar equations. To then translate everything into Fourier space seems to compound the difficulties. Moreover, it is difficult to attribute physical meaning to quantities defined in Fourier space. Turbulence consists of vortical structures—tubes, sheets, etc.—and not of waves. Earlier we tentatively suggested that you might associate $1/k$ with the size of an eddy. Of course, this is highly simplistic. If you take the Fourier transform of a compact structure of given size you generate a whole range of wave number components, albeit central around l_e^{-1} (Figure 6.5). So functions in Fourier space do not readily admit a simple physical interpretation. This is a non-trivial point. Much of the research in turbulence has centred around developing, or using, closure models. By definition, these invoke some *ad hoc* hypothesis, based on physical intuition or experimental observations, which closes the set of statistical equations. So if you are to develop or deploy such closure models, then it is important to have a sound physical intuition.

For these reasons we shall start our formal analysis of isotropic turbulence in *real space*. Later, in Chapter 8, we shall pull in the added complexity of the Fourier transform.

6.1.3 *An overview of this chapter*

The literature on the statistical theory of homogeneous turbulence is vast. The first definitive source was, perhaps, G. K. Batchelor's (1953) classic text *Theory of Homogeneous Turbulence*. With his characteristic brevity Batchelor managed to summarize the post-war developments in homogeneous turbulence in a mere 187 pages. However, by 1975 Monin and Yaglom found that 650 pages were barely sufficient to capture the key ideas! Since then many monographs have charted the various developments as they unfolded, such as the second edition of M. Lesieur's *Turbulence in Fluids* and W. D. McComb's *The Physics of Fluid Turbulence*, both published in 1990. So how can we do justice to these texts in a single chapter? The short answer is that we cannot: the best we can do is to provide a bridge from the rudimentary ideas expounded in Part I of this book to the more advanced monographs on homogeneous turbulence.

The primary aim of the chapter is to establish the vocabulary of statistical fluid mechanics, write down Newton's second law in terms of statistical quantities, and explore some of the immediate consequences of this equation. In order to simplify matters we restrict ourselves to isotropic turbulence so that there is no mean velocity field. The layout of the chapter is as shown below.

- Governing equations (Section 6.2)
  - Kinematics (the vocabulary of statistical fluid mechanics)
  - Newton's second law in statistical form (the Karman–Howarth equation)
  - The closure problem (again!)
- The dynamics of the large eddies (Section 6.3)
  - The classical theory of Landau and Kolmogorov
  - The objections of Batchelor, Birkhoff and Saffman
  - The experimental evidence
- The characteristic signature of different eddy types (Section 6.4)
- Intermittency in the small-scale eddies (Section 6.5)
- The distribution of energy and vorticity across the different eddy sizes (Section 6.6)
  - A 'real-space' version of the energy spectrum
  - Cascade dynamics in real space.

Perhaps we should say a few words about each section and how they fit together as a whole.

6.1.3.1 The governing equations

Most of the hard work comes in Section 6.2 where we establish the statistical equations which govern isotropic turbulence. The subsequent Sections 6.3–6.6, merely explore some of the consequences of these equations. Section 6.2 is broken down into three parts: (i) kinematics; (ii) dynamics; and (iii) the closure problem. In part (i) we generalize the idea, first introduced in Chapter 3, that the statistical state of a cloud of turbulence may be described in terms of correlation functions and structure functions. We also examine the kinematic relationships between these various quantities that follow from the assumption of statistical isotropy and from the conservation of mass. Next, in part (ii), we use the vocabulary of part (i) to write down Newton's second law (as applied to a turbulent fluid) in statistical form. The resulting expression, which is known as the *Karman–Howarth equation*, is the single most important equation in the theory of isotropic turbulence. It provides the starting point for nearly all dynamical theories and indeed one of the immediate consequences of the Karman–Howarth equation is Kolmogorov's celebrated four-fifth's law.

Finally, in part (iii) of Section 6.2, we remind the reader of the closure problem of turbulence. That is to say, when the Navier–Stokes equation is written in statistical form, the resulting hierarchy of equations is not closed in the sense that there are always more unknowns than equations. As emphasized in Chapter 3, this is a result of the non-linearity of the Navier–Stokes equation. The closure problem is crucial because it means that statistical models of turbulence necessarily invoke *ad-hoc* hypotheses or assumptions, which supplement the rigorous equations. There have been many such *closure schemes*, and we mention just two in Section 6.2. The simplest, and arguably the most elegant, was first put forward by Obukhov (1949). It applies only to the equilibrium range (small-to-intermediate sized eddies) and consists of assuming that the skewness of the velocity difference, Δv, is constant across the equilibrium range. This assumption is not too far out of line with the experimental data and leads to predictions which are remarkably close to the most recent computer simulations of turbulence. The second closure scheme, by contrast, is complex and involves a number of assumptions whose validity is difficult to assess. This is the famous (and widely used) EDQNM model and its ill-fated predecessor, the *quasi-normal scheme*. The EDQNM scheme is much more general than the constant skewness model of Obukhov in that it applies to all scales, large and small. Some would argue that it has notched up some notable successes over the years; others are more wary.

6.1.3.2 The dynamics of the large scales

In Sections 6.3–6.6 we explore some of the properties and consequences of the governing equations of isotropic turbulence. We start, in Section 6.3, by looking at the dynamics of the large eddies. There are two ways into this subject. One is to move into Fourier space and examine the dynamics of the low-k end of $E(k)$. The other route is to consider the implications of the momentum conservation laws for a large ensemble of vortices. This second approach, which was pioneered by Landau, is better suited to the theme of this chapter. The key point, discovered by Landau and pursued by Saffman, is that the linear and angular momenta of a large cloud of turbulence of volume V can be related to the statistical quantity $\langle \mathbf{u}(\mathbf{x}) \cdot \mathbf{u}(\mathbf{x}+\mathbf{r}) \rangle = \langle \mathbf{u} \cdot \mathbf{u}' \rangle$ as follows:

$$[\text{linear momentum}]^2 \sim V \int \langle \mathbf{u} \cdot \mathbf{u}' \rangle \, d\mathbf{r} + (\text{correction term}, C_1)$$

$$[\text{angular momentum}]^2 \sim -V \int \mathbf{r}^2 \langle \mathbf{u} \cdot \mathbf{u}' \rangle \, d\mathbf{r} + (\text{correction term}, C_2).$$

It turns out that the correction terms C_1 and C_2 vanish provided that $\langle \mathbf{u} \cdot \mathbf{u}' \rangle$ decays sufficiently rapidly with $|\mathbf{r}|$. In such cases the momentum conservation laws applied to a cloud of turbulence suggest that the integrals

$$L = \int \langle \mathbf{u} \cdot \mathbf{u}' \rangle \, d\mathbf{r}, \qquad I = -\int \mathbf{r}^2 \langle \mathbf{u} \cdot \mathbf{u}' \rangle \, d\mathbf{r}$$

should be invariants of freely evolving turbulence. Actually, when remote points in a turbulent flow are statistically independent, in the sense that $\langle \mathbf{u} \cdot \mathbf{u}' \rangle$ is transcendentally small for large $|\mathbf{r}|$, these suggestions may be confirmed directly using the Karman–Howarth equation. In particular, we shall see that the statistical independence of remote points leads to

$$L = 0, \qquad I = \text{constant}.$$

Now we would expect the angular momentum of a cloud of turbulence to be dominated by the large eddies and this is consistent with the integral expression for I given above. (The integral is dominated by the large $\mathbf{r}$ contribution to the integrand.) So it is common to estimate I as,

$$I \sim u^2 l^5$$

where l is the integral scale (the size of the large eddies). Kolmogorov used the alleged invariance to I, in conjunction with the energy decay law,

$$\frac{du^2}{dt} \sim -\frac{u^3}{l}$$

to predict the rate of decay of energy in a turbulent cloud: the result is readily shown to be $u^2 \sim t^{-10/7}$. This is known as Kolmogorov's decay law, and it is a reasonable fit to some physical and numerical experiments. So far, so good.

Things changed dramatically, however, in 1956. The key development was the discovery by Batchelor that, because of the transmission of information over large distances by the pressure field, $\langle \mathbf{u} \cdot \mathbf{u}' \rangle$ is not, in general, transcendentally small for large $|\mathbf{r}|$. It turns out that this invalidates Landau's assertion that I is proportional to the square of the angular momentum (the correction term C_2 is of order one) as well as the formal proof (via the Karman–Howarth equation) that I is an invariant of freely evolving turbulence. Batchelor went on to suggest that, in general, one would expect $L = 0$, as in the traditional theory, but that I should be time-dependent. By implication, Kolmogorov's decay law was suspect. Later, Saffman showed that, *for certain initial conditions*, even more dramatic departures from the traditional theory could arise. In particular, he showed that it is possible to produce turbulence in which L is *non-zero* and constant, while I diverges (does not exist!).

There has been much controversy over these various findings, and even today there is little agreement as to whether grid turbulence is of the Saffman type (I divergent, $L \neq 0$) or of the Batchelor type (I convergent, $L = 0$). In Section 6.3 we describe the classical theories of Landau and Kolmogorov, as well as the modern objections of

Batchelor, Birkhoff and Saffman. We also examine the experimental data. We conclude that, for certain initial conditions, the turbulence evolves to a state in which the long-range statistical correlations are weak, so that the classical theory of Landau and Kolmogorov works reasonably well. For other initial conditions, however, we must abandon the classical view and adopt Saffman's theory.

6.1.3.3 The characteristic signatures of different eddy types

In Section 6.4 we leave aside dynamics and return to kinematics. The issue in question is the following. In statistical theories we find that various quantities, such as the energy spectrum, $E(k)$, and the structure function, $\langle[\Delta v]^2\rangle$, are the object of our analysis. Closure models make predictions about the evolution of $E(k)$ or $\langle[\Delta v]^2\rangle$ and laboratory experiments or computer simulations provide data for E or $\langle[\Delta v]^2\rangle$ against which to compare the predictions. However, being statistical quantities, E and $\langle[\Delta v]^2\rangle$ tell us only about what happens, *on average*, over a large number of realizations of the flow, each realization being different in detail. In an *individual* realization, however, we expect a flow to consist of a hierarchy of eddies, and we interpret the evolution of the flow in terms of the interaction of these eddies in accordance with the laws of vortex dynamics. The problem, of course, is that our rigorous, deterministic laws and physical intuition belong to the latter view point (eddy interaction in an individual realization of the flow), whereas the statistical closure models, and the objects which they describe, belong to the former viewpoint. It is important that we establish links between the two points of view and a natural starting point is to ask whether or not we can say something about the instantaneous eddy population in a particular realization of a flow from a snapshot of $E(k)$ or $\langle[\Delta v]^2\rangle$. This is an issue that particularly exercised Townsend and the first chapter of his famous text on shear flows discusses the problem at some length. In Section 6.4 we expand on Townsend's ideas, incorporating later discoveries, such as Lundgren's stretched, spiral vortex, which gives rise to a $k^{-5/3}$ spectrum.

6.1.3.4 Intermittency in the inertial-range Eddies

In Section 6.5 we consider the significance of the observation that the distribution of $\boldsymbol{\omega}$ is somewhat patchy in a turbulent flow (Plate 8). We shall see that this necessitates a correction to Kolmogorov's theory.

As noted in Chapter 5, one of the major success stories of turbulence is Kolmogorov's universal theory of the small scales. However, from the very beginning, Landau questioned the validity of this theory. In particular, he was unconvinced that the statistical features of the small scales should be universal (independent of the manner in which the turbulence is maintained or is generated). His objection was the following. He accepted that the statistical properties of the structure function $[\Delta v]^2 = [u_x(\mathbf{x} + r\hat{\mathbf{e}}_x) - u_x(\mathbf{x})]^2$ in the inertial subrange could, as Kolmogorov claimed, be expressed as a universal function of ε (the dissipation per unit mass) and r alone. However, he observed that the relevant value of ε is the dissipation (or energy flux) at the instant considered and averaged over the small region (of size r) in which $[\Delta v]^2$ is evaluated, which we might denote $\varepsilon_{AV}(r, \mathbf{x}, t)$. In other words, the ε appearing in the unaveraged version of Kolmogorov's law is itself a random function of position and time. (Remember that $\boldsymbol{\omega}$, and hence the dissipation, is very patchy in space.) We must now consider what happens when we average $[\Delta v]^2 = f(\varepsilon_{AV}, r, \mathbf{x}, t)$ to find an expression for $\langle[\Delta v]^2\rangle$. Even though the dependence of $[\Delta v]^2$ on ε_{AV} and r might have a universal character, as in Kolmogorov's original theory, the result of the averaging may not be universal if the statistical behaviour of ε_{AV} depends on the type of flow in question, i.e. it is different for a wake, jet, or boundary layer. That is, the manner in which the dissipation varies in space, or varies with time at a particular location, would, in Landau's opinion, depend of the large-scale features of the flow, which are definitely not universal. If the dissipation in a turbulent flow were spatially uniform, or nearly uniform, this would be a non-issue, since the only relevant value of is the spatially averaged dissipation at the instant is question. The problem, though, is that it has been known since the post-war studies of Townsend and Batchelor that the vorticity, enstrophy, and dissipation are all highly intermittent in space, a result which was anticipated by G. I. Taylor as far back as 1917 and re-emphasized by Landau in 1941. It is natural, therefore, to investigate whether or not Kolmogorov's theory needs to be modified to accommodate this intermittency.

A hint as to the nature of the problem is provided by a simple, if somewhat artificial, example given in Monin and Yaglom (1975). (Actually, this example is constructed to highlight the implications of *large-scale* inhomogeneities in dissipation, rather than *small-scale* inhomogeneities. Nevertheless, it does illustrate that spatial variations in dissipation are inconsistent with Kolmogorov's original theory.) Consider a turbulent flow in which the spatially averaged dissipation, ε, averaged over a length scale of the order of the integral scale, l, varies slowly in space and on a length-scale much greater than l. That is, the turbulence is weakly inhomogeneous on a scale L, which is a scale much greater than l (Figure 6.7). Moreover, let us suppose that the nature of the weak inhomogeneity is random in space, so that when considered on a global scale L_g ($L_g \gg L$) the turbulence appears to be statistically homogeneous.

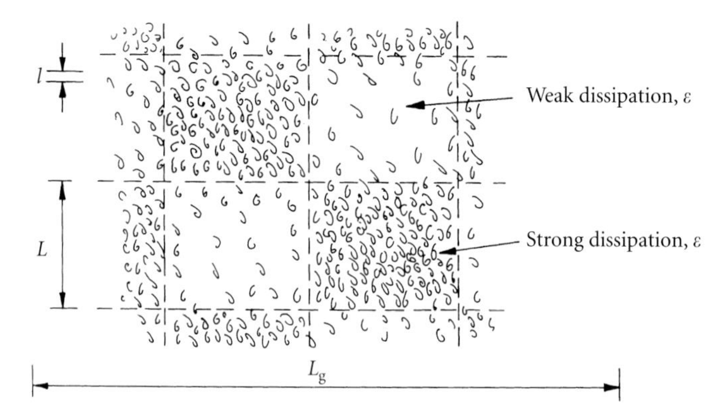

Figure 6.7 Turbulence which is weakly inhomogeneous on the scale L.

According to Kolmogorov's theory the structure function of order p, $\langle[\Delta v]^p\rangle$, is, in the inertial sub-range, given by (5.24) in the form,

$$\langle[\Delta v]^p\rangle = \beta_p(\varepsilon r)^{p/3}.$$

(Here $\langle[\Delta v]^p\rangle$, like ε, is averaged over a volume of order l^3.) Moreover, Kolmogorov's theory would have all the β_p's as universal constants, that is the same from one flow to another. Now suppose we evaluate $\langle[\Delta v]^p\rangle$ and ε at a number, N, of well separated points in our flow so that ε, and hence $\langle[\Delta v]^p\rangle$, is different at each point. We now form a super-average of ε and $\langle[\Delta v]^p\rangle$ as follows,

$$\bar{\varepsilon} = \frac{1}{N}\sum_{i=1}^{N}\varepsilon_i, \qquad \overline{\langle[\Delta v]^p\rangle} = \frac{1}{N}\sum_{i=1}^{N}\langle[\Delta v]^p\rangle_i.$$

According to Kolmogorov's theory $\bar{\varepsilon}$ and $\overline{\langle[\Delta v]^p\rangle}$ are also related by,

$$\overline{\langle[\Delta v]^p\rangle} = \beta_p(\bar{\varepsilon} r)^{p/3}$$

since the flow is homogeneous on the global scale L_g. It follows that, if Kolmogorov's theory is correct,

$$\frac{1}{N}\sum \beta_p \varepsilon_i^{p/3} r^{p/3} = \beta_p r^{p/3}\left[\frac{1}{N}\sum \varepsilon_i\right]^{p/3}$$

from which,

$$\frac{1}{N}\sum \varepsilon_i^{p/3} = \left[\frac{1}{N}\sum \varepsilon_i\right]^{p/3}.$$

This expression is satisfied if $p = 3$, which is just as well since we believe that the four-fifth's law,

$$\langle[\Delta v]^3\rangle = -\frac{4}{5}\varepsilon r$$

is (probably) exact. It is also satisfied if ε is spatially uniform so that each of the ε_i are the same. However, it is not satisfied if ε is non-uniform and $p \neq 3$, and indeed we find departures from Kolmogorov's universal equilibrium theory for $p \neq 3$. (The violation of the equality above for non-uniform ε and $p \neq 3$ is guaranteed by Hölder's inequality.)

Consider, for example, the case of $p=2$. We have already seen that there is strong evidence in favour of $\langle[\Delta v]^2\rangle \sim r^{2/3}$. If we accept the two-thirds power law, then at each point we may apply Kolmogorov's law locally to give,

$$\langle[\Delta v]^2\rangle_i = (\beta_2)_L(\varepsilon_i r)^{2/3}$$

where $(\beta_2)_L$ is the local value of β_2. Now suppose that $\varepsilon_i = (1-\gamma)\bar{\varepsilon}$ in half of the measurements, and $\varepsilon_i = (1+\gamma)\bar{\varepsilon}$ in the other half, that is the local mean dissipation can take one of two values, $(1 \pm \gamma)\bar{\varepsilon}$, with equal likelihood. (Of course, γ lies between 0 and 1.) When we form the super-average we obtain

$$\overline{\langle[\Delta v]^2\rangle} = \frac{1}{2}\left[(1-\gamma)^{2/3} + (1+\gamma)^{2/3}\right](\beta_2)_L \bar{\varepsilon}^{2/3} r^{2/3}.$$

Thus, Kolmogorov's constant appearing in the super-average is

$$(\beta_2)_{SA} = \frac{1}{2}\left[(1-\gamma)^{2/3} + (1+\gamma)^{2/3}\right](\beta_2)_L.$$

(Here the subscript SA indicates a super-average.) Since this depends on the value of γ, Kolmogorov's constant cannot be universal, but rather depends on the scale of the averaging. This suggests that strong inhomogeneities in ε at the *large scales* may lead to a variability in Kolmogorov's coefficient β_2, so that the inertial-range statistics cannot be universal after all. It is the *lack of universality* of the β_p's which Landau predicted.

A second problem is associated with the intermittency of the inertial-range vorticity, that is *small-scale* intermittency. Here the issues are a little different. It is not so much a lack of universality of β_p which comes into question, but rather the power law dependence $\langle[\Delta v]^p\rangle \sim r^{p/3}$. The point is that the $p/3$ power law comes from the assumption that $\langle[\Delta v]^p\rangle$ in the inertial sub-range depends on $\langle\varepsilon\rangle = \langle 2\nu S_{ij}S_{ij}\rangle$ and r only. (Here S_{ij} is the strain-rate tensor.) If we admit that there could be another factor, such as the extent and nature of the small-scale intermittency, then we must allow for departures from the $p/3$ power law. In short, the way in which the non space filling vorticity is distributed in space could be a relevant factor.

It should be emphasized, however, that departures from Kolmogorov's universal equilibrium theory manifest themselves only under extreme conditions. For example, in the toy problem above, a value of $\gamma = \frac{1}{2}$, which represents a strong large-scale inhomogeneity, leads to only a 3% difference between $(\beta_2)_{SA}$ and $(\beta_2)_L$. Thus very strong large-scale inhomogeneities are needed before we see significant variability in the coefficients β_p. Also, values of p greater than 4 or 5 are needed before significant departures from the $r^{p/3}$ law are observed.

Although the problem of intermittency, and the failure of Kolmogorov's theory, represents a problem for turbulence in general, and is not restricted to isotropic turbulence, the issues involved fit well into the main theme of this chapter, which is the problem of describing dynamics in real space. Consequently, in Section 6.5, we take up the issue of how best to patch up Kolmogorov's theory to allow for intermittency. In particular, we describe a popular cartoon of inertial-range intermittency, called the $\hat{\beta}$ model. This generalizes the Kolmogorov equation from

$$\langle [\Delta v]^p \rangle = \beta_p \varepsilon^{p/3} r^{p/3}$$

to

$$\langle [\Delta v]^p \rangle = \beta_p \varepsilon^{p/3} r^{p/3} (r/l)^{(3-p)(1-s)}$$

where s is a model parameter, which is less than unity for an intermittent distribution of dissipation, and equal to unity for a smooth distribution of dissipation. Note that the correction term vanishes when $p = 3$ or $s = 1$, which it should. Note also that the large scales have a direct impact on $\langle [\Delta v]^p \rangle$ through the presence of the integral scale l in the correction term. Moreover, if β_p is influenced by large-scale inhomogeneities (i.e. the degree of patchiness of the large-scale eddies) then the β_p's will not be universal, but rather depend on the type of flow. This is consistent with Landau's objection to the universality of Kolmogorov's law.

Although the $\hat{\beta}$ model produces plausible results, it does not pretend to be a truly predictive model of the effects of intermittency. Nevertheless, it does provide a convenient conceptual framework within which we can air the various issues involved.

6.1.3.4 The distribution of energy and vorticity across the different eddy sizes

In Section 6.6 we turn to the most delicate issue of Chapter 6. The problem is this. One of the enormous advantages of working in Fourier space is that the energy spectrum $E(k)$ is the central object of most spectral theories. Since we interpret $E(k)$ as the distribution of energy across the different eddy sizes this gives us an invaluable tool for describing the dynamics of cascades, in which energy is passed from scale to scale. If we wish to remain in real space then we would like some equivalent function which differentiates between the energy held at different scales. To some extent the second-order structure function does this, since we tend to interpret $\langle [\Delta v]^2 \rangle$ as,

$$\langle [\Delta v]^2 \rangle (r) \sim [\text{energy held in eddies of size } r \text{ or less}].$$

Indeed, this was the position adopted by Landau, Townsend and many others. However, as we discovered in Chapter 3, this association is rather loose; too loose to allow us to study the dynamics of cascades in the style of the spectral theories. That is, $\langle[\Delta v]^2\rangle$ contains not only information about the energy of eddies of size r or less, but also information about the enstrophy held in eddies of size greater than r.

So, if we are to find a better real-space substitute for $E(k)$, we must first ask ourselves why is it that we believe that $E(k)$ represents the distribution of energy across the range of eddies present in the flow. At a superficial level the answer seems obvious. $E(k)$ is the distribution of energy amongst the varies Fourier modes present in $\mathbf{u}(\mathbf{x})$ (see Section 6.1.2). But this is not good enough. A turbulent velocity field is not composed of waves, but of blobs of vorticity. So we need a more thoughtful justification of our interpretation of $E(k)$. On reflection, the utility of $E(k)$ in interpreting cascade dynamics lies in three properties of the function:

(1) $E(k) > 0$;
(2) $\int_0^\infty E(k)\,dk = \frac{1}{2}\langle \mathbf{u}^2\rangle$;
(3) $E(k)$ corresponding to a random array of simple eddies for fixed size l_e exhibits a sharp peak around $k \sim \pi/l_e$.

The first two properties of $E(k)$ permit us to picture $E(k)$ as the distribution of kinetic energy in k-space, while the third property allows us to associate eddy size with k^{-1}. Although this last point is rather imprecise (eddies of size l_e produce a distributed spectrum, albeit centred around $k \sim \pi/l_e$) the energy spectrum provides a convenient means of visualizing the energy distribution at a given instant.

So, if we wish to study cascade dynamics in real space we would like a function, $V(r)$, which has the properties:

(1) $V(r) > 0$;
(2) $\int_0^\infty V(r)\,dr = \frac{1}{2}\langle \mathbf{u}^2\rangle$;
(3) $V(r)$ corresponding to a random array of simple eddies of fixed size l_e exhibits a sharp peak around $r \sim l_e$.

Moreover, in order to be useful, we would like $V(r)$ to be simply related to more familiar quantities, such as $\langle[\Delta v]^2\rangle$ or $E(k)$. Unfortunately, as far as this author is aware, no such function exists! So how can we describe the energy exchange between different scales in real space? The answer, unfortunately, is 'with some difficulty'. It seems that we have to make a compromise, and one compromise is the following. We can find a function which satisfies conditions (2) and (3) above, but for which (1) has to be relaxed and replaced by the weaker condition

$$\text{(1b)} \qquad \int_0^r V(r)\,dr > 0.$$

(Actually we shall see that $V(r)$ turns out to be positive in fully developed turbulence, so the distinction between (1) and (1b) is often not important.) Moreover, $V(r)$ is simply related to $\langle[\Delta v]^2\rangle$ and, as a consequence, it is not difficult to obtain an evolution equation for $V(r)$. It has the form

$$\frac{\partial V}{\partial t} = [\text{non-linear inertial terms}] + [\text{viscous terms}]$$

which is reminiscent of the equivalent spectral equation

$$\frac{\partial E}{\partial t} = [\text{non-linear inertial terms}] + [\text{viscous terms}].$$

Thus we may loosely interpret $V(r)$ as the kinetic energy density in real space, just as $E(k)$ is the kinetic energy density in Fourier space. Moreover, we can set about investigating the transfer of energy between scales by the non-linear inertial forces.

The function $V(r)$ has one more useful property. It is usual when working in Fourier space to interpret $kE(k)$ as the kinetic energy of eddies of wave number $\sim k$. Given the correspondence $k \sim \pi/l_e$ we might write this as

$$[\text{K.E. of eddies of size } l_e] \approx [kE(k)]_{k \approx \pi/l_e}, \quad (\eta < l_e < l).$$

We shall see that $V(r)$ has the analogous property,

$$[\text{K.E. of eddies of size } l_e] \approx [rV(r)]_{r=l_e}, \quad (\eta < l_e < l).$$

Combining these two expressions we find,

$$rV(r) \approx [kE(k)]_{k=\hat{\pi}/r}, \quad (\eta < l_e < l)$$

where $\hat{\pi}$ is a number close to π. (Note that these expressions are valid only if l_e is greater than Kolmogorov's microscale, η, and less than the integral scale, l.) In fact, we shall see that, for fully developed turbulence, the correspondence between $rV(r)$ and $[kE(k)]_{k=\hat{\pi}/r}$ is remarkably good if we make the substitution $\hat{\pi} = 9\pi/8$. The two functions, $f_1(r) = [kE(r)]_{k=\hat{\pi}/r}$, and $f_2(r) = rV(r)$ are shown below for the model spectrum

$$E(k) = \hat{k}^4\left(1 + \hat{k}^2\right)^{-17/6} \exp\left[-\hat{k}\mathrm{Re}^{-3/4}\right], \quad \hat{k} = kl.$$

This spectrum exhibits a k^4 region for small k, a $k^{-5/3}$ fall off for intermediate k, and an exponential tail at large k. Such properties are thought to be typical of fully developed grid turbulence at high Re and so this model spectrum is often used for illustrative purposes. The Figure 6.8 below shows $f_1(r)$ and $f_2(r)$ corresponding to $\mathrm{Re}^{3/4} = 100$. It is clear that the curves are very similar, implying that $V(r)$ and $E(k)$ carry almost exactly the same information. This close correspondence between $V(r)$ and $E(k)$ in the range $\eta < r < l$ turns out to hold whenever $E(k)$ is a smooth function of k, which it always is in fully developed turbulence.

(a)

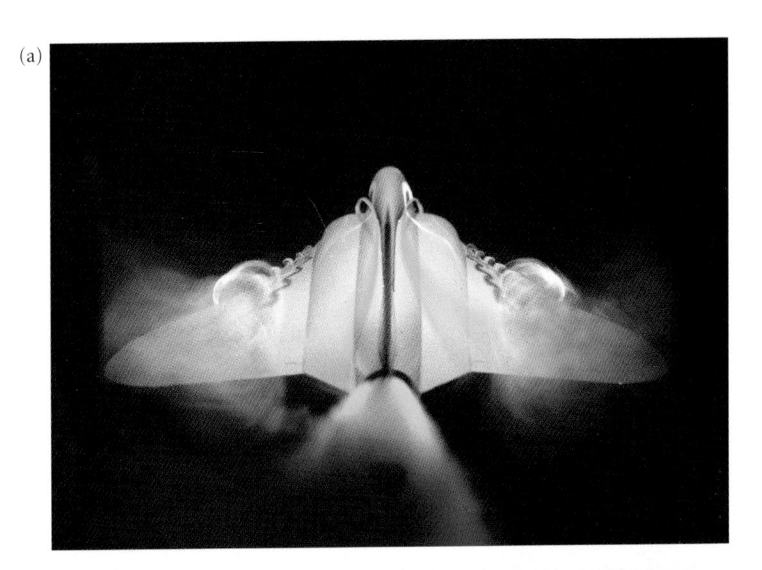

(b)

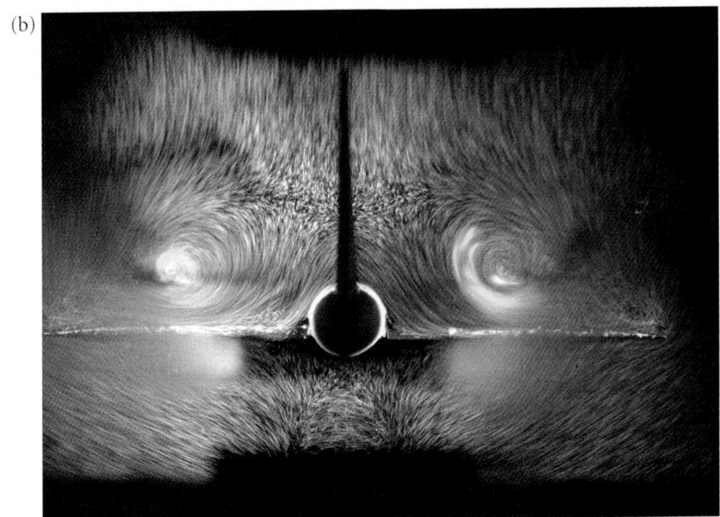

Plate 1 Turbulent vortices on the upper surface of a delta wing: (a) top view; and (b) rear view. How can we design an aircraft without a knowledge of turbulence? See the discussion in Section 1.5. [Photograph by H. Werlé of ONERA, courtesy of J. Delery.]

Plate 2 Turbulence behind a chimney. It would be difficult to simulate this numerically. See the discussion in Sections 1.5 and 1.9. [Photograph by H. Werlé of ONERA, courtesy of J. Delery.]

Plate 3 Copy of Leonardo's famous sketch of water falling into a pool. Note the different scales of motion, suggestive of the energy cascade. See the discussion in Section 1.6. [Courtesy of F. C. Davidson.]

(a)

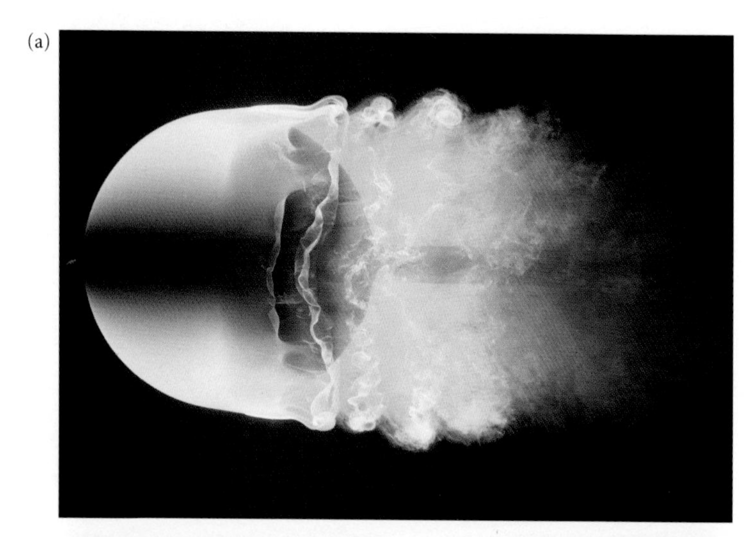

(b)

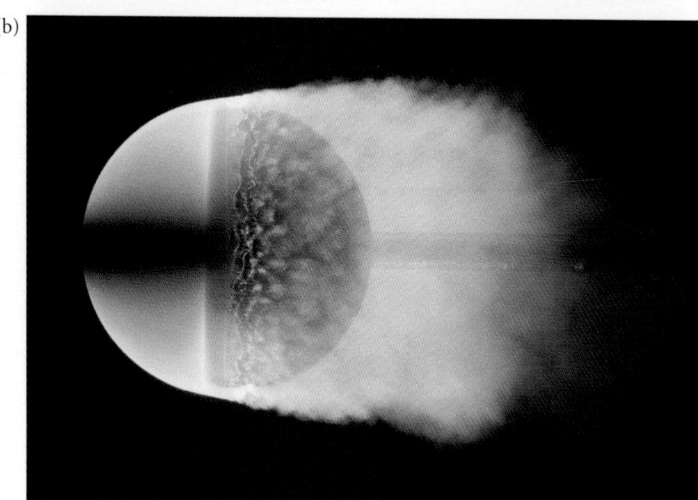

Plate 4 Flow past a sphere at (a) $\mathrm{Re} = 2 \times 10^4$ and (b) $\mathrm{Re} = 2 \times 10^5$. See the discussion in Section 1.6. [Photograph by H. Werlé of ONERA, courtesy of J. Delery.]

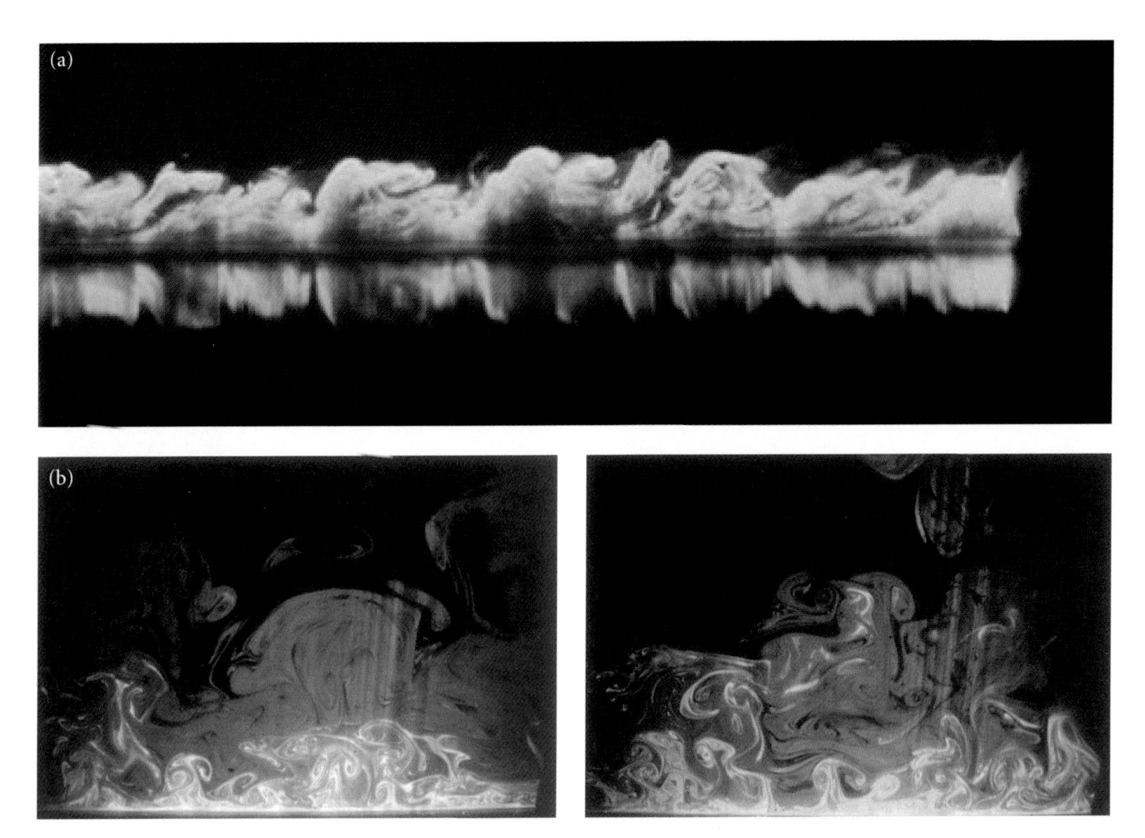

Plate 5 Two views of a boundary layer. (a) Side view of the large eddies in a turbulent boundary layer using laser induced fluorescence. The flow is from left to right. [Courtesy of efluids.com. Picture by Prof. M. Gad-el-Hak, Virginia Commonwealth University.] (b) Vortical structures in a turbulent boundary layer visualized by smoke and laser light. The picture is an oblique, transverse section of the boundary layer. Note the frequent appearance of mushroom-like structures. See the discussion in Section 4.2.6. [Courtesy of R.E. Britter, University of Cambridge].

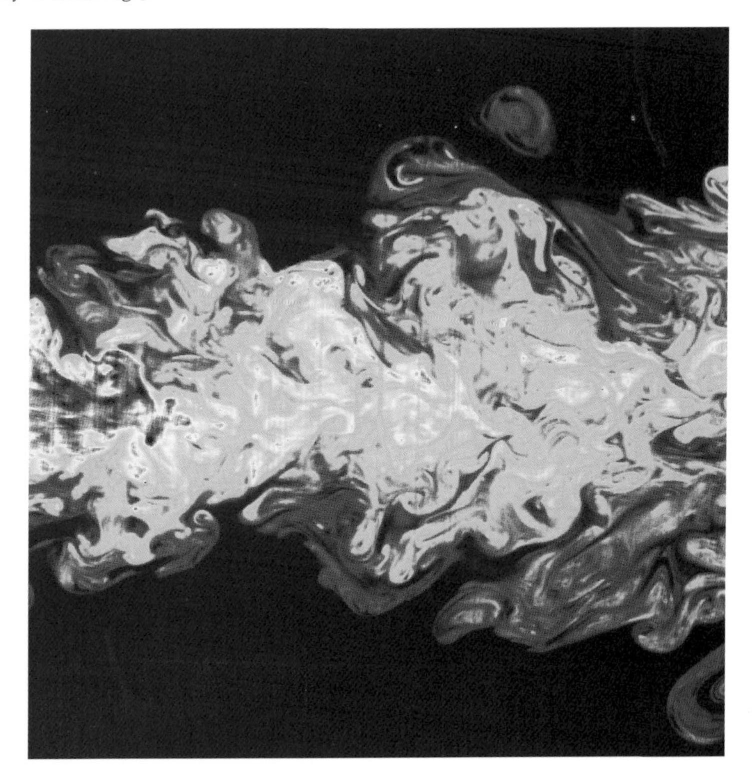

Plate 6 A turbulent jet made visible by laser-induced fluorescence. Note the convoluted outer edge of the jet. [Picture by C. Fukushima and J. Westerweel, University of Delft. Courtesy of efluids.com.]

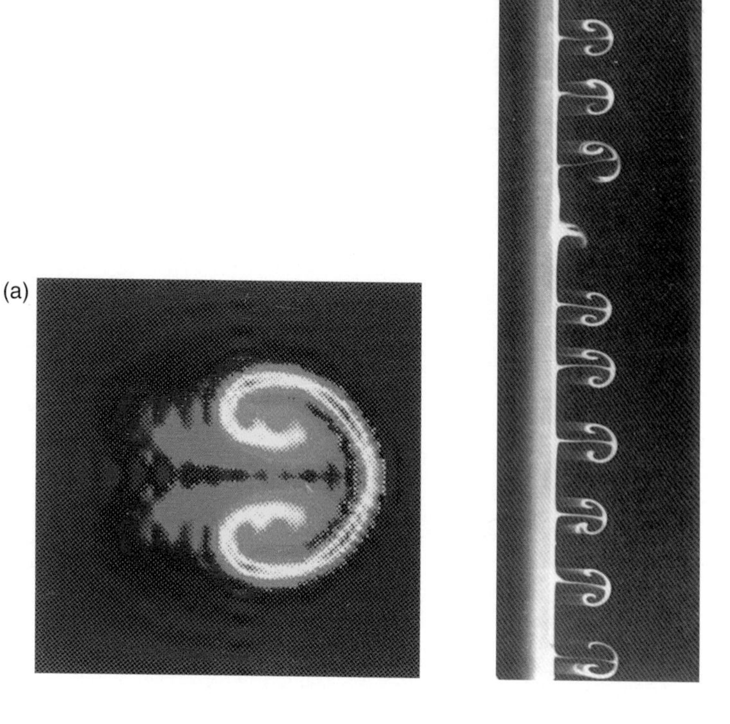

Plate 7 (a) Numerical simulation of a bursting vortex. See Section 5.1.2 and Figure 5.6 for more details. Colour contours correspond to the local intensity of angular momentum, as discussed in Section 5.1.2. As the vortex bursts a vortex sheet is created. This is an example of the transfer of energy to smaller and smaller scales by vortex stretching. (b) Instability of a swirling annulus of fluid. A brass cylinder is rotated in an oscillatory fashion at a frequency of 0.25 Hz. During each cycle a layer of swirling fluid is built up on the surface of the cylinder. This is centrifugally unstable and breaks up into an array of vortex hoops of the type shown in plate 7(a). (From Taneda *et al.* 1977; see Chapter 7 references; with permission.)

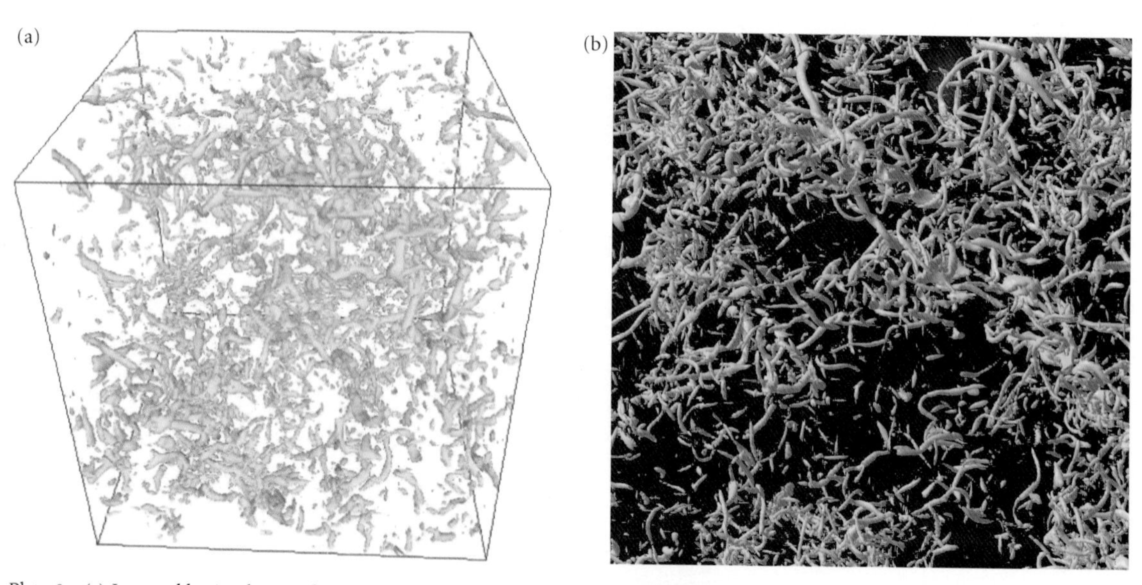

Plate 8 (a) Large-eddy simulation of isotropic turbulence in a periodic cube visualized by vorticity contours. Note that there is some evidence that the vorticity field has a spotty structure, though the very fine scales evident in the DSN adjacent are absent. See the discussion in Sections 6.1.1, 6.1.3 and 7.3. (Courtesy of M Lesieur, Institut de Mechanique de Grenoble.) (b) DNS of isotropic turbulence in a periodic cube visualized by vorticity contours. The Reynolds number based on the Taylor microscale is 1200. At the time of writing this represents the highest value of Re obtained in a simulation. Note the worm-like structure of the vorticity field (see Section 7.3). (Courtesy of Y. Kaneda, T. Ishihara, A. Uno, K. Itakura, and M. Yokokawa.)

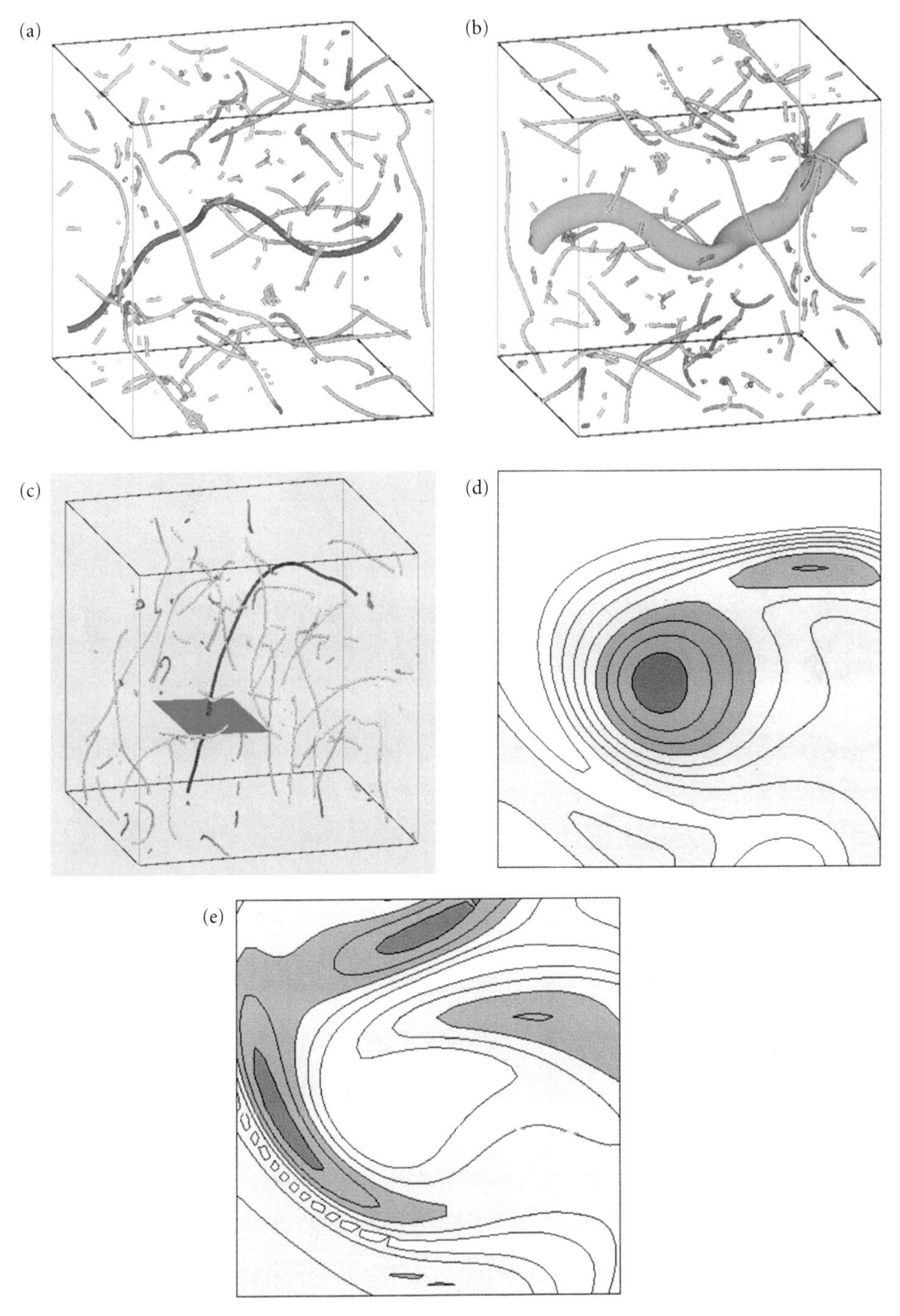

Plate 9 The DNS of almost isotropic turbulence showing the structure of the vorticity field, from Kida (2000). (See Chapter 7 references.) The axes of the vortex tubes are identified by a local minimum in pressure and the edges of the vortex tubes are determined by the point where the local value of u_θ^2/r reaches a maximum. (a) Axes of the vortex tubes. (b) One typical vortex tube. (c) Plane normal to the axis of a portion of a vortex tube. (d) Contours of axial vorticity in the plane shown in (c). (e) Contours of cross-axial vorticity in the plane shown in (c). The cross-axial vorticity is spiralled around the vortex core and this is the seat of most of the dissipation. See the discussion in Section 7.3. [Courtesy of S. Kida]

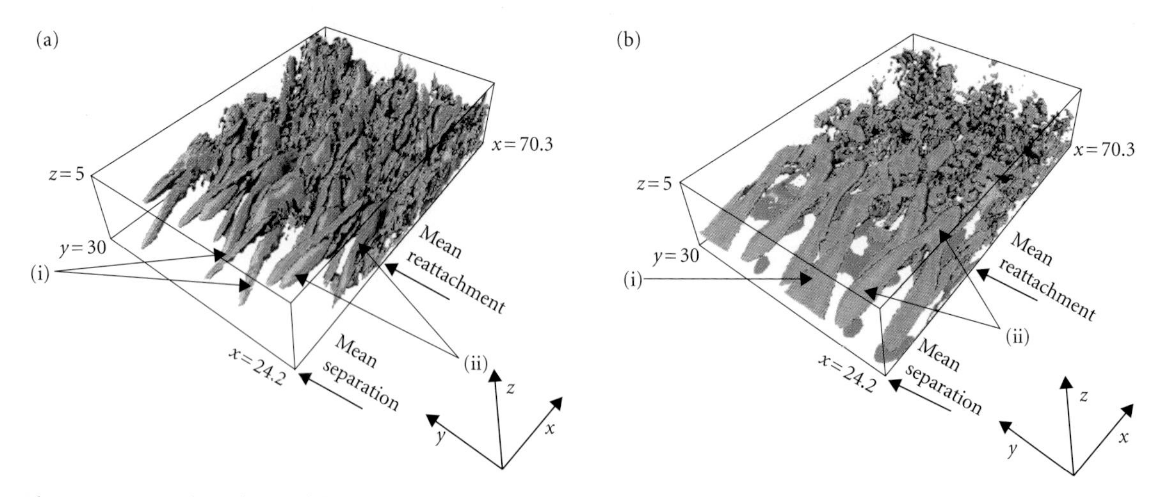

Plate 10 Numerical simulation of the separation bubble in a boundary layer undergoing transition to turbulence. See the discussion in Section 7.3. (a) Plot of ω_z showing (i) Λ-like vortices located in the separated shear layer and (ii) a distorted Λ-like vortex about to break up near the reattachment point. (b) Plot of ω_y showing the same behaviour as in (a). Red and blue represent vorticity of opposite signs. [Courtesy of M. Alam and N.D. Sandham, University of Southampton.]

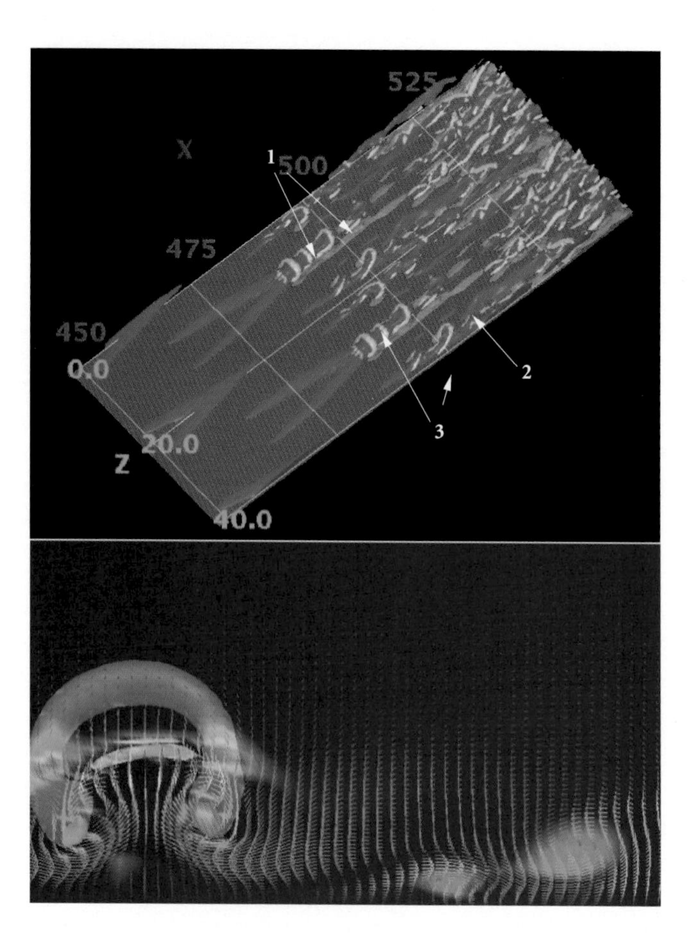

Plate 11 Numerical simulation of transition to turbulence in a boundary layer. See the discussion in Section 7.3. [Courtesy of P. Comte, University of Strasbourg.]

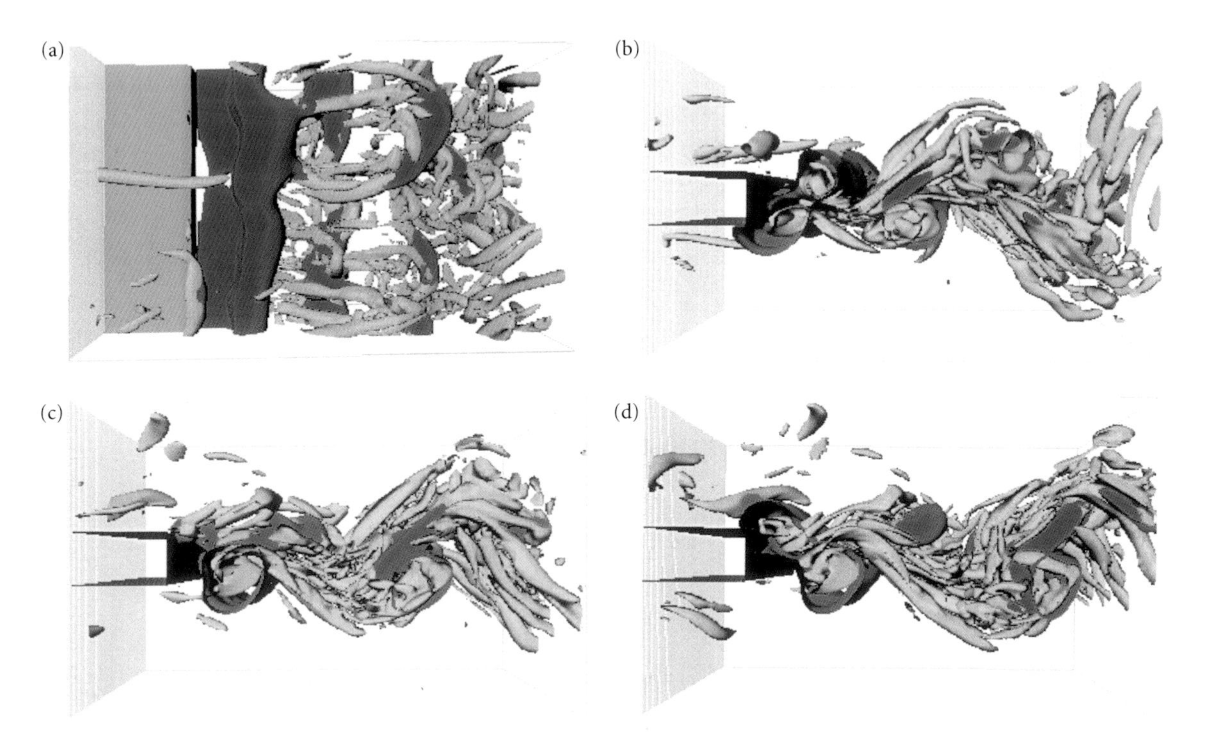

Plate 12 Numerical simulation of turbulence in the wake of a trailing edge. Red represents low pressure regions, which are associated with Karman vortices, and blue represents high vorticity levels. The figures represent instantaneous distributions at different times, increasing from (a) to (d), and show how the axial vortex tubes get wrapped around the Karman vortices. See the discussion in Section 7.3. (a) Top view. (b)–(d) Side views. [Courtesy of Y. Yao and N.D. Sandham, University of Southampton.]

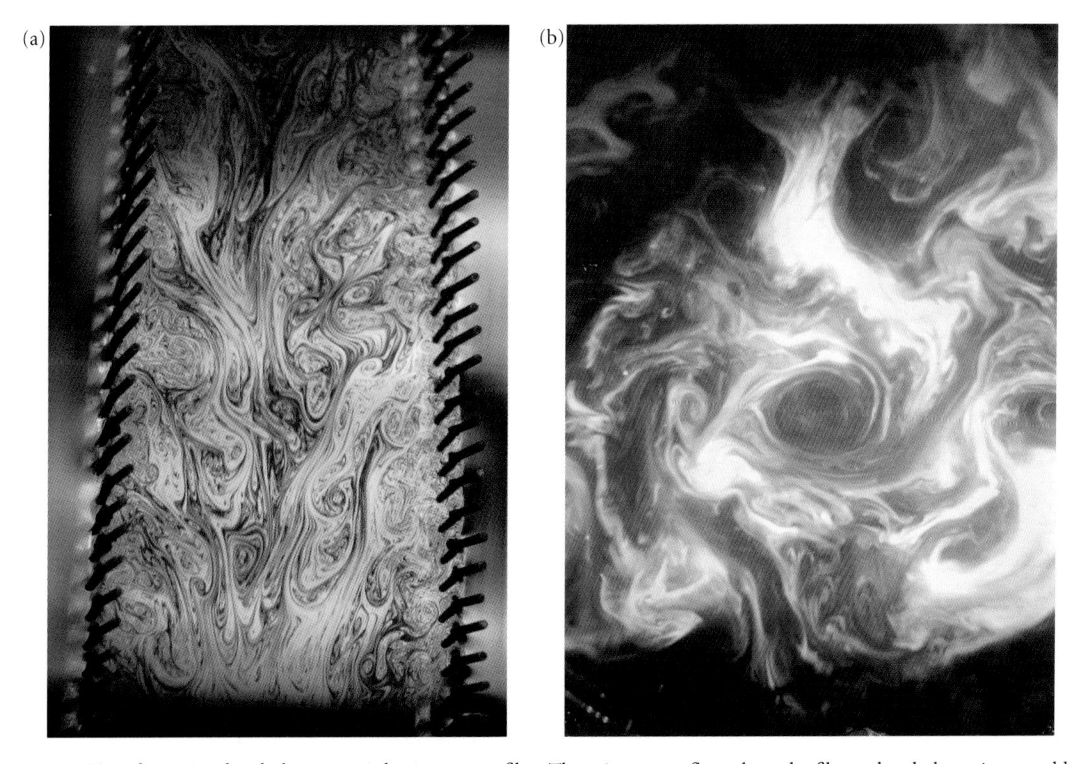

Plate 13 (a) Two-dimensional turbulence created using a soap film. There is a mean flow along the film and turbulence is created by the pegs at the sides of the film. See the discussion in Section 10.1. (b) A horizontal section of turbulence in a rapidly rotating tank. The turbulence is quasi-two-dimensional and created by an oscillating grid. ((a) Courtesy of M. Ward-Close, A. Dorn, B. Pearson, University of Nottingham, and (b) Courtesy of E. Hopfinger, Institut de Mechanique de Grenoble.)

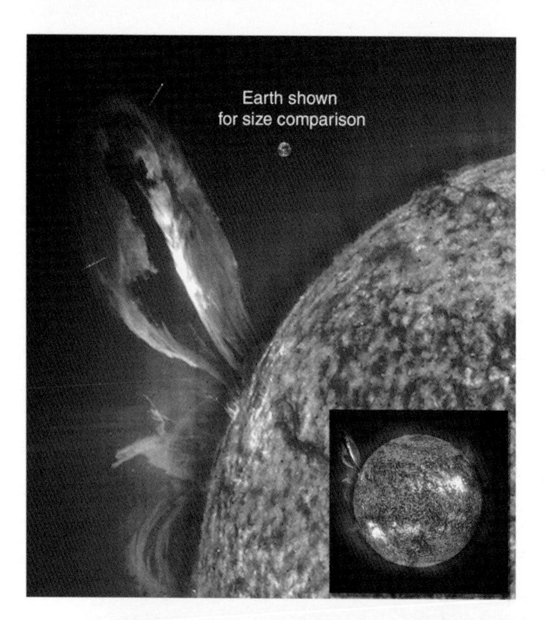

Plate 14 Turbulent activity on the surface of the sun leads to magnetic reconnection of coronal flux tubes and eventually to solar flares and coronal mass ejections (CMEs). See the discussion in Section 9.5.3. (Courtesy of the SOHO–EIT consortium. SOHO is a project of international collaboration between ESA and NASA.)

(a)

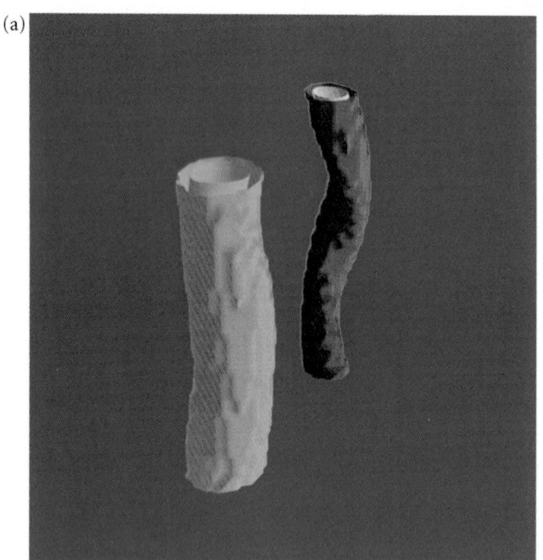

(b)

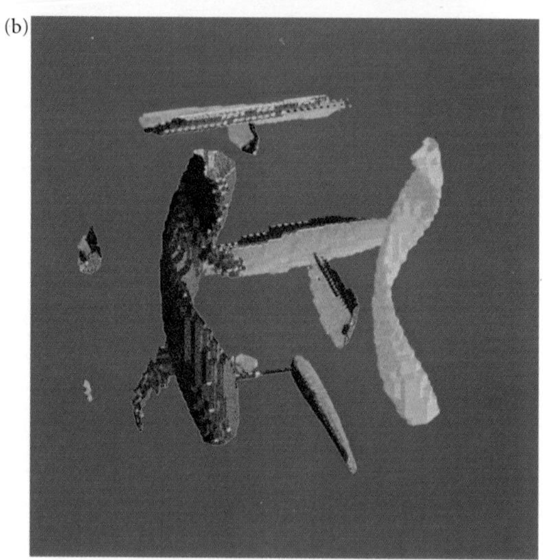

Plate 15 Numerical simulation of MHD turbulence in a periodic cube. (a and b) The figures show isocontours of vorticity at two different times for $N = 0.4$. Note the elongated sheets and tubes of vorticity. See the discussion in Section 9.4.2. (Courtesy of O. Zikanov and A. Thess.)

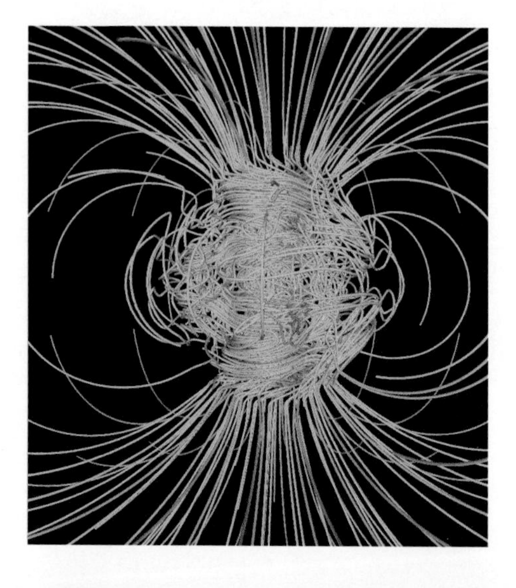

Plate 16 A numerical simulation of the earth's magnetic field which is maintained by turbulent motion in the core of the earth. The lines shown are magnetic field lines. Note the strong east–west field near the inner core of the earth. See the discussion in Section 9.5.2. (Courtesy of G. Glatzmaier, University of Santa Cruz.)

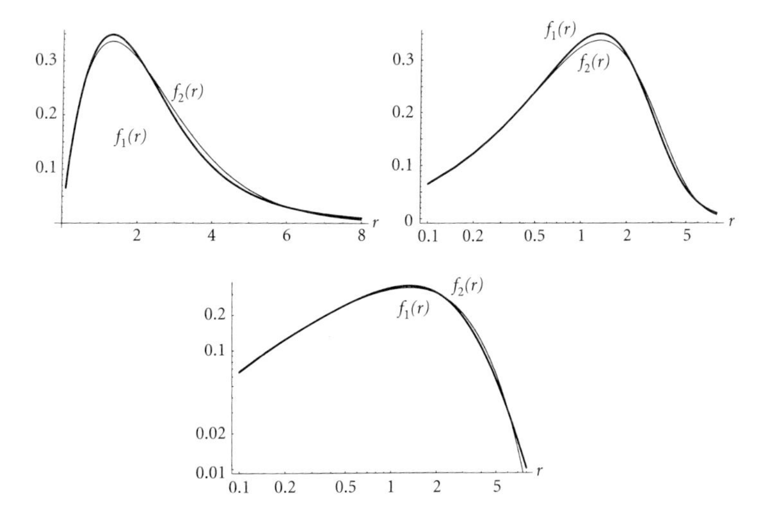

Figure 6.8 The functions $f_1(r)$ and $f_2(r)$ corresponding to the energy spectrum given above. Line–line, line–log and log–log plots are shown. The log plots emphasize the inertial sub-range.

So what is this real-space approximation to $E(k)$? It turns out to be

$$V(r) = -\frac{3}{8}r^2\frac{\partial}{\partial r}\frac{1}{r}\frac{\partial}{\partial r}\left\langle [\Delta \mathrm{v}]^2 \right\rangle.$$

The principle utility of $V(r)$ is that it allows us to discuss the dynamics of the energy cascade without moving into Fourier space. However, one must be cautious and not place too much reliance on our physical interpretation of $V(r)$. For example, it is possible to conceive of spectra (albeit unusual spectra), which give rise to small negative contributions to $V(r)$ in the range $\eta < r < l$. (See Section 6.4.1.) Clearly one cannot interpret $rV(r)$ as a kinetic energy in such cases! Moreover, we shall see that the estimates

$$[\text{K.E. of eddies of size } l_e] \approx [kE(k)]_{k=\hat{\pi}/l_e}$$

$$[\text{K.E. of eddies of size } l_e] \approx [rV(r)]_{r=l_e}$$

hold only in the range $\eta < r < l$, where η is Kolmogorov's microscale and l is the integral scale. In short, neither $E(k)$ nor $V(r)$ can be considered as representative of the kinetic energy of eddies in the ranges $r \ll \eta$ or $r \gg l$. (Actually, this has to be true as there are no eddies of size or $r \ll \eta$ or $r \gg l$!)

All-in-all, it must be conceded that $V(r)$ is very much the poor relation of $E(k)$. Nevertheless, it does allow us to extract some information about the energy cascade from the Karman–Howarth equation without reverting to Fourier analysis and perhaps that is of some value.

This concludes our overview of this chapter. As noted at the beginning of this section, we barely scratch the surface of the statistical theory of homogeneous turbulence. We hope, however, to at least provide a starting point for further study.

6.2 The governing equations of isotropic turbulence

My opinion of long standing that turbulence would finally be accounted for by integration of the Navier–Stokes equations in their complete, non-linear, form, proved wrong. . . . As in the kinetic theory of gases, statistical methods have shown their superiority. (A. Sommerfeld 1964)

6.2.1 *Kinematics*

We shall investigate the dynamics of isotropic turbulence in Section 6.2.2. First, however, we have a lot of groundwork to do. In particular, we need to introduce the various statistical quantities used to characterize the state of a cloud of turbulence and investigate the restrictions imposed on these functions by continuity and by the symmetries implied by isotropy.

There are three, related, quantities traditionally used to characterize the statistical state of a turbulent velocity field. We introduced these in Section 3.2.5: they are the velocity correlation function, the structure function, and the energy spectrum. The most fundamental of these is, perhaps, the velocity correlation function. Introduced by Taylor in 1921, the velocity correlation function is to turbulence what $\mathbf{u}(\mathbf{x}, t)$ is to a laminar flow. It is the basic building block, or common currency, of statistical theories. The energy spectrum and structure function are relatives of the velocity correlation function and, in their different ways, try to describe the manner in which energy is distributed across the different eddy sizes. The purpose of Section 6.2.1.1 is to remind you of the definitions of these quantities and describe some of their kinematic properties. In effect, we try to establish the *language* of the statistical theories and lay down some of the associated 'grammatical rules'.

Next, in Section 6.2.1.2, we explore the consequences of our assumption of isotropy and of the conservation of mass. We shall see that this greatly simplifies the form of the velocity correlation function, and it is this immense simplification that makes it possible to develop the theory of isotropic turbulence in a relatively straightforward manner.

There is a great deal of analysis in Sections 6.2.1.1 and 6.2.1.2, some of it a little tedious, but the results are of great importance and so they are summarized in Section 6.2.1.3. It is not until we reach Section 6.2.2 that we are ready for the altogether more important issue of dynamics.

6.2.1.1 Some kinematics: velocity correlation functions and structure functions

In order to simplify matters we shall restrict the discussion to homogeneous, isotropic turbulence[2] in which the mean flow is zero.

[2] See Section 3.2.5 for the definition of statistical homogeneity and isotropy.

Since, in the absence of a mean shear, there is no injection of energy into the turbulence, such a flow will always decay in the course of time. We might picture this as a fluid which is subjected to vigorous stirring and then left to itself. Of course the properties of such turbulence are time-dependent, and so we use ensemble averages, $\langle \sim \rangle$, rather than time averages, to form statistical quantities. (See Section 3.2.4). Also, since we do not need to distinguish between the mean flow and the turbulent motion we shall denote the fluctuating velocity field as simply $\mathbf{u}$, without a prime.

The concepts of the *velocity correlation tensor*, the *structure function*, and the *energy spectrum* were introduced in Section 3.2.5. The *second-order velocity correlation tensor*, for example, is defined as[3]

$$Q_{ij}(\mathbf{r}, \mathbf{x}, t) = \langle u_i(\mathbf{x}) u_j(\mathbf{x} + \mathbf{r}) \rangle \tag{6.3}$$

In homogeneous turbulence all statistical properties are, by definition, independent of $\mathbf{x}$ and so we have

$$Q_{ij}(\mathbf{r}) = \langle u_i(\mathbf{x}) u_j(\mathbf{x} + \mathbf{r}) \rangle \tag{6.4}$$

where the time dependence of Q_{ij} is understood. An alternative notation is to write

$$Q_{ij}(\mathbf{r}) = \left\langle u_i u_j' \right\rangle \tag{6.5}$$

where the prime now indicates that u_j is evaluated at location $\mathbf{x}' = \mathbf{x} + \mathbf{r}$. This correlation tensor (sometimes called correlation function) has the geometrical property

$$Q_{ij}(\mathbf{r}) = Q_{ji}(-\mathbf{r}) \tag{6.6}$$

since, in homogeneous turbulence, reversing $\mathbf{r}$ simply amounts to switching $\mathbf{x}$ and $\mathbf{x}'$ (see Figure 6.9).

Four additional properties of Q_{ij} are:

(i) $\frac{1}{2} Q_{ii}(0) = \frac{1}{2} \langle \mathbf{u}^2 \rangle =$ kinetic energy density; (6.7)

(ii) $Q_{ij}(0) = -\tau_{ij}^R / \rho = -(\text{Reynolds stress})/\rho$; (6.8)

(iii) $Q_{ij}(\mathbf{r}) \leq Q_{xx}(0)$; (6.9)

(iv) $\partial Q_{ij} / \partial r_i = \partial Q_{ij} / \partial r_j = 0$. (6.10)

The second property above gives us one interpretation of Q_{ij}. For the special case of $\mathbf{r} = \mathbf{0}$, Q_{ij} is proportional to the Reynolds stress, τ_{ij}^R. More generally, Q_{ij} tells us about the degree to which the points $\mathbf{x}$ and $\mathbf{x}'$

[3] This is sometimes called a two-point, one-time velocity correlation. Some theories work with velocity correlations relating velocities at the three points (so-called three-point correlations) while others work with the correlations between velocities measured at the same point but at different times (one-point, two-time correlations).

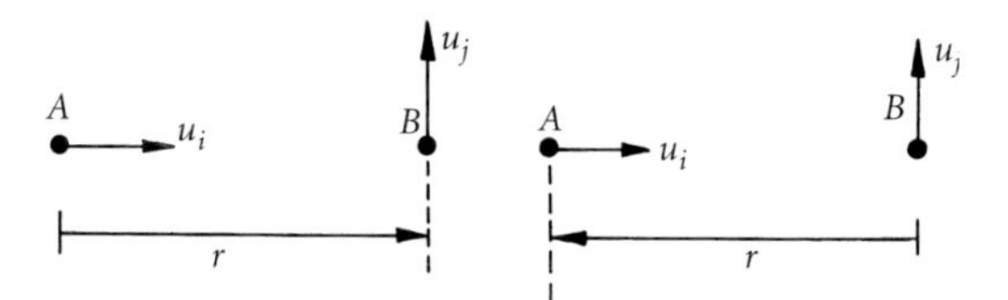

Figure 6.9 Geometrical property $Q_{ij}(\mathbf{r}) = Q_{ji}(-\mathbf{r}) = \langle (u_i)_A (u_j)_B \rangle$.

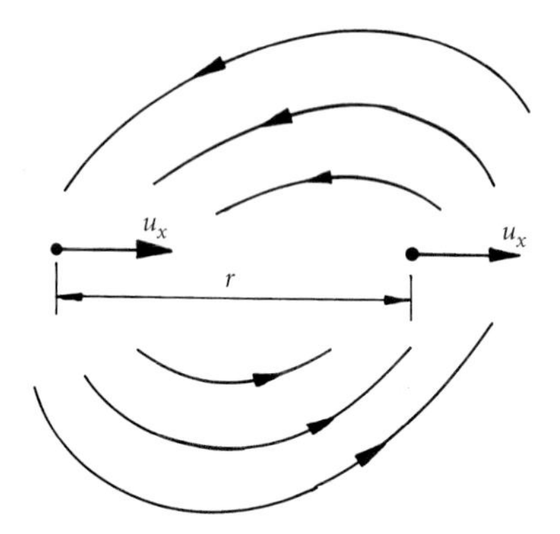

Figure 6.10 Q_{ij} is non-zero if eddies span the gap r.

are statistically correlated (see Section 3.2.5). If points $\mathbf{x}$ and $\mathbf{x}'$ are well separated, so that what happens at $\mathbf{x}$ is unrelated to events at $\mathbf{x}'$, then $Q_{ij} = 0$ (Figure 6.10). On the other hand, points that are close, relative to the integral scale l, will exhibit significant correlations between $\mathbf{x}$ and $\mathbf{x}'$ as eddies of given structure sweep through the region encompassing both $\mathbf{x}$ and $\mathbf{x}'$.

The origin of properties (iii) and (iv) is not at all obvious. Let us see where they come from. The third follows from the Schwartz inequality which states,

$$\left[\int \mathbf{A} \cdot \mathbf{B}\, dV\right]^2 \leq \int \mathbf{A}^2\, dV \int \mathbf{B}^2\, dV. \tag{6.11}$$

So, if we interpret volume averages as being equivalent to ensemble averages (see Section 3.2.4), then we have

$$Q_{ij}^2 \leq V^{-2} \int_V u_i^2\, dV \int_V u_j^2\, dV = \langle u_i^2 \rangle \left\langle u_j^2 \right\rangle = \langle u_x^2 \rangle^2. \tag{6.12}$$

The fourth relationship above follows directly from the continuity equation $\nabla \cdot \mathbf{u} = 0$, the fact that the operations $\langle \sim \rangle$ and $\partial / \partial x_i$ commute, and that $\partial / \partial x_i = -\partial / \partial r_i$, $\partial / \partial x_j' = \partial / \partial r_j$.

Let us now remind you of some of the notation introduced in Section 3.2.5. Specifically, we found it convenient to introduce three particular forms of Q_{ij}:

$$R(r) = \tfrac{1}{2} Q_{ii} = \tfrac{1}{2} \langle \mathbf{u} \cdot \mathbf{u}' \rangle \tag{6.13}$$

$$u^2 f(r) = Q_{xx}(r\hat{\mathbf{e}}_x) \tag{6.14a}$$

$$u^2 g(r) = Q_{yy}(r\hat{\mathbf{e}}_x) \tag{6.14b}$$

where u is defined as,

$$u = \langle u_x^2 \rangle^{1/2} = \left\langle u_y^2 \right\rangle^{1/2} = \langle u_z^2 \rangle^{1/2} = \left(\tfrac{1}{3} \langle \mathbf{u}^2 \rangle\right)^{1/2}.$$

The functions f and g are known as the longitudinal and lateral velocity correlation functions (e.g. Figure 6.11). They are dimensionless and satisfy $f(0) = g(0) = 1$ and f, $g \leq 1$. The shapes of f and g are shown schematically in Figure 6.12. Note that, for fully developed turbulence, f is found to be non-negative.

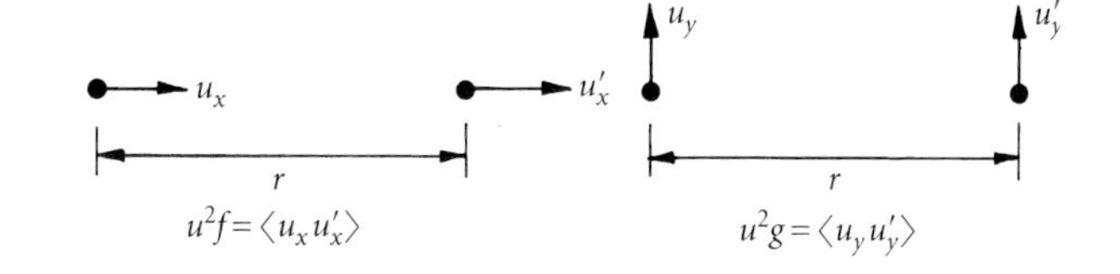

Figure 6.11 Definition of f and g, the longitudinal and lateral velocity correlation functions.

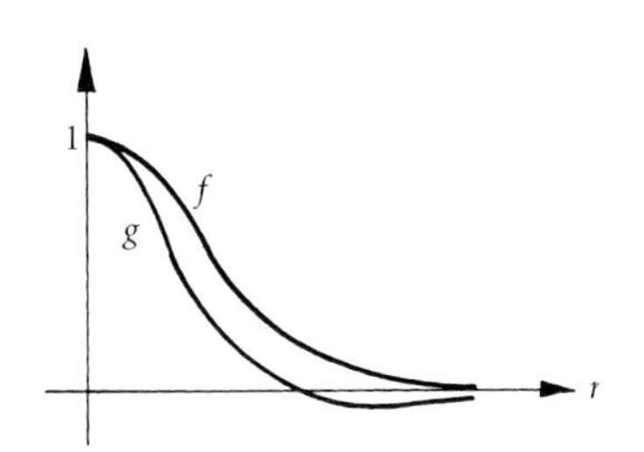

Figure 6.12 The shape of the velocity correlation functions f and g.

The integral scale, l, of the turbulence is conventionally defined as,

$$l = \int_0^\infty f(r)\, dr. \tag{6.15}$$

This definition is somewhat arbitrary, but it does provide a convenient measure of the extent of the region within which velocities are appreciably correlated. So l is representative of the size of the large, energy-containing, eddies. Actually, f and g are not independent functions. For isotropic turbulence they are related by,

$$g = f + \tfrac{1}{2} r f'(r). \tag{6.16}$$

This follows from the conservation of volume, as we shall see shortly.

The *third-order velocity correlation function* (or tensor) is defined as,

$$S_{ijk}(\mathbf{r}) = \langle u_i(\mathbf{x}) u_j(\mathbf{x}) u_k(\mathbf{x} + \mathbf{r}) \rangle \tag{6.17}$$

a special case of which is,

$$u^3 K(r) = \langle u_x^2(\mathbf{x}) u_x(\mathbf{x} + r\hat{\mathbf{e}}_x) \rangle. \tag{6.18}$$

The function $K(r)$ is known as the *longitudinal triple correlation function*. It is found to be negative and grows as $K \sim r^3$ for small r. Its general shape is shown schematically in Figure 6.13.

In Section 3.2.5 we introduced an alternative measure of the state of a field of turbulence, called *structure functions*. These are defined in terms of the longitudinal velocity increment $\Delta v = u_x(\mathbf{x} + r\hat{\mathbf{e}}_x) - u_x(\mathbf{x})$. For example, the *second-order longitudinal structure function* is defined as,

$$\langle [\Delta v]^2 \rangle = \langle [u_x(\mathbf{x} + r\hat{\mathbf{e}}_x) - u_x(\mathbf{x})]^2 \rangle. \tag{6.19}$$

Evidently, it is related to f by,

$$\langle [\Delta v]^2 \rangle = 2u^2(1 - f). \tag{6.20}$$

The *third-order longitudinal structure function* is $\langle [\Delta v]^3 \rangle$. It is readily confirmed that this is related to $K(r)$ through,

$$\langle [\Delta v]^3 \rangle = 6u^3 K(r). \tag{6.21}$$

The physical significance of $\langle [\Delta v]^2 \rangle$ is discussed in Section 3.2.5. In effect, it acts as a kind of filter, extracting information about eddies of

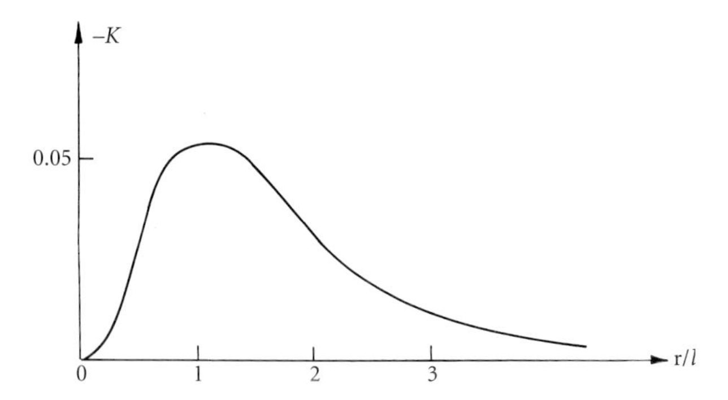

Figure 6.13 Schematic shape of $K(r)$.

size r or less. That is, all eddies of size less than or equal to r contribute directly to $\langle[\Delta \mathrm{v}]^2\rangle$, since they tend to generate different velocities at $\mathbf{x}$ and $\mathbf{x}'$. Eddies much larger than r, on the other hand, tend to have similar velocities at $\mathbf{x}$ and $\mathbf{x}'$ and so make little contribution to $\langle[\Delta \mathrm{v}]^2\rangle$. This suggests that

$$\left\langle[\Delta \mathrm{v}]^2\right\rangle \sim \frac{4}{3}[\text{all energy in eddies of size } r \text{ or less}]. \tag{6.22}$$

(The factor of $4/3$ is inserted because, as $r \to \infty$, $\left\langle(\Delta \mathrm{v})^2\right\rangle \to 2u^2 = \frac{4}{3}\left\langle\frac{1}{2}\mathbf{u}^2\right\rangle$.) However, eddies of size greater than r make a small but finite contribution to $\Delta \mathrm{v}$, of the order of $r \times$ (velocity gradient of eddy), or $r|\boldsymbol{\omega}|$ (Figure 6.14). This suggests that a more refined estimate of $\langle[\Delta \mathrm{v}]^2\rangle$ might be

$$\begin{aligned}\left\langle[\Delta \mathrm{v}]^2\right\rangle \sim\ & \frac{4}{3}[\text{all energy in eddies for size } r \text{ or less}] \\ & + \frac{4}{3}(r/\pi)^2[\text{all enstrophy in eddies of size } r \text{ or greater}].\end{aligned} \tag{6.23}$$

We shall return to this estimate shortly where we shall explain the inclusion of the factor of $4/(3\pi^2)$ in the second term on the right. In the meantime we reintroduce the energy spectrum $E(k)$, defined via the transform pair[4]

$$E(k) = \frac{2}{\pi}\int_0^\infty R(r)kr\sin(kr)\,dr \tag{6.24}$$

$$R(r) = \int_0^\infty E(k)\frac{\sin(kr)}{kr}\,dk. \tag{6.25}$$

The energy spectrum has the following useful properties:

(1) $E(k) \geq 0$;

(2) for a random array of simple eddies for fixed size l_e, $E(k)$ peaks at around $k \sim \pi/l_e$;

[4] This is, effectively, the sin-transform pair for the odd functions E/k and rR. Note that (6.24) is valid only for isotropic turbulence. In Chapter 8 we shall see how to define $E(k)$ for anisotropic turbulence.

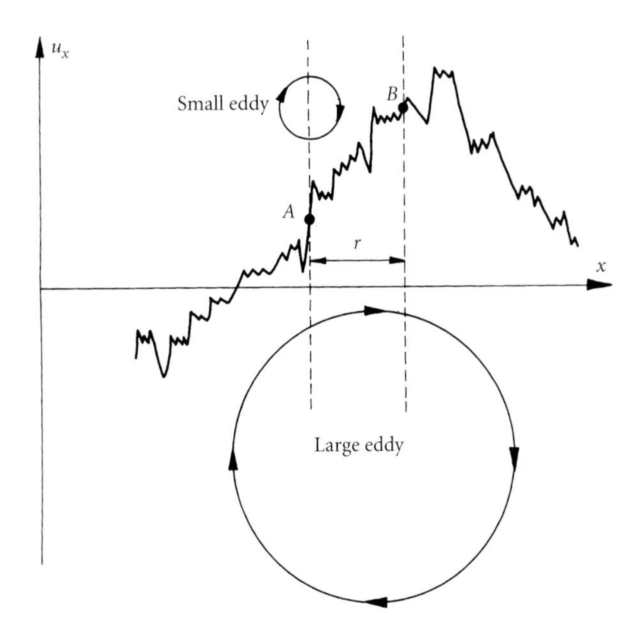

Figure 6.14 Small eddies of size l_e ($l_e \ll r$) make a contribution to $\langle[\Delta v]^2\rangle$ of the order of their kinetic energy, whereas large eddies ($l_e \gg r$) make a contribution of the order of r^2 times their enstrophy.

(3) in the limit $r \to 0$, (6.25) yields

$$\frac{1}{2}\langle \mathbf{u}^2 \rangle = \int_0^\infty E(k)\, dk. \tag{6.26}$$

(Property (i) is established in Chapter 8 while (ii) is discussed in Section 6.4.1.) In view of these properties it is conventional to interpret $E(k)\,dk$ as the contribution to $\frac{1}{2}\langle \mathbf{u}^2 \rangle$ from all eddies with wave numbers in the range $k \to k + dk$, where $k \sim \pi / l_e$. Actually we shall see in Section 6.4.1 that this is a rather naïve interpretation, particularly for $k \ll l^{-1}$ and $k \gg \eta^{-1}$, where η is the Kolmogorov microscale and l is the integral scale. Still, let us, for the meantime, stay with this simple interpretation of $E(k)$.[5] It is possible to show (see Chapter 8) that $E(k)$ has the additional property,

$$\frac{1}{2}\langle \boldsymbol{\omega}^2 \rangle = \int_0^\infty k^2 E(k)\, dk. \tag{6.27}$$

So we might think of $k^2 E(k)\,dk$ as the contribution to $\frac{1}{2}\langle \boldsymbol{\omega}^2 \rangle$ from wave numbers $k \to k + dk$.

[5] An alternative, though ultimately equivalent, interpretation of $E(k)$ is given in Chapter 8. Suppose $\hat{\mathbf{u}}(\mathbf{k})$ is the three-dimensional Fourier transform of $\mathbf{u}(\mathbf{x})$. If the turbulence is isotropic it may be shown that $2\pi k^2 \langle \hat{\mathbf{u}}^\dagger(\mathbf{k}) \cdot \hat{\mathbf{u}}(\mathbf{k}') \rangle = E(k)\delta(\mathbf{k} - \mathbf{k}')$, where $\mathbf{k}$ and $\mathbf{k}'$ are two wavevectors, $k = |\mathbf{k}|$, $\dagger$ represents a complex conjugate, and δ is the three-dimensional Dirac delta function. Thus $E(k)$ is a measure of the kinetic energy held in the *kth* mode of $\hat{\mathbf{u}}(\mathbf{k})$. Moreover, the Fourier transform of a random signal tends to concentrate information about the rapid fluctuations into its high-wave number components, and the information about the slowly varying part of the signal into its low-wave number components. (Think of the Fourier series associated with a plucked string.) Thus the high wave number part of $E(k)$ tends to be associated with the energy of the small-scale eddies and the low-wave number components with the energy of larger eddies.

The relationship between E and $\langle[\Delta v]^2\rangle$ can be established by noting that $R = u^2[g + f/2]$ which, in view of (6.16), can be rewritten as

$$R = \frac{u^2}{2r^2}\frac{d}{dr}[r^3 f].$$

Substituting $(r^3 f)'$ for R in (6.25), integrating, and using (6.20) to relate f to $\langle[\Delta v]^2\rangle$, we find,

$$\langle[\Delta v]^2\rangle = \frac{4}{3}\int_0^\infty E(k)H(kr)\,dk,$$

where,

$$H(x) = 1 + 3x^{-2}\cos x - 3x^{-3}\sin x.$$

We shall see in Chapter 8 that H acts like a filter function and that a good approximation to it is,

$$H(x) \approx \begin{cases} (x/\pi)^2, & \text{for } x < \pi \\ 1, & \text{for } x > \pi \end{cases}.$$

It follows that

$$\langle[\Delta v]^2\rangle \approx \frac{4}{3}\int_{\pi/r}^\infty E(k)\,dk + \frac{4r^2}{3\pi^2}\int_0^{\pi/r} k^2 E(k)\,dk \tag{6.28}$$

which exhibits a reassuring similarity to (6.23).

Finally we note that some authors use an alternative form of the second-order structure function, defined as,

$$\langle[\Delta \mathbf{v}]^2\rangle = \langle[\mathbf{u}(\mathbf{x}+\mathbf{r}) - \mathbf{u}(\mathbf{x})]^2\rangle \tag{6.29a}$$

(We shall distinguish between the two forms by using $\Delta\mathbf{v}$ rather than Δv in (6.29a).) Expanding the quadratic term on the right yields,

$$\tfrac{1}{4}\langle[\Delta \mathbf{v}]^2\rangle = \tfrac{3}{2}u^2 - \tfrac{1}{2}\langle\mathbf{u}\cdot\mathbf{u}'\rangle = \tfrac{1}{2}\langle\mathbf{u}^2\rangle - R(r). \tag{6.29b}$$

6.2.1.2 More kinematics: the simplifications of isotropy and the vorticity correlation function

We now investigate the consequences of: (i) the symmetries associated with isotropy; and (ii) the continuity equation. We shall see that these impose severe constraints on the general structure of Q_{ij} and S_{ijk}. Let us start with symmetry.

Let A, B, C, and D be symmetric functions of r. Then the most general form of isotropic tensors of first; second, and third-order are[6]

$$Q_i(r) = Ar_i \tag{6.30}$$

[6] Here we take isotropy to mean that Q_i, Q_{ij}, and Q_{ijk} are invariant under the full rotation group; that is, they are independent of arbitrary rigid-body rotations and reflections. A less restrictive form of isotropy, which we will not consider, is when all tensors retain spherical symmetry but not reflectional symmetry. This form of turbulence may possess mean helicity.

$$Q_{ij}(r) = Ar_i r_j + B\delta_{ij} \tag{6.31}$$

$$Q_{ijk}(r) = Ar_i r_j r_k + Br_i\delta_{jk} + Cr_j\delta_{ki} + Dr_k\delta_{ij} \tag{6.32}$$

(Consult Batchelor 1953.) Consider, for example, Q_{ij}. In this case the functions A and B are related to f and g by (6.14) and this tells us that,

$$B = u^2 g \tag{6.33}$$

$$A = u^2(f - g)/r^2. \tag{6.34}$$

Evidently, the general form of Q_{ij} can be rewritten as

$$Q_{ij}(\mathbf{r}) = u^2\left[\frac{f-g}{r^2} r_i r_j + g\delta_{ij}\right]. \tag{6.35}$$

Moreover, we know that continuity, in the form of (6.10), requires that

$$\frac{\partial Q_{ij}}{\partial r_i} = [rA'(r) + 4A + r^{-1}B'(r)]r_j = 0$$

from which

$$r^2 A'(r) + 4rA + B'(r) = 0. \tag{6.36}$$

On substituting for A and B in terms of f and g we recover (6.16). Combining these results allows us to eliminate g from (6.35) and rewrite Q_{ij} as a function of f alone:

$$Q_{ij} = \frac{u^2}{2r}\left[(r^2 f)'\delta_{ij} - f' r_i r_j\right] \tag{6.37}$$

Note that the Reynolds stresses $\langle u_x u_y\rangle$, $\langle u_y u_z\rangle$, and $\langle u_z u_x\rangle$ are all zero in isotropic turbulence.

A similar line of argument allows us to rewrite S_{ijk} as a function of $K(r)$ only. When the details are followed through we find

$$S_{ijk} = u^3\left[\frac{K - rK'}{2r^3} r_i r_j r_k + \frac{2K + rK'}{4r}\left(r_i\delta_{jk} + r_j\delta_{ik}\right) - \frac{K}{2r} r_k\delta_{ij}\right]. \tag{6.38}$$

Expressions (6.37) and (6.38) are extremely useful. They tell us that the complicated tensors Q_{ij} and S_{ijk} are simply determined by the two scalar functions $f(r)$ and $K(r)$. It is this enormous simplification which makes isotropic turbulence relatively easy to analyse.

There are a number of useful kinematic results which follow directly from (6.37). For example, from (6.13) and (6.37) we may confirm that

$$R(r) = \frac{1}{2}\langle\mathbf{u}\cdot\mathbf{u}'\rangle = \frac{u^2}{2r^2}(r^3 f)' \tag{6.39}$$

from which we see that the integral scale, l, can equally well be defined by

$$u^2 l = \int_0^\infty R\,dr = u^2 \int_0^\infty f\,dr.$$

Also, from (6.29) and (6.39), we have

$$\langle [\Delta \mathbf{v}]^2 \rangle = \frac{1}{r^2}\frac{\partial}{\partial r}\left[r^3 \langle [\Delta v]^2 \rangle \right] \tag{6.40}$$

which tells us the relationship between the two different types of structure functions introduced in the last section.

Let us now introduce the *vorticity correlation tensor* $\langle \omega_i \omega_j' \rangle$. It is readily confirmed (see Batchelor 1953) that

$$\left\langle \omega_i \omega_j' \right\rangle = \nabla^2 Q_{ij} + \frac{\partial Q_{kk}}{\partial r_i \partial r_j} - (\nabla^2 Q_{kk})\delta_{ij}. \tag{6.41a}$$

Of particular interest is the special case

$$\langle \boldsymbol{\omega} \cdot \boldsymbol{\omega}' \rangle = -\nabla^2 Q_{ii}(\mathbf{r}). \tag{6.41b}$$

In terms of $R(r)$ this becomes

$$\frac{1}{2}\langle \boldsymbol{\omega} \cdot \boldsymbol{\omega}' \rangle = -\frac{1}{r^2}\frac{\partial}{\partial r}\left(r^2 \frac{\partial R}{\partial r} \right)$$

which, when combined with (6.39), yields,

$$\langle \boldsymbol{\omega} \cdot \boldsymbol{\omega}' \rangle = -\frac{u^2}{r^2}\frac{\partial}{\partial r}\left[r^2 \frac{\partial}{\partial r}\frac{1}{r^2}\frac{\partial}{\partial r}(r^3 f) \right]. \tag{6.42}$$

This looks like a rather unpromising relationship. However, the significance of (6.42) is that the shape of f near $r = 0$ is determined by $\langle \boldsymbol{\omega}^2 \rangle$, and hence by the dissipation, $\varepsilon = \nu \langle \boldsymbol{\omega}^2 \rangle$, as we now show. Since f is even in r, $f(0) = 1$, and $f \leq 1$, we may write,

$$f(r) = 1 - \frac{r^2}{2\lambda^2} + \cdots, \quad \lambda = \text{constant}.$$

Substituting into (6.42), and recalling that the enstrophy and dissipation are related by $\varepsilon = \nu \langle \boldsymbol{\omega}^2 \rangle$, we find,

$$\lambda^2 = \frac{15u^2}{\langle \boldsymbol{\omega}^2 \rangle} = \frac{15\nu u^2}{\varepsilon} \tag{6.43}$$

from which,

$$f(r) = 1 - \frac{\varepsilon r^2}{30\nu u^2} + \cdots. \tag{6.44a}$$

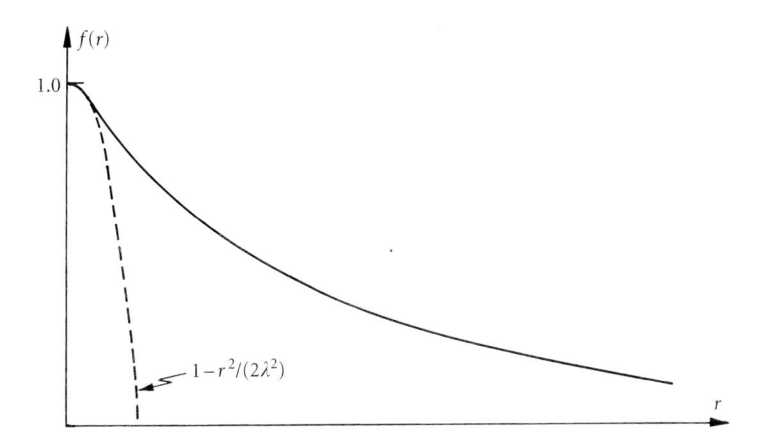

Figure 6.15 Shape of near $f(r)$ near $r = 0$.

Now we have seen that $\varepsilon \sim u^3/l$ and so it follows that, when $\mathrm{Re} = (ul/\nu) \gg 1$, the curvature of $f(r)$ near $r = 0$ is extremely large (Figure 6.15). The length scale λ is known as the *Taylor microscale*. From (6.43) we have $\lambda^2/l^2 \sim 15(ul/\nu)^{-1}$. Thus the Taylor microscale lies somewhere between the integral scale, l, and the Kolmogorov micro-scale, $\eta \sim l(\mathrm{Re})^{-3/4}$. The relationships between the three length scales are summarized below.

- (Taylor microscale)/(Integral scale) $= \lambda/l \sim \sqrt{15}\,\mathrm{Re}^{-1/2}$
- (Kolmogorov microscale)/(Taylor microscale) $= \eta/\lambda \sim \mathrm{Re}^{-1/4}/\sqrt{15}$
- (Kolmogorov microscale)/(Integral scale) $= \eta/l \sim \mathrm{Re}^{-3/4}$

Note that (6.41b) implies that

$$\tfrac{1}{2}\langle(\nabla \times \boldsymbol{\omega}) \cdot (\nabla \times \boldsymbol{\omega})'\rangle = -\nabla^2\left[\tfrac{1}{2}\langle \boldsymbol{\omega} \cdot \boldsymbol{\omega}'\rangle\right] = \nabla^4[R(r)]$$

and substituting for f yields

$$\langle(\nabla \times \boldsymbol{\omega}) \cdot (\nabla \times \boldsymbol{\omega})'\rangle = u^2\nabla^4\left[\frac{1}{r^2}\frac{\partial}{\partial r}(r^3 f)\right].$$

This allows us to determine the next term in expansion (6.44a). After a little algebra we find,

$$u^2 f = u^2 - \frac{\langle(\boldsymbol{\omega})^2\rangle}{30}r^2 + \frac{\langle(\nabla \times \boldsymbol{\omega})^2\rangle}{840}r^4 - \frac{\langle(\nabla^2\boldsymbol{\omega})^2\rangle}{45360}r^6 + \cdots. \tag{6.44b}$$

We close this section with a discussion of the sign of f. We have already noted that $f > 0$ in fully developed turbulence and it is natural to enquire as to whether or not there is some fundamental reason for this. In this respect it is useful to note that (6.25) and (6.39) combine to yield,

$$u^2 f(r) = 2\int_0^\infty E(k)\hat{H}(kr)\,dk$$

$$\hat{H}(x) = (\sin x - x\cos x)/x^3$$

The function $\hat{H}(x)$ is predominantly positive, though it has small negative regions centred around $x = 2\pi$, 4π, etc. Now suppose that we, somehow, arrange an initial condition in which $E(k)$ is a delta function centred around $k = \hat{k}$. Then,

$$u^2 f(r) = \langle \mathbf{u}^2 \rangle \hat{H}(\hat{k}r)$$

and so such an energy spectrum would give rise to negative values of f. Thus f can, in principal, become negative. However, it is important to realize that such a spectrum is not at all realistic since even if our initial conditions contained only eddies of one size, l_e, the energy spectrum would not be a δ-function but rather be of the form $E \sim k^4 \exp[-k^2 l_e^2]$ (see Section 6.4.1.). The fact that f is found to be positive in fully developed turbulence reflects the fact that $\hat{H}$ is predominantly a positive function.

6.2.1.3 A summary of the kinematic relationships

We have covered a lot of ground in the last two sections, so perhaps it is worthwhile summarizing these kinematic relationships before we move on.

1. Second-order velocity correlation function:
   - Definition: $Q_{ij} = \langle u_i(\mathbf{x})u_j(\mathbf{x}+\mathbf{r})\rangle$
   - Some properties:
     (i) $Q_{ij}(\mathbf{r}) = Q_{ji}(-\mathbf{r})$ [geometrical property]
     (ii) $\frac{1}{2}Q_{ii}(0) = \frac{1}{2}\langle \mathbf{u}^2\rangle =$ kinetic energy density
     (iii) $-Q_{ij}(0) = \tau_{ij}^R/\rho =$ Reynolds stress$/\rho$
     (iv) $Q_{ij}(\mathbf{r}) \le Q_{xx}(0)$ [Schwartz inequality]
     (v) $\partial Q_{ij}/\partial r_i = \partial Q_{ij}/\partial r_j = 0$ [continuity]
   - Special cases:
     (i) $R(r) = \frac{1}{2}Q_{ii} = \frac{1}{2}\langle \mathbf{u}\cdot\mathbf{u}'\rangle$
     (ii) $Q_{xx}(r\hat{\mathbf{e}}_x) = u^2 f(r)$ [longitudinal correlation]
     (iii) $Q_{yy}(r\hat{\mathbf{e}}_x) = u^2 g(r)$ [lateral correlation]

   See Figure 6.16.
   - General form in isotropic turbulence:

$$Q_{ij} = \frac{u^2}{2r}\left[(r^2 f)'\delta_{ij} - f' r_i r_j\right]$$

   - Relationships between f, g, and R:

     (i) $g = \dfrac{1}{2r}(r^2 f)'$

     (ii) $R = \dfrac{u^2}{2r^2}(r^3 f)'$

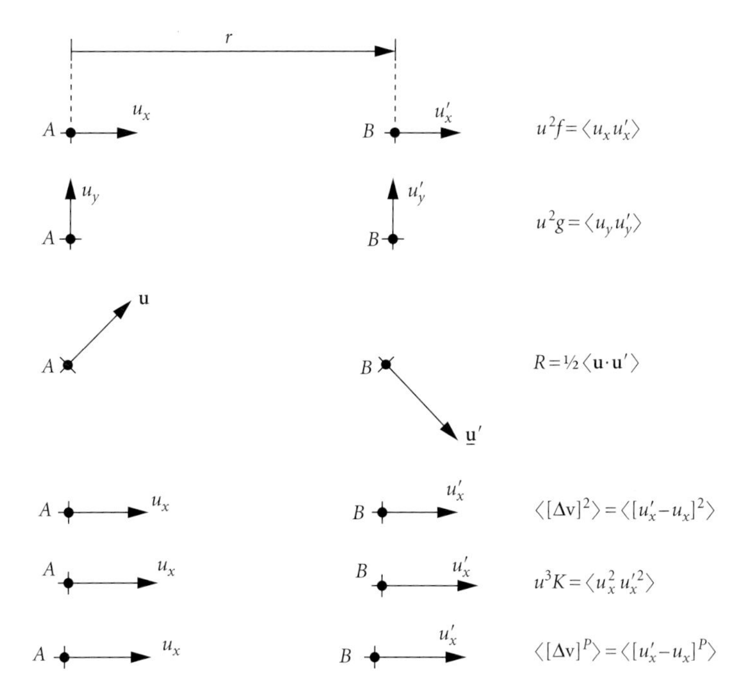

Figure 6.16 Schematic definition of second- and third-order two-point velocity correlations.

- Definition of integral scale, l:

$$l = \int_0^\infty f\,dr \quad \text{or,} \quad u^2 l = \int_0^\infty R\,dr$$

- Definition of the Taylor microscale, λ:

$$f = 1 - \frac{r^2}{2\lambda^2} + \cdots$$

- Relationship between Taylor microscale and integral scale:

$$\lambda/l \sim \sqrt{15}(ul/\nu)^{-1/2}$$

- Expansion for f:

$$u^2 f = u^2 - \frac{\langle (\boldsymbol{\omega})^2 \rangle}{30} r^2 + \frac{\langle (\nabla \times \boldsymbol{\omega})^2 \rangle}{840} r^4 - \frac{\langle (\nabla^2 \boldsymbol{\omega})^2 \rangle}{45360} r^6 + \cdots.$$

2. Third-order velocity correlation function:
   - Definition: $S_{ijk} = \langle u_i(\mathbf{x}) u_j(\mathbf{x}) u_k(\mathbf{x} + \mathbf{r}) \rangle$
   - Special case: $u^3 K(r) = \langle u_x^2(\mathbf{x}) u_x(\mathbf{x} + r\hat{\mathbf{e}}_x) \rangle$
   - General form in isotropic turbulence:

$$S_{ijk} = u^3 \left[\frac{K - rK'}{2r^3} r_i r_j r_k + \frac{2K + rK'}{4r} (r_i \delta_{jk} + r_j \delta_{ik}) - \frac{K}{2r} r_k \delta_{ij} \right].$$

3. Longitudinal structure functions
   - Definitions:

$$\left\langle [\Delta v]^2 \right\rangle = \left\langle [u_x(\mathbf{x} + r\hat{\mathbf{e}}_\mathbf{x}) - u_x(\mathbf{x})]^2 \right\rangle$$

(second order)

$$\langle [\Delta v]^p \rangle = \langle [u_x(\mathbf{x} + r\hat{\mathbf{e}}_\mathbf{x}) - u_x(\mathbf{x})]^p \rangle$$

(order p)

   - Relationships to f and K:

(i) $\left\langle [\Delta v]^2 \right\rangle = 2u^2(1 - f)$

(ii) $\left\langle [\Delta v]^3 \right\rangle = 6u^3 K$

   - Expansion in r:

$$15\left\langle [\Delta v]^2 \right\rangle = \left\langle (\boldsymbol{\omega})^2 \right\rangle r^2 - \frac{\left\langle (\nabla \times \boldsymbol{\omega})^2 \right\rangle}{28} r^4 + \frac{\left\langle (\nabla^2 \boldsymbol{\omega})^2 \right\rangle}{1512} r^6 + \cdots.$$

4. Alternative structure function
   - Definition:

$$\left\langle [\Delta \mathbf{v}]^2 \right\rangle = \left\langle [\mathbf{u}(\mathbf{x} + \mathbf{r}) - \mathbf{u}(\mathbf{x})]^2 \right\rangle$$

   - Relationship to R: $\frac{1}{4}\left\langle [\Delta \mathbf{v}]^2 \right\rangle = \frac{1}{2}\langle \mathbf{u}^2 \rangle - R(r)$
   - Relationship to longitudinal structure function:

$$\left\langle [\Delta \mathbf{v}]^2 \right\rangle = \frac{1}{r^2}\frac{d}{dr}\left[r^3 \left\langle [\Delta v]^2 \right\rangle\right].$$

5. Vorticity correlation function
   - Definition: $\langle \omega_i(\mathbf{x})\omega_j(\mathbf{x} + \mathbf{r}) \rangle = \left\langle \omega_i \omega_j' \right\rangle$
   - Relationship to Q_{ij}:

$$\langle \boldsymbol{\omega} \cdot \boldsymbol{\omega}' \rangle = -\nabla^2 Q_{ii}(\mathbf{r})$$

   - Relationship to the Taylor microscale:

$$\lambda^2 = \frac{15u^2}{\langle \boldsymbol{\omega}^2 \rangle} = \frac{15\nu u^2}{\varepsilon}.$$

6. Energy spectrum
   - Definition:

$$E(k) = \frac{2}{\pi}\int_0^\infty R(r)kr\sin(kr)\,dr$$

$$R(r) = \int_0^\infty E(k)\frac{\sin(kr)}{kr}\,dk$$

- Properties:
  (i) $E(k) \geq 0$
  (ii) $\int_0^\infty E(k)\,dk = \frac{1}{2}\langle \mathbf{u}^2 \rangle$
  (iii) $\int_0^\infty k^2 E(k)\,dk = \frac{1}{2}\langle \boldsymbol{\omega}^2 \rangle$
- Relationship to $f(r)$:

$$u^2 f(r) = 2\int_0^\infty E(k)\hat{H}(kr)\,dk, \qquad \hat{H}(x) = [\sin x - x\cos x]/x^3$$

- Relationship to the integral scale l:

$$l = \int_0^\infty f(r)dr = \frac{\pi}{2u^2}\int_0^\infty k^{-1}E(k)\,dk$$

- Relationship to $\langle [\Delta \mathrm{v}]^2 \rangle$:

$$\langle [\Delta \mathrm{v}]^2 \rangle = \frac{4}{3}\int_0^\infty E(k)H(kr)\,dk, \quad H(x) = 1 + 3x^{-2}\cos x - 3x^{-3}\sin x$$

- Approximate relationship to $\langle [\Delta \mathrm{v}]^2 \rangle$:

$$\langle [\Delta \mathrm{v}]^2 \rangle \approx \frac{4}{3}\int_{\pi/r}^\infty E(k)\,dk + \frac{4r^2}{3\pi^2}\int_0^{\pi/r} k^2 E(k)\,dk.$$

7. Miscellaneous relationships not derived here (Consult Hinze 1959)
   - Products of velocity gradients:

$$\left\langle \frac{\partial u_i}{\partial x_n}\frac{\partial u_j}{\partial x_m} \right\rangle = \frac{2u^2}{\lambda^2}\left[\delta_{ij}\delta_{mn} - \frac{1}{4}\left(\delta_{in}\delta_{jm} + \delta_{im}\delta_{jn}\right)\right]$$

   - Various expressions for the energy dissipation rate per unit mass:

$$\varepsilon = \nu\langle \boldsymbol{\omega}^2 \rangle = 15\nu\left\langle \left(\frac{\partial u_x}{\partial x}\right)^2 \right\rangle = \frac{15}{2}\nu\left\langle \left(\frac{\partial u_x}{\partial y}\right)^2 \right\rangle.$$

8. Betchov's relationships for the three principal rates of strain a, b, c. (See Section 5.3.6.)
   - $\langle (\partial u_x/\partial x)^2 \rangle = \frac{2}{15}\langle a^2 + b^2 + c^2 \rangle = \frac{1}{15}\langle \boldsymbol{\omega}^2 \rangle$
   - $\langle (\partial u_x/\partial x)^3 \rangle = \frac{24}{105}\langle abc \rangle = -\frac{2}{35}\langle \omega_i \omega_j S_{ij} \rangle$
   - $\langle (\partial u_x/\partial x)^4 \rangle = \frac{8}{105}\langle a^4 + b^4 + c^4 \rangle = \frac{4}{105}\langle (a^2 + b^2 + c^2)^2 \rangle$

6.2.2 *Dynamics*

6.2.2.1 Dynamics at last: the Karman–Howarth equation

We are finally in a position to introduce some dynamics. In particular, we can convert the Navier–Stokes equation into an evolution equation for the velocity correlation function, as shown below. First, however, perhaps we should say something about the relative roles played by the velocity and vorticity fields in turbulence.

The most obvious signature of a turbulent flow are the gusts and fluctuations associated with the velocity field. As you stand in the street in a high wind it is the various manifestations of the velocity field that you feel and see. It is the velocity field that sweeps the fallen leaves, disperses the pollutants which belch from the rear of cars, and rattles the loose window frame. In the laboratory too we are most aware of the velocity field. After all, $\mathbf{u}$ is the most natural quantity to measure, in say, a wind tunnel experiment.

At a deeper level, however, the turbulent part of the velocity field is nothing more than the instantaneous manifestation of the vorticity field. At high Re the vorticity field is virtually frozen into the fluid and advects itself in a chaotic manner in accordance with the vorticity transport equation. That is, a given distribution of $\boldsymbol{\omega}$ induces a velocity field through the Biot–Savart law, and this then advects the vorticity, like dye lines frozen into the fluid. It is this self-perpetuating, chaotic advection of the vorticity field that we call turbulence, and as the vorticity evolves, so does the velocity. Yet the man in the street cannot see or feel the vorticity field of a high wind, nor can vorticity be readily measured in the laboratory. Thus traditional theories of turbulence have tended to adopt $\mathbf{u}$, and its statistical properties, as the central object of attention. The idea is that if we can describe the dynamics of $\mathbf{u}$, then perhaps we can describe all of the essential features of the turbulence. In a formal sense this is true, since $\boldsymbol{\omega}$ is functionally related to $\mathbf{u}$ by $\boldsymbol{\omega} = \nabla \times \mathbf{u}$. Thus, if we know all about $\mathbf{u}$, we can determine the behaviour of $\boldsymbol{\omega}$. However, the focus on $\mathbf{u}$ at the expense of $\boldsymbol{\omega}$ detracts from our physical interpretation of the underlying dynamics. After all, it is $\boldsymbol{\omega}$, and not $\mathbf{u}$, which is locked into the fluid. Still, the precedent has been set and most texts couch their descriptions of turbulence in terms of the behaviour of $\mathbf{u}(\mathbf{x}, t)$ and its statistical properties. We shall not break with this tradition. Nevertheless, as you wrestle with the various dynamical theories which follow, you may find it advantageous to keep asking the question: what is the vorticity field doing?

With this word of warning, let us see if we can get an evolution equation for $\langle u_i u_j' \rangle$. Let $\mathbf{x}' = \mathbf{x} + \mathbf{r}$ and $\mathbf{u}(\mathbf{x}') = \mathbf{u}'$. Then we have,

$$\frac{\partial u_i}{\partial t} = -\frac{\partial (u_i u_k)}{\partial x_k} - \frac{\partial (p/\rho)}{\partial x_i} + \nu \nabla_x^2 u_i$$

$$\frac{\partial u_j'}{\partial t} = -\frac{\partial \left(u_j' u_k'\right)}{\partial x_k'} - \frac{\partial (p'/\rho)}{\partial x_j'} + \nu \nabla_{x'}^2 u_j'.$$

On multiplying the first of these by u'_j, and the second by u_i, adding the two and averaging, we obtain,

$$\frac{\partial}{\partial t}\left\langle u_i u'_j \right\rangle = -\left\langle u_i \frac{\partial u'_j u'_k}{\partial x'_k} + u'_j \frac{\partial u_i u_k}{\partial x_k} \right\rangle - \frac{1}{\rho}\left\langle u_i \frac{\partial p'}{\partial x'_j} + u'_j \frac{\partial p}{\partial x_i} \right\rangle$$

$$+ \nu \left\langle u_i \nabla^2_{x'} u'_j + u'_j \nabla^2_x u_i \right\rangle. \tag{6.45}$$

This somewhat messy expression can be simplified considerably if we note that:

(1) the operations of taking averages, $\langle \sim \rangle$, and differentiation commute;
(2) $\partial / \partial x_i$ and $\partial / \partial x'_j$ operating on averages may be replaced by $-\partial / \partial r_i$ and $\partial / \partial r_j$;
(3) u_i is independent of $\mathbf{x}'$ and u'_j is independent of $\mathbf{x}$.

Equation (6.45) then simplifies to the more compact result,

$$\frac{\partial Q_{ij}}{\partial t} = \frac{\partial}{\partial r_k}\left[S_{ikj} + S_{jki}\right] + 2\nu \nabla^2 Q_{ij}. \tag{6.46}$$

We have dropped the pressure terms since, in isotropic turbulence, (6.30) plus the continuity of mass demands,

$$\langle u_i p' \rangle = 0. \tag{6.47}$$

(See Exercise 6.2 at the end of the chapter.) Substituting for Q_{ij} and S_{ijk} in terms of the scalar functions $f(r)$ and $K(r)$ yields, after a little algebra,

$$\frac{\partial}{\partial t}\left[u^2 r^4 f(r,t)\right] = u^3 \frac{\partial}{\partial r}\left[r^4 K(r)\right] + 2\nu u^2 \frac{\partial}{\partial r}\left[r^4 f'(r)\right] \tag{6.48a}$$

This is probably the single most important equation in isotropic turbulence. It is known as the *Karman–Howarth equation*. In terms of R, (6.48a) may be rewritten as,[7]

$$\frac{\partial R}{\partial t} = \Gamma(r) + 2\nu \nabla^2 R, \qquad \Gamma = \frac{1}{2r^2}\frac{\partial}{\partial r}\left[\frac{1}{r}\frac{\partial}{\partial r}\left(r^4 u^3 K\right)\right]. \tag{6.48b}$$

The problem with (6.48a, b) is that we cannot predict the evolution of f or R without knowing the form of $K(r)$. Of course, the rate of change of $K(r)$ depends, in turn, on the fourth-order correlations. We have hit the closure problem of turbulence. Still, we can extract a great deal of useful information from (6.48a, b), as we shall see in the following sections.

[7] The Laplacian in this equation is acting on a spherically symmetric function and so ∇^2 represents $\frac{1}{r^2}\frac{d}{dr} r^2 \frac{d}{dr}(\sim)$.

Example 6.1 A dynamic equation for the vorticity correlation
Although a fluctuating velocity field is the most obvious manifestation of turbulence, and indeed **u** is the easiest quantity to measure, at a deeper level the more important field is $\boldsymbol{\omega}$. Vorticity, unlike velocity, is locked into, and transported by, the fluid. Like a dye, it can spread only by diffusion or advection. Moreover, at any instant the velocity field is uniquely determined by the distribution of vorticity, through the Biot–Savart law. From a dynamical perspective, then, the velocity field is only a passive manifestation of the instantaneous vorticity distribution and it is $\boldsymbol{\omega}$, and not **u**, which is of fundamental importance. Let $f_\omega(r)$ be the longitudinal correlation function for $\boldsymbol{\omega}$, and $\omega^2 = \langle \omega_x^2 \rangle$. Show that the vorticity counterpart of (6.48a) is,

$$\frac{\partial}{\partial t}[\omega^2 r^4 f_\omega(r)] = -\frac{\partial}{\partial r} r^4 \frac{\partial}{\partial r}\frac{1}{r^4}\frac{\partial}{\partial r}[r^4 u^3 K] + 2\nu\omega^2 \frac{\partial}{\partial r}\left[r^4 f_\omega'(r)\right].$$

[*Hint*: convert (6.48b) into an equation for $\langle \boldsymbol{\omega}\cdot\boldsymbol{\omega}' \rangle$ using (6.41b) and then use the vorticity analogue of (6.39) to reduce the equation to one for f_ω.]

In certain types of turbulence (i.e. turbulence with a so-called Saffman spectrum) it is found that, for large r, $f(r) \sim r^{-3}$ while K falls off as r^{-4}. Show that, in such cases,

$$\omega^2 \int_0^\infty r^4 f_\omega(r)\, dr$$

is an invariant of the motion and is proportional to the integral

$$L = \int \langle \mathbf{u}\cdot\mathbf{u}' \rangle\, d\mathbf{r}$$

The invariance of L is confirmed by numerical simulations of turbulence.

Example 6.2 The final period of decay
In the final period of decay of isotropic turbulence we have $ul/\nu \to 0$ and the non-linear term becomes unimportant (Figure 6.17). In such a case we may take $K(r) = 0$ in (6.48a). Show that *one* solution of the Karman–Howarth equations for the final period of decay is,

$$f \sim \exp[-r^2/(8\nu t)], \qquad u^2 \sim t^{-5/2}.$$

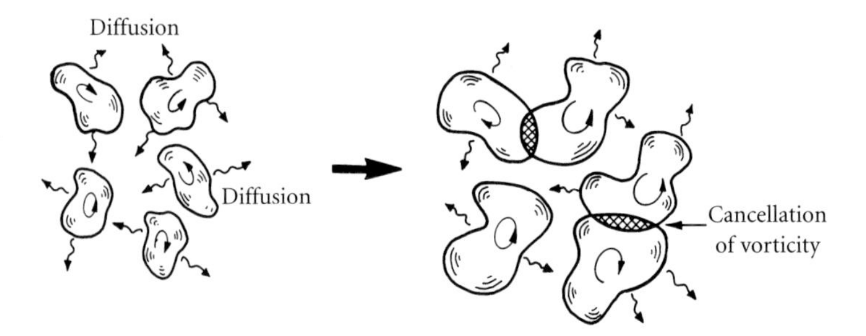

Figure 6.17 In the final period of decay the inertial forces are negligible and vorticity spreads and mixes by diffusion only.

This is a good fit to the experimental data (Monin and Yoglam 1975). Other possible solutions for the final period of decay are discussed at the end of Section 6.3.5.

6.2.2.2 Kolmogorov's four-fifths law

We shall now derive Kolmogorov's four-fifths law from the Karman–Howarth equation. This law, which is one of the landmark results of turbulence, tells us that $\langle[\Delta v]^3\rangle$ is equal to $-(4/5)\varepsilon r$ in the inertial sub-range. It is one of the few exact, but non-trivial, results in the statistical theory of isotropic turbulence. The starting point is to consider the shape of $\langle[\Delta v]^2\rangle$. For small r (6.44) shows that $\langle[\Delta v]^2\rangle = \varepsilon r^2/15\nu$. For the inertial sub-range, on the other hand, we have Kolmogorov's two-thirds law, $\langle[\Delta v]^2\rangle = \beta\varepsilon^{2/3}r^{2/3}$. So, if η is the Kolmogorov microscale, $\eta = (\nu^3/\varepsilon)^{1/4}$, we have the following estimates of $\langle[\Delta v]^2\rangle$:

$$\langle[\Delta v]^2\rangle = \varepsilon r^2/15\nu, \quad (r < \eta) \tag{6.49}$$

$$\langle[\Delta v]^2\rangle = \beta\varepsilon^{2/3}r^{2/3}, \quad (\eta \ll r \ll l) \tag{6.50}$$

$$\langle[\Delta v]^2\rangle = 2u^2, \quad (r \gg l). \tag{6.51}$$

These are shown schematically in Figure 6.18.

Now consider the Karman–Howarth equation rewritten in terms of the second- and third-order structure functions. From (6.20) and (6.21) we have

$$-\frac{2}{3}r^4\varepsilon - \frac{r^4}{2}\frac{\partial}{\partial t}\langle[\Delta v]^2\rangle = \frac{\partial}{\partial r}\left[\frac{r^4}{6}\langle[\Delta v]^3\rangle\right] - \nu\frac{\partial}{\partial r}\left[r^4\frac{\partial}{\partial r}\langle[\Delta v]^2\rangle\right]. \tag{6.52}$$

In the universal equilibrium range $\langle[\Delta v]^2\rangle \sim \varepsilon^{2/3}r^{2/3}$, or less, and so, for $r \ll l$, the second term on the left is, at most, of order $r^4\varepsilon(r/l)^{2/3}$. This is negligible by comparison with the first term and so, provided $r \ll l$, we may neglect $\partial\langle[\Delta v]^2\rangle/\partial t$ in (6.52). On integrating the remaining terms we find,

$$\langle[\Delta v]^3\rangle = -\frac{4}{5}\varepsilon r + 6\nu\frac{\partial}{\partial r}\langle[\Delta v]^2\rangle. \quad (r \ll l) \tag{6.53a}$$

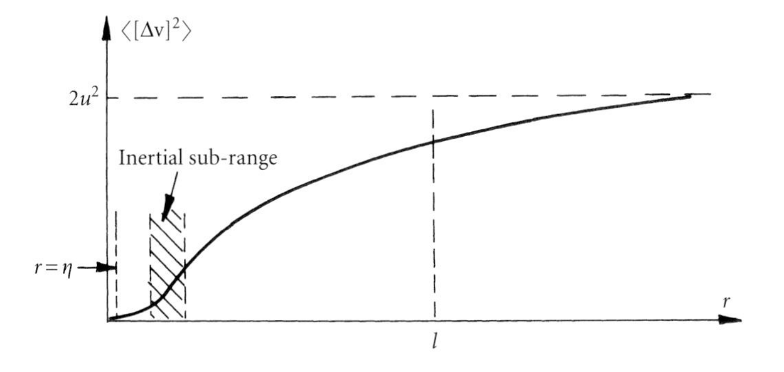

Figure 6.18 Schematic shape of $\langle[\Delta v]^2\rangle$ showing the three ranges given by equations (6.49)–(6.51).

Alternatively, introducing the skewness factor $S = \langle[\Delta v]^3\rangle / \langle[\Delta v]^2\rangle^{3/2}$, we have,

$$6\nu \frac{\partial}{\partial r}\left\langle[\Delta v]^2\right\rangle - S\left\langle[\Delta v]^2\right\rangle^{3/2} = \frac{4}{5}\,\varepsilon r. \qquad (r \ll l) \tag{6.53b}$$

These two equations, which are crucial to the dynamics of the universal equilibrium range, were first obtained by Kolmogorov. (See Kolmogorov 1941*b*.) We shall have reason to revisit (6.53a, b) time and again in this chapter. Now it turns out that $\langle[\Delta v]^3\rangle \sim r^3$ for $r \to 0$, and so (6.53a) fixes the value of $\partial\langle[\Delta v]^2\rangle / \partial r$ for small r. In fact, in the limit $r \to 0$ we recover (6.49). In the inertial sub-range, on the other hand, viscous effects are unimportant and so (6.53a) reduces to

$$\left\langle[\Delta v]^3\right\rangle = -\frac{4}{5}\,\varepsilon r \qquad (n \ll r \ll l). \tag{6.54}$$

This is Kolmogorov's celebrated four-fifths law, which we introduced in Chapter 5. It may be considered as a special case of Kolmogorov's theory of the inertial range, which predicts $\langle[\Delta v]^p\rangle = \beta_p(\varepsilon r)^{p/3}$. Evidently $\beta_3 = -4/5$. However, the four-fifths law, unlike Kolmogorov's universal equilibrium theory, is exact in the sense that the pre-factor β_3 has a unique value. Moreover, we had to make relatively few assumptions in order to reach the four-fifths law, far fewer than needed in Kolmogorov's theory of the equilibrium range. (Note, however, that there *are* assumptions, all-be-it plausible ones, lying behind the four-fifths law. In particular, we need to assume that there exists a range of eddies, which are large enough to be immune to viscous forces, but small enough to be in a state of quasi-equilibrium, that is close to being statistically steady.)

Combining (6.54) with the two-thirds law we find that the skewness factor, $S(r)$, takes a constant value in the inertial sub-range, given by,

$$S(r) = \frac{\left\langle[\Delta v]^3\right\rangle}{\left\langle[\Delta v]^2\right\rangle^{3/2}} = -\frac{4}{5}\,\beta^{-3/2} \qquad (\eta \ll r \ll l) \tag{6.55}$$

where $\beta \sim 2$ is Kolmogorov's constant. This suggests that, in the inertial sub-range, $S \sim -0.3$, which is reasonably in line with the measurements (Figure 3.19).

There is another way of looking at (6.54) and (6.55). We could take the position that the experimental data suggests that the skewness factor is constant across the inertial sub-range. We can then replace Kolmogorov's second similarity hypothesis by the hypothesis that $S(r) = \text{constant}$ $(\eta \ll r \ll l)$. Kolmogorov's two-thirds law then follows directly from the four-fifths law. One advantage of this approach is

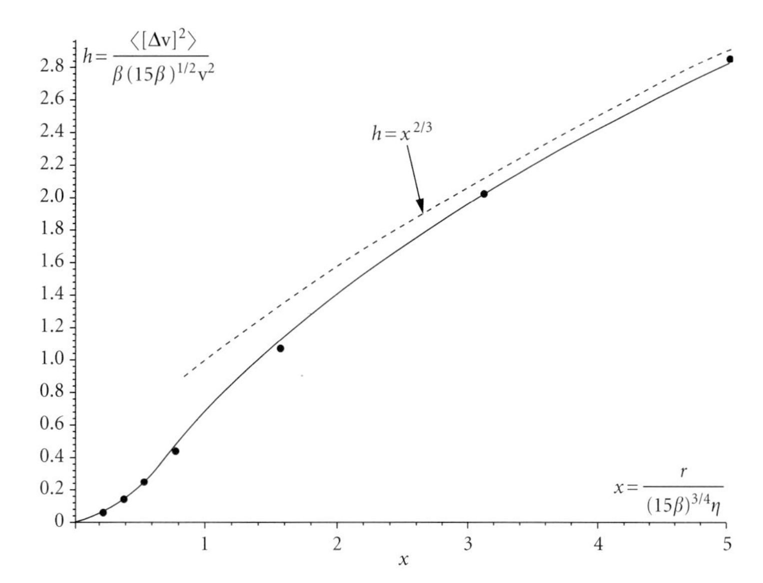

Figure 6.19 Shape of $\langle[\Delta v]^2\rangle$, normalized by $\beta(15\beta)^{1/2}\,v^2$, obtained from integrating (6.61). The results of Fukayama et al.'s (2001) DNS of turbulence ($R_\lambda = 460$) are also shown for comparison. The solid line is $h(x)$, the dashed line is $x^{2/3}$, and the dots are the DNS results.

that there is no need to regard β as a universal constant (the same for all flows), which goes some way to satisfying Landau's criticism of Kolmogorov's 2/3 law. The details are spelt out in Section 5.3.1.

6.2.2.3 The skewness factor and enstrophy production

So far we have focused on the form of $\langle[\Delta v]^3\rangle$ in the inertial subrange. However, information on the shape of $\langle[\Delta v]^3\rangle$ for $r \to 0$ can also be determined from the Karman–Howarth equation, and we shall see that this is intimately related to the rate of production of enstrophy. In fact, we shall now show that the rate of generation of enstrophy is simply dictated by the skewness $S_0 = S(r \to 0)$, which is a dimensionless measure of $\langle[\Delta v]^3\rangle_{r\to 0}$. We proceed as follows. For small r we have, from (6.44b),

$$u^2 f = u^2 - \frac{\langle \boldsymbol{\omega}^2\rangle}{30} r^2 + \frac{\langle(\nabla\times\boldsymbol{\omega})^2\rangle}{840} r^4 + \cdots.$$

On substituting this into the Karman–Howarth equation, and noting that $K \sim r^3$ for small r, we obtain,

$$10\frac{\partial}{\partial t}\left[\frac{1}{2}\langle\mathbf{u}^2\rangle\right] - \frac{\partial}{\partial t}\left[\frac{1}{2}\langle\boldsymbol{\omega}^2\rangle\right] r^2 + \cdots = 105[u^3 K/r] - 10\nu\langle\boldsymbol{\omega}^2\rangle + \nu\langle(\nabla\times\boldsymbol{\omega})^2\rangle r^2 + \cdots.$$

Of course, the terms of zero order cancel because of the energy equation

$$\frac{\partial}{\partial t}\left[\frac{1}{2}\langle\mathbf{u}^2\rangle\right] = -\nu\langle\boldsymbol{\omega}^2\rangle$$

while the terms of order r^2 yield,

$$\frac{\partial}{\partial t}\left[\frac{1}{2}\langle\boldsymbol{\omega}^2\rangle\right] = -105u^3\left[K/r^3\right]_{r\to 0} - \nu\left\langle(\nabla\times\boldsymbol{\omega})^2\right\rangle.$$

Compare this with the enstrophy equation (5.28),

$$\frac{\partial}{\partial t}\left[\frac{1}{2}\langle\boldsymbol{\omega}^2\rangle\right] = \left\langle\omega_i\omega_j S_{ij}\right\rangle - \nu\left\langle(\nabla\times\boldsymbol{\omega})^2\right\rangle.$$

Evidently the generation of enstrophy by vortex-line stretching is related to the shape of K, and hence $\langle(\Delta v)^3\rangle$, in the vicinity of $r = 0$.[8] In particular we have,

$$\left\langle\omega_i\omega_j S_{ij}\right\rangle = -105u^3\left[K/r^3\right]_0 = -(35/2)\left[\left\langle(\Delta v)^3\right\rangle/r^3\right]_0. \tag{6.56}$$

We may rewrite this in terms of the skewness factor at $r = 0$, $S_0 = S(0)$, since $\langle(\Delta v)^3\rangle = S\langle(\Delta v)^2\rangle^{3/2}$. Replacing $\langle(\Delta v)^3\rangle$ by S we obtain the remarkably useful equation,

$$\left\langle\omega_i\omega_j S_{ij}\right\rangle = -\frac{7}{6\sqrt{15}}S_0\left[\langle\boldsymbol{\omega}^2\rangle\right]^{3/2}. \tag{6.57}$$

(Here we have evaluated $\langle(\Delta v)^2\rangle$ near $r = 0$ using (6.49).) This is a particularly beautiful result as it allows us to determine the rate of generation of enstrophy from a knowledge of just one parameter, S_0. Moreover, many measurements of the skewness S_0 have been made and its value seems to be more or less independent of Re and lies in the range -0.4 to -0.5. In terms of S_0 we have

$$S_0 = -\frac{6\sqrt{15}\left\langle\omega_i\omega_j S_{ij}\right\rangle}{7\left[\langle\boldsymbol{\omega}^2\rangle\right]^{3/2}}. \tag{6.58}$$

Evidently a negative skewness is essential to the production of enstrophy. In summary, then, the third-order structure function dictates the rate of generation of enstrophy. In particular, it behaves in the following way at small r,

$$\left\langle[\Delta v]^3\right\rangle = -\tfrac{2}{35}\left\langle\omega_i\omega_j S_{ij}\right\rangle r^3 + \cdots. \tag{6.59}$$

6.2.3 *Overcoming the closure problem*

6.2.3.1 The dynamical equation for the third-order correlations and the problem of closure

One of the most striking features of the Karman–Howarth equation is that the pressure field plays no role. This is a direct result of isotropy,

[8] In a similar fashion it may be shown that the rate of generation of palinstrophy, $\langle\frac{1}{2}(\Delta\times\boldsymbol{\omega})^2\rangle$, by vortex-line stretching is equal to $\frac{3}{4}7!(u^3K_5)$, where K_5 is the coefficient which multiplies the r^5 term in an expansion of K about $r = 0$.

which demands $\langle u_i p' \rangle = 0$. This is not the case, however, when we derive dynamical equations for higher-order correlation functions. For example, when we use the Navier–Stokes equation to evaluate $\partial S_{ijk}/\partial t$ (in a manner similar to the derivation of (6.46)) we find

$$\rho \frac{\partial S_{ijk}}{\partial t} = \rho \langle uuuu \rangle - \frac{\partial}{\partial r_k} \langle u_i u_j p' \rangle - \left\langle u'_k \left(u_i \frac{\partial p}{\partial x_j} + u_j \frac{\partial p}{\partial x_i} \right) \right\rangle + \rho \nu (\sim). \tag{6.60}$$

Here $\langle uuuu \rangle$ is a symbolic representation of terms involving fourth-order velocity correlations and $\nu(\sim)$ represents the viscous term. Isotropy is now insufficient to dispense with the pressure terms. They are finite and indeed play a crucial role in determining the behaviour of the large eddies. This is discussed at some length in Section 6.3.4.

Note that (6.60) involves terms like $\langle uuuu \rangle$, so that (6.46) and (6.60) do not constitute a closed set of equations. We might try to remedy this by deriving an evolution equation for $\langle uuuu \rangle$. It has the form,

$$\rho \frac{\partial}{\partial t} \langle uuuu \rangle = \rho \langle uuuuu \rangle + \frac{\partial}{\partial r} \langle uuup \rangle + \rho \nu (\sim).$$

Evidently, we have introduced yet another set of unknowns, that is, fifth-order correlations. Of course, this is an example of the *closure problem of turbulence*. If we insist on working only with statistical quantities then we always end up with more unknowns than equations. It would appear that, even in the relatively simple case of isotropic turbulence, we cannot make definite predictions about the evolution of the motion. In this sense, all 'models of turbulence' are semi-empirical, in which the rigorous equations of turbulence are supplemented by empirical hypotheses.

We have already met one class of closure models in Chapter 4, where we discussed eddy-viscosity-type closure schemes. These models are known as *one-point closures*, since they involve turbulent quantities, such as τ_{ij}^R, evaluated at just one point in space.

Although one-point closure models dominate the engineering applications of turbulence, essentially because they are simple to implement and apply equally to homogeneous and inhomogeneous turbulence, there are other, more subtle closure models, which are favoured by physicists and mathematicians studying the fine-scale structure of turbulence. The most widely used of these are the so-called *two-point closure models*. These work with statistical quantities, such as Q_{ij} and S_{ijk}, which are evaluated at two points in space (hence the name). Two-point models are restricted, more or less, to homogeneous turbulence (unlike one-point models), but nevertheless they have proved very popular.

Perhaps the most widely used class of two-point closure models are the so-called *quasi-normal-type schemes*. We shall describe these shortly.

First, however, we give an example of a particularly simple closure scheme, which predicts Q_{ij} and S_{ijk} throughout the equilibrium range, that is, in the inertial subrange and the dissipation range.

6.2.3.2 Closure of the dynamical equations in the equilibrium range

Equation (6.53b) provides an astonishingly simple means of estimating f and K throughout the equilibrium range. We have already seen that $S = -\frac{4}{5}\beta^{-3/2} \sim -0.3$ in the inertial subrange, $\eta \ll r \ll l$, while experiments suggest $S \sim -0.4$ for $r \to 0$. The fact that S varies very little in the equilibrium range suggests that we might treat it as constant and equal to its value in the inertial subrange. This is our closure hypothesis. In this case (6.53b) reduces to

$$\frac{1}{2}\frac{dh}{dx} + h^{3/2} = x, \quad r \ll l \tag{6.61}$$

where x and h are normalized versions of r and $\langle[\Delta v]^2\rangle$,

$$h = \frac{\langle[\Delta v]^2\rangle}{\beta(15\beta)^{1/2} v^2}, \quad x = \frac{r}{(15\beta)^{3/4}\eta}.$$

Here $\eta = (\nu^3/\varepsilon)^{1/4}$ and $v = (\nu\varepsilon)^{1/4}$ are the Kolmogorov microscales and β is the Kolmogorov constant ($\beta \approx 2.0$). The boundary conditions on h are, from (6.49) and (6.50), $h \to x^2$ as $x \to 0$ and $h \to x^{2/3}$ as $x \to \infty$. This simple ordinary differential equations (ODE) may be integrated to determine $h(x)$, and hence $\langle[\Delta v]^2\rangle$, $\langle[\Delta v]^3\rangle$, f and K, throughout the equilibrium range (Obukhov 1949). The resulting shape of $\langle[\Delta v]^2\rangle$ is shown in Figure 6.19, along with the results of some recent direct numerical simulations (DNS) of turbulence.

A reasonable approximation to $h(x)$, which is continuous in $h(x)$ and its first and second derivatives (though not higher derivatives), is

$$h(x) = x^2 - \tfrac{1}{2}x^4 + \tfrac{1}{4}x^6 + \tfrac{1}{918}x^8 - \tfrac{167}{3240}x^{10}, \qquad x \le 1$$
$$h(x) = x^{2/3} - \tfrac{2}{9}x^{-2/3} - \tfrac{5}{81}x^{-2} - \tfrac{101}{6120}x^{-10/3}, \quad x \ge 1.$$

It is remarkable the such a simple closure hypothesis yields so much information and produces results reasonably close to the DNS. This might be contrasted with the equivalent (i.e. algebraic) two-point closure hypotheses formulated in spectral space, which are usually more complex and often yield less information (see Section 8.2.2.).

It has been observed that Obukhov's solution for $h(x)$ leads to the energy spectrum $E(k)$ going very slightly negative around $k \sim 0.8\eta^{-1}$. Thus the assumption of S being constant throughout the equilibrium range cannot be strictly correct. However, as noted by Monin and Yaglom (1975), this is a minor quibble as very small changes to $h(x)$ are enough to ensure $E(k)$ remains positive.[9] Also, as emphasized by Orszag (1970), it is not overly worrying if a closure hypothesis, which is after all an approximation, leads to a small negative contribution to $E(k)$. It is when the closure results in a large negative contribution, such as happens with the quasi-normal scheme, that the hypothesis has to be rejected.

6.2.3.3 Quasi-normal-type closure schemes (Part 1)

Let us now turn to a broader class of closure models, which are not restricted to the equilibrium range. We have seen that, at least symbolically, we can write evolution equations for Q_{ij}, S_{ijk}, etc. as,

$$\frac{\partial}{\partial t}\langle uu\rangle = \langle uuu\rangle + \nu\langle uu\rangle \tag{6.62a}$$

$$\frac{\partial}{\partial t}\langle uuu\rangle = \langle uuuu\rangle + \frac{1}{\rho}\langle uup\rangle + \nu\langle uuu\rangle \tag{6.62b}$$

$$\frac{\partial}{\partial t}\langle uuuu\rangle = \langle uuuuu\rangle + \frac{1}{\rho}\langle uuup\rangle + \nu\langle uuuu\rangle. \tag{6.62c}$$

Now the pressure field is related to $\mathbf{u}$ by the integral equation,

$$p(\mathbf{x}') = \frac{\rho}{4\pi}\int \frac{1}{|\mathbf{x}-\mathbf{x}'|}\frac{\partial^2 u_m u_n}{\partial x_m \partial x_n}d\mathbf{x} \tag{6.63}$$

and so terms like $\langle uup\rangle$ can be expressed as integrals over all space of $\langle uuuu\rangle$, and $\langle uuup\rangle$ as integrals of $\langle uuuuu\rangle$. So we might rewrite (6.62a) to (6.62c) as,

$$\frac{\partial}{\partial t}\langle uu\rangle = \langle uuu\rangle + \nu\langle uu\rangle \tag{6.64a}$$

$$\frac{\partial}{\partial t}\langle uuu\rangle = \langle uuuu\rangle + \int\langle uuuu\rangle + \nu\langle uuu\rangle \tag{6.64b}$$

$$\frac{\partial}{\partial t}\langle uuuu\rangle = \langle uuuuu\rangle + \int\langle uuuuu\rangle + \nu\langle uuuu\rangle \tag{6.64c}$$

$$\vdots$$

$$\frac{\partial}{\partial t}\langle u^n\rangle = \langle u^{n+1}\rangle + \int\langle u^{n+1}\rangle + \nu\langle u^n\rangle. \tag{6.64d}$$

The basis of the quasi-normal (QN) approximation is to terminate this sequence at (6.64b) and then to estimate the fourth-order correlations, $\langle uuuu\rangle$, in terms of products of the second-order correlations. Symbolically we have,

$$\frac{\partial}{\partial t}\langle uu\rangle = \langle uuu\rangle + \nu\langle uu\rangle \quad \text{(exact)} \tag{6.65}$$

$$\frac{\partial}{\partial t}\langle uuu\rangle = \langle uuuu\rangle + \int\langle uuuu\rangle + \nu\langle uuu\rangle \quad \text{(exact)} \tag{6.66}$$

$$\langle uuuu\rangle = \langle uu\rangle\langle uu\rangle \quad \text{(heuristic)}. \tag{6.67}$$

[9] We shall see in Section 6.6.2 that the energy of the eddies declines steadily with increasing k across the inertial sub-range, and then plummets to nearly zero in the range $k\eta = 0.2 \rightarrow 0.7$. The contribution from $E(k)$ to $\frac{1}{2}\langle \mathbf{u}^2\rangle$ for wave numbers in excess of $k\eta = 0.7$ is vanishingly small. Thus the small negative region of $E(k)$ in the vicinity of $k\eta = 0.8$ makes no significant contribution to $\frac{1}{2}\langle \mathbf{u}^2\rangle$.

The question then is how to relate $\langle uuuu \rangle$ to products of the type $\langle uu \rangle \langle uu \rangle$. It turns out that this can be achieved if we make certain assumptions about the velocity statistics.

We have already seen that the probability distribution of $\mathbf{u}$ measured at a single point is approximately normal (Gaussian). We have also seen that the joint-probability distribution of $\mathbf{u}$ measured at two or more points is more or less normal provided that the points are well separated ($r \gg l$). (e.g. see Van Atta and Yeh 1970.) This reflects the fact that remote points in a turbulent flow are (virtually) statistically independent and so the relationship between, say, $u_i(\mathbf{x})$ and $u_j(\mathbf{x}')$ is dictated by pure chance. Crucially, however, the joint-probability distribution of $\mathbf{u}$ measured at two or more adjacent points ($r \leq l$) is significantly non-Gaussian (Figure 3.19), reflecting the influence of the Navier–Stokes equation in determining the evolution of $\mathbf{u}$. (See Section 3.2.7.) Indeed, the non-Gaussian nature of $\mathbf{u}$ is essential to the dynamics of turbulence, since a non-zero skewness is required in order to maintain the energy cascade. Despite the acknowledged importance of departures from Gaussian behaviour, the basis of the QN scheme is to assume that, as far as the fourth-order correlations are concerned, the joint-probability distribution of $\mathbf{u}$ measured at two or more points is Gaussian. It is important to emphasize that the QN scheme makes this assumption not just for remote points, but also for adjacent points. This is clearly at variance with the measured flatness factors discussed in Section 5.5.1.

Now the attractiveness of the QN approximation lies in the fact that, if the joint-probability distribution of $\mathbf{u}$ is indeed Gaussian, then the so-called *fourth-order cumulants* of the velocity field,

$$\begin{aligned}[u_i u_j' u_k'' u_l]_{\text{cum}} = \langle u_i u_j' u_k'' u_l \rangle - \langle u_i u_j' \rangle \langle u_l u_k'' \rangle - \langle u_i u_k'' \rangle \langle u_j' u_l \rangle \\ - \langle u_i u_l \rangle \langle u_j' u_k'' \rangle \end{aligned} \tag{6.68}$$

are zero. So the QN closure scheme can be written symbolically as,

$$\frac{\partial}{\partial t} \langle uu \rangle = \langle uuu \rangle + \nu \langle uu \rangle \tag{6.69a}$$

(exact)

$$\frac{\partial}{\partial t} \langle uuu \rangle = \langle uuuu \rangle + \int \langle uuuu \rangle + \nu \langle uuu \rangle \tag{6.69b}$$

(exact)

$$\langle u_i u_j' u_k'' u_l \rangle = \langle u_i u_j' \rangle \langle u_k'' u_l \rangle + \langle u_i u_k'' \rangle \langle u_j' u_l \rangle + \langle u_i u_l \rangle \langle u_k'' u_j' \rangle \tag{6.69c}$$

(heuristic)

The simplicity of (6.9a, b, c) is rather seductive, and so QN schemes of one sort or another have been very popular ever since Millionshtchikov first suggested the procedure in 1941. Unfortunately, in its purest form the QN model rapidly runs into trouble. For example, the resulting inertial terms are time-reversible, which is inconsistent with the idea that, by virtue of inertial forces alone, turbulence possesses an 'arrow of time'. Moreover, it was shown in 1963 that, sooner or later, system (6.69) gives rise to a large, negative part of the energy spectrum, which is clearly non-physical. So a generation of theoreticians set about trying to improve the QN model with a variety of heuristic modifications. Thus, for example, around 1970 the QN scheme gave way to the eddy-damped-quasi-normal (EDQN) model, in which empirical 'damping' terms are introduced in (6.69b) in order to reduce the size of the third-order correlations (Orszag 1970). Unfortunately, this fails to ensure that $E(k)$ remains positive and so, in the mid-1970s, EDQN finally gave way to the eddy-damped-quasi-normal-Markovian (EDQNM) model. This involves the drastic step of removing a time derivative from the system (6.69). (We shall discuss this in more detail in Section 8.2.3.)

In view of the somewhat arbitrary nature of the approximations inherent in the EDQNM model, it might be thought that its predictions should be regarded with suspicion. However, it turns out that, for many purposes, it works surprisingly well, although it does run into trouble when applied to the very large eddies.[10] A nice review of QN type models and their successes is given by Lesieur (1990) who sets out the equations in some detail and shows how the algebra can be made more palatable by moving into Fourier space.[11]

Example 6.3 Use (6.68) to show that, in the QN approximation, $\langle[\Delta v]^4\rangle = 12u^4(1-f)^2$ and hence confirm that $\langle[\Delta v]^4\rangle = 3\left[\langle(\Delta v)^2\rangle\right]^2$, giving a flatness factor of 3. Is this estimate of $\langle[\Delta v]^4\rangle$ consistent with Kolmogorov's theory of the equilibrium range? Is it consistent with the log-normal or $\hat{\beta}$ models of Section 6.5?

6.3 The dynamics of the large scales

Kolmogorov's theory of the equilibrium range is considered by most researchers to be one of the success stories of turbulence. Of course, it tells us only about the small scales. This is unfortunate, since the small scales are often (but not always) of limited practical importance! The transport of momentum, and the dispersal of pollutants, is usually controlled by the large eddies.

[10] We shall see why EDQNM gives poor results for the large eddies in Section 6.3.6.

[11] We shall give a brief overview of the Fourier-space version of the QN scheme in Section 8.2.3.

Our understanding of the *large-scale* dynamics has had a rather chequered career. It started out well with a remarkable result, due to Loitsyansky, which says that isotropic turbulence possesses an integral invariant:

$$I = -\int \mathbf{r}^2 \langle \mathbf{u} \cdot \mathbf{u}' \rangle \, d\mathbf{r} = \text{constant}. \tag{6.70}$$

A little later Landau pointed out that the invariance of I is a direct consequence of the general law of conservation of angular momentum. That is to say, I is a measure of the angular momentum possessed by a cloud of turbulence and its invariance is a manifestation of the fact that the cloud maintains its angular momentum as it evolves. Now (6.70) is not just of academic interest. It is important because, as we shall see, it allowed Kolmogorov to predict the rate of decay of energy in a cloud of freely evolving turbulence. Specifically, Kolmogorov predicted

$$u^2(t) \sim t^{-10/7} \tag{6.71}$$

a law which turns out to be fairly close to some (but by no means all) observations. So, for a while, it looked like we had a sound theory of both the large and the small scales.

This comfortable world was turned on its head when, in 1956, G. K. Batchelor pointed out that, due to rather subtle effects associated with the pressure field, Loitsyansky's proof of (6.70) is flawed. The same objections also invalidate Landau's later (and quite separate) assertion that I is a measure of the angular momentum of the turbulence. Suddenly (6.70), and by implication (6.71), was in doubt.

Around the same time the QN closure scheme was gaining in popularity and, as we shall see, it too suggested that I is time dependent; and so a general consensus emerged that (6.70) and (6.71) are wrong. The final nail in the coffin came in 1967 when Saffman showed that, for certain kinds of isotropic turbulence, I does not even exist (it diverges)! Loitsyansky's invariant became Loitsyansky's integral and Kolmogorov's decay law looked destined for the scrap heap. Curiously, though, (6.71) provides a reasonably good estimate of the decay of certain types of freely evolving turbulence. So what is going on?

We shall summarize here the claims and counter-claims of Landau, Loitsyansky, Batchelor, and Saffman, ending with an overall assessment of the validity, or otherwise, of (6.70) and (6.71). It turns out that the nub of the problem lies in the ability of the pressure field to transmit information (through pressure waves) over very long distances. This means that, in principle, remote parts of a turbulent flow can be statistically correlated (they feel each other) and it is those finite, long-distance correlations, which call into question the validity of (6.70). However, we shall see that, for freely evolving turbulence, which has arisen from *certain types of initial conditions*, the long-range correlations are weak and that, to within a reasonable level of approximation, the classical view of Landau and Loitsyansky prevails. For other types of initial conditions, though, we must reject Loitsyansky and embrace an alternative theory, which is due to Birkhoff and Saffman.

6.3.1 Loitsyansky's integral

Nothing holds up the progress of science quite so much as the right idea at the wrong time. (Vincent de Vignaud)

Suppose we integrate (6.48a) from $r=0$ to $r \to \infty$. The result is

$$\frac{\partial}{\partial t}\left[u^2 \int_0^\infty r^4 f(r)\, dr\right] = [u^3 r^4 K]_\infty + 2\nu[u^2 r^4 f'(r)]_\infty. \tag{6.72}$$

Let us now make the plausible (though, as it turns out, questionable) assumption that $f(r)$ and $K(r)$ are exponentially small at large r. In effect, we are assuming that remote points in a turbulent flow ($r \gg l$) are statistically independent. If this is true then we obtain an extraordinary result. The non-linear effects, as represented by $u^3 K$, vanish and we find (Loitsyansky 1939),

$$I = 8\pi u^2 \int_0^\infty r^4 f\, dr = \text{constant}. \tag{6.73}$$

It seems that we have managed to finesse all of the difficulties normally associated with non-linearity to obtain a simple, concrete result.

I was once known as Loitsyansky's invariant, though it is now more commonly referred to as *Loitsyansky's integral*. The factor of 8π is incorporated in (6.73) so that it may be rewritten, with the aid of (6.39), in the form

$$I = -\int \mathbf{r}^2 \langle \mathbf{u} \cdot \mathbf{u}' \rangle\, d\mathbf{r} = \text{constant}. \tag{6.74}$$

This is a remarkable equation. If correct, then it is as important for the large scales as Kolmogorov's theories are for the small. It is important because it can be combined with the energy equation (3.6),

$$\frac{du^2}{dt} = -\frac{2}{3}\varepsilon \sim -\frac{u^3}{l} = -A\frac{u^3}{l} \tag{6.75}$$

to estimate the rate of decay of energy in isotropic turbulence. That is to say, (6.73) implies that, for freely evolving turbulence,

$$u^2 l^5 = \text{constant} \tag{6.76}$$

and this may be combined with (6.75) to predict the evolution of $u^2(t)$. (We are assuming here that the longitudinal correlation function, f, is approximately self-preserving in decaying, isotropic turbulence, as suggested by (5.17).) Unfortunately, the validity or otherwise of (6.74) and (6.76) has been hotly disputed, as we shall see.

The integral I has physical significance in terms of the energy spectrum $E(k)$. Consider (6.24). If we expand $\sin(kr)$ in a Taylor series in kr, and assume that $R(r)$ falls sufficiently rapidly with r, then we obtain,

$$E(k) = Lk^2/4\pi^2 + Ik^4/24\pi^2 + \cdots \tag{6.77}$$

where,

$$L = \int \langle \mathbf{u} \cdot \mathbf{u}' \rangle \, d\mathbf{r}. \tag{6.78}$$

The integral L is known as *Saffman's integral* and it is thought by many to play a key role in the dynamics of the large scales, possibly as important as I. From (6.39) we have

$$L = 4\pi u^2 [r^3 f]_\infty \tag{6.79}$$

and so, following our (tentative) assumption that $f(r)$ is exponentially small at large r, we would expect $L = 0$ and

$$E(k) = Ik^4/24\pi^2 + \cdots. \tag{6.80}$$

We shall refer to spectra of the type $E(k) \sim k^4 + \cdots$ as *Batchelor spectra* (because their key properties were established by Batchelor and Proudman in 1956), and to spectra like $E(k) \sim k^2 + \cdots$ as *Saffman spectra*. In the classical theories of Loitsyansky and Landau $[f(r)]_\infty$ and $[K(r)]_\infty$ are assumed to be exponentially small, and so all spectra are Batchelor spectra and I is an invariant.

6.3.2 *Kolmogorov's decay laws*

Kolmogorov exploited the (alleged) invariance of I to estimate the decay of $u^2(t)$ in freely evolving turbulence. He noted that (6.75) and (6.76) require,

$$u^2(t) \sim t^{-10/7}, \qquad l \sim t^{2/7}, \tag{6.81, 6.82}$$

(Kolmogorov 1941a). These are known as Kolmogorov's decay laws and, by and large, they are a reasonable fit to the experimental data and to certain (but not all) numerical simulations (see Section 6.3.6). The growth of the integral scale is, at first sight, unexpected. This is normally interpreted as follows. Equations (6.74) and (6.80) imply that

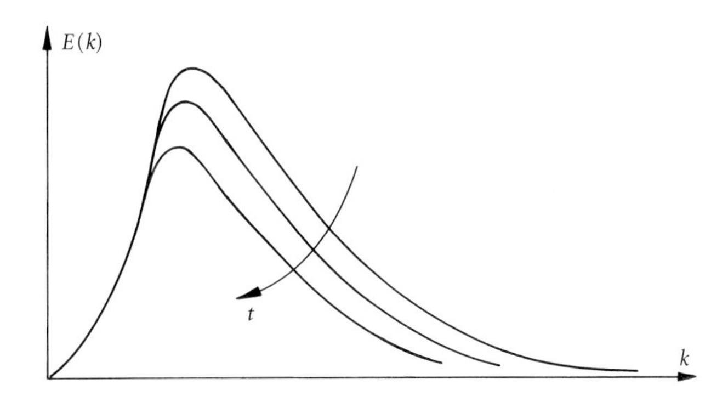

Figure 6.20 The decay of $E(k)$ in isotropic turbulence.

$E(k)$ is of fixed shape for small k. So, as the turbulence decays, $E(k)$ must evolve as shown in Figure 6.20. The spectrum collapses from the high-k end and so the average size of the eddies, as measured by $l = \int_0^\infty f\,dr$, increases.

6.3.3 *Landau's angular momentum*

The simplicity of (6.74) is striking. This suggests that there is some underlying physical principle at work and indeed there is. It was pointed out by Landau (in the first edition of Landau and Lifshitz's *Fluid Dynamics*) that the invariance of I is a direct consequence of the law of conservation of angular momentum. In fact Landau showed that

$$I = \langle \mathbf{H}^2 \rangle / V, \qquad \mathbf{H} = \int_V (\mathbf{x} \times \mathbf{u})\, dV \tag{6.83}$$

where V is the volume of some large cloud of turbulence, $V \gg l^3$, l being the integral scale. Combining this with (6.74) we obtain the *Landau–Loitsyansky equation*

$$I = -\int \mathbf{r}^2 \langle \mathbf{u} \cdot \mathbf{u}' \rangle\, d\mathbf{r} = \langle \mathbf{H}^2 \rangle / V = (\text{constant}). \tag{6.84}$$

Crucially, however, Landau had to make the same assumption as Loitsyansky, that remote points in a turbulent flow are statistically independent in the sense that f_∞ and K_∞ are exponentially small. If this is not true, then (6.83) fails.

We now consider Landau's proof of (6.83). The first question to address is: why should a cloud of turbulence possess any net angular momentum, $\mathbf{H}$? We touched on this in Chapter 3. Consider the portion of a wind tunnel grid shown in Figure 6.21. Evidently angular momentum (vorticity) is generated at the surface of the bars and then swept downstream. In fact we might suspect that the grid injects angular momentum into the flow because, if loosely suspended, it will

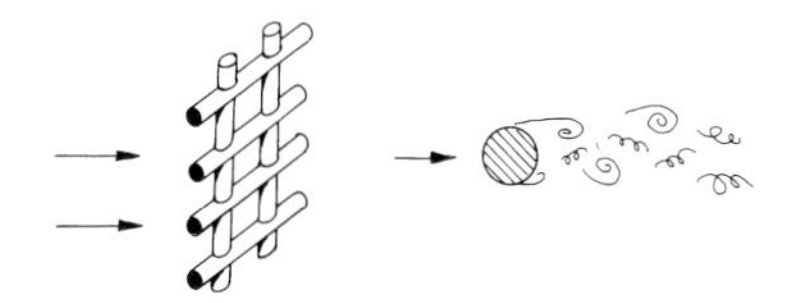

Figure 6.21 Angular momentum is injected into a turbulent flow by a grid.

shake and judder in response to the hydrodynamic forces. Now it might be thought that, in a large cloud of turbulence, ($V \gg l^3$), the total angular momentum will be very nearly zero because the vortices have random orientations. In a sense this is true; however, the central limit theorem tells us that this cancellation will be imperfect. Moreover, the theorem suggests that, if we perform many realizations of an experiment, measuring **H** in each case, then provided $V \gg l^3$ we should find

$$\langle \mathbf{H} \rangle = 0, \qquad \langle \mathbf{H}^2 \rangle \sim V.$$

(Consult Section 3.2.3.) In short, we might anticipate that $\langle \mathbf{H}^2 \rangle / V$ is finite and independent of V, as suggested by (6.84).

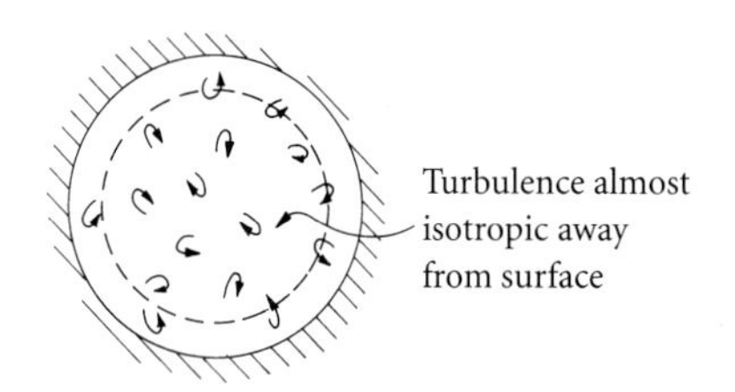

Figure 6.22 Landau's thought experiment.

The next question is: in what sense is $\langle \mathbf{H}^2 \rangle / V$ an invariant? To answer this Landau considered the following thought experiment. Consider a large cloud of turbulence freely evolving in a closed sphere of radius R, $R \gg l$, as shown in Figure 6.22. That is, we stir up the contents of the sphere and then leave the turbulence to decay. In any one realization of this experiment we have,

$$\mathbf{H}^2 = \int_V (\mathbf{x} \times \mathbf{u})\, dV \cdot \int_{V'} (\mathbf{x}' \times \mathbf{u}')\, dV'. \tag{6.85}$$

In addition, since **u** integrates to zero in a closed domain, the flow has zero linear momentum,

$$\mathbf{L} = \int \mathbf{u}\, dV = \mathbf{0}. \tag{6.86}$$

(Actually, we shall see shortly that the vanishing of **L** through the use of a closed domain is a crucial ingredient of Landau's theory. Different results are obtained in an open domain in cases where **L** is non-zero.) Combining expressions (6.85) and (6.86), it may be shown that,

$$\mathbf{H}^2 = -\iint (\mathbf{x}' - \mathbf{x})^2 \mathbf{u} \cdot \mathbf{u}'\, dV\, dV'. \tag{6.87a}$$

(See Exercise 6.5.) Now suppose we repeat the experiment many times and form ensemble averages of (6.87a) over many realizations. This yields

$$\langle \mathbf{H}^2 \rangle = -\iint \mathbf{r}^2 \langle \mathbf{u} \cdot \mathbf{u}' \rangle\, d\mathbf{r}\, dV \tag{6.87b}$$

which is beginning to look remarkably like Loitsyansky's integral. Finally Landau, like Loitsyansky before him, assumed that f_∞ is exponentially small. In such a case only those velocity correlations taken close to the boundary are aware of the presence of this surface and in this sense the turbulence is approximately homogeneous and isotropic. Also, the far-field contributions to the integral $\int \mathbf{r}^2 \langle \mathbf{u} \cdot \mathbf{u}' \rangle \, d\mathbf{r}$ are small. Integral (6.87b) then reduces to,

$$\langle \mathbf{H}^2 \rangle = -V \int \mathbf{r}^2 \langle \mathbf{u} \cdot \mathbf{u}' \rangle \, d\mathbf{r} + O[(l/R)V]$$

and in the limit $R/l \to \infty$ we have,

$$\langle \mathbf{H}^2 \rangle / V = -\int \mathbf{r}^2 \langle \mathbf{u} \cdot \mathbf{u}' \rangle \, d\mathbf{r} = I.$$

So, provided f_∞ is exponentially small, I is proportional to $\langle \mathbf{H}^2 \rangle$. However, this still does not explain why I is an invariant, since in each realization of Landau's thought experiment,

$$\frac{d\mathbf{H}}{dt} = \mathbf{T}_\nu$$

where $\mathbf{T}_\nu$ is the viscous torque exerted by the boundaries. Luckily, the central limit theorem comes to our rescue again. We can estimate $\mathbf{T}_\nu$ on the assumption that the eddies near the boundary are randomly orientated. It turns out that this suggests that $\mathbf{T}_\nu$ has negligible influence as $R/l \to \infty$. In this sense, then, $\mathbf{H}$ (and hence $\mathbf{H}^2$) is conserved in each realization and it follows that I is an invariant of the flow.

In summary, then, we have

$$I = -\int \mathbf{r}^2 \langle \mathbf{u} \cdot \mathbf{u}' \rangle \, d\mathbf{r} = \langle \mathbf{H}^2 \rangle / V = (\text{constant})$$

provided that the long-range correlations are sufficiently weak. For the case of anisotropic (but homogeneous) turbulence these arguments may be generalized to yield

$$\begin{aligned}
I_x &= -\int \left(r_y^2 + r_z^2\right) \langle \mathbf{u} \cdot \mathbf{u}' \rangle_{yz} \, d\mathbf{r} = \left\langle H_x^2 \right\rangle / V = (\text{constant}) \\
I_y &= -\int \left(r_z^2 + r_x^2\right) \langle \mathbf{u} \cdot \mathbf{u}' \rangle_{zx} \, d\mathbf{r} = \left\langle H_y^2 \right\rangle / V = (\text{constant}) \\
I_z &= -\int \left(r_x^2 + r_y^2\right) \langle \mathbf{u} \cdot \mathbf{u}' \rangle_{xy} \, d\mathbf{r} = \left\langle H_z^2 \right\rangle / V = (\text{constant})
\end{aligned} \tag{6.88}$$

where $\langle \mathbf{u} \cdot \mathbf{u}' \rangle_{xy} = \left\langle u_x u_x' + u_y u_y' \right\rangle$ and so on. One implication of (6.88) is that, if at $t = 0$ the turbulence is anisotropic, in the sense that the various components of $\mathbf{H}$ are different, then this anisotropy is preserved throughout the subsequent decay.

The fact that the invariance of I could be established by two distinct routes, and that Kolmogorov's decay law, which is based on the conservation of I, is reasonably in line with the experiments, meant that most people were, for some time, happy with (6.84). Everything changed, however, in 1956.

Example 6.4 Landau–Loitsyansky equation for a turbulent cloud in an open domain.
Landau's analysis applies to a closed domain. However, we may repeat the analysis in an open domain as follows. Consider a turbulent cloud of vorticity of radius R evolving in an infinite space. (The flow is irrotational outside the spherical volume $|\mathbf{r}| = R$.) We suppose that R is much greater than the integral scale of the turbulence, l. From Example 5.5 of Section 5.1.3 we know that the angular momentum of the turbulent cloud is conserved and can be written as

$$\mathbf{H} = \frac{1}{3}\int \mathbf{x} \times (\mathbf{x} \times \boldsymbol{\omega})\, dV = \text{constant}.$$

Let us suppose that the linear momentum, $\mathbf{L}$, of the cloud is initially zero. Then, from Example 5.3 of Section 5.1.3, we have

$$\mathbf{L} = \frac{1}{2}\int \mathbf{x} \times \boldsymbol{\omega}\, dV = \mathbf{0}, \qquad \int \boldsymbol{\omega}\, dV = \mathbf{0}$$

for all t. (The second integral equation is a direct consequence of the fact that $\boldsymbol{\omega}$ is solenoidal.) Use the vector identity

$$2[\mathbf{x} \times (\mathbf{x}_0 \times \boldsymbol{\omega})]_i = [\mathbf{x}_0 \times (\mathbf{x} \times \boldsymbol{\omega})]_i + \nabla \cdot \left[\mathbf{x} \times (\mathbf{x}_0 \times \mathbf{x})_i \boldsymbol{\omega}\right]$$

along with the integral equations above, to show that

$$\mathbf{H} = \frac{1}{3}\int (\mathbf{x} - \mathbf{x}_0) \times ((\mathbf{x} - \mathbf{x}_0) \times \boldsymbol{\omega})\, d\mathbf{x} = \text{constant}$$

where $\mathbf{x}_0$ is a constant vector. Hence confirm that the square of the angular momentum is given by

$$\begin{aligned}\mathbf{H}^2 &= \frac{1}{9}\int\!\!\int [\mathbf{r} \times (\mathbf{r} \times \boldsymbol{\omega})] \cdot [\mathbf{r} \times (\mathbf{r} \times \boldsymbol{\omega}')]\, d\mathbf{x}\, d\mathbf{x}' \\ &= \frac{1}{9}\int\!\!\int [(\boldsymbol{\omega} \cdot \boldsymbol{\omega}')r^4 - r^2(\mathbf{r} \cdot \boldsymbol{\omega})(\mathbf{r} \cdot \boldsymbol{\omega}')]\, d\mathbf{x}\, d\mathbf{x}'\end{aligned}$$

where $\mathbf{r} = \mathbf{x}' - \mathbf{x}$. If there are no long-range correlations, in the sense that $\langle \boldsymbol{\omega} \cdot \boldsymbol{\omega}' \rangle$ decays exponentially fast for large r, then the bulk of the turbulence in the cloud will be isotropic and homogeneous and integrals of the type $\int \langle \boldsymbol{\omega} \cdot \boldsymbol{\omega}' \rangle r^4\, d\mathbf{r}$ converge rapidly. In such a situation we have, in the limit $R/l \to \infty$,

$$\langle \mathbf{H}^2 \rangle / V \sim \int \left[r^4 \langle \boldsymbol{\omega} \cdot \boldsymbol{\omega}' \rangle - r^2 r_i r_j \langle \omega_i \omega_j' \rangle\right] d\mathbf{r}$$

Now use (6.41a) to confirm that this integral simplifies to

$$\langle \mathbf{H}^2 \rangle / V \sim -\int \mathbf{r}^2 \langle \mathbf{u} \cdot \mathbf{u}' \rangle\, d\mathbf{r}.$$

Since $\mathbf{H}$ is conserved for the cloud, and any change of V is a surface effect, we have

$$-\int \mathbf{r}^2 \langle \mathbf{u} \cdot \mathbf{u}' \rangle \, d\mathbf{r} \sim \langle \mathbf{H}^2 \rangle / V \approx \text{(constant)}$$

which is reminiscent of the Landau–Loitsyansky equation. Note that we had to make two crucial assumptions in this derivation: (i) the turbulence has no net linear impulse (linear momentum); and (ii) there are no long-range correlations.

6.3.4 *Batchelor's pressure forces*

'Mine is a long and a sad tale!' said the Mouse, turning to Alice and sighing. 'It is a long tail, certainly', said Alice, looking down in wonder at the mouse's tail; 'but why do you call it sad?' (Lewis Carroll)

The first sign that all was not well with the Landau–Loitsyansky equation came with the work of Proudman and Reid (1954) on isotropic turbulence. They investigated the dynamical consequences of the QN approximation. (Recall that this closes the problem at third order by assuming that the velocity statistics are Gaussian to the extent that the cumulants of the fourth-order velocity correlations are zero.) They found that, when the QN approximation is made, the triple correlations appear to decay as r^{-4} at large r, rather than exponentially fast as had traditionally been assumed. If this were also true of real turbulence it would invalidate both Loitsyansky's and Landau's proofs of the invariance of I. Indeed it turns out that the QN closure model predicts

$$\frac{d^2 I}{dt^2} = \frac{7}{5}(4\pi)^2 \int_0^\infty \frac{E^2}{k^2} dk \tag{6.89}$$

These findings were somewhat surprising and it provoked Batchelor and Proudman into revisiting the entire problem in 1956. They considered *anisotropic* (but homogeneous) turbulence and dispensed with the QN approximation. Their primary conclusion was that correlations of the form $\langle u_i u_j p' \rangle$ can decay rather slowly at large r, in fact as slowly as r^{-3}. This is a direct consequence of the slow decline in the pressure field ($p_\infty \sim r^{-3}$) associated with a local fluctuation in velocity (Figure 6.23(a)). Thus remote points in a field of turbulence are statistically correlated through the pressure field (Figure 6.23(b)).

This discovery, which lies behind the QN result, has profound implications for the way that we think about the large scales in a turbulent flow, and so it is worth taking a moment to explain its origins and significance. Long-range statistical correlations may exist

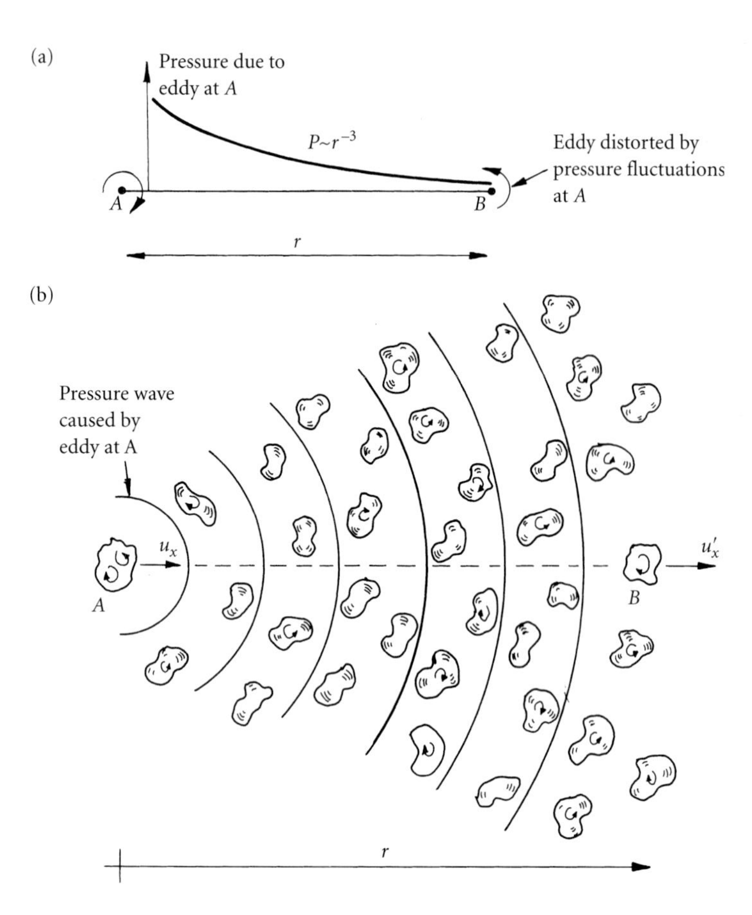

Figure 6.23 (a) The pressure field associated with a local fluctuation in velocity declines slowly with distance as r^{-3}. (b) Remote points in a field of turbulence are statistically correlated because information is transmitted over large distances by the pressure field.

between remote points because information is communicated over large distances by the pressure field, which is, of course, non-local. That is, the relationship

$$\nabla^2 p = -\rho \frac{\partial^2 u_i u_j}{\partial x_i \partial x_j} \tag{6.90}$$

may be inverted using the Biot–Savart law to give

$$p(\mathbf{x}) = \frac{\rho}{4\pi} \int \frac{\partial^2 u_i'' u_j''}{\partial x_i'' \partial x_j''} \frac{d\mathbf{x}''}{|\mathbf{x}'' - \mathbf{x}|}. \tag{6.91}$$

Thus a fluctuation in velocity at one point, say $\mathbf{x}'$, sends out pressure waves, which propagate to all parts of the flow field. Suppose, for example, that we have a single eddy located near $\mathbf{x} = \mathbf{0}$ (Figure 6.24). Then the pressure field at large distances from the eddy, due to that eddy, is

$$p = \frac{\rho}{4\pi} \frac{\partial^2}{\partial x_i \partial x_j} \left(\frac{1}{|\mathbf{x}|} \right) \int u_i'' u_j'' \, d\mathbf{x}'' + \cdots . \tag{6.92}$$

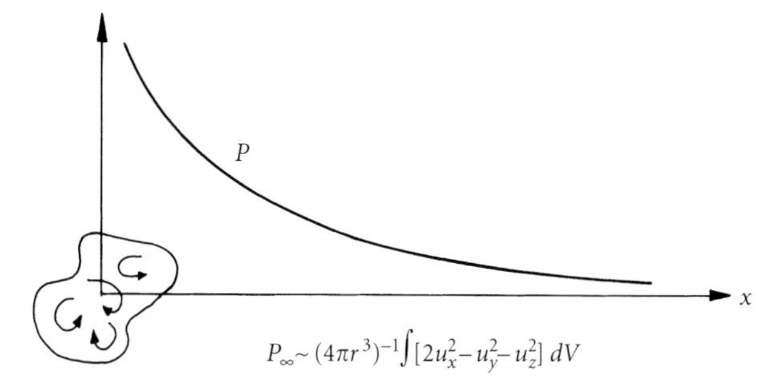

Figure 6.24 The pressure field generated by a single eddy.

(The first two terms in the Taylor expansion of $|\mathbf{x}'' - \mathbf{x}|^{-1}$ lead to integrands which integrate to zero. See Appendix II or Exercise 6.6 at the end of this chapter.) Thus the pressure field associated with our eddy falls off rather slowly, in fact as x^{-3}. This pressure field induces forces, and hence motion, throughout the fluid and of course this motion is correlated to the behaviour of the eddy at $\mathbf{x} = \mathbf{0}$.

Example 6.5 Use (6.92) to confirm that, for the isolated eddy shown in Figure 6.24, the pressure in the far field is

$$p'_\infty(r\hat{\mathbf{e}}_x) = \rho(4\pi r^3)^{-1}\int\left[2u_x^2 - u_y^2 - u_z^2\right]d\mathbf{x}$$

at a point on the x-axis a distance r from the centre of the eddy. □

Batchelor and Proudman showed that, in homogeneous turbulence, which is anisotropic, these non-local effects can give rise to long-range pressure–velocity correlations of the form,

$$\langle u_i u_j p' \rangle_\infty \sim r^{-3}, \quad r = |\mathbf{x}' - \mathbf{x}|. \tag{6.93}$$

For example, for the configuration shown in Figure 6.24 we have, using the result of the example above,

$$\langle u_x^2 p' \rangle_\infty = (4\pi r^3)^{-1}\rho\int\left\langle u_x^2\left[2(u''_x)^2 - (u''_y)^2 - (u''_z)^2\right]\right\rangle d\mathbf{x}'' \tag{6.94}$$

where p' is the pressure at $\mathbf{x} = r\hat{\mathbf{e}}_x$ and u_x^2 is measured at $\mathbf{x} = \mathbf{0}$.

Now we have already seen that the triple correlations, S_{ijk}, are governed by an equation of the type

$$\rho\frac{\partial S_{ijk}}{\partial t} = \langle uuuu' \rangle - \frac{\partial}{\partial r_k}\langle u_i u_j p' \rangle + \cdots$$

and so even if there is no algebraic tail in S_{ijk} at $t = 0$, (6.93) tells us that there will be one for $t > 0$, and in general we would expect[12]

[12] For an alternative derivation of $\langle uuu' \rangle_\infty \sim r^{-4}$ see Appendix II.

$\langle uuu' \rangle_\infty \sim r^{-4}$. It follows from (6.46) that, in anisotropic turbulence, we should expect $\langle uu' \rangle_\infty \sim r^{-5}$.

Now consider the generalized Loitsyansky-like integral

$$I_{ijmn} = \int r_m r_n \langle u_i u'_j \rangle \, d\mathbf{r}.$$

Since $\langle uu' \rangle_\infty \sim r^{-5}$, there is clearly an issue over convergence, with a possible logarithmic divergence of I_{ijmn}. Actually, it turns out that there is no divergence provided that we agree to evaluate the integral over a large sphere whose radius then recedes to infinity. The point is that the far-field form of $\langle u_i u'_j \rangle$ is such that the leading-order contribution to I_{ijmn} from a spherical annulus of large radius is zero. However, the slow decline in $\langle uuu' \rangle$ means that I_{ijmn} is, in general, time-dependent (Batchelor and Proudman 1956) and not an invariant as traditionally assumed. (e.g. See, Batchelor (1953) for the proof that, in the absence of long-range effects, I_{ijmn} is an invariant.) Many concluded that, since I_{ijmn} is time-dependent in anisotropic turbulence, then I should be time-dependent in isotropic turbulence.

So, from 1956 onwards, the word went out that Batchelor and Proudman had shown that I is time-dependent, Loitsyansky fell from grace and Kolmogorov's decay laws were rejected. However, a more careful reading of their paper shows that Batchelor and Proudman were rather more cautious. There are a number of points that should be made. First (6.93) tells us that

$$\langle u_i u_j p' \rangle_\infty \sim Cr^{-3}.$$

However, it does not tell us the magnitude of C which might, for example, be zero. It turns out that there is no rigorous way of determining the strength of the long-range correlations $\langle uuu' \rangle_\infty \sim C_3 r^{-4}$ and $\langle uu' \rangle_\infty \sim C_2 r^{-5}$, in the sense that the magnitude of C_3 and C_2 are undetermined. The only way to estimate C_2 and C_3 is either to measure them or else to invoke some closure scheme. Of course, we have no way of knowing whether or not a closure scheme can be trusted and so it is preferable to look at the experimental data. Batchelor and Proudman did just that. They looked at the shape of $f(r)$ in grid turbulence during the final period of decay and found an exponential, rather than the anticipated algebraic, decline in $\langle uu' \rangle$! This provoked them to remark '. . . it is disconcerting that the present theory [i.e. their theory] cannot do as well as the old'. The second point to note is that the positive results in their paper all relate to anisotropic turbulence. When the symmetries of isotropy are imposed, Batchelor and Proudman were unable to find any non-zero contribution to $\langle uup' \rangle_\infty$, $\langle uuu' \rangle_\infty$, etc. They concluded that, for the isotropic case, the old theory of Loitsyansky (and others) could well be valid. Of course Loitsyansky and Landau restricted their comments to isotropic turbulence, and so, based purely on a reading of Batchelor and Proudman, their claims cannot be rejected.

Actually, it was the QN result, (6.89), which probably had the bigger psychological influence and led to the widespread conviction that I is time-dependent. Of course, we now know that the QN closure scheme is fundamentally flawed, but this was not conclusively demonstrated until 1963, when computers became powerful enough for researchers to integrate the system (6.69) (and find negative energy spectra!). Prior to 1963, the QN approximation was widely regarded as reasonable, and its prediction that I is time-dependent (tentatively supported by Batchelor and Proudman) was probably regarded as a strength, rather than a weakness, of the theory. Ironically, we shall discover shortly that the QN equation (6.89) is substantially out of line with the experimental data.

So, by the early 1960s it seemed that everything was up for grabs. The old theories were regarded with suspicion and new closure schemes were providing thought-provoking results. Batchelor and Proudman had opened a can of worms, and the lid was never to be replaced.

6.3.5 *The Saffman–Birkhoff spectrum*

The final blow to Loitsyansky's invariant came in 1967. Batchelor and Proudman had shown that we have no right to pre-judge the form of f_∞ and K_∞. They could be exponentially small or else fall off algebraically. This led Saffman to return to the expansion (6.77),

$$E(k) = Lk^2/4\pi^2 + Ik^4/24\pi^2 + \cdots$$

where

$$L = \int \langle \mathbf{u} \cdot \mathbf{u}' \rangle \, d\mathbf{r} = 4\pi u^2 [r^3 f]_\infty .$$

Prior to 1967 L had usually been dismissed as zero, since it was expected that f_∞ would decline faster than r^{-3}. However Saffman, possibly inspired by the earlier work of Birkhoff (1954), showed that there is no reason why f_∞ should not be of order r^{-3}. If that were the case then we have a new form of turbulence in which

$$E(k) \sim Lk^2 + \cdots; \qquad f_\infty \sim r^{-3}$$

and Loitsyansky's integral does not even exist (it diverges)! Moreover, we can write

$$L = \left\langle \left[\int \mathbf{u} \, dV \right]^2 \right\rangle \Big/ V$$

where V is the volume of a large cloud of turbulence. So L is a measure of the square of the linear momentum held in a cloud of turbulence of size V. If the turbulence has negligible linear momentum, then $L = 0$. On the other hand, if the linear momentum is finite and of the order $V^{1/2}$ then L is non-zero. So whether or not we obtain a Saffman spectrum depends crucially on the amount of linear momentum held in the turbulence. In Section 5.1.3 we saw that this linear momentum may be represented by the net linear impulse of the turbulence, $\frac{1}{2}\int_V (\mathbf{x} \times \boldsymbol{\omega})\, dV$. Thus the linear momentum of a turbulent cloud may be considered to be the sum of the linear impulses of the individual eddies. If the individual eddies, and hence the cloud as a whole, has little linear impulse, then we obtain a $E \sim k^4$ spectrum. (Remember that in Landau's argument $\int \mathbf{u}\, dV = \mathbf{0}$ because of the use of a closed domain.) On the other hand, if the individual eddies have a significant amount of linear impulse, then the central limit theorem suggests that a cloud of turbulence composed of many such vortices will have a net linear momentum (or linear impulse), which grows as $V^{1/2}$. Such a case leads to a Saffman spectrum.

Following an argument analogous to Landau's, but with conservation of linear momentum replacing conservation of angular momentum, it may be shown that,[13] $L = \text{constant}$. A poor man's proof of the invariance of L goes something like this. (For a more detailed proof see examples 6.6 and 6.7 below.) Suppose we have a large cloud of turbulence of size R evolving in an unbounded domain. That is, the vorticity field, which constitutes the turbulent cloud, is confined to a large spherical region of radius R. (By large we mean $R \gg l$.) We know that the impulse $\frac{1}{2}\int (\mathbf{x} \times \boldsymbol{\omega})\, dV$, which is a measure of the linear momentum of the cloud, is an invariant of the motion. (See Section 5.1.3 or else Appendix II.) Thus we have

$$\int \mathbf{x} \times \boldsymbol{\omega}\, dV = \text{constant}, \qquad \int \boldsymbol{\omega}\, dV = 0.$$

This is sufficient to allow us to use (6.87a), but with $\mathbf{u}$ replaced by $\boldsymbol{\omega}$, to give,

$$\left[\int (\mathbf{x} \times \boldsymbol{\omega})\, dV\right]^2 = -\iint (\mathbf{x}' - \mathbf{x})^2 \boldsymbol{\omega} \cdot \boldsymbol{\omega}'\, dV\, dV' = \text{constant}.$$

Since R is much greater than the integral scale, l, we can rewrite this as

$$\frac{1}{V}\left\langle \left[\int (\mathbf{x} \times \boldsymbol{\omega})\, dV\right]^2 \right\rangle = -\int r^2 \langle \boldsymbol{\omega} \cdot \boldsymbol{\omega}' \rangle\, d\mathbf{r} + O(l/R) = \text{constant}.$$

[13] This result was obtained earlier by Birkhoff (1954) who anticipated a number of Saffman's findings. In particular he noted that $E \sim k^2$, $f_\infty \sim r^{-3}$ was a theoretical possibility and that L is an invariant.

and in the limit $l/R \to 0$ we obtain

$$-\int r^2 \langle \boldsymbol{\omega} \cdot \boldsymbol{\omega}' \rangle \, d\mathbf{r} = \text{constant}.$$

But the impulse $\frac{1}{2}\int (\mathbf{x} \times \boldsymbol{\omega}) \, dV$ is a measure of the linear momentum of the cloud and so we might anticipate that we can relate our vorticity invariant to L. In fact, it is readily confirmed that, provided f_∞ is no greater than r^{-3}, the relationship $\langle \boldsymbol{\omega} \cdot \boldsymbol{\omega}' \rangle = -\nabla^2 \langle \mathbf{u} \cdot \mathbf{u}' \rangle$ yields,

$$L = -\frac{1}{6}\int r^2 \langle \boldsymbol{\omega} \cdot \boldsymbol{\omega}' \rangle \, d\mathbf{r} = \text{constant},$$

as required. Thus conservation of linear impulse leads directly to the invariance of L.

Alternatively, we may note that the Karman–Howarth equation (6.48b) can be written as,

$$\frac{\partial}{\partial t}\left[r^2 \langle \mathbf{u} \cdot \mathbf{u}' \rangle\right] = \frac{\partial}{\partial r}\frac{1}{r}\frac{\partial}{\partial r}(r^4 u^3 K) + 2\nu \frac{\partial}{\partial r}\left[r^2 \frac{\partial}{\partial r}\langle \mathbf{u} \cdot \mathbf{u}' \rangle\right]$$

which, since $K_\infty \sim r^{-4}$, may be integrated to yield, once again, $L = \text{constant}$. Thus L is an invariant of Saffman's spectrum, playing a role analogous to that of I in a k^4 spectrum.

One way to understand the difference between a Saffman ($E \sim k^2$) and a Batchelor ($E \sim k^4$) spectrum is to think of a single isolated eddy (blob of vorticity). The far-field velocity induced by this eddy is (see Appendix II),

$$4\pi \mathbf{u}_\infty = (\mathbf{L} \cdot \nabla)\nabla\left(\frac{1}{r}\right) + \frac{1}{2}\frac{\partial^3}{\partial x_i \partial x_j \partial x_k}\left(\frac{1}{r}\right)\int x_i' x_j' \omega_k \, d\mathbf{x}' + \cdots,$$

$$\mathbf{L} = \frac{1}{2}\int \mathbf{x}' \times \boldsymbol{\omega}' \, d\mathbf{x}'.$$

If the eddy possesses a finite amount of linear impulse, that is $\frac{1}{2}\int \mathbf{x} \times \boldsymbol{\omega} \, dV$ is non-zero for that eddy (which is typical of an eddy in Saffman turbulence), then its influence at large distances declines as $|\mathbf{u}| \sim r^{-3}$. If it does not contain any linear impulse, that is, $\frac{1}{2}\int \mathbf{x} \times \boldsymbol{\omega} \, dV = 0$ (which is typical of an eddy in Batchelor turbulence), then the far-field is weaker, $|\mathbf{u}| \sim r^{-4}$. Thus an isolated eddy, which has linear impulse casts a longer shadow than one which does not, leading to stronger long-range correlations. (More details are given in Example 5.7 of Section 5.1.3 and Example 6.12 below.) It is this effect, which lies behind Saffman's spectrum. Put crudely, we may picture Saffman turbulence as composed of a sea of eddies, each making a finite contribution to the linear impulse (linear momentum) of the fluid (Figure 6.25). A Batchelor spectrum, on the other hand, represents a sea of eddies, each of which has a finite amount of angular

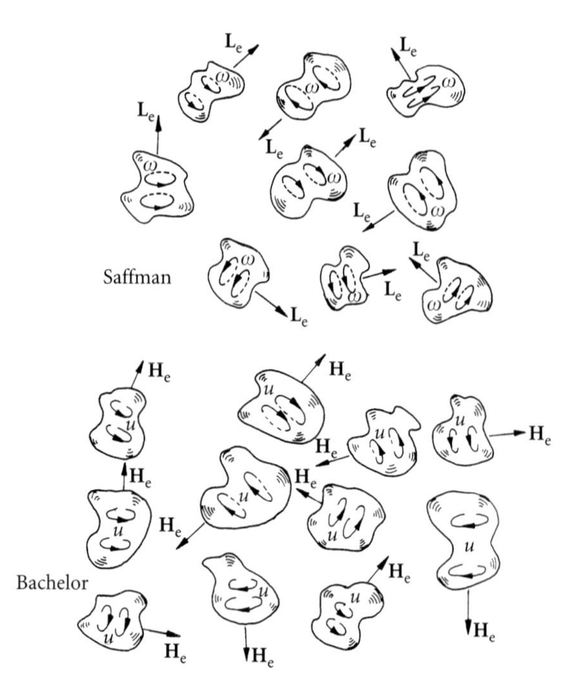

Figure 6.25 Saffman turbulence corresponds to a sea of eddies each possessing a significant amount of linear impulse, $\mathbf{L}_e$. A Batchelor spectrum, on the other hand, represents a sea of eddies which have a finite amount of angular impulse, $\mathbf{H}_e$, but negligible linear impulse.

momentum, but negligible linear impulse, that is $\frac{1}{2}\int \mathbf{x} \times \boldsymbol{\omega}\, dV \approx 0$. Of course, this does not imply that the vortices in Batchelor turbulence do not move! Each eddy is swept around by the other eddies. In a Batchelor spectrum we merely require that $\frac{1}{2}\int \mathbf{x} \times \boldsymbol{\omega}\, dV$ is near zero for a typical eddy, so that its *self-induced* translation is zero. (See Example 5.2 of Section 5.1.3.)

A third of a century has passed since Saffman's (1967) paper, yet there is still controversy over whether turbulence is of the Batchelor, $E \sim k^4$, or Saffman, $E \sim k^2$, type. It seems that both are, in principle, realizable and that either may be generated in computer simulations. The real question, of course, is which does nature prefer? It turns out that the initial conditions are crucial in this respect.

Recall that the pressure forces, which induce long-range correlations, cannot be stronger than $\langle u_i u_j p' \rangle_\infty \sim r^{-3}$ and since

$$\frac{\partial S_{ijk}}{\partial t} = -\frac{\partial}{\partial r_k}\langle u_i u_j p' \rangle + \cdots$$

the associated triple correlations cannot be greater than $\langle uuu \rangle_\infty \sim r^{-4}$. In anisotropic turbulence this yields $\langle uu \rangle_\infty \sim r^{-5}$, while in isotropic turbulence the Karman–Howarth equation tells us that $\langle uu \rangle_\infty$ cannot be stronger than r^{-6}. All of this seems incompatible with Saffman's suggestion of $f_\infty \sim r^{-3}$. However, suppose we set $f_\infty \sim r^{-3}$ at $t = 0$. There is then no necessity to rely on the pressure field to initiate long-range statistical correlations. They will be self-perpetuating. So whether we have $E \sim k^2$ or $E \sim k^4$ depends on the initial conditions. If it starts as

$E \sim k^2$ then it stays as $E \sim k^2$, and if it starts as $E \sim k^4$ then it stays as $E \sim k^4$. In particular, if turbulence is generated in such a way that

$$\left\langle \int \mathbf{u}\, dV \right\rangle \ll V^{1/2}, \quad (V \gg l^3)$$

then L will be zero at $t = 0$ and therefore remain zero. (Remember L is an invariant.) We then have a Batchelor spectrum, $E \sim k^4$, and Loitsyansky's integral exists. On the other hand if

$$\left\langle \int \mathbf{u}\, dV \right\rangle \sim V^{1/2}$$

at $t = 0$ then L will be non-zero and $E \sim k^2$. Either initial condition is readily simulated on the computer and so both k^2 and k^4 spectra are seen in numerical simulations of turbulence.

The more important question, though, is what do we find in nature. Is grid turbulence, $E \sim k^2$ or $E \sim k^4$? The answer, unfortunately, is far from clear cut. At the time of writing there is no definitive evidence either way, and indeed there is no reason to believe that all types of grid turbulence must be the same. Perhaps oscillating grids, for example, produce different results to stationary grids. While there is some indication that conventional grid turbulence *may* be of the Batchelor type, the evidence relates to the final period of decay and is not entirely convincing. (The rate of decline of u^2 in the final period of decay turns out to be approximately $u^2 \sim t^{-5/2}$, which is compatible with $E \sim k^4$, but incompatible with $E \sim k^2$. See Example 6.9 below.) So perhaps we should allow for the possibility of both $E \sim k^2$ and $E \sim k^4$ spectra.

We might note in passing that there have been a number of attempts to show that L is necessarily non-zero, implying that $E \sim k^2$ in all types of homogeneous turbulence. Typical of these is the paper of Rosen (1981). However, the analysis in Rosen's paper is flawed and no such result is established.

This more or less concludes our discussion of Saffman turbulence. In the next section we shall restrict ourselves to $E \sim k^4$ spectra and return to the controversy over the conservation (or otherwise) of Loitsyansky's integral. For the enthusiast, however, we provide in the examples below a more detailed exploration of the relationship between the linear and angular impulse (or momentum) of eddies, and the Saffman or Batchelor spectra which they produce.

Example 6.6 The linear momentum density in isotropic turbulence. Consider a spherical control volume, V, of radius R, which sits in an infinite sea of isotropic turbulence. We wish to determine $\langle \mathbf{L}^2 \rangle$ where $\mathbf{L}$ is the net linear momentum in V. Use the fact that $u_i = \nabla \cdot (\mathbf{u} x_i)$ to show that

$$\mathbf{L}^2 = \int u_i\, dV \oint_S x_i \mathbf{u} \cdot d\mathbf{S}$$

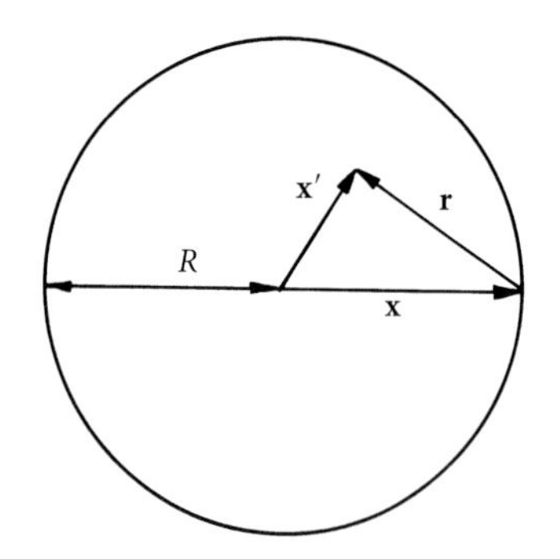

Figure 6.26 Control volume and coordinate system used in examples 6.1 and 6.2.

where S is the surface of the spherical control volume and the origin for $\mathbf{x}$ is at the centre of V. Hence confirm that

$$\langle \mathbf{L}^2 \rangle = 4\pi R^3 \int_V \langle u_x u'_x \rangle \, dV$$

where u_x is the x-component of velocity at a point on the surface $(\mathbf{x} = R\hat{\mathbf{e}}_x)$ and u'_x is measured at an interior point $\mathbf{x}'$. (See Figure 6.26.) Now evaluate the integral on the right using the general expression for Q_{ij} in isotropic turbulence. Confirm that

$$\langle \mathbf{L}^2 \rangle = 4\pi^2 R^2 u^2 \int_0^{2R} r^3 f \left[1 - (r/2R)^2\right] dr$$

where f is the usual longitudinal correlation function. For a Saffman spectrum ($f_\infty \sim r^{-3}$), this yield, in the limit $R/l \to \infty$,

$$\langle \mathbf{L}^2 \rangle / V = 4\pi u^2 (r^3 f)_\infty, \quad \text{[Saffman spectrum]}$$

(cf. 6.79). For a Batchelor spectrum ($f_\infty \sim r^{-6}$), on the other hand, we find,

$$\langle \mathbf{L}^2 \rangle = 4\pi^2 R^2 u^2 \int_0^{\infty} r^3 f \, dr, \quad \text{[Batchelor spectrum]}$$

as $R/l \to \infty$. These results are consistent with the expressions,

$$\langle \mathbf{L}^2 \rangle / V = \int \mathbf{u} \cdot \mathbf{u}' \, dV \neq 0, \qquad \langle \mathbf{L}^2 \rangle / V = 0, \quad \text{for } R \to \infty.$$

[Saffman spectrum] [Batchelor spectrum]

Example 6.7 Conservation of Saffman's integral
We shall now show that the conservation of Saffman's integral in isotropic turbulence is a direct result of the conservation of linear momentum in a large control volume.

Using the same control volume as example 6.6, show that the rate of change of $\mathbf{L}^2$ in V is,

$$\frac{d\mathbf{L}^2}{dt} = 2\mathbf{L} \cdot \frac{d\mathbf{L}}{dt} = -2 \int_V u_i \, dV \left[\oint_S u_i \mathbf{u} \cdot d\mathbf{S} + \oint_S (p/\rho) \, dS_i \right].$$

(You may ignore viscous stresses.) Now use the fact that $\langle u_i p' \rangle = 0$ in isotropic turbulence to confirm that

$$\frac{d}{dt} \langle \mathbf{L}^2 \rangle = -8\pi R^2 \int_V \langle u_i u_x u'_i \rangle \, dV$$

where, as before, $\mathbf{x}$ is located on the surface S (at $\mathbf{x} = R\hat{\mathbf{e}}_x$) and $\mathbf{x}'$ is an interior point within V. Now evaluate the integral on the right using the general expression for S_{ijk} in isotropic turbulence.

Confirm that

$$\frac{d}{dt}\langle \mathbf{L}^2 \rangle = 4\pi^2 R^2 u^3 \int_0^{2R} \left[\left[1 - (r/2R)^2 \right] \frac{1}{r}\frac{\partial}{\partial r}(r^4 K) \right] dr$$

where $K(r)$ is the longitudinal triple correlation function. Since $K_\infty \sim r^{-4}$ we have, for $R/l \to \infty$,

$$\frac{d}{dt}\langle \mathbf{L}^2 \rangle \sim R^2.$$

Compare this with the results of example 6.6 and hence confirm that, for Saffman turbulence,

$$\langle \mathbf{L}^2 \rangle / V = 4\pi u^2 (r^3 f)_\infty = \int \langle \mathbf{u} \cdot \mathbf{u}' \rangle \, dV = \text{constant}.$$

Evidently, the flux of momentum in or out of V is sufficiently small as to leave $\langle \mathbf{L}^2 \rangle / V$ unchanged in the limit $R/l \to \infty$. It is this which lies behind conservation of Saffman's integral.

Example 6.8 Kolmogorov's four-fifths law through momentum conservation

We now show how the four-fifths law can be obtained directly from the principle of linear momentum conservation, without the Karman–Howarth equation.

Let us return to examples 6.6 and 6.7, only this time we shall take R to lie in the inertial subrange, $\eta \ll R \ll l$. From example 6.6 we have

$$\langle \mathbf{L}^2 \rangle = 4\pi^2 R^2 u^2 \int_0^{2R} r^3 f \left[1 - (r/2R)^2 \right] dr.$$

Since $R \ll l$ we may take $f = 1$ and so our expression simplifies to

$$\langle \mathbf{L}^2 \rangle = 4\pi^2 R^2 u^2 \int_0^{2R} r^3 \left[1 - (r/2R)^2 \right] dr, \quad R \ll l.$$

From example 6.7, on the other hand, we know that

$$\frac{d}{dt}\langle \mathbf{L}^2 \rangle = \frac{2}{3}\pi^2 R^2 \int_0^{2R} \left[\left[1 - (r/2R)^2 \right] \frac{1}{r}\frac{\partial}{\partial r}\left(r^4 \langle [\Delta v]^3 \rangle\right) \right] dr$$

where we have substituted $\langle [\Delta v]^3 \rangle$ for $K(r)$ and it is legitimate to ignore viscous stresses since $R \gg \eta$. Show that combining these two expressions yields

$$\int_0^{2R} \left[\left[1 - (r/2R)^2 \right] \frac{1}{r}\frac{\partial}{\partial r}\left[r^4 \langle [\Delta v]^3 \rangle + \frac{4}{5}\varepsilon r^5 \right] \right] dr = 0.$$

Confirm that this is satisfied if and only if, $\langle [\Delta v]^3 \rangle = -\frac{4}{5}\varepsilon r$. This is, of course, Kolmogorov's four-fifths law.

Example 6.9 The final period of decay (reprise)
Show that one class of solutions to the Karman–Howarth equation in the final period of decay (during which the triple correlations may be ignored) is

$$f(r,t) = M\left[n, \tfrac{5}{2}, \ -r^2/8\nu t\right] = M\left[\tfrac{5}{2} - n, \tfrac{5}{2}, r^2/8\nu t\right] \exp[-r^2/8\nu t]$$

where $u^2 \sim t^{-n}$ and M is Kummer's hypergeometric function. Confirm that the corresponding energy spectrum takes the form

$$E \sim k^{2n-1} \exp[-2\nu k^2 t].$$

A Batchelor k^4 spectrum corresponds to $n = 5/2$. Verify that, in such a case, the solution above reduces to,

$$f \sim \exp[-r^2/8\nu t], \quad u^2 \sim t^{-5/2}.$$

A Saffman k^2 spectrum, on the other hand, requires $n = 3/2$, which corresponds to

$$f(r,t) \sim M[3/2, 5/2, \ -r^2/8\nu t], \quad u^2 \sim t^{-3/2}.$$

Grid turbulence data suggests $u^2 \sim t^{-5/2}$ (Monin and Yaglom 1975), which lends support to the idea that grid turbulence is of the Batchelor type.

Example 6.10 More decay laws
Show that the analogue of Kolmogorov's decay laws, (6.81) and (6.82), for a Saffman spectrum are, $u^2 \sim t^{-6/5}$, $l \sim t^{2/5}$.

Example 6.11 The global angular momentum in Saffman turbulence
In this example we show that the angular momentum in a large cloud of turbulence scales with volume V in different ways for Batchelor and Saffman turbulence.

Consider a cloud of turbulence of volume V composed of a large number of discrete vortices whose centres are located at positions $\mathbf{x}_i$. Let $\mathbf{r}_i$ be the position vector of a point measured from the centre of the i'th eddy, $\mathbf{r}_i = \mathbf{x} - \mathbf{x}_i$, and $\mathbf{L}_i$ and $\mathbf{H}_i$ be the linear and angular impulse of the i'th eddy measured relative to the centre of that eddy. Then we have,

$$\int_{V_i} \boldsymbol{\omega}\, dV = 0, \quad \mathbf{L}_i = \frac{1}{2}\int_{V_i} \mathbf{r}_i \times \boldsymbol{\omega}\, dV, \quad \mathbf{H}_i = \frac{1}{3}\int_{V_i} \mathbf{r}_i \times (\mathbf{r}_i \times \boldsymbol{\omega})\, dV.$$

Show that the net angular momentum of the cloud

$$\mathbf{H} = \frac{1}{3}\int_V \mathbf{x} \times (\mathbf{x} \times \boldsymbol{\omega})\, dV = \sum \frac{1}{3}\int_{V_i} \mathbf{x} \times (\mathbf{x} \times \boldsymbol{\omega})\, dV$$

can be written in terms of $\mathbf{L}_i$ and $\mathbf{H}_i$ as follows:

$$\mathbf{H} = \sum \mathbf{H}_i + \sum \mathbf{x}_i \times \mathbf{L}_i.$$

Hint: you may find the following vector identity useful,

$$2[\mathbf{x} \times (\mathbf{x}_0 \times \boldsymbol{\omega})]_k = [\mathbf{x}_0 \times (\mathbf{x} \times \boldsymbol{\omega})]_k + \nabla \cdot [\mathbf{x} \times (\mathbf{x}_0 \times \mathbf{x})_k \boldsymbol{\omega}],$$
$$\mathbf{x}_0 = \text{constant}.$$

In a Batchelor spectrum we have $\mathbf{L}_i = 0$ and so $\mathbf{H} = \sum \mathbf{H}_i$. In a Saffman spectrum, on the other hand, $\mathbf{H}$ is dominated by the linear momentum contribution and so $\mathbf{H} \approx \sum \mathbf{x}_i \times \mathbf{L}_i$. This leads to a difference in the way $\mathbf{H}$ scales with V in the two spectra.

Example 6.12 Why is $f_\infty \sim r^{-3}$ in turbulence composed of eddies with linear impulse?
Let us now consider the origin of Saffman's long-range correlation $f_\infty \sim r^{-3}$. Consider an eddy (a blob of vorticity) centred at $\mathbf{x} = \mathbf{0}$ and enclosed by a spherical control volume V_R of radius R. The far-field velocity, u_1', induced by this eddy at $\mathbf{x} = r\hat{\mathbf{e}}_1$ is $u_1' = L_1/2\pi r^3 + \cdots$, where $\mathbf{L} = \frac{1}{2}\int \mathbf{x} \times \boldsymbol{\omega}\, d\mathbf{x}$. Consider the product of u_1' with the velocity u_1 at some point within the eddy. Use Example 5.2 of Section 5.1.3 to show that

$$\langle u_1 u_1' \rangle = L_1^2/3\pi V_R r^3 + \cdots$$

where the average is defined as a volume average over V_R. Thus a correlation defined in this way falls off as r^{-3} if L_1 is non-zero. Now show that $\langle u_1 u_1' \rangle_\infty$ is exactly zero if $\mathbf{L} = \mathbf{0}$.

Example 6.13 Are there alternatives to $E \sim k^2$ and $E \sim k^4$ in mature turbulence?
In this example we explore alternatives to $E \sim k^2$ and $E \sim k^4$. The expansion of

$$E(k) = \frac{1}{\pi}\int_0^\infty \langle \mathbf{u} \cdot \mathbf{u}' \rangle kr \sin(kr)\, dr \tag{a}$$

to yield (6.77) in the form

$$E(k) = Lk^2/4\pi^2 + Ik^4/24\pi^2 + \cdots \tag{b}$$

is valid only if $\langle \mathbf{u} \cdot \mathbf{u}' \rangle$ decays sufficiently rapidly with r. Let us see what happens when $\langle \mathbf{u} \cdot \mathbf{u}' \rangle$ decays as an arbitrary power law. Suppose that, somehow, we arrange for $f(r)$ to fall off as

$$u^2 f(r \to \infty) = u^2 f_\infty = C_n r^{-n} \tag{c}$$

at $t = 0$. Use the Karman–Howarth equation, combined with the observation that, at most, $K_\infty \sim r^{-4}$, to show that C_n is an invariant provided that $n < 6$. Of course,

$$L = \int \langle \mathbf{u} \cdot \mathbf{u}' \rangle\, d\mathbf{r} = 4\pi u^2 [r^3 f]_\infty \tag{d}$$

and so $n=3$ is Saffman's case in which $L=4\pi C_3$. Now we would like L to be convergent since $(LV)^{1/2}$ is the net linear momentum contained in some large volume V and from the central limit theorem we would expect the total linear momentum to grow no faster than $V^{1/2}$. Thus we must restrict ourselves to $n \geq 3$ in (c). Now show that, with this restriction in place, (a) yields

$$E(k \to 0) = \frac{k^4}{3\pi}\int_0^R r^4 u^2 f\,dr - \frac{C_n k^{n-1}}{\pi}\int_{kR}^{\infty} x^{-(n-3)}\frac{d[x^{-1}\sin x]}{dx}\,dx + \cdots \tag{e}$$

where R is some length after which $u^2 f = C_n r^{-n}$ is a good approximation for f. Confirm that for $n=3$ and $n=6$ we recover the Saffman and Batchelor cases. Now consider the case where $n=4$ and show that, for small k, E grows as $E \sim C_4 k^3/4 + O(k^4)$. The implication is that if, somehow, we generate a k^3 spectrum at $t=0$, then we have a k^3 spectrum for all t. (Note, however, that a k^3 spectrum implies that the Fourier transform of Q_{ij} is non-analytic at $k=0$.)

6.3.6 A reappraisal of the long-range pressure forces in $E \sim k^4$ turbulence

Let us summarize the position so far. Both Batchelor ($E \sim k^4$) and Saffman ($E \sim k^2$) spectra can, in principle, be generated and indeed both types are observed in computer simulations. It is the initial conditions that determine which type of turbulence emerges. There is some evidence that grid turbulence is of the Batchelor ($E \sim k^4$) type, though this is far from clear cut. In this section we shall focus on turbulence with a $E \sim k^4$ spectrum.

For such spectra Loitsyansky's integral converges though it could be time dependent. That is to say, long-range pressure forces induce long-range correlations of the type $\langle u_i u_j p' \rangle \sim r^{-3}$, which in turn induce long-range triple correlations, $\langle uuu' \rangle_\infty \sim C_3 r^{-4}$. The generalized Karman–Howarth equation then tells us that, provided C_3 is non-zero, $\langle uu' \rangle_\infty \sim r^{-5}$ in anisotropic turbulence, but $\langle uu' \rangle_\infty \sim r^{-6}$ in isotropic turbulence (because symmetry kills the leading order term). More importantly, in *anisotropic* turbulence the long-range triple correlations, $\langle uuu' \rangle_\infty \sim C_3 r^{-4}$, are sufficiently strong to assure the time dependence of the Loitsyansky-like integral I_{ijmn} (Batchelor and Proudman 1956). However, Batchelor and Proudman could find no long-range triple correlations when the symmetries of isotropy are enforced, the implication being that C_3 might be zero in the isotropic case. Moreover, they found no evidence of long-range correlations in grid turbulence.

We are left, therefore, with the problem of knowing under what condition, if any, the Landau–Loitsyansky equation is valid in $E \sim k^4$

turbulence. We might dispense with Saffman's objections because these relate to $E \sim k^2$ spectra, and even point out that Batchelor and Proudman's paper is inconclusive in the case of isotropic turbulence. But there is something disconcerting about (6.89). The point is that, although the QN approximation is a flawed dynamical model, it does represent a legitimate kinematic initial condition. So, for suitable initial conditions, (6.89) is valid at $t = 0$, if not for $t > 0$. The implication is that we cannot throw out Batchelor's long-range effects in isotropic turbulence on the basis of a purely kinematic argument.

The computer simulations of turbulence are interesting in this respect.[14] Sometimes they have Gaussian initial conditions and often they start with $E \sim k^n$, $n > 4$. An example of just such a simulation by Lesieur et al. (2000) is shown in Figure 6.27. One observes the transient growth of a k^4 component in E as the turbulence recovers from its initial condition. However, once the turbulence becomes fully developed, with a full range of length-scales and a k^4 spectrum, I seems to be almost constant, as originally claimed by Loitsyansky. (We shall be more precise in our definition of 'almost' in a moment.) So it would seem that, at least for *mature* turbulence, Loitsyansky and Landau were, more or less, correct.

So what is going on? The first thing to note is that there is an apparent discrepancy between Proudman and Reid (1954) and Batchelor and Proudman (1956). The former show that, for isotropic turbulence, the QN closure model predicts

$$\frac{d^2 I}{dt^2} = 8\pi \frac{d}{dt}\left[u^3 r^4 K\right]_\infty = \frac{7}{5}(4\pi)^2 \int_0^\infty (E^2/k^2)\, dk \qquad (6.95)$$

$$\text{(QN only)}$$

(Here we have combined (6.72) and (6.89), noting that f_∞ is at most of order r^{-6} in an isotropic k^4 spectrum.) Batchelor and Proudman, on the other hand, could find no long-range effects when the symmetries of isotropy are imposed, which suggests that I might be constant.

Let us go back and repeat the analysis of Batchelor and Proudman. We shall adopt the same approach that they did and take initial conditions in which well separated points are statistically independent. In such a case f_∞, K_∞, and fourth-order cumulants $\{[u_i u_j' u_k'' u_l]_{\text{cum}}\}_{r \to \infty}$ are all exponentially small at $t = 0$. The question, then, is what happens for $t > 0$. (Note that taking $[u_i u_j' u_k'' u_l]_{\text{cum}} \sim 0$ for well separated points is *not* the same as the QN approximation, which requires $[u_i u_j' u_k'' u_l]_{\text{cum}} = 0$ for all combinations of $\mathbf{x}$, $\mathbf{x}'$, and $\mathbf{x}''$. To illustrate the difference consider the flatness factor shown in Figure 3.19. The QN

[14] Great care must be taken in interpreting such computer simulations. In order to obtain representative behaviour of the small-k part of $E(k)$ it is necessary to have a domain size of, say, 20–30 integral scales. This is rarely achieved and so many simulations should be regarded with some caution in this respect. See Section 7.2.

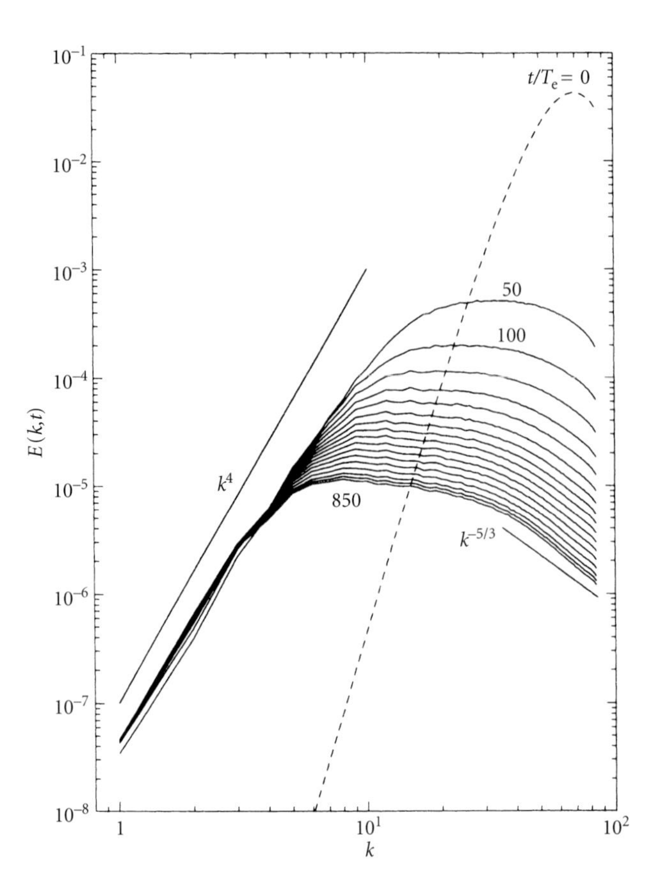

Figure 6.27 Decay simulations of Lesieur et al. (2000). This is an large eddy simulation using hyperviscosity.

approximation would require $\delta = 3$ for all r, which is clearly not true, where as here we require only that $\delta - 3$ is exponentially small for $r \to \infty$, which could well be true.)

A re-examination of the isotropic case in the spirit of Batchelor and Proudman reveals that (Davidson 2000)

$$\frac{\partial [u^3 K]_\infty}{\partial t} = \frac{3}{4\pi r^4} \int \left\langle u_x^2 \left[2u_x'^2 - u_y'^2 - u_z'^2 \right] \right\rangle d\mathbf{r} \tag{6.96a}$$

from which

$$\frac{d^2 I}{dt^2} = 8\pi \frac{d}{dt} [u^3 r^4 K]_\infty = 6 \int \langle ss' \rangle \, d\mathbf{r} = 6J \tag{6.96b}$$

where J and s are give by,

$$J = \int \langle ss' \rangle \, d\mathbf{r}, \quad s = u_x^2 - u_y^2.$$

Some hint as to where these expressions come from is given by the following argument. Consider an eddy located near $\mathbf{x} = \mathbf{0}$ as shown in Figure 6.24. We have already seen that the far-field pressure on the

x-axis induced by this eddy is (example 6.5),

$$p'(r\hat{\mathbf{e}}_x) = \rho(4\pi r^3)^{-1}\int \left[2u_x^2 - u_y^2 - u_z^2\right] d\mathbf{x}$$

and so we might expect the correlation between u_x^2 at $\mathbf{x}=0$ and $p'(r\,\hat{\mathbf{e}}_x)$ to be,

$$\langle u_x^2 p'\rangle_\infty = \rho(4\pi r^3)^{-1}\int \left\langle u_x^2\left[2\left(u_x''\right)^2 - \left(u_y''\right)^2 - \left(u_z''\right)^2\right]\right\rangle d\mathbf{x}''.$$

Now the governing equation for the triple correlations is given by (6.60) in the form,

$$\rho\frac{\partial S_{ijk}}{\partial t} + \langle uuuu\rangle = -\frac{\partial}{\partial r_k}\langle u_i u_j p'\rangle - \left\langle u_k'\left(u_i\frac{\partial p}{\partial x_j} + u_j\frac{\partial p}{\partial x_i}\right)\right\rangle + \rho\nu(\sim)$$

and in particular the longitudinal triple correlation obeys

$$\rho\frac{\partial}{\partial t}\left[u^3 K(r,t)\right] + \langle uuuu\rangle = -\frac{\partial}{\partial r}\left\langle u_x^2 p'(r\hat{\mathbf{e}}_\mathbf{x})\right\rangle - \langle uu'p\rangle + \rho\nu(\sim)$$

where $\langle uu'p\rangle$ is a symbolic representation of terms involving u, p, and u'. It turns out that the dominant long-range contribution to u^3K comes from the term involving $u_x^2 p'$ and so we have,

$$\frac{\partial}{\partial t}\left[u^3K\right]_\infty = \frac{3}{4\pi r^4}\int \left\langle u_x^2\left[2\left(u_x'\right)^2 - \left(u_y'\right)^2 - \left(u_z'\right)^2\right]\right\rangle d\mathbf{r},$$

which is (6.96a). For the isotropic turbulence this can be written in the more compact form

$$\frac{\partial}{\partial t}\left[u^3K\right]_\infty = \frac{3}{4\pi r^4}\int \langle ss'\rangle\, d\mathbf{r} \tag{6.96c}$$

and combining this with (6.72) yields

$$\frac{d^2I}{dt^2} = 8\pi\frac{d}{dt}\left[u^3r^4K\right]_\infty = 6\int \langle ss'\rangle\, d\mathbf{r} = 6J,$$

as required. (A more formal derivation of this equation is given in Appendix III.)

So it appears that there are, after all, long-range correlations in isotropic turbulence, despite the failure of Batchelor and Proudman to find them. It should be emphasized that (6.96b) is true *for all t* provided that $\{[u_i u_j' u_k'' u_l]_{\text{cum}}\}_{r\to\infty}$ remains exponentially small for $t > 0$. Now the experimental data suggests that, as far as the fourth-order correlations are concerned, the joint-probability distribution for $\mathbf{u}$ (in mature turbulence) is very close to Gaussian for well-separated points. (e.g. see Van Atta and Yeh 1970.) We might hope, therefore, that (6.96b) is a good approximation throughout most of the decay, and this is the position adopted in Davidson (2000). (It should be noted, however, that it is difficult to tell from the experiments whether

$\{[u_i u_j' u_k'' u_l]_{\text{cum}}\}_{r\to\infty}$ is exponentially or algebraically small, and so (6.96b) must be regarded with a considerable degree of caution.)

Actually, it can be shown that the QN result, (6.95), is a special case of (6.96b). That is, if we insist that $[u_i u_j' u_k'' u_l]_{\text{cum}} = 0$ for all combinations of **x**, **x**$'$, and **x**$''$, close or distant, (and it certainly is not!) then it may be shown that

$$6J_{\text{QN}} = \frac{14}{5}\int \langle \mathbf{u}\cdot\mathbf{u}'\rangle^2 \, d\mathbf{r}, \quad \text{(QN only)}.$$

Next, applying Rayleigh's theorem (see Section 8.1.1), we have

$$6J_{\text{QN}} = \frac{14}{5}\int \langle \mathbf{u}\cdot\mathbf{u}'\rangle^2 \, d\mathbf{r} = \frac{7}{5}(4\pi)^2 \int_0^\infty (E^2/k^2)\, dk, \quad \text{(QN only)}.$$

Combining this with (6.96b) yields (6.95). It should be emphasised however, that little credence should be given to the QN estimate of J.

Now J is positive since,

$$J = \int \langle ss' \rangle \, d\mathbf{r} = \left[\int s\, dV\right]^2 \bigg/ V$$

and so we may introduce a dimensionless parameter α and write $J = \alpha u^4 l^3$, $\alpha \geq 0$, from which (Davidson 2000),

$$\frac{d^2 I}{dt^2} = 8\pi \frac{d}{dt}\left[u^3 r^4 K\right]_\infty = 6\alpha u^4 l^3. \tag{6.97}$$

So I is indeed time dependent and, as suggested by Batchelor, the flaw in Loitsyansky and Landau's arguments lies in the existence of long-range pressure forces. However, the results of wind-tunnel experiments tentatively suggest that, if l is based on the estimate $l = (I/u^2)^{1/5}$, then α is rather small. In fully developed turbulence, for example, it is found that α lies in the range 0–0.03 (see Appendix III and also the data of Chasnov 1993, Lesieur et al. 1999, and Warhaft and Lumley 1978) with an average value of, perhaps, $\alpha \sim 0.01$. This might be compared with the QN estimate of $\alpha_{\text{QN}} \sim 0.6$! (See Exercise 6.7.)

So it seems likely that, in grid turbulence, the long-range effects are rather weak, and this is why I is approximately conserved in mature turbulence (Figure 6.27).

It seems also that the QN approximation greatly overestimates the strength of the long-range effects, by almost two orders of magnitude. We might note in passing that the EDQNM closure model specifies

$$\frac{dI}{dt} = 8\pi\left[u^3 r^4 K\right]_\infty \sim \theta(t) J_{\text{QN}}, \quad \text{(EDQNM only)}$$

where θ is a somewhat arbitrary model parameter with the dimensions of time (see Section 8.2.3). Evidently this simply amounts to an assertion that $[r^4K]_\infty$ is non-zero, with its magnitude being set by the arbitrary parameter θ. It is hard to reconcile the equation above with (6.96b). In particular, the arbitrary removal of a time derivative, as a result of Markovianization, seems difficult to justify.

Perhaps we should not be surprised by the weakness of the long-range correlations. This is, after all, in accord with our intuition. If the long-range correlations were strong then the influence of the boundaries of a closed domain would be felt everywhere in the interior, no matter how large the domain size relative to the integral scale. Thus, for example, the lateral boundaries of a wind tunnel could exert an influence on the decay of grid turbulence, even when the tunnel size greatly exceeds the turbulence integral scale. Yet there is little evidence that this is the case. In short, our intuition, based on experimental observations, suggests that the long-range correlations are weak in mature turbulence and this is in accordance with the observed smallness of α (See also comments in Appendix III.)

In summary, then, when we have a $E \sim k^4$ spectrum Loitsyansky's integral is time dependent and this is a result of Batchelor's long-range pressure forces. However, we have no reliable means of predicting the strength of these long-range forces and hence no reliable means of predicting the rate of change of I. All we can say with confidence is that the long-range forces appear to be weak in fully developed turbulence. In this respect the classical view of Loitsyansky and Landau is not so far from the truth. It is a beautiful but flawed theory.

It should not be forgotten, however, that the entire discussion above is restricted to $E \sim k^4$ turbulence, and the evidence that grid turbulence is of the $E \sim k^4$ type is far from clear cut. While a stationary grid may well give an $E \sim k^4$ spectrum, as suggested by the $-5/2$ decay law in the final period of decay, a grid which is vigorously shaken might be able to impart sufficient momentum to the turbulence to ensure an $E \sim k^2$ spectrum. In such a case I will not exist. The entire issue of $E \sim k^4$ versus $E \sim k^2$ spectra is still a matter of some controversy.[15]

6.4 The characteristic signature of eddies of different shape

A remarkable feature of turbulent flows is the comparative stability and permanence of the flow patterns that are usually called eddies. Use of the term implies that the whole motion can be considered to be composed of the

[15] Many researchers have suggested that we should not restrict ourselves to $E \sim k^2$ or $E \sim k^4$ spectra, but that the most general case for mature turbulence is $E(k \to 0) = C_m k^m$, $2 \le m \le 4$. In such cases the physical interpretation of C_m is unclear, though it is generally agreed that C_m is an invariant for $m < 4$. See example 6.13 at the end of Section 6.3.5.

superposition of the velocity fields of many simple elementary eddies, not much more complex in structure than the Hills spherical vortex or a vortex ring. For the concept to be useful, the duration of a single eddy must be relatively long and measurements in homogeneous turbulence show that the degree of permanence is remarkable. (A.A. Townsend (1970))

When observing the evolution of $f(r)$ or $E(k)$ in freely evolving turbulence it is natural to try and relate changes in the shape of f or E to changes in the population of eddies that make up the turbulence. This leads to the kinematic issue of relating the form of f and E to the shape and distribution of the eddies. In this respect it is useful to consider the shape of $f(r)$ or $E(k)$ associated with a sea of randomly distributed eddies of given size and structure. By using a variety of different 'model eddies' as the basic building blocks in this game we can build up an intuition as to the effect of eddy structure on f and E.

6.4.1 *Townsend's model eddy*

Consider, for example, the 'model eddy'

$$\mathbf{u} = \Omega r \exp\left[-2\mathbf{x}^2/l_e^2\right]\hat{\mathbf{e}}_\theta$$

in (r, θ, z) coordinates. This represents a blob of swirling fluid of characteristic size l_e, as shown in Figure 6.28(i). Townsend (1956) considered an ensemble of such eddies whose centres are randomly but uniformly distributed in space, but whose axes of rotation are all aligned. If we focus attention on the x–y plane the result is a kinematic representation of two-dimensional, homogeneous turbulence which is isotropic in a two-dimensional sense. Townsend showed that the longitudinal correlation function for such a field of turbulence is

$$f(r) = \exp\left[-r^2/l_e^2\right].$$

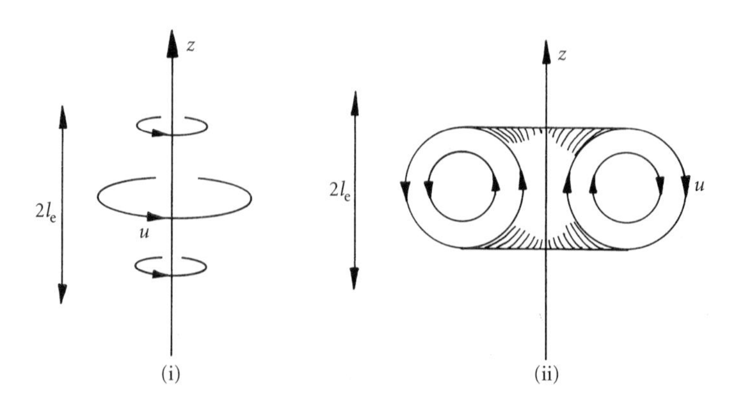

Figure 6.28 Two model eddies: (i) a swirling blob of fluid which has zero linear momentum but a finite amount of angular momentum; and (ii) a vortex ring which has zero net linear momentum and zero net angular momentum. The first eddy gives rise to a $E \sim k^4$ spectrum while the second leads to a $E \sim k^6$ spectrum.

To obtain a kinematic representation of three-dimensional turbulence, which is homogeneous and isotropic, we must consider the eddies to be randomly distributed *and* randomly orientated in space. Once again we find that f decays exponentially (see the appendix at the end of this chapter), and it is not difficult to show that the longitudinal correlation function f, the structure function $\langle[\Delta v]^2\rangle$, and the energy spectrum $E(k)$, are given by,

$$f(r) = \exp\left[-r^2/l_e^2\right] \tag{6.98a}$$

$$\tfrac{3}{4}\langle[\Delta v]^2\rangle = \tfrac{1}{2}\langle \mathbf{u}^2\rangle\left[1 - \exp\left(-r^2/l_e^2\right)\right] \tag{6.98b}$$

$$E(k) = \frac{\langle \mathbf{u}^2\rangle l_e}{24\sqrt{\pi}}(kl_e)^4 \exp\left[-l_e^2k^2/4\right] \tag{6.98c}$$

Note that $E(k)$ grows as k^4 for small k, as it must since our model eddy has a finite amount of angular momentum but no linear impulse, thus ruling out a Saffman ($E \sim k^2$) spectrum. So, in this field of turbulence, composed of eddies of fixed size l_e, the energy spectrum exhibits a sharp peak at $k = \sqrt{8}/l_e \sim \pi/l_e$. Note that the vorticity field of our model eddy is,

$$\boldsymbol{\omega} = 2\Omega\left[\left(2rz/l_e^2\right)\hat{\mathbf{e}}_r + \left(1 - 2r^2/l_e^2\right)\hat{\mathbf{e}}_z\right] \exp\left[-2\mathbf{x}^2/l_e^2\right].$$

Example 6.14 We have seen that, in isotropic turbulence, the vorticity correlation function is related to the velocity correlation by,

$$\langle \boldsymbol{\omega} \cdot \boldsymbol{\omega}'\rangle = -\nabla^2\langle \mathbf{u} \cdot \mathbf{u}'\rangle, \quad \langle \mathbf{u} \cdot \mathbf{u}'\rangle = \frac{u^2}{r^2}\frac{\partial}{\partial r}(r^3 f).$$

Confirm that the longitudinal correlation function for the vorticity, f_ω, is related to f by,

$$\omega^2 f_\omega = -\frac{u^2}{r}\frac{\partial}{\partial r}\frac{1}{r^2}\frac{\partial}{\partial r}(r^3 f), \quad \omega^2 = \tfrac{1}{3}\langle \boldsymbol{\omega}^2\rangle.$$

Thus show that, for a field of turbulence composed of a random collection of simple eddies of the type given above,

$$f_\omega = \left[1 - \frac{2}{5}\frac{r^2}{l_e^2}\right] \exp\left[-r^2/l_e^2\right]. \qquad \square$$

Let us now consider an alternative model eddy. Consider the velocity distribution,

$$\mathbf{u} = 2V\left[\left(2rz/l_e^2\right)\hat{\mathbf{e}}_r + \left(1 - 2r^2/l_e^2\right)\hat{\mathbf{e}}_z\right] \exp\left[-2\mathbf{x}^2/l_e^2\right]$$

where, once again, we use cylindrical polar coordinated (r, θ, z). This represents a poloidal velocity field reminiscent of a vortex ring, as shown in Figure 6.28. As in the previous case we now consider a field of turbulence created by randomly distributing such vortices throughout space. It is evident from the example above, and from (6.24), that the longitudinal correlation function and energy spectrum for this field of turbulence are given by,

$$f = \left[1 - \frac{2}{5}\left(\frac{r}{l_e}\right)^2\right] \exp\left[-r^2/l_e^2\right]$$

$$E(k) = \frac{\langle \mathbf{u}^2 \rangle l_e}{240\sqrt{\pi}} (kl_e)^6 \exp\left[-k^2 l_e^2/4\right].$$

Notice that, as in the previous example, E peaks at $k \sim \pi/l_e$ (actually at $k = \sqrt{12}/l_e$). This time, however, $E(k)$ grows as k^6 for small k. This is an inevitable consequence of (6.77) and of the fact that we have chosen a model eddy with zero net angular impulse and zero linear impulse (Figure 6.29).

There are two marked differences between the examples above. When our basic building block is a swirling blob of fluid we find $E \sim k^4$ at small k and $f > 0$ for all r. On the other hand, the vortex ring leads to $E \sim k^6$ and $f < 0$ for large r. In practice grid turbulence is probably of the form $E \sim k^4$, and tends to have $f > 0$ for all r. Perhaps this indicates that a swirling blob of fluid is the more plausible 'model eddy' for grid turbulence, at least at the large scales. In any event, it is clear that f can, in principle, take negative values for certain r. The fact that it is virtually always positive in mature turbulence tells us something about the dynamical processes (i.e. characteristic eddy shape) associated with the Navier–Stokes equation. In short, the observation that $f > 0$ in mature turbulence is a dynamic rather than a kinematic result.

Let us now take matters a step further. Suppose that our field of turbulence is composed of a random array of swirling eddies of size l_1, plus a random distribution of eddies of size l_2, l_3, l_4, and so on. For each size of eddy we take our basic 'building block' to be Townsend's eddy,

$$\mathbf{u}_i = \Omega_i r \exp\left[-2\mathbf{x}^2/l_i^2\right]\hat{\mathbf{e}}_\theta$$

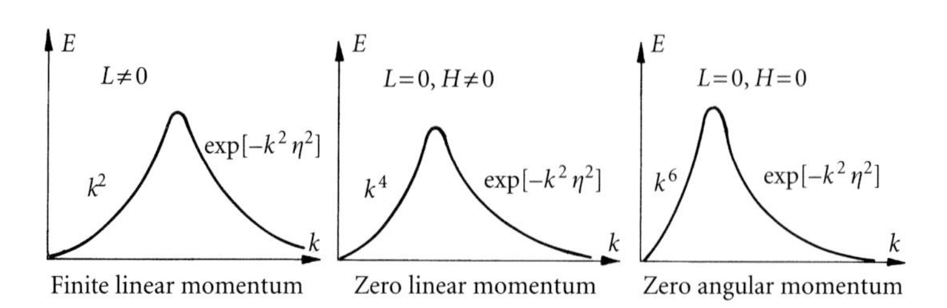

Figure 6.29 A field of turbulence composed of eddies with a significant linear impulse, $\mathbf{L} \neq \mathbf{0}$, has an energy spectrum of the form $E \sim k^2$ at small k. If the eddies have a negligible linear impulse but a significant amount of angular momentum then we find $E \sim k^4$. If both the linear and the angular impulses are negligible in a typical eddy then we find a $E \sim k^6$ spectrum.

Also, we choose $l_1, l_2, \ldots, l_N$ to be, say,

$$l_2 = 0.1l_1, \quad l_3 = 0.1l_2, \ \ldots, \quad l_N = 0.1l_{N-1}.$$

In such a case it turns out that the correlation function is composed of a sum of exponentially decaying functions of the type $\exp\left[-r^2/l_i^2\right]$ (see the appendix at the end of this chapter), and the corresponding energy spectrum takes the form,

$$E(k) = \sum_i \frac{\langle \mathbf{u}_i^2 \rangle l_i}{24\sqrt{\pi}} (kl_i)^4 \exp\left[-l_i^2 k^2/4\right].$$

(Here $\langle \mathbf{u}_i^2 \rangle$ is the contribution to $\langle \mathbf{u}^2 \rangle$ which comes from eddies of size l_i.) This is illustrated in Figure 6.30 and we can now begin to see how the energy spectrum corresponding to real turbulence, where we have a wide range of length scales, can be built up from the properties of individual eddies. (We pursue this idea in Exercise 6.12.)

We learn two more things from these kinematic models of turbulence. First, in all of the cases above E decays exponentially fast for large k. This is also true of real turbulence since it may be shown that, provided there are no singularities in the velocity field, E must be exponentially small for large k. (A finite viscosity rules out any singularities in $\mathbf{u}$.) Since the smallest eddies in a real turbulent flow have a size of the order of the Kolmogorov microscale, η, we would expect $E \sim \exp[-C(k\eta)^\alpha]$ for $k \to \infty$ where C and α are constants. In fact $\alpha = 1$, $\alpha = 4/3$, and $\alpha = 2$ (which is compatible with (6.98c)) have all been suggested at various times, though most researchers now believe that $\alpha = 1$ and that C is of the order of 5–7. (e.g. See, Saddoughi and Veeravalli 1994.) The likely explanation for the choice $\alpha = 1$ is simple.

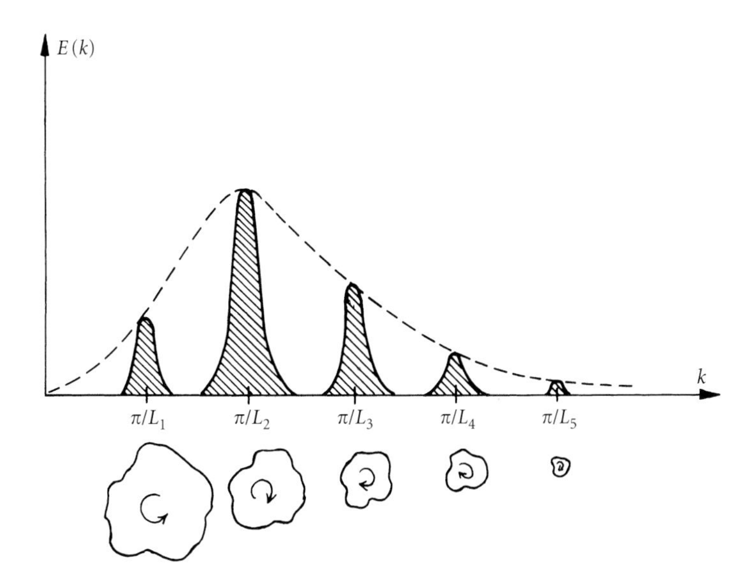

Figure 6.30 Shape of $E(k)$ for a hierarchy of well-separated eddy sizes, $l_1, l_2, l_3, \ldots, l_N$.

Eddies of size s produce a tail of the form $E \sim \exp(-k^2 s^2)$ and integrating such distributions over the range of eddy sizes typically yields $E \sim \exp(-k\eta)$. (The details are spelt out in the appendix at the end of this chapter.)

The second point to note is that (6.98c) shows that, for $k \to 0$, eddies of size l_i make a contribution to $E(k)$ of the order of $E(k) \sim \langle \mathbf{u}_i^2 \rangle l_i^5 k^4$. So when we have a hierarchy of eddy sizes we expect,

$$E(k) \sim \left\{ \sum_i \langle \mathbf{u}_i^2 \rangle l_i^5 \right\} k^4 + \cdots \quad \sim Ik^4 + \cdots$$

where I is Loitsyansky's integral. The key point is that the form of $E(k)$ at small k has nothing at all to do with eddies of size k^{-1} (Figure 6.31). Rather, it tells us something about the intensity of eddies much smaller than k^{-1}. In particular, the dominant contribution to $E(k)$ at small k comes from the energy containing eddies, and so we have, $E(k) \sim \{\langle \mathbf{u}^2 \rangle l^5\} k^4 + \cdots \sim Ik^4 + \cdots$ where l is the integral scale (size of the energy containing eddies). Evidently, the near permanence of I in freely decaying turbulence is *not* a manifestation of eddies of size k^{-1} preserving their energy. Rather, it tells us about the dynamics of the energy containing eddies. That is to say, the energy containing eddies appear to have a statistical invariant and, of course, this invariant is just a measure of their angular momentum (squared).

We now introduce a new function, $V(r)$, which we shall call the *signature function*. It is defined as,

$$V(r,t) = \langle \mathbf{u}^2 \rangle r^3 \left[\frac{\partial}{\partial r^2} \right]^2 f(r,t).$$

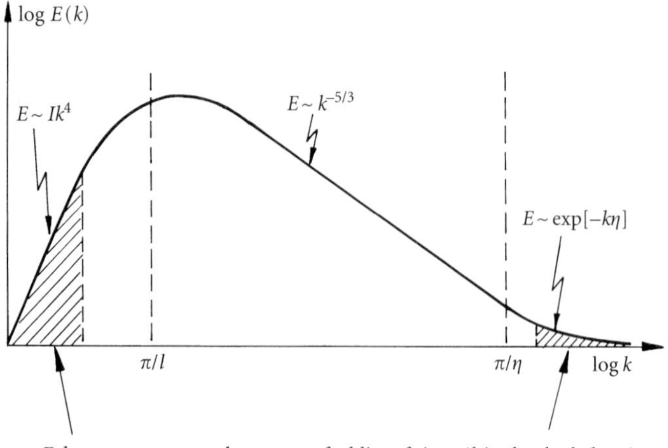

Figure 6.31 That part of the energy spectrum, which lies outside the range $\pi/l < k < \pi/\eta$ has little to do with the energy of eddies of size π/k.

For turbulence composed of one of the two model eddies discussed above we have:

$$rV(r) = \langle \mathbf{u}^2 \rangle \left(\frac{r}{l_e} \right)^4 \exp\left[-r^2/l_e^2 \right], \qquad \text{(swirling blobs of fluid)}$$

$$rV(r) = \langle \mathbf{u}^2 \rangle \left[\frac{9}{5} - \frac{2}{5} \left(\frac{r}{l_e} \right)^2 \right] \left(\frac{r}{l_e} \right)^4 \exp\left[-r^2/l_e^2 \right], \qquad \text{(vortex rings).}$$

Evidently V can take both positive and negative values, but for the case of randomly distributed swirling blobs it is strictly positive. In fact, we shall see in Section 6.6.1 that $V(r)$ is positive in fully developed turbulence provided that $f(r) > 0$, which seems to be the case in mature turbulence. So $V(r)$ is 'usually' positive for all r. It has two more properties, which are noteworthy:

(1) $V(r)$ exhibits a sharp peak around $r \sim l_e$;
(2) $\int_0^\infty V(r)\, dr = \frac{1}{2} \langle \mathbf{u}^2 \rangle$.

(This second property can be confirmed by integrating V by parts.) Now compare this with $E(k)$. It is positive, exhibits a sharp peak around $k \sim \pi / l_e$, and integrates to give $\frac{1}{2} \langle \mathbf{u}^2 \rangle$. It seems that $V(r)$ and $E(k)$ have some common characteristics, though one relates to Fourier space and the other to real space. We shall return to this issue in Section 6.6.1 where we shall see that $V(r)$, like $E(k)$, can be used to distinguish the various contributions to $\frac{1}{2} \langle \mathbf{u}^2 \rangle$, which arise from different eddy sizes.

6.4.2 *Other model eddies*

So far we have considered only very simple model eddies. Actually, there has been a long history of casting around for more sophisticated model structures which reproduce certain known features of turbulence. Perhaps the earliest such attempt was Synge and Lin (1943) who considered a random array of Hills' spherical vortices. Unlike Townsend's (1956) model eddy, a Hills' spherical vortex possesses a finite amount of linear impulse and so a Saffman ($E \sim k^2$) spectrum was inevitably found. Indeed, Synge and Lin obtain $f_\infty \sim r^{-3}$, which is the hallmark of a Saffman spectrum. However, Hills' spherical vortex is, like Townsend's (1956) model eddy, simply a compact blob of vorticity. As such, it is unlikely to be representative of the most intense turbulent vortices which are, perhaps, more likely to take the form of tubes or ribbons (Plate 8).

In 1951, Townsend explored the effect of taking different model eddy shapes. It turns out that, for large k, an ensemble of tube-like structures (Burgers vortices) gives $E \sim k^{-1} \exp[-\alpha k^2 \nu]$ where α is a model parameter, while sheet-like structures lead to $E \sim k^{-2} \exp[-\alpha k^2 \nu]$. Note that the Kolmogorov $E \sim k^{-5/3}$ lies between the tube-like and

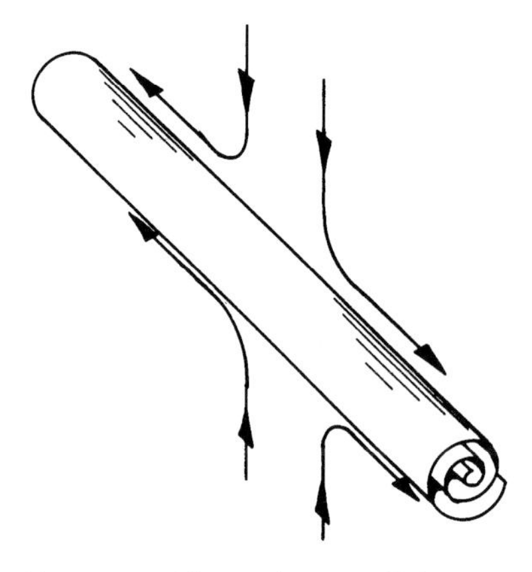

Figure 6.32 The Lundgren spiralled vortex.

sheet-like values of k^{-1} and k^{-2}. This suggests that we look for a model eddy which has both tube-like and sheet-like features. Just such a model eddy was proposed by Lundgren (1982). He considered an unsteady, Burgers-like vortex filament whose internal structure is not axisymmetric but spiralled, rather like a sheet of paper rolled up into a scroll (Figure 6.32). In such cases $E(k)$ is found to have the form

$$E \sim k^{-5/3} \exp[-\alpha k^2 \nu]$$

where, again, α is a model parameter. (The $k^{-5/3}$ spectrum arises from the spiralled internal structure of the vortex filaments.) It is remarkable that Lundgren was able to reproduce Kolmogorov's five-thirds law with this simple model eddy. Moreover, if we believe that the small scale vortex tubes in a turbulent flow have arisen from the roll-up of vortex sheets, then Lundgren's vortex tube is a natural candidate for the small-scale eddies. However, we cannot conclude from this that small-scale turbulence consists of a sea of spiralled vortex tubes. Remember that the $k^{-5/3}$ law can also be deduced on purely dimensional grounds, independent of the assumed eddy shape. That is to say, when the inertial sub-range represents a wide range of eddy sizes ($\mathrm{Re} \gg 1$), the $k^{-5/3}$ form of $E(k)$ probably tells us about the distribution of energy between the different scales, rather than about the generic shape of the eddies at a given scale. Still, Lundgren's spiralled vortex model is an intriguing result. Those interested in more details of the characteristic signature of different types of eddies could do worse than consult Pullin and Saffman (1998).

6.5 Intermittency in the inertial-range eddies

> The hypothesis concerning the local structure of turbulence at high Re, developed in the years 1939–41 by myself and Obukhov, where based physically on Richardson's idea of the existence in the turbulent flow of vortices on all possible scales $\eta < r < l$ between the external scale l and the internal scale η and of a certain uniform mechanism of energy transfer from the coarser-scaled vortices to the finer. These hypotheses were arrived at independently by a number of authors and have achieved very wide acceptance. But quite soon after they originated Landau noticed that they did not take into account a circumstance which arises directly from the assumption of the essentially accidental and random character of the mechanisms of transfer of energy from the coarser vortices to the finer. (Kolmogorov (1962))

We now turn to the phenomenon of small-scale intermittency. This is not peculiar to isotropic turbulence and indeed seems to be a universal feature of all high-Re motion. However, we discuss it here as the subject fits well with our 'real space' discussion of homogeneous turbulence. Intermittency is important because it calls into question

Kolmogorov's theory of the small scales which, for half a century, has provided the foundations of our theoretical understanding of homogeneous turbulence.

6.5.1 *A problem for Kolmogorov's theory?*

The tragedy of science—the slaying of a beautiful hypothesis by an ugly fact. (T.H. Huxley)

In Chapter 5 we saw that Kolmogorov's second similarity hypothesis led to the expression

$$\langle [\Delta \mathrm{v}]^p \rangle = \beta_p \varepsilon^{p/3} r^{p/3}, \quad \eta \ll r \ll l$$

where, according to the theory, the β_p's are universal constants, that is the same for jets, wakes, grid turbulence, and so on. However, from the very beginning Landau objected to Kolmogorov's claim that small-scale turbulence has a universal structure, pointing out that the viscous dissipation is highly intermittent and that the nature of this intermittency can depend on the large scales, and so changes from one flow to the next. (See Section 6.1.3.4.) These objections turned out to be prophetic.

Landau's remarks on intermittency are famously cryptic, and so most authors tend to present an interpretation of his comments. Of course, sometimes it is a little difficult to spot where Landau ends and the author begins. In any event, here is one interpretation of the problem. (See Monin and Yaglom 1975, or Frisch 1995, for a more complete account.) It is conventional to differentiate between intermittency at the large and small scales, which we might label integral-scale and inertial-range intermittency, respectively. The first of these relates to the clumpiness of the vorticity and dissipation when viewed on scales of order l. Think of the turbulence immediately behind a grid, or in a jet. There are active regions where the vorticity and dissipation are particularly strong, interspersed with weaker, almost irrotational, regions. In the case of grid turbulence the vorticity will be strong within the remnants of the Karman vortices, while in a jet inactive regions might correspond to external irrotational fluid, which has been engulfed by the jet. So in these flows there will exist large-scale inhomogeneities in the mean dissipation rate, and Landau noted that the precise nature of this large-scale intermittency will vary from one type of turbulence to another, being different in jets, wakes, boundary layers, and so on. The effect of this large-scale clumpiness is to moderate the local energy flux, Π, which feeds the local cascade. Since the local mean energy flux (averaged over scales of order l) has a spatial distribution which depends on the type of flow, the statistics of the small scales, which are a passive response to the local mean energy flux, might also depend on the type of flow in question. So Landau was attacking the idea that the statistics of the equilibrium range are necessarily universal (the same for jets, wakes, etc,). Thus, for example, the prefactor β_p in Kolmogorov's scaling for the structure function of order p might vary from one type of turbulence to another.

In summary, then, the key problem which Landau foresaw is that it is not the globally averaged dissipation, $\langle 2\nu S_{ij}S_{ij}\rangle$, which is important for the statistics of $[\Delta v](r)$, but rather a suitably defined local average of Π or $2\nu S_{ij}S_{ij}$. Moreover, there is not one single cascade within the turbulence, but rather many cascades at different points, all being fed at different rates by the local mean value of Π. When we average over all of these cascades to produce global statistics the non-universal form of the large-scale clumpiness will result in non-universality of the inertial-range statistics. Indeed, the simple example given in Section 6.1.3.4, in which the prefactors β_p are shown to be non-universal as a result of large-scale intermittency, is just one illustration of this.

The second problem for Kolmogorov's theory lies in the spotty structure of the small-scale vorticity field (Plate 8). This small-scale intermittency is somewhat different in character to that of the large scales. It is not associated with inhomogeneities caused by the initiation or maintenance of turbulence, that is, the large-scale clumpiness of the vorticity. Rather, it is an intrinsic feature of all forms of turbulence, and is a direct result of vortex stretching, which teases out the vorticity into finer and finer filaments. While large-scale intermittency is definitely not universal, there is at least the possibility that inertial-range intermittency may posses certain universal statistical features. Thus small-scale intermittency need not necessarily pose a problem for the universality of Kolmogorov's theory. It does, however, raise doubts as to the validity of the scaling $\langle [\Delta v]^p\rangle \sim r^{p/3}$. The point is that this scaling is a direct result of the assumption that there is only one parameter that controls the inertial-range dynamics, that is $\langle 2\nu S_{ij}S_{ij}\rangle$. Perhaps there is another relevant parameter: the degree to which the spottiness of $\nu\boldsymbol{\omega}^2$ or $2\nu S_{ij}S_{ij}$ varies with scale r within the inertial range. If this is indeed important then the scaling $\langle [\Delta v]^p\rangle \sim r^{p/3}$ becomes suspect. It is this issue of inertial-range scaling, as well as the lack of universality, which concerns us here.

The first confirmation that all is not well with Kolmogorov's theory came from a comparison of the measured exponents in $\langle [\Delta v]^p\rangle \sim r^{\varsigma_p}$ with the Kolmogorov prediction of $\varsigma_p = p/3$. For $p \leq 4$ Kolmogorov's prediction works well. However, as p increases ς_p progressively departs from $p/3$, with $\varsigma_p < p/3$. So, for example, when $p = 12$ we find measurements giving $\varsigma_p \sim 2.8$, rather than the expected value of 4. We shall see in a moment that, as suggested above, this discrepancy is due to the fact that the dissipation $2\nu S_{ij}S_{ij}$ is highly intermittent (spotty) in space.

The fact that the small scales are intermittent should not come as a surprise. Recall that we suggested a cartoon of turbulence consisting of a tangle of vortex tubes and ribbons, mutually advecting each other through their induced velocity fields. As the ribbons and tubes are teased out into finer and finer structures, so the vorticity field will become increasingly spotty, alternating between small and large values. We should expect, therefore, that $\boldsymbol{\omega}$ will be highly intermittent at the small scales, and as a consequence the dissipation $2\nu S_{ij}S_{ij}$ will be concentrated into regions surrounding thin tubes and surfaces.

There are two classes of measurements which suggest that this picture is correct. First, we note that the flatness factor of the probability distribution of $\partial u_x/\partial x$ is higher than that associated with a Gaussian distribution. It can range from $\delta_0 \sim 4$ for grid turbulence at modest Re, up to ~ 40 in the atmospheric boundary layer ($\text{Re} \sim 10^7$). In fact, in Chapter 5 we saw that the variation of δ_0 with Re is something like $\delta_0 \approx 3 + \frac{1}{2}(ul/\nu)^{1/4}$. Now high flatness factors are associated with a signal which spends a large amount of time near zero, occasionally bursting into life, and then returning to zero again (Figure 6.33). So the measured values of δ_0 are consistent with a spotty small-scale structure. The fact that δ_0 rises as Re increases is also consistent with our cartoon, since high values of Re imply that large amounts of stretching occur before the vortex tubes reach the dissipation scale, η.

The second indication that $\partial u_x/\partial x$ is spatially intermittent comes from (5.38), which suggests that, for the modest values of Re encountered in a wind tunnel,

$$\langle(\partial^n u_x/\partial x^n)^4\rangle / \langle(\partial^n u_x/\partial x^n)^2\rangle^2 \sim (n+3) \pm 20\%.$$

The higher the derivative the less Gaussian the probability distribution.

We shall now review two popular models of inertial-range intermittency. These models have their weaknesses, but they do seem to capture the essential features of the process.

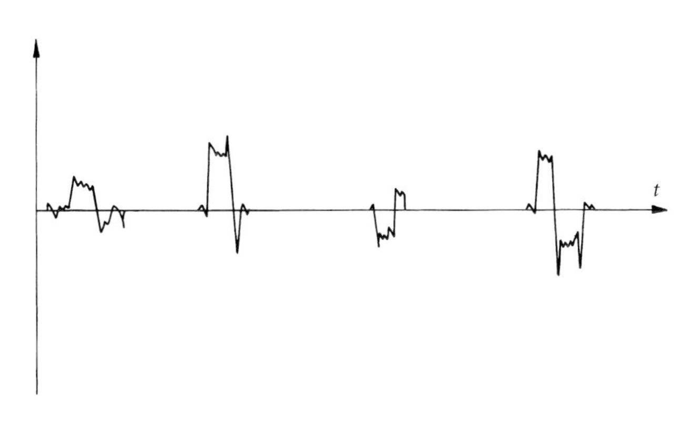

Figure 6.33 An example of an intermittent signal.

6.5.2 The $\hat{\beta}$-model of intermittency

There have been many attempts to quantify the influences of inertial-range intermittency, starting with Obukhov and Kolmogorov's work in the early 1960s. (We shall describe Kolmogorov's model in Section 6.5.3.) One popular cartoon of intermittency, called the $\hat{\beta}$ model,[16] captures the flavour of the problem (if not the details). Its starting point is Obukhov's explanation of the two-thirds law, which is reviewed in Section 5.2.1. You will recall that the argument goes something like this. Let v_r be the typical velocity associated with eddies of scale r. If these eddies break-up on a timescale of r/v_r, then the rate at which energy passes down the cascade is $\Pi(r) \sim v_r^3/r$. Moreover, if the turbulence is in statistical equilibrium, which it is in the inertial sub-range, then $\Pi(r) = \varepsilon$ where, as before, ε is the spatially averaged dissipation. It follows that $v_r^2 \sim (\varepsilon r)^{2/3}$, which yields the two-thirds law if we make the estimate $v_r^2 \sim \langle [\Delta v]^2 \rangle (r)$.

Let us now refine the argument, allowing for the fact that the eddies (vortices) probably become less space-filling as r becomes smaller (Figure 6.34). Suppose that we replace the continuous variable, r, by the discrete variable l_n, where

$$l = l_0 = 2l_1 = 2^2 l_2 = \cdots = 2^n l_n = \cdots .$$

The l_n's represent a hierarchy of eddy sizes from the integral scale, l, down to η. Now suppose that, at the scale l_n, each eddy divides into N smaller eddies of size l_{n+1}. Then the fractional reduction in volume from one generation to the next is

$$\hat{\beta} = \frac{N l_{n+1}^3}{l_n^3} = \frac{N}{2^3} = \frac{2^D}{2^3} \leq 1$$

where D is a so-called fractal dimension which characterizes the process, $D < 3$. If the largest eddies are space filling then the nth-generation vortices occupy a fraction $(\hat{\beta})^n$ of all space. This leads to a refined estimate of the spatially averaged energy flux $\Pi(l_n)$:

$$\Pi(l_n) = \Pi_n \sim (\hat{\beta})^n v_n^3 / l_n.$$

(Here v_n is a typical velocity in an eddy of scale l_n.) In statistically steady turbulence, Π_n is equal to the spatially averaged dissipation, ε, and so we may combine the estimates above to give

$$\langle (\text{K.E.})_n \rangle \sim (\hat{\beta})^n v_n^2 \sim (\hat{\beta}^n v_n^3)^{2/3} \hat{\beta}^{n/3} \sim (\varepsilon l_n)^{2/3} (2^{D-3})^{n/3}$$
$$\sim \varepsilon^{2/3} l_n^{2/3} (l_n / l_0)^{(3-D)/3}.$$

[16] The $\hat{\beta}$ that appears in this theory should not be confused with Kolmogorov's constant.

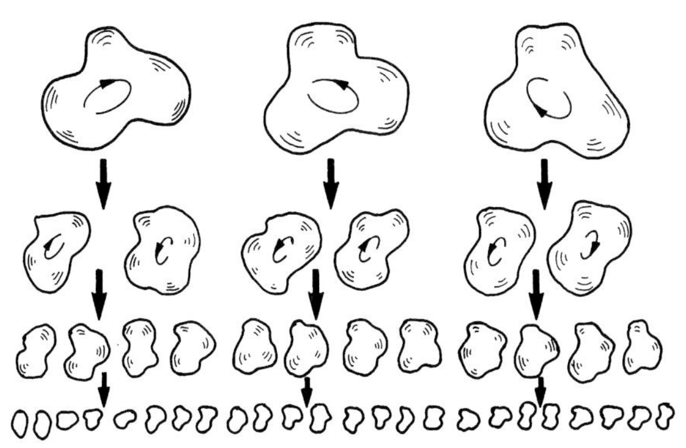
Cascade without spatial intermittency

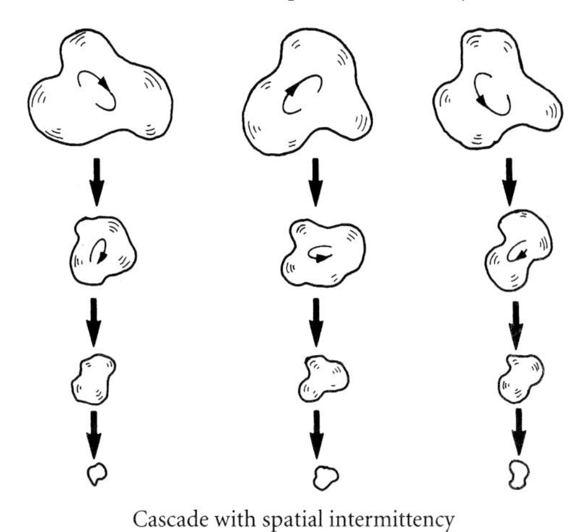
Cascade with spatial intermittency

Figure 6.34 A cascade without intermittency is space filling at every scale. In practice, vortex stretching ensures that the vorticity becomes less space filling as we move to small scales. (After Frisch 1995.)

This suggests that the second-order structure has the form

$$\langle [\Delta v]^2 \rangle \sim \varepsilon^{2/3} r^{2/3} (r/l)^{(3-D)/3}.$$

Similarly, the $\hat{\beta}$ model yields, for any positive integer p,

$$\langle [\Delta v]^p \rangle \sim \varepsilon^{p/3} r^{p/3} (r/l)^{(3-p)(3-D)/3} \tag{6.99a}$$

and from which we find the scaling exponent

$$\varsigma_p = (3 - D) + p(D - 2)/3.$$

A few comments are in order. First, this is much more of a cartoon than a predictive theory. Nevertheless, taking $D = 2.8$ gives a good fit to the experimental measurements of ς_p, up to $p = 8$. Second, the theory breaks down for $p = 0$ unless $D = 3$. Third, for $D = 3$ the eddies at all scales are space-filling and we recover Kolmogorov's 1941 theory. Fourth, for $p = 3$ the $\hat{\beta}$—correction goes to zero, which is reassuring since the four-fifths law is (probably) exact. The $\hat{\beta}$ model is compared with Kolmogorov (1941b) in Figure 6.35.

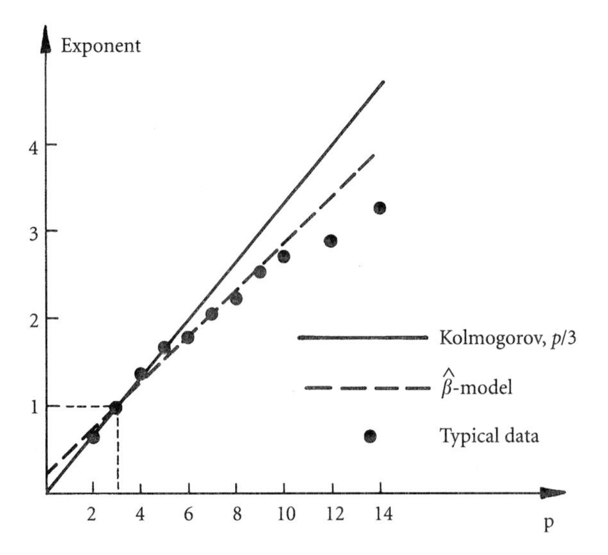

Figure 6.35 Comparison of Kolmogorov (1941b) and the $\hat{\beta}$ model.

6.5.3 *The log-normal model of intermittency*

An alternative model, called the log-normal model, was introduced by Kolmogorov and Obukhov in the early 1960s (Kolmogorov 1962). Actually, this was the *first* of many intermittency models and, somewhat surprisingly, it does a reasonable job (as good as the $\hat{\beta}$ model) of predicting the variation of ς_p with p. According to this model,

$$\varsigma_p = \frac{p}{3} + \frac{\mu}{18}(3p - p^2)$$

where μ is a free parameter, usually taken to lie in the range 0.2–0.3.

The idea behind the log-normal model is the following. In a given region of size r (where $r \ll l$) the flux of energy to the small scales, $\Pi(r)$, is equal to the dissipation averaged over that region, $\varepsilon_{AV}(r)$. Since the magnitude of $\Pi(r)$ is controlled by the dynamics of eddies of size r, we would expect the statistics of $[\Delta v](r)$, which also reflect the dynamics of eddies of size r, to depend on $\varepsilon_{AV}(r)$, the locally averaged dissipation, rather than ε, the global average. In order to correct Kolmogorov's original theory we must seek the relationship between $\varepsilon_{AV}(r)$ and r. Once this has been achieved we can refine Kolmogorov's similarity hypotheses, replacing ε by $\varepsilon_{AV}(r)$. So let us start by trying to find a model which relates to $\varepsilon_{AV}(r)/\varepsilon$ to r/l.

In isotropic turbulence the average rate of dissipation of energy, $\varepsilon = \langle 2\nu S_{ij} S_{ij} \rangle$, can be written as,

$$\varepsilon = 15\nu \left\langle \left(\frac{\partial u_x}{\partial x} \right)^2 \right\rangle.$$

Let us, therefore, consider the statistics of the dissipation-like quantity

$$\hat{\varepsilon} = 15\nu \left(\frac{\partial u_x}{\partial x} \right)^2$$

for which there exists a great deal of experimental data. In particular, we are interested in the behaviour of $\hat{\varepsilon}$ when averaged over a spherical volume of radius r:

$$\hat{\varepsilon}_{AV} = \frac{1}{V_r}\int_{V_r} \hat{\varepsilon}\, dV.$$

For large r (i.e. $r \gg l$) $\hat{\varepsilon}_{AV} = \varepsilon$, since a large volume average is equivalent to an ensemble average. For small r (i.e. $r \leq \eta$) the volume averaging has no effect and $\hat{\varepsilon}_{AV}$ is effectively the same as the point value of $\hat{\varepsilon}$. Now we expect that $\hat{\varepsilon}_{AV}$ will become increasingly intermittent as $r \rightarrow \eta$, and indeed we can use the results of Section 5.5.1 to quantify this transition to intermittency. In particular, since $\langle \hat{\varepsilon}_{AV} \rangle = \varepsilon$, and $\langle \hat{\varepsilon}^2 \rangle$ is proportional to $\langle (\partial u_x / \partial x)^4 \rangle$, we have,

$$\langle \hat{\varepsilon}_{AV}^2 \rangle / \varepsilon^2 \approx 1; \quad \text{for } r \geq l$$
$$\langle \hat{\varepsilon}_{AV}^2 \rangle / \varepsilon^2 \approx \delta_0; \quad \text{for } r \leq \eta$$

where δ_0 is the flatness factor for $\partial u_x / \partial x$ introduced in Section 5.5.1. Moreover, we have seen that δ_0 increases with Re, possibly as

$$\delta_0 \approx 3 + \tfrac{1}{2}(ul/\nu)^{0.25} \approx 3 + \tfrac{1}{2}(l/\eta)^{1/3}.$$

For large Re, therefore, we might expect,

$$\langle \hat{\varepsilon}_{AV}^2 \rangle / \varepsilon^2 \approx 1; \quad \text{for } r \geq l$$

$$\langle \hat{\varepsilon}_{AV}^2 \rangle / \varepsilon^2 \approx (l/\eta)^{1/3}; \quad \text{for } r \leq \eta.$$

If we interpolate between these two limits we have

$$\langle \hat{\varepsilon}_{AV}^2 \rangle / \varepsilon^2 \approx (l/r)^{1/3}; \quad \text{for } \eta \leq r \leq l$$

a result which has some empirical support. The question now arises: would we expect the true dissipation, $2\nu S_{ij}S_{ij}$, to behave in the same way as its one-dimensional surrogate $\hat{\varepsilon}$? To answer this we must first note that the key step in the analysis above is the identification of the flatness factor, δ_0, with $\langle \hat{\varepsilon}^2 \rangle / \varepsilon^2$. Does a similar relationship hold between δ_0 and $2\nu S_{ij}S_{ij}$? At this point Betchov's analysis comes to our rescue (see Section 5.3.6.4). Provided the turbulence is isotropic we have

$$\delta_0 = \frac{15}{7} \frac{\langle (2\nu S_{ij}S_{ij})^2 \rangle}{\langle 2\nu S_{ij}S_{ij} \rangle^2}$$

and so all of the steps above may be repeated for $2\nu S_{ij}S_{ij}$. Thus we have the empirical statement that

$$\langle \varepsilon_{AV}^2(r) \rangle / \varepsilon^2 = B(l/r)^{\mu}; \quad \text{for } \eta \leq r \leq l$$

where

$$\varepsilon_{AV}(r) = \frac{1}{V_r}\int_{V_r} 2\nu S_{ij}S_{ij}\,dV.$$

Here μ is a model parameter, called the *intermittency exponent*, which we might anticipate has a value of around 0.3. Kolmogorov adopted this approximation and assumed that μ is universal (the same for all types of turbulence) where-as he took the constant B to be non-universal, probably because he wanted to build into his model a lack of universality in order to satisfy Landau's criticisms.

The next step is to modify, or refine, Kolmogorov's second similarity hypothesis. In particular, we suppose that the statistics of the inertial-range structure function $[\Delta v](r)$ are controlled by $\varepsilon_{AV}(r)$ and r, rather than by ε and r. Specifically, Kolmogorov proposed that, for $\eta \ll r \ll l$, the probability density function for $[\Delta v](r)/[r\varepsilon_{AV}(r)]^{1/3}$ has a universal form, independent of Re and of the type of flow in question. This refined view suggests that we replace

$$\langle[\Delta v]^p\rangle = \beta_p \varepsilon^{p/3} r^{p/3}; \quad \eta \ll r \ll l$$

by

$$\langle[\Delta v]^p\rangle = \beta_p \langle\varepsilon_{AV}^{p/3}(r)\rangle r^{p/3}; \quad \eta \ll r \ll l$$

where, as before, the β_p's are universal coefficients. This is the basis of Kolmogorov's *refined similarity hypothesis* (Kolmogorov 1962). If we accept this hypothesis then the problem of predicting $\langle[\Delta v]^p\rangle$ reduces to one of estimating $\langle\varepsilon_{AV}^{p/3}(r)\rangle$. In particular, we need to estimate $\langle\varepsilon_{AV}^{p/3}(r)\rangle$ from the empirical expression,

$$\langle\varepsilon_{AV}^2\rangle/\varepsilon^2 = B(l/r)^\mu; \quad \eta \le r \le l.$$

Two special cases follow immediately:

$$p = 3\text{: } \langle\varepsilon_{AV}\rangle = \varepsilon$$
$$p = 6\text{: } \langle\varepsilon_{AV}^2\rangle = B\varepsilon^2(l/r)^\mu$$

from which

$$p = 3\text{: } \langle[\Delta v]^3\rangle = \beta_3\varepsilon r$$
$$p = 6\text{: } \langle[\Delta v]^6\rangle = B\beta_6\varepsilon^2 r^2(l/r)^\mu.$$

The first of these is, of course, the four-fifth law, which we think is exact and universal. Hence the absence of any correction term. The second is not universal because of the presence of B, whose value (according to Kolmogorov) depends on the type of flow in question.

For other values of p we need to introduce an additional hypothesis in order to predict $\langle\varepsilon_{AV}^{p/3}\rangle$. To this end Kolmogorov followed the example of Obukhov and assumed that the probability density function for ε_{AV} is a log-normal law, from which it is possible to show that

$$\langle \varepsilon_{AV}^m \rangle / \varepsilon^m = \left[\langle \varepsilon_{AV}^2 \rangle / \varepsilon^2 \right]^{m(m-1)/2}.$$

Combining this with the empirical observation that $\langle \varepsilon_{AV}^2 \rangle = B\varepsilon^2 (l/r)^{\mu}$ we have,

$$\langle \varepsilon_{AV}^m \rangle = B^{m(m-1)/2} \varepsilon^m (l/r)^{\mu m(m-1)/2}.$$

Thus Kolmogorov's refined similarity hypothesis, combined with the log-normal assumption, leads to,

$$\langle [\Delta v]^p \rangle = C_p (\varepsilon r)^{p/3} (l/r)^{\mu p(p-3)/18}$$

where $C_p = \beta_p B^{p(p-3)/18}$. This may be rewritten as

$$\langle [\Delta v]^p \rangle \sim r^{\varsigma_p}; \quad \varsigma_p = \frac{p}{3} + \frac{\mu}{18}(3p - p^2). \tag{6.99b}$$

It turns out that the choice of $\mu = 0.2$ gives a good fit to the experimental data, up to $p = 12$.

Perhaps some comments are in order at this point. First we note that, for $p \leq 4$, the correction to Kolmogorov's original theory is negligible, in the sense that it probably lies within the uncertainty associated with experimental error. (See Examples 6.22 and 6.23 at the end of Section 6.6.2.) Second, we observe that, even in his refined theory, Kolmogorov tried to hang on to a form of universality in as much as μ and the β_p's, though not the C_p's, are taken to be universal constants. Third, the power-law exponent in the log-normal model (with $\mu = 0.2$) and the $\hat{\beta}$ model (with $D = 2.8$) coincide for $p = 6$. Fourth, the intermittency correction goes to zero for $p = 3$, as it should. Fifth, unlike the $\hat{\beta}$ model, the correct limit is obtained for $p = 0$.

So, at first sight, the log-normal model seems to fair better than the $\hat{\beta}$ model. However, there is a certain amount of arbitrariness about it. For example:

- Why should μ be universal yet B be specific to the type of flow in question?
- Why should ε_{AV} be distributed according to a log-normal law?

Actually, the experimental evidence does indeed tend to support the idea that μ is universal, and so Kolmogorov's guess about the relative roles played by μ and B is possibly correct. Still, there are a number of serious deficiencies in Kolmogorov's model and these are discussed at length by Frisch (1995). For example, (6.99b) predicts that ς_p is a decreasing function of p for $p > 3/2 + 3/\mu$, which is known to be physically unacceptable. Despite it shortcomings, though, the log-normal model is of historical importance as it set the tone of the debate for the decades that followed. Readers will find an account of rival theories, of which there have been many, in Frisch (1995).

The whole problem of characterizing inertial range intermittency remains one of the most active areas of research in turbulence, although the various theories, of which the $\hat{\beta}$ and log-normal models are just two, are all somewhat speculative. Infact, not only are the various intermittency models controversial, there are those who question the very existence of inertial range intermittency at high Re (Lundgren 2003). An up-to-date and detailed review is given by Frisch (1995).

6.6 The distribution of energy and vorticity across the different eddy sizes

A great many people think they are thinking when they are merely rearranging their prejudices. (William James)

We now introduce a function, $V(r)$, which represents, approximately, the distribution of energy across the different eddy sizes in fully developed, isotropic turbulence. Unlike $E(k)$, which is normally used for this purpose, it is a function of the real-space variable r, which we may associate with eddy size. So $V(r)$, which we shall call the *signature function*, is easier to interpret than $E(k)$. A related function, $\Omega(r) \sim r^{-2}V(r)$, can be used to represent the approximate distribution of enstrophy across the different eddy sizes.

Our interpretation of $V(r)$ and $\Omega(r)$ is not restricted to kinematics. We shall derive a dynamical equation for $V(r)$ which is equivalent to the dynamical equation for $E(k)$ (see equation 6.2). This equation conforms to our cascade picture of turbulence in which energy is continually transferred from the large eddies (the large-r contribution to $V(r)$) to the small eddies, where it is destroyed by viscous forces. Despite this reassuring ring of familiarity a word of warning is, perhaps, appropriate here. It is important to bear in mind that the introduction of $V(r, t)$ and its governing equation introduces no new dynamics. We are merely repackaging the information contained in the Karman–Howarth equation. The same is true of the introduction of $E(k, t)$ and its governing equation. The most that we can hope to achieve from such an exercise is to establish a convenient framework from which to unpack the dynamical information contained in the Karman–Howarth equation. In short, we are searching for the appropriate lens through which to view the dynamics of turbulence. While such a procedure can, in principle, prove useful, there is always a danger that we give the illusion of progress through a process which amounts to no more than a re-cataloguing of terms. It is probably healthy, therefore, to keep William James' warning in mind when reading this section!

6.6.1 *A 'real-space' function which represents, approximately, the distribution of energy*

6.6.1.1 A wish list for the real-space equivalent of the energy spectrum

Many monographs on homogeneous turbulence dwell on the role of the energy spectrum $E(k)$. The reason is that it provides a useful picture of the instantaneous state of a field of turbulence. Let us recall some of the properties of $E(k)$:

(1) $E(k) \geq 0$;
(2) $\int_0^\infty E(k)\,dk = \frac{1}{2}\langle \mathbf{u}^2 \rangle$;
(3) for a random array of simple[17] eddies of fixed size l_e we find $E(k) \sim \langle \mathbf{u}^2 \rangle l_e x^4 \exp(-x^2/4)$ where $x = kl_e$. This exhibits a peak at $k \sim \pi/l_e$. (See Section 6.4.1.)

The first two properties tell us that we may regard the contribution to $\frac{1}{2}\langle \mathbf{u}^2 \rangle$ from wave number components in the range as $k \to k + dk$ as $E(k)\,dk$. Of course, we are then left with the problem of how to interpret k in terms of the compact structures (eddies) which populate a turbulent flow. Property (3) now comes to our aid. Simple eddies of size l_e produce a spectrum which peaks around $k \sim \pi/l_e$, so when we have a range of eddy sizes we may loosely associate k^{-1} with the eddy size.

Property (3) has another use. It tells us that the kinetic energy of a random array of eddies of size is l_e of the order of $l_e^{-1}E\left(l_e^{-1}\right)$. Thus we might expect that, in fully developed turbulence, the kinetic energy of eddies of size π/k_e is of the order of $k_e E(k_e)$, and that these eddies have a characteristic velocity of $[k_e E(k_e)]^{1/2}$. Indeed, such estimates are commonly made (Tennekes and Lumley 1972).[18] For example, in the inertial subrange we have $E \sim \varepsilon^{2/3} k^{-5/3}$, and so the kinetic energy of eddies of size r is, $\mathrm{v}_r^2 \sim \varepsilon^{2/3} k^{-2/3} \sim \varepsilon^{2/3} r^{2/3}$. This is consistent with Kolmogorov's two-thirds law,

$$\left\langle (\Delta \mathrm{v})^2 \right\rangle = \beta \varepsilon^{2/3} r^{2/3}.$$

[17] By a 'simple eddy' we mean a blob of vorticity characterized by a single length-scale. Of course, the exact form of $E(k)$ depends on the shape of the 'model eddy' used to construct the velocity field. However, whatever model is used $E(k)$ peaks at around $k \sim l_e^{-1}$. The details are spelt out in Section 6.4.1.

[18] In Tennekes and Lumley they approximate the energy spectrum of an eddy of size π/k_e by a smooth, localized function of width k_e centred around $k = k_e$. This has a kinetic energy of $\int_0^\infty E(k)\,dk \sim k_e E(k_e)$. The exchange of energy between eddies of different size is then interpreted in wave number space as wave number-octaves exchanging energy with adjacent octaves.

It seems that the utility $E(k)$ of lies in its ability to give an immediate impression of how energy is distributed across the different eddy sizes at each instant. Moreover, we may massage the Karman–Howarth equation into an expression of the form,

$$\partial E/\partial t = (\text{inertial effects}) + (\text{viscous effects})$$

and so we can set about trying to understand how energy is redistributed across the range of eddy sizes by inertial effects, and how it is dissipated by viscous forces. The details of how this analysis unfolds is set out in Chapter 8. However, it is natural to wonder why we invoke the Fourier transform in order to differentiate between energy held at different scales. After all, the Fourier transform is designed to decompose a signal into a hierarchy of *waves*, yet turbulence is composed of spatially compact structures (eddies) and not waves.[19] Moreover, we shall see that the dynamical equations expressed in Fourier space are rather unfamiliar, at least to the newcomer. It is natural to enquire, therefore, as to whether or not we can find a 'real-space' function, $V(r)$, which does more or less the same job as $E(k)$, r being the distance between two points in our turbulent flow. If such a function does exist then we may investigate all the usual questions about the transfer of energy between scales without the need to move into Fourier space.

Let us list some of the properties, which we would wish our function $V(r)$ to have. By analogy with the properties of $E(k)$ we require:

(1) $V(r) \geq 0$;
(2) $\int_0^\infty V(r)\,dr = \frac{1}{2}\langle \mathbf{u}^2 \rangle$;
(3) for a random array of simple eddies of fixed size l_e, $V(r)$ has a sharp peak around $r \sim l_e$ and is small for $r \ll l_e$ and $r \gg l_e$;
(4) for $\eta < r < l$ we have $[rV(r)] \approx [kE(k)]_{k \approx \pi/r} \approx$ [energy in eddies of size r].

Unfortunately, to date, no such function has been found. The best that we can do is identify a function, which we shall call the signature function, which satisfies slightly weaker conditions. In particular (1) is

[19] The fact that the Fourier transform decomposes a velocity field into a hierarchy of waves, rather than a hierarchy of eddies, has been a perennial concern in turbulence theory for many decades. It has led to more sophisticated transform techniques, such as the wavelet. However, until we can agree what is meant by an 'eddy' we are unlikely to resolve this problem. That is, we cannot design a filter or template to extract information about eddies from a turbulent signal until we agree what an eddy looks like! The purpose of this section is more humble. We aim merely to replace by $E(k)$ an equivalent real-space function, $V(r)$. Both $E(k)$ and $V(r)$ are rather blunt instruments. For example, they deal only with spatial averages and so velocity fields, which are quite different in detail can give rise to the same $E(k)$ or $V(r)$.

replaced by

(1b) $\int_0^r V(r)dr \geq 0$,

while (2)–(4) remain unchanged. In short, we have had to relax the requirement that $V(r)$ is strictly non-negative in all possible circumstances and replace it by the weaker condition (1b). In practice, however, we shall see that $V(r)$ is almost certainly positive in *fully developed*, freely evolving, isotropic turbulence.[20]

There are several potential candidates for $V(r)$. Perhaps the most obvious one was proposed by Townsend (1956), inspired by the estimate,

$$\tfrac{3}{4}\langle[\Delta v]^2\rangle(r) \sim [\text{energy contained in eddies of size } r \text{ or less}]$$

This suggests

$$V_T(r) = \frac{d}{dr}\left[\frac{3}{4}\langle[\Delta v]^2\rangle\right]$$

where the subscript indicates that this is Townsend's definition of V and the factor of $\frac{3}{4}$ is included to ensure that V_T integrates to give $\frac{1}{2}\langle \mathbf{u}^2\rangle$. Clearly conditions (1b) and (2) are satisfied by V_T. In fact it is likely that the stronger condition of $V_T > 0$ also holds. To see this we rewrite V_T as, $V_T = -\frac{3}{2}u^2 f'(r)$. Now in fully developed turbulence f is usually seen to decay monotonically, so that and $f(r) \geq 0$ and $f'(r) \leq 0$ (Figure 6.36). Thus $V(r)$ defined in this way is 'usually' positive in fully developed turbulence.

Moreover, condition (3) is (more or less) satisfied since, as we saw in Section 6.4.1, a random array of simple eddies of fixed size l_e yields,

$$\frac{d}{dr}\left[\frac{3}{4}\langle[\Delta v]^2\rangle\right] = \langle \mathbf{u}^2\rangle l_e^{-1} x e^{-x^2}, \quad x = r/l_e$$

which has a maximum at $r \sim l_e$, as required. However, there is one objection to adopting Townsend's definition of $V(r)$, despite its compliance with three of our central requirements. As explained in Chapter 3, we would expect $\langle[\Delta v]^2\rangle(r)$ to include a contribution from eddies of size greater than r. That is, our physical intuition suggests,

$$\tfrac{3}{4}\langle[\Delta v]^2\rangle \sim [\text{energy of eddies of size } r \text{ or less}] + r^2[\text{enstrophy of eddies of size } r \text{ or greater}]$$

This is given weight in the form of (6.28), which states that

$$\frac{3}{4}\langle[\Delta v]^2\rangle(r) \approx \int_{\pi/r}^{\infty} E(k)\,dk + (r/\pi)^2 \int_0^{\pi/r} k^2 E(k)\,dk.$$

[20] In particular, we shall see that $V(r) \geq 0$ whenever $f(r)$ decays monotonically from $f(0) = 1$, which is what most measurements in fully developed, isotropic turbulence suggest.

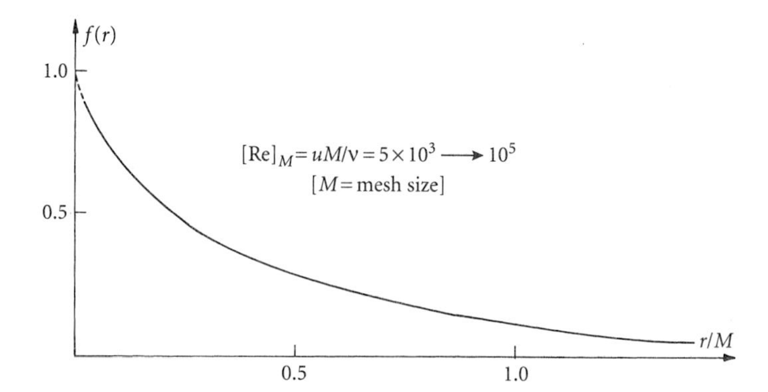

Figure 6.36 Shape of the longitudinal correlation function $f(r)$ based on the data given in Section 3.5 of Townsend (1956).

The second integral on the right may be neglected in the range of the energy containing eddies, but is non-negligible in the equilibrium range. Evidently the suggestion

$$V_T(r) = \frac{d}{dr}\left[\frac{3}{4}\langle[\Delta v]^2\rangle\right]$$

represents an imperfect candidate for $V(r)$. The question at hand, therefor, is: can we do better than $V_T(r)$? In short, we want to find a way to strip out of $V_T(r)$ the information contained in the second integral on the right.

6.6.1.2 One candidate for the real-space equivalent of the energy spectrum

All of this suggests that we might look for an alternative candidate for the signature function, $V(r)$. In this text we adopt the definition

$$V(r) = \langle \mathbf{u}^2\rangle \frac{r^2}{4}\frac{\partial}{\partial r}\frac{1}{r}\frac{\partial f}{\partial r} = -\frac{3}{8}r^2\frac{\partial}{\partial r}\frac{1}{r}\frac{\partial}{\partial r}\left[\langle[\Delta v]^2\rangle\right] \tag{6.100a}$$

or equivalently

$$\frac{3}{4}\langle[\Delta v]^2\rangle = \int_0^r V(s)\,ds + r^2\int_r^\infty \frac{V(s)}{s^2}\,ds \tag{6.100b}$$

$$\frac{1}{2}\langle \mathbf{u}^2\rangle f(r) = \int_r^\infty \left[1-(r/s)^2\right]V(s)\,ds \tag{6.100c}$$

(The equivalence of 6.100(a), 6.100(b), and 6.100(c) may be established by differentiating the integral equations above.) We shall explain the rationale behind this definition shortly. First, however, we need to establish two useful properties of V as defined above. Recall that, for small r, the structure function $\langle[\Delta v]^2\rangle$ takes the form

$$\langle[\Delta v]^2\rangle = \langle(\partial u_x/\partial x)^2\rangle r^2 + 0(r^4) = (r^2/15)\langle \boldsymbol{\omega}^2\rangle + 0(r^4)$$

Two important properties of $V(r)$ follow directly from this expression and from 6.100(b, c). Setting r to 0 in the first integral equation, and noting $V(r) \sim r^3$ for small r, we obtain

$$\frac{3}{4}\langle[\Delta v]^2\rangle = r^2 \int_0^\infty \frac{V(s)}{s^2} ds + \cdots = \frac{\langle \boldsymbol{\omega}^2 \rangle}{20} r^2 + \cdots .$$

It follows that

$$\frac{1}{2}\langle \boldsymbol{\omega}^2 \rangle = \int_0^\infty \frac{10V(s)}{s^2} ds. \tag{6.101a}$$

Similarly, setting $r = 0$ in the second integral equation yields,

$$\frac{1}{2}\langle \mathbf{u}^2 \rangle = \int_0^\infty V(s)\, ds. \tag{6.101b}$$

The integral property (6.101b) looks promising, but definition (6.100a) still seems an unlikely candidate for an energy density. Let us try to understand where it comes from. We start by re-examining our physical interpretation of $\langle[\Delta v]^2\rangle$. Recall that $\Delta v = u_x(\mathbf{x} + r\hat{\mathbf{e}}_x) - u_x(\mathbf{x}) = (u_x)_B - (u_x)_A$, where A and B are two points separated by a distance r (Figure 6.37). Now any eddy in the vicinity of A or B whose size s is much smaller than r will contribute to either $(u_x)_A$ or $(u_x)_B$, but not to both. Hence eddies for which $s \ll r$ will make a contribution to $\langle[\Delta v]^2\rangle$ of the order of their kinetic energy. On the other hand, eddies whose size is much greater than r will make a contribution to $\langle[\Delta v]^2\rangle$ of the order of $r^2(\partial u_x/\partial x)^2$. Moreover, in the limit $r \to 0$ we have, $\langle[\Delta v]^2\rangle = \langle(\partial u_x/\partial x)^2\rangle r^2 = \frac{1}{15}\langle\boldsymbol{\omega}^2\rangle r^2$, while $\langle[\Delta v]^2\rangle = 2\langle u_x^2\rangle = \frac{2}{3}\langle\mathbf{u}^2\rangle$ for $r \to \infty$. So an estimate of $\langle[\Delta v]^2\rangle$, which is compatible with our physical picture, and with the limiting cases of small and large r, is

$$\begin{aligned}\tfrac{3}{4}\langle[\Delta v]^2\rangle \approx{} & [\text{contribution to } \tfrac{1}{2}\langle\mathbf{u}^2\rangle \text{ from eddies of size } s < r] \\ & + (r^2/10)[\text{contribution to } \tfrac{1}{2}\langle\boldsymbol{\omega}^2\rangle \text{ from eddies of size } s > r].\end{aligned}$$

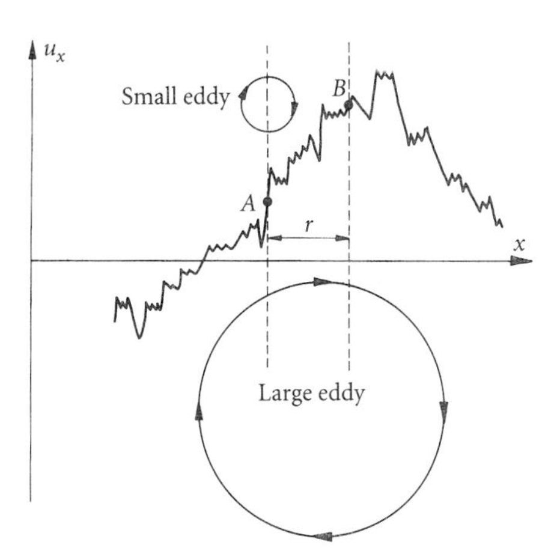

Figure 6.37 Contributions to $\langle[\Delta v]^2\rangle$ from small and large eddies.

Now suppose there exists a function $W(r)$, which has the property that its integral from 0 to r gives the contribution to $\frac{1}{2}\langle \mathbf{u}^2 \rangle$ from eddies of size r or less. In other words, $W(r)$ is the signature function we seek. Then our estimate of $\langle [\Delta v]^2 \rangle$ becomes

$$\frac{3}{4}\langle [\Delta v]^2 \rangle \approx \int_0^r W(s)\, ds + \frac{r^2}{10} \int_r^\infty \frac{\gamma W(s)}{s^2}\, ds$$

where γ is a dimensionless coefficient. Compare this with our definition of $V(r)$, (6.100b). It is natural to make the association $V(r) = W(r)$, $\gamma = 10$ and indeed this is compatible with the integral properties of $V(r)$ given by (6.101a) and (6.101b).

So it seems that there are at least tentative grounds for exploring the proposition that V, as defined by (6.100a), gives the distribution of energy across the different eddy sizes. The shape of $V(r)$, corresponding to (6.100a), and based on Townsend's data for $f(r)$, is shown below in Figure 6.38. Notice that it is positive and peaks around the integral scale, which in this case is $l \sim M/2.6$ where M is the mesh size.

Of course, the arguments above are merely suggestive and in no sense justify the assertion that definition 6.100(a) provides a suitable signature function. The real test is provided by our four acceptance criteria.

6.6.1.3 The four acceptance criteria applied to $V(r)$

Actually it is readily confirmed that definition (6.100a) satisfies all four of our requirements. Condition (4) is, perhaps, the most delicate and so we shall consider this last. Let us start, therefore, by discussing conditions (1)–(3), taking them in reverse order. Condition (3) is

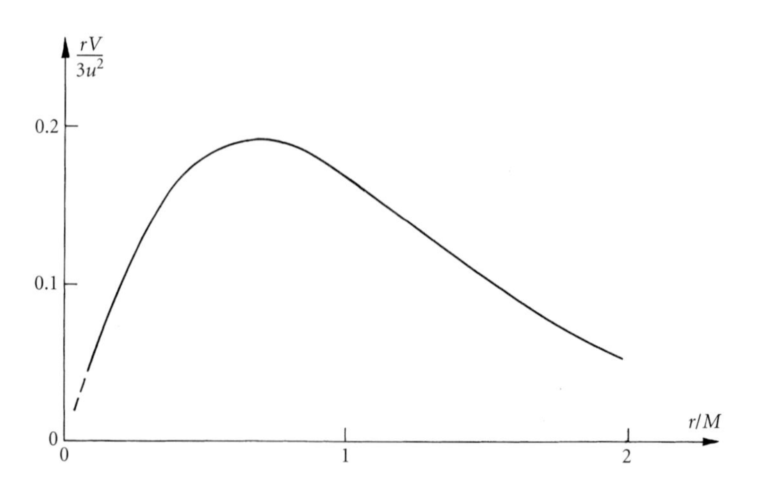

Figure 6.38 Shape of $V(r)$ corresponding to definition 6.100(a) and based on Townsend's data for the longitudinal correlation function. (See Figure 6.36.) M is the mesh size.

satisfied since, from (6.98a), a random distribution of simple eddies of size l_e yields $f(r) = \exp\lfloor -r^2/l_e^2 \rfloor$, from which,

$$rV(r) = \langle \mathbf{u}^2 \rangle (r/l_e)^4 \exp\left[-r^2/l_e^2\right]. \tag{6.102}$$

Evidently this exhibits a fairly sharp maximum around $r \sim l_e$, and indeed this is a sharper peak than that associated with Townsend's function, $V_T(r)$. Condition (2) is also satisfied since, as we have seen,

$$\int_0^\infty V(r)\, dr = \frac{1}{2}\langle \mathbf{u}^2 \rangle$$

We are left with condition (1b): we need to show that $\int_0^r V(r)\, dr \geq 0$ for all r and under all conditions. In this respect it is useful to note that we may use (6.25) and definition (6.100a) to show that $E(k)$ and $V(r)$ are related by,

$$rV(r) = \frac{3\sqrt{\pi}}{2\sqrt{2}} \int_0^\infty E(k)(rk)^{1/2} J_{7/2}(rk)\, dk$$

where $J_{7/2}$ is the usual Bessel function. Integrating this yields

$$\int_0^r V(r)\, dr = \int_0^\infty E(k) G(rk)\, dk$$

where the shape of $G(x)$ is shown below in Figure 6.39. Note that $G(x) > 0$, confirming that $\int_0^r V(r)\, dr$ is indeed positive.

Actually, it is possible to show that, in *fully developed* turbulence, the stronger condition $V(r) > 0$ holds. We proceed as follows. First we note that most measurements suggest that $f(r)$ monotonically declines in mature, isotropic turbulence. (e.g. See Comte-Bellot and Corrsin, 1971, or else Figure 6.36, which is compiled by averaging the data given in Section 3.5 of Townsend, 1956.) So we interpret our task as the need to show that $V(r)$, as defined by (6.100a), is a positive function whenever $f(r) \geq 0$ and $f'(r) \leq 0$ for all r. There are three supporting

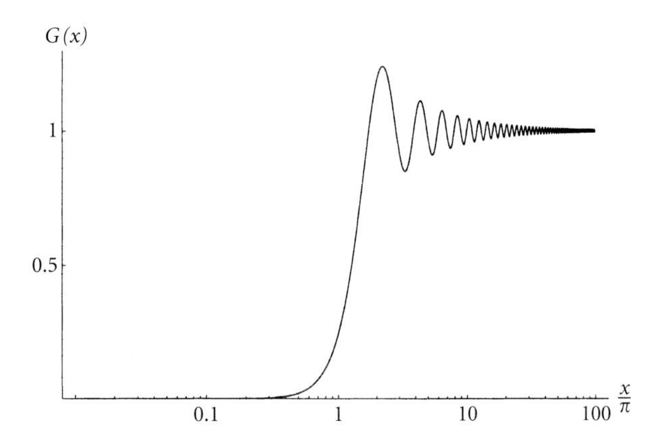

Figure 6.39 The shape of $G(x)$.

pieces of evidence that this is indeed the case. First, we know that $V \geq 0$ for small r since expansion (6.44b) combined with (6.100a) yields,

$$rV(r) = \langle(\nabla \times \boldsymbol{\omega})^2\rangle r^4/140 + \cdots, \quad r \ll \eta$$

Second, $V > 0$ in the inertial subrange, since Kolmogorov's two-thirds law requires

$$rV(r) = \tfrac{1}{3}\beta\varepsilon^{2/3}r^{2/3}, \quad \eta \ll r \ll l \tag{6.103}$$

where β is Kolmogorov's constant, η is the Kolmogorov microscale, and l is the integral scale. Third, since $f(r)$ declines monotonically and $f'(0) = 0$, $f''(r)$ must start out as negative and then turn positive. The inflection point at which $f''(r)$ changes sign lies between the dissipation range and the inertial subrange. Since $V \sim f''(r) - r^{-1}f'(r)$ it follows that V is positive at all scales greater than or equal to the inertial subrange. Thus the only possibility of a negative contribution to V comes from the dissipation range $r \sim \eta$. However, throughout the equilibrium range (6.53b) yields,

$$V = -\frac{3}{8}\frac{r^2}{\nu}\frac{\partial}{\partial r}\left[\frac{u^3K}{r}\right] = \frac{r^2}{16\nu}\frac{\partial}{\partial r}\frac{1}{r}\left[|S|\langle[\Delta v]^2\rangle^{3/2}\right] \tag{6.104}$$

where $S(r)$ is the skewness factor ($S \sim -0.3$). Moreover, we saw in Section 6.2.3.2 that a good estimate of K and $\langle[\Delta v]^2\rangle$ throughout the equilibrium range can be obtained by integrating (6.53b) on the assumption that S is constant. This integration reveals that $\langle[\Delta v]^2\rangle / r^{2/3}$ rises monotonically throughout the equilibrium range, something which is also observed in direct numerical simulations of turbulence. As a consequence we conclude, from (6.104), that $V \geq 0$ in the range $r \sim \eta$.

In summary, then, provided that $f(r)$ monotonically declines, as shown in Figure 6.36, we expect V to be positive for all r. So we might anticipate that $V \geq 0$ in fully developed turbulence.

Example 6.14 Integral properties of $V(r)$
Confirm that:

$$\int_0^\infty r^m V(r)\,dr = \frac{3}{8}|m|(2+m)\int_0^\infty r^{(m-1)}\langle[\Delta v]^2\rangle\,dr, \quad (-2 < m < 0) \tag{6.105a}$$

$$\int_0^\infty r^m V(r)\,dr = \frac{1}{4}\langle \mathbf{u}^2\rangle m(2+m)\int_0^\infty r^{(m-1)}f\,dr, \quad (m > 0). \tag{6.105b}$$

Particular cases of special interest are:

Integral scale $$l = \int_0^\infty f dr = \frac{4}{3\langle \mathbf{u}^2 \rangle} \int_0^\infty rV\, dr \qquad (6.106a)$$

Loitsyansky $$I = -\int r^2 \langle \mathbf{u} \cdot \mathbf{u}' \rangle\, d\mathbf{r} = \frac{32\pi}{105} \int_0^\infty r^5 V\, dr \qquad (6.106b)$$ □

Thus far we have shown that definition (6.100a) satisfies the first three of our acceptance criteria. It remains to show that our choice of $V(r)$ satisfies criterion (4), that is $rV(r) \approx [kE(k)]_{k \approx \pi/r}$. The idea behind this is the following. Equation (6.102) suggests that the kinetic energy associated with eddies of size l_e is,

$$\mathrm{v}_{l_e}^2 = \int_0^\infty V(l_e; r)\, dr \sim V(l_e) l_e.$$

This is analogous to the Fourier-space estimate $\mathrm{v}_k^2 \sim kE(k), k \sim \pi/l_e$ and suggests that

$$\mathrm{v}_r^2 \sim rV(r) \sim kE(k), \quad k \sim \pi/r.$$

Certainly, if this holds true then it is a useful property of $V(r)$. That is, if we accept the premise that $E(k)$ represents, approximately, the distribution of energy across the different scales, then $V(r)$ must also give an impression of the distribution of energy. We might think of $V(r)$ as a 'poor man's energy spectrum', with r interpreted as eddy size. It should be constantly born is mind, however, that this physical interpretation of $V(r)$ is imperfect. There is no one-to-one correspondence between r and eddy size. This is a deficiency, which $V(r)$ shares with $E(k)$, where the estimate $k \sim \pi/l_e$ is equally imprecise. That is to say, just as eddies of fixed size produce a distributed energy spectrum, $E(k)$, albeit peaked around $k \sim \pi/l_e$, so the same eddies give rise to a continuous distribution of $V(r)$, centred around $r \sim l_e$. One consequence of this is that we can attribute no particular physical meaning (in terms of energy) to $V(r)$, or $E(k)$, for $r < \eta$ and $r > l$, l being the integral scale. (See Figure 6.40.) Still, with these limitations in mind, let us look more closely at the relationship between $V(r)$ and $E(k)$, and see if we can arrive at the estimate $rV(r) \approx [kE(k)]_{k \approx \pi/r}$ in a less hand-waving fashion.

First we note that we may use (6.24) and (6.25), along with definition (6.100a), to show that $E(k)$ and $V(r)$ are related by the Hankel transform pair,

$$E(k) = \int_0^\infty U(r)(rk)^{1/2} J_{7/2}(rk)\, dr \qquad (6.107a)$$

$$U(r) = \int_0^\infty E(k)(rk)^{1/2} J_{7/2}(rk)\, dk \qquad (6.107b)$$

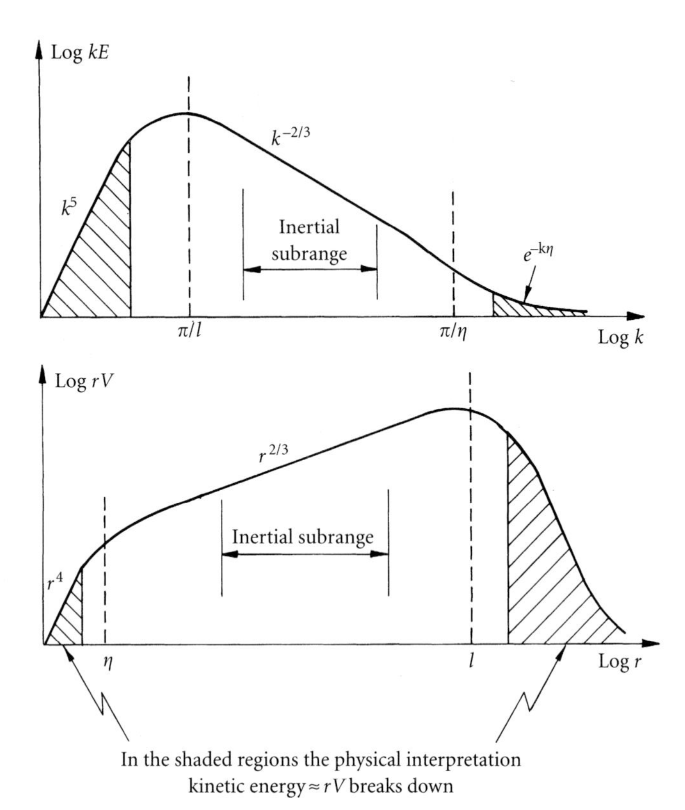

Figure 6.40 The idea that $kE(k)$ or $rV(r)$ represents the kinetic energy of eddies of size π/k or r is not meaningful for $r < \eta$ or $r > l$. See Section 6.4.1.

where,

$$U(r) = \frac{2\sqrt{2}}{3\sqrt{\pi}} rV(r)$$

and $J_{7/2}$ is the usual Bessel function. (We shall not pause to prove these results but rather leave it as an exercise for the reader.) Unfortunately, this somewhat cryptic transform pair obscures the essence of the relationship between $E(k)$ and $V(r)$. A simpler, if approximate, relationship may be established as follows. We start by integrating (6.107b) to give

$$\int_0^r V(r)\,dr = \int_0^\infty E(k)G(rk)\,dk$$

where the shape of $G(x)$ is shown in Figure 6.39. If we think of $G(x)$ as a sort of filter function then a rough approximation to $G(x)$ is: $G(x) = 0$ for $x < \hat{\pi}$, and $G(x) = 1$ for $x > \hat{\pi}$, where[21] $\hat{\pi} = 9\pi/8$. It follows that,

$$\int_0^r V(r)dr \approx \int_{\hat{\pi}/r}^\infty E(k)dk, \quad \hat{\pi} = 9\pi/8.$$

[21] Of course, we could equally use π instead of $\hat{\pi}$ as the cut-off for x in $G(x)$. The factor $9/8$ is included, however, as it ensures that rV and $[kE]_{k=\hat{\pi}/r}$ have the same mean and root-mean-square values. See Examples 6.17 and 6.18 at the end of this section.

This integral equation is satisfied provided

$$rV(r) \approx [kE(k)]_{k=\hat{\pi}/r}, \quad \eta < r < l. \tag{6.108}$$

Note that this approximate relationship is valid only if $V(r)$ and $E(k)$ are relatively smooth functions. It is not a good approximation if $E(k)$ exhibits steep gradients. (See Exercise 6.11 at the end of this chapter.) It also breaks down outside the range $\eta < r < l$. For example, $rV(r) \sim r^4$ for $r \ll \eta$, yet $kE(k)$ decays exponentially at large k, rather than as k^{-4} as (6.108) would suggest. This failure of (6.108) is indicative of the fact that neither $rV(r)$ nor $kE(k)$ can lay claim to represent the energy of eddies of size r outside the range $\eta < r < l$. Still, let us accept these restrictions and see if estimate (6.108) is consistent with the known facts.

First we note that equation (6.108) is compatible with what we know about the inertial subrange where (6.103) yields,

$$rV(r) = \tfrac{1}{3}\beta\varepsilon^{2/3}r^{2/3} = 0.667\varepsilon^{2/3}r^{2/3}$$

while the five-thirds law requires,

$$[kE(k)]_{k=\hat{\pi}/r} = \left[\alpha\varepsilon^{2/3}k^{-2/3}\right]_{k=\hat{\pi}/r} = 0.655\varepsilon^{2/3}r^{2/3}.$$

(Remember that $\beta \approx 2.0$ and $\alpha \approx 1.52$.) Evidently, in the limit $\mathrm{Re} \to \infty$, there is only a 2% difference between $rV(r)$ and $[kE(k)]_{k=\hat{\pi}/r}$ in the inertial subrange.

Another simple test of (6.108) is provided by Figure 6.41, which shows the shape of $rV(r)$ and $[kE(k)]_{k=\hat{\pi}/r}$ for the wind-tunnel data of Townsend (see Figure 6.36). As anticipated by (6.108), the shapes of the two curves are similar. Note, however, that the equilibrium range is not resolved in Figure 6.36 and hence the curves shown in Figure 6.41 relate only to the energy containing eddies.

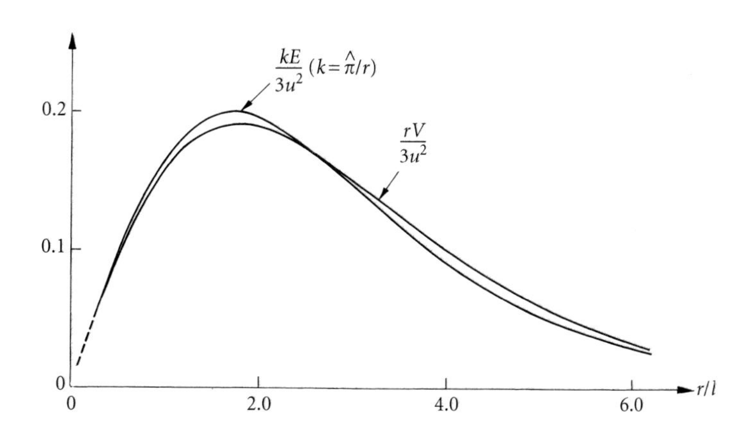

Figure 6.41 The shape of $rV(r)$ and $kE(k)$ corresponding to the wind-tunnel data shown in figure 6.36. The integral scale is denoted l.

As a final check on (6.108) we have calculated $V(r)$ corresponding to the model spectrum

$$E = \hat{k}^4\left(1+\hat{k}^2\right)^{-17/6}\exp\left[-\hat{k}/\mathrm{Re}^{3/4}\right], \quad \hat{k} = kl.$$

This exhibits a k^4 region for small k, a $k^{-5/3}$ fall-off for intermediate k, and an exponential tail at large k. In short, it possesses the properties we might expect of a realistic spectrum. A comparison of $rV(r)$ and $[kE(k)]_{k=\hat{\pi}/r}$ is shown in Figure 6.42 for the case $\mathrm{Re}^{3/4} = 100$. Three sets of curves are shown, corresponding to a linear plot, a linear-log plot and a log–log plot. (The log plots are included as they emphasize the inertial subrange.) The comparison seems favourable, lending some support for (6.108).

All-in-all it would seem that the estimate

$$rV(r) \approx [kE(k)]_{k=\hat{\pi}/r} \approx [\text{K.E. in eddies of size } r] \quad (\eta < r < l)$$

represents a reasonable approximation in fully developed, freely evolving turbulence. Additional support for (6.108) is provided by the transform pair (6.107). The details are given in the examples below. (You may find Appendix IV on Hankel transforms useful.)

Example 6.15 Power-law spectra
Use the Hankel transform relationship between $V(r)$ and $E(k)$ to show that, if E takes the form of a simple power law, $E = Ak^n$, $-5 < n < 4$, with an exponential fall-off at infinity, then,

$$[rV(r)] = \lambda_n [kE(k)]_{k=\hat{\pi}/r}$$

where the λ_n are coefficients, which depend on n. Confirm that

$$\lambda_n = \frac{3\sqrt{\pi}}{4}\left(\frac{16}{9\pi}\right)^{n+1}\frac{\Gamma\left(\frac{5}{2}+\frac{n}{2}\right)}{\Gamma\left(2-\frac{n}{2}\right)}$$

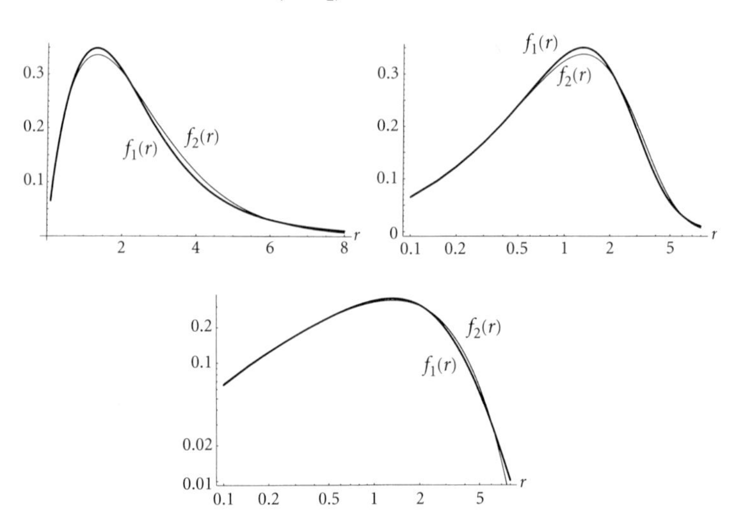

Figure 6.42 The functions $f_1(r) = [kE(k)]_{k=\hat{\pi}/r}$ and $f_2(r) = rV(r)$ corresponding to the spectrum $E = \hat{k}^4(1+\hat{k}^2)^{-17/6}\exp[-\hat{k}/100]$. r is normalized by the integral scale l.

where Γ is the gamma function. For example

n	-2	$-5/3$	-1	0	1	2
λ_n	1.04	1.02	1	1	0.96	0.80

Cases of particular interest are: $n = -1$, (tube-like eddies); $n = -5/3$, (Kolmogorov's spectrum); and $n = -2$ (sheet-like eddies).

Example 6.16 Equilibrium range spectra
Consider the equilibrium range energy spectrum

$$E(k) = \text{v}^2\eta(\eta k)^m \exp\left[-(k\eta)^2\right]$$

where v and η are the Kolmogorov microscales. Use the Hankel transform pair (6.107) and Kummer's transformation rule to show that $V(r)$ is given by

$$rV(r) = \frac{3\sqrt{\pi}}{2^6}\frac{\Gamma\left(\frac{5}{2}+\frac{m}{2}\right)}{\Gamma\left(\frac{9}{2}\right)}\frac{\text{v}^2 r^4}{\eta^4}\exp\left[\frac{-r^2}{4\eta^2}\right]M\left(2-\frac{m}{2}, \frac{9}{2}, \frac{r^2}{4\eta^2}\right)$$

where M is Kummer's hypergeometric function. Verify that Example 6.15 is a special case of this. Examples of $rV(r)$, along with $[kE(k)]_{k=\hat{\pi}/r}$, are given in Figure 6.43 corresponding to $m = -1, -5/3, -2$. Note the close correspondence between the two sets of functions, as anticipated by (6.108). Given that $M(a,b,z) > 0$ whenever a, b and z are positive, confirm that $V(r)$ is positive provided that $-5 < m \leq 4$.

Example 6.17 Equivalence of mean values
Use equations (6.106a) and (8.22) to show that

$$2u^2 l = \pi \int_0^\infty [E/k]dk = \frac{8}{9}\int_0^\infty rV\,dr,$$

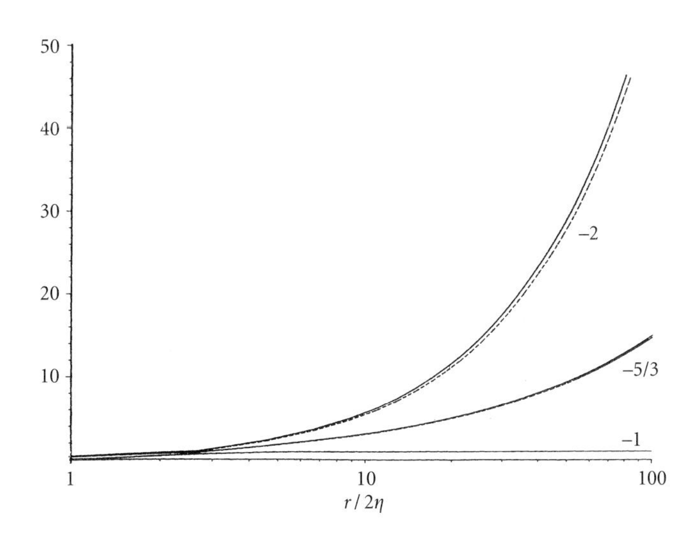

Figure 6.43 (Example 6.16) The shape of the equilibrium-range energy spectrum. The figure shows $f_1(r) = [kE(k)]_{k=\hat{\pi}/r}$ (dashed lines) and $f_2(r) = rV(r)$ (solid lines) corresponding to $E(k) \sim (\eta k)^m \exp\left[-(\eta k)^2\right]$ for $m = -1, -5/3, -2$. For the case of $m = -1$ the curves f_1 and f_2 are indistinguishable. The units of the vertical axis are arbitrary.

and hence confirm that $rV(r)$ and have $[kE(k)]_{k=\hat{\pi}/r}$ have the same mean values in the sense that

$$\int_0^\infty [rV(r)]\,dr = \int_0^\infty [kE(k)]_{k=\hat{\pi}/r}\,dr.$$

Example 6.18 Equivalence of root-mean-squares

Use the Hankel transform pair (6.107), in conjunction with Rayleigh's power theorem for transform pairs, to show that

$$\int_0^\infty E^2 dk = \frac{8}{9\pi}\int_0^\infty [r^2V^2]\,dr.$$

Hence confirm that the root-mean-square values of $rV(r)$ and $[kE(k)]_{k=\hat{\pi}/r}$ are equal in the sense that

$$\int_0^\infty [r^2V^2]\,dr = \int_0^\infty [k^2E^2]_{k=\hat{\pi}/r}dr.$$

Example 6.19 A Saffman spectrum

Consider the energy spectrum $E = k^2\exp[-k]$. Show that the corresponding signature function, $V(r)$, can be written in terms of Gauss's hypergeometric function ${}_2F_1$. Evaluate $rV(r)$ and $[kE(k)]_{k=\hat{\pi}/r}$ and confirm that they have the shape shown below in Figure 6.44.

6.6.2 *Cascade dynamics in real space*

We now introduce some dynamics. Readers who are used to thinking about cascade dynamics in Fourier space may, at first, find the following description somewhat perplexing. For example plots of, say, $E(k)$ now become plots of $V(r)$, so that small scales are to the left, and large scales to the right. However, this is simply a matter of lack of familiarity. Formally there is no difference between the 'real space' description given here and the more classical spectral approach. As we shall see, one can shift at will between the two using the appropriate transform.

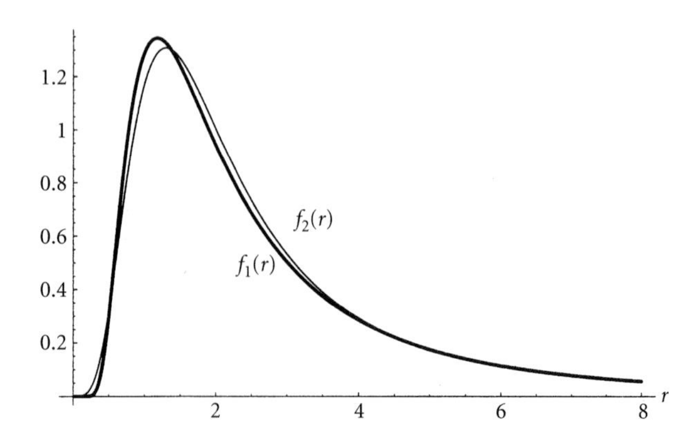

Figure 6.44 The functions $f_1(r) = [kE(k)]_{k=\hat{\pi}/r}$ and $f_2(r) = rV(r)$ corresponding to the energy spectrum $E = k^2\exp[-k]$. Note the similar shapes of f_1 and f_2, as anticipated by (6.108).

We start by integrating the Karman–Howarth equation. Invoking (6.100a) this yields an evolution equation for $V(r)$:

$$\frac{\partial V}{\partial t} = \frac{\partial \Pi_V}{\partial r} + 2\nu\left[\frac{\partial}{\partial r}\frac{1}{r^2}\frac{\partial}{\partial r}(r^2 V) - \frac{10}{r^2}V\right] \tag{6.109}$$

$$\Pi_V = \frac{3}{4}r^3\frac{\partial}{\partial r}r^{-6}\frac{\partial}{\partial r}[u^3 r^4 K]. \tag{6.110}$$

The function Π_V captures the influence of the non-linear inertial forces. It can redistribute energy from one eddy size to another, but it cannot create or destroy energy. If we are away from the dissipation scales, $r \gg \eta$, viscous forces may be neglected and equation (6.109) integrates to give,

$$\frac{d}{dt}\int_r^\infty V dr = -\Pi_V(r), \quad r \gg \eta \tag{6.111}$$

Since $\int_r^\infty V dr$ is a measure of the energy held in eddies whose size exceeds r, Π_V must represent the energy transferred by inertial forces from eddies of size r or greater to those of size r or less. We might refer to Π_V as the *real-space kinetic energy flux*. Next we integrate (6.109) from 0 to ∞. The inertial term now vanishes and we find,

$$\frac{d}{dt}\int_0^\infty V dr = -20\nu\int_0^\infty \frac{V}{r^2}dr. \tag{6.112}$$

The integral on the right can be evaluated with the aid of (6.101a) to yield the familiar statement

$$\frac{d}{dt}\left[\tfrac{1}{2}\langle \mathbf{u}^2\rangle\right] = -\nu\langle \boldsymbol{\omega}^2\rangle = -\varepsilon. \tag{6.113}$$

Let us now consider the equilibrium range, which includes the inertial subrange, $\eta \ll r \ll l$, and the dissipation scales, $r \sim \eta$. In this range we know, from (6.53a), that u^3K is related to f by,

$$u^3K(r) + 2\nu u^2 f'(r) = -\tfrac{2}{15}\varepsilon r.$$

(Remember that $\langle[\Delta v]^3\rangle = 6u^3K$.) If we substitute for K in (6.109), and integrate, we find

$$\frac{d}{dt}\int_r^\infty V dr = -\varepsilon, \quad r \ll l. \tag{6.114a}$$

Combining this with (6.111) we see that, in the inertial subrange, the real-space kinetic energy flux is,

$$\Pi_V(r) = \varepsilon, \quad \eta \ll r \ll l. \tag{6.114b}$$

These results admit a simple physical interpretation. Energy is destroyed at a rate ε by the smallest eddies. Since virtually all of this energy is held in the large eddies, ε must also be the rate of extraction of energy from the large scales (see 6.114a). The eddies in the inertial subrange have almost no energy of their own, and merely act as conduit for transferring the large-scale energy down to the dissipation scales. Thus the flux of energy through the inertial sub-range is $\Pi_V = \varepsilon$.

The general shapes of Π_V and $rV(r) \sim \mathrm{v}_r^2$ are shown in Figure 6.45. The kinetic energy v_r^2 rises with r, initially as r^4 and then as $r^{2/3}$. It peaks around the integral scale and then declines. Π_V also rises as r^4 for small r, reaches a plateau of $\Pi_V = \varepsilon$ in the inertial subrange and then falls again for $r \sim l$. It is positive, indicating that energy is transferred from large to small scales.

The precise form of V and Π_V can be determined by numerical or physical experiments. Estimates can also be obtained by closure models. Perhaps the most important region is the equilibrium range, $r \ll l$, since, according to Kolmogorov (but not Landau!), this has universal properties. The simplest closure model for the universal equilibrium range is that described in Section 6.2.3.2, in which the skewness factor for $\Delta \mathrm{v}$ is assumed constant across the range. As noted

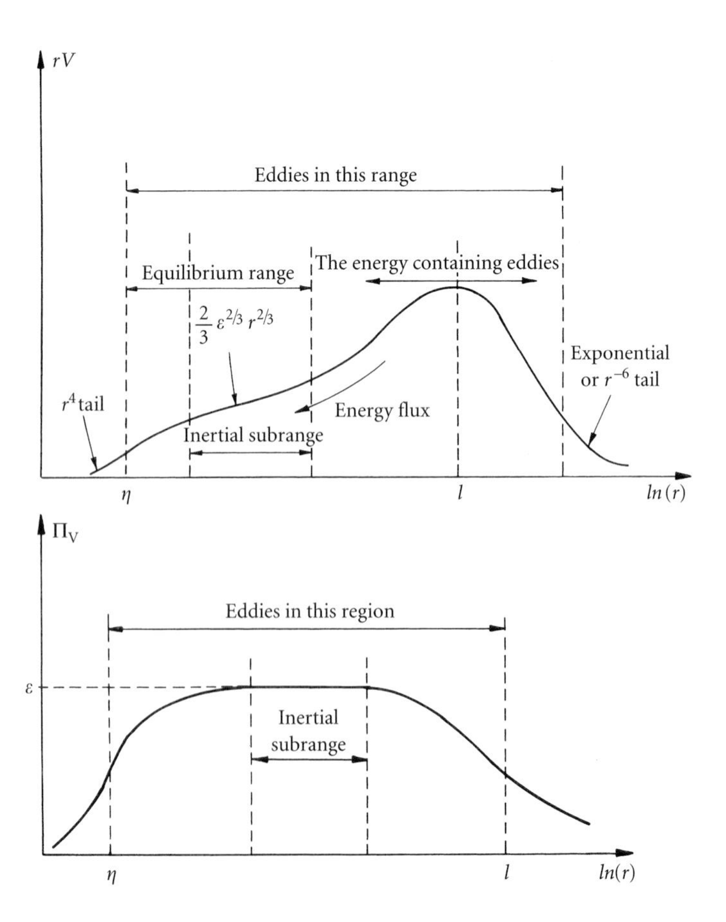

Figure 6.45 The shapes of Π_V and $rV(r)$ (schematic only).

in Section 6.2.3.2, the appeal of this closure lies partly in its simplicity, and partly because its validity is readily checked against the numerical and experimental data. (Figure 6.19, for example, shows that the results of this closure model are supported by numerical experiments.)

As we have seen, the assumption of constant skewness reduces the problem of freely decaying turbulence to that of solving the trivial equation,

$$\frac{1}{2}\frac{dy}{dx} + y^{3/2} = x \quad (r \ll l)$$

where

$$y = \frac{\langle [\Delta \mathrm{v}]^2 \rangle}{\beta (15\beta)^{1/2} \mathrm{v}^2}, \qquad x = \frac{r}{(15\beta)^{3/4}\eta}.$$

(See equation (6.61)). Here η and v are the Kolmogorov microscales of length and velocity, and β is the Kolmogorov constant, $\beta \approx 2.0$. For small and large x we have, $y_0 = x^2 + \cdots$, $y_\infty = x^{2/3}$, corresponding to the small-r expansions of $\langle [\Delta \mathrm{v}]^2 \rangle$ and Kolmogorov's two-thirds law, $\langle [\Delta \mathrm{v}]^2 \rangle = \beta \varepsilon^{2/3} r^{2/3}$, respectively. Suitably normalized versions of $rV(r)$ and $\Pi_V(r)$ are

$$z(x) = \frac{rV}{\beta\sqrt{15\beta}\mathrm{v}^2}, \qquad p(x) = \frac{\Pi_V}{\varepsilon}$$

whose asymptotic forms for small and large x are

$$z_0(x) = \tfrac{3}{2}x^4 + \cdots, \quad z_\infty(x) = \tfrac{1}{3}x^{2/3}, \quad p_0(x) = \tfrac{27}{20}x^4 + \cdots,$$
$$p_\infty(x) = 1.$$

The shape of $z(x)$ may be estimated using the polynomial approximation to $y(x)$ given in Section 6.2.3.2. This is inaccurate near $x = 1$ because, in the approximation, $y'''(x)$ is discontinuous at $x = 1$. But elsewhere the estimate is reasonable, and in particular,

$$z(x) \approx \tfrac{3}{2}x^4\left[1 - \tfrac{3}{2}x^2 - \tfrac{2}{153}x^4 + \tfrac{167}{162}x^6\right], \quad x \le 0.6$$

$$z(x) \approx \tfrac{1}{3}x^{2/3}\left[1 + \tfrac{4}{9}x^{-4/3} + \tfrac{5}{9}x^{-8/3} + \tfrac{101}{306}x^{-4}\right], \quad x \ge 1.4.$$

The precise shapes of $y(x)$, $z(x)$, and $p(x)$, predicted by the constant skewness model, are shown in Figure 6.46. The structure function $y \sim \langle [\Delta \mathrm{v}]^2 \rangle$ begins to approach the large-x asymptote rather quickly, reaching 90% of the value predicted by the two-thirds law at $x \sim 2$. (This corresponds to $r \sim 26\eta$ and $k\eta \sim 0.14$.) The subsequent convergence to $\langle [\Delta \mathrm{v}]^2 \rangle = \beta \varepsilon^{2/3} r^{2/3}$ is somewhat slower, the difference between y and y_∞ being 3% at $x \sim 5$ (i.e $r \sim 64\eta$). The signature function $z \sim rV(r)$ also approaches its large-x asymptote reasonably

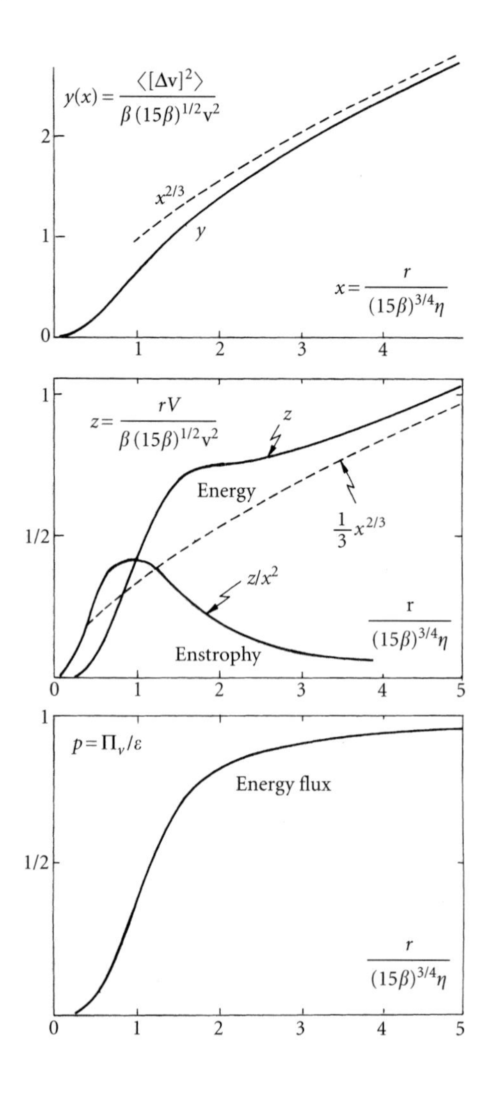

Figure 6.46 The shapes of $y \sim \langle[\Delta v]^2\rangle$, $z \sim rV(r)$, z/x^2 (normalized enstrophy) and $p = \Pi_V/\varepsilon$ predicted by the constant skewness model. The horizontal axis is $x = r/(15\beta)^{3/4}\eta$.

quickly, being within 10% of the expression $z_\infty = \frac{1}{3}x^{2/3}$ by $x \sim 4$ (i.e. $r \sim 51\eta$, $k\eta = 0.07$). Notice that the value of rV drops rapidly in the region $x \sim 0.4 \rightarrow 1.5$, which corresponds to $r \sim 5\eta \rightarrow 20\eta$ and $k\eta \sim 0.2 \rightarrow 0.7$, suggesting that the cut-off in eddy size occurs not at $r \approx \eta$, but rather at a somewhat higher value, say 6η. The approach of $rV(r)$ (or $kE(k)$) to its inertial range asymptote at $k\eta \sim 0.07$ is consistent with the experimental measurements of $E(k)$, as is the rapid fall of $rV(r)$ (or $kE(k)$) in the range $k\eta \sim 0.2 \rightarrow 0.7$. (See, for example, Frisch 1995.) The curious observation that rV, or $kE(k)$, is greater than its inertial subrange asymptote for $r/\eta > 10$ is also consistent with the experimental data (see below).

The function z/x^2, which we shall see shortly represents the normalised enstrophy density, is also shown in Figure 6.46. It exhibits a peak at $x \sim 1$ (i.e. $r \sim 13\eta$, $k\eta \sim 0.28$), which is a little smaller than the experimental data would suggest. (There is some scatter in the data, but typically the experiments suggest that z/x^2 peaks at $x \sim 2$.) Finally, the form of $p(x) = \Pi_V(r)/\varepsilon$ is also shown in Figure 6.46. It is positive, as expected, and within 10% of the inertial subrange value $\Pi_V = \varepsilon$ for $x > 3.0$ (i.e. $r > 38\eta$).

The form of the transition region from the inertial subrange to the dissipation range is best seen by plotting $(rV)/r^{2/3}$ against r/η. The top graph in Figure 6.47 shows $(rV)/r^{2/3}$ (normalized by $\text{v}^2/\eta^{2/3}$) versus r/η. Like Figure 6.46 it is based on the constant skewness model, with $\beta = 2.0$. In the inertial subrange the curve approaches the value 2/3. Based on this figure, and the enstrophy curve in Figure 6.46, we might

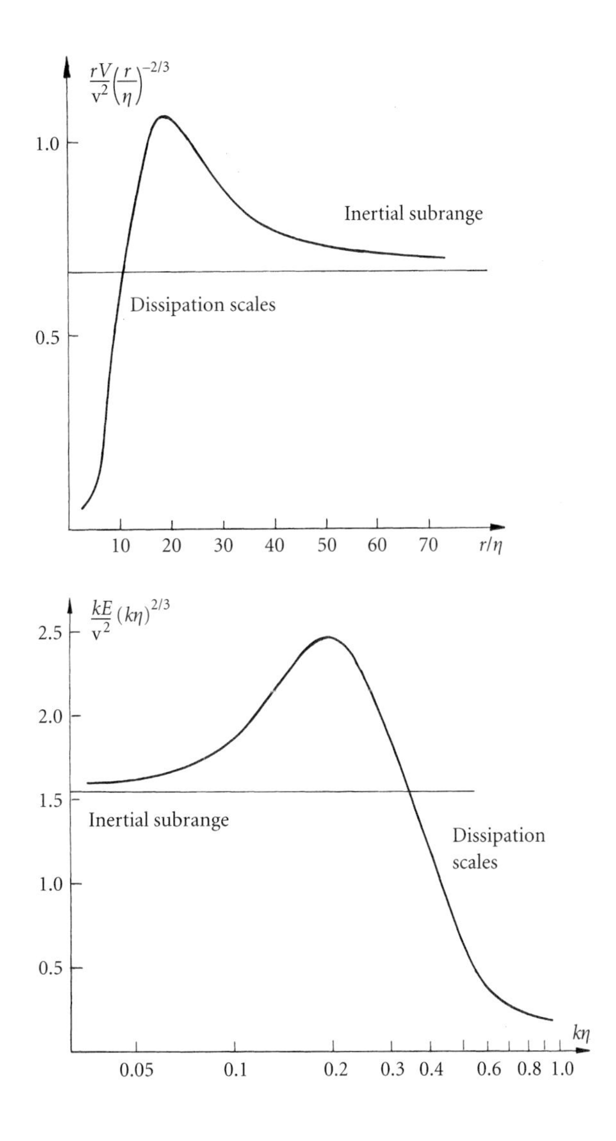

Figure 6.47 Compensated energy plots based on the constant skewness model ($\beta = 2.0$).

divide up the different size ranges as follows:

- Lower cut off in eddy size: $r \sim 6\eta$
- Dissipation range: $6\eta < r < 30\eta$
- Transition from dissipation range to inertial subrange: $30\eta < r < 70\eta$
- Inertial subrange: $r > 70\eta$.

These broad subdivisions are consistent with most of the experimental evidence. Much of the experimental data in the equilibrium range is plotted in the form $k^{5/3}E(k)$ versus $k\eta$. These are referred to as *compensated spectra*. If we recall that $rV \approx kE(k)$, where $k = \hat{\pi}/r$, we can replot $(rV)/r^{2/3}$ in this form. The result is shown in the lower graph in Figure 6.47. Note that the inertial subrange ends around $k\eta \sim 0.05$, the compensated spectrum peaks around $k\eta \sim 0.2$, and the dissipation range ends at around $k\eta \sim 0.7$. Similar compensated spectra are shown in Saddoughi and Veeravalli (1994) based on wind-tunnel measurements at very high Re. These indicate that the inertial subrange ends at $k\eta \sim 0.02$, the compensated spectra peak at $k\eta \sim 0.1$, and the dissipation range ends at $k\eta \sim 0.7$. Moreover, the characteristic overshoot of the compensated spectrum is clearly visible in the experimental data, though the magnitude of this overshoot is smaller than indicated above. (This difference may be because, at large but finite Re, there is a small correction to equation (6.53b) arising from the time dependence of the large scales, and this tends to suppress the overshoot. The details are spelt out in the four examples at the end of this section.)

In Chapter 8 we shall see that the spectral equivalent of (6.109) is

$$\frac{\partial E}{\partial t} = -\frac{\partial \Pi_E}{\partial k} - 2\nu k^2 E$$

where Π_E is called the *spectral kinetic energy flux*. In the inertial sub range this yields

$$\frac{d}{dt}\int_0^k E dk = -\Pi_E = -\varepsilon$$

since ε is the rate of loss of energy from the large eddies. Thus we expect $\Pi_V = \Pi_E = \varepsilon$ in the range $\eta \ll r \ll l$. Actually it may be shown that the general relationship between $\Pi_V(r)$ and $\Pi_E(k)$ is the same as that between $kE(k)$ and $rV(r)$; that is,

$$\frac{\Pi_E(k)}{k} = \frac{2\sqrt{2}}{3\sqrt{\pi}}\int_0^\infty \Pi_V(r)(rk)^{1/2} J_{7/2}(kr)\, dr$$

$$\frac{2\sqrt{2}}{3\sqrt{\pi}}\Pi_V(r) = \int_0^\infty \frac{\Pi_E(k)}{k}(rk)^{1/2} J_{7/2}(rk)\, dk.$$

It follows from the examples at the end of Section 6.6.1 that, provided $\Pi_V(r)$ and $\Pi_E(k)$ are relatively smooth functions, then Π_V and Π_E satisfy the approximate relationship,

$$\Pi_V(r) \approx \Pi_E(k = \hat{\pi}/r), \quad \eta \leq r \leq l.$$

Let us now consider the nature of the evolution of $V(r,t)$ in decaying turbulence. The manner in which $V(r)$ collapses with time in freely decaying turbulence takes a particularly simple form in those cases where we have a Saffman spectrum. From (6.78), (6.79), and (6.100a) we have, for a Saffman spectrum,

$$(rV)_\infty = \frac{45}{16\pi}\frac{L}{r^3}; \quad L = \int \langle \mathbf{u} \cdot \mathbf{u}' \rangle \, d\mathbf{r}$$

where L is Saffman's invariant. Evidently, in such cases, the shape of rV at large r is fixed during the decay and so the signature function evolves as shown in Figure 6.48. This is analogous to the equivalent spectral plot of $E(k,t)$ shown in Figure 6.20.

In the final period of decay inertial forces become negligible and we can solve (6.109) exactly. It is readily confirmed that, for this period, there exists a self-similar solution for rV of the form,

$$rV(r) = \frac{4\alpha(\alpha+1)}{35}\langle \mathbf{u}^2\rangle M\left(\frac{5}{2} - \alpha, \frac{9}{2}, x\right)x^2 e^{-x}, \quad \langle \mathbf{u}^2 \rangle \sim t^{-\alpha}$$

where $x = r^2/(8\nu t)$ and M is Kummer's hypergeometric function. When $\alpha = 5/2$, which corresponds to a Batchelor spectrum, we have $rV(r) = \langle \mathbf{u}^2\rangle x^2 e^{-x}$. A Saffman spectrum, on the other hand, requires $\alpha = 3/2$ (see Example 6.9 at the end of Section 6.3.5) and in such cases we have $rV(r) = \frac{3}{7}\langle \mathbf{u}^2\rangle M(1, \frac{9}{2}, x)x^2 e^{-x}$ where $\langle \mathbf{u}^2 \rangle = L/(8\pi\nu t)^{3/2}$.

We close this subsection by reminding the reader that care must be exercised in the interpretation of $E(k)$, $V(r)$, Π_E, and Π_V. We have already noted that the usual interpretation of $E(k)$, as the energy

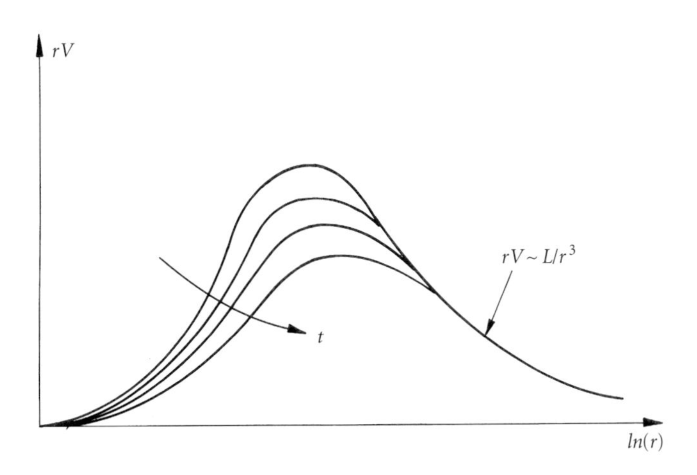

Figure 6.48 Evolution of $rV(r)$ in the free decay of a Saffman spectrum.

density of eddies of size $l_e \sim k^{-1}$, breaks down when k is very small or very large. That is, if we have a range of eddy sizes, from l down η, with each batch of eddies making a contribution to $E(k)$ of the form

$$E(k) \sim \langle \mathbf{u}^2 \rangle l_e (kl_e)^4 \exp\left[-(kl_e)^2/4\right],$$

then the energy spectrum will have a k^4 tail for $k < l^{-1}$ and an exponential tail for $k > \eta^{-1}$. Yet there are no eddies in this range. Evidently, the form of E for very small or very large k has nothing to do with the energy of eddies of size k^{-1}. By implication, Π_E does not represent an energy flux in these regions.

A similar problem arises with $V(r)$. Again imagine a range of eddy sizes from l down to η, with each batch of eddies making a contribution to $V(r)$ of the form,

$$V(r) \sim \frac{\langle \mathbf{u}^2 \rangle}{l_e} \left(\frac{r}{l_e}\right)^3 \exp\left[-(r/l_e)^2\right].$$

The signature function $V(r)$ will have an r^3 tail for $r < \eta$ and an exponential tail[22] for $r > l$, neither of which relates to the energy of eddies of size $l_e \sim r$. So our physical interpretation of both $V(r)$ and Π_V breaks down for $r \gg l$ and $r \ll \eta$.

Example 6.20 The first-order viscous correction to the two-thirds law: the bottleneck

Consider the constant-skewness model of the equilibrium range in which the governing equation is

$$y'(x) + 2y^{3/2} = 2x,$$

where $y = \langle (\Delta \mathrm{v})^2 \rangle / (\beta\sqrt{15\beta}\mathrm{v}^2)$ and $x = (15\beta)^{-3/4} r/\eta$. Show that, for large x, its solution is of the form $y = x^{2/3} - (2/9)x^{-2/3} + \cdots$ and hence confirm that, for the constant skewness model,

$$\langle [\Delta \mathrm{v}]^2 \rangle = \beta \varepsilon^{2/3} r^{2/3} \left[1 - \tfrac{10}{3}\beta (r/\eta)^{-4/3} + \cdots\right], \quad \eta \ll r \lll l$$

$$rV(r) = \tfrac{1}{3}\beta \varepsilon^{2/3} r^{2/3} \left[1 + \tfrac{20}{3}\beta (r/\eta)^{-4/3} + \cdots\right], \quad \eta \ll r \lll l.$$

Thus verify that $rV(r)$, and hence $kE(k)$, will exceed its inertial range asymptote for large r/η, as shown in Figure 6.47. This is sometimes called the *bottleneck*. However, this effect is partially masked by a second correction, arising from the time dependence of the large scales, as we shall see shortly.

Example 6.21 The first-order, large-scale correction to the two-thirds law

Use (6.52) to show that the first-order corrections to the four-fifths and two-thirds laws due to the time dependence of the large-scales are, for

[22] If there are significant long-range correlations induced by Batchelor's pressure forces then $f_\infty \sim r^{-6}$ for a Batchelor ($E \sim k^4$) energy spectrum or else $f_\infty \sim r^{-3}$ for a Saffman ($E \sim k^2$) spectrum. In such cases the exponential tail in $V(r)$ must be replaced by $V_\infty \sim r^{-7}$ or $V_\infty \sim r^{-4}$.

the constant skewness model,

$$\left\langle [\Delta v]^3 \right\rangle = -\tfrac{4}{5}\varepsilon r\left[1 + \tfrac{15}{34}\beta r^{2/3}\dot{\varepsilon}/\varepsilon^{4/3} + \cdots\right], \quad \eta \lll r \ll l$$
$$\left\langle [\Delta v]^2 \right\rangle = \beta\varepsilon^{2/3}r^{2/3}\left[1 + \tfrac{5}{17}\beta r^{2/3}\dot{\varepsilon}/\varepsilon^{4/3} + \cdots\right], \quad \eta \lll r \ll l$$

Hence confirm that,

$$\left\langle [\Delta v]^2 \right\rangle = \beta\varepsilon^{2/3}r^{2/3}\left[1 - \gamma\beta(r/l)^{2/3} + \cdots\right], \quad \eta \lll r \ll l$$
$$rV(r) = \tfrac{1}{3}\beta\varepsilon^{2/3}r^{2/3}\left[1 - \gamma\beta(r/l)^{2/3} + \cdots\right], \quad \eta \lll r \ll l$$

where l is the integral scale and γ is a positive, dimensionless coefficient. Now show that, for the case where Kolmogorov's 10/7th decay law applies, $\gamma = (3A/2)^{2/3}/3$ where A is the coefficient in the equation $du^2/dt = -A(u^3/l)$. Typically A is found to have a value of around 0.5, so that $\gamma \approx 0.2 \rightarrow 0.3$.

Example 6.22 The influence of the large-scale correction on the equilibrium range
Use the results of Example 6.21 above to show that the constant-skewness model $y'(x) + 2y^{3/2} = 2x$ can be corrected for the weak time dependence of the large scales to give,

$$y'(x) + 2y^{3/2} = 2x\left[1 - \tfrac{3}{2}\gamma\beta\sqrt{15\beta}x^{2/3}\mathrm{Re}^{-1/2}\right].$$

Solutions of this equation for $rV(r)$ are shown in Figure 6.49 for $\mathrm{Re} = 10^4, 10^5, \infty$.

Example 6.23 Finite Re corrections to the two-thirds, four-fifths, and five-thirds laws
Use the results of examples 6.20 and 6.21 above, along with (6.108), to show that, in the inertial subrange, the first-order finite-Re corrections to the four-fifths, two-thirds, and five-thirds laws are,

$$\begin{aligned}\left\langle [\Delta v]^3 \right\rangle &= -\tfrac{4}{5}\varepsilon r\left[1 - 5\beta(r/\eta)^{-4/3} + \tfrac{15}{34}\beta r^{2/3}\dot{\varepsilon}/\varepsilon^{4/3}\right]\\ &= -\tfrac{4}{5}\varepsilon r\left[1 - 5\beta(r/\eta)^{-4/3} - \tfrac{3}{2}\gamma\beta(r/l)^{2/3}\right]\\ \left\langle [\Delta v]^2 \right\rangle &= \beta\varepsilon^{2/3}r^{2/3}\left[1 - \tfrac{10}{3}\beta(r/\eta)^{-4/3} - \gamma\beta(r/l)^{2/3}\right]\\ &= \beta\varepsilon^{2/3}r^{2/3}\left[1 - \tfrac{10}{3}\beta(r/\eta)^{-4/3} - \gamma\beta(r/\eta)^{2/3}\mathrm{Re}^{-1/2}\right]\end{aligned}$$

$$E(k) \approx 0.77\beta\varepsilon^{2/3}k^{-5/3}\left[1 + 2.5(\eta k)^{4/3} - 4.6\gamma(\eta k)^{-2/3}\mathrm{Re}^{-1/2}\right].$$

The first correction term on the right of our expression for $E(k)$ is responsible for the bottleneck effect, while the second arises from the time dependence of the large scales and tends to mask the bottleneck.

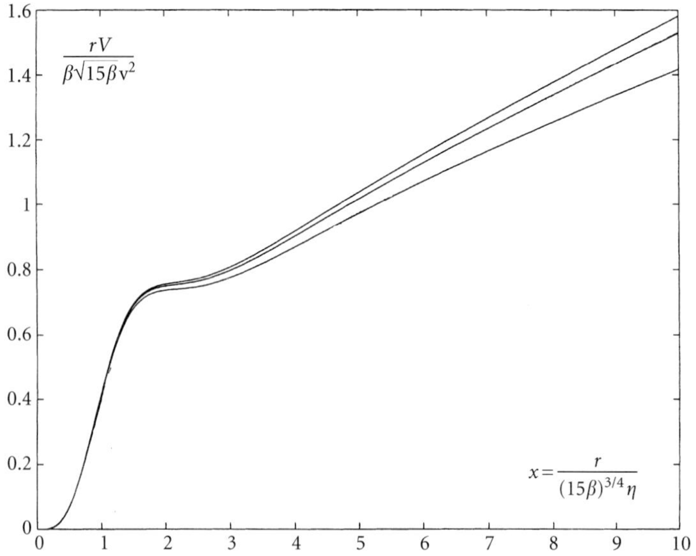

Figure 6.49 The shape of $rV(r)$ in the equilibrium range with a first-order correction for the time dependence of the large scales. $\mathrm{Re} = 10{,}000$ is the lowest curve and $\mathrm{Re} = \infty$ is the highest one. Notice that a finite value of Re changes the inertial range slope. (Constant skewness model, $\gamma = 0.21$, $\beta = 2$.)

Typically $\gamma \sim 0.21$. Now consider an experiment in which $\mathrm{Re} = 10^4$ and measurements of $\langle (\Delta v)^2 \rangle$ are made in the range $50 < r/\eta < 100$ in order to verify the two-thirds law. Show that, instead of obtaining a power-law exponent of 2/3, the data will suggest a value of ~ 0.647. This error has nothing at all to do with intermittency, and is purely a finite-Re effect. (See also the discussion in Lundgren 2003.)

Note that plotting $\langle (\Delta v)^2 \rangle$ against $\langle (\Delta v)^3 \rangle$, a rather than against r, masks the finite-Re corrections. Thus intermittency corrections to the 2/3 law are more readily detected by plotting $\langle (\Delta v)^2 \rangle$ against $\langle (\Delta v)^3 \rangle$. This is an example of what has become known as *extended self-similarity*.

6.6.3 *A 'real-space' function which represents, approximately, the distribution of the enstrophy*

It is clear from the discussion in Section 6.6.1, and (6.101a) in particular, that the real-space enstrophy density equivalent of $V(r)$ is,

$$\Omega(r) = \frac{10}{r^2} V(r) = \frac{5}{2} \langle \mathbf{u}^2 \rangle \frac{\partial}{\partial r} \frac{1}{r} \frac{\partial f}{\partial r}. \tag{6.115}$$

The relationship between $r\Omega(r)$ and its spectral equivalent, $k\left[k^2 E(k)\right]$, can be deduced from (6.25), (6.39), and (6.115). After a little algebra we find,

$$\Omega(r) = \int_0^\infty k^3 E(k) J(rk)\, dk \tag{6.116}$$

$$J(x) = -15x \left[\frac{1}{x}\frac{d}{dx}\right]^3 \left(\frac{\sin x}{x}\right). \tag{6.117}$$

It is readily confirmed that $J(x)$ has a maximum at $x \sim \pi$, and is small for $x \ll \pi$ and $x \gg \pi$. Since $\int_0^\infty J(x)\,dx = 1$ we may think of J as a sort of filter function which effectively extracts that part of $k^3E(k)$ which is centred around $r \sim \pi/k$. This gives weight to the idea that $r\Omega(r) \approx [k^3E(k)]_{k=\hat{\pi}/r}$, which is the enstrophy equivalent of (6.108). Note that a dimensionless form of $r\Omega$, based on the constant skewness model, is shown if Figure 6.46, labelled as z/x^2. In the constant skewness model $r\Omega$ peaks at $r \approx 13\eta$, which is around twice the minimum eddy size of $r \approx 6\eta$.

We can derive an evolution equation for $\Omega(r)$ from (6.109), or else directly from the Karman–Howarth equation. Either way we find,

$$\frac{\partial \Omega}{\partial t} = -\frac{\partial \Pi^*}{\partial r} + 2\nu\left[\frac{\partial}{\partial r}\frac{1}{r^6}\frac{\partial}{\partial r}(r^6\Omega)\right] \tag{6.118}$$

where

$$\Pi^* = -\frac{15}{2}\frac{1}{r}\frac{\partial}{\partial r}\frac{1}{r^4}\frac{\partial}{\partial r}(r^4u^3K). \tag{6.119}$$

If we integrate (6.118) from $r=0$ to $r \to \infty$ we obtain,

$$\frac{d}{dt}\int_0^\infty \Omega dr = \Pi_0^* - 2\nu\left[\frac{1}{r^6}\frac{\partial}{\partial r}(r^6\Omega)\right]_0, \quad \Pi_0^* = \Pi^*(0).$$

However, for small r we have $\Omega = (r/14)\langle(\nabla \times \boldsymbol{\omega})^2\rangle$, and so our equation simplifies to,

$$\frac{d}{dt}\left[\frac{1}{2}\langle \boldsymbol{\omega}^2\rangle\right] = \Pi_0^* - \nu\langle(\nabla \times \boldsymbol{\omega})^2\rangle. \tag{6.120}$$

Compare this with the enstrophy equation (5.28),

$$\frac{d}{dt}\left[\frac{1}{2}\langle \boldsymbol{\omega}^2\rangle\right] = \langle \omega_i\omega_jS_{ij}\rangle - \nu\langle(\nabla \times \boldsymbol{\omega})^2\rangle.$$

Evidently Π_0^* is the net rate of generation of enstrophy by vortex-line stretching across all of the different eddy sizes,

$$\Pi_0^* = \langle \omega_i\omega_jS_{ij}\rangle. \tag{6.121}$$

Let us, therefore, try to determine the shape of $\Pi^\star(r)$ for small r. We have $6u^3K = \langle(\Delta v)^3\rangle = S(r)\langle(\Delta v)^2\rangle^{3/2}$ and so we can rewrite (6.119) in the form,

$$\Pi^* = -\frac{5}{4}\frac{1}{r}\frac{\partial}{\partial r}\frac{1}{r^4}\frac{\partial}{\partial r}\left[S(r)\langle(\Delta v)^2\rangle^{3/2}r^4\right] \tag{6.122}$$

Moreover, near $r=0$ we know that $\langle(\Delta v)^2\rangle = (r^2/15)\langle\boldsymbol{\omega}^2\rangle$, and so (6.122) yields,

$$\Pi_0^* = -\frac{7}{6\sqrt{15}}S_0\langle\boldsymbol{\omega}^2\rangle^{3/2} \tag{6.123}$$

where S_0 is the skewness factor at $r=0$. Evidently the rate of generation of enstrophy by vortex stretching is related to the skewness by

$$\langle \omega_i \omega_j S_{ij} \rangle = \Pi_0^* = -\frac{7}{6\sqrt{15}} S_0 \langle \boldsymbol{\omega}^2 \rangle^{3/2}. \tag{6.124}$$

We have arrived back at (6.57).

6.6.4 *A footnote: can we capture Richardson's vision with our mathematical analysis?*

We end this chapter by returning to Richardson's picture of the energy cascade as a hierarchy of eddies of different sizes exchanging energy by vortex stretching. We ask: to what extent does our mathematical analysis capture this vision? In this respect the dynamic equation for $V(r)$ looks encouraging, as energy appears to pass from large to small scales. However, we might question the extent to which we have been successful in decomposing the velocity field, and its associated kinetic energy, into a hierarchy of 'eddies' of different size. Does $V(r)$, or $E(k)$ for that matter, really represent the distribution of energy across the eddies?

A moment's thought is sufficient to confirm that, to a large extent, we have failed. The key point is that we have not yet agreed on what we mean by an *eddy*. Is it a sheet, a tube, or a blob of vorticity? Unless we know what our basic building block is, how can we construct an algorithm which decomposes $\mathbf{u}$ or $\langle \mathbf{u}^2 \rangle$ into a distribution of such entities? Evidently, we cannot. The problem is made worse by the fact that the geometry of the 'generic eddy' may be different in different size ranges. For example, perhaps a typical eddy resembles a vortex blob at the large scales, a vortex sheet at the intermediate scales, and a vortex tube at the small scales. Or perhaps there is no such thing as a 'generic eddy'. Perhaps the vorticity field is just a shapeless mess!

Consider, as an example of our failure, the traditional strategy of deploying Fourier analysis. (Similar deficiencies arise with our use of a real-space function $V(r)$.) Here we divide up the velocity field into a hierarchy of mutually orthogonal Fourier modes (waves): $\mathbf{u} = \sum \mathbf{u}_i$. The orthogonality of the Fourier modes means that the kinetic energy may be likewise divided between the various modes:

$$\frac{1}{2}\langle \mathbf{u}^2 \rangle = \sum \frac{1}{2}\langle \mathbf{u}_i^2 \rangle.$$

Now the amplitudes of the individual modes is given by the Fourier transform of $\mathbf{u}(\mathbf{x})$, $\hat{\mathbf{u}}(\mathbf{k})$, while the energy spectrum, $E(k)$, provides a measure of the energy contained in the ith mode. That is, $E(k)$ is related to $\hat{\mathbf{u}}(\mathbf{k})$ by,

$$E(k)\delta(\mathbf{k}-\mathbf{k}') = 2\pi k^2 \left\langle \hat{\mathbf{u}}(\mathbf{k}) \cdot \hat{\mathbf{u}}^\dagger(\mathbf{k}') \right\rangle, \quad k = |\mathbf{k}|.$$

(Here δ is the three-dimensional Dirac delta function and $\mathbf{k}$ and $\mathbf{k}'$ are distinct wavevectors—see Chapter 8.) Now, if a single Fourier mode (wave) happened to represent a sea of eddies of given size, then this would be an ideal means of realizing Richardson's vision. That is, all eddies of a given size would be represented by their contribution to a single value of k, with eddy size $\sim k^{-1}$. However, we saw in Section 6.4.1 that a random distribution of simple eddies (vortex blobs) of fixed size l_e gives rise to a Fourier transform, and associated energy spectrum, which is distributed in $\mathbf{k}$-space, albeit peaked around $|\mathbf{k}| = \pi/l_e$. Thus, we cannot equate k^{-1} to eddy size. To illustrate the point, suppose we had an imaginary field of turbulence composed of a sea of eddies of two sizes, l_1 and l_2, in which $\mathbf{u} = \mathbf{u}_1 + \mathbf{u}_2$ and $\hat{\mathbf{u}} = \hat{\mathbf{u}}_1 + \hat{\mathbf{u}}_2$. From the expression above we have,

$$[E(k) - E_1(k) - E_2(k)]\delta(\mathbf{k} - \mathbf{k}') = 2\pi k^2 \left[\left\langle \hat{\mathbf{u}}_1(\mathbf{k}) \cdot \hat{\mathbf{u}}_2^\dagger(\mathbf{k}') + \hat{\mathbf{u}}_2(\mathbf{k}) \cdot \hat{\mathbf{u}}_1^\dagger(\mathbf{k}') \right\rangle \right]$$

where E_1 and E_2 are the energy spectra for $\mathbf{u}_1$ and $\mathbf{u}_2$ acting alone. Evidently, $E = E_1 + E_2$ if, and only if, $\hat{\mathbf{u}}_1(\mathbf{k})$ and $\hat{\mathbf{u}}_2(\mathbf{k})$ do not overlap in $\mathbf{k}$-space. Yet we have seen that, despite the difference in size of the two groups of eddies, $\hat{\mathbf{u}}_1$ and $\hat{\mathbf{u}}_2$ inevitably overlap. We conclude that the total energy spectrum, $E(k)$, is not simply the sum of $E_1(k)$ and $E_2(k)$. (Only in cases where l_1 and l_2 are markedly different, so that the overlap of $\hat{\mathbf{u}}_1(\mathbf{k})$ and $\hat{\mathbf{u}}_2(\mathbf{k})$ is slight, can we write. $E(k) \approx E_1(k) + E_2(k)$.) This is an inevitable consequence of the fact that the Fourier transform decomposes $\mathbf{u}$ into a hierarchy of *waves*, not *eddies*.

We might try to improve matters by using a different transform whose basis functions are a better match to a random sea of eddies. However, since we cannot agree as to what an eddy is, and we suspect that its generic shape changes as we pass down the energy cascade, such a strategy does not look promising.

All in all, we are a long way from realizing Richardson's vision. The energy spectrum, $E(k)$, is a blunt instrument, and the signature function, $V(r)$, an even cruder device. Moreover, since we cannot agree on what we mean by an eddy, it is far from clear how we might improve upon the situation. And that is not our only problem. Suppose that we did agree on the types of eddies to be found, and fixed on a satisfactory way of measuring the energy distribution across a hierarchy of such eddies. What then? It is unlikely that any statistical closure model could mimic, say, the intricate details of a field of vortex ribbons rolling up into a tangle of vortex tubes. In short, statistical models cannot reproduce the detailed dynamics of individual vortex events. At best they might capture the broad consequences of a multitude of such events. We should not expect too much from these models.

Exercises

6.1 Use (6.40) to show that $\langle \omega_i \omega_i' \rangle$ and $\langle [\Delta \mathbf{v}]^2 \rangle$ are related by

$$\langle \omega_i \omega_i' \rangle = \frac{1}{2r^2} \frac{\partial}{\partial r} \left[r^2 \frac{\partial}{\partial r} \langle [\Delta \mathbf{v}]^2 \rangle \right].$$

6.2 From (6.30) we have $\langle p' u_i \rangle = A(r) r_i$. Conservation of mass also requires $\partial \langle p' u_i \rangle / \partial r_i = 0$. Show that $rA' + 3A = 0$, and hence deduce that, for isotropic turbulence, $\langle p' u_i \rangle = 0$.

6.3 Use (6.40) to show that, for small r, $\langle [\Delta \mathbf{v}]^2 \rangle = \varepsilon r^2 / 3\nu$.

6.4 Derive Kolmogorov's decay laws from (6.75) and (6.76).

6.5 Confirm that, for flow evolving in a closed sphere,

$$\mathbf{H}^2 = -\iint (\mathbf{x}' - \mathbf{x})^2 \mathbf{u} \cdot \mathbf{u}' dV \, dV'$$

where $\mathbf{H}$ is the angular momentum in the sphere. Hint: first show that

$$\begin{aligned}(\mathbf{x} \times \mathbf{u}) \cdot (\mathbf{x}' \times \mathbf{u}') &= (\mathbf{x} \cdot \mathbf{x}')(\mathbf{u} \cdot \mathbf{u}') - (\mathbf{x} \cdot \mathbf{u}')(\mathbf{x}' \cdot \mathbf{u}) \\ &= 2(\mathbf{x} \cdot \mathbf{x}')(\mathbf{u} \cdot \mathbf{u}') + \nabla \cdot (\sim)\end{aligned}$$

where the divergence vanishes on integration.

6.6 Show that, for large $|\mathbf{x}|$,

$$|\mathbf{x}' - \mathbf{x}|^{-1} = |\mathbf{x}|^{-1} - \frac{\partial}{\partial x_i} \left(\frac{1}{|\mathbf{x}|} \right) x_i' + \frac{1}{2} \frac{\partial^2}{\partial x_i \partial x_j} \left(\frac{1}{|\mathbf{x}|} \right) x_i' x_j' + \cdots$$

and hence confirm that, for a localized vorticity field, the far-field pressure is

$$4\pi p(\mathbf{x})/\rho = \int \frac{\partial^2 u_i' u_j'}{\partial x_i' \partial x_j'} \frac{d\mathbf{x}'}{|\mathbf{x}' - \mathbf{x}|} = \frac{\partial^2}{\partial x_i \partial x_j} \left(\frac{1}{|\mathbf{x}|} \right) \int u_i' u_j' d\mathbf{x}' + \cdots.$$

6.7 Estimate the QN value of α in (6.97) by taking $f(r)$ to be of the form $\exp\left[-r^2/L^2\right]$ for some L. [Hint: first find I in terms of L and hence find the relationship between $l = (I/u^2)^{1/5}$ and L.]

6.8 Give a dynamical reason, based on the method of formation of the turbulence, why you might expect Saffman's integral to be zero in grid turbulence. [Hint: consult Saffman 1967.]

6.9 Show that the signature function and second-order structure function are related by,

$$\int_0^r V(r) \, dr = -\frac{3}{8} r^3 \frac{d}{dr} \left[\langle [\Delta v]^2 \rangle / r^2 \right]$$

Now use the integral relationship below (6.107)

$$\int_0^r V(r) \, dr = \int_0^\infty E(k) G(rk) \, dk, \quad G(rk) \geq 0$$

to show that $\langle [\Delta v]^2 \rangle / r^2$ is a monotonically decreasing function of r.

6.10 Consider the Lundgren energy spectrum (see Section 6.4.2)

$$E = \alpha\varepsilon^{2/3}k^{-5/3}\exp\left[-(k\eta)^2\right], \quad \alpha = 1.52$$

where η is the Kolmogorov microscale. Use the Hankel transform pair (6.107) to show that the corresponding signature function is,

$$rV(r) = \frac{3\sqrt{\pi}\Gamma(5/3)}{2^6\Gamma(9/2)}\frac{\alpha\varepsilon^{2/3}r^4}{\eta^{10/3}}M\left(\frac{5}{3},\frac{9}{2},-\frac{r^2}{4\eta^2}\right)$$

where Γ is the gamma function and M is Kummer's hypergeometric function. Use the asymptotic form of $M(a,b,z)$ for large z to show that, for $r \gg \eta$, we recover Kolmogorov's two-thirds law in the form,

$$rV(r) = \tfrac{1}{3}\beta\varepsilon^{2/3}r^{2/3}.$$

6.11 From (6.107b) we have

$$V(r) = \frac{3}{2}\int_0^{\infty}\left[k^2E(k)x^{-1}j_3(x)\right]dx, \quad x = kr$$

where $j_3(x)$ is the usual spherical Bessel function of the first kind. The function $x^{-1}j_3(x)$ peaks around $x = \hat{\pi} = 9\pi/8$ and is small for $x < 1$ and $x > 10$. Consequently, the primary contribution to $V(r)$ come $k^2E(k)$ from in the range $k = r^{-1} \to 10r^{-1}$. Let us suppose that $k^2E(k)$ varies sufficiently slowly in the range $k = r^{-1} \to 10r^{-1}$ that it may be approximated by a linear function of k matched to $k^2E(k)$ at $k = \hat{\pi}/r$, that is,

$$k^2E(k) \approx \left[k^2E(k)\right]_{k_0} + \left[\partial(k^2E)/\partial k\right]_{k_0}(k - k_0), \quad k_0 = \hat{\pi}/r.$$

Verify that substituting this estimate into our expression for $V(r)$ yields

$$V(r) \approx \left[k^2E(k)\right]_{k_0}(3\pi/32) + k_0\left[\partial(k^2E)/\partial k\right]_{k_0}(1/rk_0 - 3\pi/32),$$

and show that this can be rearranged to give

$$\frac{rV(r)}{\left[kE(k)\right]_{k_0}} = \lambda \approx \left[1 - 0.0409(\partial kE/\partial k)_{k_0}/E(k_0)\right].$$

Evidently, when $E(k)$ is a slowly varying function we have $rV(r) \cong [kE(k)]_{k_0}$ in accordance with (6.108). On the other hand, when $\partial(kE)/\partial k$ greatly exceeds $E(k)$ the estimate (6.108) becomes inaccurate. A comparison of the estimate of λ given above with the exact value of λ_n for spectra of the form $E = Ak^n$ is shown in Figure 6.50. (See Example 6.15 of Section 6.6.1 for the exact values of λ_n.)

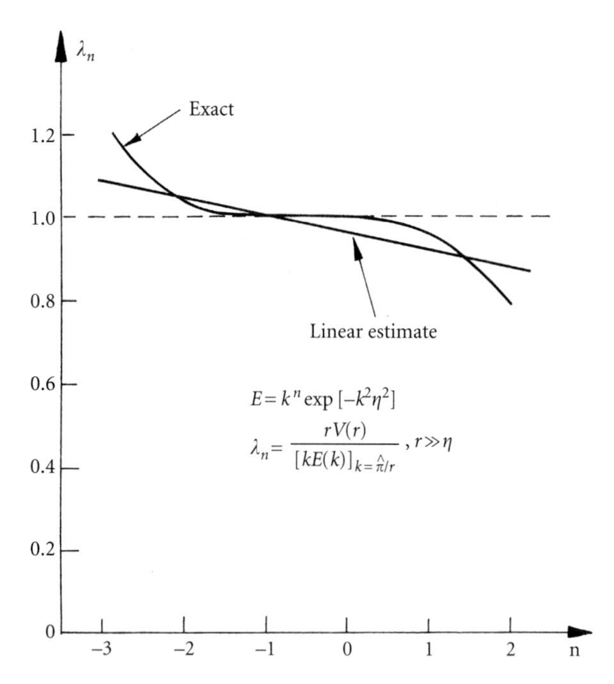

Figure 6.50 A comparison of the linear estimate of λ above with the exact solution for power law spectra given in example 6.15 of Section 6.6.1.

6.12 In Section 6.4.1 we saw that a random distribution of Townsend 'model eddies' (spherical blobs of vorticity) of fixed size l_e gives rise to the longitudinal correlation function $f(r) = \exp[-r^2/l_e^2]$. The corresponding spectrum is,

$$E(k) = \frac{\langle \mathbf{u}^2 \rangle l_e}{24\sqrt{\pi}} (kl_e)^4 \exp\left[-l_e^2 k^2/4\right].$$

Suppose that an artificial field of turbulence is composed of a random array of Townsend eddies of size l_1, plus a random distribution of eddies of size l_2, l_3, l_4, and so on. (All eddies are statistically independent.) Also, we choose $l_1, l_2, \ldots, l_N$ to be, $l_2 = 0.1 l_1, \ldots, l_N = 0.1 l_{N-1}$. In such a case the correlation function will be the sum of exponentially decaying functions of the type $f \sim \exp[-r^2/l_i^2]$, and the corresponding energy spectrum will be of the form (see the Appendix),

$$E(k) = \sum_i \frac{\langle \mathbf{u}_i^2 \rangle l_i}{24\sqrt{\pi}} (kl_i)^4 \exp\left[-l_i^2 k^2/4\right].$$

Here $\langle \mathbf{u}_i^2 \rangle$ is the contribution to $\langle \mathbf{u}^2 \rangle$, which comes from eddies of size l_i. Find the corresponding form of the signature function, $V(r)$, defined by (6.100). Confirm that, for the case $N = 5$ and $\langle \mathbf{u}_i^2 \rangle / \langle \mathbf{u}^2 \rangle = (0.05, 0.25, 0.4, 0.25, 0.05)$, $rV(r)$ has the form shown in Figure 6.51. Also shown in the figure is $[kE(k)]_{k=\hat{\pi}/r}$ which, according to (6.108), should have a similar form.

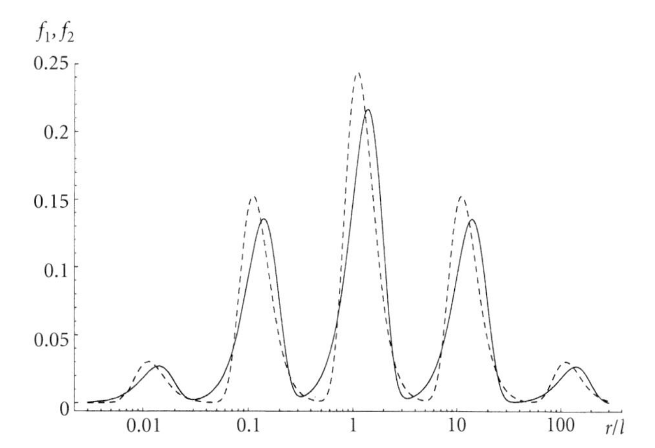

Figure 6.51 A comparison of $f_1(r) = [kE(k)]_{k=\hat{\pi}/r}$ (chain line) and $f_2(r) = rV(r)$ (solid line) for 'turbulence' composed of a random distribution of Townsend eddies of five different sizes and varying intensity.

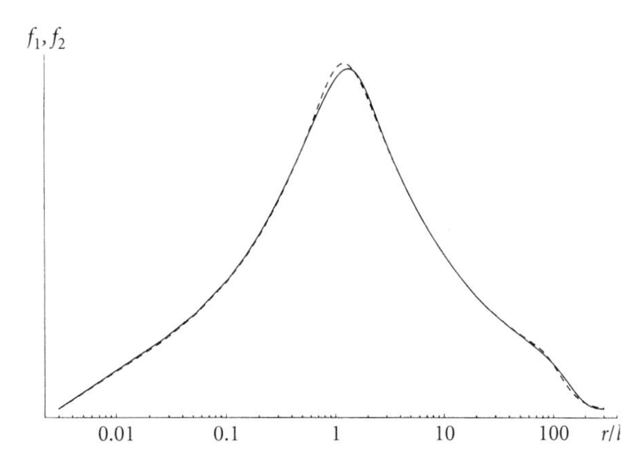

Figure 6.52 A comparison of $f_1(r) = [kE(k)]_{k=\hat{\pi}/r}$ (chain line) and $f_2(r) = rV(r)$ (solid line) for 'turbulence' composed of a random distribution of Townsend eddies of a 100 different sizes and varying intensity.

Finally, Figure 6.52 shows the case where the range of eddy sizes is the same but there are a hundred different eddy sizes rather than five, that is, $N = 100$ and $l_N = (0.1)^{1/25} l_{N-1}$. Note that the individual signatures all blend to form a smooth spectrum. Note also that the correspondence between $rV(r)$ and $[kE(k)]_{k=\hat{\pi}/r}$ is better for the smooth spectrum than for that shown in Figure 6.51, as anticipated in Example 6.11 above.

Appendix: Turbulence composed of Townsend's model eddy

We wish to explore the kinematic properties of an artificial field of turbulence created by randomly sprinkling eddies of given shape throughout all space. In particular, consider the vector potential

$$\mathbf{A} = \tfrac{1}{4}\Omega l_e^2 \exp\left[-2\mathbf{x}^2/l_e^2\right]\hat{\mathbf{e}}_z. \tag{A.1}$$

The corresponding velocity field is, in (r,θ,z) coordinates,

$$\mathbf{u} = \nabla \times \mathbf{A} = \Omega r \exp\left[-2\mathbf{x}^2/l_e^2\right]\hat{\mathbf{e}}_\theta. \tag{A.2}$$

Of course, this is Townsend's model eddy of Section 6.4.1, where l_e is the eddy size. Now suppose that we create a field of turbulence by distributing such eddies randomly yet uniformly in space. Then we have

$$\mathbf{A} = \frac{1}{4}\Omega l_e^2 \sum_m \exp\left[-2(\mathbf{x}-\mathbf{x}_m)^2/l_e^2\right]\hat{\mathbf{e}}_\mathbf{m} \tag{A.3}$$

where $\hat{\mathbf{e}}_m$ and $\mathbf{x}_m$ give the orientation and position of the *mth* eddy. The components of $\hat{\mathbf{e}}_m$ and $\mathbf{x}_m$ constitute a set of independent random variables. Now consider the product of $\mathbf{A}$ at $\mathbf{x}=\mathbf{0}$ with that at $\mathbf{x}=\mathbf{r}$:

$$\mathbf{A}(0)\cdot\mathbf{A}(\mathbf{r}) = \left(\tfrac{1}{4}\Omega l_e^2\right)^2 \sum_m \exp\left[-2(\mathbf{x}_m)^2/l_e^2\right]\hat{\mathbf{e}}_m \cdot \sum_n \exp\left[-2(\mathbf{r}-\mathbf{x}_n)^2/l_e^2\right]\hat{\mathbf{e}}_n.$$

Since the components of $\hat{\mathbf{e}}_m$ and $\hat{\mathbf{e}}_n$ are independent random variables with zero mean we have $\langle\hat{\mathbf{e}}_m\cdot\hat{\mathbf{e}}_n\rangle = 0$ for $m \neq n$. It follows that

$$\langle\mathbf{A}(0)\cdot\mathbf{A}(\mathbf{r})\rangle = \frac{1}{16}\Omega^2 l_e^4 \sum_m \left\langle \exp\left[-2\left(\mathbf{x}_m^2 + (\mathbf{r}-\mathbf{x}_m)^2\right)/l_e^2\right]\right\rangle$$

which, after a little algebra, simplifies to

$$\langle\mathbf{A}(0)\cdot\mathbf{A}(\mathbf{r})\rangle = \frac{1}{16}\Omega^2 l_e^4 \exp\left[-r^2/l_e^2\right] \sum_m \left\langle \exp\left[-4\mathbf{y}_m^2/l_e^2\right]\right\rangle.$$

The quantity $\mathbf{y}_m = \mathbf{x}_m - \frac{1}{2}\mathbf{r}$ is a new random variable obtained from $\mathbf{x}_m$ by a shift of origin. Since the summation on the right is simply a coefficient whose value depends on l_e, we have

$$\langle\mathbf{A}(0)\cdot\mathbf{A}(\mathbf{r})\rangle = \langle\mathbf{A}\cdot\mathbf{A}'\rangle = A_0^2 \exp\left[-r^2/l_e^2\right] \tag{A.4}$$

for some constant A_0. We can now find $\langle\mathbf{u}\cdot\mathbf{u}'\rangle$ and the longitudinal correlation function, $f(r)$, from the relationships

$$\frac{1}{r^2}\frac{\partial}{\partial r} r^3 u^2 f(r) = \langle\mathbf{u}\cdot\mathbf{u}'\rangle = -\nabla^2\langle\mathbf{A}\cdot\mathbf{A}'\rangle.$$

It is readily confirmed that, since $f(0)=1$,

$$f(r) = \exp\left[-r^2/l_e^2\right]. \tag{A.5}$$

The corresponding energy spectrum can be found from (6.24) and may be shown to be

$$E(k) = \frac{\langle\mathbf{u}^2\rangle l_e}{24\sqrt{\pi}}(kl_e)^4 \exp\left[-l_e^2k^2/4\right]. \tag{A.6}$$

Now suppose that the turbulence is composed of a sea of Townsend eddies of size l_1 (randomly but evenly distributed in space) plus a sea of eddies of size l_2, l_3, l_4 and so on. The analysis above may be adapted to yield,

$$f(r) = \sum_i \frac{\langle \mathbf{u}_i^2 \rangle}{\langle \mathbf{u}^2 \rangle} \exp\left[-r^2/l_i^2\right] \tag{A.7}$$

$$E(k) = \sum_i \frac{\frac{1}{2}\langle \mathbf{u}_i^2 \rangle l_i}{12\sqrt{\pi}} (kl_i)^4 \exp\left[-l_i^2 k^2/4\right] \tag{A.8}$$

where $\frac{1}{2}\langle \mathbf{u}_i^2 \rangle$ is the contribution to $\frac{1}{2}\langle \mathbf{u}^2 \rangle$, which comes from eddies of size l_i:

$$\frac{1}{2}\langle \mathbf{u}^2 \rangle = \sum \tfrac{1}{2}\langle \mathbf{u}_i^2 \rangle.$$

Examples of such spectra are given in Exercise (6.12) above. In the limit where there is a continuous distribution of eddy sizes we can replace l_i by the continuous variable s and $\frac{1}{2}\langle \mathbf{u}_i^2 \rangle$ by the energy density $\hat{E}(s)$, which has the property

$$\frac{1}{2}\langle \mathbf{u}^2 \rangle = \int_0^\infty \hat{E}(s)\, ds = \int_0^\infty E(k)\, dk. \tag{A.9}$$

Then (A.8) becomes

$$E(k) = \int_0^\infty \frac{\hat{E}(s)s}{12\sqrt{\pi}} (ks)^4 \exp\left[-(ks)^2/4\right] ds \tag{A.10}$$

which shows the relationship between the energy spectrum, $E(k)$, and the 'real space' energy density, $\hat{E}(s)$.

Of course, this is all a little artificial because real turbulence is not composed of a collection of Townsend model eddies. Nevertheless, (A.3) is admissible in a kinematic sense and represents a legitimate initial condition. The important thing about (A.10) is that it is $\hat{E}(s)$, and *not* $E(k)$, which represents the distribution of energy across the different eddy sizes. So (A.10) give us an opportunity to see just how good (or poor) a job $E(k)$ does in representing the distribution of energy. By way of illustration, consider the energy distribution given by

$$\hat{E}(s) = \frac{\langle \mathbf{u}^2 \rangle \exp(2\eta/L)}{\sqrt{\pi}L} \exp\left[-(s/L)^2 - (\eta/s)^2\right], \quad \eta \ll L.$$

The exponential above is more or less constant and equal to unity for $(\pi\eta) < s < (L/\pi)$, and is extremely small for $s < (\eta/\pi)$ and $s > (L\pi)$. So this represents a random velocity field in which energy is distributed relatively evenly across the size range $\eta < s < L$, but $\hat{E}(s)$ is exponentially small outside this range. The corresponding energy spectrum is, from (A.10),

$$E(k) = \frac{\langle \mathbf{u}^2 \rangle \eta^2 e^{2\eta/L}}{12\pi L} (\eta k)^4 \frac{K_3\left(2\sqrt{(\eta/L)^2 + (\eta k/2)^2}\right)}{\left((\eta/L)^2 + (\eta k/2)^2\right)^{3/2}}$$

where K_3 is the usual modified Bessel function. It is instructive to look at the limits of small and large k. For small η/L we find that

$$E(k \to 0) = \frac{\langle \mathbf{u}^2 \rangle L^5 k^4}{12\pi}, \qquad E(k \to \infty) = \frac{\sqrt{2}\langle \mathbf{u}^2 \rangle \eta^2}{3\sqrt{\pi} L} (k\eta)^{1/2} \exp[-k\eta].$$

We observe two interesting features of these expressions:

1. $E(k)$ falls off as $\exp(-k\eta)$, rather than as $\exp(-k^2\eta^2)$. This is precisely what is observed in practice.
2. If we adopt the rule of thumb that $k \sim \pi/s$, then $E(k)$ gives the misleading impression that there is a significant amount of energy in the range $s > L$, despite the fact that there are virtually no eddies there. This is a common failing of the energy spectrum. Treating $E(k)$ as an energy distribution makes physical sense only if the turbulence is composed of waves. But it is not! Numerical simulations show that it is composed of lumps of vorticity (blobs, sheets, tubes . . .).

Suggested reading

Books

Batchelor, G.K. (1953) *The Theory of Homogeneous Turbulence*. Cambridge University Press. [Still one of the best accounts of homogeneous turbulence.]

Frisch, U. (1995) *Turbulence*. Cambridge University Press. [Discusses intermittency at length.]

Hinze, J.O. (1959) *Turbulence*. McGraw–Hill.

Landau, L.D. and Lifshitz, E.M. (1959) *Fluid Mechanics*, 1st edition. Pergamon.

Lesieur, M. (1990) Turbulence in Fluids, 2nd edition. Kluwer Academic Publishers. [Gives one of the best discussions of two-point closures of the EDQNM type.]

McComb, W.D. (1990) *The Physics of Fluid Turbulence*. Oxford University Press. [This gives a detailed discussion of two-point closures.]

Monin, A.S. and Yaglom, A.M. (1975) *Statistical Fluid Mechanics II*. MIT Press.

Tennekes, H. and Lumley J.L. (1972) *A First Course in Turbulence*. MIT Press.

Townsend, A.A. (1956) *The Structure of Turbulent Shear Flow*. Cambridge University Press. [See chapter 1 for a discussion of the physical interpretation of $E(k)$.]

Journal References

Batchelor, G.K. and Proudman, I. (1956) *Phil. Trans. Roy. Soc. A*, **248**, 369–405.

Birkhoff, G. (1954) *Comm. Pure Appl. and Math.*, 7,19–44.

Chasnov, J.R. (1993) *Phys. Fluids*, **5**, 2579–2581

Comte–Bellot, G. and Corrsin, S. (1966) *J. Fluid Mech.*, **25**, 657–682.

Davidson, P.A. (2000) *J. Turbulence*, **1**.

Fukayama, D. et al. (2001) *Phys. Rev E*, **64**, 016304

Kolmogorov, A.N. (1941*a*) *Dokl. Akad. Nauk SSSR*, **31**(6), 538–541.

Kolmogorov, A.N. (1941*b*) *Dokl. Akad. Nauk SSSR*, **32**(1), 19–21.

Kolmogorov, A.N. (1962) *J. Fluid Mech.*, **13**, 82–85.

Lesieur, M., Ossia, S., and Metais, O. (1999) *Phys. Fluids*, **11**(6), 1535–1543.

Lesieur, M. et al. (2000) *European Congress on Computational Methods in Science and Engineering*, Barcelona, September 2000.

Loitsyansky, L.G. (1939) *Trudy Tsentr. Aero.-Giedrodin Inst.*, **440**, 3–23.

Lundgren, T.S. (1982) *Phys. Fluids*, **25**, 2193–2203.

Lundgren, T.S. (2003) *Phys. Fluids*, **15**(4), 1074–1081.

Obukhov, A.M. (1949) *Dokl. Akad. Nauk SSSR*, **67**(4), 643–646.

Orszag, S.A. (1970) *J. Fluid Mech.*, **41**(2), 363–386.

Proudman, I. and Reid, W.H. (1954) *Phil. Trans. Roy. Soc., A*, **247**, 163–189.

Pullin, D.I. and Saffman, P.G. (1998) *Ann. Rev. Fluid Mech.*, **30**, 31–51.

Rosen, G. (1981) *Phys. Fluids*, **24**(3), 558–559.

Saddoughi, S.G. and Veeravalli, S.V. (1994) *J. Fluid Mech.*, **268**, 333–372.

Saffman, P.G. (1967) *J. Fluid Mech.*, **27**, 581–593.

Synge, J.L. and Lin, C.C. (1943) *Trans. Roy. Soc. Canada*, **37**, 45–79.

Townsend, A.A. (1951) *Proc. Roy. Soc. London. A*, **208**, 534–542.

Van Atta, C.W. and Yeh, T.T. (1970) *J. Fluid Mech.*, **41**(1), 169–178.

Warhaft, Z. and Lumley, J.L. (1978) *J. Fluid Mech.*, **88**, 659–684.

CHAPTER 7

The role of numerical simulations

> It seems that the surge of progress which began immediately after the war has now largely spent itself . . . we have got down to the bedrock difficulty of solving non-linear partial differential equations.
>
> G.K. Batchelor (1953)

In this chapter we discuss the insights into the nature of turbulence which have resulted from the application of numerical methods to the Navier–Stokes equation. These computerized integrations, sometimes called numerical experiments, have become extremely popular in the last 30 years, largely as a result of the growth in computer power. There is now a vast literature devoted to the do's and dont's of implementing these numerical simulations.[1] We shall discuss none of these technical issues here, but rather focus on the basic ideas and the end results. We are interested primarily in homogeneous turbulence, and in the insights into homogeneous turbulence which the computations have yielded.

7.1 What is DNS or LES?

7.1.1 *Direct numerical simulations*

In 1972 a new chapter in the theory of turbulence began: Orszag and Patterson demonstrated that it was possible to perform computer simulations of a fully developed turbulent flow. It is important to understand that these simulations do not require some turbulence model to parameterize the influence of the turbulent eddies. Rather, every eddy, from the largest to the smallest, is computed. Starting from specified initial conditions the Navier–Stokes equation, $\partial \mathbf{u}/\partial t = (\sim)$, is integrated forward in time in some specified domain. It is like carrying out an experiment, only on the computer rather than in a wind tunnel. Indeed, such simulations are called *numerical experiments*.

[1] Those readers interested in numerical issues might consult Canuto et al. (1987) or Hirsch (1988).

The potential benefits were immediately clear. Initial conditions can be controlled in such simulations in a way which is just not possible in the laboratory. Moreover, the amount of data which can be recovered is overwhelming: in effect, the entire history of the velocity field $\mathbf{u}(\mathbf{x},t)$ is available for inspection. The possibilities seemed endless and many fluid dynamicists were attracted to the growing field of *direct numerical simulation* (DNS).

However, there was, and is, a catch. We have already seen that the Kolmogorov microscale (the approximate size of the smallest eddies) is,

$$\eta = (ul/\nu)^{-3/4} l = \mathrm{Re}^{-3/4} l \tag{7.1}$$

where l is the size of the large, energy-containing eddies (the integral scale). When Reynolds number (Re) is large, which it always is, η is small. It is easy to estimate from (7.1) the number of points at which $\mathbf{u}$ must be calculated in order to resolve every eddy in the turbulence. The spatial separation of the sampling points, Δx, cannot be greater than η. So, at a minimum, we require,[2]

$$\Delta x \sim \eta \sim \mathrm{Re}^{-3/4} l.$$

The number of data points required at any instant for a three-dimensional simulation is therefore,

$$N_x \sim \left(\frac{L_{\mathrm{BOX}}}{\Delta x}\right)^3 \sim \left(\frac{L_{\mathrm{BOX}}}{l}\right)^3 \mathrm{Re}^{9/4}. \tag{7.2}$$

Here L_{BOX} is a typical linear dimension of the computational domain (flow field size). The factor $\mathrm{Re}^{9/4}$ is referred to as the number of degrees of freedom of the turbulence, a phrase coined (in this respect) by Landau and Lifshitz (1959). We can rewrite (7.2) as

$$\mathrm{Re} \sim \left(\frac{l}{L_{\mathrm{BOX}}}\right)^{4/3} N_x^{4/9} \tag{7.3}$$

and we see immediately that there is a problem. To obtain a large Reynolds number we need a vast number of data points. Of course, we want to have more than just one large eddy in our simulation and so (7.3) can be written as

$$\mathrm{Re} \ll N_x^{4/9}.$$

In the pioneering (1972) work of Orszag and Patterson the Reynolds number, based on the Taylor microscale,[3] was limited to $R_\lambda \sim 35$.

[2] In Section 6.6.2 we saw that the smallest eddies have a size of around 6η and so it has become common to take $\Delta x \sim 2\eta$ giving a crude, but probably adequate, representation of the smallest eddies.

[3] It has become conventional to use a Reynolds number based on the Taylor microscale, $u\lambda/\nu$, rather than ul/ν, to characterize DNS. This is because most simulations focus attention on the small-to-intermediate scales, which are characterized by λ or η.

(Recall that $R_\lambda \sim \sqrt{15}\text{Re}^{1/2}$). To increase this by, say, a factor of 2 requires increasing N_x by a factor of around 20. So, over the years, the rate of rise of R_λ has been depressingly slow. Despite the advent of parallel processors, the rate of rise of R_λ has been a mere factor of ~ 2 every decade, reaching around 80 by 1985 and around 160 by the mid 1990s.[4] By comparison, the values of R_λ of interest to the engineer (e.g. flow over an aircraft) or the physicist (e.g. atmospheric flows) are vast. There is a huge chasm between what can be realized by DNS and what the engineer needs to know.

We might estimate the computational cost of a numerical simulation as follows. The maximum permissible time step in a simulation is of the order of $\Delta t \sim \Delta x/u \sim \eta/u$ since, in order to maintain numerical stability and accuracy, we cannot let a fluid particle move by more than one grid spacing in each time step.[5] If T is the total duration of the simulation then the minimum number of time steps needed to complete the simulation is,

$$N_t \sim \frac{T}{\Delta t} \sim \frac{T}{\eta/u} \sim \frac{T}{l/u}\text{Re}^{3/4}.$$

Now it happens that the number of computer operations required for a simulation is approximately proportional to $N_x N_t$, and so the computing time needed scales as,

$$\text{Computer time} \propto N_x N_t \sim \left(\frac{T}{l/u}\right)\left(\frac{L_{\text{BOX}}}{l}\right)^3 \text{Re}^3. \tag{7.4}$$

Of course, the constant of proportionality depends on the speed of the computer but typically, even with the fastest of machines currently available, we find that simulations in which $\text{Re} \sim 10^3$ require several days to complete, while simulations in which $\text{Re} \sim 10^4$ may take weeks. This is indicated in Table 7.1 below.

It is evident from Table 7.1 that large values of R_λ can be achieved only if L_{BOX} is restricted to $10l$ or less. Even then, values of R_λ much above ~ 700 are difficult to realise with current computers.[6] The same information is shown in Figure 7.1 for a teraflop computer.

[4] At the time of writing the most ambitious simulations tend to have $R_\lambda \sim 400$, (with one exception which is discussed below). However, such simulations are possible only in the simplest of geometries and model only small regions of space, that is, regions of size comparable with the integral scale.

[5] The optimal value of Δt depends on the form of time stepping used. In practice, many researchers use a fourth-order Runge–Kutta scheme, in which case it is found necessary to take the *Courant number*, $u\Delta t/\Delta x$, to be no greater than ~ 0.1.

[6] At the time of writing the most ambitious simulation corresponds to $R_\lambda = 1200$ (see Plate 8) and was carried out by Kaneda et al. (2003). This required a massive 16.4 teraflop computer, and even then only two eddy turn-over times were achieved and the computational domain was not much larger than the integral scale (one large eddy size). Of course, things will improve with time as computers become faster. It is an empirical observation that the maximum computer speed doubles every 18 months or so. This is

Table 7.1 Estimated computer run times for (i) a gigaflop computer and (ii) a teraflop computer, based on equation (7.4)

Gigaflop computer $L_{BOX}(l)$	Re = 100 $R_\lambda = 39$	Re = 500 $R_\lambda = 87$	Re = 1000 $R_\lambda = 122$	Re = 5000 $R_\lambda = 274$
10	2 h	9 days	2 months	24 years
20	1 day	2 months	2 years	2 centuries
50	9 days	3 years	24 years	A long time
100	2 months	24 years	2 centuries	A very long time
Teraflop computer $L_{BOX}(l)$	Re = 1000 $R_\lambda = 122$	Re = 5000 $R_\lambda = 274$	Re = 10,000 $R_\lambda = 386$	Re = 50,000 $R_\lambda = 866$
10	2 h	9 days	2 months	24 years
20	1 day	2 months	2 years	2 centuries
50	9 days	3 years	24 years	A long time
100	2 months	24 years	2 centuries	A very long time

Note: We have used the rule of thumb that a simulation for $R_\lambda = 90$ and $L_{BOX} = 3l$ takes around half a day at a computing rate of 1 gigaflop. The simulation is assumed to take place in a periodic cube. The total run time is five eddy turn-over times, $T = 5(l/u)$, and the resolution is assumed to be $\Delta x = 2\eta$. At the time of writing there are relatively few computers which can approach a speed of 1 teraflop. Note that $\text{Re} = ul/\nu$ where l is the integral scale.

Equation (7.4) has been the curse of DNS, overshadowing all of the developments for the past three decades. It has had a profound effect on the way that we view these simulations. There is no pretence that DNS can provide a brute-force method of solving engineering problems: it clearly cannot. Instead, it is regarded as a scientific tool. Like a laboratory experiment it can be used to investigate the development of turbulent flows at modest Re and in very simple geometries. Perhaps its most striking success in this respect has been the identification of generic structures in turbulence. For example, we are now in a better position to answer the old and much debated question: what is an eddy? We shall return to this in Section 7.3.

In order to raise the value of R_λ investigators have, by and large, gone for the simplest possible geometries. For those interested in homogeneous turbulence these often take the form of the so-called periodic cube. This is a cubic domain which has the special property that whatever happens at one face of the cube happens at the opposite face. This is clearly unphysical but the rationale is that, if $L_{BOX} \gg l$, then the

known as Moore's law. From (7.4) we see that this translates to a doubling of the maximum achievable R_λ every 10 years or so, which is more or less what has happened. So the rate of improvement is likely to be slow.

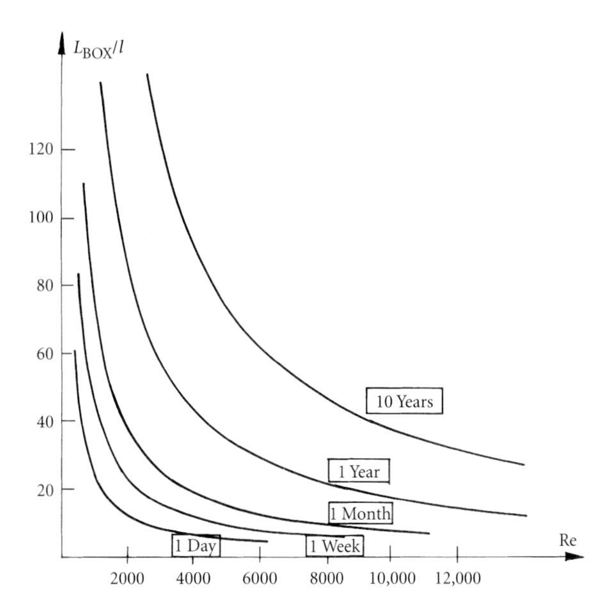

Figure 7.1 Estimated computation times for simulations in a periodic cube at a computing rate of 1 teraflop. The total run time is taken to be five eddy turn-over times, $T = 5(l/u)$, and the resolution is $\Delta x \approx 2\eta$. Note that $\mathrm{Re} = ul/\nu$ where l is the integral scale.

bulk of the turbulence does not care about the idiosyncratic boundary conditions. The great advantage of the periodic cube is that it lends itself to particularly efficient numerical algorithms for solving the Navier–Stokes equations (so-called pseudo-spectral methods).[7] Many (possibly most) DNS studies have been performed in periodic cubes, or else in domains which are periodic in at least one direction.

7.1.2 *Large eddy simulations*

There has been another assault on (7.4), called *large eddy simulation* (LES). This is a sort of half-way house between DNS and turbulence closure schemes. The idea of LES is to compute both the mean flow and the large, energy-containing eddies exactly. The small-scale structures are not simulated, but their influence on the rest of the flow is parameterized by some heuristic model. The success of LES rests on the fact that, by and large, energy and information tends to travel down the energy cascade to small scales, but not in the reverse direction. In short, the small scales are somewhat passive, simply mopping up whatever energy is passed down to them. So, if we cut the spectrum at some intermediate point, and provide a dust-bin for the energy flux, then

[7] In this method the velocity field is represented by a truncated Fourier series with time-dependent coefficients. The Navier–Stokes equation yields an evolution equation for each Fourier coefficient and the scheme works by marching forward in time from some specified initial condition. The most expensive part of the procedure is the evaluation of the non-linear terms in Fourier space which involves N_x^2 operations at each time step. To avoid this it is normal to transform to physical space at each step, evaluate the non-linear terms in real space, and then transform back to Fourier space. This involves $N_x \log N_x$ operations. (See Canuto et al. 1987.)

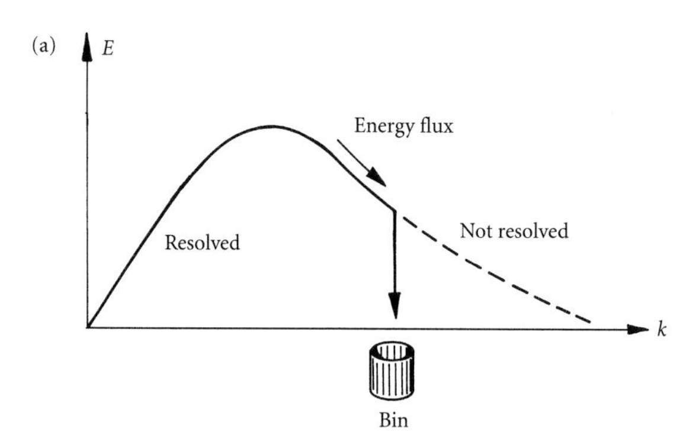

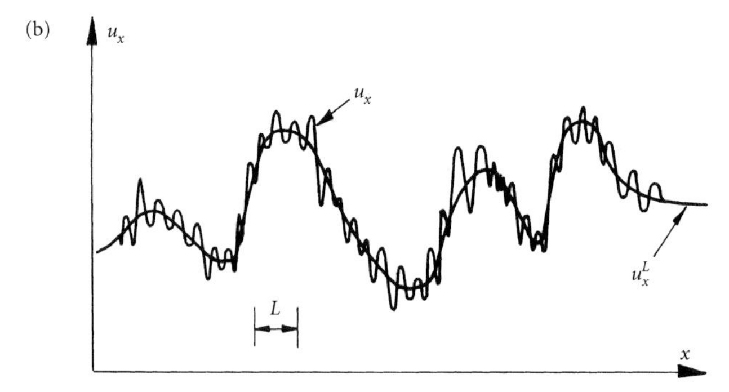

Figure 7.2 (a) Schematic of LES, (b) the effect of filtering a signal.

perhaps the large scales will not notice the absence of the small scales (Figure 7.2(a)).[8]

The attraction of LES lies in the fact that often (though not always) it is the large scales which are the most important. For example, they dominate the transfer of momentum, heat, and chemical pollutants. Yet in DNS virtually all of the effort goes into computing the small-to-intermediate scales. This can be seen from the following simple example. Suppose that $k_{\max}$ is the maximum wave number used in a spectral simulation. To obtain resolution at the Kolmogorov scale we need $k_{\max}\eta \sim \pi$. However, the wave numbers which characterize the large, energy-containing scales, k_E, satisfy $k_E l \sim \pi$. It follows that $k_E \sim k_{max}(\eta/l) \sim k_{max}\mathrm{Re}^{-3/4}$. Suppose, for example, that $\mathrm{Re} = 10^3$ (i.e. $R_\lambda \sim 120$). Then $k_E^3 \sim 2 \times 10^{-7} k_{\max}^3$ and we conclude that only 0.01% of the modes used in the simulation are required to calculate the behaviour of the all-important large eddies (defined, say, by $k_E < k < 10k_E$), the other 99.99% of the modes going to simulate the small-to-intermediate scales. Clearly, for those interested in the effects

[8] Actually they probably do! See Lesieur (1990), or Hunt et al. (2001). Lesieur, for example, notes that random fluctuations in the small scales will, eventually, lead to random fluctuations in the large scales, over and above those generated by the large-scale dynamics themselves. These are not captured by an LES as the information contained in the small scales is suppressed. In this sense LES is ill-posed. Nevertheless, we might still expect an LES to capture the statistical trends and typical coherent structures of the large-scale motion.

of the large eddies, LES is a much more attractive proposition than DNS.

The entire subject of LES is becoming increasingly important and so perhaps it is worth taking a moment to explore some of the details. The first step in formalizing LES is to introduce the idea of *filtering*. Consider one component of velocity, say u_x, measured along a straight line in a turbulent flow, say $u_x(x)$ measured along $y = z = 0$. This signal will contain both small and large-scale fluctuations and we wish to distinguish between these different scales. Consider the new function,

$$u_x^L(x) = \int_{-\infty}^{\infty} u_x(x - r)G(r)dr$$

where $G(r)$ is a *filter function* defined by

$$G(r) = 1/L, \quad |r| < L/2$$
$$G(r) = 0, \quad |r| > L/2$$

Evidently $u_x^L(x)$ is the local average of u_x in the neighbourhood of x, the average being performed over the length L. Thus $u_x^L(x)$ is a smoothed out, or filtered, version of u_x in which fluctuations of scales much less than L are absent (Figure 7.2(b)). An alternative filter function, called the *Gaussian filter*, is

$$G(r) = \frac{\exp(-r^2/L^2)}{\pi^{1/2}L}.$$

Like the *box filter* above, it is even in r, has an integral equal to unity, is of the order of $1/L$ at the origin, and is small for $r \gg L$. In fact, almost any smooth function with these four properties constitutes a suitable filter. (See Figure 7.3, or else Table 8.1 in Section 8.1.2, for examples of common filters.) This idea of smoothing the velocity field by forming a *convolution* with a filter probably dates back to Leonard (1974), although the essential idea of LES is much earlier, going back to Smagorinsky (1963) in meteorology and to Deardorff (1970) in engineering.[9]

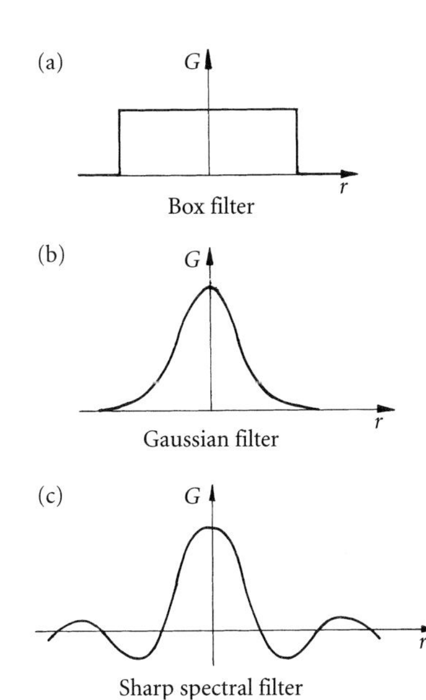

Figure 7.3 Three common filters: (a) the *box filter*; (b) the *Gaussian filter*; and (c) the *sinc filter* (also known as the *sharp spectral filter*), defined as $\sin(\pi r/L)/\pi r$.

Let us now introduce the notation $\bar{u}_x(x) = u_x^L(x)$, so that $\bar{u}_x$ represents the smoothed-out signal. It is readily confirmed that the operations of filtering and differentiation commute, in the sense that

$$\frac{\partial \bar{u}_x}{\partial t} = \overline{\left(\frac{\partial u_x}{\partial t}\right)}, \qquad \frac{\partial \bar{u}_x}{\partial x} = \overline{\left(\frac{\partial u_x}{\partial x}\right)}.$$

All of these ideas extend in an obvious way to three dimensions, where it is conventional to introduce the following notation. We define $\mathbf{u}'$ as $\mathbf{u} - \bar{\mathbf{u}}$ and write

$$\mathbf{u} = \bar{\mathbf{u}} + \mathbf{u}'$$

[9] Those unfamiliar with the idea of convolution integrals will find a discussion in Section 8.1.1.

where $\bar{\mathbf{u}}$ is referred to as the *filtered velocity* and $\mathbf{u}'$ the *residual velocity*. This is reminiscent of the Reynolds decomposition. However, there are important differences. For example, in general $\bar{\bar{\mathbf{u}}} \neq \bar{\mathbf{u}}$ and hence $\overline{\mathbf{u}'} \neq 0$, that is, the filtered residual velocity is non-zero.[10] Also, $\bar{\mathbf{u}}$ does not represent the mean motion, but rather the sum of the mean motion plus the large scales in the turbulence.

Next we turn to the Navier–Stokes equation, which we write as,

$$\frac{\partial u_i}{\partial t} + \frac{\partial}{\partial x_j}(u_i u_j) = -\frac{1}{\rho}\frac{\partial p}{\partial x_i} + \nu \nabla^2 u_i.$$

Applying our filter to this equation, and remembering that the operations of filtering and differentiation commute, we obtain,

$$\frac{\partial \bar{u}_i}{\partial t} + \frac{\partial}{\partial x_j}(\overline{u_i u_j}) = -\frac{1}{\rho}\frac{\partial \bar{p}}{\partial x_i} + \nu \nabla^2 \bar{u}_i$$

where $\bar{\mathbf{u}}$ is solenoidal. This can be rewritten in the more familiar form

$$\frac{\partial \bar{u}_i}{\partial t} + \frac{\partial}{\partial x_j}(\bar{u}_i \bar{u}_j) = -\frac{1}{\rho}\frac{\partial \bar{p}}{\partial x_i} + \frac{1}{\rho}\frac{\partial \tau_{ij}^R}{\partial x_j} + \nu \nabla^2 \bar{u}_i$$

$$\tau_{ij}^R = \rho\left[\bar{u}_i \bar{u}_j - \overline{u_i u_j}\right]$$

which looks remarkably like the Reynolds-averaged momentum equation. Thus the effect of filtering is to introduce fictitious stresses, called *residual stresses*, which are analogous to the Reynolds stresses introduced by time-averaging. Since the filtered equation of motion contains only scales (eddies) of size L and above, it may be integrated on a mesh of size $\sim L$. Thus, provided we have some means of estimating the residual stress tensor, we have removed the troublesome small scales. Typically the scale L is chosen so that $\bar{\mathbf{u}}$ contains the bulk of the energy-containing eddies and L lies in the inertial subrange. It only remains to find a means of characterizing τ_{ij}^R.

The most popular method of accounting for the unresolved scales in LES is to use an eddy-viscosity model. That is, we write the residual stress tensor as

$$\tau_{ij}^R = 2\rho\nu_R \bar{S}_{ij} + \tfrac{1}{3}\delta_{ij}\tau_{kk}^R$$

where ν_R is the eddy viscosity of the residual motion. This yields,

$$\frac{\partial \bar{u}_i}{\partial t} + \frac{\partial}{\partial x_j}(\bar{u}_i \bar{u}_j) = -\frac{1}{\rho}\frac{\partial \bar{p}^*}{\partial x_i} + 2\frac{\partial}{\partial x_j}\left[(\nu + \nu_R)\bar{S}_{ij}\right]$$

[10] For the Gaussian filter, for example, the double filtered velocity, $\bar{\bar{\mathbf{u}}} = (\mathbf{u}^L)^L$, is equal to $\mathbf{u}^{\sqrt{2}L}$, that is, double filtering is equivalent to filtering once with a filter of greater width. (Readers may wish to confirm this for themselves.) On the other hand, for the so-called *sharp-spectral filter* the double-filtered velocity is equal to the single-filtered velocity, $\bar{\bar{\mathbf{u}}} = \bar{\mathbf{u}}$, and so the filtered residual velocity is zero, $\overline{\mathbf{u}'} = \mathbf{0}$. (See Exercise 7.2.)

where $\bar{p}^*$ is a modified pressure. The final, and crucial, step is to prescribe ν_R. Physical arguments suggest that ν_R should be determined by the most energetic of the unresolved eddies, that is, eddies of a scale a little less than L. So, on dimensional grounds, perhaps a natural candidate for ν_R should be

$$\nu_R \sim L(\mathrm{v}_L^2)^{1/2}$$

where v_L^2 is the kinetic energy of eddies of size L. In the *Smagorinsky model*, which was developed in the meteorological community in the 1960s, v_L^2 is taken to be of the order of $L^2(\bar{S}_{ij}\bar{S}_{ij})$. The prescription for ν_R is then

$$\nu_R = C_S^2 L^2 \left(2\bar{S}_{ij}\bar{S}_{ij}\right)^{1/2}.$$

The dimensionless constant C_s, which is usually given a value of ~ 0.1, is called the *Smagorinsky coefficient*.[11] This model, which is extremely popular, has a number of shortcomings (it is too dissipative near walls)[12] but seems to work well for isotropic turbulence, free-shear flows, and channel flows. (See Lesieur 1990, for more details.) Various refinements of the model have been proposed, aimed particularly at improving the near-wall treatment of turbulence. However, we shall not detail these modifications here. We merely note that one important refinement was introduced by Germano et al. (1991). They proposed the so-called *dynamic model* in which C_s is allowed to be a function of position and time. This employs two filters of different widths, the grid filter and a test filter, which allows one to estimate optimum values of C_s depending on the state of the local flow.

There have been many other suggestions for ν_R. For example, from the discussion in Chapter 6, we might expect v_L^2 to be proportional to the second-order structure function evaluated at a separation of L. This suggests

$$\mathrm{v}_L^2 \sim \left\langle [\Delta\bar{\mathbf{v}}(\mathbf{x}, \mathbf{r}, t)]^2 \right\rangle_{|\mathbf{r}|=L},$$

where

$$\Delta\bar{\mathbf{v}}(\mathbf{x}, \mathbf{r}, t) = [\bar{\mathbf{u}}(\mathbf{x} + \mathbf{r}, t) - \bar{\mathbf{u}}(\mathbf{x}, t)].$$

[11] One of the curious features of the filtered momentum equation and the accompanying Smagorinsky model is that, once C_s is chosen, the model is completely independent of the choice of filter. That is, the net effect of filtering is to produce a new version of the Navier–Stokes equation in which we replace $\mathbf{u}$ by $\bar{\mathbf{u}}$ and ν by $\nu + \nu_R$, the prescription of $\nu_R(\mathbf{x}, t)$ being independent of the type of filter used.

[12] See Exercise 7.3.

(Here the angled brackets indicate an angular average over a spherical surface of radius L.) The resulting expression for ν_R is then (Lesieur 1990),

$$\nu_R = CL\sqrt{\left\langle [\Delta\bar{\mathbf{v}}(\mathbf{x}, L, t)]^2 \right\rangle}$$

where C is yet another dimensionless coefficient. This yields good results for isotropic turbulence and free-shear flows. However, like the Smagorinsky model, it is overly dissipative near boundaries.

Many other sub-grid closure models have been suggested, though we shall not pause to describe them here. Some of these models, along with their strengths and weaknesses, are discussed in detail in Lesieur (1990).

Large eddy simulations is, in the eyes of some, the future of computational fluid dynamics (CFD). It avoids the need to rely on some (ultimately flawed) closure scheme to parameterize the all-important large eddies, while simultaneously circumventing the restrictions imposed by (7.4). Indeed, it is now possible to simulate flows of engineering interest using LES (e.g. see the simulation of flow over a cube described in Section 4.6.3). However, although LES is substantially less demanding than DNS, it is still orders of magnitude more difficult to implement than, say, the k–ε model, and so it is unlikely to displace eddy-viscosity models in cases where high accuracy is not required. Moreover, the Achilles' heel of LES is the presence of boundaries where the dynamically important eddies are very small. To incorporate a boundary layer into an LES one must either make the mesh so small near the surface that the LES effectively becomes a DNS, or else separately model the boundary layer, say with a suitable eddy-viscosity model, and then patch this to an LES of the outer flow. The former strategy is practical only for relatively low values of Re, while the latter is problematic to the extent that the eddy-viscosity model provides only statistically averaged information as a boundary condition for the LES.

Perhaps one way to put the limitations of LES in perspective is to reflect on the insightful observations of Sherman (1990), who introduced his review of turbulence thus, 'Viscosity is both midwife and executioner in the life story of turbulence. As midwife, it delivers the vorticity from its birthplace at the wall. Thus, the flow receives something without which it cannot be unstable, and hence possibly turbulent. As executioner, it applies torques that rub out the local concentrations of vorticity.' Well, in LES viscosity is neither midwife nor executioner; indeed, it plays almost no role at all. Both the creation of vorticity at surfaces and the destruction of enstrophy within the fluid interior have to be modelled. LES serves merely as a guide to middle-age. But then there are many flows for which this is the most important period.

There is no doubt that both DNS and LES have contributed greatly to our understanding of turbulence and of turbulent flows. We shall give some examples of these successes in Section 7.3. First, though, we take a rather cautious approach, and discuss one of the limitations of certain DNS and LES. In particular, we start with a word of warning about the periodic cube, which has found such favour in some circles.

Example 7.1 It is required to perform a DNS of a simple, low-Re grid turbulence experiment. The elapsed time of the experiment is 20 s, the working section is $1\,\text{m} \times 1\,\text{m} \times 1\,\text{m}$, the rms fluctuating velocity is $u = 1\,\text{m/s}$, and the integral scale $l = 1\,\text{cm}$. The entire working section is to be modelled and there is a 10 gigaflop computer dedicated to the purpose. Use Table 7.1 to confirm that the expected run time is ~ 8000 yrs.

Example 7.2 Show that the Smagorinsky model ensures that kinetic energy is transferred from the large-scale motion to the unresolved scales.

Example 7.3 Show that the filtered momentum equation and the residual stress becomes identical to the Reynolds-averaged momentum equation and Reynolds stress when we interpret the overbar as a time average.

7.2 On the dangers of periodicity

A religious man said to a whore, 'You're drunk,
Caught every minute in a snare.'
She replied, 'Oh Shaikh, I am what you say,
Are you what you seem?' (Omar Khayyam)

Like Omar Khayyam's shaikh, there is more to a periodic cube than meets the eye. The idiosyncratic, yet seemingly benign, boundary conditions turn out to be somewhat of a problem. Let us go back to (7.3)

$$\text{Re} \sim \left(\frac{l}{L_{\text{BOX}}}\right)^{4/3} N_x^{4/9}. \tag{7.5}$$

The major concern of many researchers has been to achieve as high a value of Re as possible. As a result, the ratio of L_{BOX} to l has not always been as large as one would have liked. (Many researchers take $L_{\text{BOX}} \sim \pi l$.) This is important because, as we have seen, the boundary conditions for a periodic cube are unphysical. If we wish to pretend that the turbulence in the cube mimics, say, turbulence in a wind tunnel,

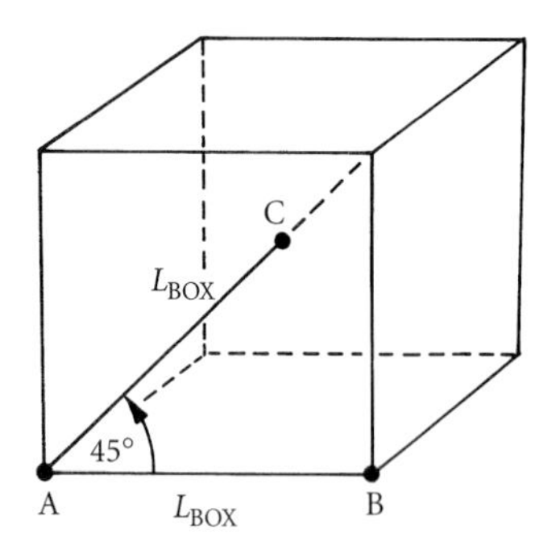

Figure 7.4 A periodic cube. Points A and B are perfectly correlated while A and C are not. Evidently, there is anisotropy at the scale of the box.

then it is essential that $L_{BOX} \gg l$ so that the bulk of the turbulence does not 'feel' the imposed periodicity. This is particularly important if information is required about the low-to-intermediate k end of the energy spectrum, that is, the behaviour of the large eddies. The indications are that, to obtain results at low-to-intermediate k which are not influenced by the imposed periodicity we require, perhaps,[13] $L_{BOX} > 20l \rightarrow 40l$. This is rarely achieved in DNS and if it is, then it is at the cost of having an extremely low value of R_λ. (See Table 7.1.)

There are two distinct problems which arise from the boundary conditions. The first is that anisotropy is imposed at the large scales by virtue of the periodicity. Consider Figure 7.4. Points A and B are perfectly correlated, while A and C, which have the same separation, are not. Evidently, we have anisotropy on the scale of L_{BOX}. This anisotropy may not be important if $L_{BOX} \gg l$, but it could influence the large-scale dynamics if L_{BOX} is only a few multiples of l. The second, and more serious, problem is that artificial long-range correlations are established on the scale L_{BOX}. That is to say, events on one face of the cube are perfectly correlated to events on the opposite face. Consider a region of space much greater than L_{BOX}. Because of the assumed periodicity, it will consist of a set of periodic cubes as shown in Figure 7.5. Of course, the flow in each cube will be identical. Moreover, each cube communicates with every other cube via the pressure field. Thus we have a set of discrete flows, all forced to be identical, and all influencing each other. Clearly we cannot pretend that turbulence in a periodic cube is a small part of a larger, isotropic field of turbulence.

Now consider the longitudinal correlation function for this extended region of space. It looks something like that shown in Figure 7.6(a). There is a periodic spike of height 1 and width $\sim 2l$. This is a direct result of the artificial long-range correlations established by the imposed periodicity. Now we saw in Chapter 6 that the large-scale dynamics are heavily influenced by the form of f and K at $r \rightarrow \infty$. That is to say, we get entirely different behaviour depending on whether $f_\infty \sim e^{-r^2}$, $f_\infty \sim r^{-6}$ or $f_\infty \sim r^{-3}$. For example, if f_∞ is exponentially small then the turbulent energy decays as $u^2 \sim t^{-10/7}$, while $f_\infty \sim r^{-3}$ gives $u^2 \sim t^{-1.2}$, (see Sections 6.3.2 and 6.3.5). Now compare Figure 7.6(a) with Figure 7.6(b). Evidently it is meaningless to talk about f_∞ or K_∞ in a periodic simulation. So how can we interpret the results of periodic simulations in terms of real-world turbulence? Clearly, as far as the large-scale dynamics are concerned, the answer is: with some difficulty! Perhaps the most we can do is to insist that $L_{BOX} \sim 20l \rightarrow 40l$ and hope for the best.

[13] The precise restriction on L_{BOX}/l in three dimensions has not been established. However, see Lowe (2001) for a discussion of the effects of periodicity in two dimensions, where pollution from the symmetry planes is still evident at $L_{BOX}/l = 100$. It should be said, however, that long-range correlations are known to be stronger in two dimensions than in three.

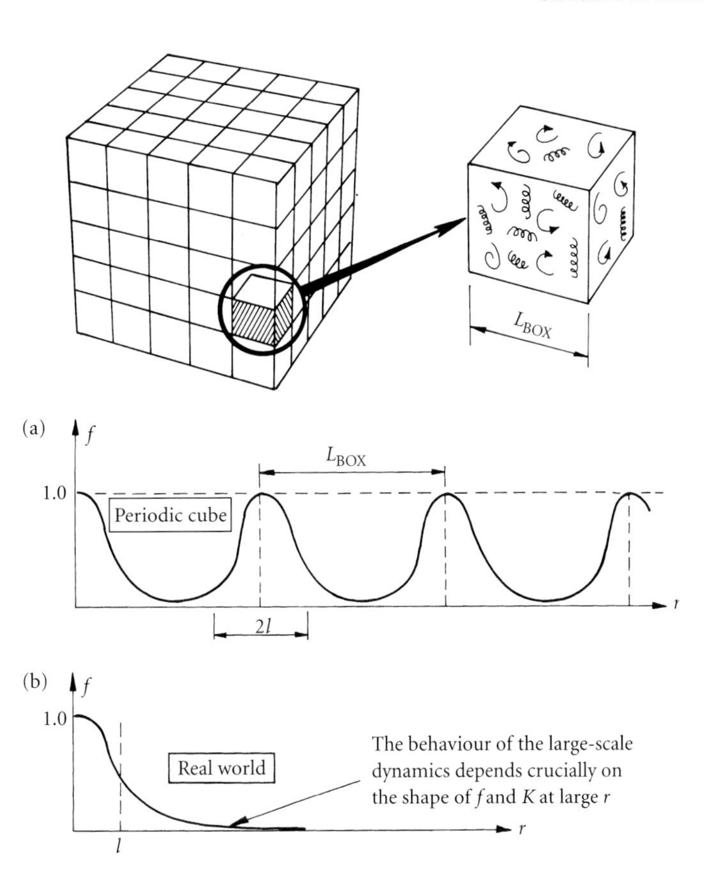

Figure 7.5 A simulation in a periodic cube is, in fact, equivalent to a flow evolving in a multitude of periodic cubes. The flow in each cube is identical and the various cubes communicate with each other via the pressure field. In what sense is this representative of isotropic turbulence?

Figure 7.6 (a) The longitudinal correlation function for an array of periodic cubes. (b) The shape of f for isotropic turbulence. The large-scale dynamics of the turbulence depend crucially on the shape of f and K at large r, so what does flow in a periodic cube tell us about the large scales?

Certainly, a simulation in which $L_{BOX} \leq 8l$, say, is unlikely to produce any meaningful results as far as the large scales are concerned.

So how can we tell if our box is big enough? Perhaps a useful check in this respect is to start a simulation with a k^4 energy spectrum at small k and then observe the subsequent behaviour of $E(k \to 0)$. If the k^4 spectrum appears to evolve to, say, a k^2 spectrum (which it can if L_{BOX} is too small) then there is something badly wrong. That is to say, the behaviour of $E(k \to 0)$ depends critically on the form of the long-range correlations (see Section 6.3.4) and in isotropic turbulence one finds either $E \sim Ik^4$, where I is almost a constant (see Section 6.3.6), or else $E \sim Lk^2$ where L is strictly constant (see Section 6.3.5). However, it is not possible, in isotropic turbulence, for a k^4 spectrum to spontaneously convert to a k^2 spectrum.

7.3 Structure in chaos

We used to have lots of questions to which there were no answers. Now with the computer there are lots of answers to which we haven't thought up the questions. (Peter Ustinov)

We shall now give a few examples of the kind of intriguing results which DNS and LES have produced. There is no attempt here to give a systematic coverage of the more important studies. That would fill an entire text in its own right. Rather, we present just a few typical studies, chosen almost at random, to illustrate the allure of numerical experiments. We are interested primarily in homogeneous turbulence.

7.3.1 *Tubes, sheets, and cascades*

We start by considering 'almost' isotropic turbulence in a periodic cube. Most of these simulations have focused on the small scales since it is extremely difficult to simultaneously achieve $R_\lambda \gg 1$ and $L_{BOX} > 20l$, and so reliable data for the large scales is hard to obtain. The primary interest in these simulations lies in: (i) probing the authenticity of the Richardson–Kolmogorov view of turbulent cascades; and (ii) determining the generic structure of the small-scale vorticity field. Since there is no pretence to reproduce the large-scale behaviour, these studies often take $L_{BOX} \sim \pi l$. They then appeal to the multistage cascade picture of turbulence to suggest that the small-scales will not know that the large scales are unphysical. There are two categories of such simulations: freely decaying turbulence and statistically steady turbulence. In the latter case a random stirring force is applied at the large scales and once again one appeals to the cascade picture to argue that the precise details of the forcing does not influence the small eddies.

Typical of these simulations is Vincent and Meneguzzi (1991), Ruetsch and Maxey (1992), and Vincent and Meneguzzi (1994). The first of these is a study of statistically steady flow at $R_\lambda \approx 150$. Their primary findings are broadly in line with our intuitive view of turbulence, based on the experimental evidence. They found that:

(1) the peak vorticity is largely organized into a sparse network of thin, long tubes of diameter somewhere between η and λ and of a length comparable with the integral scale, l;
(2) there exists a $k^{-5/3}$ inertial range;
(3) Kolmogorov's constant is $\alpha \sim 2$ (they found a value $\sim 25\%$ higher than most experiments suggest);
(4) the probability distribution of $\partial u'_x / \partial x$ is highly non-Gaussian;
(5) the probability distribution of $\Delta \upsilon = u_x(\mathbf{x}_0 + r\hat{\mathbf{e}}_x) - u_x(\mathbf{x}_0)$ becomes increasingly Gaussian as $r \to \infty$;
(6) the skewness of $(\partial u_x / \partial x)$ has a value of $S_0 \sim -0.5$;
(7) the flatness factor is much larger than that suggested by a Gaussian distribution, consistent with the notion that the vorticity field is highly intermittent;
(8) the exponent of ζ_n in the structure–function law $\langle [\Delta \upsilon]^n \rangle \sim r^{\zeta_n}$ is considerably lower, for $n \geq 5$, than both the Kolmogorov (1941) and $\hat{\beta}$-model estimates of ζ_n;
(9) the vorticity is, on average, aligned with the intermediate principal rate of strain.

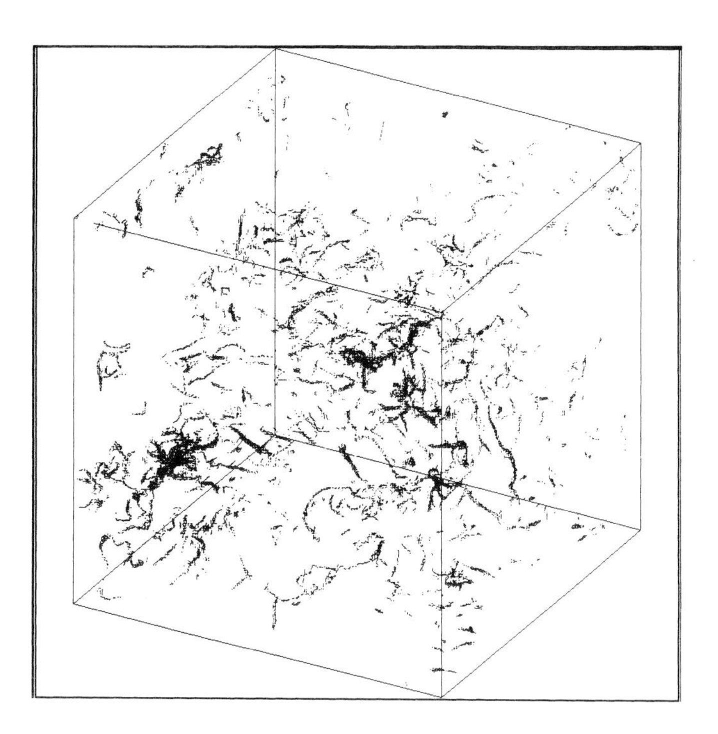

Figure 7.7 Vorticity contours in the simulations of Vincent and Meneguzzi (1991) (with permission). See also Plate 8.

Most of this is broadly in line with the experimental evidence summarized in Chapters 3 and 5, which gives us some confidence in the accuracy of the simulations. The novelty of the DNS largely resides in observation (1), which is not inconsistent with the experiments, but tips the balance between tubes versus sheets, as the favoured structure for $\boldsymbol{\omega}$, firmly towards the former. (This conclusion is consistent with the earlier simulations of Siggia (1981) and those of She et al. (1991).) These thin vortex tubes are often called *worms* (Figure 7.7). Notice that tube-like vortices are also evident in Plate 8 which shows high-resolution LES and DNS of homogeneous turbulence.

Interestingly, within a year, the pendulum had swung the other way. In another forced simulation, Ruetsch and Maxey (1992) noted that, while the most intense vorticity seems to reside in tubes, moderate levels of $\boldsymbol{\omega}$ tend to take the form of sheets. These sheets form in the manner foreseen by Betchov in 1956 (see Section 5.3.2) and then roll up via a Kelvin–Helmholtz instability, as anticipated by Townsend (1976).[14] Suddenly we had reverted to a rather classical view of the intermediate-to-small scales.[15]

[14] Betchov anticipated that sheet formation would be important because of his prediction that the principal strains in a turbulent flow would generally consist of one large compressive strain plus two weaker extensional ones. Such a situation favours sheet formation. His prediction about the strains was later confirmed by the DNS of Ashurst et al. (1987), who computed the relative magnitudes of the principal strains to be (3, 1, −4).

[15] This picture of the intermediate vorticity being sheet-like, while the very intense vorticity resides in slender tubes, is also seen in the DNS of Jiménez et al. (1993).

Vincent and Meneguzzi revisited the subject in 1994 in an unforced decay simulation. They confirmed the view of Ruetsch and Maxey that vortex sheets appear first and that the tubes are essentially the debris resulting from a succession of Kelvin–Helmholtz instabilities. They noted that vortex tubes are more stable (long-lived) structures than sheets, and this is why they are more frequently observed in DNS. However, in their view, one of the main mechanisms responsible for the energy cascade is the formation, and subsequent disintegration, of vortex sheets.

It is encouraging that these simulations are broadly in line with the earlier hypotheses of Betchov, Townsend, Corrsin, and others.[16] However, it must always be borne in mind that the value of R_λ in these numerical experiments is rather low. Moreover, there are many who regard the debate over the relative importance of sheets and tubes in the energy cascade as still unresolved, a significant number of whom still see the vortex tube as the crucial structure. The jury is still out on this issue.

7.3.2 On the taxonomy of worms

Whether or not vortex tubes (worms) dominate the energy cascade, they appear to be a ubiquitous feature of the DNS of isotropic turbulence and it is natural to try and catalogue some of their features. There have been a multitude of such studies. The three main themes have been:

(1) what is their structure (diameter, length, shape, number density, etc.)?
(2) how do these features vary with R_λ?
(3) do the tubes play an important dynamical role or are they 'passive debri', as suggested by Vincent and Meneguzzi (1994)?

Typical of these studies are Jiménez et al. (1993), Jiménez and Wray (1998) and Kida (2000). Let us start with Jiménez and Wray. These authors performed a sequence of forced, statistically steady simulations with Re in the range $R_\lambda = 40 \rightarrow 170$. They focused attention on regions of particularly intense enstrophy, identified by the threshold $\boldsymbol{\omega}^2 \approx \langle \boldsymbol{\omega}^2 \rangle R_\lambda$. Such strong vorticity is found only at the dissipation scale η. In line with other authors they conclude that this peak vorticity resides in a sparse network of thin, long tubes. The radius of these tubes is around $\delta \sim 5\eta$, which is comparable with the smallest structures seen in high-Re turbulence (see Section 6.6.2). This radius does not appear to vary with R_λ. The vortex tubes themselves

[16] An informative cartoon of the disintegration of a vortex sheet into an ensemble of vortex tubes is given by Lin and Corcos (1984).

Table 7.2 DNS results of Kida (2000). See also Plate 9

R_λ	% of flow volume occupied by vortex cores	% of net enstrophy in vortex cores	% of net dissipation in vortex cores	Mean core radius in multiples of η
86	22	46	19	8
120	16	39	14	6
170	13	36	14	4

resemble inhomogeneous Burgers-like vortices, whose length may be as much as the integral scale l. They are stretched by a fluctuating strain field whose magnitude is of the order of $\alpha \sim \langle \boldsymbol{\omega}^2 \rangle^{1/2}$. Although the tubes are long, the straining is coherent only over relatively short portions of the vortices, perhaps of the order of a few multiples of η. The fractional volume of the flow occupied by these Burgers-like vortices is small, falling off as R_λ^{-2}. So, as Re rises, the turbulence becomes more intermittent, which is precisely what we would expect.

By virtue of their choice of a high threshold for $\boldsymbol{\omega}^2$, Jiménez and Wray have catalogued the properties of only the most intense vortices. However, the bulk of the vorticity lies outside these tubes and so they contribute very little to the energy cascade.[17] Kida (2000), by contrast, was interested in cataloguing the overall structure of the vorticity field, and so abandoned the idea of a high threshold in enstrophy as a means of visualizing vortex tubes. Instead Kida looked for a local minimum in pressure and used this to identify the axes of the vortex tubes (Plate 9(a)). The 'edges', or radii, of the vortex tubes were determined by looking for the local maximum in u_θ^2/r, where u_θ is the local swirl velocity in the plane normal to the vortex axis. One such vortex tube is shown in Plate 9(b). These simulations were statistically steady and spanned the range $R_\lambda = 80 \to 170$. The advantage of Kida's visualization method is that it is not restricted to the most intense vortices. Its disadvantage, however, is that it overemphasizes the importance of tubes, as distinct from vortex sheets or ribbons, as only vortex tubes are identified. Some of Kida's findings are tabulated in Table 7.2.

The fractional volume occupied by the vortex tubes, as well as their mean radius, appears to fall off with R_λ. The first of theses observations is consistent with Jiménez and Wray (1998), while the second appears to be contradictory. However, it should be remembered that Jiménez and Wray were looking at only the most intense vortices, whereas Kida is averaging over a range of vortex strengths. Note that

[17] This was demonstrated in a rather dramatic way by Jiménez et al. (1993) who showed that artificially removing these intense vortex tubes from a numerical simulation makes almost no difference to the subsequent rate of loss of energy.

the precise values of the mean core radii are probably less important than the general variation of radius with R_λ, since Kida's definition of core radius is a little arbitrary.

One of the more interesting features of Table 7.2 is that, although much of the enstrophy resides in the vortex cores, very little of the dissipation, $2\nu S_{ij}S_{ij}$, is found there. This is entirely consistent with Figure 5.28, which relates to a conventional Burgers vortex and shows that the bulk of the dissipation in such a vortex lies outside the vortex core. In this respect it is also interesting to look at Plate 9(d) and 9(e) which show vorticity contours in a plane normal to the axis of a worm-like vortex. (The plane in question is shown in Plate 9(c).) Plate 9(d) shows the contours of axial vorticity while Plate 9(e) shows the vorticity components normal to the vortex axis. It is evident that stray vortex filaments are wrapped around the central vortex core, presumably by the strong local swirling motion. It turns out that most of the local dissipation is found not in the vortex core itself but rather in the surrounding annular region where the cross-axial vorticity is strong. Note, however, that this does not necessarily imply that it is the cross-axial vorticity which causes the dissipation, since the bulk of the dissipation in a conventional Burgers vortex also lies in an annulus surrounding the vortex core. (See Figure 5.28.)

This annular distribution of dissipation is emphasized in Table 7.3 which shows the fractional volume, enstrophy and dissipation contained within tubular regions surrounding the vortex tubes. Three cases are considered, corresponding to

$$R = R_{\text{core}}, \quad R = 2R_{\text{core}}, \quad \text{and} \quad R = 3R_{\text{core}}.$$

It is intriguing that three-quarters of the total dissipation lies within a distance of $2R_{\text{core}}$ from the axis of the vortex tubes, and 96% lies within a distance of $3R_{\text{core}}$. On the other hand, there is a close correspondence between the second (% volume) and fourth (% dissipation) columns of Table 7.3, so perhaps we should not read too much into this result.

The results of Kida (2000) tentatively suggest that much of the dissipation of energy is located in annular regions surrounding vortex tubes. However, it should be emphasized that Re here is modest,

Table 7.3 DNS of Kida (2000), $R_\lambda = 86$

R/R_{core}	% of flow volume occupied	% of enstrophy	% of dissipation
1	22	46	19
2	70	82	76
3	93	96	96

and that the visualization scheme adopted by Kida biases our view of events towards vortex tubes and away from vortex sheets or ribbons.

7.3.3 *Structure and intermittency*

Let us now turn to the inertial range and ask: 'what are the typical structures which characterize the energy cascade?'. The DNS of Kida (2000) (and many others) emphasize the importance of fine-scale vortex tubes in turbulence, and so it is tempting to assume that these are the structures which dominate the energy cascade and are responsible for inertial-range intermittency. The question arises, therefore, as to whether or not it is the emergence of slender vortex filaments which cause departures from Kolmogorov's original theory, that is, departures from the inertial-range scaling $\langle[\Delta v(r)]^p\rangle \sim r^{p/3}$. Certainly She, Jackson, and Orszag (1991) and Jiménez and Wray (1998) expressed the view that slender vortex filaments are largely responsible for intermittency, though it is possible that Jiménez and Wray had in mind intermittency in the dissipation range, rather than inertial-range intermittency.[18] Moreover Chorin, in a sequence of papers, explores the possibility that inertial-range dynamics could be modelled as a tangle of vortex tubes (e.g. see Chorin 1990).

On the other hand, Ruetsch and Maxey (1992) and Vincent and Meneguzzi (1994) were of the opinion that the energy cascade can be characterized by the formation and subsequent disintegration of vortex sheets, while Boratav and Peltz (1997) found that the structures which most influence inertial-range intermittency are ribbon-like and not tubular. Boratav and Peltz also suggested that it might be intermittency in the *strain field*, rather than in the vorticity field, which is important for departures from Kolmogorov's theory. (For vortex sheets the distinction is not so important since the two fields are more or less coincident. However, for tubes, ribbons, or more complex structures the distinction could be significant.) Tsinober (2002), like Boratav and Peltz, emphasizes the different roles played by the strain and vorticity fields. He also argues against the dominance of vortex filaments in the inertial subrange, suggesting that intermittency and anomalous scaling in the inertial range is unrelated to the vortex filaments.

The suggestion by Boratav and Peltz that due attention should be paid to the strain field, as well as the ongoing debate about the

[18] In an earlier paper (Jiménez et al. 1993) Jiménez had already nailed his colours to the mast by noting that the bulk of the vorticity and dissipation resides in sheet-like structures and that the very intense vorticity found in worms is of limited dynamical significance. She, Jackson, and Orszag also noted the existence of large-scale vortex sheets.

structure of the inertial-range eddies, raises the subtle question of how we should visualize the results of DNS. Different visualization methods emphasize different facets of the flow field and can give qualitatively different pictures. Moreover, a bewildering range of schemes have been adopted at one time or another by various researchers. The simplest, and most obvious, strategy is to plot iso-surfaces of $\boldsymbol{\omega}^2$ or $S_{ij}S_{ij}$, depending on which field is of interest. However, the results obtained depend greatly on the threshold set for $\boldsymbol{\omega}^2$ or $S_{ij}S_{ij}$. Some investigators, such as Kida, prefer to search for regions of low pressure, which tend to mark the axes of vortex tubes. This is particularly effective at picking out intense vortex filaments, but tells us little about vortex sheets, or about the strain field. A related strategy is to plot iso-surfaces of Q, the second-order invariant of the velocity-gradient tensor. In Section 5.3.6 we saw that this is defined as

$$Q = -\frac{1}{2}\frac{\partial u_i}{\partial x_j}\frac{\partial u_j}{\partial x_i} = -\frac{1}{2}S_{ij}S_{ij} + \frac{1}{4}\boldsymbol{\omega}^2 = \frac{1}{2}\nabla^2(p/\rho).$$

A region where vorticity dominates over strain can be identified by plotting surfaces of positive Q, while surfaces of negative Q mark regions where strain is dominant. Those looking for coherent vortices sometimes use a search strategy which combines a positive Q with low pressure. Note, however, that $Q = 0$ in a vortex sheet, and so any visualization method based on Q cannot differentiate between regions of low activity (low $\boldsymbol{\omega}^2$ and $S_{ij}S_{ij}$) and intense vortex sheets or ribbons. Note also that

$$\nabla^2(p/\rho) = 2Q$$

and so there is a close relationship between the low-pressure technique of searching for the axes of vortices and the strategy of looking for regions of large positive Q.

There are many other visualization techniques which have been advocated from time to time, such as the Q–R criterion of Chong, Perry, and Cartwell (1990) (see Section 5.3.6) or the combined enstrophy and strain method of Jeong and Hussain (1995). (The various schemes are nicely summarized in Jeong and Hussain.) Given the diversity of techniques, it is hardly surprising that different researchers tend to promote rather different pictures of the inertial-range eddies.

All in all it seems that we are a long way from having a unified picture (or even a naive cartoon) of the structures which characterize the energy cascade. While the irreversible flux of energy towards small scales is certainly driven by vortex stretching, it is by no means clear as to whether anomalous scaling in the inertial range is caused by intermittency in the strain field or by intermittency in the vorticity field.

Nor do we have a clear idea of the geometrical relationship between these two fields. (Are they largely coincident, as in vortex sheets, or somewhat distinct, as in vortex filaments?) Finally, we do yet not know if the slender vortex tubes which are so evident in the numerical simulations are really typical of the inertial-range eddies; many would say not.

To resolve these issues we require simulations in larger boxes and at higher Re, so that we can achieve a true separation of scales between η, λ, l, and L_{BOX}. As we saw in Section 7.1.1, this is hard to achieve, and so progress may be slow. But perhaps the fact that the picture is still incomplete is not so important. The key point is that it is the evidence generated through the use of DNS which is fuelling this debate, and as the DNS gets better, which it surely will, it is likely that we shall converge on an agreed cartoon of the energy cascade. While such a cartoon will be imperfect, because labels such as tube-like or sheet-like are subjective and imprecise, a simple picture of the cascade (if there is a simple picture!) would be of considerable conceptual value.

7.3.4 Shear flows

Although we are primarily concerned with homogeneous turbulence, we cannot leave the topic of DNS without a discussion of shear layers, where DNS has contributed so much. Let us, therefore, turn to the DNS (and LES) of boundary layers. This is particularly interesting in view of the fact that there was a substantial body of research in the 1980s and 1990s devoted to the study of coherent structures in boundary layers (see Section 4.2.6). Since one of the great strengths of DNS is the identification of robust structures, it is natural to see how the results of the DNS stack up against the received wisdom of the experimentalists. The results are encouraging. In an early LES Moin and Kim (1985) confirmed the predominance of vortical structures (hairpin vortices) inclined at 45° to the wall, precisely as observed in the experiments (see Section 4.2.6). In Chapter 4 we suggested that these hairpin vortices are initiated by a horizontal perturbation of the mean spanwise vorticity. In boundary layers this spanwise vorticity is most intense near the wall and so it is not so surprising that many hairpins in a boundary layer should originate in the near-wall region. However, the mechanism does not itself require the presence of a wall: it merely requires a mean shear. Thus hairpin vortices should be seen in any planar shear flow, with or without a wall. This was confirmed by Rogers and Moin (1987) who performed a DNS of homogeneous turbulent shear flow. As in a boundary layer, hairpin vortices are the dominant structure.

There have been many other numerical simulations of fully developed, turbulent boundary-layers, and these are nicely summarized in Moin and Mahesh (1998). The main findings are:

(1) there are many streamwise vortex rolls located near the wall, though unlike the cartoon shown in Figure 4.13(c), the rolls do not always appear in pairs;
(2) turbulent 'bursts' in the near-wall region may not be as dramatic nor as important as was once thought, and could simply represent the passage of a vortex over the measuring station;
(3) low-speed streaks occur near the wall and are often longer than individual vortex rolls;
(4) hairpin vortices often appear in streamwise arrays, one hairpin triggering another (Figure 7.8).

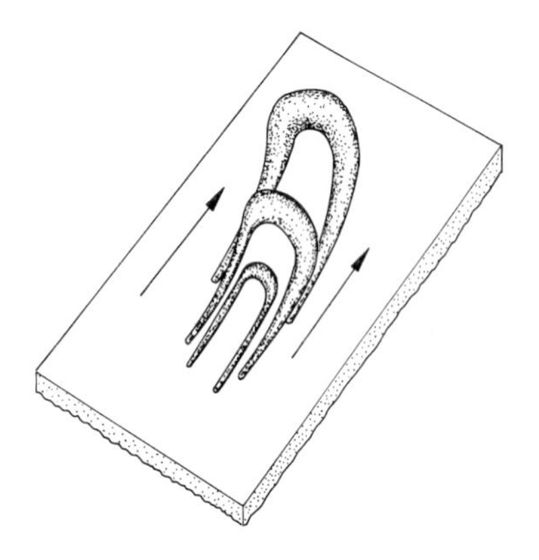

Figure 7.8 Hairpin vortices are often found to appear in streamwise arrays, one hairpin triggering another.

Direct numerical simulations has also been applied to boundary layers which are in a state of transition, say a laminar boundary layer becoming turbulent, or a turbulent boundary layer leaving the trailing edge of a plate. As with fully developed boundary layers, the DNS has proven to be particularly good at identifying the dominant vortical structures. A couple of examples will be mentioned. Alam and Sandham (2000) used DNS to study a laminar boundary layer which is disturbed, causing laminar separation followed by turbulent reattachment. They focused on the so-called separation bubble which encompasses that part of the boundary layer which lies between the initial separation point and the subsequent reattachment. Plate 10 shows the vorticity contours for: (a) the normal vorticity component; and (b) the spanwise vorticity. (Red and blue represent vorticity of opposite signs.) It is clear that separation initially takes the form of a few Λ-shaped vortices which subsequently break-up into the disorganized mess we call turbulence. Similar results were obtained in an earlier study by Ducros Comte and Lesieur (1996), as shown in Plate 11. Once again we see the power of DNS in identifying structure in chaos.

Another DNS study by Sandham et al. (2001) focused on the structure of turbulence in a trailing edge flow, such as the rear of a plate or aerofoil (Plate 12). The Re, based on the mean flow and trailing edge thickness, is around 1000. Here we find coherent spanwise vorticity, in the form of Karman vortices (shown red), interacting with small, intense, longitudinal vortex tubes (shown blue) which are generated in the boundary layer upstream of the trailing edge.[19] The axial vortex tubes are strained and deformed as they interact with the quasi-two-dimensional Karman vortices. The axial vortex tubes become wrapped around the Karman vortices and intensified by strain, generating a flow pattern which is somewhat reminiscent of a

[19] The Karman vortices (shown red) are visualized using the low-pressure method, while the axial vorticity (shown blue) is visualized using the positive Q criterion.

mixing layer (e.g. see Comte, Lesieur, and Lamballais (1992)). Eventually the axial vortices become powerful enough to destroy the Karman vortex street.

As in the previous example, DNS offers a unique insight into the structure of the turbulent vorticity field, an insight which would be hard to obtain by any other means. Many more examples of DNS and LES studies of mixing layers may be found in Lesieur (1995).

7.4 Postscript

This concludes our brief survey of numerical simulations. We have barely scratched the surface, but the interested reader will find a more detailed discussion of DNS in Moin and Mahesh (1998), and of LES in Lesieur (1990).

Perhaps the one message you should take away from this chapter is the following. At present, and indeed for the foreseeable future, our ability to simulate turbulent flows by DNS is rather feeble. We can achieve only modest Re and squeeze only a handful of energy-containing eddies into the simulations. Nevertheless, those simulations which have been performed are already providing fascinating new insights, and it is certain that matters can only improve as computers become more powerful. Love it or loath it, DNS is here to stay. Half a century after his call for a concerted numerical attack on turbulence von Neumann's dream has finally been realised.

Exercises

7.1 Consider the filtered signal

$$u_x^L(x) = \int_{-\infty}^{\infty} u_x(x-r)G(r)dr$$

where $G(r)$ is a filter function. Let and $\hat{u}_x(k)$ and $\hat{G}(k)$ be the one-dimensional transforms of u_x and G. Use the convolution theorem to show that

$$\hat{u}_x^L(k) = 2\pi\hat{u}_x(k)\dot{G}(k).$$

Now consider the filter function

$$G(r) = \frac{\sin(\pi r/L)}{\pi r}.$$

Confirm that $\hat{u}_x^L(k) = \hat{u}_x(k)$ for $k < \pi/L$ and $\hat{u}_x^L(k) = 0$ for $k > \pi/L$. Evidently this particular filter is equivalent to truncating the Fourier transform in spectral space. It is called the *sharp spectral filter*.

7.2 Show that the residual stress tensor can be written as

$$\tau_{ij}^R/\rho = (\overline{u}_i\overline{u}_j - \overline{\overline{u}_i\overline{u}_j}) - (\overline{\overline{u}_i u_j'} + \overline{\overline{u}_j u_i'}) - \overline{u_i' u_j'}.$$

The first term on the right

$$(\bar{u}_i\bar{u}_j - \overline{\bar{u}_i\bar{u}_j})$$

is, at first sight, unexpected. It is sometimes called the *Leonard stress*. Show that the Leonard stress is zero for a sharp spectral filter

7.3 Show that, if L is much greater than the integral scale, l, the Smagorinsky model reduces to a conventional eddy-viscosity model of the mean flow. Show also that the specification of eddy viscosity is inappropriate for flow near a surface.

7.4 Show that, if we set $L = \eta$ in the Smagorinsky model, then the model gives

$$\nu + \langle \nu_R^2 \rangle^{1/2} = (1 + C_S^2)\nu \approx 1.01\nu.$$

Thus reasonable behaviour is obtained in the limit of fine resolution. That is to say, an LES using the Smagorinsky model blends smoothly into a DNS as the resolution becomes finer.

Suggested reading

Alam, M. and Sandham, N.D. (2000) *J. Fluid Mech.*, **410**, 1–28.
Ashurst, W.T. et al. (1987) *Phys. Fluids*, **30**, 2343–2353.
Betchov, R. (1956) *J. Fluid Mech.*, **1**, 497–504.
Boratav, O.N. and Peltz, R.B. (1997) *Phys. Fluids*, **9**, 1400–1415.
Canuto, C. et al. (1987) *Spectral Methods in Fluid Dynamics*. Springer.
Chong, M.S., Perry, A.E. and Cantwell, B.J. (1990) In: *Topological Fluid Mechanics*, Eds. H.K. Moffatt & A. Tsinober, Cambridge University Press.
Chorin, A.J. (1990) In: *Topological Fluid Mechanics* (eds. H.K. Moffatt and A. Tsinober). Cambridge University Press.
Comte, P., Lesieur, M. and Lamballais, E. (1992) *Phys. Fluids*, **4**, 2761–2778.
Deardorff, J.W. (1970) *J. Fluid Mech.*, **41**, 453–480.
Ducros, F., Comte, P. and Lesieur, M. (1996) *J. Fluid Mech.*, **326**, 1–36.
Germano et al. (1991) *Phys. Fluids A*, **3**, 1760–1765.
Hirsch, C. (1988) *Numerical Computation of Internal and External Flows*. Wiley.
Hunt, J.C.R. et al. (2001) *J. Fluid Mech.*, **436**, 353–379.
Jeong, J. and Hussain, F. (1995) *J. Fluid Mech.*, **285**.
Jiménez, J. and Wray, A.A. (1998) *J. Fluid Mech.*, **373**, 255–285.
Jiménez, J. et al. (1993) *J. Fluid Mech.*, **225**, 65–90.
Kaneda, Y. et al. (2003) *Phys. Fluids*, **15**(2), L21–L24.
Kida, S. (2000) In: *Mechanics for a New Millenium*. (eds. H. Aref and J.W. Phillips) Kluwer Academic Publishers.
Landau, L.D. and Lifshitz, L.M. (1959) *Fluid Mechanics*. Pergamon Press, New York.
Leonard, A. (1974) *Adv. Geophys*, **18**A, 237–248.
Lesieur, M. (1990) *Turbulence in Fluids*, 2nd edition. Kluwer Academic Publishers, Netherlands.

Lesieur, M. (1995) Mixing layer vortices. In: *Fluid Vortices*. (ed. S.I. Green). Kluwer Academic Publishers, Netherlands.

Lin, S.J. and Corcos, G.M. (1984) *J. Fluid Mech.*, **141**, 139–178.

Lowe, A.J. (2001) The direct numerical simulation of isotropic two-dimensional turbulence in a periodic square. Ph.D. Thesis, University of Cambridge.

Moin, P. and Kim, J. (1985). *J. Fluid Mech.*, **155**, 441–462.

Moin, P. and Mahesh, K. (1998) *Ann. Rev. Fluid Mech.*, **30**, 539–578.

Pope, S.B. (2000) *Turbulent Flows*. Cambridge University Press.

Rogers, M.M. and Moin, P. (1987) *J. Fluid Mech.*, **176**, 33–66.

Ruetsch, G.R. and Maxey, M.R. (1992) *Phys. Fluids A*, **4**(12), 2747–2760.

Sandham, N.D. et al. (2001) *Theor. and Comput. Fluid Dyn.*, **14**(5).

She, Z.-S., Jackson, E. and Orszag, S.A. (1991) *Proc Roy. Soc. London A*, **434**.

Sherman, F.S. (1990) *Viscous Flow*. McGraw Hill, New York.

Siggia, E.D. (1981) *J. Fluid Mech.*, **107**, 375–406.

Smagorinsky, J. (1963). *Mon. Weath. Rev.*, **91**, 99–164.

Taneda, S., Honji, H. and Tatsuno, M. (1977) *Int. Symp. Flow Vis.*, Tokyo.

Townsend, A.A. (1976) *The Structure of Turbulent Shear Flow*. Cambridge University Press.

Tsinober, A. (2002) *An Informal Introduction to Turbulence*. Kluwer Academic Publishers.

Vincent, A. and Meneguzzi, M. (1991) *J. Fluid Mech.*, **225**, 1–20.

Vincent, A. and Meneguzzi, M. (1994) *J. Fluid Mech.*, **258**, 245–254.

CHAPTER 8

Isotropic turbulence (in spectral space)

> Nothing puzzles me more than time and space; and yet nothing troubles me less, as I never think about them.
>
> Charles Lamb

We now move from real space to Fourier space, recasting the equations of turbulence in terms of $\mathbf{k}$ and t, rather than $\mathbf{r}$ and t. It is important to note that this produces no new information; it simply represents a rearrangement of what we know (or think we know) already. However, it does have the great advantage of placing the all-important energy spectrum, $E(k)$, centre stage.

8.1 Kinematics in spectral space

In the first half of this chapter we focus on kinematics. In particular, we seek to establish the kinematic properties of the energy spectrum and its various relatives. However, rather than jump directly to a discussion of three-dimensional velocity fields, we start by describing how the Fourier transform can be used to distinguish between the different scales present in a one-dimensional, random signal. The intention is to show that the three-dimensional energy spectrum, as used in turbulence theory, is a natural extension of the use of the Fourier transform in the decomposition of one-dimensional signals.

Perhaps we should say something about the way that Section 8.1 is organized. It is divided up into the following subsections:

(1) the properties of the Fourier transform;
(2) the Fourier transform as a filter;
(3) the autocorrelation function and one-dimensional spectrum;
(4) the three-dimensional energy spectrum, $E(k)$;
(5) one-dimensional spectra in three-dimensional turbulence;
(6) the relationship between $E(k)$ and the second-order structure function;
(7) footnote 1—singularities in the spectrum;
(8) footnote 2—the transform of the velocity field;
(9) footnote 3—the physical significance of one-dimensional and three-dimensional spectra in turbulence.

We start, in Section 8.1.1, by summarizing the relevant properties of the Fourier transform. Next, we discuss the way in which the Fourier transform acts like a sort of filter, sifting out the different scales which are present in a fluctuating signal. It is this ability to differentiate between scales which makes the Fourier transform so useful in turbulence. Section 8.1.2, on filtering, has much in common with Section 7.1.2 on LES.

When dealing with a one-dimensional random signal, say $g(x)$, it is common to combine the Fourier transform with the idea of the autocorrelation function, defined as $\langle g(x)g(x+r)\rangle$. In Section 8.1.3 we shall see that the transform of the autocorrelation function is proportional to $|G(k)|^2$, where $G(k)$ is the transform of $g(x)$. The transform of $\langle g(x)g(x+r)\rangle$ is referred to as the *one-dimensional energy spectrum* of $g(x)$, and it provides a convenient measure of the relative amplitudes of the different Fourier modes (scales) present in the original signal. The simple ideas embodied in the one-dimensional energy spectrum provide the basis for the development of the three-dimensional spectrum, $E(k)$.

We turn to turbulence in Section 8.1.4. Here we extend the idea of filtering (via the Fourier transform) to a three-dimensional, turbulent velocity field. This is achieved by taking the transform of the velocity correlation tensor, Q_{ij}. The end result is the all-important three-dimensional energy spectrum, $E(k)$, which provides a rough measure of the distribution of kinetic energy amongst the different eddy sizes. In practice, however, one rarely measures $E(k)$ in an experiment. Instead, one ends up with a variety of one-dimensional spectra representing different facets of the three-dimensional velocity field. So, while most theoretical papers on turbulence talk about $E(k)$, their experimental counterparts tend to discuss certain one-dimensional spectra. We discuss the relationship between $E(k)$ and its various one-dimensional surrogates in Section 8.1.5.

We close Section 8.1 with three footnotes. First, in Section 8.1.7, we warn readers that singularities can occur in the spectrum at $\mathbf{k} = \mathbf{0}$, and that these singularities are related to the lack of absolute convergence of integral moments of Q_{ij}. This can be particularly problematic in anisotropic turbulence. Our second footnote, in Section 8.1.8, describes an alternative, though equivalent, way of defining $E(k)$. Here one works directly with the transform of the velocity field, rather than the transform of the velocity correlation Q_{ij}. This more direct approach has become increasingly popular in recent years. Our final footnote constitutes a word of caution. Energy spectra provide a powerful means of distinguishing between scales. However, one must be very careful in attributing physical meaning to $E(k)$ and its one-dimensional surrogates. There are a number of pitfalls lying in wait for the unwary, and these are discussed in Section 8.1.9.

Let us start, then, by reminding you of some of the properties of the Fourier transform. (Those who are already familiar with the Fourier transform may wish to go directly to Section 8.1.2.)

8.1.1 *The Fourier transform and its properties*

I had worked on the theory of Fourier integrals under Hardy's guidance for a good many years before I discovered that this theory has applications in applied mathematics. (Titchmarsh)

The Fourier transform of some function $f(x)$ is defined as

$$F(k) = \frac{1}{2\pi} \int_{-\infty}^{\infty} f(x) e^{-jkx} dx. \tag{8.1}$$

This integral exists, and hence $F(k)$ is well defined, provided the integral of $|f(x)|$ from $-\infty$ to ∞ exists.[1] Fourier's integral theorem tells us that the inverse transform is

$$f(x) = \int_{-\infty}^{\infty} F(k) e^{jkx} dk. \tag{8.2}$$

If $f(x)$ is an even function then it is readily confirmed that

$$F(k) = \frac{1}{\pi} \int_0^{\infty} f(x) \cos(kx) dx. \tag{8.3}$$

This implies that $F(k)$ is also even and so, from (8.2),

$$f(x) = 2 \int_0^{\infty} F(k) \cos(kx) dk. \tag{8.4}$$

The transform pair (8.3) and (8.4) represent a so-called cosine transform. If $f(x)$ is odd, on the other hand, we have

$$F(k) = -\frac{j}{\pi} \int_0^{\infty} f(x) \sin(kx) dx, \qquad f(x) = 2j \int_0^{\infty} F(k) \sin(kx) dk.$$

In such cases it is conventional to absorb the factor of j into the definition of F and we obtain the sine transform pair,

$$F(k) = \frac{1}{\pi} \int_0^{\infty} f(x) \sin(kx) dx \tag{8.5}$$

$$f(x) = 2 \int_0^{\infty} F(k) \sin(kx) dk. \tag{8.6}$$

In fact we already met such a pair in the form of (6.24) and (6.25),

$$[E/(2k)] = \frac{1}{\pi} \int_0^{\infty} [rR] \sin(kr) dr \tag{8.7}$$

[1] There are additional restrictions on $f(x)$ to do with discontinuities and with the boundedness of the number of maxima and minima of $f(x)$. We are not concerned with such special cases.

$$[rR] = 2\int_0^\infty [E/(2k)]\sin(kr)dk. \tag{8.8}$$

The principal utility of the Fourier transform arises form the property

$$\text{Transform}[f'(x)] = jkF(k)$$

which is readily confirmed by differentiating (8.2). Thus linear, ordinary differential equations (ODE), such as $f''(x) + \omega^2 f(x) = g(x)$ transform into linear algebraic equations, in this case $-k^2F + \omega^2F = G(k)$. It is the ability to transform a linear differential equation into an algebraic equation which makes the Fourier transform such a useful device. Additional properties of the Fourier transform include the following. (See Bracewell, 1986, for more details)

1. Rayleigh's theorem:

$$\int_{-\infty}^{\infty} |f(x)|^2 dx = 2\pi \int_{-\infty}^{\infty} |F(k)|^2 dk. \tag{8.9}$$

2. Power theorem (for real functions of f and g):[2]

$$\int_{-\infty}^{\infty} f(x)g(x)dx = 2\pi \int_{-\infty}^{\infty} F(k)G^\dagger(k)dk. \tag{8.10}$$

3. Convolution theorem:

$$2\pi F(k)G(k) = \text{Transform}\left[\int_{-\infty}^{\infty} f(u)g(x-u)du\right]. \tag{8.11}$$

4. Autocorrelation theorem (for a real function f):

$$2\pi|F(k)|^2 = \text{Transform}\left[\int_{-\infty}^{\infty} f(u)f(u+x)du\right]. \tag{8.12}$$

Theorems 1, 2, and 4 are, in effect, special cases of 3.

The concept of the Fourier transform is readily generalized to functions of more than one variable. Thus, for example, we have the transform pair,

$$F(k_x, k_y) = \frac{1}{(2\pi)^2}\int_{-\infty}^{\infty}\int_{-\infty}^{\infty} f(x,y)e^{-jk_x x}e^{-jk_y y}dxdy$$
$$f(x,y) = \int_{-\infty}^{\infty}\int_{-\infty}^{\infty} F(k_x,k_y)e^{jk_x x}e^{jk_y y}dk_x dk_y.$$

This simply represents the process of transforming first with respect to one variable, say x, and then the other variable, y. A more compact notation is,

$$F(\mathbf{k}) = \frac{1}{(2\pi)^2}\int f(\mathbf{x})e^{-j\mathbf{k}\cdot\mathbf{x}}d\mathbf{x} \tag{8.13}$$

[2] †represents a complex conjugate.

$$f(\mathbf{x}) = \int F(\mathbf{k}) e^{j\mathbf{k}\cdot\mathbf{x}} d\mathbf{k} \tag{8.14}$$

When written in this way we may regard $\mathbf{k}$ as a vector. Of course, this generalizes to three dimensions, with the factor of $(2\pi)^2$ in (8.13) replaced by $(2\pi)^3$.

We can also take the Fourier transform of a vector by simply transforming each of its components one by one. Thus, for example, the Fourier transform, $\hat{\mathbf{u}}$, of a three-dimensional velocity field, $\mathbf{u}$, satisfies the transform pair,

$$\hat{\mathbf{u}}(\mathbf{k}) = \frac{1}{(2\pi)^3} \int \mathbf{u}(\mathbf{x}) e^{-j\mathbf{k}\cdot\mathbf{x}} d\mathbf{x} \tag{8.15}$$

$$\mathbf{u}(\mathbf{x}) = \int \hat{\mathbf{u}}(\mathbf{k}) e^{j\mathbf{k}\cdot\mathbf{x}} d\mathbf{k} \tag{8.16}$$

We note in passing that the continuity equation gives us,

$$\nabla \cdot \mathbf{u} = \int j\mathbf{k} \cdot \hat{\mathbf{u}} e^{j\mathbf{k}\cdot\mathbf{x}} d\mathbf{k} = 0$$

and since this must be true for all $\mathbf{x}$ we have

$$\mathbf{k} \cdot \hat{\mathbf{u}} = 0. \tag{8.17}$$

Also, the vorticity is related to $\hat{\mathbf{u}}$ by,

$$\boldsymbol{\omega} = \nabla \times \mathbf{u} = \int (j\mathbf{k}) \times \hat{\mathbf{u}} e^{j\mathbf{k}\cdot\mathbf{x}} d\mathbf{k} \tag{8.18}$$

Now suppose we have a function $g(\mathbf{x})$ which is defined in three dimensions, but which has spherical symmetry: $g = g(|\mathbf{x}|)$. Then it is readily confirmed that $G(\mathbf{k})$ also has spherical symmetry. Thus we have,

$$G(k) = \frac{1}{(2\pi)^3} \int g(r) e^{-j\mathbf{k}\cdot\mathbf{x}} d\mathbf{x}, \quad k = |\mathbf{k}|$$

in (r, θ, ϕ) coordinates. Integrating over θ and ϕ we find, after a little algebra,

$$G(k) = \frac{1}{2\pi^2} \int_0^\infty r^2 g(r) \frac{\sin(kr)}{kr} dr \tag{8.19}$$

Similarly, it may be shown that,

$$g(r) = 4\pi \int_0^\infty k^2 G(k) \frac{\sin(kr)}{kr} dk. \tag{8.20}$$

Actually we have come across this transform pair already. For example $R(r)$ and $E/(4\pi k^2)$ are related by just such a transform:

$$\left[\frac{E}{4\pi k^2}\right] = \frac{1}{2\pi^2} \int_0^\infty r^2 R(r) \frac{\sin(kr)}{kr} dr \tag{8.21a}$$

$$R(r) = 4\pi \int_0^\infty k^2 \left[\frac{E}{4\pi k^2}\right] \frac{\sin(kr)}{kr} dk \tag{8.21b}$$

Thus $R = \frac{1}{2}\langle \mathbf{u} \cdot \mathbf{u}' \rangle$ and $E/(4\pi k^2)$ represent a three-dimensional transform pair. We shall explain the significance of this shortly. Finally we note that (8.21b) may be integrated to give the relationship between $E(k)$ and the integral scale l:

$$l = \int_0^\infty f dr = \frac{\pi}{2u^2} \int_0^\infty \frac{E}{k} dk. \tag{8.22}$$

(The details are spelt out in Exercise 8.1.)

8.1.2 *The Fourier transform as a filter*

So far we have merely defined the Fourier transform and summarized some of its more common properties. However, we have said nothing about its ability to differentiate between different scales in a fluctuating signal which is, after all, the primary reason for its use in turbulence theory.

The key point is that the Fourier transform may be thought of as a sort of *filter* which sifts out the various scales which are present in a turbulent signal. The manner in which this is achieved is, perhaps, best explained in one dimension. Consider a function $f(x)$ which is composed of fluctuations of different characteristic lengths. For example, $f(x)$ might be the instantaneous velocity component $u_x(x, y=0, z=0)$ in a turbulent flow, so that eddies of different sizes cause fluctuations in $u_x(x)$ of different scales. We now introduce the so-called *box function*,

$$H(r) = \begin{cases} L^{-1}, & |r| < L/2 \\ 0, & |r| > L/2 \end{cases}$$

and consider the new function $f^L(x)$ defined by,

$$f^L(x) = \int_{-\infty}^{\infty} H(r) f(x-r) dr.$$

Physically, f^L is constructed from f as follows. For any one value of x, we evaluate the average value of f in the vicinity of x by integrating f from $x - L/2$ to $x + L/2$ and dividing by L. Thus $f^L(x)$ is a filtered or 'smoothed out' version of $f(x)$, the smoothing operation being carried out over the length L. In short, $f^L(x)$ is the local average of $f(x)$, the averaging being defined in terms of the length-scale L. The effect of this smoothing operation is to filter out those parts of $f(x)$ whose characteristic length-scale is significantly smaller than L. (See Figure 8.1(a)).

(a)

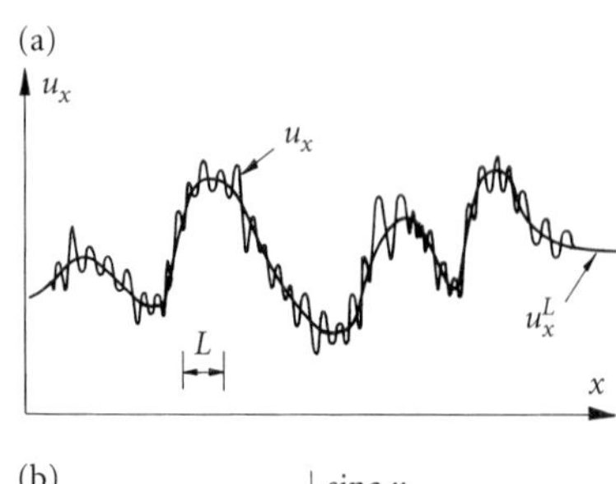

(b)

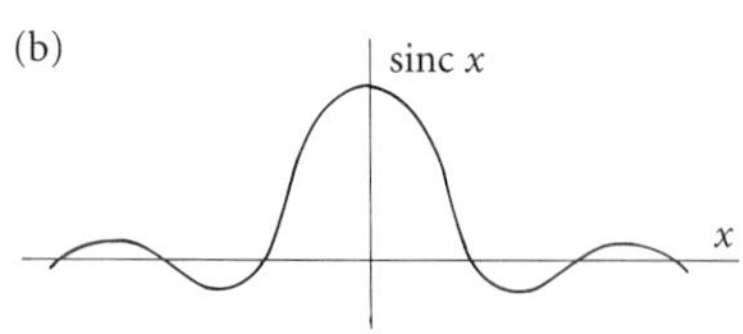

Figure 8.1 (a) Smoothing of a turbulent signal using the box filter. (b) The shape of $\text{sinc}(x) = \sin x / x$.

Let us now take the Fourier transform of $f^L(x)$, invoking the convolution theorem in the process. The result is

$$F^L(k) = \frac{\sin(kL/2)}{kL/2} F(k)$$

since the Fourier transform of $H(r)$ is $\sin(\eta)/2\pi\eta$, where $\eta = kL/2$. The function $\sin(x)/x$, sometimes written $\text{sinc}(x)$, is known as the filtering function. It has the properties

$$\text{sinc}(0) = 1, \qquad \int_{-\infty}^{\infty} \text{sinc}x dx = \pi$$

Thus $\text{sinc}(kL/2)$ peaks at the origin, falls to zero at $k = \pi/L$, and oscillates with a diminishing amplitude thereafter (Figure 8.1(b)). If we ignore the weak oscillations for $k > \pi/L$ we may regard $F^L(k)$ as a truncated version of $F(k)$, with contributions to $F^L(k)$ for $k > \pi/L$ suppressed by $\text{sinc}(kL/2)$.

So the operation of smoothing $f(x)$ over a length-scale L, suppressing contributions of scale less than L, is more or less equivalent to throwing out the high-wave number contribution to $F(k)$, the cut-off being around $k = \pi/L$. The implication is that the short length-scale information contained in $f(x)$ is held in the high wave number part of $F(k)$. In this sense the Fourier transform allows us to differentiate between scales, though it is an imperfect filter because the cut-off in k-space is not abrupt.

Of course there is no reason why we should restrict ourselves to the box function, $H(r)$, if we wish to smooth out $f(x)$. We might equally use, say, the *Gaussian filter*

$$G(r) = \frac{\exp[-r^2/L^2]}{\pi^{1/2}L}.$$

Like $H(r)$, this has unit area, $\int_{-\infty}^{\infty} G(r)dr = 1$, and so the function

$$f^L(x) = \int_{-\infty}^{\infty} G(r)f(x-r)dr$$

is also a locally averaged, or filtered, version of $f(x)$, in which fluctuations on a scale much less than L are suppressed. This time the Fourier transform of $f^L(x)$ is,

$$F^L(k) = \exp[-k^2L^2/4]F(k).$$

As before, the operation of filtering the function in real space corresponds to suppressing the high wave numbers in the Fourier transform of $f(x)$. However, as in the previous example, there is not an abrupt cut-off in wave number, but rather a gradual suppression of high wave number components.

Table 8.1 Different filters and their Fourier transforms

Name	Filter	Fourier transform	Properties
Box filter	$H(r) = \begin{cases} L^{-1}, & \|r\| < L/2 \\ 0, & \|r\| > L/2 \end{cases}$	$\frac{1}{2\pi}\mathrm{sinc}(kL/2)$	Sharp in real space but oscillatory in spectral space
Gaussian filter	$G(r) = \frac{\exp[-r^2/L^2]}{\sqrt{\pi}L}$	$\frac{1}{2\pi}\exp[-(kL/2)^2]$	Gaussian in both real space and spectral space
Sinc filter or sharp spectral filter	$S(r) = \frac{\sin(\pi r/L)}{\pi r}$	$\frac{1}{2\pi}\begin{cases} 1, & \|k\|L < \pi \\ 0, & \|k\|L > \pi \end{cases}$	Sharp in spectral space but oscillatory in real space

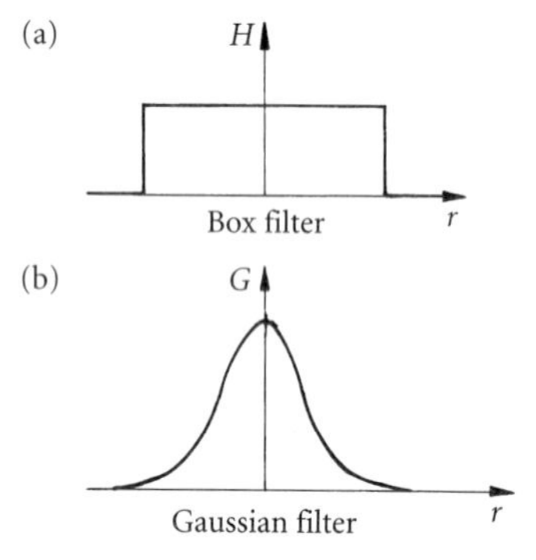

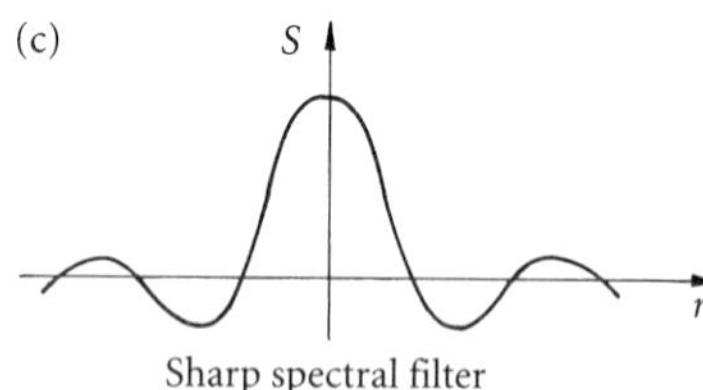

Figure 8.2 The shape of the box, Gaussian, and sinc filter.

In order to obtain a sharp cut-off in Fourier space we need to adopt the real-space filter function,

$$S(r) = \frac{\sin(\pi r/L)}{\pi r}$$

which has unit area and a Fourier transform proportional to the box function. (See Table 8.1.) However, we have paid a heavy price for the sharp cut-off in k-space: a smoothing operation involving sinc($\pi r/L$) admits scales less than L since sinc(x) has an oscillatory tail extending to infinity (Figure 8.2). So the idea of the Fourier transform as a mechanism of distinguishing between scales is an imperfect one, and this deficiency will come back to haunt us time and again.

8.1.3 *The autocorrelation function*

The ability of the Fourier transform to differentiate between scales is the primary reason it is introduced into turbulence theory. That is, we need a means of distinguishing between small and large eddies. As suggested above, the Fourier transform represents an imperfect filter. Nevertheless, it is almost universally adopted as the filter of choice in turbulence.

In practice, when working with random signals, the Fourier transform is usually applied in conjunction with the idea of the *autocorrelation function.* In fact, the two concepts (Fourier transform and autocorrelation function) invariably go hand-in-hand in turbulence.

The idea of the autocorrelation is, perhaps, best introduced in terms of a one-dimensional function. Consider, for example, the velocity component $u_x(x)$ measured along the line $y=0$, $z=0$ at the instant $t=0$. If the flow is laminar then u_x might have a simple, smooth shape as shown below.

The *autocorrelation function* of such a signal is defined as,

$$\mathrm{v}(r) = \int_{-\infty}^{\infty} u_x(x)u_x(x-r)dx = \int_{-\infty}^{\infty} u_x(x)u_x(x+r)dx.$$

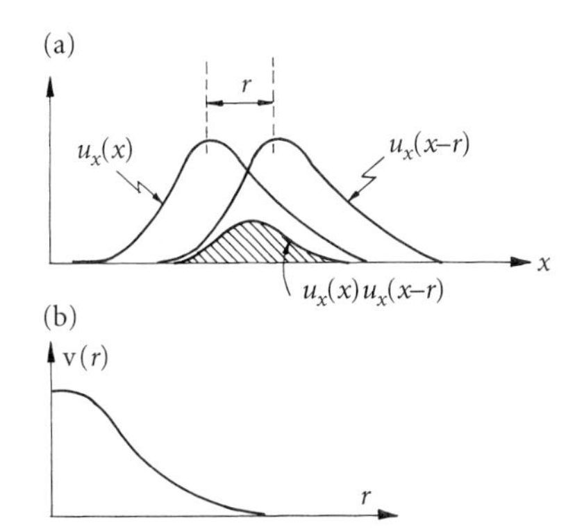

Figure 8.3 The autocorrelation function for $u_x(x)$ in a laminar flow.

(It is evident from Figure 8.3 that v(r) is even in r, and hence either of the integrals above may be used to define v(r).) It is possible to prove, using the Schwartz inequality, that v(r) has a maximum value at $r = 0$, and so it is often convenient to normalize v(r) by v(0), which yields the so-called *autocorrelation coefficient*:

$$\rho(r) = \frac{\int_{-\infty}^{\infty} u_x(x)u_x(x+r)dx}{\int_{-\infty}^{\infty} u_x^2(x)dx}, \quad \rho(r) \le 1.$$

When the flow is turbulent, statistically homongeneous, and has no mean velocity component, u_x will be a random function with zero mean and no particular beginning or end. In such a case the autocorrelation function is redefined in terms of the convergent integral

$$\mathrm{v}(r) = \frac{1}{2X}\int_{-X}^{X} u_x(x)u_x(x+r)dx$$

and the autocorrelation coefficient becomes,

$$\rho(r) = \frac{\int_{-X}^{X} u_x(x)u_x(x+r)dx}{\int_{-X}^{X} u_x^2(x)dx}.$$

Here X is taken to be much greater then any characteristic length-scale associated with the fluctuations in $u_x(x)$. Since spatial averages are equivalent to ensemble averages we see that, for a turbulent flow, the autocorrelation coefficient of $u_x(x)$ is equal to the longitudinal correlation function, $f(r)$:

$$\rho(r) = \langle u_x(x)u_x(x+r)\rangle / u^2 = f(r).$$

Of course, the autocorrelation function is simply the two-point velocity correlation: $\mathrm{v}(r) = \langle u_x(x)u_x(x+r)\rangle$.

So why is the autocorrelation function so useful and what has it to do with the Fourier transform? This is best illustrated using a trivial example. Suppose that $u_x(x)$ takes a particularly simple form, say,

$$u_x(x) = A_1 \sin(k_1 x + \phi_1) + A_2 \sin(k_2 x + \phi_2) + A_3 \sin(k_3 x + \phi_3).$$

Let us calculate v(r). Since $u_x(x)$ does not die out at large $|x|$ we take the approach of integrating over a large distance $2X$, divide by $2X$ and let $X \to \infty$. It is readily verified that

$$\mathrm{v}(r) = \left[A_1^2 \cos(k_1 r) + A_2^2 \cos(k_2 r) + A_3^2 \cos(k_3 r)\right].$$

Now suppose that we take the Fourier transform of v(r). This consists of spikes at $k = k_1, k_2$, and k_3, the amplitude of the spikes being proportional to A_1^2, A_2^2, and A_3^2 respectively.

More generally, the Fourier transform of the autocorrelation of some one-dimensional function $g(x)$ is referred to as the *one-dimensional energy spectrum* of $g(x)$. From the autocorrelation theorem (see theorem 4 of Section 8.1.1) the one-dimensional energy spectrum is always a positive quantity. Moreover, the example above suggests that the one-dimensional energy spectrum picks out the different Fourier components contained in $g(x)$ and that the magnitude of the energy spectrum (at a given k) is proportional to the square of amplitude of the kth Fourier mode. For a well-behaved function which dies out at infinity these statements are all formalized by the autocorrelation theorem:

$$\text{Transform}\left[\int_{-\infty}^{\infty} g(x)g(x+r)dx\right] = 2\pi|G(k)|^2$$
$$= \text{one-dimensional energy spectrum of } g(x). \qquad (8.23a)$$

For a random function with no particular beginning or end the Fourier transform is not well defined and so we have to do a little more work. Suppose that $g(x)$ is just such a statistically homogeneous random function, with zero mean. To make use of the autocorrelation theorem we introduce $g_X(x)$, define as,

$$g_X(x) = \begin{cases} g(x), & |x| < X \\ 0, & |x| > X \end{cases}$$

Unlike $g(x)$, the 'clipped function' $g_X(x)$ has a well defined transform, $G_X(k)$. We now invoke the autocorrelation theorem for $g_X(x)$:

$$\text{Transform}\left[\int_{-\infty}^{\infty} g_X(x)g_X(x+r)dx\right] = 2\pi|G_X(k)|^2.$$

Since $g_X(x)$ is zero outside the range $-X < x < X$, this reduces to

$$\text{Transform}\left[\int_{\max[-X,\,-X-r]}^{\min[X,\,X-r]} g(x)g(x+r)dx\right] = 2\pi G_X(k)G_X^{\dagger}(k).$$

We now ensemble average the result, divide by $2X$, and let X become large. If $\langle g(x)g(x+r)\rangle$ falls off rapidly as $|r|$ increases, and we shall assume that it does, then the limits on the integral can be replaced by $\pm X$. The end result is

$$\text{Transform}\left[\frac{1}{2X}\int_{-X}^{X}\langle g(x)g(x+r)\rangle dx\right] = \frac{\pi}{X}\left\langle G_X(k)G_X^{\dagger}(k)\right\rangle.$$

Since $g(x)$ is statistically homogeneous, $\langle g(x)g(x+r)\rangle$ is independent of x and this simplifies to,

$$\text{Transform}[\langle g(x)g(x+r)\rangle] = \lim_{X\to 0}[(\pi/X)\langle G_X(k)G_X^{\dagger}(k)\rangle].$$

Note that the right-hand side remains finite as $X\to\infty$ since $G_X(k)$, being the integral of a random function, grows on average as $X^{1/2}$ for large X. (This is a consequence of the central limit theorem.) Once again we find that the transform of the autocorrelation function is positive and that it picks out the different Fourier modes present in the original signal, with the magnitude of the one-dimensional energy spectrum being proportional to the square of the kth Fourier mode.

All of these ideas extend in an obvious way to three dimensions. If $g(\mathbf{x})$ is a well behaved, non-random function of x, y, and z, and $G(\mathbf{k})$ is its three-dimensional transform, then the three-dimensional autocorrelation function for $g(\mathbf{x})$ is,

$$\mathrm{v}(\mathbf{r}) = \int g(\mathbf{x})g(\mathbf{x}+\mathbf{r})d\mathbf{x}$$

and the autocorrelation theorem tells us that

$$\mathrm{Transform}[\mathrm{v}(\mathbf{r})] = (2\pi)^3|G(\mathbf{k})|^2. \tag{8.23b}$$

When $g(\mathbf{x})$ is a random function with zero mean we follow the same procedure as in one dimension. The autocorrelation is redefined as

$$\mathrm{v}(\mathbf{r}) = \frac{1}{V}\int_V g(\mathbf{x})g(\mathbf{x}+\mathbf{r})d\mathbf{x} = \langle g(\mathbf{x})g(\mathbf{x}+\mathbf{r})\rangle$$

where V is some large volume, say a cube of side $2X$. It is readily confirmed that the autocorrelation theorem then yields

$$\mathrm{Transform}[\mathrm{v}(\mathbf{r})] = \lim_{X\to 0}\left[(\pi/X)^3\langle G_X(\mathbf{k})G_X^{\dagger}(\mathbf{k})\rangle\right]. \tag{8.23c}$$

Clearly, the right-hand sides of (8.23b) and (8.23c) represent a form of energy spectrum for the three-dimensional autocorrelation $\mathrm{v}(\mathbf{r})$, analogous to the one-dimensional energy spectrum in (8.23a). However, we shall refrain from calling this a three-dimensional energy spectrum since, as we shall see in a moment, this name is reserved for a different, though related, quantity.

In summary then, the procedure of forming an autocorrelation (be it one dimensional or three-dimensional) and then taking its Fourier transform provides a convenient means of extracting the relative magnitudes of the different Fourier modes present in the original signal. This, in turn, provides a means of differentiating between large and small scales (low and high frequencies) in the original function. The tradition in turbulence theory has been to use just this sort of procedure to differentiate between the small- and large-scale eddies. However, rather than work with one-dimensional, scalar functions, such as $u_x(x)$, it is conventional to take the Fourier transform of the velocity correlation tensor, $Q_{ij}(\mathbf{r})$. The general idea, though, is the same since the diagonal components of Q_{ij} represent autocorrelation functions for the velocity components u_i.

It has to be said, however, that this procedure is imperfect. There are two reasons for this. First we might note that, in the simple one-dimensional example above, the phases ϕ_1, ϕ_2, and ϕ_3 of the three wave components contained in $u_x(x)$ do not appear in the autocorrelation function v(r). Thus we lose information in the process of forming the autocorrelation, and this information is also missing in the one-dimensional energy spectrum. In short, an infinite number of different signals can give rise to the same energy spectrum. Second, in turbulence the flow does not consist of a set of waves, but rather of spatially compact structures, that is, eddies, and it is the distribution of eddy sizes which is important, not the Fourier components of **u**. Nevertheless, our interpretation of the Fourier transform as a sort of filter (see Section 8.1.2) means that small eddies tend to give rise to high-frequency components in the energy spectrum, while large eddies tend to be associated with low-frequency components. So we can infer something about the distribution of eddy sizes from examining the energy spectrum. However, making the connection between eddy size and wave number is not a straightforward matter, as we found out in Section 6.4.1. We shall return to this issue in Section 8.1.9.

So far we have defined the autocorrelation and energy spectrum only for simple scalar functions. In a moment we shall see how these ideas generalize to a three-dimensional velocity field. Here the details are rather intricate, involving some fairly tedious algebra. For example, defining the energy spectrum of a three-dimensional velocity field is a more complex procedure than defining the one-dimensional energy spectrum associated with, say, u_x. Nevertheless, the main point does not change. The combined use of autocorrelation function and the Fourier transform is an attempt to differentiate between the various scales which arise in a turbulent velocity field.

8.1.4 *The transform of the correlation tensor and the three-dimensional energy spectrum*

Let us now return to turbulence. It is conventional to introduce the *spectrum tensor*, $\Phi_{ij}(\mathbf{k})$, defined as the transform of Q_{ij}:

$$\Phi_{ij}(\mathbf{k}) = \frac{1}{(2\pi)^3}\int Q_{ij}(\mathbf{r})e^{-j\mathbf{k}\cdot\mathbf{r}}d\mathbf{r} \qquad (8.24a)$$

$$Q_{ij}(\mathbf{r}) = \int \Phi_{ij}(\mathbf{k})e^{j\mathbf{k}\cdot\mathbf{r}}d\mathbf{k} \qquad (8.24b)$$

and observe that the incompressibility condition, (6.10), requires

$$k_i\Phi_{ij}(\mathbf{k}) = k_j\Phi_{ij}(\mathbf{k}) = 0. \qquad (8.25)$$

Note that, since Q_{xx}, Q_{yy}, and Q_{zz} are three-dimensional autocorrelation functions for u_x, u_y, and u_z, the diagonal components of Φ_{ij} are all non-negative, and in particular, $\Phi_{ii} \geq 0$. Now for isotropic turbulence Φ_{ij} must be an isotropic tensor and (6.31) tells us that it is of the form

$$\Phi_{ij} = A(k)k_i k_j + B(k)\delta_{ij}$$

where A and B are even functions of $k = |\mathbf{k}|$. However, the incompressibility condition (8.25) demands

$$(Ak^2 + B)k_j = 0$$

and so Φ_{ij} simplifies to,

$$\Phi_{ij} = B(k)\left[\delta_{ij} - \frac{k_i k_j}{k^2}\right]. \tag{8.26}$$

Let us now consider two special cases of Φ_{ij} and Q_{ij}, namely, Φ_{ii} and Q_{ii}. These are of particular interest since Q_{xx}, Q_{yy}, and Q_{zz} represent three-dimensional autocorrelations for u_x, u_y, and u_z, and so terms such as Φ_{xx} represent a sort of energy spectrum for u_x, in the sense of (8.23c). From (8.26) and (6.39) we see that Φ_{ii} and Q_{ii} are spherically symmetric functions, given by,

$$\tfrac{1}{2}\Phi_{ii} = B(k) \tag{8.27}$$

$$\tfrac{1}{2}Q_{ii} = \tfrac{1}{2}\langle \mathbf{u} \cdot \mathbf{u}' \rangle = R(r). \tag{8.28}$$

Moreover (8.24b) yields

$$\frac{1}{2}\langle \mathbf{u}^2 \rangle = \frac{1}{2}\int \Phi_{ii} d\mathbf{k} = \int_0^\infty 2\pi k^2 \Phi_{ii} dk$$

and so integrating $\frac{1}{2}\Phi_{ii}$ over all $\mathbf{k}$-space gives the kinetic energy density of the turbulence. Sometimes we talk about Φ_{ii} representing the 'distribution' of kinetic energy in spectral space.

We now reintroduce the *three dimensional energy spectrum* of the velocity field, $E(k)$. This first appeared in Chapter 3 defined in terms of the transform of $R(r)$. This time, however, we define it in an alternative, but equivalent, way. Noting that the diagonal components of Φ_{ij} represent energy spectra for the different velocity components (see (8.23c)), it seems natural to focus on Φ_{ii}. We write

$$E(k) = 2\pi k^2 \Phi_{ii}, \quad E(k) > 0 \tag{8.29}$$

from which

$$\frac{1}{2}\langle \mathbf{u}^2 \rangle = \int_0^\infty E dk. \tag{8.30}$$

Returning to (8.26) we now see that Φ_{ij} may be expressed as,

$$\Phi_{ij} = \frac{E(k)}{4\pi k^2}\left[\delta_{ij} - \frac{k_i k_j}{k^2}\right]. \tag{8.31}$$

Let us now check that our new definition of $E(k)$ is consistent with the old. We note that Φ_{ii} and Q_{ii} are three-dimensional transform pairs which have spherical symmetry. We may therefore apply (8.19) and (8.20) to give,

$$\Phi_{ii} = \frac{1}{2\pi^2}\int_0^\infty r^2 Q_{ii}\frac{\sin(kr)}{kr}dr, \qquad Q_{ii} = 4\pi\int_0^\infty k^2\Phi_{ii}\frac{\sin(kr)}{kr}dk.$$

Substituting for Φ_{ii} and Q_{ii} in terms of $E(k)$ and $R(r)$ we recover the old definitions

$$E(k) = \frac{2}{\pi}\int_0^\infty R(r)kr\sin(kr)dr \tag{8.32}$$

$$R(r) = \int_0^\infty E(k)\frac{\sin(kr)}{kr}dk. \tag{8.33}$$

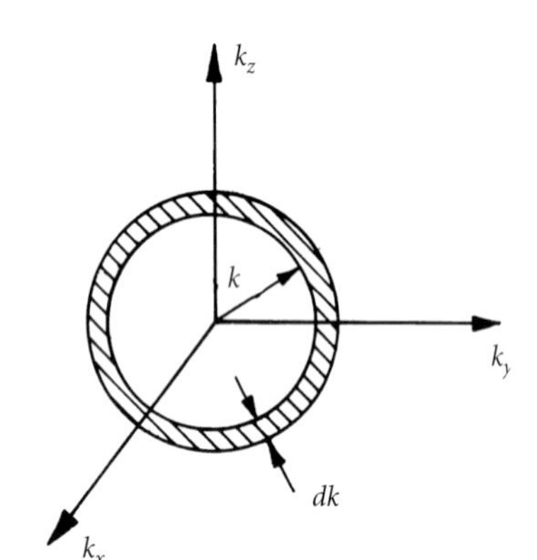

Figure 8.4 $E(k)dk$ represents the contribution to $\frac{1}{2}\langle \mathbf{u}^2\rangle$ from Φ_{ii} which lies within a spherical annulus in **k**-space.

We have come full circle. However, the advantage of our new definition is that the meaning of $E(k)$ (in spectral terms) is now clear. In particular:

(1) $\Phi_{ii} = E(k)/2\pi k^2$ is equal to the sum of the energy spectra of the three velocity components (in the sense of 8.23c);
(2) $E(k)dk$ represents the contribution of Φ_{ii} to $\frac{1}{2}\langle \mathbf{u}^2\rangle$ which is contained in a spherical annulus in **k**-space of thickness dk (Figure 8.4).

We note in passing that care must be exercised when examining the form of Φ_{ij} for small $|\mathbf{k}|$. For example, we may combine (6.77) and (8.31) to give

$$\Phi_{ij}(k) = \frac{6L + Ik^2}{96\pi^3}\left(\delta_{ij} - \frac{k_i k_j}{k^2}\right) + \ldots$$

For a Batchelor spectrum ($L = 0$, $E \sim k^4$) the form of Φ_{ij} at small k is well behaved. For a Saffman spectrum, on the other hand, where $E \sim k^2$ and $L \neq 0$, we find,

$$16\pi^3\Phi_{ij}(k) = L\left(\delta_{ij} - k_i k_j/k^2\right) + \cdots$$

The spectrum tensor, Φ_{ij}, is now non-analytic at $k = 0$ (Figure 8.5), its value depending on the manner in which we approach the origin. For example, $16\pi^3\Phi_{xx} = L(1 - k_x^2/k^2) + \cdots$. This is equal to zero if $k_x = k$, or equal to L if $k_x \ll k_y$, k_z. Moreover, (8.24a) tells us that

$$8\pi^3\Phi_{ij}(\mathbf{0}) = \int Q_{ij}(\mathbf{r})d\mathbf{r}.$$

Evidently, in a Saffman spectrum, the non-analytic behaviour of $\Phi_{ij}(\mathbf{k} \to \mathbf{0})$ corresponds to the lack of absolute convergence of the

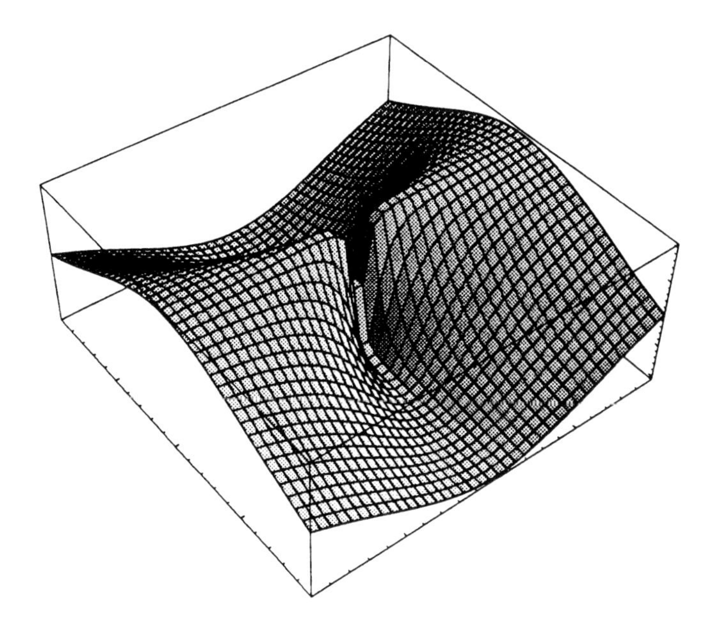

Figure 8.5 The form of $\Phi_{xx}(k_x, k_y, k_z = 0)$ near $k = 0$ in a Saffman spectrum. Note the latent singularity at the origin which, obviously, is not resolved in this plot.

integral of $Q_{ij}(\mathbf{r})$. (Remember $f_\infty \sim r^{-3}$ in such a spectrum.) This is the first hint that great care must be exercised when examining Φ_{ij} in the region of the origin. We shall return to this matter in Section 8.1.7.

Finally we consider the vorticity field. We have already seen that $\langle \omega_i \omega_i' \rangle$ is related to Q_{ii} via

$$\langle \omega_i \omega_i' \rangle = -\nabla^2 Q_{ii}(r)$$

and from (8.24b) we have,

$$\langle \omega_i \omega_i' \rangle = \int \Phi_{ii} k^2 e^{j\mathbf{k}\cdot\mathbf{r}} d\mathbf{k}.$$

In particular, as suggested in Chapter 6, we find,

$$\frac{1}{2}\langle \boldsymbol{\omega}^2 \rangle = \int_0^\infty k^2 E(k) dk. \tag{8.34}$$

Given that E is non-negative, and that

$$\frac{1}{2}\langle \mathbf{u}^2 \rangle = \int_0^\infty E(k) dk, \qquad \frac{1}{2}\langle \boldsymbol{\omega}^2 \rangle = \int_0^\infty k^2 E(k) dk$$

it is customary to interpret $E(k)$ and $k^2 E(k)$ as the distributions of kinetic energy and enstrophy in wave number space, with eddies of size l_e being associated (roughly) with wave number $k \sim l_e^{-1}$.

8.1.5 *One-dimensional energy spectra*

At the risk of confusing the reader, two alternative energy spectra are sometimes introduced via the cosine transform pair (8.3) and (8.4).

They appear particularly in experimental papers as the quantities most commonly measured in experiments. They are[3]

$$F_{11}(k) = \frac{1}{\pi}\int_0^\infty u^2 f(r)\cos(kr)dr \tag{8.35a}$$

$$F_{22}(k) = \frac{1}{\pi}\int_0^\infty u^2 g(r)\cos(kr)dr \tag{8.35b}$$

with inverse transforms,[4]

$$u^2 f(r) = 2\int_0^\infty F_{11}(k)\cos(kr)dk \tag{8.35c}$$

$$u^2 g(r) = 2\int_0^\infty F_{22}(k)\cos(kr)dk \tag{8.35d}$$

Here g and f are the usual transverse and longitudinal correlation functions. Of course, $F_{11}(k)$ and $F_{22}(k)$ are simply the one-dimensional energy spectra of $u_x(x, 0, 0)$ and $u_y(x, 0, 0)$ of the type introduced in Section 8.1.3. It is readily confirmed, by integrating by parts, that

$$\frac{d}{dk}\left[\frac{1}{k}\frac{dF_{11}}{dk}\right] = \frac{1}{\pi}\int_0^\infty \left(u^2 r^3 f\right)' \frac{\sin kr}{k^2 r} dr.$$

Substituting for f in terms of R, and using (8.32), we find a simple relationship between $E(k)$ and $F_{11}(k)$:

$$E(k) = k^3 \frac{d}{dk}\left[\frac{1}{k}\frac{dF_{11}}{dk}\right]. \tag{8.36a}$$

It is also possible to show that F_{11} and F_{22} are related by (see Exercise 8.2)

$$\frac{d^2 F_{11}}{dk^2} = -\frac{2}{k}\left[\frac{dF_{22}}{dk}\right]. \tag{8.36b}$$

A third one-dimensional spectrum is defined by the transform pair

$$E_1(k) = \frac{1}{\pi}\int_0^\infty \langle \mathbf{u}\cdot\mathbf{u}'\rangle \cos(kr)dr \tag{8.37a}$$

$$\langle \mathbf{u}\cdot\mathbf{u}'\rangle = 2\int_0^\infty E_1(k)\cos(kr)dk. \tag{8.37b}$$

Since $\langle \mathbf{u}\cdot\mathbf{u}'\rangle = u^2(f+2g)$ we see that E_1 is related to F_{11} and F_{22} by

$$E_1(k) = F_{11}(k) + 2F_{22}(k).$$

[3] There is no uniformity of notation for one-dimensional energy spectra. Batchelor (1953), Hinze (1975), Tennekes and Lumley (1972), and Monin and Yaglom (1975) all use different symbols. We shall adopt the same convention as Tennekes and Lumley for the one-dimensional spectra of $u^2 f$ and $u^2 g$, that is, F_{11} and F_{22}, and introduce E_1 for the one-dimensional spectrum of $\langle \mathbf{u}\cdot\mathbf{u}'\rangle$. Readers should be warned that Hinze uses E_1 for what we call $2F_{11}$.

[4] Of course, k represents k_x here rather that the magnitude of $\mathbf{k}$.

(a)

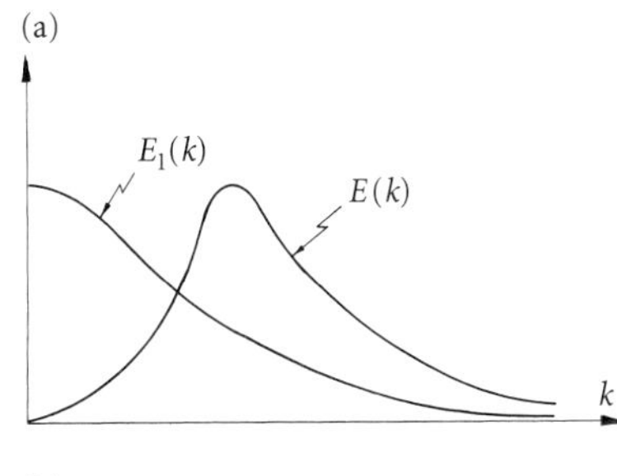

(b)

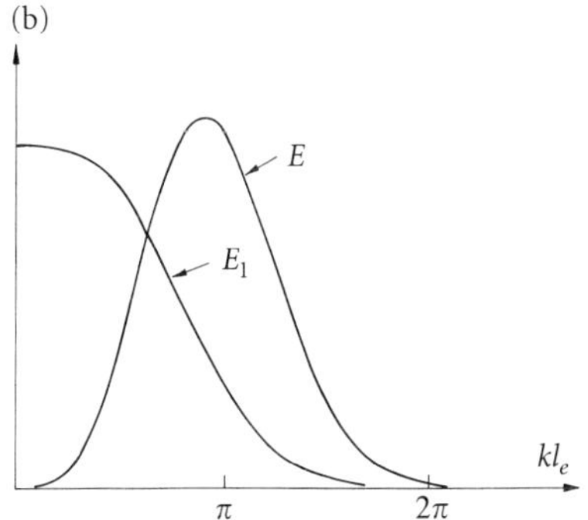

Figure 8.6(a) Schematic of the shapes of $E(k)$ and $E_1(k)$ for fully developed turbulence. (b) The spectra $E(k)$ and $E_1(k)$ for a random array of spherical eddies of size l_e. See Section 6.4.1 for more details.

Also, from (8.36a, b), one may show that

$$\frac{dE_1}{dk} = -\frac{E}{k}, \qquad E_1(k) = \int_k^\infty \frac{E(p)}{p} dp \tag{8.37c}$$

which defines the relationship between $E(k)$ and $E_1(k)$. Note that, like $E(k)$, $E_1(k)$ integrates to give the total kinetic energy:

$$\frac{1}{2}\langle \mathbf{u}^2 \rangle = \int_0^\infty E_1(k)dk.$$

Clearly $E_1(k)$ is just the one-dimensional energy spectrum of the autocorrelation function $\langle \mathbf{u} \cdot \mathbf{u}' \rangle$. In this respect it is somewhat reminiscent of $E(k)$ which comes from the three-dimensional transform of the same autocorrelation function. However, we shall see in a moment that $E(k)$ and $E_1(k)$ are rather different functions. The properties of F_{11}, F_{22}, and E_1 are summarized in Table 8.2.

The reason why F_{11} and F_{22} are commonly discussed is that they are easier to measure than $E(k)$ in an experiment. Thus most experimental papers will show measurements of F_{11} or F_{22}, rather than $E(k)$. However, one must be very cautious when it comes to the interpreting of the physical significance of F_{11}, F_{22}, or $E_1(k)$. For example, although $E(k)$ and $E_1(k)$ have similar shapes at large k (e.g. $E(k) \approx \frac{5}{3}E_1(k)$ in the inertial subrange), they have very different shapes for small to intermediate k (Figure 8.6(a)). Let us compare $E(k)$, $F_{11}(k)$, and $E_1(k)$ at small k. We have already seen that expanding (8.32) yields,

$$E = \frac{L}{4\pi^2}k^2 + \frac{I}{24\pi^2}k^4 + \cdots, \qquad L = 4\pi u^2 [r^3 f]_\infty.$$

In grid turbulence we might anticipate that $L = 0$ since f probably decays faster than r^{-3} at large r (see Section 6.3.5). Thus normally we expect,

$$E = \frac{I}{24\pi^2}k^4 + \cdots.$$

Table 8.2 Properties of one-dimensional energy spectra

Symbol	$F_{11}(k)$	$F_{22}(k)$	$E_1(k)$
One-dimensional transform partner	$u^2 f(r)$	$u^2 g(r)$	$\langle \mathbf{u} \cdot \mathbf{u}' \rangle$
Relationship to $E(k)$	$E = k^3 \frac{d}{dk}\left[\frac{1}{k}\frac{dF_{11}}{dk}\right]$	—	$\frac{dE_1}{dk} = -\frac{E}{k}$
Relationship to other one-dimensional spectra	$\frac{d^2 F_{11}}{dk^2} = -\frac{2}{k}\left[\frac{dF_{22}}{dk}\right]$	—	$E_1 = F_{11} + 2F_{22}$
Integral value	$\int_0^\infty F_{11}(k)dk = \frac{1}{2}u^2$	$\int_0^\infty F_{22}(k)dk = \frac{1}{2}u^2$	$\int_0^\infty E_1(k)dk = \frac{1}{2}\langle \mathbf{u}^2 \rangle$
Value at $k = 0$	$u^2 l/\pi$	$u^2 l/2\pi$	$2u^2 l/\pi$

On the other hand, expanding (8.35a) and (8.37a) gives,

$$F_{11} = \frac{u^2 l}{\pi} - \left[\frac{u^2}{2\pi}\int_0^\infty r^2 f dr\right]k^2 + \left[\frac{I}{24\pi^2}\right]\frac{k^4}{8} + \cdots$$

$$E_1 = \frac{2u^2 l}{\pi} - \left[\frac{I}{24\pi^2}\right]\frac{k^4}{4} + \cdots$$

which are not at all like $E(k)$! Now $E_1(k)$ is non-negative, since it is proportional to the one-dimensional energy spectrum of $\langle \mathbf{u} \cdot \mathbf{u}' \rangle$. It also satisfies,

$$\frac{1}{2}\langle \mathbf{u}^2 \rangle = \int_0^\infty E_1(k) dk.$$

It is tempting, therefore, to interpret $E_1(k)$ as the energy density of eddies measured in **k**-space, just like we interpret $E(k)$ as an energy density. However, this is clearly inappropriate for small to intermediate k as E and E_1 have very different shapes in this region. This raises a number of questions, such as: how do we know that it is $E(k)$, and not $E_1(k)$, which gives the best impression of the distribution of energy amongst the different eddy sizes, and if it does, what property does $E(k)$ have that $E_1(k)$ does not?

Well it turns out that it is indeed E, and not E_1, which represents, approximately, the kinetic energy distribution in spectral space. The point is that a random array of eddies of fixed size l_e tends to produce a three-dimensional spectrum $E(k)$ which is sharply peaked around $k \sim \pi / l_e$ (see Section 6.4.1). However, the same eddies produce a fairly broad one-dimensional spectrum, $E_1(k)$, which does not exhibit a peak at $k \sim \pi / l_e$, but rather at $k = 0$. The difference is shown in Figure 8.6(b). Thus the properties of $E(k)$ and $E_1(k)$ are somewhat different, as shown in Table 8.3.

It is the third property of $E(k)$ in Table 8.3, which allows us to associate the different parts of the three-dimensional spectrum with eddies of size π / k. Put crudely, we may think of eddies of a given size l_e contributing a spike in the energy spectrum at $k = \pi / l_e$, with the total energy spectrum $E(k)$ being the sum of many such spikes. $E_1(k)$, on the other hand, does not possess this property and so, despite the fact that E_1 integrates to give $\frac{1}{2}\langle \mathbf{u}^2 \rangle$, we do not know how the various eddy sizes present in a turbulent flow contribute to the different parts of $E_1(k)$. This curious and misleading property of the one-dimensional spectrum (when applied to a three-dimensional process) is due to something called *aliasing*, which we will discuss in Section 8.1.9.

It should also be born in mind that even $E(k)$ has its limitations as a representation of the spread of energy. This point was emphasized in Chapter 6, but it may be worth repeating here. While it is true that eddies of size l_e produce a peak in $E(k)$ around $k \sim \pi / l_e$, other wave numbers are also excited. So eddies of a given size, l_e, produce a

Table 8.3 Properties of $E(k)$ and $E_1(k)$

Properties of $E(k)$	Properties of $E_1(k)$
(1) $E(k) \geq 0$	(1) $E_1(k) \geq 0$
(2) $\int_0^\infty E(k)dk = \frac{1}{2}\langle \mathbf{u}^2 \rangle$	(2) $\int_0^\infty E_1(k)dk = \frac{1}{2}\langle \mathbf{u}^2 \rangle$
(3) Eddies of fixed size l_e produce a spectrum $E(k)$ which is sharply peaked at $k \sim \pi/l_e$.	(3) Eddies of fixed size l_e produce a spectrum $E_1(k)$ which is peaked at $k = 0$, and not at $k \sim \pi/l_e$.

spread of wave number contributions to $E(k)$, typically of the form $E(k) \sim k^4 \exp\left[-k^2 l_e^2\right]$. This causes particular difficulties when it comes to interpreting the low-k and high-k ends of a turbulent spectrum. Suppose, for example, that we have a turbulent flow which instantaneously consists of well-defined, compact structures (eddies) of sizes which range from l down to η. When we take the transform of $\langle \mathbf{u} \cdot \mathbf{u}' \rangle$ to determine $E(k)$ we find that E is a continuous function with a significant contribution to $E(k)$ in the range $k = 0 \rightarrow \pi/l$, that is, $E \sim Ik^4$. It also has a finite contribution in the range $k > \pi/\eta$, of the form $E \sim \exp[-k\eta]$. Yet, if we interpret k as $\sim\pi/(\text{eddy size})$, there are no eddies in these ranges! The shape of $E(k)$ at small k is controlled by structures of size much smaller than k^{-1}, while that at very large k is controlled by eddies of size η. In either case we are just detecting the tails of energy spectra of the form $E(k) \sim k^4 \exp\left[-k^2 l_e^2\right]$. We shall return to these issues in Section 8.1.9.

8.1.6 *Relating the energy spectrum to the second-order structure function*

In Chapters 3 and 6 we noted that the second-order structure function, $\langle [\Delta v]^2 \rangle$, provides an alternative means of distinguishing between the energy held at different scales. In particular, it is common to associate $\langle [\Delta v]^2 \rangle$ with the energy of eddies of size r or less. However, we have already noted that this is a little simplistic in that eddies of size greater than r also make a contribution to $\langle [\Delta v]^2 \rangle$, of the order of $r^2 \times$ (enstrophy of eddies). Let us explore the relationship between $E(k)$ and $\langle [\Delta v]^2 \rangle$ in a little more detail. We start with (8.33) in the form,

$$R(r) = \frac{u^2}{2r^2}\left(r^3 f\right)' = \int_0^\infty E(k)\frac{\sin kr}{kr}dk.$$

Integrating once to find f, and noting that $\langle [\Delta v]^2 \rangle = 2u^2(1 - f)$, yields

$$\left\langle [\Delta v]^2 \right\rangle = \frac{4}{3}\int_0^\infty E(k)H(kr)dk \tag{8.38}$$

$$H(\chi) = 1 + \frac{3\cos\chi}{\chi^2} - \frac{3\sin\chi}{\chi^3}. \tag{8.39}$$

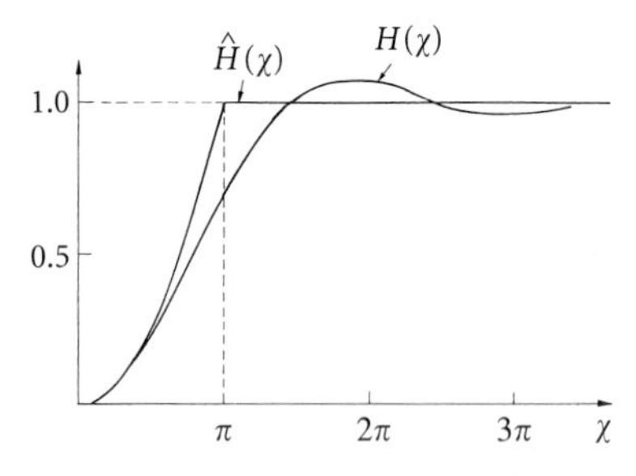

Figure 8.7 The shape of $H(\chi)$ and $\hat{H}(\chi)$.

The function $H(\chi)$ has the shape shown in Figure 8.7. For small χ it grows slowly as $H(\chi) \sim \chi^2/10$. For large χ, on the other hand, H approaches the asymptote $H = 1$ in an oscillatory manner. Also shown in Figure 8.7 is a rough approximation to $H(\chi)$, given by,

$$\hat{H}(\chi) = \begin{cases} (\chi/\pi)^2, & \chi < \pi \\ 1, & \chi > \pi \end{cases}$$

Now if we regard $H(kr)$ as a weighting function, filtering $E(k)$, then we should get a reasonable estimate of the general shape of $\langle[\Delta v]^2\rangle$ if we replace H by $\hat{H}$. Making this substitution in (8.38) yields

$$\frac{3}{4}\langle[\Delta v]^2\rangle \approx \int_{\pi/r}^{\infty} E(k)dk + \frac{r^2}{\pi^2}\int_0^{\pi/r} k^2 E dk. \tag{8.40}$$

In words this says,

$$\langle[\Delta v]^2\rangle \approx \tfrac{4}{3}[\text{all energy in eddies of size } r \text{ or less}]$$
$$+ \tfrac{4}{3}(r/\pi)^2[\text{all enstrophy in eddies of size } r \text{ or greater}].$$

As discussed in Section 6.2.1, this fits well with our physical interpretation of $\langle[\Delta v]^2\rangle$.

This more or less completes all we want to say about the kinematics of spectral space. We conclude, however, with a couple of footnotes. First, we wish to emphasize that great care must be exercised when evaluating Φ_{ij} or its derivatives near $\mathbf{k} = \mathbf{0}$. In particular, this tensor is often non-analytic at the origin. Second, many authors introduce the spectrum tensor Φ_{ij} in a different, though equivalent, way. It seems prudent to review briefly this alternative approach. Finally, we note that caution must be exercised when attributing physical meaning to $E(k)$ or $E_1(k)$. The idea that these functions represent some sort of distribution of kinetic energy amongst the different eddy sizes should not be pushed too far. There are many pitfalls lying in wait for the unwary. We start, however, with the warning about singularities in the derivatives of Φ_{ij}.

8.1.7 *A footnote: singularities in the spectrum arising from anisotropy*

We have already noted in Section 8.1.4 that $\Phi_{ij}(\mathbf{k})$ ceases to be analytic at $\mathbf{k} = \mathbf{0}$ when we have a Saffman spectrum, $E \sim k^2$. This is related to the lack of absolute convergence of the integral

$$\int Q_{ij}(\mathbf{r})d\mathbf{r}.$$

However, in isotropic turbulence there is no singularity in a Batchelor ($E \sim k^4$) spectrum. Matters are somewhat worse if we move from isotropic to anisotropic turbulence. (We shall retain the assumption of homogeneity.) In anisotropic turbulence we find that, for large r, the triple correlations decay as r^{-4}, as in the isotropic case. However, in a Batchelor spectrum the anisotropic version of the Karman–Howarth equation yields $Q_{ij} \sim r^{-5}$ at large r, rather than the isotropic r^{-6}. In fact a detailed kinematic analysis reveals (Batchelor and Proudman 1956),

$$Q_{ij} \sim \pi^2 C_{lmnp} \left[\delta_{il} \nabla^2 - \frac{\partial^2}{\partial r_i \partial r_l} \right] \left[\delta_{jm} \nabla^2 - \frac{\partial^2}{\partial r_j \partial r_m} \right] \frac{\partial^2 r}{\partial r_n \partial r_p} + 0(r^{-6}).$$

(Batchelor spectrum, anisotropic turbulence)

The r^{-5} behaviour of Q_{ij} in anisotropic turbulence,[5] leads to problems over the convergence of integrals of the form

$$I_{mnij} = \int r_m r_n Q_{ij}(\mathbf{r}) d\mathbf{r}$$

though these problems are not as severe as it might seem at first sight. A careful analysis reveals that the integrals are in fact convergent, because the r^{-5} terms cancel when integrated over the volume $|\mathbf{r}| > R$ (for some large R). However, I_{mnij} is not absolutely convergent.

Now in an analytic spectrum we know from the definition of Φ_{ij} that,

$$I_{mnij} = -8\pi^3 \left(\partial^2 \Phi_{ij} / \partial k_m \partial k_n \right)_{k=0}.$$

Thus the lack of absolute convergence of I_{mnij} raises the question of just how well defined the second-order derivatives of Φ_{ij} are at $\mathbf{k} = \mathbf{0}$. It turns out that they are not well defined, having a value which depends upon the way in which we approach the origin. In fact, for small k, the anisotropic form of the spectrum tensor in Batchelor turbulence is, (Batchelor and Proudman 1956),

$$\Phi_{ij} = C_{lmnp} \left[\delta_{il} - \frac{k_i k_l}{k^2} \right] \left[\delta_{jm} - \frac{k_j k_m}{k^2} \right] k_n k_p + 0(k^3 \ln k)$$

$$E(k) = Ck^4 + 0(k^5 \ln k)$$

(Batchelor spectrum, anisotropic turbulence).

The equivalent results for a Saffman–Birkhoff spectrum are (Saffman 1967),

$$Q_{ij} \sim -\pi^2 C_{lm} \left[\delta_{il} \nabla^2 - \frac{\partial^2}{\partial r_i \partial r_l} \right] \left[\delta_{jm} \nabla^2 - \frac{\partial^2}{\partial r_j \partial r_m} \right] r + \cdots$$

$$\Phi_{ij} = C_{lm} \left[\delta_{il} - \frac{k_i k_l}{k^2} \right] \left[\delta_{jm} - \frac{k_j k_m}{k^2} \right] + \cdots$$

$$E(k) = \tfrac{4}{3}\pi C_{ii} k^2 + 0(k^3)$$

(Saffman–Birkhoff spectrum, anisotropic turbulence).

[5] We note in passing that, for isotropic turbulence, Batchelor and Proudman showed that $C_{lmnp} = A\delta_{lm}\delta_{np} + B(\delta_{ln}\delta_{mp} + \delta_{lp}\delta_{mn})$, where A and B are scalars. It follows that, in the isotropic case, the leading-order term in our expansion for Q_{ij} is identically zero and we revert to $Q_{ij} \sim r^{-6}$. This r^{-6} dependence in isotropic turbulence can also be derived from the Karman–Howarth equation by noting that $K_\infty(r) \sim ar^{-4} + br^{-5} + \cdots$.

Evidently, when the turbulence is anisotropic, both spectra are non-analytic at $\mathbf{k} = \mathbf{0}$ since the value of Φ_{ij} (or its second derivatives in the case of a Batchelor spectrum) depends on the way in which we approach the origin. One consequence of this non-analytic behaviour is that great care must be exercised when trying to extract information about the evolution of I_{mnij} from DNS data by examining the behaviour of $(\partial^2 \Phi_{ij} / \partial k_m \partial k_n)_{k=0}$. (Remember that turbulence in a periodic cube is always anisotropic at the scale of the box.)

8.1.8 *Another footnote: the transform of the velocity field*

In Sections 8.1.4 and 8.1.5 we applied the Fourier transform to statistically averaged quantities, such as Q_{ij}. We have not yet applied it to the chaotic velocity field itself, which is a little surprising since one of the great assets of the Fourier transform is that it can filter a chaotic signal to provide information about scales (see Section 8.1.2). We shall now show what happens when we follow such a procedure, and how this leads us back to Φ_{ij}, but by a different route.

Suppose that $\mathbf{u}(\mathbf{x}, t)$ is statistically homogeneous with zero mean, though not necessarily isotropic. We define the Fourier transform of $\mathbf{u}(\mathbf{x})$ in the usual way:

$$\hat{\mathbf{u}}(\mathbf{k}) = \frac{1}{(2\pi)^3} \int \mathbf{u}(\mathbf{x}) e^{-j\mathbf{k}\cdot\mathbf{x}} d\mathbf{x}.$$

Since the integral does not converge we must regard $\hat{\mathbf{u}}$ as a generalized function. However, it turns out that this does not prevent us from formally manipulating the integral according to the usual rules.

The complex conjugate of $\hat{\mathbf{u}}(\mathbf{k})$ has the property $\hat{\mathbf{u}}^\dagger(-\mathbf{k}) = \hat{\mathbf{u}}(\mathbf{k})$ and so we may form the product,

$$\hat{u}_i^\dagger(\mathbf{k})\hat{u}_j(\mathbf{k}') = \frac{1}{(2\pi)^6} \iint u_i(\mathbf{x}) u_j(\mathbf{x}') e^{j(\mathbf{k}\cdot\mathbf{x} - \mathbf{k}'\cdot\mathbf{x}')} d\mathbf{x} d\mathbf{x}'.$$

We now let $\mathbf{r} = \mathbf{x}' - \mathbf{x}$ and ensemble average our equation. Since the only random terms on the right are u_i and u_j' we have,

$$\left\langle \hat{u}_i^\dagger(\mathbf{k})\hat{u}_j(\mathbf{k}') \right\rangle = \frac{1}{(2\pi)^6} \iint \left\langle u_i u_j' \right\rangle e^{j(\mathbf{k}-\mathbf{k}')\cdot\mathbf{x}} e^{-j\mathbf{k}'\cdot\mathbf{r}} d\mathbf{x} d\mathbf{r}.$$

Next we note that, because of homogeneity, $\langle \mathbf{u} \cdot \mathbf{u}' \rangle$ depends on $\mathbf{r}$ but not on $\mathbf{x}$ and so we rearrange the double integral as follows:

$$\left\langle \hat{u}_i^\dagger(\mathbf{k})\hat{u}_j(\mathbf{k}') \right\rangle = \frac{1}{(2\pi)^6} \int \left\langle u_i u_j' \right\rangle e^{-j\mathbf{k}'\cdot\mathbf{r}} \left[\int e^{j(\mathbf{k}-\mathbf{k}')\cdot\mathbf{x}} d\mathbf{x} \right] d\mathbf{r}.$$

Finally we observe that the inner integral is simply $(2\pi)^3\ \delta(\mathbf{k}-\mathbf{k}')$ where δ is the three-dimensional Dirac delta function. It follows that,

$$\left\langle \hat{u}_i^{\dagger}(\mathbf{k})\hat{u}_j(\mathbf{k}')\right\rangle = \delta(\mathbf{k}-\mathbf{k}')\frac{1}{(2\pi)^3}\int Q_{ij}e^{-j\mathbf{k}'\cdot\mathbf{r}}d\mathbf{r}.$$

Evidently $\hat{u}_i^{\dagger}(\mathbf{k})$ and $\hat{u}_j(\mathbf{k}')$ are uncorrelated unless, $\mathbf{k}=\mathbf{k}'$. From definition (8.24a) we can rewrite this as,

$$\left\langle \hat{u}_i^{\dagger}(\mathbf{k})\hat{u}_j(\mathbf{k}')\right\rangle = \delta(\mathbf{k}-\mathbf{k}')\Phi_{ij}(\mathbf{k}). \tag{8.41a}$$

Thus we see the relationship between the spectrum tensor and the transform of the velocity field. The filtering property of the Fourier transform tells us that small eddies are associated with the high wave number components of $\hat{\mathbf{u}}(\mathbf{k})$, and so small scales will contribute most to the large-k part of Φ_{ij}. For isotropic turbulence (8.41a) may be rewritten in terms of $E(k)$,

$$2\pi k^2\left\langle \hat{\mathbf{u}}^{\dagger}(\mathbf{k})\cdot\hat{\mathbf{u}}(\mathbf{k}')\right\rangle = \delta(\mathbf{k}-\mathbf{k}')E(k) \tag{8.41b}$$

which confirms that $E(k)\geq 0$.

So what have we gained from this alternative approach? One advantage of this procedure is that the rule of thumb,

small eddies $\leftrightarrow$ high wave number part of Φ_{ij} and E
large eddies $\leftrightarrow$ small wave number part of Φ_{ij} and E

is now seen to be an immediate consequence of the filtering property of the Fourier transform, as discussed in Section 8.1.2. Another benefit is that we discover that $\hat{u}_i^{\dagger}(\mathbf{k})$ and $\hat{u}_j(\mathbf{k}')$ are uncorrelated unless $\mathbf{k}=\mathbf{k}'$. The disadvantage, of course, is that great care must be exercised in interpreting generalized functions like $\hat{\mathbf{u}}(\mathbf{k})$.

8.1.9 *Definitely the last footnote: what do $E(k)$ and $E_1(k)$ really represent?*

A very common pitfall when using any kind of transform is to forget the presence of the analysing function in the transformed field, which may lead to severe misinterpretations, the structure of the analysing function being interpreted as characteristic of the phenomena under study. (M. Farge 1992)

We conclude our review of spectral kinematics by returning to a theme initially developed in Section 6.4.1. The question at issue is: what physical meaning should we attribute to $E(k)$ and $E_1(k)$? One of the points which we wish to emphasize is that turbulence consists of eddies, not waves. Moreover, an eddy (i.e. a spatially compact vortical structure) gives rise to a range of wave numbers in $E(k)$, and not to a 'spike' at $k \sim (\text{eddy size})^{-1}$. So we cannot strictly associate k^{-1} with eddy size. For example, in Section 6.4.1 we found that a sea of randomly distributed spherical eddies of fixed size l_e gives rise to the spectrum

$$E(k) \sim \langle \mathbf{u}^2 \rangle l_e (kl_e)^4 \exp\left[-(kl_e/2)^2\right]. \tag{8.42a}$$

There are several implications of this:

(1) Even if eddy interactions in the energy cascade are, by and large, restricted to eddies of similar size (which they may or may not be), the associated interactions need not be localized in k-space.

(2) For small k (i.e. $k \ll l^{-1}$) the energy spectrum of fully developed turbulence will be of the form,

$$E(k) \sim (u^2 l^5) k^4 + \cdots$$

where l is the integral scale. This part of the spectrum has nothing at all to do with spatial structures of size k^{-1}, but rather represents the low-k tails of the spectra associated with the energy-containing eddies.

(3) For large k (i.e. $k \gg \eta^{-1}$) the energy spectrum of fully developed turbulence will be of the form,

$$E(k) \sim \exp[-(\eta k)^\alpha]$$

where η is the Kolmogorov microscale. This part of $E(k)$ represents the high-k tails of the spectra associated with the dissipation-scale eddies.

Let us remind ourselves of the origin of (8.42a). In Section 6.4.1 we noted that a sea of randomly distributed eddies of simple form and fixed size, l_e, gives (Townsend, 1976)

$$f(r) = \exp\left[-r^2/l_e^2\right], \qquad E(k) = \frac{\langle \mathbf{u}^2 \rangle l_e}{24\pi^{1/2}} (kl_e)^4 \exp\left[-(kl_e/2)^2\right]. \tag{8.42b, c}$$

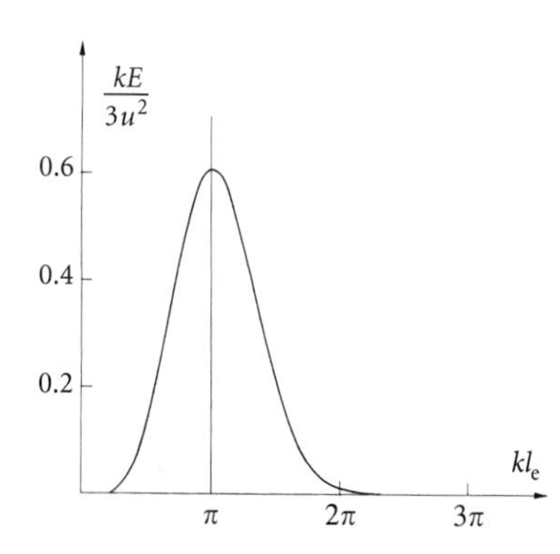

Figure 8.8 Energy spectrum $E(k)$ produced by a sea of spherical eddies of fixed size l_e.

By a 'simple' eddy we mean a spherical blob of swirling fluid which has some angular momentum but no net linear momentum. If we allow the 'model eddy' to have linear momentum then we find $E(k \to 0) \sim k^2$. Alternatively, if we remove any net linear or angular momentum from the model eddy then we have $E(k \to 0) \sim k^6$. (See Section 6.4.1 for the details.) However, grid turbulence is probably of the form $E \sim k^4$ and so we shall stay with (8.42b) and (8.42c). Noting that (8.42c) exhibits a sharp peak at around $k \sim \pi/l_e$ (Figure 8.8), we may summarize the main physical properties of $E(k)$ as:

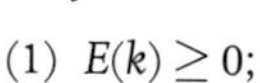

(1) $E(k) \geq 0$;

(2) $\frac{1}{2}\langle \mathbf{u}^2 \rangle = \int_0^\infty E dk$;

(3) eddies of size l_e contribute to $E(k)$ primarily in the region $k = \pi/l_e$.

Properties (1) and (2) allow us to interpret Edk as the energy associated with wave numbers $k \rightarrow k + dk$, and property (3) allows us to think of k as being, approximately, π/(eddy size). This is the usual interpretation of $E(k)$. However, as noted above, this interpretation goes badly wrong if $k \ll l^{-1}$ or $k \gg \eta^{-1}$. In these ranges of the spectrum $E(k)$ has nothing at all to do with the energy contained is structures of size k^{-1}. We are merely detecting the tails of the distributions associated with energy-containing eddies (if $k \ll l^{-1}$) or dissipation-scale eddies (if $k \gg \eta^{-1}$).

Consider now the one-dimensional spectrum, $E_1(k)$. Like $E(k)$ it has the properties:

(1) $E_1(k) \geq 0$;

(2) $\int_0^\infty E_1(k)dk = \frac{1}{2}\langle \mathbf{u}^2 \rangle$.

Crucially, however, it does not possess the third property of $E(k)$. Let us calculate $E_1(k)$ for our sea of randomly dispersed spherical eddies. Combining (8.37c) with (8.42c) we find,

$$E_1(k) = \frac{\langle \mathbf{u}^2 \rangle l_e}{3\pi^{1/2}} \left[1 + k^2 l_e^2/4\right] \exp\left[-k^2 l_e^2/4\right].$$

Evidently, E_1 peaks at $k = 0$ rather than $k \sim \pi/l_e$ (see Figure 8.6(b)). This is why the one-dimensional energy spectrum fails to give a reasonable estimate of the distribution of energy in fully developed turbulence. Eddies of all sizes contribute to $E_1(k)$ primarily at $k = 0$, rather than at $k \sim \pi/l_e$. The physical reason for this peculiar behaviour is discussed in Tennekes and Lumley (1972). We shall give only a brief summary of their argument.

The tendency for one-dimensional spectra to give a misleading impression of three-dimensional processes is known as *aliasing* (see Figure 8.9). Consider a wave of length λ propagating at an angle θ to the x-axis. Now suppose we measure $\mathbf{u}$ at different positions along the x-axis. We will obtain a sinusoidal function of wavelength $\lambda/\cos\theta$. When we take the one-dimensional energy spectrum of $\mathbf{u}$, as described in Section 8.1.5, we get a spike in the spectrum corresponding to

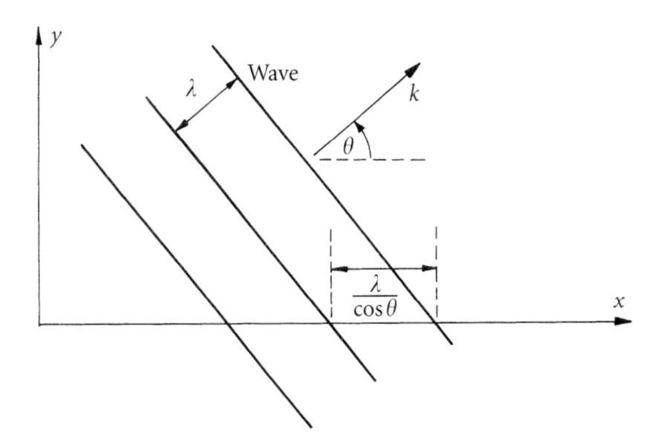

Figure 8.9 The problem of aliasing.

wavelength $\lambda/\cos\theta$. Thus the one-dimensional spectrum has misinterpreted a wave of wave number k as being one of wave number $k\cos\theta$, the energy of the wave appearing closer to the origin in **k**-space than it should have. We talk of energy being aliased towards the origin $k=0$. This is why $E_1(k)$ is peaked around $k=0$ while $E(k)$ is peaked at $k \sim \pi/l_e$. In short, one-dimensional spectra cannot distinguish between waves of wave number k moving parallel to the line of measurement and waves of wave number larger than k propagating obliquely to the line of measurement.

The tendency for $E_1(k)$ to aliase energy towards $k=0$ is captured by equation (8.37c):

$$E_1(k) = \int_k^\infty E(p)p^{-1}dp.$$

This confirms that all wave numbers greater than or equal to k contribute to $E_1(k)$.

In summary, then, some caution must be exercised when it comes to attributing physical meaning to either $E(k)$ or $E_1(k)$. $E(k)$ gives an impression of the energy distribution only in the range $\eta < k^{-1} < l$, while $E_1(k)$ gives a misleading impression for all k.

With this word of caution let us now leave behind kinematics and discuss the altogether more important issue of dynamics.

8.2 Dynamics in spectral space

8.2.1 *An evolution equation for E(k)*

It is important to emphasize at this point that, so far, we have discovered nothing new about turbulence. We have simply established an alternative language for describing turbulent flows. Let us now see if we can use this new language in a constructive way. Our starting point is the Karman–Howarth equation (6.48a)[6]

$$\frac{\partial}{\partial t}[u^2 r^4 f] = \frac{\partial}{\partial r}[u^3 r^4 K] + 2\nu\frac{\partial}{\partial r}[u^2 r^4 f'] \tag{8.43}$$

which, using the relationship,

$$R = \frac{1}{2}\langle \mathbf{u}\cdot\mathbf{u}'\rangle = \frac{u^2}{2r^2}(r^3 f)'$$

can be rewritten as,[7]

$$\frac{\partial R}{\partial t} = \Gamma(r,t) + 2\nu\nabla^2 R; \quad \Gamma(r,t) = \frac{1}{2r^2}\frac{\partial}{\partial r}\frac{1}{r}\frac{\partial}{\partial r}[u^3 r^4 K]. \tag{8.44}$$

[6] Here f is the longitudinal correlation function and K is the triple correlation function as defined in Chapter 6.

[7] The Laplacian in this equation represents $\frac{1}{r^2}\frac{\partial}{\partial r}r^2\frac{\partial}{\partial r}(\sim)$ since it is acting on a spherically symmetric function.

We now recall that $E(k)$ and $R(r)$ are related by

$$E(k) = \frac{2}{\pi}\int_0^\infty R(r)kr\sin(kr)dr \tag{8.45}$$

which allows us to convert (8.44) into an evolution equation for $E(k)$. After a little algebra we find,

$$\frac{\partial E}{\partial t} = T(k,t) - 2\nu k^2 E \tag{8.46}$$

$$T(k,t) = \frac{k}{\pi}\int_0^\infty \frac{1}{r}\frac{\partial}{\partial r}\frac{1}{r}\frac{\partial}{\partial r}[r^4 u^3 K]\sin(kr)dr. \tag{8.47}$$

This is the spectral equivalent of the Karman–Howarth equation. $T(k, t)$, which we normally abbreviate to $T(k)$, is called the *spectral kinetic energy transfer function*. In the language of the energy cascade we interpret $T(k)$ as representing the removal of energy from the large scales and the deposition of energy at the small scales. We expect, therefore, that T is negative for small k (removal of energy from large eddies) and positive for large k (deposition of energy in small eddies).

Notice the symmetry in the relationship between $E(k)$ and $R(r)$ and between $T(k)$ and $\Gamma(r)$:

$$E(k) = \frac{2}{\pi}\int_0^\infty R(r)kr\sin(kr)dr, \qquad R(r) = \int_0^\infty E(k)\frac{\sin(kr)}{kr}dk$$

$$T(k) = \frac{2}{\pi}\int_0^\infty \Gamma(r)kr\sin(kr)dr, \qquad \Gamma(r) = \int_0^\infty T(k)\frac{\sin(kr)}{kr}dk$$

Now the non-linear inertial terms, which are represented by $T(k)$, transfer energy from scale to scale without destroying energy, and so we might anticipate that

$$\int_0^\infty T(k)dk = 0. \tag{8.48}$$

In fact this is readily confirmed by integrating (8.47). (Consult Exercise 8.3.) The implication is that

$$\frac{d}{dt}\int_0^\infty E dk = -2\nu\int k^2 E dk \tag{8.49}$$

which we recognize as the familiar statement

$$\frac{d}{dt}\left[\frac{1}{2}\langle \mathbf{u}^2\rangle\right] = -\varepsilon = -\nu\langle \boldsymbol{\omega}^2\rangle \tag{8.50}$$

The energy equation (8.46) is sometimes written in the alternative form,

$$\frac{\partial E}{\partial t} = -\frac{\partial \Pi_E}{\partial k} - 2\nu k^2 E \tag{8.51}$$

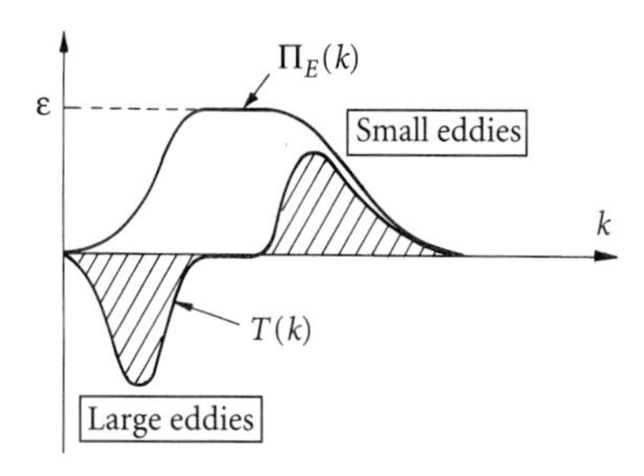

Figure 8.10 Schematic shape of $T(k)$ and $\Pi_E(k)$.

where

$$\Pi_E = -\int_0^k T(k)dk = \int_k^\infty T(k)dk.$$

The function $\Pi_E(k)$ is called the *spectral kinetic energy flux* and represents the net transfer of energy from all eddies of wave number less than k to those of wave number greater than k. The shapes of $T(k)$ and $\Pi_E(k)$ are shown in schematic form (Figure 8.10). $T(k)$ is negative for small k (removal of energy from the large eddies) and positive for large k (inertial transfer of energy to the small eddies). Π_E, on the other hand, is positive since the flux of energy is towards the small scales. In the inertial subrange both $\partial E/\partial t$ and viscous effects are negligible, so that $T = 0$ (from 8.46) and $\Pi_E = \varepsilon$.

Let us now consider the shape of $T(k)$ and $\Pi_E(k)$ in a little more detail. It is possible to show, using (8.47), that $\Gamma(r)$ and $\Pi_E(k)$ are related by the transform pair:

$$\Pi_E(k) = -\frac{2}{\pi}\int_0^\infty \frac{1}{r}\frac{\partial}{\partial r}(r\Gamma)\sin(kr)dr \tag{8.52a}$$

$$\frac{1}{r}\frac{\partial}{\partial r}[r\Gamma] = -\int_0^\infty \Pi_E(k)\sin(kr)dk. \tag{8.52b}$$

Moreover, integrating (8.47) by parts yields,

$$T(k) = \frac{k^4}{3\pi}\int_0^\infty \frac{d}{dr}[u^3r^4K(r)]G(kr)dr \tag{8.53}$$

where

$$G(\chi) = 3[\sin\chi - \chi\cos\chi]\chi^{-3}.$$

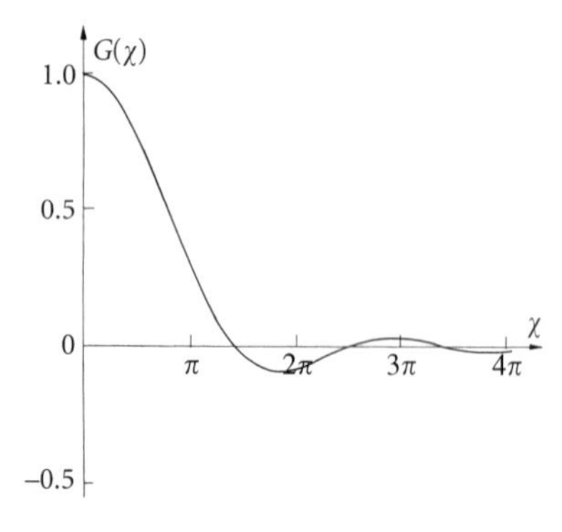

Figure 8.11 Shape of $G(\chi)$ in equation (8.53).

This shows, in a simple way, the relationship between $T(k)$ and the triple correlation function $K(r)$. The general shape of $G(\chi)$ is shown below (Figure 8.11). It is positive for $\chi < 5\pi/4$ and oscillates about $G = 0$ as χ increases, the oscillations diminishing rapidly in amplitude. For small k, $G(\chi) \approx 1 - (\chi^2/10) + \cdots$ and so we have,

$$T(k) = \frac{k^4}{3\pi}[u^3r^4K]_\infty - \frac{k^6}{30\pi}\int_0^\infty r^2\frac{d}{dr}[u^3r^4K]dr + \cdots. \tag{8.54}$$

When combined with (8.46) and the expansion,

$$E = \frac{L}{4\pi^2}k^2 + \frac{I}{24\pi^2}k^4 + \cdots$$

this yields the familiar results

$$L = \text{constant}, \qquad \frac{dI}{dt} = 8\pi[u^3r^4K]_\infty.$$

This shows the relationship between Loitsyansky's integral and $[r^4 K]_\infty$. In the absence of long-range triple correlations the k^4 term in the expansion for T vanishes and $T \sim k^6$ at small k.

8.2.2 *Closure in spectral space*

James was being teased by a theological colleague who said to him 'A philosopher is like a blind man in a dark cellar, looking for a black cat that isn't there.' 'Yes,' said William James, 'and the difference between philosophy and theology is that theology finds the cat.' (A. J. Ayer on William James)

When working in real space, it is natural to frame the closure question in the form: can I relate $\langle uuu \rangle$ to $\langle uu \rangle$ using some plausible physical hypothesis, thus closing the system? In spectral space the equivalent question is: can I evoke some plausible closure hypothesis relating $T(k)$ (or its time derivative) to $E(k)$, thus allowing (8.51) to be integrated? There have been many stabs at this, and several of them have managed to reproduce the $k^{-5/3}$ form of E in the inertial subrange. However, just like their analogues in real space, they all run into trouble sooner or later and, after a while, one wonders whether we have been looking for James' non-existent cat! In short, it is probable that there is no single closure scheme which is reliable for all k.[8]

Broadly speaking there are two classes of closure hypotheses in spectral space. Simple, algebraic schemes postulate that the spectral energy flux, $\Pi_E(k)$, can be related to $E(k)$ by an expression of the form $\Pi_E = \Pi_E(E, k)$. The functional relationship between $\Pi_E(k)$ and $E(k)$ is then chosen to ensure that the known characteristics of the inertial subrange are reproduced, that is, $E = \alpha\varepsilon^{2/3}k^{-5/3}$, $\Pi_E = \varepsilon$. The hope is that, when the expression $\Pi_E = \Pi_E(E, k)$ is applied outside the inertial range, say in the dissipation range, results are obtained which are compatible with the experimental data. In effect, this is an exercise in interpolation. The second class of closure schemes postulate a dynamic equation for $T(k, t)$ of the form $\partial T/\partial t = (\sim)$. The quasi-normal approximation and its relatives (e.g. EDQN) are examples of such an approach. This is a bolder and more complex strategy. Let us start, therefore, with the simple algebraic models, of which there have been many. These apply only to the equilibrium range (small scales). Some of the more common proposals are listed below.

- Obukhov's hypothesis (1941):

$$\Pi_E(k) = \alpha_1 \int_k^\infty E dk \left\{ \int_0^k k^2 E dk \right\}^{1/2}. \tag{8.55}$$

[8] In brief, the situation in spectral space is broadly similar to that in real space. It is possible to come up with a reasonable closure scheme for the universal equilibrium range (see Section 6.2.3.2 for the real space version of such a closure scheme) but there is no satisfactory scheme for the small-k end of the spectrum.

- Ellison's modification of Obukhov's hypothesis (1961):

$$\Pi_E(k) = \alpha_2 kE(k)\left\{\int_0^k k^2 E dk\right\}^{1/2}. \tag{8.56}$$

- Heisenberg's hypothesis (1948):

$$\Pi_E(k) = \alpha_3 \int_k^\infty k^{-3/2} E^{1/2} dk \int_0^k k^2 E dk. \tag{8.57}$$

- Kovasznay's hypothesis (1948):

$$\Pi_E(k) = \alpha_4 E^{3/2} k^{5/2}. \tag{8.58}$$

- Pao's (1965) hypothesis:

$$\Pi_E(k) = \alpha_5 \varepsilon^{1/3} k^{5/3} E. \tag{8.59}$$

It is readily confirmed that, with appropriate choices of α_i, all five suggestions yield $\Pi_E = \varepsilon$ when $E = \alpha\varepsilon^{2/3} k^{-5/3}$, as required. The physical basis of the first two proposals for Π_E is a sort of mixing length argument. Recall that, in a shear flow, energy is transferred from the mean flow to the turbulence at a rate $\tau_{ij}^R \bar{S}_{ij}$. By analogy, we might imagine that the flux of energy down the cascade at wave number k is the product of: (i) a sort of Reynolds stress associated with the eddies smaller than k^{-1}; and (ii) the mean strain of eddies larger than k^{-1}. In short, we suppose that, $\Pi_E(\hat{k}) = \tau_{ij}^R(k > \hat{k}) S_{ij}(k < \hat{k})$. (See Example 5.10) In (8.55) and (8.56) we take $S_{ij}^2(k < \hat{k})$ to be proportional to the enstrophy of eddies of size k^{-1} and larger, while $\tau_{ij}^R(k > \hat{k})$ is taken to be of the order of the energy contained in eddies of size k^{-1} and smaller.

Heisenberg's suggestion, on the other hand, is essentially an eddy-viscosity concept. Energy is destroyed at the small scales at a rate $2\nu\int_0^\infty k^2 E dk$. Heisenberg supposed that the energy flux, Π_E, at wave number k is the product of an eddy viscosity, ν_t, times the enstrophy of the eddies of size greater than k^{-1}. The eddy viscosity parameterizes the influence of eddies smaller than k^{-1} and on dimensional grounds is set equal to $\alpha_3 \int_k^\infty k^{-3/2} E^{1/2} dk$.

One problem with (8.55) $\rightarrow$ (8.57) is that, in a sense, they contravene our intuitive ideas about the energy cascade. That is, we would expect the energy flux $\Pi_E(k)$ to depend primarily (though not exclusively) on eddies whose size is of the order of k^{-1}, since eddies of greatly different size interact only weakly.[9] This is the basis of Kovasznay's suggestion. If we hypothesize that $\Pi_E(k)$ is a function

[9] This idea should not be pushed too far. We have already seen that eddies of size l_e produce a continuous spectrum with contributions from all k, albeit peaked around $k \sim \pi/l_e$. Thus, even if interactions are more or less localized in terms of eddy sizes, the same interactions need not be localized in spectral space.

of $E(k)$ and k only, and that $\Pi_E(k)$ depends only on eddies of size $\sim k^{-1}$, then (8.58) follows on dimensional grounds. Pao's suggestion is also a local one, but one in which we allow Π_E to depend on ε and request that Π_E is proportional to E.

Now all five suggestions above for $\Pi_E(k)$ get things more or less right in the inertial subrange, which is hardly surprising since they are designed to do just that. They are useful only if they make predictions outside this range which are in accord with the experimental data. If we restrict ourselves to the equilibrium range (which is the range these closures address) then $\partial E/\partial t$ is negligible and we have,

$$0 = -\partial \Pi_E/\partial k - 2\nu k^2 E.$$

Substituting for $\Pi_E(k)$ using any one of (8.55) $\rightarrow$ (8.59) allows us to estimate $E(k)$ in the dissipation range and this provides the acid test for these closure hypotheses. For example, one thing to look out for is the fact that E should decay exponentially fast as $k \rightarrow \infty$, since there are no singularities in the velocity field. It turns out that only (8.56) and (8.59) satisfy this criterion, all other proposals yielding anomalous results for $k \geq \eta^{-1}$.[10] (η is Kolmogorov's microscale.) In fact, it is readily confirmed that (8.56) predicts $E \sim (\eta k)^{-1} \exp[-\sqrt{2}(\eta k)^2/\alpha_2]$ for $\eta k \rightarrow \infty$, while (8.59) yields

$$E = \alpha \varepsilon^{2/3} k^{-5/3} \exp\left[-(3\alpha/2)(\eta k)^{4/3}\right] \tag{8.60}$$

(Pao Spectrum)

throughout the equilibrium range. Actually, it turns out that (8.60) provides a quite reasonable description of $E(k)$ for $k \leq (2\eta)^{-1}$, though it somewhat overestimates $E(k)$ for $\eta k > 0.5$.

It is clear from the wide range of hypotheses that have been proposed, and the failure of most of them in the dissipation range, that this kind of heuristic appeal to a 'spectral eddy viscosity', or a 'spectral Reynolds stress', is deeply unsatisfactory. In fact, only Pao's hypothesis really withstands scrutiny. We can justify it on the basis of just two assumptions:

(1) the transfer of energy is local in k-space;

(2) Π_E depends on $E(k)$ in a power-law fashion, that is, $\Pi_E \sim E^m$.

The argument goes something like this. In the equilibrium range E depends on k, ε, and ν only. (This is the essence of Kolmogorov's theory.) If the energy transfer is localized in k-space then Π_E will also depend on only k, ε, and ν:

$$E = E(k, \varepsilon, \nu), \qquad \Pi_E = \Pi_E(k, \varepsilon, \nu).$$

[10] Both Obukhov's and Kovasznay's schemes break down as k approaches η, while Heisenberg's scheme leads to $E \sim k^{-7}$ in the dissipation range. The details are spelt out in Monin and Yaglom (1975). See also Exercise 8.6.

The only dimensionally consistent possibility is,

$$E = \alpha\varepsilon^{2/3}k^{-5/3}\hat{E}(\eta k), \qquad \Pi_E = \varepsilon\hat{\Pi}_E(\eta k)$$

where η is Kolmogorov's microscale, $\eta = (\nu^3/\varepsilon)^{1/4}$, and $\hat{E}$ and $\hat{\Pi}_E$ are dimensionless functions. Now in the equilibrium range we have

$$\frac{\partial \Pi_E}{\partial k} = -2\nu k^2 E$$

Substituting for E and Π_E in terms of $\hat{E}$ and $\hat{\Pi}_E$, and introducing $\chi = (\eta k)^{4/3}$, this yields

$$\frac{d\hat{\Pi}_E}{d\chi} = -\frac{3}{2}\alpha\hat{E}(\chi).$$

So far we have used only approximation (1) above. We now invoke the second assumption. Since $\hat{\Pi}_E$ and $\hat{E}$ are functions only of χ, we have $\hat{\Pi}_E = \hat{\Pi}_E(\hat{E})$. We shall assume that $\hat{\Pi}_E$ depends on $\hat{E}$ in a power-law fashion: $\hat{\Pi}_E \sim \hat{E}^m$. (This is equivalent to the assertion that $\Pi_E \sim E^m$.) It is readily confirmed, using our dynamic equation for $\hat{\Pi}_E$, that $m = 1$ is the only choice of m which leads to an exponential decay of $\hat{E}$ in the dissipation range. Thus we require that $\hat{\Pi}_E$ is proportional to $\hat{E}$: $\hat{\Pi}_E \sim \hat{E}$. Moreover, in the inertial subrange we have

$$\Pi_E = \varepsilon, \qquad E = \alpha\varepsilon^{2/3}k^{-5/3}$$

and so the constant of proportionality is $\hat{\Pi}_E/\hat{E} = 1$, that is, $\hat{\Pi}_E$ and $\hat{E}$ are equal. If we now revert back to Π_E and E we obtain

$$\Pi_E = \frac{1}{\alpha}\varepsilon^{1/3}k^{5/3}E(k).$$

This is the Pao (1965) hypothesis. The beauty of Pao's closure scheme is that it gives a simple prediction of E and Π_E within the equilibrium range, that is,

$$E = \alpha\varepsilon^{2/3}k^{-5/3}\exp\left[-(3\alpha/2)(\eta k)^{4/3}\right],$$

based on just two plausible assumptions. Moreover, its predictions are a reasonable fit to the experimental data down to $\eta k \approx 0.5$, which corresponds roughly to the smallest eddies present in the flow. Of the five simple closure schemes listed above it is the one which has best withstood the test of time.

Example 8.1 The Pao closure model applied to the entire energy spectrum

One of the appealing features of Pao's closure scheme is that it produces passable results for the entire spectrum, as we now show. The spectral evolution equation for $E(k, t)$, combined with the Pao closure estimate of $\Pi_E(k, t)$, yields,

$$\frac{\partial E}{\partial t} = -\frac{\partial}{\partial k}\left[\frac{\varepsilon^{1/3}k^{5/3}E}{\alpha}\right] - 2\nu k^2 E.$$

In the equilibrium range, which includes both the inertial subrange and the dissipation scales, we may neglect the term on the left-hand side. We have already seen that the resulting equation integrates to yield

$$E = \alpha\varepsilon^{2/3}k^{-5/3}\exp\left[-(3\alpha/2)(\eta k)^{4/3}\right], \quad kl \gg 1$$

We shall now consider the opposite end of the spectrum, to which the Pao closure hypothesis is not normally applied. We restrict ourselves to the region $\eta k \ll 1$, which includes both the large scales and the inertial subrange. Here the viscous term is negligible and the Pao closure scheme reduces to

$$\frac{\partial E}{\partial t} = -\frac{\partial}{\partial k}\left[\frac{\varepsilon^{1/3}k^{5/3}E}{\alpha}\right], \quad k\eta \ll 1$$

Let us introduce the length-scale $\hat{l}(t)$ defined by

$$\hat{l}^{2/3} = \hat{l}_0^{2/3} + \int_0^t \varepsilon^{1/3}dt.$$

It is readily confirmed that, in freely decaying turbulence, $\hat{l}$ is of the order of the integral scale, l. (This follows from the observation that, in freely decaying turbulence, $\varepsilon \sim u^3/l$ and $l \sim ut$.) Show that our evolution equation for $E(k,t)$ can be rewritten in terms of $\hat{l}(t)$ as follows:

$$\left[\frac{\partial}{\partial \hat{l}^{2/3}} - \frac{2}{3\alpha}\frac{\partial}{\partial k^{-2/3}}\right]\left(k^{5/3}E\right) = 0; \quad k\eta \ll 1.$$

Hence confirm that, for $k\eta \ll 1$, the Pao closure model admits the solution,

$$E = \alpha k^{-5/3}F(\chi), \qquad \Pi_E = \varepsilon^{1/3}F(\chi); \qquad \chi = \hat{l}^{2/3}(t) + (3\alpha/2)k^{-2/3}.$$

Here F is an arbitrary function of χ whose form is fixed by the initial conditions. The only constraint on $F(\chi)$ is that set by the requirement that $\Pi_E = \varepsilon$ for $k\hat{l} \to \infty$. This demands $F(\hat{l}^{2/3}) = \varepsilon^{2/3}$. In fully developed turbulence we expect either $E \sim k^4$ for $k \to 0$ (a Batchelor spectrum) or else $E \sim k^2$ (a Saffman spectrum). It is the initial conditions that dictate which form is found (see Section 6.3.5). Show that the choice of $F(\chi) \sim \chi^{-17/2}$ yields,

$$E = \frac{I}{24\pi^2}\frac{k^4}{\left[1 + (2/3\alpha)\left(k\hat{l}\right)^{2/3}\right]^{17/2}}, \quad k\eta \ll 1$$

where,

$$I = \text{constant}, \; \hat{l} \sim t^{2/7}, \langle \mathbf{u}^2 \rangle \sim t^{-10/7}.$$

(You will need to exploit the fact that $F(\hat{l}^{2/3}) = \varepsilon^{2/3}$ and $d\langle \mathbf{u}^2 \rangle / dt = -2\varepsilon \sim -u^3/\hat{l}$.) This is a Batchelor spectrum with a constant Loitsyansky integral, I, and a ten-seventh's Kolmogorov decay law (see Section 6.3.2). Now confirm that the choice of $F \sim \chi^{-11/2}$ yields a Saffman spectrum,

$$E \sim \frac{k^2}{\left[1 + (2/3\alpha)\left(k\hat{l}\right)^{2/3}\right]^{11/2}}, \qquad k\eta \ll 1$$

with a constant Saffman integral, $\int \langle \mathbf{u} \cdot \mathbf{u}' \rangle d\mathbf{r}$, and a Saffman decay law of $\langle \mathbf{u}^2 \rangle \sim t^{-6/5}$. All of this is roughly in accord with the theoretical discussion in Section 6.3.5. Thus we see that the Pao closure scheme yields passable results for both large and small k in fully developed turbulence. Finally, confirm that, for $\mathrm{Re} \to \infty$, we can combine the small and large-k forms of $E(k)$ to yield

$$E = \frac{\alpha \varepsilon^{2/3} k^{-5/3} \exp\left[-(3\alpha/2)(\eta k)^{4/3}\right]}{\left[1 + (3\alpha/2)\left(k\hat{l}\right)^{-2/3}\right]^m}$$

$$\varepsilon \hat{l}^m = \text{constant}, \qquad \langle \mathbf{u}^2 \rangle \hat{l}^{2(m-1)/3} = \text{constant}.$$

Here $m = 17/2$ for a Batchelor ($E \sim k^4$) spectrum and $m = 11/2$ for a Saffman ($E \sim k^2$) spectrum. Many model spectra of this type have been proposed at one time or another based on an interpolation between different known results. What makes the spectrum above interesting is that we have obtained it from a single closure hypothesis. No interpolation is involved.

Example 8.2 The Pao closure model applied to hyper-viscous fluids
In some numerical simulations there is insufficient resolution to resolve the Kolmogorov scales. In such cases a crude form of LES can be performed by replacing the real fluid by a so-called hyper-viscous fluid. This involves replacing viscous force $\nu \nabla^2 \mathbf{u}$ by the term $\nu_m \nabla^{2m} \mathbf{u}$ where m is a positive integer. We can gain some insight into the spectral implications of this modification using the Pao closure model. Confirm that, for a hyper-viscous fluid, this closure predicts,

$$E = \alpha \varepsilon^{2/3} k^{-5/3} \exp\left[-\frac{3\alpha}{3m-1}(\eta_m k)^{2(3m-1)/3}\right],$$

$$\eta_m = \left(\nu_m^{3/2} / \varepsilon^{1/2}\right)^{1/(3m-1)}$$

in the equilibrium range. Evidently, the effect of introducing hyper-viscosity is to produce a faster fall-off of $E(k)$ in spectral space.

Example 8.3 The skewness prediction of the Pao closure model
The rate of generation of enstrophy by vortex stretching is equal to $\int_0^\infty k^2 T(k) dk$. Use (6.58) to show that the skewness of $\partial u_x / \partial x$, S_0,

is equal to

$$S_0 = \frac{\langle (\partial u_x/\partial x)^3 \rangle}{\langle (\partial u_x/\partial x)^2 \rangle^{3/2}} = -\frac{3\sqrt{30}}{14} \int_0^\infty k^2 T dk \left[\int_0^\infty k^2 E dk \right]^{-3/2}.$$

In the equilibrium range we have $T = 2\nu k^2 E$ and since the dominant contribution to the integrals above come from this range we have

$$S_0 = -\frac{3\sqrt{30}\nu}{7} \int_0^\infty k^4 E dk \left[\int_0^\infty k^2 E dk \right]^{-3/2}.$$

Show that the Pao closure model predicts $S_0 = -1.28$. (You may take $\alpha = 1.52$ and ignore any contribution to the integrals outside the equilibrium range.) This estimate of $|S_0|$ is considerably larger than the measured value of $|S_0| \approx 0.4$.

Example 8.4 The small-k behaviour of the Pao closure model
Use the results of Example 8.1 to calculate the form of $T(k)$ predicted by the Pao model for $k \to 0$. Show that it is incompatible with the existence of long-range, pressure-induced velocity correlations of the sort predicted by Batchelor (see Section 6.3.4).

8.2.3 *Quasi-Normal type closure schemes (Part 2)*

Simple algebraic spectral models have now largely given way to more complex dynamical models for $T(k)$ whose validity is not, in principle, restricted to the equilibrium range. As noted in Section 6.2.3.3, the earliest two-point closure model goes back to Millionshtchikov's 1941 quasi-normal (QN) hypothesis. However, since then there has been a multitude of proposals. Indeed, there are several monographs largely devoted to the subject (e.g. McComb, 1990).

There was a particularly active period throughout the 1960s and 1970s, most notably in the physics community. In the early days there was great optimism: an almost religious conviction that the might of modern physics would finally tame the beast of turbulence. However, as the decades passed, the number of positive results remained depressingly small. As with the application of chaos theory to turbulence, initial enthusiasm progressively gave way to a sense of disappointment. Nevertheless, there were some lasting successes, and at least two closure schemes which were developed in this era and are still with us today: the Eddy-Damped-Quasi-Normal-Markovian (EDQNM) and Test Field models. We shall focus on the former.

The EDQNM model is usually described as a variant of Millionshtchikov's QN scheme, and we shall stay with this convention. However, as you will shortly discover, the EDQNM closure model contains a great deal of ED and M, but comparatively little QN. Still, we shall follow the usual route of describing how EDQNM evolved as an attempt to patch up the worst excesses of the QN model.

So let us start with Millionshtchikov's elegant but flawed QN hypothesis. This was discussed in a qualitative fashion in Section 6.2.3.3. Here we will fill in some of the gaps. In this model (8.46) is retained in its exact form,

$$\frac{\partial E}{\partial t} = T(k,t) - 2\nu k^2 E(k,t).$$

It is also possible to derive an exact equation for $T(k,t)$. It has the form

$$\frac{\partial T}{\partial t} = \text{(Non-linear terms)} + \text{(Viscous terms)}.$$

The problem, of course, is that the non-linear terms on the right consist of fourth-order correlations. This is where the QN scheme seems made to measure. We assume that the fourth-order correlations, and only the fourth-order correlations, can be modelled on the assumption that the joint probability distribution for the velocity field at two or more points is Gaussian. This allows us to rewrite the fourth-order correlations in terms of products of second-order correlations, which is a particular feature of Gaussian statistics. That is, if the probability distribution for $\mathbf{u}$ is normal then we have

$$\left\langle u_i u_j' u_k'' u_l''' \right\rangle = \left\langle u_i u_j' \right\rangle \left\langle u_k'' u_l''' \right\rangle + \left\langle u_i u_k'' \right\rangle \left\langle u_j' u_l''' \right\rangle + \left\langle u_i u_l''' \right\rangle \left\langle u_j' u_k'' \right\rangle$$

where $\mathbf{u}$, $\mathbf{u}'$, $\mathbf{u}''$, and $\mathbf{u}'''$ represent velocities at $\mathbf{x}$, $\mathbf{x}'$, $\mathbf{x}''$, and $\mathbf{x}'''$.[11] Thus, in the QN model, the right-hand side of our $T(k)$ equation is rewritten in terms of $E(k)$, and closure is achieved.

Perhaps the most important paper in the development of the QN model was that of Proudman and Reid (1954). They were the first to systematically explore the dynamical implications of this closure hypothesis, thus initiating much controversy and debate. We shall outline briefly the central results of their analysis, without making any attempt to reproduce the rather tortuous algebra which lies behind them.

It turns out that one cannot readily apply the QN hypothesis directly to an evolution equation for $T(k)$. Rather, it is necessary to start with the Fourier transform of a generalized version of the two-point triple correlation introduced in Chapter 6. An evolution equation for this triple correlation can be obtained from the Navier–Stokes equation and this, in turn, can be greatly simplified with the help of the QN hypothesis.

[11] This is sometimes referred to as a *cumulant discard* hypothesis since terms of the form $\langle uuuu \rangle - \langle uuuu \rangle_{QN}$, where $\langle uuuu \rangle_{QN}$ is the QN estimate of $\langle uuuu \rangle$, are known as fourth-order cumulants.

Let us start by introducing some notation. We use the symbol $S_{i,j,k}$ to represent the third-order, three-point velocity correlation tensor:

$$S_{i,j,k}(\mathbf{r},\mathbf{r}') = \langle u_i(\mathbf{x})u_j(\mathbf{x}')u_k(\mathbf{x}'')\rangle = \left\langle u_i u_j' u_k'' \right\rangle$$

where $\mathbf{r} = \mathbf{x}' - \mathbf{x}$ and $\mathbf{r}' = \mathbf{x}'' - \mathbf{x}$. The six-dimensional Fourier transform of $S_{i,j,k}$ is defined as,

$$\Phi_{ijk}(\mathbf{k},\mathbf{k}') = j(2\pi)^{-6} \iint S_{i,j,k}\, e^{-j(\mathbf{k}\cdot\mathbf{r}+\mathbf{k}'\cdot\mathbf{r}')} d\mathbf{r} d\mathbf{r}'$$

and it turns out that $T(k)$ can be written in terms of Φ_{ijk} as follows,

$$T(k) = 4\pi k^2 k_j \int \Phi_{iij}(\mathbf{k},\mathbf{k}') d\mathbf{k}'.$$

The governing equation for Φ_{ijk}, which may be deduced from the Navier–Stokes equation, takes the form,

$$\left[\frac{\partial}{\partial t} + \nu\left(\mathbf{k}^2 + \mathbf{k}'^2 + \mathbf{k}''^2\right)\right]\Phi_{ijk} = \text{Transform}[\langle uuuu\rangle]$$

where $\langle uuuu\rangle$ is a symbolic representation of the fourth-order correlations and $\mathbf{k}$, $\mathbf{k}'$, and $\mathbf{k}''$ constitute a *triad* of wave vectors, related by,

$$\mathbf{k} + \mathbf{k}' + \mathbf{k}'' = \mathbf{0}.$$

The fourth-order correlations are now estimated using the QN hypothesis, which allows us to express $\langle uuuu\rangle$ in terms of products of $\langle uu\rangle$. The right-hand side of our dynamical equation for Φ_{ijk} can then be evaluated in terms of Φ_{ij} and, after a great deal of algebra, we obtain,

$$\begin{aligned}\left[\frac{\partial}{\partial t} + \nu(\mathbf{k}^2 + \mathbf{k}'^2 + \mathbf{k}''^2)\right]\Phi_{ijk}(\mathbf{k},\mathbf{k}') &= P_{i\alpha l}(\mathbf{k}'')\Phi_{\alpha j}(\mathbf{k})\Phi_{lk}(\mathbf{k}') \\ &\quad + P_{j\alpha l}(\mathbf{k})\Phi_{\alpha k}(\mathbf{k}')\Phi_{li}(\mathbf{k}'') \\ &\quad + P_{k\alpha l}(\mathbf{k}')\Phi_{\alpha i}(\mathbf{k}'')\Phi_{lj}(\mathbf{k})\end{aligned}$$

where

$$P_{i\alpha l}(\mathbf{k}) = k_l \Delta_{i\alpha}(\mathbf{k}) + k_\alpha \Delta_{il}(\mathbf{k}), \quad \Delta_{i\alpha}(\mathbf{k}) = \delta_{i\alpha} - k_i k_\alpha / \mathbf{k}^2.$$

In the interests of brevity, we shall rewrite this equation in symbolic form:

$$\left[\frac{\partial}{\partial t} + \nu(\mathbf{k}^2 + \mathbf{k}'^2 + \mathbf{k}''^2)\right]\Phi_{ijk}(\mathbf{k},\mathbf{k}') = \sum_{\text{QN}} \Phi_{ij}\Phi_{ij}. \qquad (8.61\text{a})$$

(QN model)

Note that the non-linear interactions involve only Fourier modes which satisfy the triad relationship $\mathbf{k} + \mathbf{k}' + \mathbf{k}'' = \mathbf{0}$. This is not peculiar to the QN model, but is a generic feature of a Fourier description of the problem.

The final step is to integrate over $\mathbf{k}'$ to obtain an evolution equation for $T(k,t)$. This takes the form

$$\frac{\partial T}{\partial t} = 4\pi k^2 \int \left[G(\mathbf{k}, \mathbf{k}') \frac{E(k'')}{k''^2} \left\{ \frac{E(k')}{k'^2} - \frac{E(k)}{k^2} \right\} \right] d\mathbf{k}' + \nu(\sim) \tag{8.61b}$$

where $G(\mathbf{k}, \mathbf{k}')$ is a purely geometric function which depends only on $\mathbf{k}$ and $\mathbf{k}'$ and whose exact form need not concern us.[12] If we combine our expression for $T(k)$ with the energy equation

$$\frac{\partial E}{\partial t} = T(k,t) - 2\nu k^2 E(k,t)$$

then we have a closed system which may be integrated to yield $E(k,t)$.

Note that (8.61b) seems to suggest that a whole spectrum of wave numbers contribute to $T(k)$, and hence $\Pi_E(k)$, at wave number k. This goes against our intuition that, in the energy cascade, the energy flux at wave number k, $\Pi_E(k)$, should depend primarily on wave numbers in the vicinity of k. In fact, for the inertial range, it may be shown that the integral in (8.61b) is indeed dominated by contributions in which $k \sim k' \sim k''$.

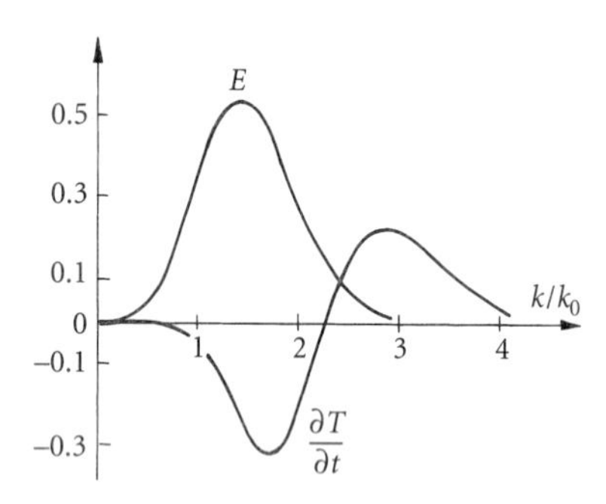

Figure 8.12 The shape of $\partial T/\partial t$ calculated using the QN equation (8.61b) for $E(k) \sim k^4 \exp[-(k/k_0)^2]$. (The units of the vertical axis are arbitrary.)

It is sometimes instructive to take test spectra, $E(k)$, and calculate the resulting form of $\partial T/\partial t$ predicted by (8.61b). One such case is discussed in Proudman and Reid (1954), corresponding to $E(k) \sim k^4 \exp[-(k/k_0)^2]$. The results, which are shown in Figure 8.12, are plausible, with the triple correlations transferring energy from $k \sim k_0$ to higher wave numbers.

One of the first applications of the QN closure model was to investigate the form of $E(k)$ for small k. For example, Proudman and Reid (1954) showed that (8.61b) predicts

$$\frac{\partial T}{\partial t} = \frac{14}{15} k^4 \int_0^\infty \frac{E^2(p)}{p^2} dp + O(k^6).$$

When combined with (8.46), and the expansion

$$E = \frac{I}{24\pi^2} k^4 + \cdots,$$

we obtain

$$\frac{d^2 I}{dt^2} = \frac{7}{5} (4\pi)^2 \int_0^\infty \frac{E^2(p)}{p^2} dp \tag{8.62}$$

[12] Those interested in the details of (8.61b) will find them in Monin and Yaglom (1975), Section 19.3. The function $G(\mathbf{k}, \mathbf{k}')$ is given by the expression $8\pi^2 G = [\mathbf{k} \cdot \mathbf{k}' + k^2 k'^2 / k''^2][1 - (\mathbf{k} \cdot \mathbf{k}')^2 / k^2 k'^2]$.

where I is Loitsyansky's integral. We discussed the significance and deficiencies of this equation at length in Chapter 6.

Proudman and Reid also used (8.61b) to show that the QN hypothesis predicts,

$$\frac{d^2}{dt^2}\left\langle \frac{1}{2}\boldsymbol{\omega}^2 \right\rangle = \frac{2}{3}\left[\left\langle \frac{1}{2}\boldsymbol{\omega}^2 \right\rangle\right]^2 + \text{(viscous term)}.$$

This may be integrated to yield

$$\frac{d}{dt}\left\langle \frac{1}{2}\boldsymbol{\omega}^2 \right\rangle = \frac{2}{3}\left[\left\langle \frac{1}{2}\boldsymbol{\omega}^2 \right\rangle\right]^{3/2} + \text{(viscous term)} \tag{8.63}$$

which might be compared with the exact result (6.57):

$$\frac{d}{dt}\left\langle \frac{1}{2}\boldsymbol{\omega}^2 \right\rangle = \frac{7|S_0|}{6\sqrt{15}}\left[\langle \boldsymbol{\omega}^2 \rangle\right]^{3/2} - \nu\langle (\nabla \times \boldsymbol{\omega})^2 \rangle.$$

(Here S_0 is the skewness factor of $\partial u_x/\partial x$.) Evidently, the QN model predicts a skewness of

$$|S_0| = \sqrt{30}/7 = 0.782.$$

This is considerably higher than the measured values of $|S_0| \approx 0.4$ and is the first hint that the QN hypothesis overestimates the strength of the triple correlations. Note also that the QN model violates Betchov's inequality (see Section 5.5.1),

$$|S_0| \leq [4\delta_0/21]^{1/2} = 0.756$$

since the flatness factor for a Gaussian distribution is $\delta_0 = 3$. The problem, of course, is the artificial nature of assuming Gaussian statistics for $\mathbf{u}$ when evaluating the fourth-order correlations, yet insisting on non-Gaussian statistics as far as the third-order correlations are concerned.

In its day (8.61a) was thought to be a significant breakthrough, particularly as wind-tunnel data showed that the neglect of fourth-order cumulants is indeed valid for well separated points $\mathbf{x}$, $\mathbf{x}'$, and $\mathbf{x}''$. However, it was soon discovered that all was not well. For example, Ogura (1963) showed that integration of the QN equations leads to a significant negative contribution to $E(k)$, which is physically unacceptable (Figure 8.13). As Orszag (1970) noted, there is nothing overly worrying about a closure scheme producing weak negative values of E for uninteresting regions of k. A closure scheme is, after all, an approximation. The problem with the results of Ogura's calculation (and many subsequent ones) is that the violation of the condition $E(k) > 0$ is not weak, and it occurs in the middle of the dynamically most important range of eddy sizes. This renders the QN scheme useless. Perhaps we should not be overly surprised by this failure. For example, in Chapter 5 we stressed that two-point statistics are decidedly non-Gaussian for small separation r. Moreover, in Chapter 6

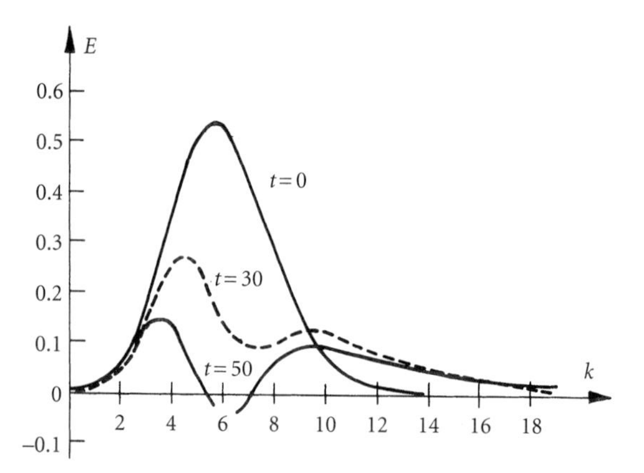

Figure 8.13 Ogura's calculation of $E(k, t)$ using the QN model. Note the eventual appearance of a negative contribution to $E(k, t)$.

we saw that (8.62) overestimates the rate of growth of I by one or two orders of magnitude, essentially because the QN approximation grossly overestimates the strength of the long-range correlations. (We shall return to this point shortly.)

In the late 1960s theoreticians began to offer explanations for the failure of the QN model and to look for alternatives. In an influential paper Orszag (1970) argued that part of the problem lay in the fact that the QN closure fails to build in an arrow of time. His starting point was the entirely plausible (but unproven) suggestion that an *inviscid* fluid would, like its viscous counterpart, lead on average to greater mixing as time progresses. That is to say, even though the governing equations of motion for inviscid flow are time reversible, the statistical behaviour of this 'inviscid turbulence' should be *irreversible*. The implication is that the statistical effects of non-linearity posses an arrow of time and so any closure scheme for the non-linear inertial terms should be non-symmetric in t. In short, Orszag argued that time reversibility of a closure and irreversible mixing are inconsistent. The inviscid QN equations, however, are symmetric in time. Indeed, the only irriversibilty in the QN model arises from the viscous term on the left of (8.61a). This led Orszag to embrace an idea which had been around for several years: that of introducing 'eddy damping' into the QN model. This has the effect of replacing νk^2 in (8.61a) by $\nu_t k^2$ where ν_t is an eddy viscosity which might, for example, be of the order of $\nu_t \sim [E(k)/k]^{1/2}$. Later, it became conventional to supplement, rather than replace, ν by ν_t. Thus, in the conventional Eddy-Damped-Quasi-Normal (EDQN) model we replace (8.61a) by

$$\left[\frac{\partial}{\partial t} + \nu(\mathbf{k}^2 + \mathbf{k}'^2 + \mathbf{k}''^2) + \mu(\mathbf{k}) + \mu(\mathbf{k}') + \mu(\mathbf{k}'')\right]\Phi_{ijk}(\mathbf{k}, \mathbf{k}') = \sum_{\text{QN}} \Phi_{ij}\Phi_{ij} \tag{8.64}$$

where,

$$\mu(\mathbf{k}) \sim [k^3 E(k)]^{1/2}$$

or, as in some later schemes,

$$\mu(\mathbf{k}) \sim \left[\int_0^k p^2 E(p) dp\right]^{1/2}.$$

Note that the EDQN equations are irreversible, even when $\nu = 0$. An additional motivation for the introduction of eddy damping, over and above the need to introduce irreversibility, is the following. If ν is small then for all but the largest of wave numbers the QN scheme gives,

$$\Phi_{ijk}(\mathbf{k}, \mathbf{k}') = \int_0^t \sum_{\text{QN}} \Phi_{ij} \Phi_{ij} dt.$$

Thus, in the QN model, the instantaneous value of the triple correlation depends on the entire history of the product of the double correlations. This, argued Orszag, is physically implausible. We would expect the turbulence to have a short memory and that the instantaneous value of Φ_{ijk} should depend on only the immediate history of $\Phi_{ij}\Phi_{ij}$.[13] Now consider the EDQN model, rewritten as,

$$\left[\frac{\partial}{\partial t} + \frac{1}{\theta}\right] \Phi_{ijk} = \sum_{\text{QN}} \Phi_{ij} \Phi_{ij}$$

where,

$$\theta = [\mu(\mathbf{k}) + \mu(\mathbf{k}') + \mu(\mathbf{k}'') + \nu(\mathbf{k}^2 + \mathbf{k}'^2 + \mathbf{k}''^2)]^{-1}.$$

This may be integrated to give,

$$\Phi_{ijk} = \int_0^t \left[\exp\left[(\tau - t)/\theta\right] \sum_{\text{QN}} \Phi_{ij} \Phi_{ij}\right] d\tau.$$

The instantaneous value of Φ_{ijk} now depends on the immediate history of $\Phi_{ij}\ \Phi_{ij}$ only, with θ playing the role of a relaxation time. In the QN scheme this memory becomes very large as $\nu \to 0$, whereas in the EDQN scheme θ is always of the order of the eddy turn-over time. So the essential argument underlying (8.64) is that the instantaneous state of the triple and double correlations are always closely allied, as if the turbulence is close to a state of statistical equilibrium.

[13] This argument is plausible only for the equilibrium range where eddies adjust rapidly to their global environment. It is less convincing for the energy containing eddies whose memory (turn-over time) is comparable with the decay time of the turbulence.

Another way of viewing the damping terms in (8.64) is to regard them as a simple model for the fourth-order cumulants, which are neglected in (8.61a). (You will recall that the fourth-order cumulants, $[uuuu]_{\text{CUM}}$, are the difference between $\langle uuuu\rangle$ and the QN estimate of $\langle uuuu\rangle$: $[uuuu]_{\text{CUM}} = \langle uuuu\rangle - \langle uuuu\rangle_{\text{QN}}$.) So some see terms of the form $\mu\Phi_{ijk}$ in (8.64) as representing the influence of the neglected cumulants in (8.61a).

In any event, the introduction of 'eddy damping' into (8.64) has the enormous practical advantage of reducing the predicted rate of growth of Φ_{ijk}. This is important since we know from our discussion of Loitsyansky's integral (see Section 6.3.6) that the QN model grossly overestimates the rate of growth of the triple correlations.

Whatever the merits or faults of the EDQN scheme its fate was sealed by the discovery that, like the QN model, it fails to guarantee that $E(k)$ remains positive. (See Lesieur, 1990, for a discussion of this.) It seemed that more drastic action was required. So in the mid 1970s, EDQN finally gave way to the EDQNM scheme. The so-called Markovianization of EDQN involves the drastic step of removing the time derivative from (8.64), converting it into a simple algebraic relationship between Φ_{ijk} and $\sum_{\text{QN}} \Phi_{ij}\Phi_{ij}$. In particular, for large times, $(k^3E)^{1/2}t \gg 1$, EDQNM specifies (Lesieur 1990),

$$[\nu(\mathbf{k}^2 + \mathbf{k}'^2 + \mathbf{k}''^2) + \mu(\mathbf{k}) + \mu(\mathbf{k}') + \mu(\mathbf{k}'')]\Phi_{ijk}(\mathbf{k},\mathbf{k}') = \sum_{\text{QN}} \Phi_{ij}\Phi_{ij}$$

$$\text{(EDQNM model)} \qquad (8.65)$$

The removal of the derivative from (8.64) is, perhaps, justifiable for the small eddies where the timescale μ^{-1} is rapid by comparison with the timescale for the evolution of large eddies, and hence by comparison with $E(k,t)$. However, it is difficult to conceive of any rational justification for the removal of the time derivative when dealing with the all-important energy-containing eddies. Still it turns out that this procedure does at least ensure that $E(k)$ remains positive.

If we compare (8.61a) and (8.65) it becomes clear that we have all but lost the simple elegance of the QN scheme (Table 8.4). Equation (8.61a) rests on the simple hypothesis that, as far as the fourth-order correlations are concerned, the turbulence statistics are Gausian and so fourth-order correlations may be expressed as the product of second-order correlations. This plausible (but flawed) hypothesis can, and has, been checked time and again in wind-tunnel experiments and we know the limits of its validity. In (8.65), on the other hand, we have given up any hope of justifying the closure on the basis of a simple statistical principle. In fact the role of the fourth-order correlations have all but vanished in (8.65). Instead, we simply prescribe an algebraic relationship between third and second-order correlations.

Table 8.4 A comparison of different QN-type closures

Model	Characteristic equation	Rationale
Exact	$\left[\dfrac{\partial}{\partial t}+\nu(\mathbf{k}^2+\mathbf{k}'^2+\mathbf{k}''^2)\right]\Phi_{ijk}(\mathbf{k},\mathbf{k}')=\text{Transform}\langle uuuu\rangle$	—
QN	$\left[\dfrac{\partial}{\partial t}+\nu(\mathbf{k}^2+\mathbf{k}'^2+\mathbf{k}''^2)\right]\Phi_{ijk}=\text{Transform}\langle uuuu\rangle=\sum_{\text{QN}}\Phi_{ij}\Phi_{ij}$	Experimental data show that the QN approximation works for well separated points (though not adjacent points).
EDQN	$\left[\dfrac{\partial}{\partial t}+\nu(\mathbf{k}^2+\mathbf{k}'^2+\mathbf{k}''^2)+\mu(\mathbf{k})+\mu(\mathbf{k}')+\mu(\mathbf{k}'')\right]\Phi_{ijk}=\sum_{\text{QN}}\Phi_{ij}\Phi_{ij}$	• QN model does not work • The need for irreversibility • The need to suppress the growth of Φ_{ijk} in the QN model
EDQNM (for $\mu t\gg 1$)	$[\nu(\mathbf{k}^2+\mathbf{k}'^2+\mathbf{k}''^2)+\mu(\mathbf{k})+\mu(\mathbf{k}')+\mu(\mathbf{k}'')]\Phi_{ijk}=\sum_{\text{QN}}\Phi_{ij}\Phi_{ij}$	$\mu(\mathbf{k})\gg\partial/\partial t$ for the small (but not the large) eddies.

Note: The QN model postulates an algebraic relationship between Transform $\langle uuuu\rangle$ and Φ_{ij}. By contrast, the EDQNM model (at large times) postulates an algebraic relationship between Φ_{ijk} and Φ_{ij}. So the EDQNM model closes the problem at third, rather than fourth, order.

Moreover, the precise nature of that relationship is partly at the disposal of the modeller who is free to choose his preferred version of $\mu(\mathbf{k})$, subject to certain obvious constraints, such as reproducing the five-thirds law.

The EDQNM model is still popular today and seems to perform well in a variety of circumstances (Lesieur 1990; Cambon 2002). However, its reliability in predicting the evolution of the large eddies is questionable. For example, we have already seen that the QN equation (8.62)

$$\frac{d^2I}{dt^2}=\frac{7}{5}(4\pi)^2\int_0^\infty\frac{E^2(p)}{p^2}dp$$

overestimates the strength of the long-range correlations by one or two orders of magnitude. In the EDQNM model (8.62) is replaced by (Lesieur 1990)

$$\frac{dI}{dt}=\frac{7}{5}(4\pi)^2\int\theta(p,t)\frac{E^2(p)}{p^2}dp$$

where θ is a model parameter which has the dimensions of time. There is little reason to have confidence in this equation.

There have been other, more sophisticated, spectral models. In 1959 Kraichnan introduced his direct-interaction-approximation (DIA) scheme. However, this does not reproduce the $k^{-5/3}$ behaviour in the inertial subrange (Hinze 1975). A more satisfactory approach is Kraichnan's test-field model (TFM) which is rather similar to EDQNM, but where the Markovianization step is less drastic. In particular, the procedure for determining the relaxation time θ is more sophisticated than that adopted in EDQNM, with θ being evaluated by examining the behaviour of the triple correlations in an auxiliary field (the test-field of the name).

Although different spectral closure models have their enthusiasts, it is probably fair to say there is still no satisfactory closure scheme which encompasses both the large and the small scales. It seems the hunt for William James' cat is still on, but there is no guarantee that we shall ever find it.[14]

Exercises

8.1 The integral scale, l, is usually defined in terms of the longitudinal correlation function as,

$$l = \int_0^\infty f dr = \frac{1}{u^2} \int_0^\infty R dr.$$

Use (8.21b) to show that l is related to $E(k)$ by

$$l = \frac{\pi}{2u^2} \int_0^\infty \frac{E}{k} dk.$$

8.2 Derive expression (8.36b) from (8.35a, b) and find the forms of the one-dimensional spectra F_{11}, F_{22}, E_1 in the inertial subrange.

8.3 Use equation (8.47) to show that,

$$T(k) = \frac{d}{dk}\left[\frac{2}{\pi}\int_0^\infty \frac{1}{r}\frac{\partial}{\partial r}(r\Gamma)\sin(kr)dr\right]$$

and hence confirm that $\int_0^\infty T(k)dk = 0$.

8.4 Use (8.36a) to confirm that

$$F_{11}(k^*) = \frac{1}{2}\int_{k^*}^\infty \left[1 - (k^*/k)^2\right]\frac{E(k)}{k} dk$$

Evidently $F_{11}(k)$ contains energy from all wave numbers greater than k.

8.5 Energy transfer in the cascade is due to vortex stretching. At any one scale, $k \to k + \Delta k$, three physically important quantities are the energy, $E\Delta k$, the enstrophy, $k^2 E\Delta k$, and the strain-rate which we might take as of order (enstrophy)$^{1/2}$. Thus we might expect $\Pi_E(k)$ to be a function of k, $E\Delta k$, and $k^2 E\Delta k$. The latter quantity might be approximated by

$$k^2 E\Delta k \sim \int_0^k k^2 E dk$$

[14] See quote at the beginning of Section 8.2.2.

since the integral is dominated by contributions close to the upper limit. Suppose that

$$\Pi_E = \Pi_E\left(k, E, \int_0^k k^2 E dk\right).$$

If Π_E is proportional to the square root of the enstrophy, show that Π_E must be of the form

$$\Pi_E \sim kE\left\{\int_0^k k^2 E dk\right\}^{1/2}.$$

This is Ellison's correction to Obukhov's closure scheme. If Π_E is to be determined only by the local properties of the cascade, on the other hand, we have $\Pi_E = \Pi_E(k,E)$. Show that the only dimensionally consistent possibility is Kovasznay's estimate,

$$\Pi_E = \alpha_4 E^{3/2} k^{5/2}.$$

8.6 In the equilibrium range of the spectrum $\partial E/\partial t$ is negligible and so

$$\partial \Pi_E/\partial k = -2\nu k^2 E.$$

Show that, in this range, Kovasznay's estimate yields,

$$E \sim \varepsilon^{2/3} k^{-5/3}\left[1 - (k\hat{\eta})^{4/3}\right]^2, \quad k\hat{\eta} < 1$$

where $\hat{\eta} \sim \eta$.

8.7 Show that the one-dimensional energy spectrum, F_{11}, and the signature function, V, are related by

$$\frac{1}{k}\frac{d}{dk}(k^3 F_{11}) \sim \text{sin transform}[V].$$

8.8 Confirm that, if $f(r) > 0$ for all r, as appears to be the case, then F_{11} is a maximum at $k = 0$.

Suggested reading

Batchelor, G.K. (1953) *The Theory of Homogeneous Turbulence*. Cambridge University Press. (The dynamical equations in spectral form are discussed in chapter 5.)

Batchelor, G.K. and Proudman, I. (1956) *Phil. Trans. Roy. Soc.* A, **248**, 369–405. (Gives the form of $\Phi_{ij}(\mathbf{k})$ and $Q_{ij}(\mathbf{r})$ for small k and large r in a Batchelor spectrum.)

Bracewell, R.N. (1986) *The Fourier Transform and Its Applications*. McGraw–Hill. (This gives a careful introduction to the Fourier transform.)

Cambon, C. (2002) In: *Closure Strategies for Turbulent and Transitional Flows* (eds. B. Launder and N. Sandham). Cambridge University Press. (Discusses the status of EDQNM-like models.)

Hinze, J.O. (1975) *Turbulence*. McGraw–Hill. (An excellent reference text. Consult Chapter 3 for isotropic turbulence.)

Kovasznay, L.S. (1948) *J. Aeronaut. Sci.*, **15**(12), 745–753.

Kraichnan, R.H. (1959) *J. Fluid Mech.*, **5**(4), 497–543.

Ellision, T.H. (1961) *Coll. Intern. de CNRS à Marseille*, Paris, CNRS, 113–121.

Heisenberg, W. (1948) *Z. Physik*, **124**, 628–657.

Lesieur, M. (1990) *Turbulence in Fluids*. Kluwer Academic Publishers. (This provides an accessible yet detailed account of QN-type closure schemes.)

McComb, W.D. (1990) *The Physics of Fluid Turbulence*. Clarendon Press. (Provides a useful compliment to Lesieur's discussion of two-point closure models.)

Monin, A.S. and Yaglom, A.M. (1975) *Statistical Fluid Mechanics II*. MIT Press. (Singularities in the spectrum are discussed in Chapter 15 and the QN model in Chapter 18.)

Obukhov, A.M. (1941) *Dokl. Akad. Nauk SSSR*, **32**(1), 22–24.

Oguru, Y. (1963) *J. Fluid Mech.*, **16**(1), 38–40.

Orszag, S.A. (1970) *J Fluid Mech.*, **41**(2), 363–386. (This paper discusses the EDQN closure scheme.)

Pao, Y.-H. (1965) *Phys. Fluids*, **8**(6), 1063. (This describes the most successful of the algebraic spectral closure models.)

Proudman, I. and Reid, W. (1954) *Phil. Trans. Roy. Soc.* A, **247**, 163–189. (A key work in the development of QN closure schemes.)

Saffman, P.G. (1967) *J. Fluid Mech.*, **27**, 581–593. (Gives the form of $\Phi_{ij}(\mathbf{k})$ and $Q_{ij}(\mathbf{r})$ for small k and large r in a Saffman spectrum.)

Tennekes, H. and Lumley, J.L. (1972) *A First Course in Turbulence*. MIT Press. (Chapter 8 discusses in detail the relationship between $E(k)$ and one-dimensional spectra.)

Townsend, A.A. (1976) *The Structure of Turbulent Shear Flow*, 2nd edition. Cambridge University Press. (Chapter 1 discusses the physical significance of $E(k)$.)

PART III

Special topics

1
2
3
4

CHAPTER 9

The influence of rotation, stratification, and magnetic fields on turbulence

9.1 The importance of body forces in geophysics and astrophysics

By and large, engineers do not have to worry too much about the influence of body forces on turbulence. Perhaps a little buoyancy crops up from time to time, but that is about it. The primary concern of the engineer is the influence of complex boundaries and the way in which these generate and shape the turbulence. The physicist, on the other hand, generally has to contend with flows in which body forces are the dominant factor. Astrophysicists, for example, might be concerned with the formation and evolution of stars, or perhaps with violent activity on the surface of the sun (solar flares, sun spots, coronal mass ejections, etc.). In either case turbulence plays a crucial role, transferring heat from the interior of a star to its surface, and triggering solar flares and coronal mass ejections. Moreover, this is a special kind of turbulence, shaped and controlled by intense magnetic fields. Geophysicists, on the other hand, might be interested in the motion of the earth's liquid core, and in particular, the manner in which turbulence in the core stretches and twists the earth's magnetic field in a way which prevents it from being extinguished through the natural forces of decay. Here the dominant forces acting on the turbulence are the Coriolis and Lorentz forces, arising from the earth's rotation and the terrestrial magnetic field, respectively. Indeed, the non-linear inertial force, $\mathbf{u} \cdot \nabla \mathbf{u}$, which has been the obsession of Chapters 1–8, is almost completely unimportant in geodynamo theory! Large-scale atmospheric and oceanic flows are also heavily influenced by body forces, in this case buoyancy and, at the very large scales, the Coriolis force.

Turbulence on the surface of the sun (Encyclopaedia Britannica).

In view of the difficulty of making predictions about conventional turbulence, it might be thought that the task of incorporating gravitational, Coriolis, and Lorentz forces into some coherent statistical model is so overwhelming as to be quite impractical. In a sense this is true. The equations of turbulence incorporating these forces are extremely complex. Curiously though, there are aspects of these

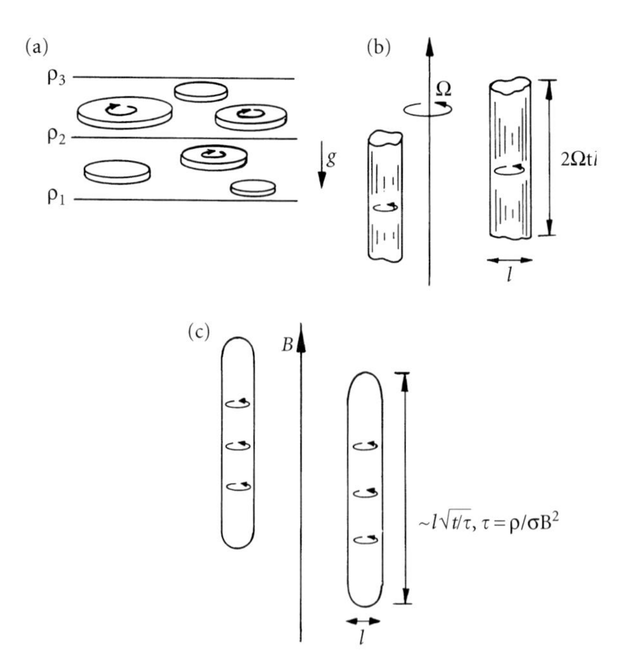

Figure 9.1 The type of large-scale eddies found in: (a) strongly stratified turbulence, (b) rapidly rotating fluid, and (c) a conducting fluid threaded by a magnetic field, **B**.

complex flows, that are easier to understand than conventional turbulence. The point is that buoyancy, Coriolis, and Lorentz forces all tend to organize and shape the turbulence, promoting vortices of a particular structure at the expense of other eddies. For example, it turns out that turbulence in a strongly stratified medium is dominated (at the large scales) by flat 'pancake' vortices (Figure 9.1(a)). A rapidly rotating fluid, on the other hand, tends to extrude vortices along the rotation axis forming columnar eddies (Plate 13; Figure 9.1(b)). Finally, a magnetic field causes vortices to diffuse along the magnetic field lines giving rise, once again, to columnar or sheet-like structures (Figure 9.1(c)). So, while the governing equations for these flows are messy and complex, the flows themselves tend to look more organized than conventional turbulence. The key to understanding turbulence in the presence of a body force is to isolate the mechanism by which that force organizes and shapes the motion. If this can be done, a great deal of useful information can be extracted from the analysis.

We shall look at the influence of rotation, stratification, and magnetic fields in turn, taking rotation and stratification together as they share many common characteristics. The discussion is brief, but the interested reader will find a more comprehensive discussion in the following texts and papers:

Rotational effects: Greenspan (1968), Cambon et al. (1997), and Iida and Nagano (1999).

Stratification: Panchev (1971), Monin and Yaglom (1971), Turner (1973), and Riley and Lelong (2000).

MHD turbulence: Moffatt (1978), Biskamp (1993), and Davidson (2001).

9.2 The influence of rapid rotation and stable stratification

We shall discuss the structure of turbulence in a rapidly rotating system in Sections 9.2.4 and 9.2.5, and turbulence in a stratified fluid in Section 9.2.6. First, however, we summarize some of the properties of the Coriolis force. In particular, we shall see that it tends to promote a form of internal wave motion, called an *inertial wave*. It is these waves which so dramatically shape the turbulent eddies in a rapidly rotating fluid. It turns out that inertial waves have a structure closely related to that of internal gravity waves, and it is this similarity in wave structure that underpins the close analogy between rotating and stratified turbulence. So let us start with the Coriolis force.

9.2.1 *The Coriolis force*

A frame of reference, which rotates at a constant rate Ω relative to an inertial frame is not inertial. The accelerations of a particle measured in the two frames are related by[1]

$$(d\mathbf{u}/dt)_{\text{inertial}} = (d\hat{\mathbf{u}}/dt)_{\text{rot}} + 2\boldsymbol{\Omega} \times \hat{\mathbf{u}} + \boldsymbol{\Omega} \times (\boldsymbol{\Omega} \times \hat{\mathbf{x}}) \tag{9.1}$$

where $\hat{}$ indicates quantities measured in the rotating frame. The terms $2\boldsymbol{\Omega} \times \hat{\mathbf{u}}$ and $\boldsymbol{\Omega} \times (\boldsymbol{\Omega} \times \hat{\mathbf{x}})$ are referred to as the Coriolis and centripetal accelerations, respectively. Equation (9.1) arises from applying the operator $(d/dt)_{\text{inertial}} = (d/dt)_{\text{rot}} + \boldsymbol{\Omega}\times$ to the radius vector $\mathbf{x}$, which gives $\mathbf{u} = \hat{\mathbf{u}} + \boldsymbol{\Omega} \times \hat{\mathbf{x}}$. Differentiating once more gives (9.1). If we multiply both sides of (9.1) by the mass of the particle, m, we have

$$m(d\mathbf{u}/dt)_{\text{inertial}} = m(d\hat{\mathbf{u}}/dt)_{\text{rot}} + m[2\boldsymbol{\Omega} \times \hat{\mathbf{u}}] + m[\boldsymbol{\Omega} \times (\boldsymbol{\Omega} \times \hat{\mathbf{x}})].$$

Since the left-hand side of this equation is equal to $\mathbf{F}$, the sum of the forces acting on the particle, we can rewrite this as

$$m(d\hat{\mathbf{u}}/dt)_{\text{rot}} = \mathbf{F} - m[2\boldsymbol{\Omega} \times \hat{\mathbf{u}}] - m[\boldsymbol{\Omega} \times (\boldsymbol{\Omega} \times \hat{\mathbf{x}})]$$

and we see that Newton's second law does not apply in the rotating frame. But we can 'fix' things if we add to the real forces $\mathbf{F}$, the fictitious forces $\mathbf{F}_{\text{cor}} = -m[2\boldsymbol{\Omega} \times \hat{\mathbf{u}}]$ and $\mathbf{F}_{\text{cen}} = -m[\boldsymbol{\Omega} \times (\boldsymbol{\Omega} \times \hat{\mathbf{x}})]$. These are known as the Coriolis and centrifugal forces, respectively.

[1] For simplicity we shall take the two frames of reference to have a common origin so that $\mathbf{x} = \hat{\mathbf{x}}$. For a detailed discussion of rotating frames of reference, see, for example, Symon (1960).

Note that the centrifugal force is irrotational and may be written as $\nabla[\frac{1}{2}m(\boldsymbol{\Omega} \times \hat{\mathbf{x}})^2]$, that is,

$$\begin{aligned}\nabla[\tfrac{1}{2}(\boldsymbol{\Omega} \times \mathbf{x})^2] &= (\boldsymbol{\Omega} \times \mathbf{x}) \cdot \nabla(\boldsymbol{\Omega} \times \mathbf{x}) + (\boldsymbol{\Omega} \times \mathbf{x}) \times [\nabla \times (\boldsymbol{\Omega} \times \mathbf{x})] \\ &= \boldsymbol{\Omega} \times (\boldsymbol{\Omega} \times \mathbf{x}) + 2(\boldsymbol{\Omega} \times \mathbf{x}) \times \boldsymbol{\Omega} = -\boldsymbol{\Omega} \times (\boldsymbol{\Omega} \times \mathbf{x}).\end{aligned}$$

This is important in the context of fluid mechanics since the centrifugal force may be simply absorbed into the pressure term, $-\nabla p$, to form a modified pressure gradient. In the absence of a free surface, such forces produce no motion. So, introducing the modified pressure, $\hat{p} = p - \frac{1}{2}\rho(\boldsymbol{\Omega} \times \hat{\mathbf{x}})^2$, the Navier–Stokes equation in a rotating frame of reference becomes

$$\partial\hat{\mathbf{u}}/\partial t + \hat{\mathbf{u}} \cdot \nabla\hat{\mathbf{u}} = -\nabla(\hat{p}/\rho) + 2\hat{\mathbf{u}} \times \boldsymbol{\Omega} + \nu\nabla^2\hat{\mathbf{u}}. \tag{9.2}$$

From now on we shall omit the ˆ on $\hat{\mathbf{u}}$, on the understanding that $\mathbf{u}$ is measured in the rotating frame, and the ˆ on $\hat{p}$, on the understanding that p refers to the modified pressure. Note that, by necessity, the fictitious Coriolis force $2\mathbf{u} \times \boldsymbol{\Omega}$ cannot create or destroy energy, as evidenced by the fact that $(2\mathbf{u} \times \boldsymbol{\Omega}) \cdot \mathbf{u} = 0$. Also, the relative strength of the non-linear inertial force, $\mathbf{u} \cdot \nabla\mathbf{u}$, and the Coriolis force, $2\mathbf{u} \times \boldsymbol{\Omega}$, is given by the so-called *Rossby number*, $\text{Ro} = u/l\Omega$, where l is a typical scale of the motion.

In the next few pages we shall omit the viscous term in (9.2) since the effect of the Coriolis force does not depend on viscosity. Also, to focus thoughts, we shall take $\boldsymbol{\Omega}$ to point in the z-direction and assume that Ro is small, so that the fluid is primarily in a state of rigid body rotation. Equation (9.2) becomes

$$\frac{D\mathbf{u}}{Dt} = 2\mathbf{u} \times \boldsymbol{\Omega} - \nabla(p/\rho). \tag{9.3}$$

Note that the Coriolis force tends to deflect a fluid particle in a direction normal to its instantaneous velocity, as illustrated in Figure 9.2. Thus, a fluid particle travelling radially outward experiences a force which tends to induce rotation in a sense opposite to that of $\boldsymbol{\Omega}$, so that its angular velocity measured in an inertial frame is reduced. Conversely, a particle moving radially inward will start to rotate (in the non-inertial frame) in the same sense as $\boldsymbol{\Omega}$. We might anticipate that, when viewed in an inertial frame, this curious behaviour is a direct consequence of the law of conservation of angular momentum and this is indeed more or less true. (Note, however, that individual fluid particles can exchange angular momentum via the pressure force and so this interpretation is a little simplistic. See Example 9.1.)

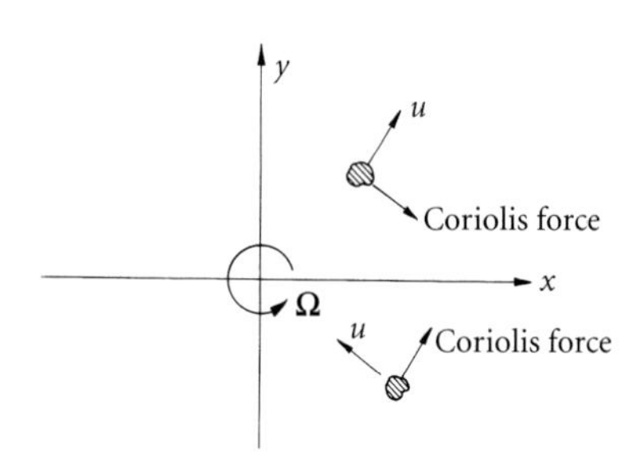

Figure 9.2 The influence of the Coriolis force on motion in the x–y plane.

We shall see shortly that the Coriolis force has an extraordinary effect on a rotating fluid. In particular, it endows the fluid with a kind of elasticity, which allows it to propagate internal waves, called *inertial waves*.[2] We can gain some insight into the origin of this phenomenon if we restrict ourselves to axisymmetric motion; that is, in our rotating frame of reference, we take $\mathbf{u}$ to be of the form $\mathbf{u}(r, z) = (u_r, u_\theta, u_z)$ in (r, θ, z) coordinates. Now suppose that, in our rotating frame, we have poloidal motion in the r–z plane, as shown in Figure 9.3(a). Initially the fluid has no relative rotation, $u_\theta = 0$. Fluid at A is swept inward to A' while fluid at B is carried outward to B'. This radial movement gives rise to a Coriolis force, $-2u_r\Omega\hat{\mathbf{e}}_\theta$, which induces positive relative rotation at A', $u_\theta > 0$, and negative relative rotation at B' (Figure 9.3(b)). Note that the direction of this induced rotation is such as to conserve angular momentum in an inertial frame of reference. The induced swirl itself now gives rise to a Coriolis force, $2u_\theta\Omega\hat{\mathbf{e}}_r$. This force opposes the original motion, tending to move the fluid at A' radially outward and the fluid at B' inward (Figure 9.3(c)). The whole process now begins in reverse and since energy is conserved in an inviscid fluid, we might anticipate that oscillations are set up, in which fluid particles oscillate about their equilibrium radii. These oscillations, which are the hallmark of a rapidly rotating fluid, are a manifestation of inertial wave propagation.

We shall return to inertial waves in Section 9.2.3, where we shall analyse their properties in detail. In the meantime, we consider another, closely related, consequence of rigid body rotation: the tendency for the Coriolis force to produce two-dimensional motion.

Example 9.1 Angular momentum conservation in inertial and non-inertial frames

Let us temporarily return to the use of the ˆ to indicate variables in the rotating frame. Show that (9.3) yields

$$(D(\hat{\mathbf{x}} \times \hat{\mathbf{u}})/Dt)_{\text{rot}} = 2\hat{\mathbf{x}} \times (\hat{\mathbf{u}} \times \boldsymbol{\Omega}) + \nabla \times [(\hat{p}/\rho)\hat{\mathbf{x}}]. \tag{9.4}$$

The second term on the right-hand side integrates to zero for a localized disturbance, but the first need not. Evidently, angular momentum, as measured in the rotating frame, is not conserved. The first term on the right may be transformed using the identity:

$$\{2\mathbf{x} \times [\mathbf{v} \times \mathbf{K}]\}_i = \{(\mathbf{x} \times \mathbf{v}) \times \mathbf{K}\}_i + \nabla \cdot [\mathbf{x} \times (\mathbf{x} \times \mathbf{K})_i \mathbf{v}] \tag{9.5}$$

where $\mathbf{v}$ is any solenoidal vector field and $\mathbf{K}$ is a constant vector. By equating $\hat{\mathbf{u}}$ to $\mathbf{v}$ and $\boldsymbol{\Omega}$ to $\mathbf{K}$, show that (9.4) can be rearranged to give

$$\begin{aligned}(D/Dt)_{\text{rot}}[\hat{\mathbf{x}} \times (\hat{\mathbf{u}} + (\boldsymbol{\Omega} \times \hat{\mathbf{x}}))] + \boldsymbol{\Omega} \times [\hat{\mathbf{x}} \times (\hat{\mathbf{u}} + (\boldsymbol{\Omega} \times \hat{\mathbf{x}}))] \\ = \nabla \times [(p/\rho)\hat{\mathbf{x}}].\end{aligned} \tag{9.6}$$

[2] A particularly simple and beautiful explanation of inertial waves, which is different from ours and relies on an analogy between rotation and density stratification, is given by Rayleigh (1916).

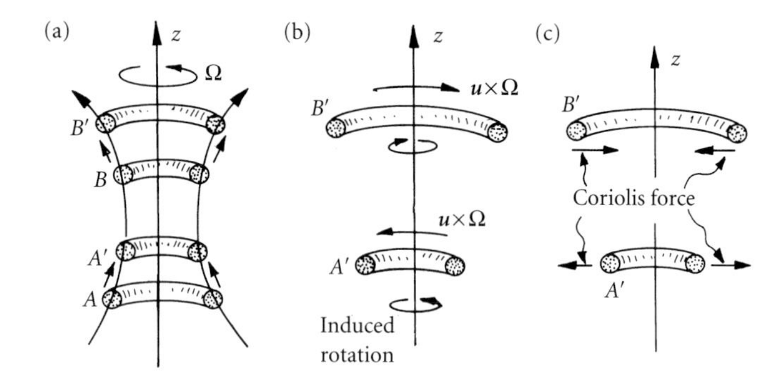

Figure 9.3 Sequence of events that leads to inertial waves through the action of the Coriolis force.

Since $\mathbf{u} = \hat{\mathbf{u}} + \mathbf{\Omega} \times \hat{\mathbf{x}}$, this simplifies to

$$[(D/Dt)_{\text{rot}} + \mathbf{\Omega}\times](\hat{\mathbf{x}} \times \mathbf{u}) = \nabla \times [(p/\rho)\hat{\mathbf{x}}]. \tag{9.7}$$

Finally show that this reduces to the inertial frame equation:

$$[(D/Dt)_{\text{inertial}}](\mathbf{x} \times \mathbf{u}) = \nabla \times [(p/\rho)\mathbf{x}] \tag{9.8}$$

which, unlike (9.4), does conserve angular momentum. (The term on the right integrates to zero for a localized disturbance.)

9.2.2 *The Taylor–Proudman theorem*

In a rotating frame of reference, our inviscid equation of motion is,

$$\frac{D\mathbf{u}}{Dt} = 2\mathbf{u} \times \mathbf{\Omega} - \nabla(p/\rho), \quad \mathbf{\Omega} = \Omega\hat{\mathbf{e}}_z. \tag{9.9}$$

We are particularly interested in cases where the departures from rigid-body rotation are slight, so the Rossby number, $u/l\Omega$, is small. In such cases the inertial term $\mathbf{u} \cdot \nabla\mathbf{u}$ may be neglected by comparison with the Coriolis force and we have

$$\frac{\partial\mathbf{u}}{\partial t} = 2\mathbf{u} \times \mathbf{\Omega} - \nabla(p/\rho). \tag{9.10}$$

We may eliminate pressure by taking the curl of (9.10). This provides us with a linearized vorticity equation:

$$\frac{\partial\boldsymbol{\omega}}{\partial t} = 2(\mathbf{\Omega} \cdot \nabla)\mathbf{u}. \tag{9.11}$$

If the motion is steady, or quasi-steady, we may neglect $\partial\boldsymbol{\omega}/\partial t$, which yields,

$$(\mathbf{\Omega} \cdot \nabla)\mathbf{u} = 0. \tag{9.12}$$

We have arrived at the *Taylor–Proudman theorem*. In cases where $u \ll \Omega l$ and $\partial\mathbf{u}/\partial t$ is small, the motion must be purely two dimensional,

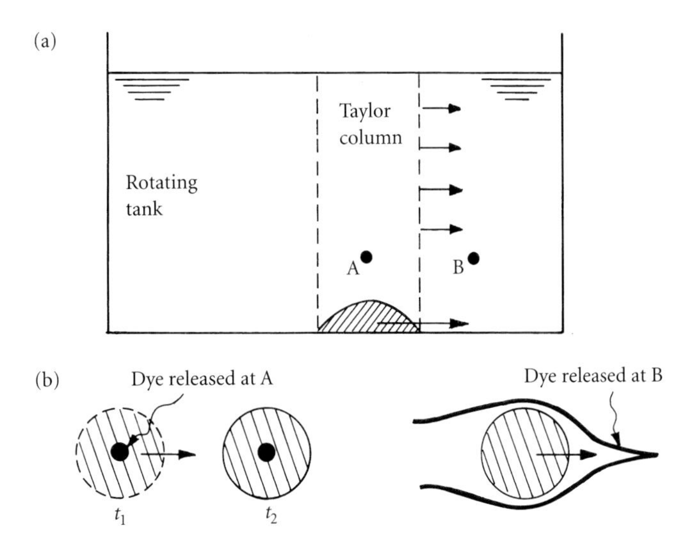

Figure 9.4 The effect of moving a small object slowly across the bottom of a rotating tank. (a) Side view. (b) Plan view.

in the sense that **u** is not a function of z. Note that this need not exclude a finite value of u_z. Many remarkable results follow from this theorem. Consider, for example, Figure 9.4 which shows water in a rapidly rotating tank in which a small object is slowly towed across the base of the tank. As the object moves, the column of fluid located between the object and the upper surface also moves. It is as if this column of fluid is rigidly attached to the obstacle at the base of the tank. This effect may be visualized by releasing dye into the water. A blob of dye released at A moves as a coherent blob, always located above the instantaneous centre of the object. On the other hand, dye released at point B ahead of the column divides into two streaks as the object passes below point B. This column of fluid is called a *Taylor column* after G.I. Taylor who first demonstrated the effect in 1921.

This bizarre behaviour can be explained using (9.12) and indeed, the existence of such columns was anticipated by Proudman some 5 years before Taylor's demonstration. The key point is this. Equation (9.12) demands that $\partial u_z / \partial z = 0$ at all points in the fluid and so the axial straining of any fluid element is zero. It follows that a vertical column of fluid cannot be stretched or compressed. Consequently, there can be no flow over the object shown in Figure 9.4 since this would entail a change in column height. Instead, as the object moves, the fluid flows around the cylinder which circumscribes the object. It is as if the Taylor column were solid.

There are many other examples of two-dimensional motion resulting from the Taylor–Proudman theorem. Evidently (9.12) is a very powerful constraint. One obvious question which arises from Figure 9.4 is: 'how does the fluid within the Taylor column know that it must move with the object?' It turns out that the answer to this question takes us back to the ability of rotating fluids to sustain waves.

9.2.3 *Properties of inertial waves*

We now quantify the wave motion anticipated in Section 9.2.1. Our starting point is (9.11). On differentiating this with respect to time, and taking its curl, we obtain the wave-like equation,

$$\frac{\partial^2}{\partial t^2}(\nabla^2 \mathbf{u}) + 4(\mathbf{\Omega} \cdot \nabla)^2 \mathbf{u} = \mathbf{0}. \tag{9.13}$$

This equation is satisfied by plane waves of the form:

$$\mathbf{u} = \hat{\mathbf{u}} \exp[j(\mathbf{k} \cdot \mathbf{x} - \varpi t)] \tag{9.14}$$

with a dispersion relationship and phase velocity of

$$\varpi = \pm 2(\mathbf{k} \cdot \mathbf{\Omega})/|\mathbf{k}| \tag{9.15}$$

$$\mathbf{C}_\mathrm{p} = 2(\mathbf{k} \cdot \mathbf{\Omega})\mathbf{k}/|\mathbf{k}|^3. \tag{9.16}$$

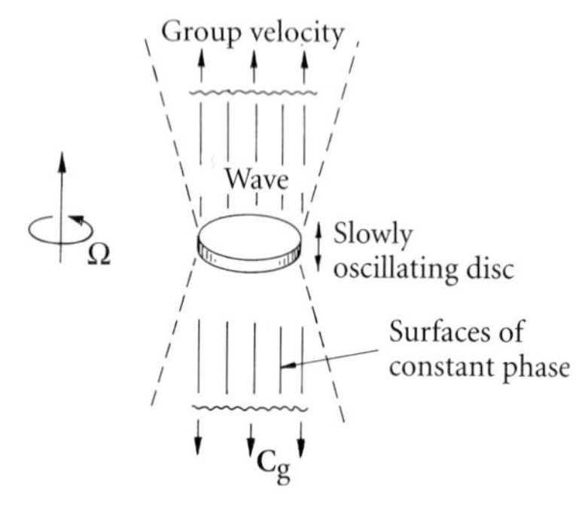

Figure 9.5 An inertial wave. Note that energy propagates parallel to the wave crests.

Note that the angular frequency, ϖ, depends only on the direction and not the magnitude of $\mathbf{k}$. More importantly, the group velocity (i.e. the velocity at which energy propagates via wave packets) is given by

$$\mathbf{C}_\mathrm{g} = \frac{\partial \varpi}{\partial k_i} = \pm\left[\frac{2\mathbf{\Omega}}{|\mathbf{k}|} - \mathbf{C}_\mathrm{p}\right] = \pm\, 2\mathbf{k} \times (\mathbf{\Omega} \times \mathbf{k})/|\mathbf{k}|^3. \tag{9.17}$$

It is evident from (9.17) that these are somewhat unusual waves in which the group velocity is perpendicular to the phase velocity (i.e. $\mathbf{C}_\mathrm{p} \cdot \mathbf{C}_\mathrm{g} = 0$). Thus, a wave appearing to travel in one direction, according to the surfaces of constant phase, is actually propagating energy in a perpendicular direction, as illustrated in Figure 9.5.

Note that $\mathbf{C}_\mathrm{g}$ is co-planar with $\mathbf{\Omega}$ and $\mathbf{k}$ and that the wave frequency varies from $\varpi = 0$ to $\varpi = 2\Omega$, depending on the relative orientation of $\mathbf{k}$ and $\mathbf{\Omega}$. Low-frequency waves have a group velocity aligned with the rotation axis and of magnitude $2\Omega/|\mathbf{k}|$. This is the situation shown in Figure 9.5. So energy propagates away from a slowly oscillating object at a speed of $2\Omega l$, where l is its characteristic size, and in the directions of $\pm\hat{\mathbf{e}}_z$. High-frequency waves, on the other hand, have $\mathbf{k}$ aligned with $\mathbf{\Omega}$ and have negligible group velocity. In general, then, energy propagates fastest in the low-frequency waves and for those disturbances which have the largest wavelength.

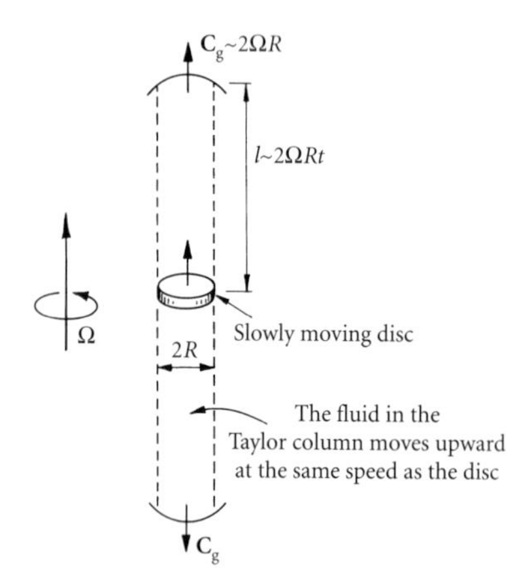

Figure 9.6 The formation of a Taylor column by an inertial wave.

This sort of behaviour can be seen if a small object, say a penny, is suddenly given an axial velocity as shown in Figure 9.6. If the speed V of the penny is much less than ΩR, R being the radius of the penny, then low-frequency waves propagate along the rotation axis in the $\pm\mathbf{\Omega}$ directions, carrying energy away from the disc. The largest wavelengths travel fastest so there is a wavefront a distance $2\Omega Rt$ above and below the penny. The fluid which lies in the region $|z| > 2\Omega Rt$

remains quiescent (in our rotating frame) since the inertial waves have not yet reached these outer regions. For $|z| < 2\Omega Rt$, on the other hand, we find a column of fluid which moves along the axis at exactly the same speed as the penny. Of course, this is a Taylor column, whose length is increasing at a rate $l \sim C_g t$.

It is now clear how Taylor columns form in the rotating tank shown in Figure 9.4. As the object is slowly towed across the base of the tank, it continually emits fast inertial waves which span the depth of the fluid on a timescale which is infinitesimally small compared with the timescale of movement of the object. The waves continually establish and re-establish a Taylor column, which drifts across the tank attached to the object. By suppressing the time derivative in (9.11), we filtered out the dynamics of these waves, but the quasi-steady equation (9.12) still captures the long-term, or smoothed out, effect of the wave propagation, that is, the formation of a Taylor column. The reader interested in more details of the formation of Taylor columns in confined domains will find a comprehensive discussion in Greenspan (1968).

It is instructive to consider one last model problem. Consider an isolated eddy sitting in an infinite rotating fluid. Let us suppose that the eddy has size l and typical velocity u, such that $u \ll \Omega l$. The dominant force acting on the eddy is then the Coriolis force. For simplicity we suppose that the eddy consists of a blob of swirling fluid of angular momentum $\mathbf{H}$ whose axis of rotation is inclined at an angle θ relative to $\boldsymbol{\Omega}$, as shown in Figure 9.7(a). We imagine that (somehow) this eddy is established at $t = 0$ and we ask the question: what happens for $t > 0$? Of course the eddy starts to propagate inertial waves, so the energy of the eddy disperses with a group velocity of $\mathbf{C}_g$.

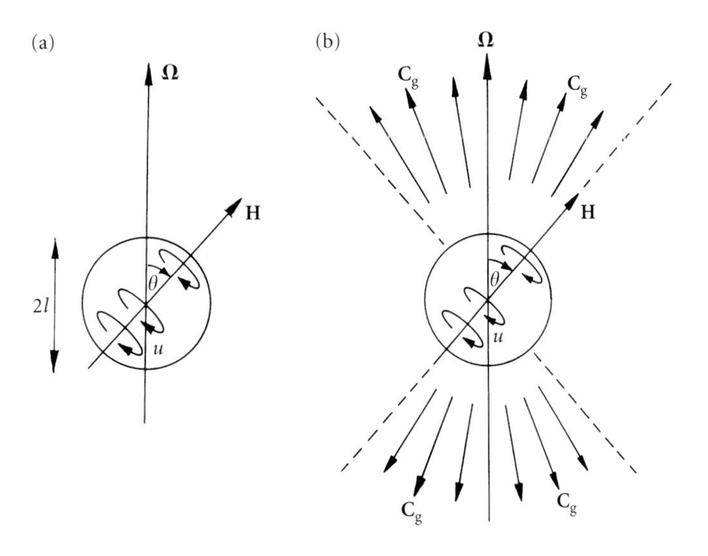

Figure 9.7 (a) Eddy with angular momentum $\mathbf{H}$ in a rotating frame. (b) The propagation of inertial waves from the eddy.

The key to determining $\mathbf{C}_g$ is to examine the Fourier transform of the initial velocity field, which is tantamount to decomposing the initial velocity field into a myriad of plane waves. It turns out that the wave-vector, $\mathbf{k}$, of the transform is more or less confined to the plane normal to $\mathbf{H}$. Thus, we can imagine a spectrum of waves with phase velocities distributed evenly throughout the plane perpendicular to $\mathbf{H}$. The corresponding group velocity for each of the wave components can be calculated from (9.17), and so we find energy radiating in a variety of directions, with the bulk of the energy flux being confined to cones defined by $\mathbf{H}$ and $\mathbf{\Omega}$, as shown in Figure 9.7(b). Thus, the energy of the eddy propagates primarily in the $\pm\mathbf{\Omega}$ directions. We shall see shortly that this has important consequences for turbulence.

Example 9.2 Consider a small blob of buoyant fluid immersed in a rapidly rotating liquid of finite depth. Let the gravitational acceleration, $\mathbf{g}$, be anti-parallel to $\mathbf{\Omega}$, and $\delta\rho$ be the small perturbation in density within the buoyant blob. Show that, in the Boussinesq approximation, the linearized equation of motion is

$$\frac{\partial \boldsymbol{\omega}}{\partial t} = 2(\mathbf{\Omega} \cdot \nabla)\mathbf{u} + \nabla\Psi \times \mathbf{g}, \quad \Psi = \delta\rho/\bar{\rho} < 0.$$

Consider quasi-steady solutions of this equation in which the transient phenomenon of inertial waves is filtered out. Show that a Taylor column of swirling fluid is formed above and below the blob, with anti-cyclonic rotation ($u_\theta < 0$) above the blob, and cyclonic rotation ($u_\theta > 0$) below the blob.

9.2.4 *Turbulence in rapidly rotating systems*

It should be clear by now that the large (though not necessarily the small) eddies in a turbulent flow will be greatly influenced by bulk rotation. In particular, we might expect the large eddies to form columnar structures, the reshaping of eddies being driven by low-frequency inertial waves. In the laboratory, the effects of rotation can be seen by spinning a tank of water while simultaneously stirring the fluid (Plate 13). Whenever $u \ll \Omega l$, l being the integral scale, the turbulence rapidly adopts a quasi-two-dimensional structure, in which the large-scale turbulent eddies are extruded into columnar vortices, presumably by inertial waves (Ibbetson and Tritton 1975; Hopfinger et al. 1982). The same behaviour is seen in numerical simulations (Bartello et al. 1994, Godeferd et al, 1999), where initially isotropic turbulence evolves towards a two-dimensional state. One way to interpret this evolution is to picture the initial state as a random distribution of eddies of the type shown in Figure 9.7. While energy disperses from these eddies in a variety of directions, there is a preference for energy to propagate along the rotation axis.

We shall return to the idea of inertial waves shaping eddies shortly. First, however, let us see where a classical one-point statistical analysis takes us. The simplest mathematical model of rotating turbulence relates to the case where there is no mean flow (in the rotating frame of reference) and in which the turbulence is homogeneous. In such cases the evolution equation for the Reynolds stresses, adapted from (4.8), becomes,

$$\frac{d}{dt}\langle u_i u_j \rangle = 2\Omega\beta_{ij} + \left\langle \frac{p}{\rho}\left(\frac{\partial u_i}{\partial x_j} + \frac{\partial u_j}{\partial u_i}\right)\right\rangle - 2\nu\left\langle \frac{\partial u_i}{\partial x_k}\frac{\partial u_j}{\partial x_k}\right\rangle \tag{9.18}$$

where the Coriolis effects are tied up in the symmetric matrix β_{ij},

$$\beta_{ij} = \begin{bmatrix} 2\langle u_x u_y \rangle & \langle u_y^2 - u_x^2 \rangle & \langle u_y u_z \rangle \\ \langle u_y^2 - u_x^2 \rangle & -2\langle u_x u_y \rangle & -\langle u_x u_z \rangle \\ \langle u_y u_z \rangle & -\langle u_x u_z \rangle & 0 \end{bmatrix}. \tag{9.19}$$

Now, while the Rossby number based on the large scales is often small, it is almost invariably large when based on the Kolmogorov microscales, η and v. In short, the Coriolis force is relatively weak at the small scales since Ω is usually negligible by comparison with the vorticity of the smallest eddies. In such a situation we might hope that the small scales will be approximately isotropic and that, as in conventional turbulence, we can replace the dissipation tensor by its isotropic counterpart,

$$2\nu\left\langle \frac{\partial u_i}{\partial x_k}\frac{\partial u_j}{\partial x_k}\right\rangle = \frac{2}{3}\varepsilon\delta_{ij}. \tag{9.20}$$

The Reynolds stress equation then simplifies to

$$\frac{d}{dt}\langle u_i u_j \rangle = \left\langle \frac{p}{\rho}\left(\frac{\partial u_i}{\partial x_j} + \frac{\partial u_j}{\partial x_i}\right)\right\rangle + 2\Omega\beta_{ij} - \frac{2}{3}\varepsilon\delta_{ij}. \tag{9.21}$$

Note that $\beta_{ii} = 0$ and so the Coriolis force does not, of course, appear in the turbulence energy equation, which is the trace of (9.21).

This is about as far as rigorous (or near rigorous) statistical arguments will take us. If we are to make progress beyond this point, we must enter the uncertain world of turbulence closure models. A Reynolds stress model, for example, might try to estimate the pressure-rate-of-strain term in (9.21) as a function of Ω, ε, and $\langle u_i u_j \rangle$, thus allowing $\langle u_i u_j \rangle(t)$ to be calculated (see Examples 9.4 and 9.5). We shall not, however, pursue this route. We merely note that, when trying to understand the behaviour of rotating turbulence, one-point closure models (i.e. models which work only with the Reynolds stresses) are of limited help. To emphasize the point, suppose that the large scales in our homogeneous, rotating turbulence are statistically axisymmetric, which is not an unreasonable situation. Then the results of Appendix V tell us that

$$\langle u_x u_y \rangle = \langle u_y u_z \rangle = \langle u_x u_z \rangle = 0, \quad \langle u_x^2 \rangle = \langle u_y^2 \rangle.$$

In this case $\beta_{ij} = 0$ and all the effects of rotation seem to have disappeared from (9.21)! There is little hint here of the power of inertial waves to create columnar eddies. In fact, all suggestion of the key physical processes seems to have been eliminated. So a one-point closure model can readily miss the crucial fact that, due to the axial elongation of eddies, the turbulent diffusion of heat and the rate of dissipation of energy are very different in rotating and non-rotating turbulence.

Example 9.3 Use the Reynolds stress equation (9.21) to show that $\langle u_x u_y \rangle$ and $\langle u_y^2 - y_x^2 \rangle$ are related by

$$\frac{d}{dt}\langle u_x u_y \rangle = 2\Omega\left\langle \left(u_y^2 - u_x^2\right)\right\rangle + \left\langle \frac{p}{\rho}\left(\frac{\partial u_x}{\partial y} + \frac{\partial u_y}{\partial x}\right)\right\rangle$$

$$\frac{d}{dt}\left\langle \left(u_y^2 - u_x^2\right)\right\rangle = -8\Omega\langle u_x u_y \rangle + \left\langle 2\frac{p}{\rho}\left(\frac{\partial u_y}{\partial y} - \frac{\partial u_x}{\partial x}\right)\right\rangle.$$

Now show that the same expressions follow directly from the linear equation (9.10).

Example 9.4 In Section 4.6.2 we introduced Rotta's 'return to isotropy' model for the pressure-rate-of-strain term:

$$\langle 2pS_{ij} \rangle = -c_R(\rho\varepsilon/k)\left[\langle u_i u_j \rangle - \tfrac{1}{3}\langle \mathbf{u}^2 \rangle \delta_{ij}\right], \quad c_R > 1.$$

Here k is the kinetic energy, $\frac{1}{2}\langle \mathbf{u}^2 \rangle$, and c_R is an adjustable coefficient. Use this model to estimate the pressure terms in Example 9.3 and show that the resulting equations are

$$\frac{db_{xy}}{dt} + \frac{\varepsilon}{k}(c_R - 1)b_{xy} = 2\Omega\left[b_{yy} - b_{xx}\right]$$

$$\frac{d}{dt}\left[b_{yy} - b_{xx}\right] + \frac{\varepsilon}{k}(c_R - 1)\left[b_{yy} - b_{xx}\right] = -8\Omega b_{xy}$$

where b_{ij} is the anisotropy tensor

$$b_{ij} = \langle u_i u_j \rangle / \langle \mathbf{u}^2 \rangle - \tfrac{1}{3}\delta_{ij}.$$

Now eliminate $[b_{yy} - b_{xx}]$ to show that, in this approximation,

$$\left[\frac{d}{dt} + \frac{\varepsilon}{k}(c_R - 1)\right]^2 b_{xy} + (4\Omega)^2 b_{xy} = 0.$$

So this model predicts an oscillatory relaxation to an axisymmetric state, in which $\beta_{ij} = 0$.

Example 9.5 In Section 4.6.2 we saw that, in an inertial frame of reference, $\langle 2pS_{ij}\rangle$ depends on both the mean flow and the turbulence. In a rotating frame, therefore, we would expect Rotta's expression in Example 9.4 to be supplemented by one involving Ω. A common estimate for $\langle 2pS_{ij}\rangle$ is

$$\langle 2pS_{ij}\rangle = -2c_R\rho\varepsilon b_{ij} - 2\hat{c}_R\rho\Omega\beta_{ij}, \qquad \hat{c}_R = \text{constant}, \quad \hat{c}_R < 1.$$

Show that, in such a model, the conclusions of Example 9.4 remain valid, but with Ω replaced by $\Omega(1 - \hat{c})$. □

We have suggested that one-point closure models have some difficulty in capturing the phenomena which underlie rotating turbulence. In order to build up a mathematical picture of what happens in rapidly rotating turbulence, it is, perhaps, more fruitful to return to the unaveraged equations of motion. Consider the case where $\text{Ro} = u/\Omega l$ is small, l, being the integral scale. Then the short-term evolution of the large eddies is more or less governed by the linear equation (9.13). Consider, for example, the case of turbulence forced by a slowly oscillating grid. The low frequency of the forcing will initiate inertial waves which propagate parallel to $\mathbf{\Omega}$, extruding the turbulent eddies in the axial direction. In such a situation we can might tentatively neglect $\partial^2/\partial z^2$ in the Laplacian of (9.13), which then becomes

$$\frac{\partial^2}{\partial t^2}\left(\nabla_\perp^2 \mathbf{u}\right) + 4(\mathbf{\Omega}\cdot\nabla)^2\mathbf{u} = (\text{source term}).$$

Here $\nabla_\perp^2$ represents the two-dimensional Laplacian defined in a plane normal to $\mathbf{\Omega}$. If we take the Fourier transform of this equation in the transverse plane (but not in the direction of $\mathbf{\Omega}$), we have

$$\frac{\partial^2\hat{\mathbf{u}}}{\partial t^2} = \left(\frac{4\Omega^2}{k_\perp^2}\right)\frac{\partial^2\hat{\mathbf{u}}}{\partial z^2} + (\text{source term})$$

where $\hat{\mathbf{u}}$ is the transformed velocity. Evidently $\hat{\mathbf{u}}$, and hence turbulent energy, spreads by wave propagation along the $\mathbf{\Omega}$-axis, creating columnar structures in the process. Thus, we might picture the large scales as an ensemble of columnar eddies immersed in a sea of small-amplitude inertial waves (Figure 9.8). In the linear approximation, these waves exist without interaction. However, even when Ro is

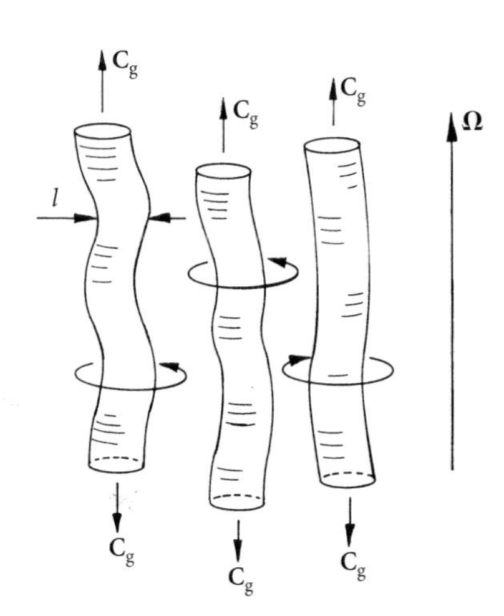

Figure 9.8 We might picture the large scales in rotating turbulence as an ensemble of columnar eddies immersed in a sea of inertial waves and smaller scale debris.

small, we cannot completely neglect the small but finite, non-linear interactions of the inertial waves. The reason is that two small-amplitude waves with distinct frequencies can combine, via the weak non-linear term $\mathbf{u} \cdot \nabla \mathbf{u}$, to produce an inertial force at a third frequency which, in turn, can excite a third wave (Cambon et al. 1997). Under the right conditions, a so-called *resonant triad* can occur in which the natural frequency of the third wave matches that of the inertial forcing. The weak non-linearities then produce a significant response. Nevertheless, the general trend for forced turbulence at small Ro is towards a quasi-two-dimensional flow, dominated by large, columnar eddies (Godeferd et al 1999, Hopfinger et al 1982).[3] Within the large scales two sets of dynamics coexist. The columnar eddies interact with each other on the slow timescale of l/u, while the inertial waves propagate on the fast timescale of Ω^{-1}, enforcing the quasi-two-dimensionality. Of course, these large columnar eddies are immersed in a multitude of smaller, disorganized structures whose small size leaves them immune to the direct influence of the Coriolis force. All in all, we would expect turbulence in a rapidly rotating system to be very different from conventional turbulence.

9.2.5 *Turbulence with moderate rotation*

When the Rossby number has a value of the order of unity, we end up with a more complex situation, with the large scales exhibiting some three-dimensional features as well as certain quasi-two-dimensional characteristics. This case is discussed in some detail by Bartello et al. (1994) who performed numerical simulations in a periodic cube. One of the more striking features of their simulations is that, when $u \sim l\Omega$, cyclonic columnar vortices are much longer lived than anti-cyclonic columnar vortices. The former are quite stable while the latter, which may be introduced into the simulation as a initial condition, are unstable and rapidly fragment into three-dimensional vortices. Indeed, quite high values of $\Omega l/u$ (greater than 10) are required before long-lived anti-cyclonic columns are observed in accordance with the Taylor–Proudman theorem.

The relative stability of cyclonic vortices, and instability of anti-cyclonic vortices, can be understood in terms of Rayleigh's circulation

[3] The situation is less clear when we have freely evolving turbulence at small Ro. As with forced turbulence, the flow evolves towards a quasi-two-dimensional state, (Bartello et al. 1994, Godeferd et al 1999). However, unlike slowly forced turbulence, there is no obvious reason why low-frequency inertial waves should be generated in preference to high-frequency waves. One way to understand how this type of flow evolves is to go back to the model problem given at the end of Section 9.2.3, which describes how a single eddy behaves.

criterion.[4] Consider an axisymmetric vortex $\mathbf{u}(r) = (0, u_\theta, 0)$ in (r, θ, z) coordinates. Let $\Gamma_r = u_\theta r$ be the angular momentum measured in the rotating frame. Then the vorticity of the vortex, also measured in a rotating frame, is $\omega_z = r^{-1}\partial\Gamma_r/\partial r$. Of course, the total angular momentum in an inertial frame of reference is $\Gamma = \Omega r^2 + \Gamma_r$, and so Rayleigh's stability criterion, in the form $(\Gamma/r)\partial\Gamma/\partial r \geq 0$, yields

$$(\Omega r^2 + \Gamma_r)(2\Omega + \omega_z) \geq 0$$

as a necessary and sufficient condition for the stability of the vortex to small-amplitude, axisymmetric disturbances. It is readily confirmed that cyclonic vortices ($\Gamma_r > 0$) generally satisfy this relationship (provided the velocity profile has a simple, smooth shape). On the other hand, anti-cyclonic vortices ($\Gamma_r < 0$) fail to satisfy our stability criterion when $|\omega_z| > 2\Omega$, that is, $|u_\theta| > 2\Omega l$. This instability condition is easily met. The tendency to observe cyclonic rather than anti-cyclonic vortices is consistent with the laboratory experiments of Hopfinger et al. (1982).

9.2.6 *From rotation to stratification (or from cigars to pancakes)*

Let us now turn from rotation to stratification. We claimed in Section 9.1 that rapid rotation and strong stratification both lead to a form of quasi-two-dimensional turbulence, but that the structure of the turbulence is very different in the two cases. Rapid rotation tends to promote columnar eddies while stratification favours pancake-like structures. Why should this be? Since we believe that columnar vortices are the result of inertial wave propagation, it is natural to suppose that the pancake eddies are caused by inertial gravity waves. Let us see if this is so.

Consider a stratified, incompressible fluid subject to small perturbations. In the unperturbed state, the density varies as

$$\rho_0(z) = \bar{\rho} + \Delta\rho_0(z) = \bar{\rho} + (d\rho_0/dz)z$$

where $\bar{\rho}$ and $d\rho_0/dz$ are both constant. For simplicity we shall assume that $\Delta\rho_0 \ll \bar{\rho}$ and adopt the Boussinesq approximation in which the small variations in density are important only to the extent that they contribute to the buoyancy force, $\rho\mathbf{g} = -\rho g\hat{\mathbf{e}}_z$. The equations governing a disturbance in our stratified fluid are then,

$$\frac{D\rho}{Dt} = 0, \qquad \nabla \cdot \mathbf{u} = 0 \quad \text{(incompressibility, continuity)}$$

[4] This states that, in an inertial frame of reference, the axisymmetric vortex $(0, u_\theta(r), 0)$ in (r, θ, z) coordinates is stable if and only if the radial gradient in $(ru_\theta)^2$ is everywhere non-negative.

$$\bar{\rho}\frac{D\mathbf{u}}{Dt} = -\nabla p + \rho \mathbf{g} \quad \text{(momentum)}$$

where viscous forces have been neglected. For small-amplitude disturbances, in which $\rho = \rho_0(z) + \delta\rho$, these equations may be linearized by neglecting terms which are quadratic in the small quantities $|\mathbf{u}|$ and $\delta\rho$. The result is

$$\frac{\partial}{\partial t}\delta\rho + u_z\frac{d\rho_0}{dz} = 0, \qquad \nabla \cdot \mathbf{u} = 0$$

$$\bar{\rho}\frac{\partial \mathbf{u}}{\partial t} = -\nabla(\delta p) + \delta\rho\mathbf{g}.$$

The unknown pressure perturbation may be eliminated by taking the curl of the linearized momentum equation. It is also convenient to take a second time derivative at the same time. This yields

$$\bar{\rho}\frac{\partial^2 \boldsymbol{\omega}}{\partial t^2} = \nabla\left(\frac{\partial \delta\rho}{\partial t}\right) \times \mathbf{g} = -\nabla\left(u_z\frac{d\rho_0}{dz}\right) \times \mathbf{g}.$$

To obtain a wave equation, we take the curl once more. The end result is

$$\frac{\partial^2}{\partial t^2}\nabla^2\mathbf{u} + N^2[\nabla^2(u_z\hat{\mathbf{e}}_z) - \nabla(\partial u_z/\partial z)] = 0 \tag{9.22}$$

where,

$$N^2 = -\frac{g}{\bar{\rho}}\frac{d\rho_0}{dz}. \tag{9.23}$$

The quantity N is referred to as the *Väisälä–Brunt frequency*. As we shall see, it provides an upper limit on the frequency of internal gravity waves. It is conventional to work with the z-component of this wave equation

$$\frac{\partial^2}{\partial t^2}\nabla^2 u_z + N^2\nabla_\perp^2 u_z = 0. \tag{9.24}$$

We see immediately that there is a similarity to the governing equation for inertial waves (9.13), the z-component of which is

$$\frac{\partial^2}{\partial t^2}\nabla^2 u_z + (2\Omega)^2\nabla_{||}^2 u_z = 0.$$

(Here $\perp$ and $||$ indicate components in the x–y plane and along the z-axis, respectively.) This is the first hint that there is a close relationship between inertial waves and internal gravity waves. In any event, we can obtain the dispersion relationship for gravity waves by looking for plane-wave solutions of the type $u_z = \hat{u}_z \exp[j(\mathbf{k}\cdot\mathbf{x} - \varpi t)]$, which yields

$$\varpi^2 = N^2 k_\perp^2 / k^2. \tag{9.25}$$

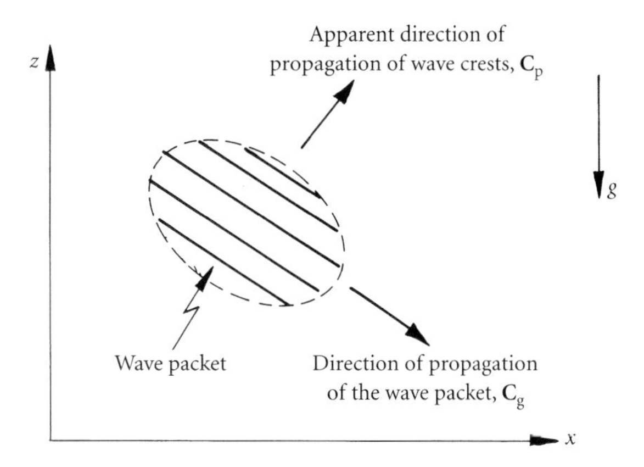

Figure 9.9 The relative orientation of the group and phase velocities in a gravity wave.

Evidently, the maximum possible frequency occurs when $k_{\|}=0$ and this corresponds to $\varpi_{\max}=N$. The phase and group velocities are readily shown to be

$$\mathbf{C}_p = \frac{Nk_\perp \mathbf{k}}{k^3} \tag{9.26}$$

$$\mathbf{C}_g = \mp \frac{N}{k^3 k_\perp}[\mathbf{k} \times (\mathbf{k}_{\|} \times \mathbf{k})] = \pm\left[\frac{N\mathbf{k}_\perp}{kk_\perp} - \mathbf{C}_p\right]. \tag{9.27}$$

We see immediately that the phase and group velocities are perpendicular (Figure 9.9), so that a wave packet propagates at right angles to the apparent direction of advancement of the wave crests. This is an astonishing property which gravity waves share with inertial waves. Moreover, like inertial waves, the wave frequency depends only on the direction of **k** and not on its magnitude. The main difference between the two types of waves is that $\varpi \sim k_\perp/k$ for gravity waves, whereas $\varpi \sim k_{\|}/k$ for inertial waves, so the direction of propagation of energy is different for the two cases.

Consider, for example, the situation where we have low-frequency gravity waves, $\varpi \ll N$, $k_\perp \ll k$. Then the phase velocity is vertical while the group velocity (the direction of propagation of energy) is horizontal. So low-frequency disturbances propagate in the x–y plane. This contrasts with low-frequency inertial waves which propagate energy in the z-direction (Figure 9.10). So, in a strongly stratified fluid (a fluid in which N^{-1} is much smaller that the eddy turn-over time), the large eddies spread in the horizontal plane, forming pancake-like structures (Figure 9.11).

Thus, eddies in a strongly stratified fluid adopt a quasi-two-dimensional form, but it is very different to the quasi-two-dimensional turbulence associated with a rapidly rotating fluid. The eddies are flat and disc-like, rather than long and columnar.

A review of stratified turbulence is given by Riley and Lelong (2000). When the stratification is strong, $u/Nl < 1$, the turbulence

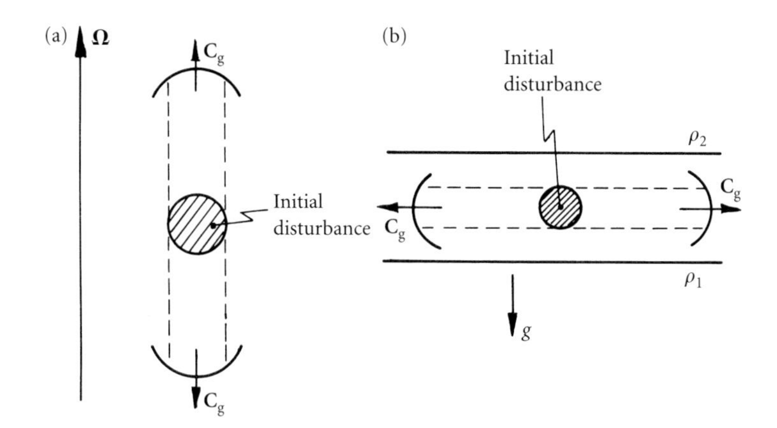

Figure 9.10 A comparison of the propagation of energy in low-frequency (a) inertial and (b) internal gravity waves.

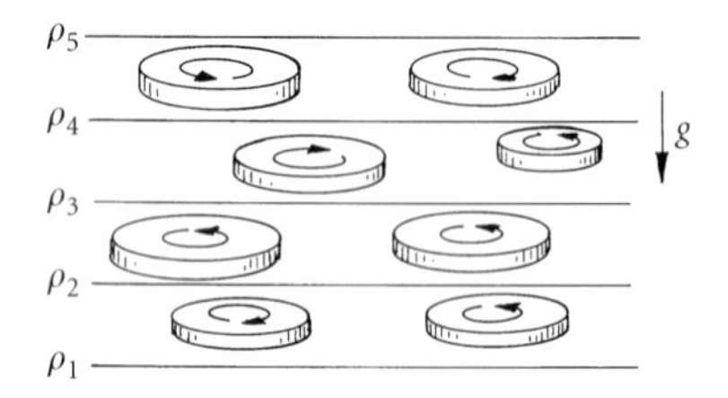

Figure 9.11 Schematic representation of turbulence in a strongly stratified fluid.

organizes itself into partially decoupled horizontal strata (layers of pancakes). The vertical velocity component is small and the flow in each layer is approximately two dimensional. However, the decoupling of the horizontal strata is incomplete, since intense vertical gradients in velocity form as pancake-like eddies slide over each other. This generates an intense shear between layers, which in turn promotes small-scale turbulence and mixing between horizontal strata. Although the primary velocity components are horizontal, which are associated with a vertical component of vorticity, the strong shear between horizontal strata leads to an intense horizontal component of vorticity. Indeed, it is the horizontal vorticity which dominates the mean dissipation. That is, most of the dissipation of energy arises from pancake eddies sliding over each other. All in all, this quasi-two-dimensional turbulence is very different to that associated with rapid rotation.

Example 9.6 A Loitsyansky-type invariant for stratified turbulence
Consider stratified turbulence evolving in a spherical domain, V, whose radius is very much larger than the integral scale. Show that, when viscous stresses on the boundary are ignored, the vertical component of global angular momentum, H_z, is conserved. Now use Landau's argument of Section 6.3.3, in conjunction with equation (6.88), to show that, when long-range interactions are weak, and the bulk of the turbulence is homogeneous,

$$\langle H_z^2 \rangle / V = -\int (r_x^2 + r_y^2)\left[\langle u_x u_x' \rangle + \left\langle u_y u_y' \right\rangle\right] d\mathbf{r} = \text{constant}.$$

Thus, homogeneous, stratified turbulence has a Loitsyansky-like invariant in cases where the long-range interactions are weak.

Now show that exactly the same result follows directly from (6.48b), generalized to include buoyancy forces. In cases where the large scales evolve in a self-similar fashion, this invariant implies that

$$u^2 l_\perp^4 l_\| = \text{constant}.$$

9.3 The influence of magnetic fields I—the MHD equations

Gladstone: What use is this electricity?
Faraday: Why sir, there is every possibility that you will soon be able to tax it!

We now set aside buoyancy and Coriolis forces and turn our attention to the Lorentz force. In particular, we consider the turbulent motion of an electrically conducting (but non-magnetic) fluid in the presence of a magnetic field. The study of the interaction of a conducting fluid and a magnetic field is called magnetohydrodynamics (MHD). It is a vast subject in its own right and we can give only a hint of the issues involved in the few pages available to us here. For a more detailed discussion, see Shercliff (1965), Moffatt (1978), Biskamp (1993), or Davidson (2001).

Our survey of MHD turbulence comes in three parts. First, in Section 9.3, we sketch out the basic equations and theorems of MHD. Next, in Section 9.4, we return to turbulence. Here we shall see that, as in stratified and rotating flows, eddies are shaped by wave motion, the wave in question being the so-called Alfvén wave. We close in Section 9.5 with a discussion of the role of MHD turbulence in solar phenomena and in geodynamo theory. The recurring theme throughout is that a magnetic field (usually just referred to as *the field*) acts to dissipate energy and to shape the turbulence by extruding eddies along the magnetic field lines.

9.3.1 *The interaction of moving conductors and magnetic fields: a qualitative overview*

The mutual interaction of a magnetic field, **B**, and a moving conducting fluid arises for the following reason. Relative movement of the magnetic field and the fluid causes an e.m.f., of the order $\mathbf{u} \times \mathbf{B}$, to develop. Ohm's law tells us that this e.m.f. drives a current, whose density in space is of the order of $\sigma\mathbf{u} \times \mathbf{B}$, where σ is the electrical conductivity of the fluid. This current has two effects.

Effect 1: the action of **u** *on* **B**.
The induced current gives rise to a second magnetic field. This adds to the original magnetic field and the change is such that the fluid appears to 'drag' the magnetic field lines along with it.

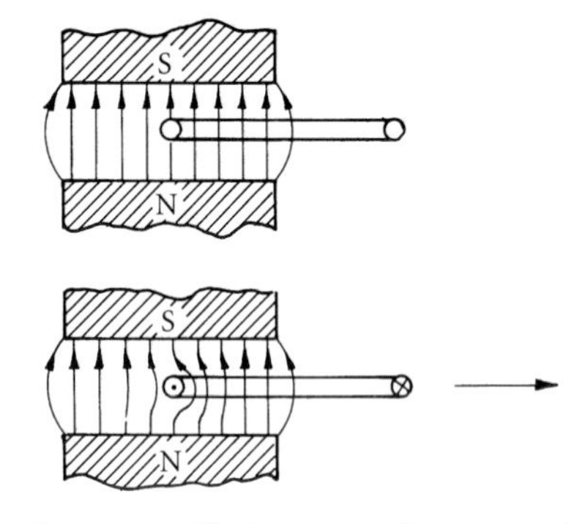

Figure 9.12 The interaction of a magnet and a moving wire loop.

Effect 2: the back reaction of **B** *on* **u**.
The combined magnetic field interacts with the induced current density, **J**, to give rise to a Lorentz force per unit volume, **J** × **B**. This acts on the fluid and is such as to oppose the relative motion.

These two effects are also evident in conventional electrodynamics. Consider, for example, the situation shown in Figure 9.12 where a wire loop is being dragged through a magnetic field. Current is induced in the wire by the relative movement of the wire and magnet. This current distorts the magnetic field as shown (the wire seems to drag the magnetic field with it) and interacts with **B** to produce a Lorentz force which opposes the relative motion.

It is conventional to divide MHD into two sub-categories: the study of highly conducting fluids and the study of poorly conducting fluids. This subdivision is not arbitrary. It turns out that the dominant physical processes are very different in the two cases. Of course, one immediate question is: 'highly conducting relative to what?' The convention, which we will see has a sound physical basis, is the following. The product of the electrical conductivity, σ, with the permeability of free space, μ, has the dimensions of $\mathrm{m}^{-2}\,\mathrm{s}$. It follows that $\lambda = (\sigma\mu)^{-1}$ has the dimension of a diffusivity and indeed λ is referred to as the *magnetic diffusivity*. We can form a dimensionless group containing λ which is analogous to the Reynolds number:

$$R_\mathrm{m} = ul/\lambda = \mu\sigma\mu l. \tag{9.28}$$

This is referred to as the *magnetic Reynolds number* and it provides a convenient dimensionless measure of σ. The convention, therefore, is to divide MHD into the study of high- and low-R_m phenomena. Virtually all terrestrial MHD is of the low-R_m type. This includes most laboratory experiments and metallurgical processes. In astrophysics, on the other hand, R_m is invariably vast, largely because of the enormous length-scales involved.

We can gain some appreciation of the significance of R_m from the dynamics of the wire loop shown in Figure 9.12. First, however, we need to introduce some elementary concepts from electromagnetism. They are the laws of Ohm, Faraday, and Ampere. Let us take these one at a time.

Ohm's equation is an empirical law which relates the current density, **J**, in a conducting medium to the electric field, **E**. For a stationary conductor it states that $\mathbf{J} = \sigma\mathbf{E}$, where σ is the electrical conductivity. We may interpret this as **J** being proportional to the Coulomb force $\mathbf{f} = q\mathbf{E}$, which acts on the free charge carriers, q being their charge. If, however, the conductor is moving with velocity **u** in a magnetic field **B**, the free charges experience an additional force $\mathbf{f} = q\mathbf{u} \times \mathbf{B}$, and so Ohm's law is modified to read,

$$\mathbf{J} = \sigma(\mathbf{E} + \mathbf{u} \times \mathbf{B}). \tag{9.29}$$

The quantity $\mathbf{E} + \mathbf{u} \times \mathbf{B}$, which is the net electromagnetic force per unit charge, is often called the *effective electric field*, since it is the electric field which would be measured in a frame of reference moving with velocity $\mathbf{u}$. We use the symbol $\mathbf{E}_r$ to represent $\mathbf{E} + \mathbf{u} \times \mathbf{B}$. Thus, the force law and Ohm's law may be rewritten as

$$\mathbf{f} = q(\mathbf{E} + \mathbf{u} \times \mathbf{B}) = q\mathbf{E}_r, \qquad \mathbf{J} = \sigma(\mathbf{E} + \mathbf{u} \times \mathbf{B}) = \sigma\mathbf{E}_r.$$

Faraday's law, on the other hand, tells us about the e.m.f. which is generated in a conducting medium as a result of: (i) a time-dependent magnetic field; or (ii) motion of the conductor within the magnetic field. In either case, it takes the form:

$$\oint_C \mathbf{E}_r \cdot d\mathbf{r} = \oint_C \left[\mathbf{E} + \mathbf{v} \times \mathbf{B}\right] \cdot d\mathbf{r} = -\frac{d}{dt}\int_S \mathbf{B} \cdot d\mathbf{S}. \tag{9.30}$$

Here C is a closed curve composed of line elements $d\mathbf{r}$ and S is any surface which spans the curve C. (The relative sense of $d\mathbf{r}$ and $d\mathbf{S}$ is determined by the right-hand convention.) The velocity $\mathbf{v}$ in (9.30) is the velocity of the line element $d\mathbf{r}$, so that $\mathbf{E}_r$ is the effective electric field as measured in a frame of reference moving with $d\mathbf{r}$. Now the e.m.f. around a closed curve C is defined as $\oint_C \mathbf{E}_r \cdot d\mathbf{r}$, and so Faraday's law tells us that an e.m.f. in induced around a curve C whenever there is a net rate of change of magnetic flux, $\Phi = \int \mathbf{B} \cdot d\mathbf{S}$, through any surface which spans C; e.m.f. $= -\,d\Phi/dt$. Since $\mathbf{B}$ is solenoidal, it is immaterial which surface is considered. Note that C may be stationary, move with the conducting medium, or execute some other motion. It does not matter. For the particular case where C is a material curve (i.e. it moves with the medium), Faraday's and Ohm's laws combine to tell us that the induced e.m.f. will drive a current through the conductor according to

$$\oint_{C_m} \mathbf{J} \cdot d\mathbf{r} = -\sigma\frac{d}{dt}\int_S \mathbf{B} \cdot d\mathbf{S} = -\sigma\frac{d\Phi}{dt}. \tag{9.31}$$

Here the subscript 'm' on C_m indicates a material curve.

Finally, we require Ampere's law. This tells us about the magnetic field associated with a given distribution of current, $\mathbf{J}$. It may be written,

$$\oint_C \mathbf{B} \cdot d\mathbf{r} = \mu\int_S \mathbf{J} \cdot d\mathbf{S}. \tag{9.32a}$$

From Stoke's theorem we see that this integral equation is equivalent to the differential expression:

$$\nabla \times \mathbf{B} = \mu\mathbf{J}. \tag{9.32b}$$

Thus, Ampere's law may be converted into an equation of the form $\mathbf{B}=f(\mathbf{J})$ using the Bio–Savart law (2.28):

$$\mathbf{B}(\mathbf{x})=\frac{\mu}{4\pi}\int\frac{\mathbf{J}'\times\mathbf{r}}{r^3}d\mathbf{x}',\quad \mathbf{r}=\mathbf{x}-\mathbf{x}'. \tag{9.32c}$$

This reveals the true nature of Ampere's law: given the distribution of $\mathbf{J}'=\mathbf{J}(\mathbf{x}')$, it allows us to calculate $\mathbf{B}(\mathbf{x})$.

Let us now return to the wire loop shown in Figure 9.13. In particular, we are interested in the behaviour of the loop when subject to an impulsive force. We can estimate the magnitude of the current and magnetic field induced by movement of the loop as follows. From Ohm's law we have $|\mathbf{J}|\sim\sigma|\mathbf{u}\times\mathbf{B}_0|$, where $\mathbf{B}_0$ is the irrotational magnetic field set up by the magnet and $\mathbf{u}$ is the instantaneous velocity of the loop. (We know that $\mathbf{B}_0$ is irrotational since $\nabla\times\mathbf{B}=\mu\mathbf{J}$ and, in the absence of the wire loop, there are no currents in the air gap between the magnet poles.) The magnetic field associated with the current density $\mathbf{J}$ can be estimated using Ampere's law, or equivalently the Biot–Savart law. Noting that $\mathbf{B}_0$ does not contribute to either (9.32a) or (9.32c), we have $|\mathbf{B}_{\text{IN}}|\sim\mu|\mathbf{J}|l$, where $\mathbf{B}_{\text{IN}}$ is the induced magnetic field and l is some characteristic geometric length-scale. Combining this with our estimate of $\mathbf{J}$, we have,

$$|\mathbf{B}_{\text{IN}}|\sim\sigma\mu ul|\mathbf{B}_0|=R_{\text{m}}|\mathbf{B}_0|. \tag{9.33}$$

The significance of R_{m} now becomes clear. If R_{m} is small, the induced magnetic field is negligible by comparison with the imposed one. In effect, this arises from the fact that a low electrical conductivity implies a weak induced current, and hence a weak induced magnetic field. When R_{m} is large, $\mathbf{B}_{\text{IN}}$ is non-negligible. Actually, for large R_{m} it turns out that (9.33) is invalid because there is a large degree of cancellation between $\mathbf{E}$ and $\mathbf{u}\times\mathbf{B}$ in Ohm's law, and so the estimate $|\mathbf{J}|\sim\sigma|\mathbf{u}\times\mathbf{B}_0|$ is incorrect. For large conductivities we usually replace estimate (9.33) by $|\mathbf{B}_{\text{IN}}|\sim|\mathbf{B}_0|$, so that the induced magnetic field cannot be neglected. More importantly, (9.31) tells us that, as

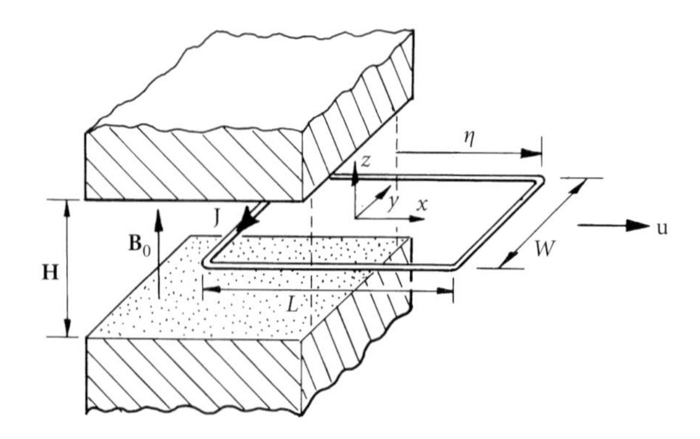

Figure 9.13 The geometry of the wire loop.

$\sigma \to \infty$, $d\Phi/dt \to 0$, since near-infinite currents are physically unacceptable. So, if the wire loop is made of a superconducting material, the magnetic flux passing through the loop cannot change as it moves. This is why, in Figure 9.12, the wire appears to drag the magnetic field lines along with it.

Let us now consider the cases of high and low R_m separately. We start with low R_m. Suppose the wire loop sits on a frictionless horizontal surface. At $t = 0$ we give it a tap and we wish to determine its subsequent motion. Since R_m is low, $\mathbf{B}_{IN}$ is small and so, from (9.29) or (9.31), the current induced in the wire is of order $J \sim \sigma B_0 u$, where u is the speed of the wire loop. The resulting Lorentz force per unit volume, $\mathbf{F} = \mathbf{J} \times \mathbf{B}$, acts to retard the motion and is of order $F \sim \sigma B_0^2 u$. It is now a simple task to show that u declines according to

$$\frac{du}{dt} \sim -\frac{u}{\tau}, \quad \tau = \left(\sigma B_0^2/\rho\right)^{-1} \tag{9.34a}$$

where ρ is the density of the wire. Evidently, if we give the wire loop a tap, it will move forward. However, its linear momentum decreases exponentially on a timescale of τ. This timescale is extremely important in low-R_m MHD and is called the *magnetic damping time*. Clearly the induced Lorentz force acts to oppose the relative motion of the conductor and the magnet and it is a simple matter to show that the rate at which the wire loop loses kinetic energy is exactly matched by the rate of rise of thermal energy caused by Ohmic heating. So, at low R_m, the primary role of the Lorentz force is to convert mechanical energy into heat.

Consider now the case where R_m is large ($\sigma \to \infty$). The key point to note here is that, from (9.31), Φ is conserved as the loop moves forward. If η marks the position of the front of the wire and the loop has length L and width W (Figure 9.13), then

$$\Phi = B_0 W(L - \eta) + \bar{B}_{IN} WL = \Phi_0 = B_0 W(L - \eta_0).$$

Here $\bar{B}_{IN}$ is the mean vertical component of $\mathbf{B}_{IN}$ within the loop, and Φ_0 and η_0 are the initial values of Φ and η. Now we know from Ampere's law that $\bar{B}_{IN} = \mu l J$, where l is some geometric length-scale whose exact value need not concern us. It follows that

$$\bar{B}_{IN} = \mu l J = \frac{\eta - \eta_0}{L} B_0.$$

We now assume that $(\eta - \eta_0) \ll L$, use the expression above to evaluate the Lorentz force $F_x = -JB_0$, and apply Newton's second law to the loop. After a little algebra we find that,

$$\frac{d^2}{dt^2}(\eta - \eta_0) + \frac{\mathrm{v}_a^2}{h^2}(\eta - \eta_0) = 0$$

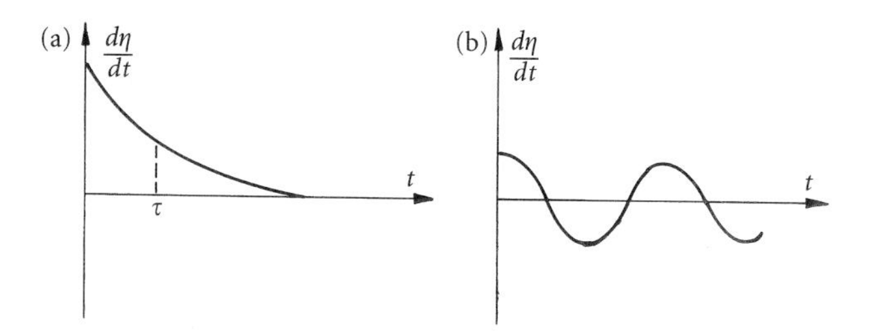

Figure 9.14 Response of the wire loop to an impulsive force, (a) low R_m, (b) high R_m.

where $h^2 = 2lL(1 + L/W)$ and v_a is defined as

$$v_a = B_0/(\rho\mu)^{1/2}. \tag{9.34b}$$

The quantity v_a is extremely important in high-R_m MHD. It is called the *Alfvén velocity*.

It seems that the behaviour of our wire loop is quite different when R_m is large. Rather than grind to a halt on a timescale of τ, the loop oscillates, with a frequency proportional to the Alfvén velocity. There is no dissipation of kinetic energy this time. Instead, the magnetic field behaves as if it is an elastic spring holding the wire loop in place. This is shown in Figure 9.14.

Let us summarize our findings. We have seen that three important quantities in MHD are the magnetic Reynolds number, the Alfvén velocity, and the magnetic damping time. These are tabulated below.

Magnetic Reynolds number	$R_m = \mu\sigma ul = ul/\lambda$
Alfvén velocity	$v_a = B_0/(\rho\mu)^{1/2}$
Magnetic damping time	$\tau = (\sigma B_0^2/\rho)^{-1}$

The behaviour of a conducting body in a magnetic field depends critically on the value of R_m. When R_m is low, the magnetic field is dissipative. If the wire loop is given a tap, it slides forward as if it is immersed in treacle. The induced Lorentz force opposes the motion, converting kinetic energy into heat via Ohmic dissipation. Also, at low R_m, the induced magnetic field is negligible.

When R_m is large, on the other hand, the field is non-dissipative and acts like an elastic spring anchoring the wire loop in place. When given a tap, the loop oscillates back and forth with a frequency proportional to the Alfvén velocity. Also, the magnetic flux passing through the wire does not change with time.

We shall see shortly that the response of a conducting fluid to an imposed magnetic field is similar to that of the wire. When R_m is large, disturbances in the fluid set up elastic oscillations which propagate as waves at a speed v_a. When R_m is low, on the other hand, the magnetic field is dissipative in nature. It opposes the motion while converting kinetic energy into heat, the heating rate per unit volume being J^2/σ.

9.3.2 *From Maxwell's equations to the governing equations of MHD*

We now set out the governing equations of MHD. These consist of the Navier–Stokes equation plus a reduced form of Maxwell's equations. As before, we restrict the discussion to incompressible fluids. Let us start with Maxwell's equations.

The governing equations of electrodynamics for materials where polarization or magnetization is not a factor are:

1. Ohm's law, $\mathbf{J} = \sigma(\mathbf{E} + \mathbf{u} \times \mathbf{B})$. (9.35)
2. Charge conservation, $\nabla \cdot \mathbf{J} = -\partial\rho_e/\partial t$. (9.36)
3. Force law for a point charge, $\mathbf{F} = q(\mathbf{E} + \mathbf{u} \times \mathbf{B})$. (9.37)
4. Gauss' law, $\nabla \cdot \mathbf{E} = \rho_e/\varepsilon_0$. (9.38)
5. Solenoidal nature of $\mathbf{B}$, $\nabla \cdot \mathbf{B} = 0$. (9.39)
6. Faraday's law in differential form, $\nabla \times \mathbf{E} = -\dfrac{\partial \mathbf{B}}{\partial t}$. (9.40)
7. Ampere–Maxwell equation, $\nabla \times \mathbf{B} = \mu\left[\mathbf{J} + \varepsilon_0 \dfrac{\partial \mathbf{E}}{\partial t}\right]$. (9.41)

Here ρ_e is the charge density and ε_0 is the permitivity of free space. The last four equations are known collectively as Maxwell's equations. Not all of these equations are independent. For example, taking the divergence of (9.40) gives $\nabla \cdot (\partial\mathbf{B}/\partial t) = 0$, from which (with suitable initial conditions) we can deduce (9.39). Similarly, taking the divergence of (9.41) and invoking Gauss' law, we have

$$\nabla \cdot \mathbf{J} = -\varepsilon_0 \nabla \cdot [\partial\mathbf{E}/\partial t] = -\frac{\partial\rho_e}{\partial t}$$

which just expresses the law of conservation of charge. It turns out that, in MHD, we can simplify these equations considerably. In particular, ρ_e, when put into appropriate dimensionless form, turns out to be minute. So, in MHD, we work with a reduced set of equations that are equivalent to assuming a small but finite charge density.

The way in which (9.35)–(9.41) are simplified in the limit of small ρ_e is rather subtle. The details are spelt out in Shercliff (1965) or Davidson (2001). Here we merely note the end result. It transpires that (9.36) simplifies to $\nabla \cdot \mathbf{J} = 0$, which perhaps is not so surprising. Also, the second term on the right of (9.41), the so-called displacement current, is omitted. (We can see that this must be so by taking the divergence of (9.41), invoking Gauss' law, and comparing the result with our simplified version of (9.36).) The final simplification comes from the force law (9.37). When this is integrated over a unit volume of conductor, it becomes, $\mathbf{F} = \rho_e\mathbf{E} + \mathbf{J} \times \mathbf{B}$. Since ρ_e is negligible, the net electromagnetic body force simplifies to $\mathbf{F} = \mathbf{J} \times \mathbf{B}$. We are now in a position to write down the governing equations of MHD. They are:

1. Ohm's law plus the Lorentz force:

$$\mathbf{J} = \sigma(\mathbf{E} + \mathbf{u} \times \mathbf{B}) \tag{9.42}$$

$$\mathbf{F} = \mathbf{J} \times \mathbf{B}. \tag{9.43}$$

2. Faraday's law in differential form plus the solenoidal constraint on $\mathbf{B}$:

$$\nabla \times \mathbf{E} = -\frac{\partial \mathbf{B}}{\partial t} \tag{9.44}$$

$$\nabla \cdot \mathbf{B} = 0. \tag{9.45}$$

3. Ampere's law plus charge conservation:

$$\nabla \times \mathbf{B} = \mu \mathbf{J} \tag{9.46}$$

$$\nabla \cdot \mathbf{J} = 0. \tag{9.47}$$

To these, of course, we must add the Navier–Stokes equation in which $\mathbf{J} \times \mathbf{B}$ appears as a body force per unit volume. Note that (9.45) and (9.47) follow from Faraday's and Ampere's laws, since $\nabla \cdot \nabla \times (\sim) = 0$. Note also that Gauss' law is omitted in the MHD approximation, since it merely specifies ρ_e, which is a small quantity whose distribution is of no interest to us. In MHD, the divergence of $\mathbf{E}$ is determined from (9.42).

Often Ampere's law and Faraday's law appear in integral form and indeed this is how we introduced them in the previous section. Clearly the integral version of (9.46) is simply (9.32a). The relationship between (9.30) and (9.44) is, however, a little less obvious. It may be established as follows. In Section 2.3.3 we saw that if $\mathbf{G}$ is any vector field which permeates a fluid, and S_m is any open material surface embedded in that fluid, then a kinematic analysis tells us that

$$\frac{d}{dt}\int_{S_m} \mathbf{G} \cdot d\mathbf{S} = \int_{S_m} \left[\partial \mathbf{G}/\partial t - \nabla \times (\mathbf{u} \times \mathbf{G})\right] \cdot d\mathbf{S}. \tag{9.48}$$

(Here $\mathbf{u}$ is the velocity of the fluid which is also the velocity of any point on the material surface S_m.) However, the differential form of Faraday's law tells us that:

$$\nabla \times [\mathbf{E} + \mathbf{u} \times \mathbf{B}] = -\{\partial \mathbf{B}/\partial t - \nabla \times (\mathbf{u} \times \mathbf{B})\}. \tag{9.49}$$

Combining these two expressions (with $\mathbf{G} = \mathbf{B}$), and integrating (9.49), yields

$$\oint_{C_m} [\mathbf{E} + \mathbf{u} \times \mathbf{B}] \cdot d\mathbf{r} = -\frac{d}{dt}\int_{S_m} \mathbf{B} \cdot d\mathbf{S}$$

where C_m is the bounding curve for the material surface S_m. Finally we introduce the effective electric field $\mathbf{E}_r = \mathbf{E} + \mathbf{u} \times \mathbf{B}$, and define the e.m.f. to be the closed line integral of $\mathbf{E}_r$ around C_m. This yields the integral version of Faraday's law:

$$\text{e.m.f.} = \oint_{C_m} \mathbf{E}_r \cdot d\mathbf{r} = -\frac{d}{dt}\int_{S_m} \mathbf{B} \cdot d\mathbf{S}. \tag{9.50}$$

Actually (9.50) applies to any surface, S, with bounding curve C. The surface S could be stationary, move with the fluid, or execute some motion different to that of the fluid. It does not matter. It is necessary only to ensure that $\mathbf{E}_r$ is evaluated using the velocity $\mathbf{v}$ of the line element $d\mathbf{r}$, $\mathbf{E}_r = \mathbf{E} + \mathbf{v} \times \mathbf{B}$.

Returning now to the differential form of Faraday's law, we may eliminate $\mathbf{E}$ using Ohm's law and then $\mathbf{J}$ using Ampere's law. The end result is rather beautiful. We obtain an evolution equation for $\mathbf{B}$ (sometimes called the induction equation) which has a familiar form:

$$\frac{\partial \mathbf{B}}{\partial t} = \nabla \times [\mathbf{u} \times \mathbf{B}] + \lambda \nabla^2 \mathbf{B}. \tag{9.51}$$

Here λ is the magnetic diffusivity $(\mu\sigma)^{-1}$. It seems that vorticity and magnetic fields evolve in identical ways, except that they have different diffusivities. This suggests that, for a perfectly conducting fluid, there should be MHD analogous of: (i) Kelvin's theorem; and (ii) the 'frozen-in' property of vortex-lines (Figure 9.15). It turns out that there are. For a perfect conductor ($\lambda = 0$), we may show that:

Theorem 1:

$$\frac{d}{dt}\int_{S_m} \mathbf{B} \cdot d\mathbf{S} = 0 \tag{9.52a}$$

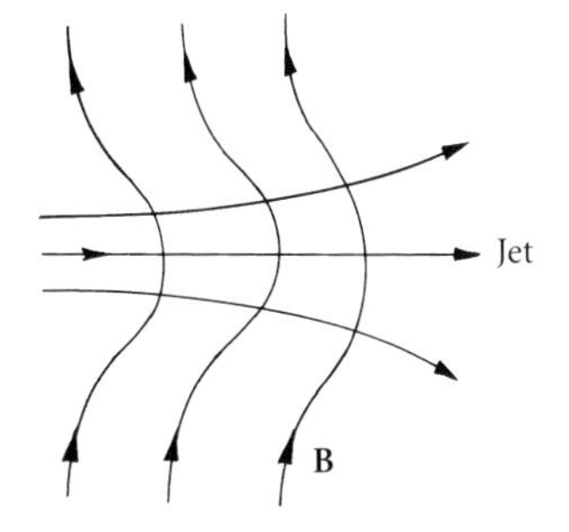

Figure 9.15 An illustration of Theorem 2. Flow through a magnetic field causes the field lines to bow out.

Theorem 2:

The **B**-lines are 'frozen' into the fluid (9.52b)

where, as always, S_m is a material surface (a surface which moves with the fluid).

Theorem 1, which is the analogue of Kelvin's theorem, follows directly from Faraday's law in integral form:

$$\oint_{C_m} \mathbf{E}_r \cdot d\mathbf{r} = \frac{1}{\sigma}\oint_{C_m} \mathbf{J} \cdot d\mathbf{r} = -\frac{d}{dt}\int_{S_m} \mathbf{B} \cdot d\mathbf{S}. \tag{9.53}$$

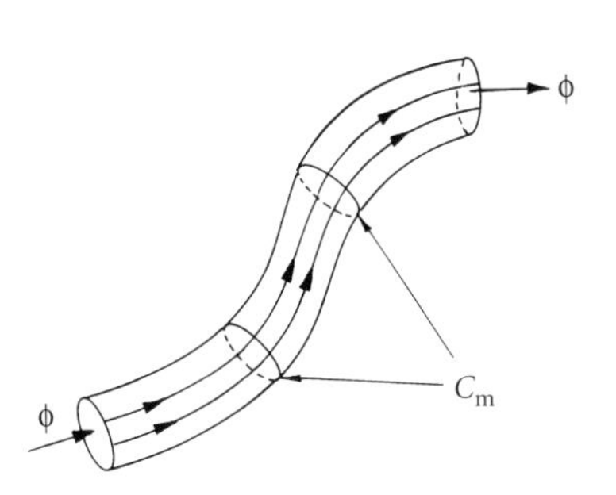

Figure 9.16 A flux tube sitting in a perfectly conducting fluid.

As $\sigma \to \infty$ (with $|\mathbf{J}|$ finite), the line integral on the left tends to zero, which yields (9.52a). Theorem 2 follows from the first if we consider Figure 9.16. This shows a magnetic flux tube (the analogue of a streamtube) sitting in a perfectly conducting fluid. Since $\mathbf{B}$ is solenoidal, the flux of $\mathbf{B}$ along the tube, Φ, is constant. Now consider a material curve C_m, which at some initial instant, $t = 0$, encircles the flux tube. The flux enclosed by C_m is equal to Φ at $t = 0$ and from

Theorem 1, this is true at all subsequent times. That is, as C_m is swept around by the flow, being stretched and twisted by the motion, it always encircles the same magnetic flux. This is also true of each and every material curve which encloses the tube at $t=0$. The only way this can be realized for all possible velocity fields is if the flux tube, like the material curve C_m, is frozen into the fluid, so that they execute the same motion. If we now let the flux tube have vanishingly small cross-section, we recover Theorem 2. (The more mathematically inclined reader may wish to try and devise a direct proof of Theorem 2 from (2.39) and (9.51).)

We close this section by deriving an energy equation for MHD. If we take the dot product of **u** with the Navier–Stokes equation, we obtain

$$\frac{D}{Dt}\left(\frac{\rho\mathbf{u}^2}{2}\right) = -\nabla\cdot(p\mathbf{u}) + \nabla\cdot(u_i\tau_{ij}) - 2\rho\nu S_{ij}S_{ij} + (\mathbf{J}\times\mathbf{B})\cdot\mathbf{u}$$

where S_{ij} is the strain-rate tensor and $\tau_{ij}=2\rho\nu S_{ij}$ is the viscous stress. However, the rate of working of the Lorentz force may be rewritten (with the help of Ohm's law) as

$$(\mathbf{J}\times\mathbf{B})\cdot\mathbf{u} = -\mathbf{J}\cdot(\mathbf{u}\times\mathbf{B}) = \mathbf{J}\cdot(\mathbf{E}-\mathbf{J}/\sigma)$$

and so the rate of change of kinetic energy is,

$$\frac{D}{Dt}\left(\frac{\rho\mathbf{u}^2}{2}\right) = -\nabla\cdot(p\mathbf{u}) + \nabla\cdot(u_i\tau_{ij}) - 2\rho\nu S_{ij}S_{ij} + \mathbf{J}\cdot\mathbf{E} - \mathbf{J}^2/\sigma.$$

We shall return to this in a moment. In the meantime we note that the dot product of **B** with Faraday's law yields,

$$\frac{\partial}{\partial t}\left(\frac{\mathbf{B}^2}{2}\right) = -\mathbf{B}\cdot\nabla\times\mathbf{E} = -\mathbf{E}\cdot\nabla\times\mathbf{B} - \nabla\cdot(\mathbf{E}\times\mathbf{B}).$$

Substituting for $\nabla\times\mathbf{B}$ using Ampere's law gives us an expression for the rate of change of magnetic energy density, $\mathbf{B}^2/2\mu$,

$$\frac{\partial}{\partial t}\left(\frac{\mathbf{B}^2}{2\mu}\right) = -\mathbf{J}\cdot\mathbf{E} - \nabla\cdot[(\mathbf{E}\times\mathbf{B})/\mu].$$

Finally we add our magnetic and kinetic energy equations, eliminating $\mathbf{J}\cdot\mathbf{E}$ in the process:

$$\begin{aligned}\frac{\partial}{\partial t}\left[\frac{\rho\mathbf{u}^2}{2}+\frac{\mathbf{B}^2}{2\mu}\right] = &-\nabla\cdot[p\mathbf{u}] + \nabla\cdot[u_i\tau_{ij}] - \nabla\cdot\left[\left(\tfrac{1}{2}\rho\mathbf{u}^2\right)\mathbf{u}\right] \\ &- \nabla\cdot[\mathbf{E}\times\mathbf{B}/\mu] - 2\rho\nu S_{ij}S_{ij} - \mathbf{J}^2/\sigma.\end{aligned} \tag{9.54a}$$

To interpret the terms on the right-hand side of (9.54a), it is convenient to integrate the equation over some fixed control volume V

with bounding surface S. The four divergences then represent: (i) the rate of working of the pressure force on S; (ii) the rate of working of the viscous stresses on S; (iii) the rate of transport of kinetic energy across S; and (iv) the flux of magnetic energy across S, the quantity $(\mathbf{E} \times \mathbf{B})/\mu$ being called the Poynting flux vector. When V extends to include all space, the surface integrals vanish and we have:

$$\frac{dE}{dt} = \frac{d}{dt}\int_V \left[\frac{\rho \mathbf{u}^2}{2} + \frac{\mathbf{B}^2}{2\mu}\right] dV = -\int_V (\mathbf{J}^2/\sigma)\, dV - 2\rho\nu \int S_{ij}S_{ij}\, dV \tag{9.54b}$$

showing that the total electro-mechanical energy declines due to Ohmic heating (sometimes called Joule heating) and due to viscous dissipation.

This concludes our brief overview of the governing equations of MHD. Those unfamiliar with electrodynamics may feel the need for a more gentle exposition, in which case Shercliff (1965) or Davidson (2001) is recommended. We now consider the simplifications which arise in the MHD equations when R_m is particularly small or particularly large. We start with small R_m.

9.3.3 *Simplifying features of low magnetic Reynolds number MHD*

In Section 9.3.1 we saw that, when R_m is small, the induced magnetic field is negligible by comparison with the imposed magnetic field. This suggests that we can simplify the governing equations of MHD when $R_m \to 0$, and this is indeed the case. In this respect it is useful to think of $R_m \to 0$ as the limit of very small velocity.

Let $\mathbf{B}_0$ represent the steady, imposed magnetic field which would exist in a given region if $\mathbf{u} = 0$, and let $\mathbf{e}$, $\mathbf{j}$, and $\mathbf{b}$ be the infinitesimal perturbations to the fields $\mathbf{E}_0 = 0$, $\mathbf{J}_0 = 0$, and $\mathbf{B}_0$ due to a small but finite velocity field. We have, for small $|\mathbf{u}|$,

$$\nabla \times \mathbf{e} = -\partial \mathbf{b}/\partial t, \qquad \mathbf{j} = \sigma(\mathbf{e} + \mathbf{u} \times \mathbf{B}_0).$$

Faraday's equation tells us that $|\nabla \times \mathbf{e}| \sim |\mathbf{u}||\mathrm{b}|$ and so $|\nabla \times \mathbf{e}|$ is a second-order quantity. It follows that, to leading order in $|\mathbf{u}|$,

$$\mathbf{J} = \mathbf{J}_0 + \mathbf{j} = \sigma(\mathbf{e} + \mathbf{u} \times \mathbf{B}_0), \qquad \nabla \times \mathbf{e} = \mathbf{0}.$$

Thus, $\mathbf{e} = -\nabla V$, V being an electrostatic potential, and Ohm's law and the Lorentz force reduce to,

$$\mathbf{J} = \sigma(-\nabla V + \mathbf{u} \times \mathbf{B}_0), \tag{9.55a}$$

$$\mathbf{F} = \mathbf{J} \times \mathbf{B}_0. \tag{9.55b}$$

Equations (9.55a) and (9.55b) are all that we need in order to evaluate the Lorentz force in low-R_m MHD. There is no need to calculate **b** since it does not appear in the Lorentz force, and so Ampere's law is entirely redundant. Moreover, **J** is completely determined by (9.55a) since,

$$\nabla \cdot \mathbf{J} = 0, \qquad \nabla \times \mathbf{J} = \sigma \nabla \times (\mathbf{u} \times \mathbf{B}_0) \tag{9.56a}$$

and a vector field is uniquely determined if its divergence and curl are known.

Note that (9.56a) allows us to estimate the magnitude of $\mathbf{F} = \mathbf{J} \times \mathbf{B}_0$. This, in turn, leads us to a useful dimensionless group called the *interaction parameter*. Consider the case where $\mathbf{B}_0$ is uniform and let $l_{\|}$ and $l_{\perp}$ be characteristic length scales for **u** normal and perpendicular to $\mathbf{B}_0$. Sometimes $l_{\|} \sim l_{\perp}$ but more often $l_{\|} > l_{\perp}$ since, as we shall see, the magnetic field tends to elongate eddies in the direction of $\mathbf{B}_0$. Equation (9.56a) in the form,

$$\nabla \times \mathbf{J} = \sigma (\mathbf{B}_0 \cdot \nabla)\mathbf{u} \tag{9.56b}$$

tells us that $|\mathbf{J}| \sim \sigma B_0 u \, l_{\perp}/l_{\|}$ and so the curl of the Lorentz force is of the order of

$$|\nabla \times (\mathbf{J} \times \mathbf{B}_0)| = |(\mathbf{B}_0 \cdot \nabla)\mathbf{J}| \sim \frac{\sigma B_0^2 u}{l_{\perp}} (l_{\perp}/l_{\|})^2.$$

We might compare this with the curl of the inertial force $\nabla \times [\rho(\mathbf{u} \cdot \nabla)\mathbf{u}] \sim \rho u^2/l_{\perp}^2$. The ratio of the two estimates is

$$\frac{\nabla \times (\text{Lorentz force})}{\nabla \times (\text{inertial force})} \sim \frac{(l_{\perp}/u)}{\tau} \left(\frac{l_{\perp}}{l_{\|}}\right)^2$$

where τ is the magnetic damping time defined by (9.34a). In cases where $l_{\|} \sim l_{\perp} = l$, this ratio reduces to

$$N = \frac{\sigma B_0^2 l}{\rho u} = \frac{l/u}{\tau}. \tag{9.57}$$

The dimensionless parameter N (which should not be confused with the Väisälä–Brunt frequency) is referred to as the *interaction parameter* and it may be thought of as the ratio of the eddy turn-over time to the magnetic damping time. Actually, when $\mathbf{B}_0$ is uniform, $\mathbf{F} = \mathbf{J} \times \mathbf{B}_0$ has a particularly simple form. It is readily confirmed that (9.55a) and (9.55b) yield,

$$\nabla \times (\mathbf{F}/\rho) = -\frac{1}{\tau} \nabla^{-2} [\partial^2 \boldsymbol{\omega} / \partial x_{\|}^2] \tag{9.58}$$

where $x_{\|}$ is the coordinate aligned with $\mathbf{B}_0$ and ∇^{-2} is an operator defined via the Biot–Savart law. Finally we note that, in the limit

$R_m \to 0$, the energy equation (9.54b) reduces to

$$\frac{d}{dt}\int \frac{1}{2}\mathbf{u}^2 dV = -\frac{1}{\rho\sigma}\int \mathbf{J}^2 dV - 2\nu \int S_{ij}S_{ij}\, dV \tag{9.59}$$

since the magnetic energy associated with the induced field, **b**, can be ignored when R_m is small.

9.3.4 *Simple properties of high magnetic Reynolds number MHD*

We now turn to high-R_m MHD. The two topics we wish to touch on are: (i) integral invariants of ideal MHD and (ii) Alfvén waves. Let us start with the invariants. When $\lambda = 0$ (a so-called perfect conductor), the evolution equation for **B** reduces to

$$\frac{\partial \mathbf{B}}{\partial t} = \nabla \times (\mathbf{u} \times \mathbf{B}). \tag{9.60}$$

We have already seen that two properties of this equation are: (i) $(d/dt)\int_{S_m} \mathbf{B} \cdot d\mathbf{S} = 0$, and (ii) the **B**-lines are frozen into the fluid.

These are the MHD analogues of (i) Kelvin's theorem and (ii) the 'frozen-in' property of $\boldsymbol{\omega}$-lines in an inviscid fluid. It is property (ii) which gives high-R_m MHD its distinctive character. There is one other result we can borrow from conventional hydrodynamics. In Exercise 2.7 of Chapter 2, we saw that, when $\nu = 0$, an ideal fluid possesses an integral invariant called helicity. This is defined as

$$H_\omega = \int_{V_\omega} \mathbf{u} \cdot \boldsymbol{\omega}\, dV$$

where V_ω is any material volume which encloses the vortex lines. The equivalent result in electrodynamics is that, when $\lambda = 0$,

$$\frac{d}{dt}\int_{V_B} \mathbf{B} \cdot \mathbf{A}\, dV = 0 \tag{9.61}$$

where **A** is the vector potential for **B** which satisfies $\nabla \times \mathbf{A} = \mathbf{B}$, $\nabla \cdot \mathbf{A} = 0$. (Here V_B is any material volume which encloses the flux tubes.) The proof of (9.61) follows from (9.60) in the form

$$\frac{\partial \mathbf{A}}{\partial t} = \mathbf{u} \times \mathbf{B} + \nabla\phi$$

from which

$$\frac{D}{Dt}(\mathbf{A} \cdot \mathbf{B}) = \nabla \cdot [(\phi + \mathbf{A} \cdot \mathbf{u})\mathbf{B}].$$

Integrating over V_B provides the desired result: $H_B = \int \mathbf{B} \cdot \mathbf{A} dV$ = constant. The conservation of magnetic helicity, H_B, has the

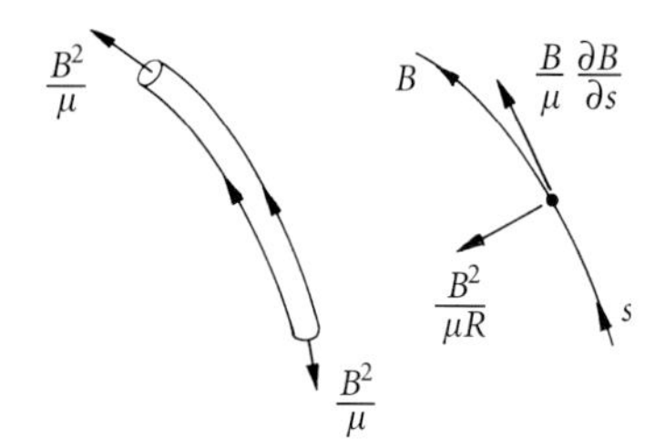

Figure 9.17 The force $(\mathbf{B}\cdot\nabla)(\mathbf{B}/\mu)$ may be thought of as being caused by a tension $T = B^2/\mu$ in the field lines.

same topological significance as the conservation of kinetic helicity, H_ω. (See Exercise 2.8 in Chapter 2.) That is to say, H_ω and H_B represent the degree of linkage of the vortex tubes and **B**-tubes respectively.

Example 9.7 Cross-helicity
Confirm that there is a second helicity invariant of ideal (i.e. diffusionless) MHD of the form $H_c = \int_{V_B} \mathbf{u}\cdot\mathbf{B}dV$. This is called *cross-helicity*.

We now turn to our second topic: that of Alfvén waves. It is the existence of these waves which gives high-R_m MHD its special character. To understand how they arise, we need to introduce the idea of *Faraday tensions*. Faraday pictured magnetic field lines as it they were elastic bands held in tension. Given the opportunity they will tend to straighten out. To appreciate where this idea comes from we need to think about the form of the Lorentz force. Using Ampere's law, this may be rewritten as

$$\mathbf{F} = \mathbf{J}\times\mathbf{B} = (\mathbf{B}\cdot\nabla)[\mathbf{B}/\mu] - \nabla[\mathbf{B}^2/2\mu].$$

The second term on the right-hand side is usually unimportant as it may be simply absorbed into the pressure gradient. Indeed, $\mathbf{B}^2/2\mu$ is referred to as the *magnetic pressure*. The more important term is the first one, which may be written in terms of curvilinear coordinates as,

$$(\mathbf{B}\cdot\nabla)[\mathbf{B}/\mu] = \frac{B}{\mu}\frac{\partial B}{\partial s}\hat{\mathbf{e}}_t - \frac{B^2}{\mu R}\hat{\mathbf{e}}_n.$$

Here $B = |\mathbf{B}|$, $\hat{\mathbf{e}}_t$ and $\hat{\mathbf{e}}_n$ are unit vectors tangential and normal, respectively, to a magnetic field line, R is the radius of curvature of the field line, and s is a coordinate measured along the field line.

It is now clear where Faraday's notion of tension comes from. If a magnetic field line is curved, it exerts a force on the medium directed towards the centre of curvature, of magnitude $B^2/\mu R$. This is exactly what would happen if the field line carried a tension of $T = B^2/\mu$, as shown in Figure 9.17.

We have already seen that, when R_m is large, the magnetic field lines are frozen into the fluid. We now see that they also behave as if they are in tension. These two features combine to give us Alfvén waves. Consider what happens if we apply an impulsive force to a fluid which is threaded by a uniform magnetic field. If the medium is highly conducting, then the **B**-lines are frozen into the fluid, and so the field lines will start to bow out (Figure 9.18). However, the resulting curvature of the lines creates a back reaction, $B^2/\mu R$, on the fluid. So as the field lines become more distorted, the Lorentz force increases

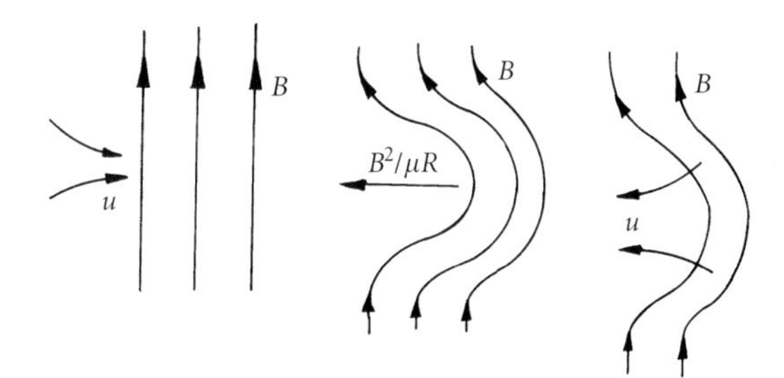

Figure 9.18 Magnetic field lines behave like elastic bands frozen into the fluid. This results in Alfvén waves.

and eventually the fluid will come to rest. The Faraday tensions then reverse the flow and push the fluid back to its starting point. However, the inertia of the fluid ensures that it will overshoot the neutral point and the entire process now starts in reverse. Oscillations result, just as they did in the loop of wire in Section 9.3.1.

The properties of these waves are readily established. Suppose we have a uniform, steady magnetic field, $\mathbf{B}_0$, which is perturbed by an infinitesimally small velocity field, $\mathbf{u}$. Let $\mathbf{j}$ and $\mathbf{b}$ be the resulting perturbations in current and magnetic field. Then, from (9.51),

$$\frac{\partial \mathbf{b}}{\partial t} = \nabla \times [\mathbf{u} \times \mathbf{B}_0] + \lambda \nabla^2 \mathbf{b}, \quad \nabla \times \mathbf{b} = \mu \mathbf{j}$$

which yields,

$$\frac{\partial \mathbf{j}}{\partial t} = \frac{1}{\mu}(\mathbf{B}_0 \cdot \nabla)\boldsymbol{\omega} + \lambda \nabla^2 \mathbf{j}. \tag{9.62}$$

Now consider the vorticity equation. Since $\nabla \times (\mathbf{u} \times \boldsymbol{\omega})$ is quadratic in the small quantity $\mathbf{u}$, this simplifies to

$$\rho \frac{\partial \boldsymbol{\omega}}{\partial t} = (\mathbf{B}_0 \cdot \nabla)\mathbf{j} + \rho \nu \nabla^2 \boldsymbol{\omega}. \tag{9.63}$$

The next step is to eliminate $\mathbf{j}$ from (9.62) and (9.63). This gives us the governing equation for Alfvén waves,

$$\frac{\partial^2 \boldsymbol{\omega}}{\partial t^2} = \frac{1}{\rho \mu}(\mathbf{B}_0 \cdot \nabla)^2 \boldsymbol{\omega} + (\lambda + \nu)\nabla^2 \left(\frac{\partial \boldsymbol{\omega}}{\partial t}\right) - \lambda \nu \nabla^4 \boldsymbol{\omega}. \tag{9.64}$$

It is readily confirmed that (9.64) supports plane-wave solutions of the form $\boldsymbol{\omega} \sim \hat{\boldsymbol{\omega}} \exp[j(\mathbf{k} \cdot \mathbf{x} - \varpi t)]$ and that the corresponding dispersion relationship is,

$$\varpi = \pm \left[\mathrm{v}_a^2 k_\parallel^2 - (\lambda - \nu)^2 k^4/4\right]^{1/2} - j[(\lambda + \nu)k^2/2].$$

Here $k_\parallel$ is the component of $\mathbf{k}$ in the direction of $\mathbf{B}_0$, and $\mathbf{v}_a$ is the Alfvén velocity $\mathbf{B}_0/(\rho\mu)^{1/2}$. When $\nu = 0$ and λ is small, which is a good

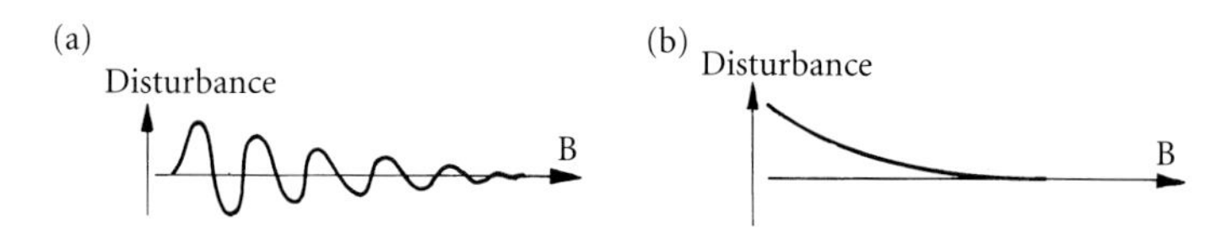

Figure 9.19 Alfvén waves at (a) high- and (b) low-R_m.

approximation for most high-R_m flows, we find,

$$\varpi = \pm v_a k_\parallel - (\lambda k^2/2)j. \tag{9.65}$$

This represents the propagation of transverse inertial waves, with a group velocity equal to $\pm\mathbf{B}_0/(\rho\mu)^{1/2}$. So, as we might have expected, these waves carry energy in the $\pm\mathbf{B}_0$ direction, the magnetic field lines acting like plucked harp strings, guiding the wave motion. Now consider the limit $\nu = 0$ and $\lambda \to \infty$, which is characteristic of many low-R_m flows. We find,

$$\varpi = -j\lambda k^2, \quad -j\left(k_\parallel/k\right)^2\tau^{-1} \tag{9.66a, b}$$

where τ is the magnetic damping time $\left(\sigma B_0^2/\rho\right)^{-1}$. The first root is unsurprising and of little interest. It is a heavily damped wave which is rapidly eradicated by Ohmic dissipation. The second root, however, is a surprise. We naturally think of Alfvén waves as a strictly high-R_m phenomenon, since they rely on the 'frozen-in' property of **B**. It turns out, however, that this is not strictly true, as can be seen from (9.66b). This represents a non-oscillatory disturbance which decays rather slowly, on a timescale of τ. As we shall see, (9.66b) represents the slow diffusion of disturbances along the **B**-lines. The two extremes of high and low R_m are shown in Figure 9.19.

Example 9.8 Magnetostrophic waves
Since both rotation and magnetic fields lead to internal wave propagation, it is natural to ask what happens when we have rapid rotation and a mean magnetic field present in a conducting fluid. Such a situation arises frequently in astrophysics where, in addition to inertial waves and Alfvén waves, an entirely new class of wave manifests itself. These are called *magnetostrophic waves*, and they are characterized by having an extremely low frequency. Consider a rapidly rotating fluid threaded by a uniform magnetic field $\mathbf{B}_0$. In the absence of viscous and Ohmic dissipation, small-amplitude disturbances, **b** and $\boldsymbol{\omega}$, are governed by the linearized equations:

$$\frac{\partial \mathbf{b}}{\partial t} = \nabla \times [\mathbf{u} \times \mathbf{B}_0], \quad |\mathbf{b}| \ll |\mathbf{B_0}|$$

$$\rho\frac{\partial \boldsymbol{\omega}}{\partial t} = 2\rho(\boldsymbol{\Omega} \cdot \nabla)\mathbf{u} + (\mathbf{B}_0 \cdot \nabla)\mathbf{j}, \quad |\boldsymbol{\omega}| \ll \Omega.$$

Show that these combine to give the wave-like equation

$$\left[\frac{\partial^2}{\partial t^2} - \frac{1}{\rho\mu}(\mathbf{B}_0 \cdot \nabla)^2\right]^2 \nabla^2\mathbf{u} + 4(\boldsymbol{\Omega} \cdot \nabla)^2\left(\frac{\partial^2\mathbf{u}}{\partial t^2}\right) = \mathbf{0}$$

and that the corresponding dispersion relationship is $\varpi^2 \pm \varpi_\Omega\varpi - \varpi_B^2 = 0$. Here ϖ is the angular frequency, and ϖ_Ω and ϖ_B are the inertial wave and Alfvén wave frequencies, $2(\boldsymbol{\Omega} \cdot \mathbf{k})/k$ and $(\mathbf{B}_0 \cdot \mathbf{k})/\sqrt{\rho\mu}$. It frequently happens that $\varpi_\Omega \gg \varpi_B$ (the weak field limit), in which case the dispersion equation yields two sets of solutions: $\varpi = \mp\varpi_\Omega$ and $\varpi = \pm\varpi_B^2/\varpi_\Omega$. The first of these is simply an inertial wave. The second has a frequency much lower than either ϖ_Ω or ϖ_B and it is this branch of the dispersion relationship which gives rise to magnetostrophic waves. These waves are important in astrophysics because they are very long lived. For example, in the core of the earth they have a period of $\sim 10^3$ years.

9.4 The influence of magnetic fields II—MHD turbulence

Let us now turn to turbulence. Traditionally MHD turbulence has been studied by two rather distinct communities. On the one hand, engineers have studied low-R_m turbulence motivated largely by the need to understand the flow of liquid metal in technological devices. On the other hand, plasma physicists and astrophysicists tend to study turbulence at high magnetic Reynolds number, $ul/\lambda \gg 1$. Much of the astrophysical work has focused on homogeneous turbulence and is motivated by a need to understand the behaviour of accretion discs, the dynamics of the solar atmosphere, and motion deep within the interior of the sun and the planets (in particular those motions which are thought to control the evolution of the solar and planetary magnetic fields).

Some problems require a knowledge of high-R_m *and* low-R_m turbulence. For example, in the liquid core of the earth, the largest motion occurs at a magnetic Reynolds number of ~ 100, while the smallest eddies correspond to $R_m < 1$ (Figure 9.20). In geodynamo theory, both motions are thought to be important.

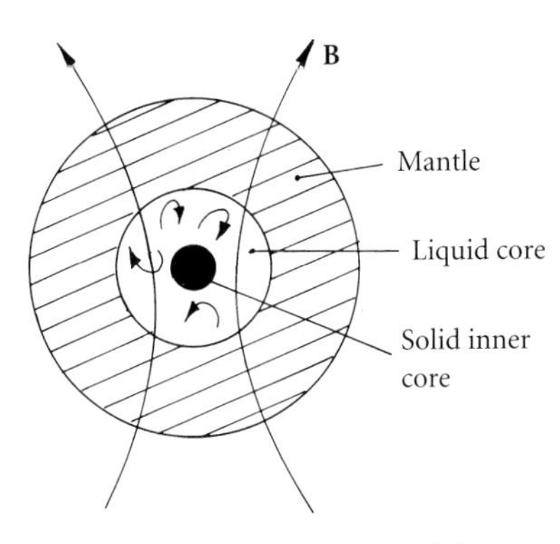

Figure 9.20 Motion in the core of the earth maintains the terrestrial field. See also Plate 16.

Here we shall consider both high- and low-R_m turbulence. In the case of low R_m, we have in mind the need to characterize turbulent motion in the presence of a uniform, imposed magnetic field. This is relevant to small-scale motion in the core of the earth, and also to liquid-metal turbulence in the many metallurgical operations where magnetic fields are used to suppress motion in the melt. In the case of high-R_m turbulence, we are interested in: (i) the evolution of a field of turbulence in the presence of an imposed large-scale magnetic field; and (ii) the evolution of a small random seed magnetic field in the

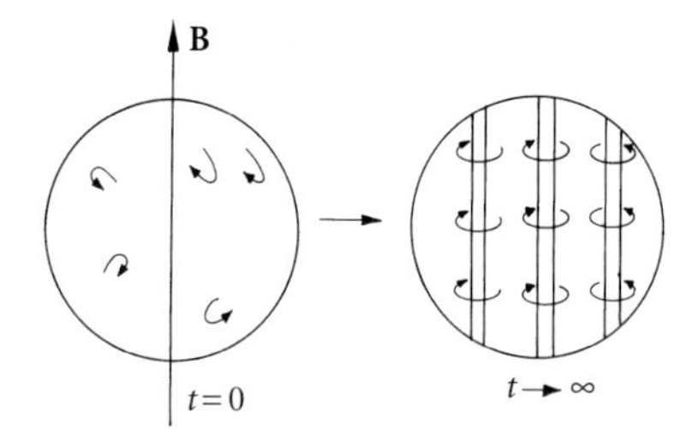

Figure 9.21 A magnetic field organizes turbulence into columnar eddies.

presence of forced turbulence. Both problems are relevant to, say, turbulence in the sun (Plate 14).

In Section 9.4.3 we shall see that the ideas of Landau and Kolmogorov can be redeveloped in the context of MHD turbulence. This provides a unified view of freely decaying homogeneous turbulence, valid for arbitrary magnetic Reynolds number. The arguments follow those of Davidson (1997, 2000). First, however, in order to develop some physical intuition, we shall give a qualitative overview of the influence of an imposed magnetic field on the large-scale eddies. We are particularly interested in the manner in which the imposed field organizes and shapes an initially isotropic cloud of turbulence. In a sense, this kind of initial value problem is a little artificial (how do we get such an isotropic cloud in the first place?) but it is well suited to exposing the various phenomena associated with MHD turbulence.

9.4.1 *The growth of anisotropy in MHD turbulence*

Let us start by considering a somewhat contrived thought experiment, first proposed by Davidson (1995, 1997), which is designed to bring out the crucial role played by angular momentum conservation in MHD turbulence. For simplicity, we temporarily leave aside viscous forces.

Suppose that a conducting fluid is held in a large insulated sphere of radius R (Figure 9.21). The sphere sits in a uniform, imposed field $\mathbf{B}_0$, so that the total magnetic field is $\mathbf{B} = \mathbf{B}_0 + \mathbf{b}$, $\mathbf{b}$ being associated with the currents induced by the motion $\mathbf{u}$ within the sphere. We place no restriction on R_m, nor on the interaction parameter which we define here to be $N = \sigma B_0^2 l / \rho u$, l being the initial integral scale of turbulence. When R_m is small, we have $|\mathbf{b}| \ll |\mathbf{B}_0|$, but in general $|\mathbf{b}|$ may be as large as $|\mathbf{B}_0|$. At $t = 0$, the fluid is vigorously stirred and then left to itself. We wish to characterize the anisotropy introduced into the turbulence by $\mathbf{B}_0$.

We attack the problem as follows. The global torque exerted on the fluid by the Lorentz force is

$$\mathbf{T} = \int \mathbf{x} \times (\mathbf{J} \times \mathbf{B}_0) dV + \int \mathbf{x} \times (\mathbf{J} \times \mathbf{b}) dV. \tag{9.67}$$

However, a closed system of currents produces zero net torque when it interacts with its self-field, $\mathbf{b}$,[5] and it follows that the second integral

[5] This is because an isolated system must conserve angular momentum, so that any isolated system of currents interacting with their associated magnetic field cannot produce a net torque. Alternatively, we can write $\mathbf{J} \times \mathbf{b}$ in terms of Maxwell stresses (see Exercise 9.2) so the volume integral becomes a surface integral which tends to zero as the surface recedes to infinity.

on the right is zero. The first integral, on the other hand, can be transformed using identity (9.5),

$$2\mathbf{x} \times [\mathbf{v} \times \mathbf{B}_0] = [\mathbf{x} \times \mathbf{v}] \times \mathbf{B}_0 + \nabla \cdot [(\mathbf{x} \times (\mathbf{x} \times \mathbf{B}_0))\mathbf{v}] \tag{9.68}$$

where $\mathbf{v}$ is any solenoidal field. Setting $\mathbf{v} = \mathbf{J}$, we obtain

$$\mathbf{T} = \left\{ \frac{1}{2} \int (\mathbf{x} \times \mathbf{J}) dV \right\} \times \mathbf{B}_0 = \mathbf{m} \times \mathbf{B}_0 \tag{9.69}$$

where $\mathbf{m}$ is the net dipole moment of the current distribution within the sphere. Evidently, the global angular momentum evolves according to

$$\rho \frac{d\mathbf{H}}{dt} = \mathbf{T} = \mathbf{m} \times \mathbf{B}_0, \quad \mathbf{H} = \int (\mathbf{x} \times \mathbf{u}) dV. \tag{9.70}$$

We see immediately that $H_{||}$, the component of $\mathbf{H}$ parallel to $\mathbf{B}_0$, is conserved. This, in turn, places a lower bound on the total energy of the flow,

$$E = E_{\mathbf{b}} + E_{\mathbf{u}} \geq E_{\mathbf{u}} \geq \rho \mathbf{H}_{||}^2 \left(2 \int \mathbf{x}_{\perp}^2 dV \right)^{-1} \tag{9.71}$$

where

$$E_{\mathrm{b}} = \int (\mathbf{b}^2 / 2\mu) dV, \quad E_{\mathrm{u}} = \int (\rho \mathbf{u}^2 / 2) dV.$$

(This follows from the Schwarz inequality in the form $\mathbf{H}_{||}^2 \leq \int \mathbf{u}_{\perp}^2 dV \int \mathbf{x}_{\perp}^2 dV$.) However, the energy declines due to Joule dissipation and so we also have

$$\frac{d}{dt} \int_{V_{\mathrm{R}}} \tfrac{1}{2} \rho \mathbf{u}^2 dV + \frac{d}{dt} \int_{V_{\infty}} (\mathbf{b}^2 / 2\mu) dV = -\frac{1}{\sigma} \int_{V_{\mathrm{R}}} \mathbf{J}^2 dV. \tag{9.72}$$

Evidently, one component of angular momentum is conserved, requiring that E is non-zero, yet energy is dissipated as long as $\mathbf{J}$ is finite. The implication is that the turbulence evolves to a state in which $\mathbf{J} = \mathbf{0}$, yet E_{u} is non-zero, to satisfy (9.71). However, if $\mathbf{J} = \mathbf{0}$, then Ohm's law reduces to $\mathbf{E} = -\mathbf{u} \times \mathbf{B}_0$, while Faraday's law requires that $\nabla \times \mathbf{E} = \mathbf{0}$. It follows that, at large times, $\nabla \times (\mathbf{u} \times \mathbf{B}_0) = (\mathbf{B}_0 \cdot \nabla)\mathbf{u} = \mathbf{0}$, and so $\mathbf{u}$ becomes independent of $\mathbf{x}_{||}$ as $t \to \infty$. The ultimate state is therefore two dimensional of the form $\mathbf{u}_{||} = \mathbf{0}$, $\mathbf{u}_{\perp} = \mathbf{u}_{\perp}(\mathbf{x}_{\perp})$. In short, the turbulence eventually approaches a state consisting of one or more columnar eddies, each aligned with $\mathbf{B}_0$ (see Figure 9.21). Note that all of the components of $\mathbf{H}$, other than $\mathbf{H}_{||}$, are destroyed during this evolution. As we shall see, this behaviour, which is reminiscent of the Taylor–Proudman theorem, is a result of energy propagation by Alfvén waves.

When R_m is small, this transition will occur on the timescale of $\tau = (\sigma B_0^2/\rho)^{-1}$, the magnetic damping time. The proof of this is straightforward. At low R_m, the current density is governed by (9.55a),

$$\mathbf{J} = \sigma(-\nabla V + \mathbf{u} \times \mathbf{B}_0), \tag{9.73}$$

and so the dipole moment becomes

$$\mathbf{m} = \frac{1}{2}\int_V \mathbf{x} \times \mathbf{J} dV = (\sigma/2)\int_V \mathbf{x} \times (\mathbf{u} \times \mathbf{B}_0) dV - (\sigma/2)\oint_S (V\mathbf{x}) \times d\mathbf{S}.$$

The surface integral vanishes while the volume integral transforms, with the aid of (9.68), to yield

$$\mathbf{m} = (\sigma/4)\mathbf{H} \times \mathbf{B}_0.$$

Substituting into (9.70) we obtain

$$\frac{d\mathbf{H}}{dt} = -\frac{\mathbf{H}_\perp}{4\tau}, \quad \tau^{-1} = \sigma B_0^2/\rho \tag{9.74}$$

and so $\mathbf{H}_{||}$ is conserved while $\mathbf{H}_\perp$ declines as $\mathbf{H}_\perp = \mathbf{H}_{\perp 0} \exp(-t/4\tau)$.

In summary, whatever the initial condition, and for any R_m or N, our confined flow evolves towards the two-dimensional state,

$$\mathbf{u}_\perp = \mathbf{u}_\perp(\mathbf{x}_\perp), \qquad \mathbf{H}_{||} = \mathbf{H}_{||}(0), \qquad \mathbf{H}_\perp = \mathbf{0}, \qquad \mathbf{u}_{||} = \mathbf{0}. \tag{9.75}$$

The simplicity of this result is surprising, particularly since we are dealing with the evolution of a fully non-linear system. This is the first hint that the conservation of angular momentum plays a crucial role in MHD turbulence.

9.4.2 *The evolution of eddies at low magnetic Reynolds number*

The precise manner in which these eddies elongate along the **B**-lines is readily investigated in the special case where inertia, $\mathbf{u} \cdot \nabla \mathbf{u}$, is weak relative to $\mathbf{J} \times \mathbf{B}$ and the conductivity, σ, is low, that is, $N \gg 1$ and $R_m \ll 1$. (This kind of low-R_m turbulence is relevant to small-scale motion in the core of the earth.) When R_m is small, the Lorentz force is linear in **u** and so, if N is large, the equation of motion for the fluid is itself linear. In particular, (9.58) may be 'uncurled' to yield,

$$\frac{\partial \mathbf{u}}{\partial t} = -\nabla(p/\rho) - \frac{1}{\tau}\nabla^{-2}[\partial^2\mathbf{u}/\partial x_{\|}^2], \qquad R_m \ll 1, \quad N \gg 1 \tag{9.76}$$

where ∇^{-2} is a symbolic operator defined via the Biot–Savart law. For localized disturbances in the infinite domains we have $p = 0$, and so (9.76) reduces to

$$\frac{\partial \mathbf{u}}{\partial t} = -\frac{1}{\tau}\nabla^{-2}[\partial^2\mathbf{u}/\partial x_{\|}^2], \qquad R_m \ll 1, \quad N \gg 1. \tag{9.77}$$

Because of the linearity of (9.77), we may think of MHD turbulence at high N and low R_m as the superposition of many vortices of different sizes and orientation, each evolving independently. It makes sense, therefore, to focus on individual vortices and ask what happens to them. Note that, if (9.77) is Fourier transformed in the perpendicular plane, and if $l_{\perp} \ll l_{\|}$, then we obtain a diffusion equation for the transform of $\mathbf{u}$,

$$\frac{\partial \hat{\mathbf{u}}}{\partial t} = \frac{1}{\tau k^2}[\partial^2\hat{\mathbf{u}}/\partial x_{/\!/}^2], \qquad R_m \ll 1, \quad N \gg 1. \tag{9.78}$$

This suggests a sort of pseudo-diffusion of momentum along the magnetic field lines. Of course, this is simply the last vestige of Alfvén wave propagation at low R_m, as discussed in Section 9.3.4 and described by (9.66b).

Let us now return to our large sphere of radius R. Suppose that, instead of turbulence, we start with a single vortex whose size is much less than R. Then the arguments of Section 9.4.1 still apply. If the axis of the vortex is parallel to $\mathbf{B}_0$, the vortex retains its angular momentum as its energy declines. On the other hand, perpendicular vortices lose their angular momentum exponentially fast on a timescale of 4τ. In both cases the vortices will elongate in the direction of $\mathbf{B}_0$. Moreover, it is clear that this behaviour should not depend on the existence of the remote spherical boundary. This suggests that we look at isolated vortices in an infinite domain, and that we should consider two particular cases: vortices parallel and perpendicular to $\mathbf{B}_0$. So now we remove the boundary and consider a single isolated vortex at $t = 0$. The discussion is restricted to inviscid fluids, and to high N (small inertia) and low R_m (low conductivity). Also, we choose the initial distribution of $\mathbf{u}$ to be such that $\mathbf{H}$ is a convergent integral. The analysis of Section 9.4.1 then reduces to the three integral statements:

$$\mathbf{H}_{\|} = \text{constant} \tag{9.79a}$$

$$\mathbf{H}_{\perp}(t) = \mathbf{H}_{\perp}(0)\exp[-t/4\tau] \tag{9.79b}$$

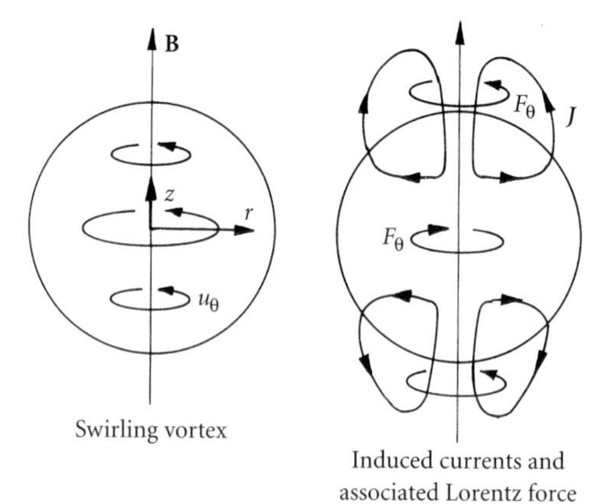

Figure 9.22 Magnetic damping of an axisymmetric vortex aligned with **B**. The current pattern demonstrates the mechanism of elongation of the vortex.

$$\frac{dE}{dt} = -\frac{1}{\sigma}\int J^2 dV.$$

We shall see that these tell us a great deal about the evolution of individual vortices. We start with a vortex whose angular momentum is parallel to $\mathbf{B}_0$.

Consider a vortex whose axis is aligned with $\mathbf{B}_0$. For simplicity, we restrict ourselves to an axisymmetric vortex, described in terms of cylindrical polars (r, θ, z), with $\mathbf{B}_0$ parallel to z. Let δ be the initial radius of the vortex, and subscripts θ and p indicate azimuthal and poloidal components of $\mathbf{u}$ and $\mathbf{J}$. We shall take an initial condition in which $\mathbf{u}$ is azimuthal, $\mathbf{u} = (0, u_\theta, 0)$. The interaction of this velocity with $\mathbf{B}_0$ produces an e.m.f., $\mathbf{u} \times \mathbf{B}_0$, which is radial. This, in turn, drives a poloidal current, $\mathbf{J}_p = (J_r, 0, J_z)$, in accordance with Ohm's law (Figure 9.22). Our first task is to find the relationship between u_θ and the induced current.

It is convenient to introduce the angular momentum, Γ, and the Stokes streamfunction, Ψ, defined through the expression

$$\mathbf{u} = \mathbf{u}_\theta + \mathbf{u}_p = (\Gamma/r)\hat{\mathbf{e}}_\theta + \nabla \times [(\Psi/r)\hat{\mathbf{e}}_\theta]. \qquad (9.80)$$

It is also convenient to introduce a Stokes streamfunction, say $\sigma B_0 \phi$, for $\mathbf{J}_p$,

$$\mathbf{J}_p = \sigma B_0 \nabla \times [(\phi/r)\hat{\mathbf{e}}_\theta].$$

The curl of Ohm's law (9.73) then yields,

$$\nabla_*^2 \phi = \left(\frac{\partial^2}{\partial z^2} + r\frac{\partial}{\partial r}\frac{1}{r}\frac{\partial}{\partial r}\right)\phi = -\frac{\partial \Gamma}{\partial z}. \qquad (9.81)$$

Moreover, the azimuthal component of the Lorentz force per unit mass is simply,

$$F_\theta = -\frac{1}{\tau}\frac{J_r}{\sigma B_0} \qquad (9.82)$$

$$= \frac{1}{r\tau}\frac{\partial \phi}{\partial z} \qquad (9.83)$$

and so the governing equation for Γ is,

$$\frac{\partial \Gamma}{\partial t} = -\frac{1}{\tau}\frac{\partial^2}{\partial z^2}[\nabla_*^{-2}\Gamma] \qquad (9.84)$$

where ∇_*^2 is defined via (9.81) and ∇_*^{-2} is the inverse operator. (We have ignored the inertial term $\mathbf{u}\cdot\nabla\mathbf{u}$ here since N is taken to be large.) Note that the pseudo-diffusion term, which first appeared in (9.77) and (9.78), is also present on the right-hand side of (9.84), and so we might anticipate that angular momentum propagates along the magnetic field lines. Note also that (9.83) confirms that the global angular momentum is conserved since $\partial\phi/\partial z$ integrates to zero. Of course, this is a special case of (9.79a). In fact, we can use conservation of angular momentum to determine the manner in which the flow evolves. From (9.56b) and (9.59), we have the estimates

$$\frac{dE}{dt} \sim -\left(\frac{\delta}{l_z}\right)^2 \frac{E}{\tau}, \quad E \sim \rho u_\theta^2 \delta^2 l_z \qquad (9.85)$$

where E is the kinetic energy of the vortex, δ is the radius of the vortex, and l_z is a characteristic length-scale for the eddy in the axial direction. This may be integrated to yield

$$E \sim E_0 \exp\left[-\frac{1}{\tau}\int_0^t (\delta/l_z)^2 dt\right]. \qquad (9.86)$$

Since angular momentum must be conserved, there is only one way in which this decrease in energy can be accommodated. The length-scale l_z must increase with time to reduce the dissipation, thus avoiding the exponential decline in energy. In fact, conservation of $H \sim u_\theta \delta^3 l_z$, combined with (9.85), yields the scaling laws

$$l_z \sim \delta(t/\tau)^{1/2}, \qquad u_\theta \sim (t/\tau)^{-1/2}. \qquad (9.87,\ 9.88)$$

These simple scalings may be confirmed by exact analysis. Let $\hat{u}$ be the first-order Hankel-cosine transform of u_θ,

$$\hat{u}(k_r,\ k_z) = 4\pi \int_0^\infty \int_0^\infty \Gamma(r,\ z) J_1(k_r r)\cos(k_z z)\ dr dz. \qquad (9.89)$$

Then (9.84) transforms to give $\partial\hat{u}/\partial\hat{t} = -(k_z/k)^2\hat{u}$ and so $\hat{u}$ evolves as $\hat{u} = \hat{u}_0 \exp\left(-(k_z/k)^2\hat{t}\right)$. Here $\hat{t}$ is the dimensionless time t/τ, $\hat{u}_0$ represents the initial condition, and k is the magnitude of $\mathbf{k}$. We can now determine Γ by taking the inverse transform. For large values of t, this yields (Davidson 1997),

$$\Gamma(\mathbf{x}, t) = (t/\tau)^{-1/2} G\left(r, z/(t/\tau)^{1/2}\right) \qquad (9.90)$$

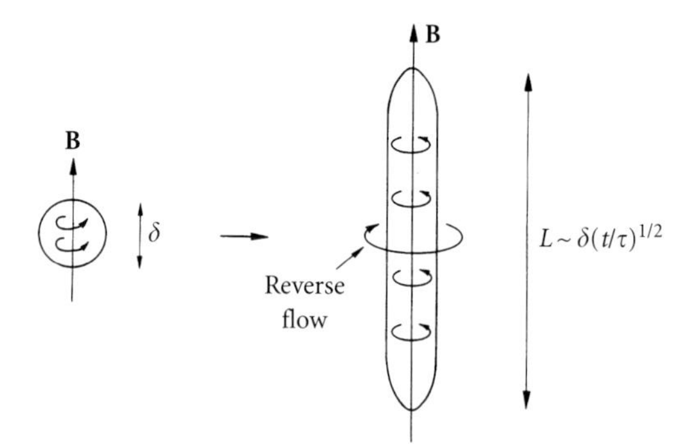

Figure 9.23 Magnetic damping of a parallel vortex for $R_m \ll 1$, $N \gg 1$.

where the form of G is determined by the initial conditions. We have arrived back at (9.87, 9.88). As expected, the angular momentum propagates along the z-axis according to $l_z \sim (t/\tau)^{1/2}$ and decays in magnitude as $\Gamma \sim (t/\tau)^{-1/2}$. The total kinetic energy of the vortex therefore declines at a rate[6] $E \sim (t/\tau)^{-1/2}$.

The mechanism for the propagation of angular momentum is described in Davidson (1995) and shown in Figure 9.22. The term $\mathbf{u}_\theta \times \mathbf{B}_0$ tends to drive a radial current J_r. Near the centre of the vortex, where the axial gradient in Γ is small, this is counter-balanced by an electrostatic potential, V, and so almost no current flows. However, near the top and bottom of the vortex, the current can return through regions of small or zero swirl. The resulting inward flow of current above and below the vortex gives rise to a positive azimuthal torque which, in turn, creates positive angular momentum in previously stagnant regions. Notice also that regions of reverse flow form in an annular zone surrounding the initial vortex where F_θ is negative. Thus, the generic shape of a parallel vortex is as shown in Figure 9.23.

Example 9.9 Consider the initial condition of a spherical vortex in which $\Gamma_0 = \Omega r^2 \exp[-(r^2 + z^2)/\delta^2]$. Confirm that

$$\Gamma(\hat{t} \to \infty) = \tfrac{3}{4}\pi^{1/2}\Omega\delta r\hat{t}^{-1/2}(\delta/r)^4\zeta^{5/2}H(\zeta), \quad \zeta = r^2/(\delta^2 + z^2/\hat{t}) \tag{9.91}$$

where $H(\zeta)$ is the hypergeometric function $M(5/2, 2, -\zeta)$ and $\hat{t} = t/\tau$. The function $H(\zeta)$ is negative for large ζ, confirming the reverse flow shown in Figure 9.23.

We now turn to transverse vortices. In the interest of simplicity, we shall consider a two-dimensional vortex whose axis is normal to the imposed magnetic field. $\mathbf{B}_0$ is taken to be in the z-direction and the flow is confined to the (x, z)-plane and independent of y. We shall take the vortex to be initially axisymmetric (Figure 9.24). Once again

[6] The laws $E \sim (t/\tau)^{-1/2}$, $l_z \sim (t/\tau)^{1/2}$ were first proposed by Moffatt (1967).

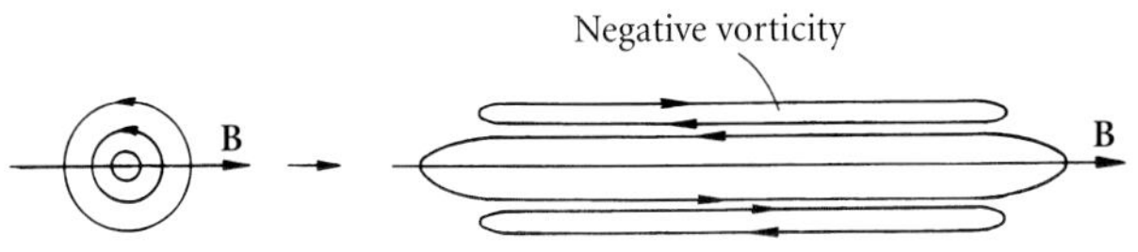

Figure 9.24 Generic shape of a transverse vortex at large times for $R_m \ll 1$, $N \gg 1$.

we shall find that angular momentum provides the key to determining the motion.

It is readily confirmed that the electrostatic potential is zero for this geometry (this follows from the divergence of (9.73)) and so $\mathbf{J} = -\sigma u_x B_0 \hat{\mathbf{e}}_y$. The Lorentz force per unit mass is then $-(u_x/\tau)\hat{\mathbf{e}}_x$ and so the global magnetic torque acting on the fluid is

$$T_y = -\tau^{-1} \int z u_x \, dV = -H_y/2\tau. \tag{9.92}$$

Here H_y is the angular momentum

$$H_y = \int (z u_x - x u_z) dV = 2 \int \psi dV \tag{9.93}$$

where ψ is the two-dimensional streamfunction for $\mathbf{u}$. Evidently, the angular momentum of the vortex decays as

$$H_y(t) = H_y(0) e^{-t/2\tau} \tag{9.94}$$

which is the two-dimensional counterpart of (9.79b). It is tempting to conclude, therefore, that the vortex decays on a timescale of 2τ. However, this appears to contradict (9.77) which, in the present context, 'uncurls' to give

$$\frac{\partial \psi}{\partial t} = -\frac{1}{\tau} \nabla^{-2} \frac{\partial^2 \psi}{\partial z^2}. \tag{9.95}$$

We may rewrite (9.95) in Fourier space (transforming only in the x-direction) to give

$$\frac{\partial \hat{\psi}}{\partial t} \sim \frac{1}{\tau k^2} \frac{\partial^2 \hat{\psi}}{\partial z^2} \tag{9.96}$$

which suggests that the cross-section of the vortex distorts from a circle to a sheet on a timescale of τ. If this picture is correct, the distortion should proceed in accordance with

$$l_z \sim \delta (t/\tau)^{1/2} \tag{9.97}$$

$$u_z \sim (t/\tau)^{-1/2}. \tag{9.98}$$

We appear to have a contradiction. On the one hand, (9.94) suggests that the flow is destroyed exponentially fast on a timescale of 2τ. On the other, (9.98) suggests that the energy declines only algebraically, as $(t/\tau)^{-1/2}$. Thus, the disappearance of the angular momentum cannot be a manifestation of the decay of the vortex. In fact, if estimates (9.97) and (9.98) are correct, and we shall see that they are, then the only possibility is that the flow evolves into a multi-cellular structure in which the total angular momentum is exponentially small due to the cancellation of **H** in adjacent cells (Figure 9.24). We may confirm this is so by introducing the Fourier transform

$$\Psi(k_x, k_z) = 4\int_0^{\infty}\int_0^{\infty} \psi(x,z)\cos(xk_x)\cos(zk_z)dxdz \tag{9.99}$$

and applying this transform to (9.95). Let $\hat{t}$ be the dimensionless time t/τ, k the magnitude of $\mathbf{k}$, and Ψ_0 the transform of ψ at $t=0$. Then the transformed version of (9.95), $\partial\Psi/\partial\hat{t} = -(k_z/k)^2\Psi$, is readily integrated to give $\Psi = \Psi_0 \exp\left[-(k_z/k)^2\hat{t}\right]$. Taking the inverse transform and looking for solutions at large values of t, we find (Davidson 1997),

$$\psi(\mathbf{x}, t) \sim \hat{t}^{-1/2}F(z/\hat{t}^{1/2}, x) \tag{9.100}$$

where the form of F depends on the initial conditions. It would appear, therefore, that the arguments leading to (9.97) and (9.98) are essentially correct. An initially axisymmetric vortex progressively distorts into a sheet-like structure, with a longitudinal length-scale given by (9.97), while $|\mathbf{u}|$ declines as $u_z \sim \hat{t}^{-1/2}$. As with the parallel vortex, the kinetic energy of the eddy falls off as $E \sim \hat{t}^{-1/2}$.

To demonstrate the multi-cellular structure of the vortex, we must consider a specific example. Suppose, for example, that the initial eddy structure is described by

$$\psi_0(r) = \Phi_0 e^{-r^2/\delta^2}, \quad r^2 = x^2 + z^2. \tag{9.101}$$

Then the inverse transform of Ψ may be integrated to give, at large t/τ,

$$\psi(\mathbf{x}, t) = \frac{\Phi_0}{(\pi\hat{t})^{1/2}}\frac{\zeta}{x^2}G(\zeta), \quad \zeta = \frac{x^2}{\delta^2 + z^2/\hat{t}} \tag{9.102}$$

where G is Kummer's hypergeometric function, $G(\zeta) = M(1, \frac{1}{2}, -\zeta)$. This solution has the form shown in Figure 9.24. The vorticity diffuses along the **B**-lines in accordance with (9.97), while simultaneously adopting a layered structure with zero net angular momentum. That is, for $t \gg \tau$, the global angular momentum, H_y, is

$$H_y = \frac{4\Phi_0\delta^2}{\pi^{1/2}}\int_0^{\infty}(1+x^2)^{-1/2}\int_0^{\infty}\zeta^{-1/2}G(\zeta)d\zeta dx \tag{9.103}$$

which integrates to zero since $\int_0^{\infty}\zeta^{-1/2}G(\zeta)d\zeta = 0$.

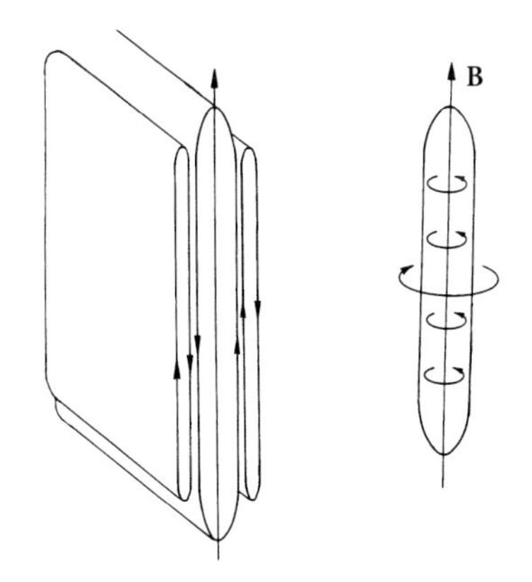

Figure 9.25 Typical structures in low-R_m turbulence.

In summary, then, there are two sorts of structure which we expect to emerge in low-R_m turbulence: sheets and columns (Figure 9.25, Plate 15). The sheets arise when **H** and **B** are mutually perpendicular. They necessarily consist of a set of platelets of oppositely signed vorticity.

9.4.3 *The Landau invariant for homogeneous MHD turbulence*

The arguments above, involving angular momentum, are reminiscent of Landau's derivation of the Loitsyansky invariant for isotropic turbulence (see Section 6.3.3). It is natural to see if Landau's arguments can be generalized to MHD turbulence. We shall see that they can and that, in the absence of long-range statistical correlations, MHD turbulence possesses an integral invariant.

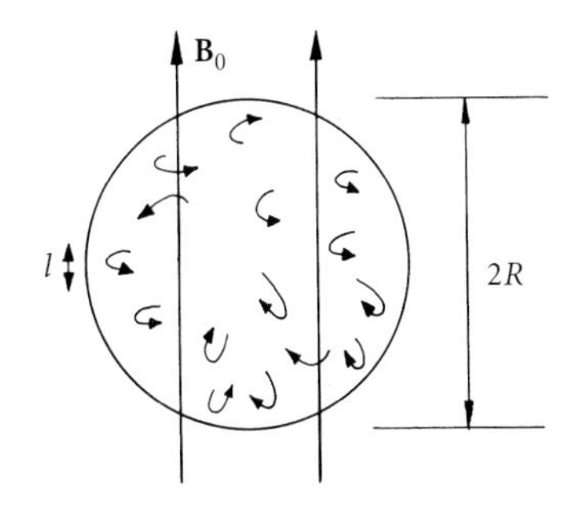

Figure 9.26 MHD turbulence evolving in a large sphere and subject to a mean field.

Let us repeat Landau's thought experiment, adapted now to MHD turbulence (Figure 9.26). As in Section 9.4.1, our conducting fluid is held in a large sphere of radius R, $R \gg l$, which sits in a uniform, imposed magnetic field, $\mathbf{B}_0$. At $t = 0$, the fluid is set into turbulent motion and then left to itself. We are interested in the behaviour of the turbulence for periods in which the integral scale, l, remains much smaller than R. Since $R \gg l$, we may follow the strategy of Section 6.3.3 and ignore the viscous torque exerted by the fluid on the boundary $r = R$. It follows from (9.70) that $\mathbf{H}_{\|}$ is conserved during this period, and following Landau's arguments (see equation (6.88)) we have, for any value of R_m and any N,

$$\left\langle \mathbf{H}_{\|}^2 \right\rangle = -\int\int \mathbf{r}_{\perp}^2 \langle \mathbf{u}_{\perp} \cdot \mathbf{u}'_{\perp} \rangle d\mathbf{r} d\mathbf{x} = \text{constant}$$

where $\mathbf{r} = \mathbf{x}' - \mathbf{x}$. If we now ignore Batchelor's long-range pressure forces, we have, in the spirit of Landau,

$$I_{\|} = \left\langle \mathbf{H}_{\|}^2 \right\rangle / V = -\int \mathbf{r}_{\perp}^2 \langle \mathbf{u}_{\perp} \cdot \mathbf{u}'_{\perp} \rangle d\mathbf{r} = \text{constant} \tag{9.104}$$

This applies to homogenous turbulence at value any N. It was first noted in the context of low-R_m turbulence, by Davidson (1997), and then later extended to high-R_m turbulence in Davidson (2000).

As in conventional turbulence, this invariant may also be derived from the Karman–Howarth equation. The argument proceeds as follows. Consider the equation of motion

$$\frac{\partial u_i}{\partial t} = -\frac{\partial}{\partial x_k}[u_k u_i - b_k b_i/\rho\mu] - \frac{\partial}{\partial x_i}\left[\frac{p}{\rho}\right] + (\mathbf{J} \times \mathbf{B}_0)_i/\rho + \nu\nabla^2 u_i \tag{9.105}$$

in which $\mathbf{b}$ is the induced magnetic field associated with the current $\mathbf{J}$ and p includes the magnetic pressure, $\mathbf{b}^2/2\mu$. This yields the generalized Karman–Howarth equation:

$$\begin{aligned}\frac{\partial}{\partial t}\left\langle u_i u'_j\right\rangle &= \frac{\partial}{\partial r_k}\left[\left\langle u_i u_k u'_j - b_i b_k u'_j/\rho\mu\right\rangle - \left\langle u'_j u'_k u_i - b'_j b'_k u_i/\rho\mu\right\rangle\right] \\ &\quad + 2\nu\nabla^2\left\langle u_i u'_j\right\rangle + \frac{1}{\rho}\left[\frac{\partial}{\partial r_i}\langle p u'_j\rangle - \frac{\partial}{\partial r_j}\langle p' u_i\rangle\right] \\ &\quad + \frac{1}{\rho}\left[\langle(\mathbf{J}\times\mathbf{B}_0)_i u'_j + (\mathbf{J}'\times\mathbf{B}_0)_j u_i\rangle\right].\end{aligned} \tag{9.106}$$

Consider first the case where $\mathbf{B}_0$ and $\mathbf{b}$ are both zero, that is, conventional hydrodynamic turbulence. Then, following the arguments of Batchelor (1953), it is readily shown that (9.106) yields

$$I_{ijmn} = \int r_m r_n \left\langle u_i u'_j\right\rangle d\mathbf{r} = \text{constant} \tag{9.107}$$

provided, of course, that there are no long-range correlations. This is a generalization of Loitsyansky's integral. When $\mathbf{b}$ is finite, but $\mathbf{B}_0$ remains zero (no mean field), Batchelor's arguments may be repeated and again we find that I_{ijmn} is an invariant. This was first noted by Chandrasekhar (1951) in the context of isotropic turbulence. Let us turn now to the case where $\mathbf{B}_0$ is finite. In the absence of long-range correlations, only the final term in (9.106) can contribute to the rate of change of integrals of the type I_{ijmn} and so

$$\frac{d}{dt}\int \mathbf{r}_\perp^2\langle \mathbf{u}_\perp\cdot\mathbf{u}'_\perp\rangle d\mathbf{r} = \frac{1}{\rho}\int [\mathbf{r}_\perp^2\langle(\mathbf{J}\times\mathbf{B}_0)_\perp\cdot\mathbf{u}'_\perp + (\mathbf{J}'\times\mathbf{B}_0)_\perp\cdot\mathbf{u}_\perp\rangle] d\mathbf{r}. \tag{9.108}$$

The integrand on the right consists of terms of the form $r_\perp^2(J_y u'_x - J_x u'_y)$ and $r_\perp^2(J'_y u_x - J'_x u_y)$. (We take $\mathbf{B}_0$ to point in the z direction.) Such terms can be converted into surface integrals since $3y^2 J_y = \nabla\cdot(y^3\mathbf{J})$, $2yu_y = \nabla\cdot(y^2\mathbf{u})$, etc. Moreover, in the absence of long-range correlations, these surface integrals are zero, and so (9.108) yields,

$$I_\parallel = -\int \mathbf{r}_\perp^2\langle \mathbf{u}_\perp\cdot\mathbf{u}'_\perp\rangle d\mathbf{r} = \text{constant} \quad (\text{any } N, \text{ any } R_m). \tag{9.109}$$

We have arrived back at (9.104), but by a different route. However, the Landau-like derivation is to be preferred since it exposes the physical origin of the invariant (9.109). Of course (6.109) comes with all the usual caveats discussed in Sections 6.3.4–6.3.5. In particular, it applies only if the turbulence has negligible linear momentum, and so has a low-k spectrum of the form $E \sim k^4$. Moreover, it applies only if the long-range statistical correlations are weak, as seems to be the case for fully developed hydrodynamic turbulence. Despite these limitations, it is an important result as it extends the earlier studies of Landau and Chandrasekhar to MHD turbulence in the presence of a mean magnetic field. We shall now show how (6.109) may be used to predict the rate of decay of energy in low-R_m turbulence.

9.4.4 *Decay laws at low magnetic Reynolds number*

Let us now repeat Kolmogorov's arguments of Section 6.3.2, adapted to homogeneous, low-R_m MHD turbulence. The aim is to determine how a field of initially isotropic turbulence evolves in an imposed magnetic field (Figure 9.27).

We start with the curl of Ohm's law, $\nabla \times \mathbf{J} = \sigma \mathbf{B}_0 \cdot \nabla \mathbf{u}$, from which the Joule dissipation can be estimated as,

$$\frac{\langle \mathbf{J}^2 \rangle}{\rho\sigma} \sim \left(\frac{l_{\min}}{l_{\|}}\right)^2 \frac{u^2}{\tau}, \qquad \tau = \left(\sigma B_0^2/\rho\right)^{-1}.$$

(Here $l_{\min}$ and $l_{\|}$ are suitably defined integral scales.) Now we know that the effect of $\mathbf{B}_0$ is to introduce anisotropy into the turbulence, with $l_{\|} > l_{\perp}$. Thus we have

$$\frac{\langle \mathbf{J}^2 \rangle}{\rho\sigma} = \frac{\beta}{2}\left(\frac{l_{\perp}}{l_{\|}}\right)^2 \frac{\langle \mathbf{u}^2 \rangle}{\tau} \tag{9.110}$$

where β is of order unity. (In fact it can be shown that $\beta = 2/3$ when the turbulence is isotropic.) We can use (9.110) to estimate the rate of decay of kinetic energy. That is, the energy equation,

$$\frac{d}{dt}\frac{1}{2}\langle \mathbf{u}^2 \rangle = -\nu\langle \boldsymbol{\omega}^2 \rangle - \langle \mathbf{J}^2 \rangle/\rho\sigma \tag{9.111}$$

can be written as

$$\frac{du^2}{dt} = -\alpha\frac{u^3}{l_{\perp}} - \beta\left(\frac{l_{\perp}}{l_{\|}}\right)^2 \frac{u^2}{\tau}. \tag{9.112}$$

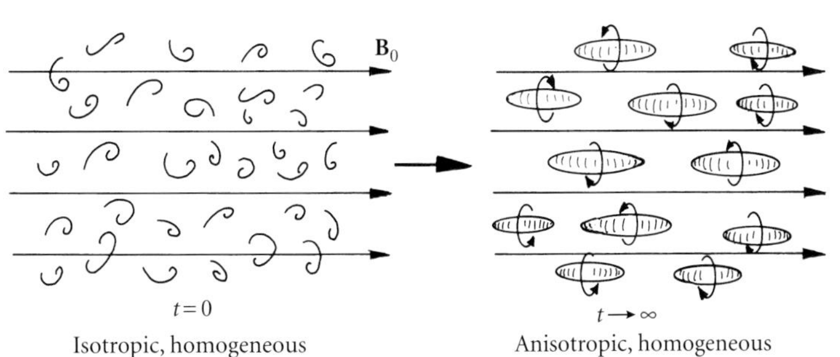

Figure 9.27 Homogeneous MHD turbulence at low R_m.

Here we define $u^2 = \frac{1}{3}\langle \mathbf{u}^2 \rangle$ and we have made the usual estimate of the viscous dissipation. (In conventional turbulence, α is of the order of unity.) Now our energy equation might be combined with (9.109) in the form

$$u^2 l_\perp^4 l_\| = \text{constant} \tag{9.113}$$

which offers the possibility of predicting $u^2(t)$ as well as $l_\perp$ and $l_\|$. In low-R_m turbulence it is conventional to categorize the flow according to the value of the interaction parameter, $N = \sigma B_0^2 l_\perp / \rho u$. When N is small (negligible magnetic effects), (9.112) and (9.113) reduce to

$$\frac{du^2}{dt} = -\alpha \frac{u^3}{l}, \quad u^2 l^5 = \text{constant} \tag{9.114}$$

which yields the familiar Kolmogorov law $u^2 \sim t^{-10/7}$. When N is large, on the other hand, inertia is unimportant and we have,

$$\frac{du^2}{dt} = -\beta \left(\frac{l_\perp}{l_\|} \right)^2 \frac{u^2}{\tau}, \quad u^2 l_\perp^4 l_\| = \text{constant}. \tag{9.115}$$

We have already seen that $l_\perp$ remains constant during the decay of high-N turbulence and in such a case our equations predict

$$u^2 \sim u_0^2 (t/\tau)^{-1/2}, \qquad l_\| = l_0 [1 + 2\beta t/\tau]^{1/2}, \quad l_\perp = l_0 \tag{9.116}$$

which we also know to be correct. For intermediate values of N, however, we have a problem. Equations (9.112) and (9.113) between them contain three unknowns u^2, $l_\perp$, and $l_\|$. To close the system we might tentatively introduce the heuristic equation

$$\frac{d}{dt} (l_\| / l_\perp)^2 = 2\beta/\tau \tag{9.117}$$

which has the merit of being exact for $N \to 0$ and $N \to \infty$ but cannot be justified for intermediate N. (Essentially the same equation was proposed by Widlund et al. (1998) in their one-point closure model of MHD turbulence.) Integrating (9.112), (9.113), and (9.117) yields (Davidson 2000, 2001),

$$u^2/u_0^2 = \hat{t}^{-1/2} [1 + (7/15)(\hat{t}^{3/4} - 1) N_0^{-1}]^{-10/7} \tag{9.118}$$

$$l_\perp / l_0 = [1 + (7/15)(\hat{t}^{3/4} - 1) N_0^{-1}]^{2/7} \tag{9.119}$$

$$l_\| / l_0 = \hat{t}^{1/2} [1 + (7/15)(\hat{t}^{3/4} - 1) N_0^{-1}]^{2/7} \tag{9.120}$$

where N_0 is the initial value of N and $\hat{t} = 1 + 2(t/\tau)$. (For simplicity we have taken $\alpha = \beta = 1$.) The high- and low-N results above are special cases of (9.118)–(9.120). For the case of $N_0 = 7/15$ we obtain the power laws,

$$u^2/u_0^2 \sim \hat{t}^{-11/7}, \qquad l_{\|}/l_0 \sim \hat{t}^{5/7} \tag{9.121}$$

and indeed these power laws are reasonable approximations to (9.118) and (9.120) for all values of N_0 around unity. Experiments of low-R_m, homogeneous turbulence were carried out by Alemany et al. (1979) and they suggest $u^2 \sim t^{-1.6}$ for $N_0 \sim 1$. This compares favourably with (9.121), which predicts $u^2 \sim t^{-1.57}$.

9.4.5 *Turbulence at high magnetic Reynolds number*

We now turn to high-R_m turbulence. There are two canonical problems of interest here. The first, which is a continuation of the previous section, is the influence of a uniform, imposed magnetic field on homogeneous turbulence. The second relates to the case in which $\mathbf{B}_0 = \mathbf{0}$, where we are interested in the fate of a small, random, seed magnetic field in a turbulent flow.

Let us start with the influence of an imposed field. When the long-range correlations are weak, $I_{\|}$ in (9.109) is an invariant of high-R_m turbulence, and it is natural to explore the implications of this. Now for large R_m, the mean field $\mathbf{B}_0$ tends to promote an equipartition of energy between $\mathbf{b}$ and $\mathbf{u}$. This is known as the Alfvén effect and arises because small-scale disturbances tend to convert their energy into Alfvén waves, which distributes the energy equally between the induced field, $\mathbf{b}$, and the velocity field, $\mathbf{u}$ (e.g. see Oughton et al. 1994). Thus, for large R_m, we might anticipate:

$$u^2 l_\perp^4 l_{\|} = \text{constant}, \qquad \langle \mathbf{u}^2 \rangle \sim \langle \mathbf{b}^2 \rangle / (\rho\mu)$$

$$\frac{d}{dt}\left[\frac{\rho\langle \mathbf{u}^2 \rangle}{2} + \frac{\langle \mathbf{b}^2 \rangle}{2\mu}\right] = -\rho\nu\langle \boldsymbol{\omega}^2 \rangle - \langle \mathbf{J}^2 \rangle/\sigma.$$

These combine to yield,

$$E l_\perp^4 l_{\|} = \text{constant}, \qquad \frac{dE}{dt} = -\rho\nu\langle \boldsymbol{\omega}^2 \rangle - \langle \mathbf{J}^2 \rangle/\sigma,$$

where E is the energy per unit volume. This suggest that, as E falls, $l_\perp^4 l_{\|}$ must increase and the indications are that $l_{\|}$ grows faster than $l_\perp$, as in low-R_m turbulence (see Oughton et al. 1994).

So far we have considered the influence of an imposed, uniform magnetic field on freely evolving turbulence. When R_m is large, there is a second problem of interest: that of the influence of a prescribed field[7] of turbulence on a small, random, 'seed' magnetic field. In a

[7] Prescribed in a statistical sense only.

sense this represents a change of emphasis: in this second problem we focus on the way in which **u** shapes **B**, rather than the influence of **B** on **u**. The main question here is whether or not the energy of the seed field grows by field-line stretching, or is extinguished by Ohmic dissipation. An intriguing argument, generally attributed to Batchelor, suggests that the fate of the seed field is completely determined by the *magnetic Prandtl number*, $P_m = \nu/\lambda$.

The argument goes something like this. Batchelor noted that there are two competing effects at work in high-R_m turbulence. On the one hand, random stretching of the magnetic flux tubes by **u** will tend to increase $\langle \mathbf{B}^2 \rangle$. On the other hand, the turbulent generation of small-scale magnetic structures will tend to enhance the Ohmic dissipation, thus reducing $\langle \mathbf{B}^2 \rangle$. The fate of a small seed field is determined by the relative magnitude of these two effects. For homogeneous turbulence, this competition is captured by the magnetic energy equation,

$$\frac{\partial}{\partial t}\left\langle \frac{B^2}{2\mu} \right\rangle = \left\langle \frac{B_i B_j}{\mu} S_{ij} \right\rangle - \frac{1}{\sigma}\langle \mathbf{J}^2 \rangle$$

which can be obtained by taking the dot product of **B** with (9.51). The starting point for Batchelor's analysis lies in the evolution equations for $\boldsymbol{\omega}$ and **B**:

$$\partial \boldsymbol{\omega}/\partial t = \nabla \times (\mathbf{u} \times \boldsymbol{\omega}) + \nu \nabla^2 \boldsymbol{\omega}, \qquad \partial \mathbf{B}/\partial t = \nabla \times (\mathbf{u} \times \mathbf{B}) + \lambda \nabla^2 \mathbf{B}.$$

Here we have ignored the Lorentz force, $\mathbf{J} \times \mathbf{B}$, in the $\boldsymbol{\omega}$-equation on the assumption that **B** is very weak, so there is no back-reaction on **u**. Evidently, when $\lambda = \nu$, there exists a solution for the seed field of the form $\mathbf{B} = \text{constant} \times \boldsymbol{\omega}$. This suggests that if $\boldsymbol{\omega}$ is statistically steady, then so is **B**. The implication is that, if $\lambda = \nu$, flux-tube stretching and Ohmic dissipation exactly balance. If λ exceeds ν, however, we would expect enhanced Ohmic dissipation and a decline in $\langle \mathbf{B}^2 \rangle$, while $\lambda < \nu$ should led to a spontaneous growth in the seed field, a growth which is curtailed only when $\mathbf{J} \times \mathbf{B}$ is sufficiently large to suppress the turbulence. Thus, according to Batchelor, the condition for the amplification of this random, seed field is $P_m > 1$. Such a situation is never realized in liquid metals, where $P_m \sim 10^{-6}$, but might be realized in the astrophysical context, perhaps in the solar corona or the interstellar gas.

This idea, that a magnetic field can be continually intensified by a velocity field, is closely related to the topic of geodynamo theory, which we shall discuss in Section 9.5.2. (Remember that geodynamo theory tries to explain the maintenance of the earth's magnetic field in terms of flux-tube stretching in the liquid core of the earth.) There are, however, some crucial differences between Batchelor's analysis and conventional dynamo theory. In dynamo theory we are usually interested in the generation of a *large-scale field* from small-scale turbulence, so that, on the local scale of the turbulence, there is a mean component of **B**, that is, $\langle \mathbf{B} \rangle \neq 0$. In Batchelor's analysis, however, **B** is random and, like **u**, has zero mean, $\langle \mathbf{B} \rangle = 0$. Moreover, in dynamo theory we almost always find it necessary to endow the turbulence with a large amount of helicity, $\langle \mathbf{u} \cdot \boldsymbol{\omega} \rangle$, if the dynamo is to be efficient. In Batchelor's analysis, there is no requirement for the turbulence to be helical, and indeed the analysis does not distinguish between helical and non-helical turbulence.

In any event, intriguing as they are, Batchelor's arguments are flawed. The problems are two-fold. First, the analogy between **B** and $\boldsymbol{\omega}$ is not exact: $\boldsymbol{\omega}$ is functionally related to **u** in a way that **B** is not. Thus, there are many more possible solutions of the induction equation than of the vorticity equation. Second, if the turbulence is to be statistically steady we need a forcing term in the vorticity equation, representing some kind of mechanical stirring. Since the corresponding term is absent in the induction equation, the analogy between **B** and $\boldsymbol{\omega}$ is again broken. However, despite these weaknesses, there seems to be some element of truth in Batchelor's hypothesis. Numerical simulations of forced, non-helical MHD turbulence at high R_m tend to show a growth in $\langle \mathbf{B}^2 \rangle$ when P_m is of order unity or greater, and a fall in $\langle \mathbf{B}^2 \rangle$ when P_m is smaller than ~ 1. So, although the analysis is flawed, the underlying ideas do seem to have some merit.

If we accept the argument that a seed field is amplified for sufficiently small λ, it is natural to ask what the spatial structure of this field might be. Will it form an intricate, fine-scale pattern due to flux tube stretching, or a coarse, large-scale structure due to flux tube mergers? In this context it is interesting to note that there are reasons to believe that there is an *inverse cascade* of the magnetic helicity, and hence of magnetic energy, in freely evolving, high-R_m turbulence. That is to say, the integral scale for **B** grows as the flow evolves, with magnetic energy being transferred from small to large scales. The arguments are rather tentative and rest on the approximate conservation of magnetic helicity, which, in turn, relies on the equation (see Section 9.3.4):

$$\frac{D}{Dt}(\mathbf{A} \cdot \mathbf{B}) = \nabla \cdot [(\phi + \mathbf{u} \cdot \mathbf{A})\mathbf{B}] - \sigma^{-1}[2\mathbf{J} \cdot \mathbf{B} + \nabla \cdot (\mathbf{J} \times \mathbf{A})]. \tag{9.122}$$

When averaged, this becomes[8]

$$\frac{d}{dt}[\langle \mathbf{A} \cdot \mathbf{B} \rangle] = -2\langle \mathbf{J} \cdot \mathbf{B} \rangle / \sigma. \tag{9.123}$$

[8] We are assuming here that the turbulence is statistically homogeneous, so that divergences disappear on averaging. However, we explicitly exclude isotropic turbulence (turbulence which is invariant under rotations and reflections) since reflectionally symmetric turbulence has no mean helicity.

(Note that, for a perfect conductor, the magnetic helicity, $H_B = \langle \mathbf{A} \cdot \mathbf{B} \rangle$, is conserved, as shown in Section 9.3.4.) Combining (9.123) with the energy equation, we have

$$\frac{dE}{dt} = -\rho\nu\langle \boldsymbol{\omega}^2 \rangle - \langle \mathbf{J}^2 \rangle / \sigma, \qquad \frac{dH_B}{dt} = -2\langle \mathbf{J} \cdot \mathbf{B} \rangle / \sigma.$$

The next step is to show that, as $\sigma \to \infty$, dE/dt remains finite while dH_B/dt tends to zero. We proceed as follows. The Schwarz inequality tells us $\langle \mathbf{J} \cdot \mathbf{B} \rangle^2 \le \langle \mathbf{J}^2 \rangle \langle \mathbf{B}^2 \rangle$. This may be written as

$$|\langle \mathbf{J} \cdot \mathbf{B} \rangle| / \sigma \le (2\mu/\sigma)^{1/2} [|\dot{E}|E]^{1/2} \tag{9.124}$$

and so we can place an upper bound on the rate of destruction of magnetic helicity:

$$|\dot{H}_B| / \mu \le (8\lambda)^{1/2} |\dot{E}|^{1/2} E^{1/2}. \tag{9.125}$$

We now let $\sigma \to \infty$. In the process, however, we assume that $\dot{E}$ remains finite. We might try to justify this as follows. We expect that, as $\sigma \to \infty$, more and more of the Joule dissipation is concentrated into thin current sheets. However, by analogy with viscous dissipation at small ν, we might expect $\langle \mathbf{J}^2 \rangle / \sigma$ to remain finite in the limiting process. If this is true, it follows that, in the limit $\lambda \to 0$, H_B is conserved. Thus, for small λ, we have the destruction of energy subject to the approximate conservation of magnetic helicity. In a finite domain this presents us with a well-defined variational problem. Minimizing E subject to the conservation of H_B in a bounded domain gives us $\nabla \times \mathbf{B} = \alpha \mathbf{B}$ and $\mathbf{u} = \mathbf{0}$, where α is an eigenvalue of the variational problem. The implication is that $\mathbf{B}$ ends up with a large length-scale, comparable with the domain size.

In summary, then, the assumption that $\dot{E}$ remains finite as $\sigma \to \infty$ leads to the conservation of helicity, and minimizing energy subject to the invariance of H_B gives, for a finite domain, a large-scale static magnetic field with $\mathbf{J}$ and $\mathbf{B}$ aligned.

9.5 The combined effects of Coriolis and Lorentz forces

9.5.1 *The shaping of eddies by Coriolis and magnetic forces*

The structure of turbulence is more complex when both Lorentz and Coriolis forces act together. Some hint as to how Coriolis forces might influence the decay of a low-R_m eddy is furnished by the following simple model problem. Suppose that, in a rotating frame of reference, we have a small, localized, low-R_m eddy sitting in a locally uniform field $\mathbf{B}_0 = B\hat{\mathbf{e}}_x$. It has finite angular momentum and is subject to

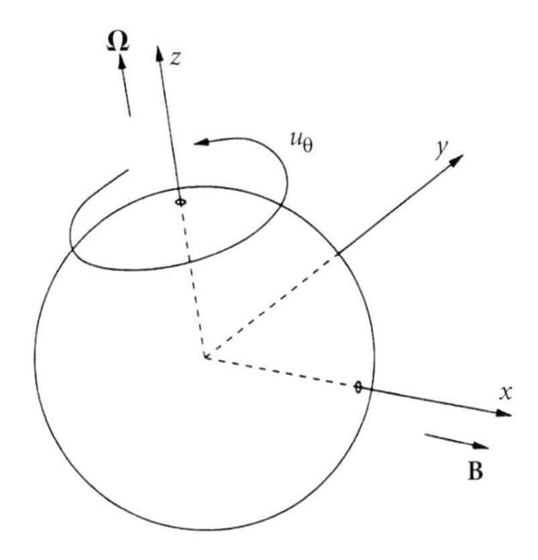

Figure 9.28 An eddy sitting in a rapidly rotating fluid and an ambient magnetic field.

Coriolis and Lorentz forces (with $\mathbf{\Omega} = \Omega\hat{\mathbf{e}}_z$). However, viscous forces are neglected (Figure 9.28).

The momentum equation in a frame of reference rotating at $\mathbf{\Omega}$ is,

$$\frac{D\mathbf{u}}{Dt} = 2\mathbf{u} \times \mathbf{\Omega} - \nabla(p/\rho) + \rho^{-1}\mathbf{J} \times \mathbf{B}_0 \tag{9.126}$$

from which we have

$$\frac{D(\mathbf{x} \times \mathbf{u})}{Dt} = 2\mathbf{x} \times (\mathbf{u} \times \mathbf{\Omega}) - \mathbf{x} \times \nabla(p/\rho) + \rho^{-1}\mathbf{x} \times (\mathbf{J} \times \mathbf{B}_0). \tag{9.127}$$

Using (9.68) to rearrange the Coriolis and Lorentz torques, we find

$$\frac{D(\mathbf{x} \times \mathbf{u})}{Dt} = (\mathbf{x} \times \mathbf{u}) \times \mathbf{\Omega} + \nabla \times (p\mathbf{x}/\rho) + (2\rho)^{-1}(\mathbf{x} \times \mathbf{J}) \times \mathbf{B}_0 + \nabla \cdot (\sim\mathbf{u}) + \nabla \cdot (\sim\mathbf{J}).$$

Next we integrate over all space, or else assume that the flow is confined to a large spherical volume and insist that $\mathbf{u} \cdot d\mathbf{S}$ and $\mathbf{J} \cdot d\mathbf{S}$ are zero on the remote boundary. Either way we find,

$$\rho\frac{d\mathbf{H}}{dt} = \rho\mathbf{H} \times \mathbf{\Omega} + \mathbf{m} \times \mathbf{B}_0 \tag{9.128}$$

where $\mathbf{H}$ is the angular momentum, $\int \mathbf{x} \times \mathbf{u}\, dV$, and $\mathbf{m}$ is the dipole moment induced by the interaction of the eddy with $\mathbf{B}_0$. Finally, following the arguments leading up to (9.74), and on the assumption that the low-R_m form of Ohm's law applies, we recast $\mathbf{m}$ in terms of $\mathbf{H}$:

$$\mathbf{m} = \frac{1}{2}\int \mathbf{x} \times \mathbf{J}\, dV = (\sigma/4)\mathbf{H} \times \mathbf{B}_0. \tag{9.129}$$

This gives,

$$\frac{d\mathbf{H}}{dt} = \mathbf{H} \times \mathbf{\Omega} - \frac{\mathbf{H}_\perp}{4\tau}, \quad \mathbf{H}_\perp = (0, H_y, H_z) \tag{9.130}$$

where τ is the Joule damping time, $\tau = (\sigma B^2/\rho)^{-1}$. From this we may show that H_z declines exponentially on a timescale of 4τ. H_y and H_x, on the other hand, decay in a fashion that depends on the value of the so-called Elsasser number $\Lambda = \sigma B^2/(2\rho\Omega) = (2\Omega\tau)^{-1}$. If Λ is greater than four, then H_y and H_x decay exponentially, while if Λ is less than four they oscillate as they decay. In either case the characteristic decay time is 4τ and the decay of $|\mathbf{H}|$ is exponentially fast. However, the Coriolis force does no work on the flow, so the energy equation

$$\frac{dE}{dt} = -\frac{1}{\sigma}\int J^2\, dV \sim -\left(\frac{l_{\min}}{l_B}\right)^2 \frac{E}{\tau} \tag{9.131}$$

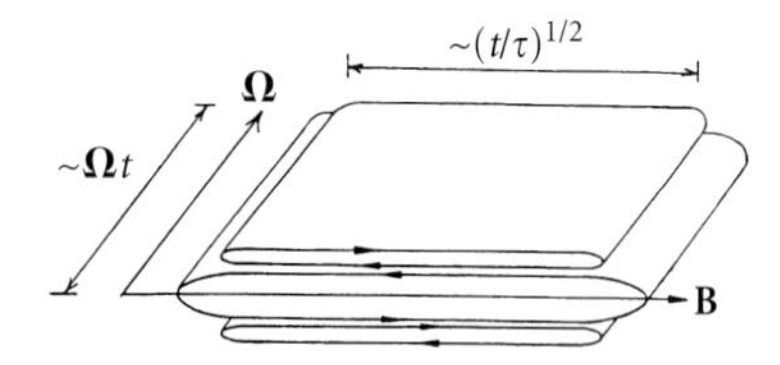

Figure 9.29 Evolution of a low-R_m vortex subject to both Coriolis and Lorentz forces.

holds as before. (l_B is a typical length-scale parallel **B**.) Therefore, if we repeat the arguments of Section 9.4.4, we would expect $E \sim (t/\tau)^{-1/2}$.

We have the makings of a paradox. On the one hand, (9.130) suggests that the flow is destroyed exponentially fast on a timescale of 4τ. On the other hand, the energy appears to decline only algebraically, as $(t/\tau)^{-1/2}$. Thus, the disappearance of the angular momentum cannot be a manifestation of the decay of the vortex. We met this paradox in Section 9.4.2, along with its resolution. The flow evolves into a multi-cellular structure in which the total angular momentum is exponentially small due to the cancellation of **H** between adjacent cells. The situation is as depicted in Figure 9.29.

So it would seem that, whatever the value of Λ, plate-like structures will emerge. Moreover, when Λ is small or of order unity, the eddy presumably undergoes a substantial elongation before being destroyed, the Coriolis force extruding the eddy into a partial Taylor column. Thus, we have extension along the rotation axis by inertial wave propagation progressing hand in hand with pseudo-diffusion along the **B**-lines. The axial length-scale will increase as $l_\Omega \sim l_0 \Omega t$, while the length-scale parallel to $\mathbf{B}_0$ grows as $l_B \sim l_0 (t/\tau)^{1/2}$ (see Section 9.4.2). The ratio of the timescales for these two processes is simply the Elsasser number Λ.

When N is large, this behaviour may be confirmed by exact analysis. Combining (9.58) with (9.10) yields, after a little algebra,

$$\left[\frac{\partial}{\partial t}\nabla^2 + \frac{1}{\tau}\left[\partial^2/\partial x_{||}^2\right]\right]^2 \mathbf{u} + 4(\boldsymbol{\Omega}\cdot\nabla)^2(\nabla^2\mathbf{u}) = \mathbf{0}, \quad N \gg 1 \tag{9.132}$$

where $x_{||}$ is the coordinate parallel to $\mathbf{B}_0$. We now consider initial conditions that are axisymmetric about z and look for solutions in which ω_z remains even in x. (Reversing the magnetic field does not change the dynamics.) It is a simple matter to solve (9.132) by taking the Fourier transform in y and z, and the Fourier-cosine transform in x. This yields a second-order equation for the transform of ω_z:

$$\left[\frac{\partial}{\partial t} + \frac{1}{\tau}\left[k_B/k\right]^2\right]^2 W_z + (2\boldsymbol{\Omega}\cdot\mathbf{k}/k)^2 W_z = 0.$$

This is readily integrated and the inverse transform then yields, for large times (Davidson and Siso-Nadal 2002; Siso-Nadal and Davidson 2003),

$$\omega_z(\mathbf{x},\ t) \sim (t/\tau)^{-3/2} F\left(x/(t/\tau)^{1/2}, y, z/\Omega t\right) \tag{9.133}$$

where the form of F depends on the initial conditions. This confirms that l_B and l_Ω grow as, $l_B \sim (t/\tau)^{1/2}$, $l_\Omega \sim \Omega t$. We conclude that the situation is indeed as shown in Figure 9.29. The eddy is extruded in the Ω-direction at a rate $2\Omega t$, forming a sort of Taylor column. Simultaneously, it diffuses along the **B**-lines by low-R_m Alfvén wave propagation. This requires $l_B \sim (t/\tau)^{1/2}$. In the meantime, the eddy sheds its angular momentum by fragmenting into a set of platelets of oppositely signed vorticity.

Evidently, when both Coriolis and Lorentz forces act, the turbulence is highly anisotropic. These forces are especially important in the astrophysical context, such as in the violent motion on the surface of the sun or the turbulent convection which occurs in many planetary interiors.

9.5.2 *Turbulence in the core of the earth*

Parker considered fairly irregular motions of thermal convection in the earth's core, together with a non-uniform rotation. Like Bullard before him (and like Elsasser before him) he supposed the non-uniform rotation, acting on a dipole field, to extend the lines of force in the azimuthal direction and to create an azimuthal field. The novel feature in his approach was when he came to consider the interaction of convective motions with the azimuthal field ... Parker pointed out that Coriolis forces acting on horizontally moving material will produce vortex effects ... so matter rising in a convective cell would not simply lift the lines of force of the azimuthal field: it would also twist them to create a north-south field. Hence loops of magnetic force are produced, which in general reinforce the original dipole field. (T.G. Cowling 1957)

The physical insight crammed into these few words is breathtaking and we shall spend the rest of this section explaining the implications of Parker's theory of the geomagnetic field.

It is now generally agreed that the earth's magnetic field owes its existence to turbulent motion in the molten core of the earth, motion which stretches and twists the field, maintaining it against the natural forces of decay. The temperature in the earth's interior is well above the Curie point at which ferromagnetic material loses its permanent magnetism, and so we cannot picture the earth's field as arising from some sort of giant magnet. Nor can the earth's magnetic field be the relic of some primordial field trapped within its interior during its formation, as this would have long ago decayed through Ohmic dissipation. (See Example 9.9.) The only plausible explanation is that **B** is maintained by *dynamo action*, in which mechanical energy is converted into magnetic energy via the Lorentz force.

From the perspective of turbulence theory this is not such a surprising concept. The evolution equation for **B** is precisely the same as that for $\boldsymbol{\omega}$ (except we must replace λ by ν) and we know that turbulence tends to increase $\boldsymbol{\omega}^2$ by vortex-line stretching. It would not be surprising, therefore, if forced turbulence tends to stretch the magnetic

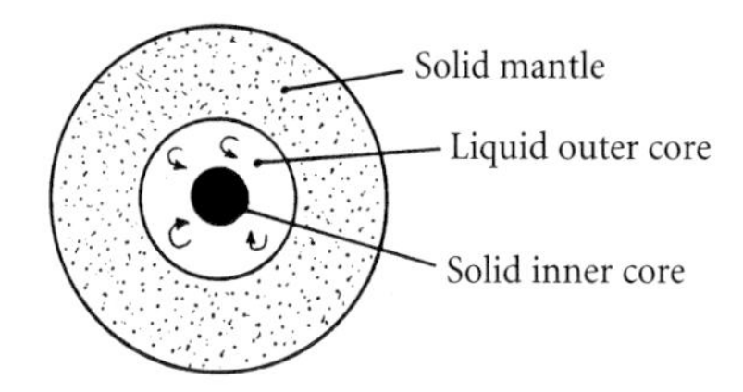

Figure 9.30 Structure of the earth. The solid inner core has radius $\sim 10^3$ km, the liquid core has a radius of $\sim 3 \times 10^3$ km, and the outer radius of the mantle is $\sim 6 \times 10^3$ km.

field lines, increasing $\mathbf{B}^2$. Of course, R_m must be large enough for field-line stretching to outway Ohmic dissipation. (See Example 9.10.) Also, we need an energy source to maintain the turbulence in the face of viscous and Lorentz forces, which tend to suppress it. The likely energy source is natural convection, driven partly by temperature differences in the liquid core and partly by compositional buoyancy which arises from the slow solidification of the solid inner core. (As the solid inner core grows by solidification it releases iron, which is rich in an admixture of lighter elements.) At present, the radius of the solid inner core is roughly one-sixth of the earth's outer radius (Figure 9.30).

There is one other important factor in dynamo theory: rotation. It is thought that several of the planets have magnetic fields maintained by dynamo action, and it is striking that the average planetary field strength tends to be proportional to the rate of rotation of the planet in question. Moreover, in some cases, such as that of the earth, the magnetic axis is more or less aligned with the rotation axis. So it seems likely, therefore, that the Coriolis force is important in dynamo action. Indeed, order-of-magnitude estimates suggest that the dominant forces acting on the liquid interior of the earth are the Lorentz force, the Coriolis force, and buoyancy. The viscous and non-linear inertial forces are almost completely negligible outside boundary layers since both $\Omega l/u$ and Re are vast.

So it seems plausible that turbulence, modified in some sense by rotation, maintains the earth's magnetic field by flux tube stretching. However, there is at least one problem with this line of thought. Turbulence is, by its very nature, random. The earth's magnetic field, on the other hand, appears to be more or less steady, though it is subject to occasional reversals. It is natural, therefore, to question whether or not turbulence is really required, or even desirable, in a geodynamo theory. Perhaps some steady-on-average convective flow could stretch the flux tubes in the desired manner? However, it is readily confirmed that, if this mean flow is steady and axisymmetric, no dynamo action is possible. The proof is as follows. We employ cylindrical polar coordinates (r, θ, z) and divide $\mathbf{B}$ into azimuthal, $\mathbf{B}_\theta = (0, B_\theta, 0)$, and poloidal, $\mathbf{B}_p = (B_r, 0, B_z)$, components (Figure 9.31). Both $\mathbf{B}$ and $\mathbf{u}$ are assumed to be axisymmetric.

If the velocity field is similarly divided, $\mathbf{u} = \mathbf{u}_\theta + \mathbf{u}_p$, the induction equation yields evolution equations for $\mathbf{B}_\theta$ and $\mathbf{B}_p$:

$$\frac{\partial \mathbf{B}_p}{\partial t} = \nabla \times (\mathbf{u}_p \times \mathbf{B}_p) + \lambda \nabla^2 \mathbf{B}_p \tag{9.134a}$$

$$\frac{\partial \mathbf{B}_\theta}{\partial t} = \nabla \times (\mathbf{u}_p \times \mathbf{B}_\theta) + \nabla \times (\mathbf{u}_\theta \times \mathbf{B}_p) + \lambda \nabla^2 \mathbf{B}_\theta. \tag{9.134b}$$

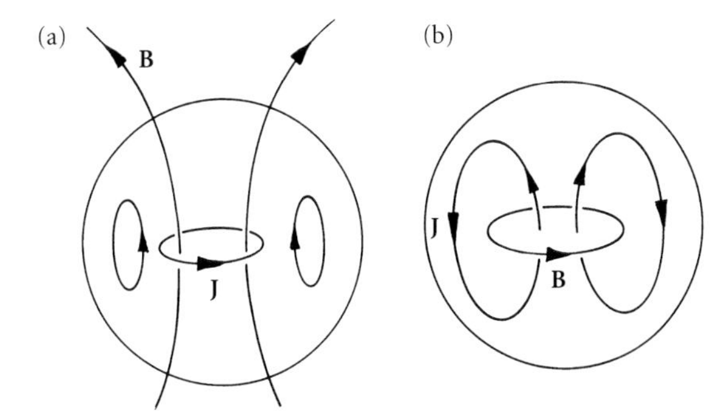

Figure 9.31 (a) Poloidal and (b) azimuthal magnetic fields. The poloidal field is maintained by azimuthal currents and the azimuthal field by poloidal current.

The first of these can be 'uncurled' to yield an equation for the vector potential, $\mathbf{A}_\theta$, of the poloidal field, $\mathbf{B}_p$. (Remember that the vector potential, $\mathbf{A}$, for a field $\mathbf{B}$, is defined by $\nabla \times \mathbf{A} = \mathbf{B}$, $\nabla \cdot \mathbf{A} = 0$.) So, after a little algebra, (9.134a,b) yield

$$\frac{D}{Dt}(rA_\theta)\hat{\mathbf{e}}_\theta = \lambda[r\nabla^2 \mathbf{A}_\theta] \tag{9.135a}$$

$$\frac{D}{Dt}\left(\frac{B_\theta}{r}\right)\hat{\mathbf{e}}_\theta = \mathbf{B}_p \cdot \nabla\left(\frac{u_\theta}{r}\right)\hat{\mathbf{e}}_\theta + \lambda[r^{-1}\nabla^2 \mathbf{B}_\theta]. \tag{9.135b}$$

The key point is that there is no source term in (9.135a) which might maintain $\mathbf{A}_\theta$, and hence $\mathbf{B}_p$. In fact, it is readily confirmed that, whatever the velocity field, $\mathbf{B}_p$ simply decays by Ohmic dissipation. When $\mathbf{B}_p$ is zero, (9.135b) reduces to a simple advection-diffusion equation for B_θ/r. Again, there is no source term for $\mathbf{B}_\theta$ and so, as $\mathbf{B}_p$ falls away, so does $\mathbf{B}_\theta$.

So it seems that the geodynamo cannot be maintained by a steady, axisymmetric flow. The motion in the core must be non-axisymmetric and almost certainly unsteady.

The most popular cartoon for dynamo action is the so-called *α–Ω dynamo*. The idea is the following. Convection in the core of the earth systematically transports angular momentum from place to place and it would not be surprising if, due to the principle of conservation of angular momentum, there are systematic differences in angular velocity throughout the earth's core. It is believed by some that the solid inner core of the earth, and its surrounding fluid, rotates slightly faster than the mantle. There are simple dynamical arguments to suggest that this is so (see Example 9.11) and indeed some seismic studies lend tentative support to the idea. Now a typical estimate of u and λ in the earth's core are $u \sim 2 \times 10^{-4}$ m/s and $\lambda \sim 2$ m^2/s. Thus, R_m based on the width of the liquid annulus, $l \sim 2 \times 10^6$ m, is around $ul/\lambda \sim 200$. At the large scale, then, we might consider $\mathbf{B}$ as almost

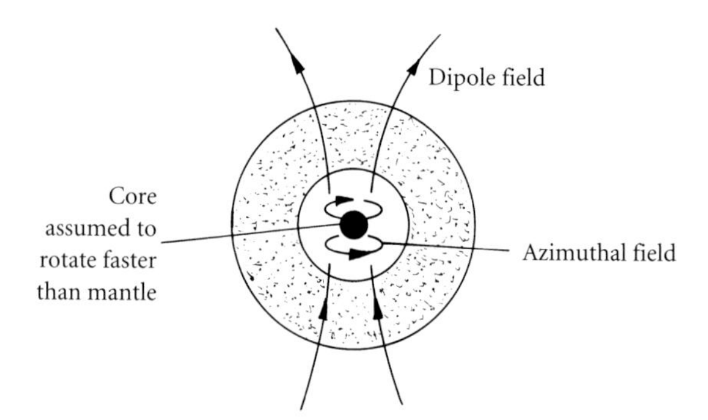

Figure 9.32 The Ω-effect relies on differential rotation sweeping out an azimuthal field from the observed dipole.

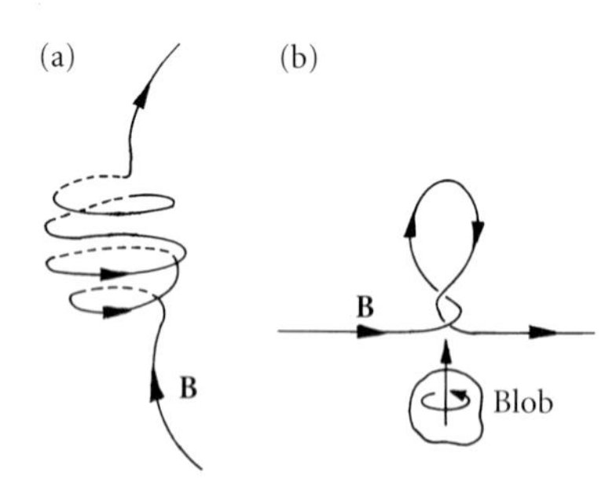

Figure 9.33 The Ω and α effects as envisaged by Parker in 1955. (a) Ω-effect (b) α-effect.

frozen into the liquid core, so that differential rotation (if it exists) will tend to spiral out an azimuthal field, $\mathbf{B}_\theta$, from the observed dipole one, $\mathbf{B}_\text{d}$ (Figure 9.32). Indeed, order of magnitude estimates based on (9.135b) give

$$B_\theta \sim R_\text{m} B_\text{d} \gg B_\text{d}$$

so the dominant field in the interior of the earth is probably azimuthal (east–west) rather than dipole (north–south). This process, whereby systematic variations in angular velocity sweep out an azimuthal field from the dipole field, is referred to as the *Ω-effect*. It is evident in Plate 16, which illustrates a crude numerical simulation of the geodynamo. Note that the Ω-effect produces an azimuthal field which is skew-symmetric about the equator.

To get a regenerative cycle, however, we need some way of converting the azimuthal field back into a dipole field, thus completing the cycle, $\mathbf{B}_\text{d} \to \mathbf{B}_\theta \to \mathbf{B}_\text{d}$. This is where the so-called α-effect comes in. In the α–Ω dynamo, the *α-effect* converts B_θ back into a dipole field. The idea is that small-scale turbulence in the core, which is driven by buoyancy, tends to tease out a small-scale magnetic field from the large-scale azimuthal one (Figure 9.33). This is a quasi-random process, but if conditions are dynamically favourable then this random, small-scale field can reorganise itself so as to reinforce the original dipole field.[9] Note that, since R_m has the modest value of $\sim$200 at the largest scales, it is likely to be small (say $R_\text{m} \sim 1$) at the scale of the α-effect.

The name α-effect arises as follows. Suppose that in some region of the liquid core we divide $\mathbf{u}$ and $\mathbf{B}$ up into mean and fluctuating parts: $\mathbf{u} = \mathbf{u}_0 + \mathbf{u}'$, $\mathbf{B} = \mathbf{B}_0 + \mathbf{B}'$. Then the induction equation (9.51) can be ensemble-averaged to give,

[9] For a detailed account of the α-effect, see Moffatt (1978). The original idea of an α–Ω dynamo was due to Parker (1955). It was later placed on a firm mathematical footing by Steenbeck et al. (1966).

$$\frac{\partial \mathbf{B}_0}{\partial t} = \nabla \times (\mathbf{u}_0 \times \mathbf{B}_0) + \lambda \nabla^2 \mathbf{B}_0 + \nabla \times (\langle \mathbf{u}' \times \mathbf{B}' \rangle).$$

Evidently, the turbulent fluctuations have given rise to a mean e.m.f., $\langle \mathbf{u}' \times \mathbf{B}' \rangle$. This is reminiscent of the Reynolds stresses which occur in conventional turbulence when the Navier–Stokes equation is averaged. In the α-effect it is supposed that $\mathbf{B}'$ is a linear function of $\mathbf{B}_0$, since it arises from the interaction of $\mathbf{u}'$ with $\mathbf{B}_0$. So the turbulence-induced e.m.f. is modelled as, $\langle \mathbf{u}' \times \mathbf{B}' \rangle_i = \alpha_{ij}(\mathbf{B}_0)_j$, for some α_{ij}. Often it is assumed that the turbulence is quasi-isotropic,[10] so that the assumed form of $\langle \mathbf{u}' \times \mathbf{B}' \rangle$ simplifies to

$$\langle \mathbf{u}' \times \mathbf{B}' \rangle = \alpha \mathbf{B}_0 \tag{9.136}$$

where α is a pseudo-scalar,[11] which has the dimensions of velocity. In effect, this is a closure hypothesis which asserts that the turbulent e.m.f. gives rise to a mean current: $\langle \mathbf{J} \rangle = \sigma\alpha\mathbf{B}_0$. The end result is,

$$\frac{\partial \mathbf{B}_0}{\partial t} = \nabla \times (\mathbf{u}_0 \times \mathbf{B}_0) + \lambda \nabla^2 \mathbf{B}_0 + \nabla \times (\alpha \mathbf{B}_0). \tag{9.137}$$

It is the appearance of α in (9.137) which gives rise to the term α-effect. Let us see where (9.137) leads. For simplicity we assume that the mean fields, $\mathbf{u}_0$ and $\mathbf{B}_0$, possess axial symmetry and so we divide $\mathbf{u}_0$ and $\mathbf{B}_0$ into azimuthal and poloidal components (Figure 9.31). Moreover, it is convenient to introduce a vector potential for the poloidal field $\mathbf{B}_\text{p}$, defined by, $\mathbf{B}_\text{p} = \nabla \times (\mathbf{A}_\theta)$, $\nabla \cdot \mathbf{A}_\theta = 0$. If we likewise divide (9.137) into azimuthal and poloidal components, we obtain evolution equations for B_θ and A_θ. They turn out to be reminiscent of (9.135a,b), but with extra terms associated with α:

$$\frac{D}{Dt}(rA_\theta) = \alpha r B_\theta + \lambda \nabla_*^2 (rA_\theta) \tag{9.138}$$

$$\frac{D}{Dt}\left(\frac{B_\theta}{r}\right) = \mathbf{B}_\text{p} \cdot \nabla \left(\frac{u_\theta}{r}\right) - \alpha \left[r^{-2} \nabla_*^2 (rA_\theta) \right] + \lambda \left[r^{-2} \nabla_*^2 (rB_\theta) \right]. \tag{9.139}$$

[10] By quasi-isotropic we mean turbulence which is statistically independent of rotations though not reflections.

[11] A pseudo-scalar is a scalar which changes sign as we move from a right-handed to a left-handed coordinate system. Pseudo-scalars are discussed in Appendix I. The fact that α is a pseudo-scalar means that only turbulence which lacks reflectional symmetry can give rise to an α-effect. This rules out fully isotropic turbulence but not quasi-isotropic turbulence, which is statistically independent of rotations but not reflections (Moffatt 1978).

(Here ∇_*^2 is the Stokes operator defined by (9.81).) We recognize the first term on the right of (9.139) as the Ω-effect, whereby differential rotation acting on the dipole field, $\mathbf{B}_p$, provides a source of azimuthal field. This is thought to be the primary mechanism for generating B_θ. The first term on the right of (9.138), on the other hand, is the α-effect which converts B_θ back into a dipole field, thus completing the cycle $B_p \to B_\theta \to B_p$. Integration of equations (9.138) and (9.139), with suitable specifications of α and u_θ/r, do indeed yield self-sustaining magnetic fields provided that the *dynamo number*, $(\alpha l/\lambda)(u_\theta l/\lambda)$, is large enough.

This raises the question: how big is α? It seems plausible that it depends on $|\mathbf{u}'|$, l, and λ, where l and $|\mathbf{u}'|$ are integral length and velocity scales of the turbulence. Dimensional analysis then suggests $\alpha = |\mathbf{u}'| f(|\mathbf{u}'| l/\lambda)$. However, this cannot be correct since α is a pseudo-scalar which changes sign when we convert from a right-handed coordinate system to a left-handed one. $|\mathbf{u}'|$, on the other hand, is a true scalar. Since helicity, $\mathbf{u} \cdot \boldsymbol{\omega}$, is the most common pseudo-scalar encountered in turbulence theory, a more promising estimate of α might be

$$\alpha = \frac{\langle \mathbf{u}' \cdot \boldsymbol{\omega}' \rangle l}{|\mathbf{u}'|} f(|\mathbf{u}'| l/\lambda)$$

and indeed such estimates are frequently used. When $R_m = |\mathbf{u}'| l/\lambda$ is large, it is thought that λ ceases to be a relevant parameter, whereas at low R_m, it turns out that α is inversely proportional to λ. Thus, in the limit of small and large R_m, we have,

$$\alpha \sim -\frac{\langle \mathbf{u}' \cdot \boldsymbol{\omega}' \rangle l}{|\mathbf{u}'|}, \quad (R_m \gg 1), \qquad \alpha \sim -\frac{\langle \mathbf{u}' \cdot \boldsymbol{\omega}' \rangle l^2}{\lambda}, \quad (R_m \ll 1).$$

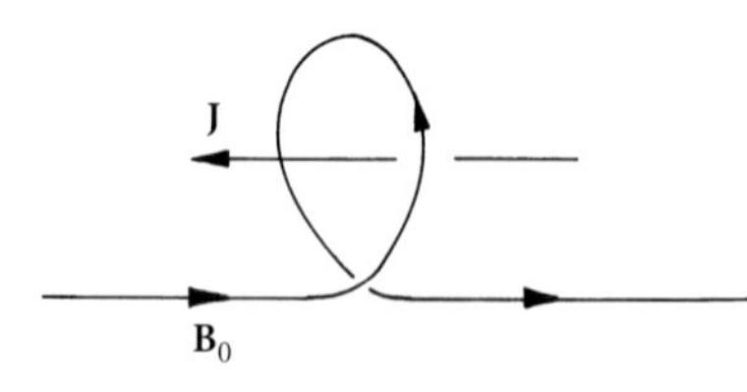

Figure 9.34 A positive helicity tends to give rise to an induced current which is anti-parallel to $\mathbf{B}_0$, and hence to a negative value of α.

Note the minus sign. This is because positive helicity tends to induce a **B**-loop whose associated current density, with which we associate $\sigma\alpha\mathbf{B}_0$, is anti-parallel to $\mathbf{B}_0$ (Figure 9.34).

Actually it may be shown that, when $R_m \ll 1$, the α-effect can be formally justified. Moreover, for homogeneous turbulence, the value of α_{ij} is (Davidson 2001),

$$\alpha_{ij} = -\lambda^{-1}\left[\langle \mathbf{a}' \cdot \mathbf{u}' \rangle \delta_{ij} - \left\langle a_i' u_j' + a_j' u_i' \right\rangle\right], \quad R_m \ll 1$$

where $\mathbf{a}'$ is the vector potential for $\mathbf{u}'$. For quasi-isotropic turbulence, this simplifies to $\alpha = -\langle \mathbf{a}' \cdot \mathbf{u}' \rangle/(3\lambda)$. Since $|\mathbf{a}'| \sim |\boldsymbol{\omega}'| l^2$, this is consistent with the estimate above.

So far we have considered only kinematic aspects of the α-effect. There is also the issue of why small-scale motion in the core should be *dynamically* predisposed to generate a mean dipole field from the internal east–west field. Here there are two popular cartoons, which we might label as *blobs* and *waves*. Let us start with blobs. Consider a small blob of light material which has been released from the inner core and is floating up towards the mantle. It is shown in Example 9.2 in Section 9.2.3 that, if we neglect the Lorentz forces acting on the blob, the balance between the Coriolis, pressure, and buoyancy forces causes the blob (and its surrounding fluid) to spin about a vertical axis. Now suppose that the earth's dipole field points from south to north. (At present it is in the opposite direction.) Then the Ω-effect will produce an azimuthal field which is negative in the north and positive in the south (Figure 9.32). As the blob passes up through the east–west field, its spin will ensure that it spirals out a **B**-loop as shown in Figure 9.34. There is an azimuthal current associated with this loop and if many such blobs float out towards the mantle, each inducing an azimuthal current, there is the possibility that the small-scale currents add to produce a global azimuthal current, which in turn supports a dipole field. For this cartoon to work, we need the helicity, $\mathbf{u}' \cdot \boldsymbol{\omega}'$, of the buoyant blobs to be positive in the north, so that they sweep out a positive azimuthal current from the negative azimuthal field, and negative in the south, so that the same positive azimuthal current is generated from a positive azimuthal field. In this cartoon, then, the turbulence is associated with a multitude of small spiralling, buoyant blobs whose helicity is skew-symmetric about the equator.

The other popular cartoon concerns waves. Here the small-scale motion is pictured as a sea of small helical waves, generated by buoyancy forces. Both inertial waves and magnetostrophic waves are potential candidates since both possess mean helicity. Indeed it is readily confirmed that a helical wave passing through a mean magnetic field produces an α-effect. (See Example 9.12.) So the idea is that small-scale inertial or magnetostrophic waves spiral through the east–west field, inducing azimuthal current, which in turn supports a net dipole field. Of course, a sea of entirely random waves will produce no net effect, since the small-scale currents induced by the waves will tend to cancel. As with the blob cartoon, we require some global organization of the waves so that their individual effects are additive. In short, to support a northward pointing dipole, we require the waves to have positive helicity in the north and negative helicity in thesouth.

The α–Ω model has its supporters and its detractors. Perhaps the weakest link in the chain of arguments is the α-effect itself. It presupposes a two-scale structure for **u**, with a significant amount of energy at the smaller scales. Can such small eddies, blobs, or waves survive the intense Ohmic dissipation associated with B_θ? Let us consider this question in a little more detail. At the scale at which α is intended to operate, we have negligible viscous and non-linear inertial forces, and R_m is of the order of unity. The governing equation for a small-scale eddy is therefore,

$$\frac{\partial \mathbf{u}}{\partial t} = 2\mathbf{u} \times \mathbf{\Omega} - \nabla(p/\rho) + \rho^{-1}\mathbf{J} \times \mathbf{B}_0 + (\text{buoyancy forces}). \tag{9.140}$$

(Here $\mathbf{B}_0$ is the local large-scale azimuthal field.) Substituting for the Lorentz force using (9.58) we have,

$$\rho \frac{\partial}{\partial t}(\nabla^2 \mathbf{u}) + \sigma(\mathbf{B}_0 \cdot \nabla)^2 \mathbf{u} = -2\rho(\mathbf{\Omega} \cdot \nabla)\boldsymbol{\omega} + (\text{buoyancy}). \tag{9.141}$$

However, we have already studied the dynamics of eddies governed by this equation since (9.141) is effectively the same as (9.132). As we have seen, such eddies rapidly fragment into a network of platelets (Figure 9.29) and, in the absence of buoyancy, their energy decays on a timescale of $\tau \sim (\sigma B_0^2/\rho)^{-1}$. For typical estimates of σ, ρ, and B_θ, we find that τ is of the order of days in the core of the earth. This is extremely rapid by comparison with the other relevant timescales. For example, the time taken to convect material from the solid inner core to the mantle is $\sim$300 years, while the large-scale diffusion time for $\mathbf{B}$, l^2/λ, is of the order of 10^5 years. Evidently, the Ohmic dissipation acting on the small scales is extremely potent, and it is by no means clear that it is plausible to conceive of an energetic small-scale motion, as required by the α-effect.

There is a growing feeling, therefore, that classical dynamo models, such as the α–Ω model, are, at best, pedagogical idealizations. They capture some of the key physical processes in a simple mathematical framework, but are too idealized to be realistic models. In all probability there is no formal separation of scales with an energetic small-scale motion. Rather, the dynamo processes probably all operate at the large scale only. Thus, for example, the Ω-effect might be combined with a large-scale helical motion, say large-scale inertial or magnetostrophic waves, to produce an α–Ω like model but without the need for any small-scale motion. Such a cartoon is discussed in Example 9.13.

Example 9.9 Decay time for the earth's field in the absence of dynamo action
Show that, in the absence of any motion in the interior of the earth, the energy of the terrestrial magnetic field will decay as

$$\frac{dE_B}{dt} = -\int (\mathbf{J}^2/\sigma)\, dV$$

where $\mathbf{J}$ is the terrestrial current density. Confirm that the decay time for the field, in the absence of core motion, is of the order of

$\tau_d \sim R_c^2/\lambda$, where R_c is the radius of the conducting core. Actually, a more accurate estimate is $R_c^2/\pi^2\lambda$. Show that, for $\lambda \sim 2\,m^2/s$, we have $\tau_d \sim 10^4$ years (H. Lamb 1889). The earth's magnetic field, on the other hand, has been around for at least 10^8 years.

Example 9.10 A large magnetic Reynolds number is needed for dynamo action
Let us now allow for core motion. Show that the energy of the terrestrial magnetic field varies according to

$$\frac{dE_B}{dt} = \frac{1}{\mu}\int \mathbf{u} \cdot [\mathbf{B} \times (\nabla \times \mathbf{B})] dV - \frac{1}{\sigma}\int \mathbf{J}^2\, dV = P - D.$$

Use the Schwarz inequality to show that the production of integral, P, is bounded from above according to

$$\mu^2 P^2 \leq u_{\max}^2 \int \mathbf{B}^2 dV \int (\nabla \times \mathbf{B})^2\, dV$$

where $u_{\max}$ is the maximum velocity (in a rotating frame of reference) in the core. Also use the calculus of variations to bound the dissipation integral, D, from below according to $D \geq 2\pi^2(\lambda/R_c^2)E_B$. Combining the inequalities we have

$$\frac{dE_B}{dt} \leq (2E_B D/\lambda)^{1/2}[u_{\max} - \pi\lambda/R_c]$$

which shows that a necessary condition for dynamo action is $R_m = (u_{\max}R_c/\lambda) > \pi$.

Example 9.11 Driving the Ω-effect by natural convection
Consider the equation of motion for the core in which viscous and non-linear terms are neglected by comparison with the Coriolis force. We have,

$$\frac{\partial \mathbf{u}}{\partial t} = 2\mathbf{u} \times \boldsymbol{\Omega} - \nabla(p/\bar{\rho}) + (\delta\rho/\bar{\rho})\mathbf{g} + \mathbf{J} \times \mathbf{B}/\bar{\rho}$$

where $\delta\rho$ is the density perturbation which drives the natural convection, $\bar{\rho}$ is the mean density, $\mathbf{g}$ the gravitational acceleration, and we have used the Boussinesq approximation to accommodate variations in density. Show that the equivalent angular momentum equation is,

$$\frac{\partial}{\partial t}(\mathbf{x} \times \mathbf{u}) = 2\mathbf{x} \times (\mathbf{u} \times \boldsymbol{\Omega}) + \nabla \times (p\mathbf{x}/\bar{\rho}) + (\text{Lorentz torque}).$$

Now consider the cylindrical volume V_1 shown in Figure 9.35. It is concentric with the rotation axis, symmetric about the equator, and has radius R. Show that our angular momentum equation integrates to give

$$\frac{d}{dt}\int_{V_1} (\mathbf{x} \times \mathbf{u})_z\, dV = -\Omega \oint_{S_1} r^2 \mathbf{u} \cdot d\mathbf{S} + (\text{Lorentz torque})$$

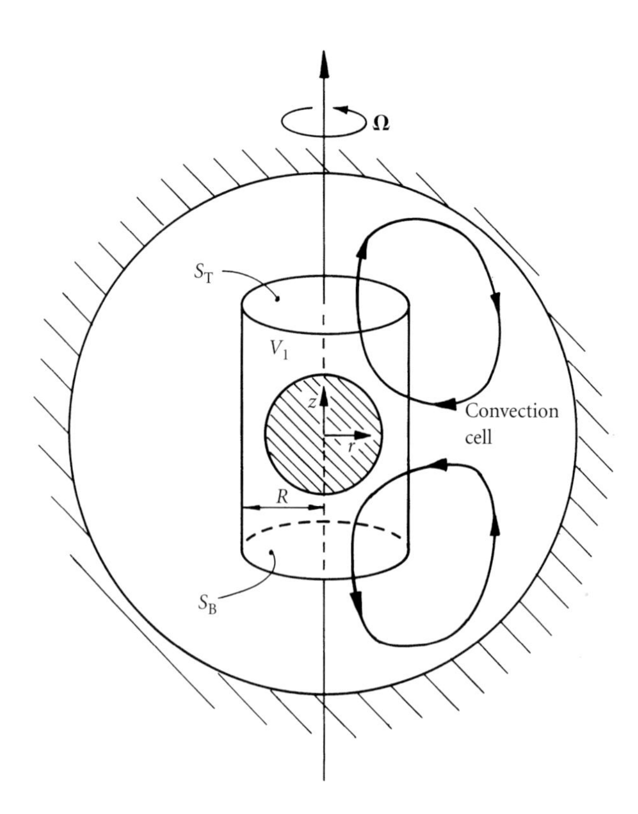

Figure 9.35 Natural convection in the core tends to increase the angular momentum $(\mathbf{x} \times \mathbf{u})_z$ near the inner core at the expense of the angular momentum near the mantle.

where S_1 encloses V_1 and we are using cylindrical polar coordinates (r, θ, z). Now use mass conservation to show that this integral equation can be rewritten as

$$\frac{d}{dt}\int_{V_1} (\mathbf{x} \times \mathbf{u})_z \, dV = \Omega \int_{S_T+S_B} (R^2 - r^2)\mathbf{u} \cdot d\mathbf{S} + (\text{Lorentz torque})$$

where S_T and S_B are the top and base of the cylinder. When V_1 is relatively short there is a net axial flow out through S_T and S_B, driven by convection. The integral on the right will then be positive and so the angular momentum within V_1 will tend to increase as a result of natural convection. On the other hand, if the cylinder spans the liquid core, then the first integral on the right is zero. By implication, the buoyant convection shown in Figure 9.35 tends to increase the angular momentum of the fluid near the inner core at the expense of the angular momentum of the fluid near the mantle. This is the dynamical basis of the Ω-effect. Of course, in an inertial frame of reference, this mechanism is associated with the spin up of fluid which moves radially inward and spin down of fluid which moves outward.

Example 9.12 The α-effect produced by helical waves
Consider a small amplitude wave of the form $\mathbf{u} = \text{Re}[\hat{\mathbf{u}} \exp(j(\mathbf{k} \cdot \mathbf{x} - \varpi t))]$, where $\hat{\mathbf{u}} = u_0(-j, 1, 0)$ and $\mathbf{k} = (0, 0, k)$ in (x, y, z) coordinates. Confirm that the wave is helical in the sense that $\nabla \times \mathbf{u} = k\mathbf{u}$, so that the helicity density is $\boldsymbol{\omega} \cdot \mathbf{u} = k\mathbf{u}^2$. Suppose that the wave travels through a uniform field $\mathbf{B}_0$. Show that the induced magnetic field has amplitude

$$\hat{\mathbf{b}} = \frac{\mathbf{B}_0 \cdot \mathbf{k}}{\varpi^2 + k^4\lambda^2}(j\lambda k^2 - \varpi)\hat{\mathbf{u}}$$

and confirm that the mean e.m.f. is

$$\langle \mathbf{u} \times \mathbf{b} \rangle = -\frac{\lambda k^2(\mathbf{B}_0 \cdot \mathbf{k})u_0^2}{\varpi^2 + k^4\lambda^2}(0, 0, 1).$$

Example 9.13 A one scale α–Ω cartoon of the geodynamo
It is possible to conceive of a multitude of α–Ω-like cartoons for the geodynamo, and if there is no requirement to add mathematical 'flesh' to the physical 'bare bones', then these can be conceived as one-scale models. Of course, with little in the way of hard evidence, it is difficult to distinguish between those which are hopelessly wrong and those which contain a germ of truth. Here is one such model. There are many others.

Let us suppose that differential rotation exists within the core, as anticipated by the Ω-effect. Then an intense subsurface azimuthal field will be swept out from the observed dipole field, being strongest near the inner core and skew-symmetric about the equator. For the sake of argument let us take the dipole field to point from south to north, as shown in Figure 9.36. Then the azimuthal field will be positive in the south and negative in the north. It seems plausible that the turbulent convection near the inner core is partly suppressed by the strong azimuthal field, but that the convection is less inhibited in regions closer to the mantle. Let us suppose, therefore, that the most active motion occurs in the vicinity of the mantle. The Rossby number is small and so the motion will be quasi-two-dimensional. This two-dimensionality is maintained by the continual propagation of inertial waves along the rotation axis. That is, if a blob or eddy moves from left to right, its associated Taylor column moves with it, the translation of the Taylor column being accomplished by the propagation of fast inertial waves. Thus, as the turbulence churns away in the outer regions of the core, it continually sends a stream of inertial waves down into the inner core, where the intense azimuthal field sits. Because these waves are helical, they will produce an α-effect as they

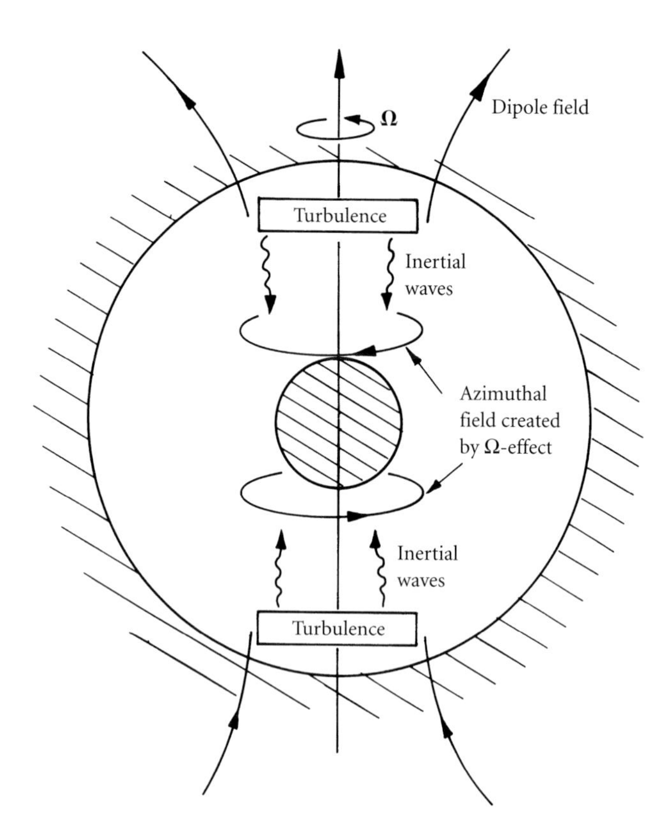

Figure 9.36 A one-scale α–Ω-like cartoon of the geodynamo.

spiral through the azimuthal field. Consider inertial waves of the form $\mathbf{u} = \hat{\mathbf{u}} \exp[j(\mathbf{k} \cdot \mathbf{x} - \varpi t)]$. Use the results of Section 9.2.3 to show that the frequency, helicity density, and group velocity of these waves are: $\varpi = \pm 2(\boldsymbol{\Omega} \cdot \mathbf{k})/k$, $\hat{\boldsymbol{\omega}} \cdot \hat{\mathbf{u}} = \mp k\hat{\mathbf{u}}^2$, and $\mathbf{c}_g = \pm 2\boldsymbol{\Omega}/k$, where $\mathbf{k}$ is almost perpendicular to $\boldsymbol{\Omega}$. Now use the results of Example 9.12 to show that the mean e.m.f. induced by the inertial waves spiralling through the azimuthal field is

$$\langle \mathbf{u} \times \mathbf{b} \rangle = \pm \frac{(\mathbf{B}_0 \cdot \mathbf{k}) u_0^2 \lambda k}{\varpi^2 + \lambda^2 k^4} \mathbf{k}$$

where u_0 is the amplitude of the wave. Since $\mathbf{c}_g$ is parallel to $\boldsymbol{\Omega}$ in the south and anti-parallel to $\boldsymbol{\Omega}$ in the north, we must take the top sign in the south and the bottom sign in the north. Show that this results in a mean current, $\sigma\langle \mathbf{u} \times \mathbf{b} \rangle$, which is azimuthal and in the positive θ direction in both hemispheres. In short, the direction of the induced current is exactly that required to support the dipole field. We have a self-consistent cartoon. Now construct a similar argument based on magnetostrophic waves.

9.5.3 *Turbulence near the surface of the sun*

Turbulence plays a central role in astrophysics, influencing the dynamics of the interstellar medium, accretion discs, and the 10^{23} stars which inhabit the universe. Here we focus on the role of turbulence in stars, and in particular our own star. We shall see that turbulence controls the transport of heat from the interior of the sun to its surface, plays a crucial role in the 22-year solar cycle during which the sun's magnetic field comes and goes, and triggers the violent explosions which are all too evident on the surface of the sun (Plate 14).

The interior structure of a star depends, amongst other things, on its mass. Roughly speaking, there are two types. Stars whose mass is similar to, or smaller than, our own sun have quiescent, radiative cores surrounded by a turbulent, convective envelope. Larger stars tend to have a convective core embedded in a radiative envelope. In either case, the Reynolds number in the convective region is invariably very large ($\text{Re} > 10^{10}$) and so the motion is highly turbulent.

Let us focus on our own star. Its interior is conventionally divided into three regions (Figure 9.37). The inner core of radius $\sim 2 \times 10^5$ km is where the thermonuclear reactions are concentrated. This is surrounded by a radiative zone of thickness $\sim 3 \times 10^5$ km. Here heat is transported diffusely outward by radiation. The outermost region of the sun's interior is convectively unstable and so heat is carried to the surface by large-scale convective motions. This *convection zone* is $\sim 2 \times 10^5$ km deep. The surface of the sun is called the photosphere. This is a thin, transparent layer of relatively dense material, about 500 km deep. Above this we have the solar atmosphere, which is divided into the chromosphere and the corona. There is no clear upper boundary to the corona, which extends out to the planets in the form of the solar wind.

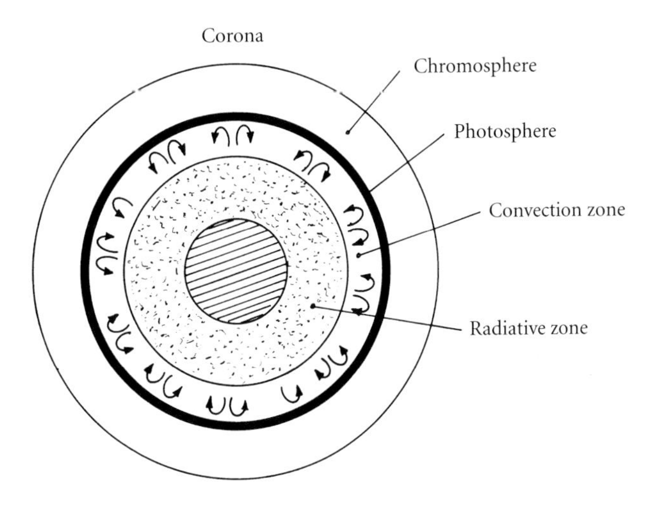

Figure 9.37 The structure of the sun.

Some hint as to the nature of the motion in the convective zone can be obtained from photographs of the photosphere. It has a granular appearance which is strongly suggestive of Bénard convection. The convection cells (granules) appear to evolve on a timescale of minutes and are relatively bright at their centre, where hot plasma rises to the surface, and dark at their edges, where cooler plasma falls back into the interior. The cells have a diameter of $\sim 10^3$ km.

The turbulent motion in the convection zone is important because it controls the rate of transfer of heat to the surface. There have been many attempts to model this convective heat transfer, ranging from crude mixing-length models to more complex EDQNM or Reynolds stress type schemes. These are nicely summarized in Canuto and Christensen-Dalsgaard (1998). It must be said, though, that turbulence in the convective zone is hard to characterize because, in addition to all the usual difficulties associated with turbulent heat transfer, there are extremely large variations in density as well as strong Lorentz and Coriolis forces.

The convection zone is also important as it is the seat of the solar dynamo. The decay time for the solar dipole field, in the absence of convective motion, is around 10^9 years, which is similar to the age of the sun itself. At first sight, therefore, there is no need to invoke dynamo action in order to explain the origin of the solar field. However, the sun's magnetic field is continually changing in a way which is inconsistent with the notion that it is a relic of some primordial field, trapped within the sun at its time of formation. In particular, the dipole field seems to reverse every 11 years, suggesting some sort of cycle with a period of around 22 years. It is ironic that, while a geodynamo theory is needed to explain the unexpected persistence of the earth's magnetic field, dynamo action is required in the solar context in order to explain the rapid fluctuations of the sun's field. There are other differences. While one of the problems of constructing a model of the geodynamo lies in the modest value of R_m, the magnetic Reynolds number in the solar convective zone is vast, exceeding 10^7. So the magnetic field lines which pervade the solar interior and atmosphere are virtually frozen into the plasma.

At one time it was thought that the entire convective zone contributed equally to dynamo action. However, it is now agreed that dynamo action is restricted to a relatively thin layer at the interface of the radiative and convective zones. In large part this view has arisen from measurements of rotation within the sun. The sun does not rotate as a rigid body. The surface rotates faster at the equator than near the poles, while the radiative zone rotates more or less like a rigid body at a rate somewhere between the equatorial and polar surface rates. The difference in angular velocity between the radiative and convection zones leads to strong differential rotation (azimuthal shear) at the base of the convection zone. Thus, an intense azimuthal (east–west) magnetic field builds up near the bottom of the convection zone as the dipole field is spiralled out by the differential rotation. This has lead to speculation that the solar dynamo is of the α–Ω type, operating in a thin, spherical shell at the base of the convection zone. In this picture strong azimuthal fields build up near the base of the convective region. These fields support low-frequency magnetostrophic waves which, when combined with buoyancy-driven motion, regenerate the dipole field from the local azimuthal one. However, as with the various theories of the geodynamo, the α–Ω solar dynamo is a little idealistic, more of a cartoon than a model.

Turbulence is also important in the solar atmosphere, where it triggers violent explosions (Plate 14). This atmosphere has an extremely complex structure. It is threaded with vast magnetic flux tubes which arch up from the photosphere and which are constantly evolving, being jostled by turbulent convection in the photosphere. There are also so-called prominences which extend from the chromosphere up into the corona. These are arch-like, tubular structures of length $\sim 10^5$ km and thickness $\sim 10^4$ km. They contain relatively cold, chromospheric gas, and are threaded by a magnetic field of ~ 10 G. A prominence is itself surrounded by thinner flux tubes which arch up from the photosphere, criss-crossing the prominence. Some flux tubes sit below the prominence, providing a magnetic cushion. Others lie above, forming a so-called magnetic arcade (Figure 9.38).

Quiescent prominences are stable, long-lived structures which survive for many weeks, while eruptive prominences give rise to spectacular releases of mass and energy. Mass which is propelled from the sun in this way is called a *coronal mass ejection* (Plate 14), and the sudden release of energy is called a *solar flare*.

Flares are powered by magnetic energy stored in the solar atmosphere. This energy is released when atmospheric turbulence triggers magnetic reconnection, allowing the local magnetic field to relax to a lower energy state. The energy of the coronal flux loops is subsequently replenished by turbulent jostling of the flux-tube footpoints

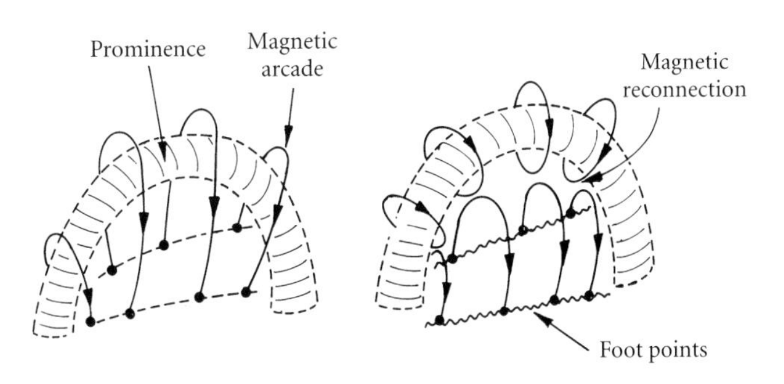

Figure 9.38 A cartoon of a two-ribbon solar flare.

in the photosphere, which stretches and twists the overlying flux tubes. The largest flares are called *two-ribbon flares* and they are thought to arise as follows. Consider a prominence which is supported by a magnetic cushion and has a magnetic arcade overlying it. Now suppose the prominence starts to rise as a result of a build-up of magnetic pressure in the magnetic cushion. The field lines in the overlying arcade, which have their roots in the photosphere, will become increasingly stretched. Eventually magnetic reconnection will occur, in which the arcade flux tubes pinch off, releasing magnetic energy. When this occurs, the restraining force associated with flux tubes is removed and the prominence is propelled explosively upward. Some of this energy is propagated down the arcade field lines to their footpoints in the chromosphere and photosphere. The footpoints of these field lines then appear as two highly energetic 'ribbons' in the chromosphere, from which the name of the flare derives.

There can be no more spectacular manifestation of turbulence than a solar flare or coronal mass ejection. They are vast in scale, extending over a length of $\sim 10^5$ km, and they release prodigious amounts of energy, of the order of 10^{25} J. The mass released by these events spirals through the solar system, enhancing the solar wind, and one or two days after a large flare the earth is buffeted by magnetic storms, often with dramatic effects.

This concludes our brief overview of geodynamo and solar turbulence. We have hardly scratched the surface of these important topics, but interested readers will find more details in Moffatt (1978) (for the geodynamo) and Canuto and Christensen-Dalsgaard (1998) (for solar turbulence).

Exercises

9.1 Use (9.35), (9.36), and (9.38) to show that

$$\frac{\partial \rho_e}{\partial t} + \frac{\sigma \rho_e}{\varepsilon_0} + \sigma \nabla \cdot (\mathbf{u} \times \mathbf{B}) = 0.$$

The quantity $\tau_e = \varepsilon_0/\sigma$ is called the charge relaxation time and has a value of $\sim 10^{-18}$ s in liquid metals. For flows evolving on typical MHD timescales, $\partial \rho_e / \partial t$ is negligible by comparison with ρ_e/τ_e, so that the charge distribution is given by $\rho_e = -\varepsilon_0 \nabla \cdot (\mathbf{u} \times \mathbf{B})$. Use this estimate of ρ_e to confirm that the terms involving ρ_e in (9.36), (9.41), and $\mathbf{F} = \rho_e \mathbf{E} + \mathbf{J} \times \mathbf{B}$ are all negligible.

9.2 Show that the Lorentz force is entirely equivalent to the action of an imaginary set of stresses of the form $\tau_{ij} = (B_i B_j/\mu) - (B^2/2\mu)\delta_{ij}$. These are called Maxwell stresses. [Hint: use Ampere's law to eliminate **J**.]

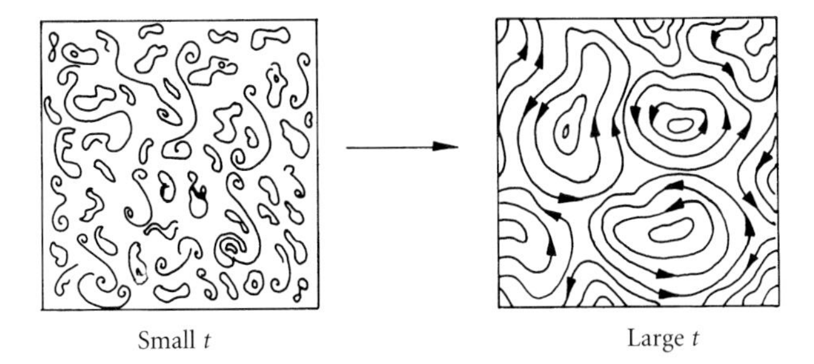

Figure 9.39 The evolution of two-dimensional, isotropic MHD turbulence showing a schematic of the flux lines.

9.3 Construct a proof, along the lines of the derivation of the 'frozen-in' property of $\boldsymbol{\omega}$-lines given in Chapter 2, to show that **B**-lines are frozen into a perfectly conducting fluid.

9.4 Show that finite-amplitude solutions of the ideal induction equation and Euler's equation exist in the form, $\mathbf{u} = \mathbf{f}(\mathbf{x} \pm \mathbf{h}_0 t)$, $\mathbf{h} = \mathbf{h}_0 \pm \mathbf{f}(\mathbf{x} \pm \mathbf{h}_0 t)$, where $\mathbf{h} = \mathbf{B}/(\rho\mu)^{1/2}$, $\mathbf{h}_0 =$ constant and $\mathbf{f}$ is an arbitrary solenoidal vector field.

9.5 Consider a stratified, rapidly rotating fluid in which the bulk rotation rate is $\boldsymbol{\Omega} = \Omega\hat{\mathbf{e}}_z$ and the Brunt buoyancy frequency is N. Show that small perturbations to equilibrium are governed by

$$\frac{\partial^2}{\partial t^2}\nabla^2 u_z + (2\Omega)^2\nabla_{\|}^2 u_z + N^2\nabla_{\perp}^2 u_z = 0$$

and that the corresponding dispersion relationship and group velocity are

$$\varpi^2 = (2\Omega)^2\frac{k_{\|}^2}{k^2} + N^2\frac{k_{\perp}^2}{k^2}, \qquad \mathbf{C}_g = \pm\frac{(2\Omega)^2 - N^2}{\varpi k^4}\left[\mathbf{k} \times \left(\mathbf{k}_{\|} \times \mathbf{k}\right)\right].$$

Confirm that the frequency, ϖ, always lies between N and 2Ω. (Note that the group velocity is zero when $N = 2\Omega$.)

9.6 Consider the two-dimensional turbulent MHD flow $\mathbf{u}(x, y) = (u_x, u_y, 0)$, $\mathbf{B}(x, y) = (B_x, B_y, 0)$. Let us suppose that there is no mean field, $\mathbf{B}_0 = 0$, and that the motion is statistically homogeneous. Let $\mathbf{A}(x, y) = (0, 0, A)$ be the vector potential for $\mathbf{B}$, that is, $\mathbf{B} = \nabla \times \mathbf{A}$. Show that the induction equation uncurls to yield $DA/Dt = \lambda\nabla^2 A$, from which,

$$d\left\langle\frac{1}{2}A^2\right\rangle/dt = -\lambda\langle\mathbf{B}^2\rangle = -2E_B/\sigma,$$

where E_B is the magnetic energy density. Now consider the limit of $\sigma \to \infty$. E_B is always bounded from above by the value of the total energy density E, which falls due to Ohmic and viscous dissipation. It follows that, in the limit $\sigma \to \infty$, we have $\langle A^2\rangle =$ constant. So, as E_B falls, it does so subject to the constraint that $\langle A^2\rangle$ is constant. Show that the variational problem of minimizing E_B subject to the conservation of $\langle A^2\rangle$ leads to a progressive transfer of magnetic energy from small to large scales (Figure 9.39).

Suggested reading

Books

Turner, J.S. (1973) *Buoyancy Effects in Fluids*. Cambridge University Press. (Chapter 4 discusses the influence of buoyancy forces on turbulence.)

Symon, K. (1960) *Mechanics*, 2nd edn. Addison-Wesley.

Greenspan, H.P. (1968) *The Theory of Rotating Fluids*. Cambridge University Press. (This is *the* book on rotating fluids.)

Shercliff, J.A. (1965) *A Textbook of Magnetohydrodynamics*. Pergamon Press. (Although rather dated, this is still one of the best introductions to MHD.)

Batchelor, G.K. (1953) *The Theory of Homogeneous Turbulence*. Cambridge University Press. (Consult chapter 5 for the derivation of invariant (9.107).)

Monin, A.S. and Yaglom, A.M. (1971) *Statistical Fluid Mechanics*. MIT Press. (See chapter 4 for the influence of buoyancy on atmospheric flows.)

Biskamp, D. (1993) *Non-linear Magnetohydrodynamics*. Cambridge University Press.

Panchev, S. (1971) *Random Functions and Turbulence*. Pergamon Press. (See chapters 8–10 for a discussion of atmospheric turbulence.)

Moffatt, H.K. (1978) *Magnetic Field Generation in Electrically Conducting Fluids*. Cambridge University Press.

Davidson, P.A. (2001) *An Introduction to Magnetohydrodynamics*. Cambridge University Press.

Journal references

Bartello, P., Metais, O., and Lesieur, M. (1994) *J. Fluid Mech.*, **273**.

Rayleigh (1916) *Proc., Royal Soc.*, **XCIII**.

Davidson P.A. (1995) *J. Fluid Mech.*, **299**.

Davidson P.A. (1997) *J. Fluid Mech.*, **336**.

Chandrasekhar, S. (1951) *Proc. Royal Soc. Lond.*, **A**, 204.

Widland, O. et al. (1998) *Phys. Fluids*, **10**(8).

Oughton, S. et al. (1994) *J. Fluid Mech.*, **280**.

Alemany, A. et al. (1979) *J. Méc.*, **18**.

Moffatt, H.K. (1967) *J. Fluid Mech.*, **28**, 571–592.

Hopfinger, E.J. et al. (1982) *J. Fluid Mech.*, **125**, 505.

Parker, E.N. (1955) *Astrophys. J.*, **122**.

Davidson, P.A. (2000) Was Loitsyansky correct? *J. Turbulence*.

Davidson, P.A. and Siso-Nadal, F. (2002) *Geophys. Astrophys. Fluid Dyn.*, **96**(1).

Ibbetson, A. and Tritton, D.J. (1975) *J. Fluid Mech.*, **68**(4).

Canuto, V.M. and Christensen-Dalsgaard (1998) *Ann. Rev. Fluid Mech.*, **30**.

Cambon, C. et al. (1997) *J. Fluid Mech.*, **337**.

Riley, N. and Lelong, M.-P. (2000) *Ann. Rev. Fluid Mech.*, **32**, 613–651.

Iida, O. and Nagano, Y. (1999) *Phys. Fluids*, **11**(2).

Steenbeck M., Krause F., and Radler, K.-H. (1966) *Z Naturforsch*, **21a**, 369–376.

Siso-Nadal, F. and Davidson, P. (2003) *J. Fluid Mech.*, **493**.

Godeferd, F.S. et al. (1999) *J. Fluid Mech.*, **393**, 257–280.

Lamb, H. (1889) *Roy. Soc. Phil. Trans.*, 513.

CHAPTER 10

Two-dimensional turbulence

> Common sense is the collection of prejudices acquired by age eighteen
>
> (A. Einstein)

For one brought up in the tradition of Richardson and Kolmogorov, the idea of the energy cascade seems entirely natural. Of course vorticity is teased out into finer and finer filaments, carrying its energy to smaller and smaller scales. The 'grinding down' of eddies is an entirely natural process, or so one might think. Well, life is not always that simple. In two-dimensional turbulence, for example, it is observed that energy propagates in the reverse direction, from the small scales to the large!

There are two implications of this observation. First, the Richardson–Kolmogorov hypothesis of an energy cascade (in three-dimensional turbulence) is a non-trivial conjecture, not to be passed over lightly. Second, two-dimensional turbulence is, in some ways, more subtle and counter-intuitive than its three-dimensional counterpart. To understand such flows we must first follow Einstein's advice and re-examine our prejudices. It is ironic that, while the governing equations are inevitably simpler in two dimensions, some of the dynamical processes are, perhaps, harder to pin down.

Maybe we should start with a note of caution. It is often said that there is no such thing as two-dimensional turbulence, and indeed it is true that all real flows are three dimensional. Nevertheless, certain aspects of certain flows are 'almost' two-dimensional. For example, large-scale atmospheric flows might have a horizontal length-scale of, say, 10^3 km, yet the troposphere is only around 10 km deep. Also, in the laboratory, a strong magnetic field or intense rotation tends to suppress one component of motion. Thus approximate two-dimensionality can be maintained either by virtue of geometry, or else by internal wave propagation (inertial waves or Alfven waves).[1] But this process is never perfect and there is always some aspect of the motion which is non-planar. For example, small-scale atmospheric phenomena are clearly three-dimensional. Moreover, it is in the nature of turbulence that the small and large scales interact, perhaps through some

[1] The tendency for magnetic fields or rapid rotation to promote two-dimensional motion (through wave propagation) is discussed in Chapter 9.

kind of cascade. Thus a great deal of caution is required when interpreting the results of two-dimensional theory. Nevertheless, it seems natural to investigate the behaviour of chaotic, two-dimensional flow, if not in the hope that it sheds some light on 'almost' two-dimensional turbulence, then in the interests of scientific curiosity.

The subject of two-dimensional turbulence has received a great deal of attention in the last 30 years and great strides have been made in our understanding of the phenomena involved. We cannot possibly do justice to all of this work here. So our aims are rather more modest, in that we hope to provide only a simple introduction to the subject, intended as a stepping stone to more serious study.

10.1 The classical picture of two-dimensional turbulence: Batchelor's self-similar spectrum

Two-dimensional and three-dimensional turbulence have different properties, but both contain the two basic ingredients of randomness and convective non-linearity .... In spatially homogeneous two-dimensional turbulence, the mean-square vorticity is unaffected by convection and can only decrease under the action of viscosity. Consequently the rate of dissipation of energy tends to zero with the viscosity. On the other hand, the mean-square vorticity gradient is increased by convective mixing and it seems likely that the rate of decrease of mean-square vorticity tends to a non-zero limit as $\nu \rightarrow 0$. This suggests the existence of a **cascade of mean-square vorticity**. (G.K. Batchlor 1969)

Formal theories of two-dimensional turbulence began to appear in the sixties, culminating in Batchelor's (1969) paper on freely evolving turbulence. In this seminal work he suggested that fully developed, two-dimensional turbulence has: (i) a direct cascade of enstrophy from large to small scales; (ii) a self-similar energy spectrum; and (iii) an inverse flow of energy from small to large scales. That is to say, as the flow evolves the enstrophy is continually passed down to small-scale structures, where it is destroyed (hypothesis I). Moreover, the way in which energy is distributed across the different eddy sizes does not change with time, provided the energy spectrum and characteristic length-scales are suitably normalized (hypothesis II). Finally, in contrast to three-dimensional turbulence, kinetic energy is passed from small-scale structures up to large-scale eddies (hypothesis III). It is now thought that Batchelor's theory is flawed. Nevertheless, in conjunction with the related work of Kraichnan (1967), it has set the scene for debate for some 30 years, and so it seems a natural starting point for our discussion. Later, in Section 10.2, we shall see how the Bachelor–Kraichnan picture needs to be modified to allow for the existence of coherent vortices.

10.1.1 *What is two-dimensional turbulence?*

We shall consider two-dimensional motion, $\mathbf{u}(x, y) = (u_x, u_y, 0) = \nabla \times (\psi \hat{\mathbf{e}}_z)$, in which a suitably defined Reynolds number (Re) is high. It is observed that, as in three dimensions, motion in two dimensions becomes chaotic in time and space when Re exceeds some modest value. This random motion shares many of the features of its three-dimensional counterpart, such as self-induced chaotic mixing of the vorticity field. It seems appropriate, therefore, to follow the example of Batchelor (and others) and refer to this random, two-dimensional motion as *turbulent*.

In order to simplify the discussion we shall consider turbulence which is statistically homogeneous and isotropic (in a two-dimensional sense) and in which there are no body forces. In short, we consider freely evolving, isotropic turbulence. The velocity and vorticity fields are governed by

$$\frac{D\mathbf{u}}{Dt} = -\nabla\left(\frac{p}{\rho}\right) - \nu\nabla \times \boldsymbol{\omega} \tag{10.1}$$

$$\frac{D\omega}{Dt} = \nu\nabla^2\omega \tag{10.2}$$

Note that there is no vortex stretching in (10.2). This is the hallmark of two-dimensional turbulence, in the sense that the absence of vortex stretching gives planar motion its distinctive character. Expressions (10.1) and (10.2) can be manipulated into energy, enstrophy, and palinstrophy equations of the form[2]

$$\frac{D}{Dt}\left[\frac{1}{2}\mathbf{u}^2\right] = -\nabla \cdot \left[\frac{p\mathbf{u}}{\rho}\right] - \nu\left[\omega^2 + \nabla \cdot (\boldsymbol{\omega} \times \mathbf{u})\right] \tag{10.3a}$$

$$\frac{D}{Dt}\left[\frac{1}{2}\omega^2\right] = -\nu\left[(\nabla\omega)^2 - \nabla \cdot (\omega\nabla\omega)\right] \tag{10.3b}$$

$$\frac{D}{Dt}\left[\frac{1}{2}(\nabla\omega)^2\right] = -S_{ij}(\nabla\omega)_i(\nabla\omega)_j - \nu\left[(\nabla^2\omega)^2 - \nabla \cdot ((\nabla^2\omega)\nabla\omega)\right] \tag{10.3c}$$

We now average these equations, noting that an ensemble average is equivalent to a spatial average. Statistical homogeneity of the turbulence ensures that all divergences integrate to zero and we obtain,

$$\frac{d}{dt}\left[\frac{1}{2}\langle\mathbf{u}^2\rangle\right] = -\nu\langle\omega^2\rangle \tag{10.4}$$

[2] Palinstrophy is defined as $\frac{1}{2}(\nabla \times \boldsymbol{\omega})^2$, which in two-dimensions is $\frac{1}{2}(\nabla\omega)^2$. The etymology of the word is given in Lesieur (1990). It was introduced by Pouquet et al. (1975) and is constructed from *palin* and *strophy*, which are the Greek for *again* and *rotation* respectively. Thus Palinstrophy is 'again rotation' or 'curl curl'.

$$\frac{d}{dt}\left[\tfrac{1}{2}\langle\omega^2\rangle\right] = -\nu\left\langle(\nabla\omega)^2\right\rangle \tag{10.5}$$

$$\frac{d}{dt}\left[\tfrac{1}{2}\left\langle(\nabla\omega)^2\right\rangle\right] = -\left\langle S_{ij}(\nabla\omega)_i(\nabla\omega)_j\right\rangle - \nu\left\langle(\nabla^2\omega)^2\right\rangle \tag{10.6}$$

These three equations form the basis of most phenomenological theories of two-dimensional turbulence. The key point to note is that $\langle\omega^2\rangle$ declines monotonically as the flow evolves and so the enstrophy, $\frac{1}{2}\langle\omega^2\rangle$, is bounded from above by its initial value. Now consider the limit $\text{Re} \to \infty$. That is, imagine that we perform a sequence of experiments with similar initial conditions but in which ν is made progressively smaller. Since $\langle\omega^2\rangle$ is always finite it follows that, as $\nu \to 0$, we have conservation of energy in the sense that

$$\frac{d}{dt}\left[\frac{1}{2}\langle\mathbf{u}^2\rangle\right] \to 0 \quad \text{as} \quad \nu \to 0.$$

Thus, if we observe the flow for a finite time, $\langle\frac{1}{2}\mathbf{u}^2\rangle$ is conserved to within a margin of order Re^{-1}. (Note that we let $\nu \to 0$ while retaining t as finite. We cannot guarantee conservation of energy if we let $t \to \infty$ as $\nu \to 0$, since a vanishingly small dissipation integrated over a very long period of time can give rise to a finite loss of energy.)

It is the near conservation of energy which gives high-Re, two-dimensional turbulence its special character. It implies that the turbulence is long-lived. Moreover, it lies in stark contrast with three-dimensional turbulence in which $\lim_{\nu\to 0}\{d\langle\frac{1}{2}\mathbf{u}^2\rangle/dt\}$ is finite, independent of ν, and of order u^3/l. (Here l is the size of the large eddies, that is, the integral scale.) The difference, of course, is that we have no vortex stretching in two dimensions. In conventional turbulence vortex stretching intensifies the small-scale vorticity until the dissipation, $\nu\langle\omega^2\rangle$, is large enough to mop up all of the energy which is released by the break-up of the large-scale structures. We talk of energy cascading down to the small scales. If ν is made very small then $\langle\omega^2\rangle$ simply increases in such a way that the dissipation remains finite and of order u^3/l. In short, the rate of dissipation of energy (in three dimensions) is set by the rate at which the large-scale eddies break-up, which is an inviscid process. In two dimensions, on the other hand, $\langle\omega^2\rangle$ is fixed by the initial conditions and cannot grow (by vortex stretching) to compensate for a small value of ν.

Now consider the vorticity field. From (10.2) we see that it is advected and diffused rather like a passive scalar. In the limit of $\text{Re} \to \infty$ diffusion is small (except in regions where large gradients build up) and so ω is materially conserved and the isovortical lines act like material lines. They are continually teased out by the flow, rather like cream being stirred into coffee. Thus, after a while, the vorticity field adopts the structure of thin, sinuous sheets, as shown schematically in Figure 10.1. Plate 13 also illustrates two-dimensional turbulence: Plate 13(a)

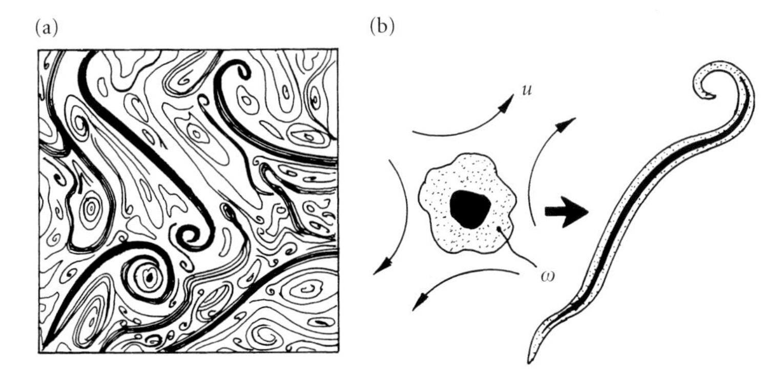

Figure 10.1 (a) Schematic of isovorticity lines in two-dimensional turbulence. The pattern is reminiscent of cream being stirred into coffee. (b) Filamentation of a blob of vorticity.

shows a horizontal section through a tank of rapidly rotating fluid, while Plate 13(b) shows two-dimensional turbulence on a soap film.

The appearance of vortex sheets, while intuitively plausible, may be rationalized as follows. It is commonly believed that material lines are, on average, continually extended in two-dimensional turbulence, just as they are in three dimensions. Now suppose that the turbulence is generated initially with velocity and vorticity gradients of similar length-scales. The process of extension of material lines, and hence of the isovortical lines, will amplify the vorticity gradients as shown above. As a blob of vorticity is teased out into a filament, $(\nabla\omega)^2$ rises and the enstrophy is increasingly associated with a fine-scale structure. This continual filamentation of the vorticity is sometimes referred to as an *enstrophy cascade*, by analogy with the energy cascade in three dimensions. That is, vorticity is continually teased out to finer and finer scales, so that ω^2 is passed down from the large to the small scales. Note, however, that there is limited justification for the use of the word cascade here. In three dimensions the term is invoked to suggest a multistage process in which big eddies break-up into smaller ones, which in turn produce even smaller ones, and so on. But in two dimensions it is less clear that the transfer of enstrophy to small scales is a multistage process (see Section 10.1.7). Still, the term enstrophy cascade is firmly established in the literature and we shall conform to this convention.

Now the rate of filamentation of vorticity is controlled by the inviscid, large-scale eddies and is halted only when the vortex sheets are thin enough for viscosity to act, destroying enstrophy and diffusing vorticity. This suggests that, as in three dimensions, viscosity plays a passive role, mopping up enstrophy at the small scales at a rate determined by the large-scale processes. (In three dimensions it is energy which is mopped up at the small scales.) If this picture is correct, and there is mounting evidence that it is, then we are led to the non-trivial conclusion that $d\langle\omega^2\rangle/dt$ is finite and independent of ν when Re is large. This is the analogue of the three-dimensional result

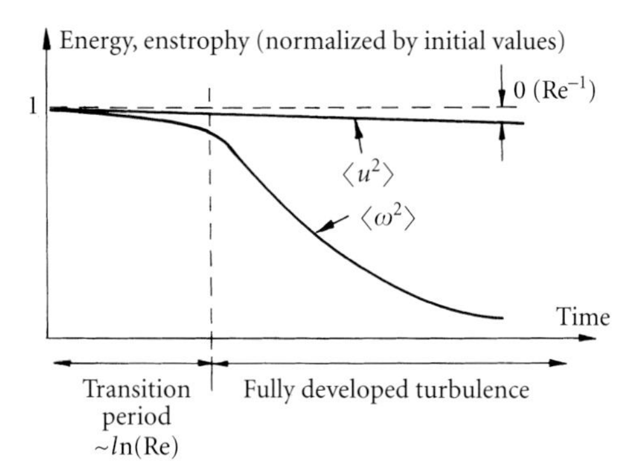

Figure 10.2 Schematic representation of the variation of energy and enstrophy with time. The transition period is thought to be of order ln(Re). (See Example 10.2.)

$\lim_{\nu\to 0}\{d\langle \mathbf{u}^2\rangle/dt\} \sim u^3/l$. Of course, if our initial condition consists of a random collection of large-scale structures, then it will take a finite time for the thin vortex filaments to form. (Again, think of blobs of cream being stirred into coffee.) Thus, initially, $\langle\omega^2\rangle$ will be (almost) conserved. However, after some transition period, which is thought to be of order ln(Re), the turbulence becomes fully developed, with a wide range of scales, right down to the dissipation scale. At this point $d\langle\omega^2\rangle/dt$ becomes finite and adopts a value which is independent of ν (provided Re is large enough). This is the picture proposed by Batchelor and Kraichnan, and illustrated in Figure 10.2.

Example 10.1 Consider the term $G = -S_{ij}(\nabla\omega)_i(\nabla\omega)_j$ in (10.6), which represents the rate of generation of palinstrophy. Suppose that we adopt coordinates aligned with the local principal rates of strain, with the compressive strain orientated along the y-axis. Show that

$$G = -S_{ij}(\nabla\omega)_i(\nabla\omega)_j = a\left[\left(\frac{\partial\omega}{\partial y}\right)^2 - \left(\frac{\partial\omega}{\partial x}\right)^2\right]$$

where $a = \partial u_x/\partial x$. Sketch the flow and confirm that the generation of palinstrophy corresponds to a situation where the vorticity is being teased out into a thin strip aligned with the principal axis of positive strain (the x-axis in this example).

10.1.2 *What does the turbulence remember?*

Recall Batchelor's (1953) comment about the influence of initial conditions on three-dimensional turbulence: 'We put our faith in the tendency for dynamical systems with a large number of degrees of freedom, and with coupling between those degrees of freedom, to approach a statistical state which is independent (partially, if not wholly) of the initial conditions.' When expressed in this way it seems likely that particular details of the initial conditions are equally unimportant in two dimensions. That is to say, when the turbulence becomes fully developed, it remembers little about its initial conditions, though perhaps it remembers a little more than its three-dimensional counterpart. For example, for $\nu \to 0$, the flow remembers how much energy it has, $\frac{1}{2}\langle \mathbf{u}^2 \rangle$, and indeed this is the basis of many of the theories of two-dimensional turbulence. However, there are other, more subtle, possibilities. For example, it is observed that in some turbulent flows, such as shear layers, certain types of structure retain their general shape and intensity for periods much greater than the typical eddy turn-over time. These robust entities are called *coherent structures*. Perhaps, in two dimensions, there are coherent structures embedded in the sea of filamentary vorticity, in which the peak vorticity is preserved for long periods of time. If so, this could endow the turbulence with a kind of memory not foreseen in Batchelor's comment.

In his 1969 paper Batchelor made a bold assumption. He argued that the flow remembers only $\frac{1}{2}\langle \mathbf{u}^2 \rangle$. Based on this hypothesis, he used dimensional arguments to show that the energy spectrum should have a self-similar form, valid at all times other than the transition period. In recent years there has been a multitude of papers pointing out that Batchelor's self-similar theory is flawed. There *are* coherent vortices and their peak vorticity *is* preserved for long periods. However, despite its many faults, Batchelor's theory still provides a convenient starting point for understanding *certain features* of the turbulence. We start, therefore, with the classical self-similar theory. Our discussion of coherent vorticity, and its implications for Batchelor's theory, is left until Section 10.2.

10.1.3 *Batchelor's self-similar spectrum*

Let us introduce some notation. Let

$$u^2 = \langle u_x^2 \rangle = \langle u_y^2 \rangle$$

as in our discussion of three-dimensional turbulence. Also, we re-introduce the second- and third-order *longitudinal structure functions*, defined in terms of the longitudinal velocity increment, Δv:

$$\langle [\Delta v(r)]^2 \rangle = \langle [u_x(\mathbf{x} + r\hat{\mathbf{e}}_x) - u_x(\mathbf{x})]^2 \rangle = \langle [u_x' - u_x]^2 \rangle \tag{10.7a}$$

$$\langle [\Delta v(r)]^3 \rangle = \langle [u_x(\mathbf{x} + r\hat{\mathbf{e}}_x) - u_x(\mathbf{x})]^3 \rangle = \langle [u_x' - u_x]^3 \rangle. \tag{10.7b}$$

It is natural to assume that eddies of size much larger that r make little contribution to $\Delta v(r)$, the velocity at $\mathbf{x}$ and $\mathbf{x}' = \mathbf{x} + \mathbf{r}$ being almost the same. Thus only structures of size r (or smaller) make a significant contribution to Δv. This suggests that $\langle (\Delta v(r))^2 \rangle$ is an indication of the energy per unit mass contained in eddies of size less than or equal to r.[3] In terms of the energy spectrum, $E(k)$,

$$\tfrac{1}{2}\left\langle (\Delta \mathrm{v}(r))^2 \right\rangle \sim \int_{\pi/r}^{\infty} E(k)\,dk \sim (\text{energy contained in eddies of size smaller than } r \sim \pi/k).$$

(Recall that $E(k)$ is defined such that $E(k)\,dk$ gives all of the energy contained in eddies whose size lies in the range $k^{-1} \to (k+dk)^{-1}$, k being a wave number.)[4]

Now we have already suggested that viscosity plays a somewhat passive role, in the sense that it mops up the enstrophy which cascades down from above, but does not influence any of the large- to intermediate-scale dynamics. (The viscous forces are negligible at these larger scales.) If we assume, as Batchelor did, that mature turbulence remembers only u^2, then the form of $\langle(\Delta \mathrm{v})^2\rangle$ outside the dissipation range can depend on only u^2, r, and t, there being no other relevant physical parameter. It follows immediately that, except at the smallest scales, $\Delta \mathrm{v}(r)$ must be of the form,

$$\left\langle (\Delta \mathrm{v}(r))^2 \right\rangle = u^2 g(r/ut). \tag{10.8}$$

The same arguments applied to the energy spectrum yields

$$E(k) = u^3 t\, h(kut). \tag{10.9}$$

Batchelor went further and suggested that g and h are universal functions, valid for any initial condition, a claim which would now be regarded as flawed. Actually, it turns out that (10.8) and (10.9) are often poor representations of $\langle(\Delta \mathrm{v})^2\rangle$ and $E(k)$, (Bartello and Warn 1996). However, just for the moment, we shall stay with Batchelor's model and see where it leads. We shall explore its deficiencies in Section 10.2.

In this picture, then, the characteristic scale of the turbulence grows as $l \sim ut$. That is, if we normalize $\langle(\Delta \mathrm{v})^2\rangle$ by u^2, and r by ut, we obtain

[3] We shall see in Section 10.3.1 that, as in three dimensions, a better approximation is $\tfrac{1}{2}\langle[\Delta \mathrm{v}(r)]^2\rangle \approx \int_{\pi/r}^{\infty} E(k)\,dk + (r/\pi)^2 \int_0^{\pi/r} k^2 E(k)\,dk$.

The second integral on the right reflects the fact that eddies of size greater than r make a contribution to $(\Delta \mathrm{v})^2$ of the order of $r^2 \times$ (enstrophy of eddy). In fact, as we shall see, this second integral often dominates $(\Delta \mathrm{v})^2$. However, for the present purposes this is not important.

[4] The energy spectrum in two-dimensions is defined in the same way as in three-dimensional turbulence: $E(k) = \tfrac{1}{2}\int_S \Phi_{ii}(k)\,dS$. Here S is the circle (sphere in three dimensions) of radius k in $\mathbf{k}$-space and Φ_{ij} is the Fourier transform of the velocity correlation function. (The details are spelt out in Chapter 8.) $E(k)$ has the property that $\tfrac{1}{2}\langle \mathbf{u}\cdot\mathbf{u}\rangle = \int_0^{\infty} E(k)\,dk$.

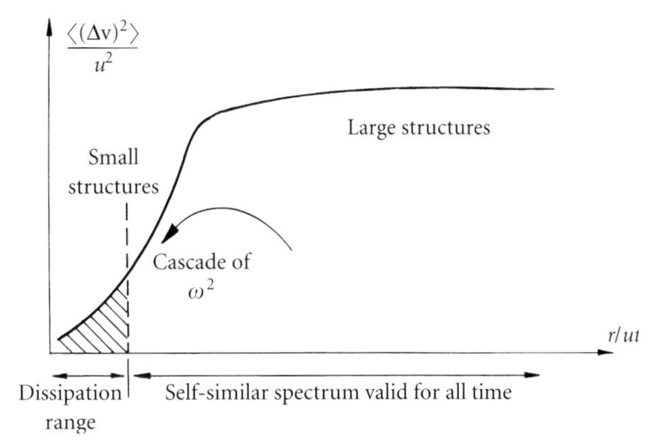

Figure 10.3 Batchelor's universal energy spectrum.

a self-similar energy spectrum valid throughout the evolution of the flow. So, in the Batchelor model the characteristic size of the energy-containing eddies grows as ut (see Figure 10.3).

10.1.4 *The inverse energy cascade of Batchelor and Kraichnan*

When a number of vortices having a similar sense of rotation exist in proximity to one another, they tend to approach one another, and to amalgamate into one intense vortex. (D. Brunt 1929, paraphrasing the findings of Ayrton 1919)

We have now reached an extraordinary conclusion. It would seem that the energy of the turbulence is held in eddies whose characteristic size grows continually as the flow develops. In particular, according to Batchelor, the integral scale grows as $l \sim ut$. This tendency for energy to move upscale is certainly seen in numerical simulations of decaying, two-dimensional turbulence. It is also seen in forced, two-dimensional turbulence; that is, in turbulence where the motion is continually stimulated by energy which is 'injected' into the flow at some intermediate length-scale through the action of some prescribed body force. This is precisely the opposite of three-dimensional turbulence where energy is passed down from large to small scales. In the context of forced turbulence this two-dimensional phenomenon is called the *inverse cascade of energy* (Kraichnan 1967), a term not generally applied to freely evolving, two-dimensional turbulence. However, since energy moves to larger and larger scales in both forced and free turbulence, we shall break with convention and apply the same term to both.[5]

[5] As already noted, the term cascade must be applied with care in two-dimensional turbulence since it is by no means clear that the transfer of energy (to large scale) or enstrophy (to small scale) is really a multistage process.

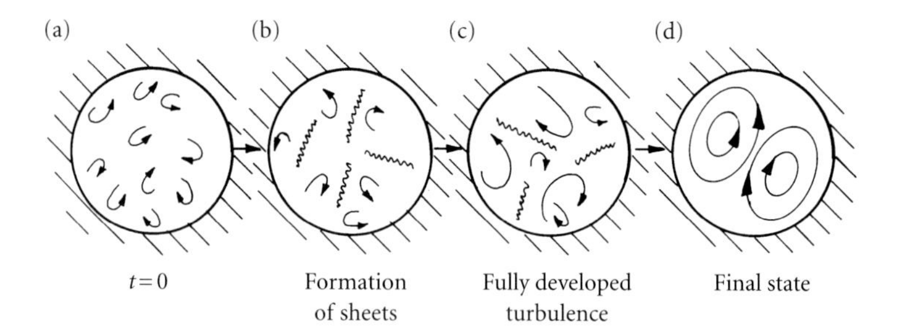

Figure 10.4 Evolution of two-dimensional turbulence in a confined domain.

The accumulation of energy in the large scales means that confined, two-dimensional turbulence will eventually evolve into a large-scale structure which fills the domain, as illustrated above. Figure 10.4(a) represents some initial condition at $t = 0$. There then follows some transition period in which the vorticity begins to form filaments and the turbulence adjusts to its fully developed state, which includes both large-scale eddies and thin, dissipative vortex sheets. The end of this transition is represented by Figure 10.4(b). Batchelor's theory now applies and we have a direct cascade of enstrophy from the large scales down to the small, accompanied by a continual growth in the integral scale (the inverse energy cascade). The size of eddies now increases, as indicated by Figure 10.4(c), until eventually we are left with only one or two eddies which fill the domain. The flow is now quasi-steady, in the sense that the rapid, cascade-enhanced evolution of the flow has ceased and any subsequent change occurs on the slow viscous time-scale, l^2/ν (Figure 10.4(d)).

The key question, of course, is why do the eddies continually grow in size. There are many schools of thought here, although the difference between some theories may just be a matter of semantics. We give four typical explanations below.

One school talks about vortex merger. That is, they attribute the growth of the integral scale, and indeed the scale of all the eddies except the smallest, to the merger of like-signed vortices. Actually this is hardly a recent discovery. The tendency for like-signed vortices to amalgamate is discussed by Ayrton (1919). There were also two papers by Fujiwhara (1921, 1923) in which it was suggested that like-signed vortices tend to merge and that this could be the mechanism by which cyclones develop.

An alternative cartoon of the combined inverse energy cascade and the 'direct' enstrophy cascade is provided by Figure 10.5. A blob of vorticity (Figure 10.5(a)) is teased out into a strip of thickness δ and length l. Area is preserved by the vortex patch since ω is materially conserved (except at the smallest scales). Thus, as δ falls, l increases. The process of teasing and twisting the vortex patch continues until δ becomes so small that diffusion begins to set in (Figure 10.5(c)). The direct cascade of enstrophy is associated with the reduction of δ, while

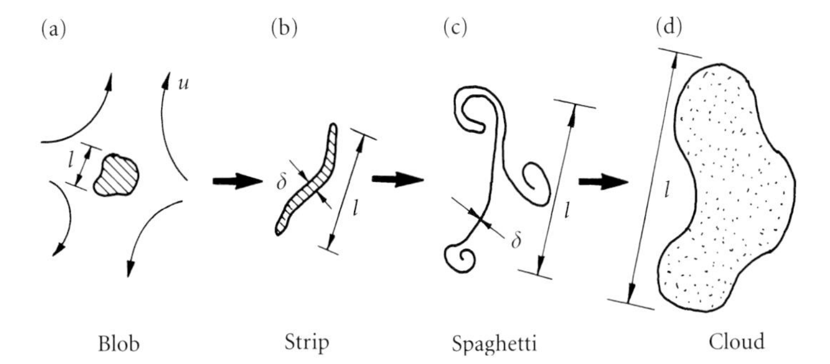

Figure 10.5 Destruction of a blob of vorticity.

the inverse cascade of energy is associated with the growth of l, since it is l which characterizes the eddy size associated with the vortex patch.[6,7] Note that this suggests that the growth in l is, in some way, related to the rate at which two material points tend to separate under the influence of turbulent mixing. We shall return to this issue later.

A third explanation for (or perhaps interpretation of) the growth of l relies on the use of the energy spectrum $E(k)$. We have (see Chapter 8),

$$\frac{1}{2}\langle \mathbf{u}^2 \rangle = u^2 = \int_0^\infty E(k)\, dk \tag{10.10}$$

$$\frac{1}{2}\langle \omega^2 \rangle = \int_0^\infty k^2 E(k)\, dk. \tag{10.11}$$

Now u^2 is (almost) conserved while $\langle \omega^2 \rangle$ decreases at a finite rate. It follows that the area under the curve $E(k)$ is constant, while the integral of $k^2E(k)$, which is weighted towards larger values of k, falls. Thus $E(k)$ must continually grow at small k and become depleted at large k, as shown in Figure 10.6. We interpret this as a build up of energy at the large scales. Actually, in those situations where Batchelor's self-similar spectrum holds, this third view is nothing more than a reinterpretation of Figure 10.3. It also follows directly from (10.9). That is, the growth of E at small k, and the corresponding decline at large k, is an inevitable consequence of (10.9) in the form

$$\frac{\partial E}{\partial t} = u^3 \frac{\partial}{\partial \chi}[\chi h], \quad \chi = kut.$$

Clearly E declines in regions where $E \sim k^{-n}$, $n > 1$ (i.e. to the right of the spectrum) and grows where $n < 1$ (to the left of the spectrum).

[6] Note that, if one considers a multitude of vortex patches simultaneously undergoing such a process, eventually overlapping, then it is easy to see how this might be interpreted as 'vortex merger'.

[7] Think of the Biot–Savart law applied to the patch.

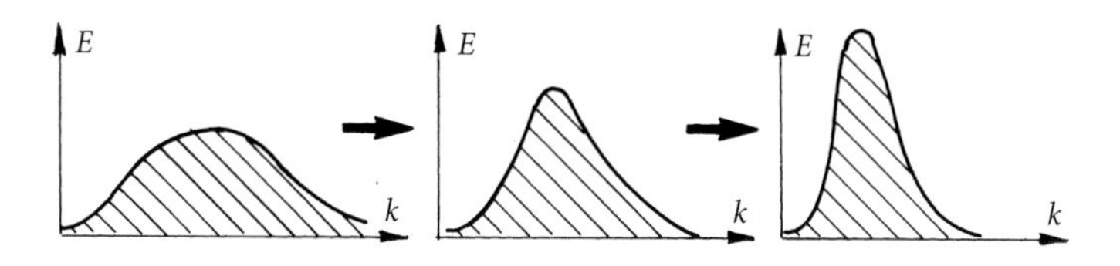

Figure 10.6 Evolution of the energy spectrum.

Note that, in terms of Batchelor's self-similar spectrum, as expressed by (10.9), we have

$$u^2 = \int_0^\infty E(k)\,dk \approx u^2 \int_0^{\chi_\nu} h(\chi)\,d\chi \qquad (10.12)$$

$$\frac{1}{2}\langle\omega^2\rangle = \int_0^\infty k^2E(k)\,dk \approx t^{-2} \int_0^{\chi_\nu} \chi^2 h(\chi)\,d\chi \qquad (10.13)$$

where $\chi = kut$ and χ_ν is the viscous cut-off for χ based on the smallest scales in the flow.[8] The first of these expressions simply normalizes h. The second suggests that, according to Batchelor's theory, the enstrophy declines as t^{-2}. We interpret this simply as $\omega \sim u/l$, where l is the integral scale, $l \sim ut$. Actually we shall see in Section 10.1.6 that χ_ν is time dependant and that a more refined analysis of (10.13) gives a logarithmic correction to the t^{-2} law, of the form $\omega^2 \sim t^{-2} \ln t$.[9]

We now turn to our fourth, and final, explanation for the growth in l. It is motivated by the fact that there is, perhaps, a weakness in the argument that a decline in $\int k^2E(k)\,dk$, subject to the conservation of $\int E(k)\,dk$, forces a shift of E to small k. The problem is that the decline of $\int k^2E(k)\,dk$ is caused by viscous forces, and these do not influence the bulk of the eddies. Ideally, we would like an explanation for the inverse cascade which does not depend on viscosity. Now when $\nu = 0$ we have conservation of both $\int k^2E(k)\,dk$ and $\int E(k)\,dk$, and so a natural way to frame the question is: would we expect the mean eddy size to grow when we have chaotic mixing of the vorticity field subject to the conservation of both $\langle\omega^2\rangle$ and $\langle\mathbf{u}^2\rangle$? The answer turns out to be 'probably'.

Let us start by defining the centroid of $E(k)$ by the expression

$$k_c = \frac{\int kE(k)\,dk}{\int E(k)\,dk}.$$

[8] Note that we use '$\approx$' rather than '$=$' in (10.12) and (10.13) since the self-similar equation (10.9) is not valid all the way down to the dissipation scale.

[9] Numerical experiments suggest that the enstrophy decays more slowly than this, perhaps as t^{-n} where n lies between 0.8 and 1.2 (e.g. Ossai and Lesieur 2001; Bartello and Warn 1996; and Chasnov 1997, found n to lie in the range 1.0 to 1.2, $\sim$1.2, and $\sim$0.8, respectively). This discrepancy is often attributed to the fact that some of the vorticity (that which lies within coherent vortices) is protected from Batchelor's filamentation process and so the rate of destruction of enstrophy is less than that predicted by the self-similar theory. See Section 10.2.

Next we note that, because of the conservation of $\langle \omega^2 \rangle$ and $\langle \mathbf{u}^2 \rangle$,

$$\frac{d}{dt}\int (k_c - k)^2 E(k)\, dk = u^2 \frac{dk_c^2}{dt} - 2\frac{d}{dt}\left[k_c \int kE(k)\, dk\right] = -u^2 \frac{dk_c^2}{dt}$$

But we would expect that, in an Euler flow, chaotic mixing will continually spread $E(k)$ over a greater range of wave numbers, so that $\int (k_c - k)^2 E(k) dk$ continually grows. *If* this is true then k_c must decline, which represents the expected shift of energy to small k (large scales).[10]

10.1.5 *Different scales in two-dimensional turbulence*

Let l be the integral scale of the turbulence (i.e. the size of the large, energy containing eddies) and η the smallest scale (i.e. the thickness of the vortex sheets). Similarly, let u be the typical large-scale velocity and v be the characteristic velocity of the smallest scales, which have a characteristic time $\tau = \eta / \mathrm{v}$. In three dimensions we saw that the small scales, called the Kolmogorov scales, are related to the large scales as shown in Table 10.1. Let us now determine the equivalent relationships in two-dimensional turbulence. We know that $\eta \mathrm{v}/\nu \sim 1$, since scales smaller than this cannot exist (they would be eliminated by viscosity). Also, since ω is materially conserved (except at the smallest scales) we have $u/l \sim \mathrm{v}/\eta$. It follow that $\eta/l \sim \mathrm{v}/u \sim \mathrm{Re}^{-1/2}$, where $\mathrm{Re} = ul/\nu$. These relationships are tabulated below. This table also shows the time evolution of the small scales based on the estimate $l \sim ut$. It should be remembered, however, that Batcherlor's self-similar spectrum is now thought to be flawed. (See Section 10.2.) Thus the estimate $l \sim ut$, as well as the expressions $\eta \sim (\nu t)^{1/2}$ and $\mathrm{v} \sim (\nu/t)^{1/2}$ given below, must be regarded with a degree of caution. If we compare Tables 10.1 and 10.2 we see that one of the main differences between three- and two-dimensional turbulence is that, in three dimensions, the small scales evolve extremely rapidly, while in two dimensions the small scales evolve no faster than the large eddies. There is no quasi-equilibrium range in two-dimensional turbulence.

Table 10.1 Small scales in three-dimensional turbulence. Note that $\mathrm{Re} = ul/\nu$

Dimension	Ratio of Kolmogorov scale to large scale
Length	$\eta/l \sim \mathrm{Re}^{-3/4}$
Velocity	$\mathrm{v}/u \sim \mathrm{Re}^{-1/4}$
Time	$u\tau/l \sim \mathrm{Re}^{-1/2}$

10.1.6 *The shape of the energy spectrum: the k^{-3} law*

The relationship $\mathrm{v}/u \sim \mathrm{Re}^{-1/2}$ suggests that the large eddies are much more energetic than the smallest structures. The next question is: can

[10] Actually, the problem with this argument is that we have no proof that energy does continually spread over a greater range of wave numbers in an Euler flow, although it certainly seems likely. The most that we can deduce from this line of reasoning is that such a spread is consistent with the growth of the integral scale.

Table 10.2 Different scales in two-dimensional turbulence. Note that the final two columns are valid only for Batchelor's self-similar spectrum in which $l \sim ut$

Dimension	Ratio of small to large scales (Always true)	True only for Batchelor's spectrum	
		Time dependence of large scales	Time dependence of small scales
Length	$\eta/l \sim \mathrm{Re}^{-1/2}$	$l \sim ut$	$\eta \sim (\nu t)^{1/2}$
Velocity	$v/u \sim \mathrm{Re}^{-1/2}$	$u \sim$ constant	$v \sim (\nu/t)^{1/2}$
Time	$u\tau/l \sim 1$	$l/u \sim t$	$\tau \sim t$

we predict how energy is distributed across the full range of eddy sizes? In short, we wish to determine the shape of the energy spectrum.

10.1.6.1 The shape at small k

Let us start by reminding you of the shape of $E(k)$ in three dimensions. For small k we have,

$$E(k) = (I/24\pi^2)k^4 + \cdots, \qquad I = -\int \mathbf{r}^2 \langle \mathbf{u} \cdot \mathbf{u}' \rangle \, d\mathbf{r} \tag{10.14}$$

where $\mathbf{u}' = \mathbf{u}(\mathbf{x} + \mathbf{r}) = \mathbf{u}(\mathbf{x}')$. This comes from expanding the expression for the spectral tensor in terms of a Taylor series in k (see Chapter 8). The integral I is, of course, Loitsyansky's integral. In the early theories of turbulence I was thought to be an invariant of freely decaying turbulence. In practice, however, it appears to exhibit a slight time dependence.

In Chapter 6 we refer to equation (10.14) as Batchelor's (three-dimensional) spectrum in order to distinguish it from the so-called Saffman spectrum,

$$E \sim Lk^2 + \cdots, \qquad L = \int \langle \mathbf{u} \cdot \mathbf{u}' \rangle \, d\mathbf{r} \tag{10.15}$$

However, you will recall that Saffman's spectrum requires special initial conditions, conditions which are possibly not met in, say, wind-tunnel turbulence. Thus Batchelor's three-dimensional spectrum (10.14) is, perhaps, the norm in three dimensions.

In two-dimensional turbulence the analogue of (10.14) is

$$E(k) = (I/16\pi)k^3 + \cdots \tag{10.16}$$

$$I = -\int \mathbf{r}^2 \langle \mathbf{u} \cdot \mathbf{u}' \rangle \, d\mathbf{r}. \tag{10.17}$$

Note that Batchelor's two-dimensional self-similar spectrum, if it is valid, requires that $I \sim t^4$, which lies in stark contrast with the three-dimensional situation where I is only a very weak function of time. We shall return to this issue in Section 10.3.3 where we shall see that the numerical experiments suggest that $I \sim t^2$. Note also that (10.16) applies only when the turbulence has reached a mature state, in which the details of the initial conditions are largely forgotten and the full range of length-scales has developed. It is perfectly possible to specify $E \sim k^n$, $n > 3$, as an initial condition, though such a condition will not, in general, persist.

10.1.6.2 The k^{-3} law for the inertial range

Let us now consider the shape of the spectrum at points where $\eta \ll k^{-1} \ll l$. That is to say, we consider how the energy is distributed between eddies which are much larger than the dissipative scale, but much smaller than the large-scale structures. In effect we are looking for the two-dimensional analogue of Kolmogorov's five-thirds law,

$$E \sim \varepsilon^{2/3} k^{-5/3}, \quad \eta \ll k^{-1} \ll l. \tag{10.18}$$

(See Chapter 5 for the derivation of (10.18).) This question was first addressed by Kraichnan (1967) in the context of forced turbulence, and a little later by Batchelor (1969) in the context of freely evolving turbulence. We follow Batchelor's arguments.

It has been suggested there is a direct enstrophy cascade form large to small scales. Let β be the rate at which enstrophy is removed form the large-scale structures and deposited in the dissipation range,

$$\beta = \nu \langle (\nabla \omega)^2 \rangle = -\frac{d}{dt} \frac{1}{2} \langle \omega^2 \rangle.$$

Since $\langle \omega^2 \rangle$ has the dimensions of t^{-2}, the dimensions of β must be t^{-3}. Indeed, according to Batchelor's self-similar spectrum, $\langle \omega^2 \rangle \sim t^{-2}$ (see 10.13) and so $\beta \sim t^{-3}$.[11] Now, if we believe that the destruction of $\langle \omega^2 \rangle$ is genuinely a cascade process, in the sense $\langle \omega^2 \rangle$ that is passed down, step by step, through a hierarchy of eddies, then the eddies in the range $\eta \ll k^{-1} \ll l$ should know nothing of the large-scale structures, nor of the dissipation scales. That is, these eddies are the offspring of larger structures which, in turn, have come from yet bigger eddies, and so on. So perhaps they do not 'feel' the strain field of the largest eddies, but only that of the eddies of comparable size. *If* this is all true (and we shall see shortly that it is not!), we might argue that, by analogy with Kolmogorov's theory in three dimensions, E is a function only of β, k, t, and ν. But the spectrum should not be a function of ν in the range $k^{-1} \gg \eta$, since viscous stresses influence

[11] Actually, we noted in Section 10.1.4 that, in practice, the enstrophy decays more like t^{-1}. The significance of this is discussed in Section 10.2.2.

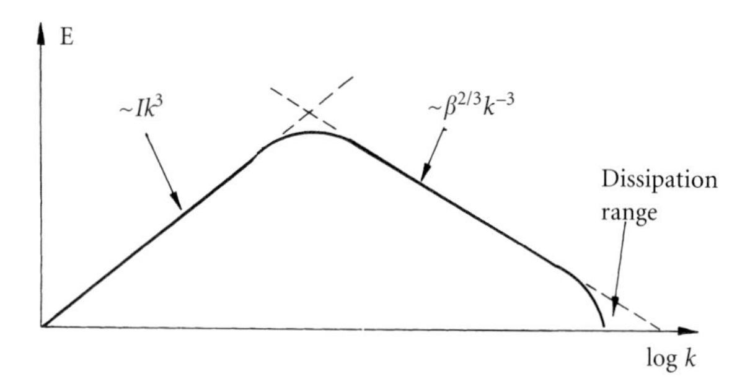

Figure 10.7 Shape of the energy spectrum for fully developed turbulence.

only the smallest eddies. Thus we have $E = E(\beta, k, t)$. Given the dimensions of E, β, and k, the only possibility is,

$$E = \mathrm{F}(\beta^{1/3}t)\beta^{2/3}k^{-3}, \quad \eta \ll k^{-1} \ll l \tag{10.19a}$$

where F is a dimensionless function of $\beta^{1/3}t$. Moreover, if we accept Batchelor's self-similar spectrum, then $\beta^{1/3}t = \text{constant}$. It then follows that, throughout the *inertial range*, we have

$$E \sim \beta^{2/3}k^{-3} \sim t^{-2}k^{-3}, \quad \eta \ll k^{-1} \ll l \tag{10.19b}$$

(Batchelor's self-similar spectrum only).
This k^{-3} law is illustrated in Figure 10.7.[12] When similar cascade-like arguments are applied to the second and third-order structure functions we find

$$\langle [\Delta \mathrm{v}]^2 \rangle = c_2 \beta^{2/3} r^2, \quad \eta \ll r \ll l \tag{10.20a}$$

$$\langle [\Delta \mathrm{v}]^3 \rangle = c_3 \beta r^3, \quad \eta \ll r \ll l \tag{10.20b}$$

Here c_2 and c_3 are, in general, dimensionless functions of $\beta^{1/3}t$, but for the particular case of Batchelor's self-similar spectrum they are constants. These expressions are the counterpart of Kolmogorov's five-thirds, two-thirds, and four-fifths laws, respectively. Equations (10.20a) and (10.20b) are reasonably well supported by the numerical evidence (Lindborg and Alvelius 2000). However, unlike Kolmogorov's five-thirds law, (10.19a) is far from being universally accepted.[13] In fact, numerical simulations suggest $E \sim k^{-n}$, where n is typically a little

[12] Note that, in those cases where Batchelor's self-similar spectrum is valid, the net enstrophy can be estimated by integrating (10.19b) across the inertial range, say from k_1 to k_2. This gives

$$\langle \omega^2 \rangle \sim t^{-2} \ln(k_2/k_1) \sim t^{-2} \ln(l/\eta) \sim t^{-2} \ln(t).$$

[13] We shall see shortly that $\langle [\Delta \mathrm{v}]^2 \rangle$ often scales as r^2 even in the absence of a k^{-3} regime in the energy spectrum. This is a subtle issue which we shall explore in detail in Section 10.3.1.

larger than 3. Some of the more recent numerical evidence is reviewed in Lindborg and Alvelius (2000).

10.1.7 *Problems with the k^{-3} law*

The underlying problem with the k^{-3} law is that it relies on the idea that the transfer of $\langle \omega^2 \rangle$ from large to small scales is a multistage cascade, in which eddy interactions are localized in Fourier space. That is, it assumes that the evolution of the vortex patch shown in Figure 10.5 is controlled only by vortical structures whose sizes are similar to that of the patch, so that the patch is oblivious to the large-scale eddies. At first sight this seems reasonable. Large-scale vortical structures produce a velocity field which is almost uniform on the scale of the eddy shown in Figure 10.5(a). Surely such a velocity field simply advects the blob around without changing its shape. However, this argument does not withstand scrutiny. The mean-square strain associated with eddies lying in the wave number range $k_{\min} \ll k \ll k_{\max}$ is of the order of their enstrophy, $\int k^2 E\, dk$. In the inertial range, where we have suggested that (10.19a) applies, this mean-square strain is of the order of $\beta^{2/3} \ln(k_{\max}/k_{\min})$. Now suppose we are interested in the strain field experienced by an eddy of size k_0^{-1}. The eddies slightly larger than k_0^{-1}, say $0.1k_0 \to k_0$, make a contribution to the total mean-square strain of the order of $\beta^{2/3}$ ln10. However, the next octave of wave numbers, $0.01k_0 \to 0.1k_0$, makes precisely the same contribution. Thus our eddy of size k_0^{-1} feels the strain of different wave number octaves with equal intensity. This contravenes the idea of a cascade which is local in **k**-space. So we have a contradiction. If we assume that interactions are localized in **k**-space, that is, we have a true cascade, then dimensional analysis gives us $E \sim k^{-3}$, which in turn tells us that interactions are *not* local! On the other hand, non-local interactions undermine the dimensional argument for the k^{-3} law.[14]

The fact that, in practice, E is found to be close to $E \sim k^{-3}$ tells us that, in two dimensions, vortical structures of rather different size interact, disrupting our notion of a simple enstrophy cascade. These 'non-local' interactions have been extensively studied and some researchers have proposed a weak logarithmic correction to the k^{-3} law. (e.g. see the review of Lesieur 1990.) However, it is difficult to verify (or otherwise) such a correction through numerical experiments since limitations in computing power severely limit the extent of the inertial range which can be realized.

[14] Actually, we shall see later that there are other arguments in favor of a k^{-3} law which do not depend on the interactions being local in **k**-space, that is, on the existence of a multistage cascade. This is discussed over page in Examples 10.1 and 10.3.

Despite the controversy over the reason for, or even the existence of, an $E \sim k^{-3}$ law, it seems that $E \sim k^{-3}$ is not too far out of line with the numerical simulations. Such simulations tend to indicate an inertial-range scaling of $E \sim k^{-n}$ where n is close to, but often slightly greater than, 3. Brachet et al. (1988), for example, find that $n \approx 4$ during the initial transition period and then $n \approx 3$ for fully developed turbulence. (See, also, Lindborg 1999; Lindborg and Alvelius 2000.) Interestingly, it seems that the exponent in the $E \sim k^{-n}$ relationship tends to depend on the initial conditions, which casts some doubt on Batchelor's central assumption that the turbulence remembers only u^2. This is the first hint that two-dimensional turbulence is more complex than suggested by the early theories.

10.1.8 A Richardson-type law for the inertial range

Before closing our review of the classical theories of two-dimensional turbulence we return to the issue, raised in Section 10.1.4, of the rate of growth of the length-scale shown in Figure 10.5. This is closely related to the question of how rapidly two material points move apart in two-dimensional turbulence.[15]

Let us see if we can estimate the average rate of separation of two points in the inertial range. Suppose that $\delta\mathbf{x}$ represent the instantaneous separation of two material points whose initial separation lies in the range $\eta \ll |(\delta\mathbf{x})_0| \ll l$. Consider a sequence of experiments in statistically identical flows in which, at $t = t_0$, we tag two points separated by a distance $|(\delta\mathbf{x})|_0$. We can then form ensemble averages of quantities such as $(\delta\mathbf{x})^2$. The question is, how rapidly does $\langle(\delta\mathbf{x})^2\rangle$ grow? If $|\delta\mathbf{x}|$ lies in the inertial range we might anticipate that its rate of growth will depend on only $\langle(\delta\mathbf{x})^2\rangle$ and the mean velocity difference at the scale $|\delta\mathbf{x}|$. If this is so then dimensional considerations tells us that

$$\frac{d}{dt}\langle(\delta\mathbf{x})^2\rangle \sim \langle(\Delta\mathrm{v})^2\rangle^{1/2}\langle(\delta\mathbf{x})^2\rangle^{1/2}$$

Let us now make the assumption that (10.20a) holds true. Substituting for the velocity difference then yields,

$$\frac{d}{dt}\langle(\delta\mathbf{x})^2\rangle \sim \beta^{1/3}\langle(\delta\mathbf{x})^2\rangle$$

This expression was first proposed by Lin (1971) for forced, statistically steady turbulence, where it predicts an exponential growth in $\langle(\delta\mathbf{x})^2\rangle$. Lin did not attempt to apply this equation to freely evolving turbulence,

[15] See Section 5.4.2 for the equivalent discussion of two-particle dispersion in three dimensions.

though the argument above seems equally applicable to both types of flow. (Note, however, that Lesieur 1990, has questioned the validity of Lin's equation for freely decaying turbulence.)

Example 10.1 A speculative model of inviscid turbulence
Consider two-dimensional, chaotic motion of an unbounded, inviscid fluid. We suppose that the flow is statistically homogeneous. From the discussion above we would expect the motion to satisfy the following conditions:

(1) $\frac{d}{dt}\int_0^\infty E(k)\,dk = 0,$

(2) $\frac{d}{dt}\int_0^\infty kE(k)\,dk < 0,$

(3) $\frac{d}{dt}\int_0^\infty k^2E(k)\,dk = 0,$

(4) $\frac{d}{dt}\int_0^\infty k^4E(k)\,dk > 0.$

The first and third of these simply represent conservation of energy and enstrophy, while the second was shown in Section 10.1.4 to be an inevitable consequence of the tendency for chaotic motion to spread energy across an ever greater range of scales. The fourth is associated with equation (10.6) and Figure 10.1, in which we expect $\langle(\nabla\omega)^2\rangle$ to continually increase as filaments of vorticity are teased out by the turbulent motion. Consider an energy spectrum of the form

$$E(k,t) = A(t)k^{-3}, \qquad k_1(t) < k < k_2(t), \quad k_1 \ll k_2$$

with a rapid fall-off of $E(k)$ outside the range $k_1 < k < k_2$. Show that all four conditions above are met provided that

$$A \sim t^{-2}, \qquad k_1 \sim (ut)^{-1}, \qquad k_2 \sim (ut)^{-1}\exp\left[\tfrac{1}{4}\langle\omega^2\rangle t^2\right]$$

where u and $\langle\omega^2\rangle$ are, of course, constants. (This is shown in Figure 10.8.) Now consider a fixed wave number k^* such that $k_1 < k^* < k_2$. Confirm that the rate of transfer of enstrophy from wave numbers less than k^* to those greater than k^* is of order $\beta \sim t^{-3}$.

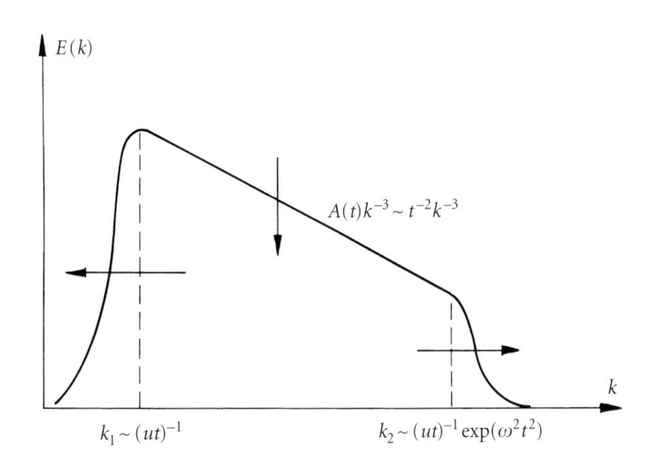

Figure 10.8 Log–log plot of one candidate for $E(k)$ in two-dimensional, inviscid turbulence.

Evidently, the lower end of our spectrum is that predicted by Batchelor's self-similar theory, with $k_1 \sim (ut)^{-1}$ and $\beta \sim t^{-3}$. Moreover, the behaviour of k_2 is consistent with the idea that enstrophy cascades down to smaller and smaller scales, there being no viscous cut-off in an Euler flow.

Example 10.2 The transition from quasi-inviscid to fully developed turbulence
In the inviscid model of Example 10.1, we found that $\langle \omega^2 \rangle t^2 \sim \ln(k_2 ut)$. Since $\langle \omega^2 \rangle$ is constant in an Euler flow, this equation effectively determines the evolution of $k_2(t)$. Now suppose that the fluid is viscous and that the initial value of k_2^{-1} is much greater than the dissipation scale, $\eta \sim \sqrt{\nu t}$. Confirm that the time, t_c, taken to reach the viscous cut-off is given by

$$\langle \omega^2 \rangle t_c^2 \sim \ln(ut_c/\sqrt{\nu t_c}) \sim \tfrac{1}{4}\ln(\langle \omega^2 \rangle t_c^2) + \tfrac{1}{2}\ln(\mathrm{Re})$$

where $\mathrm{Re} = u^2/\langle \omega^2 \rangle^{1/2} \nu$ is determined by the initial condition. For large Re a good approximation to this is $\langle \omega^2 \rangle t_c^2 \sim \ln(\mathrm{Re})$.

Example 10.3 Why k^{-3} in the inertial range?
Repeat the analysis of Example 10.1 for a spectrum of the form $E \sim A(t)\, k^{-n}$, $n \geq 3$. Show that no choice of n, other than $n = 3$, allows us to satisfy all four integral conditions for $E(k)$. So perhaps there are reasons for believing in a k^{-3} inertial range even in the absence of a multistage cascade.

Example 10.4 Palinstrophy generation during the transition period
Show that the inviscid model of Example 10.1 implies a palinstrophy generation rate of

$$G = -\left\langle S_{ij}(\nabla\omega)_i(\nabla\omega)_j \right\rangle \sim \langle \omega^2 \rangle \langle (\nabla\omega)^2 \rangle t.$$

An alternative estimate of the palinstrophy generation rate has been suggested by Herring et al. (1974) and Pouquet et al. (1975). It is

$$G = -\left\langle S_{ij}(\nabla\omega)_i(\nabla\omega)_j \right\rangle = S_2 \langle \omega^2 \rangle^{1/2} \tfrac{1}{2} \langle (\nabla\omega)^2 \rangle$$

where S_2 is a positive, dimensionless coefficient of the order of unity.

Example 10.5 More on the generation of palinstrophy
A simple model problem, which illustrates the generation of palinstrophy, is the following. Suppose that, at $t = 0$, we have a region of space where ω varies linearly with y, say $\omega = Sy$. We now place a point vortex of strength Γ at the origin, which starts to wind up the vorticity field (Figure 10.9). For simplicity we shall assume that the velocity associated with $\omega(\mathbf{x}, t)$ is much weaker than that of the point vortex, so that $\mathbf{u} = (\Gamma/r)\hat{\mathbf{e}}_\theta$ in polar coordinates. Show that ω evolves

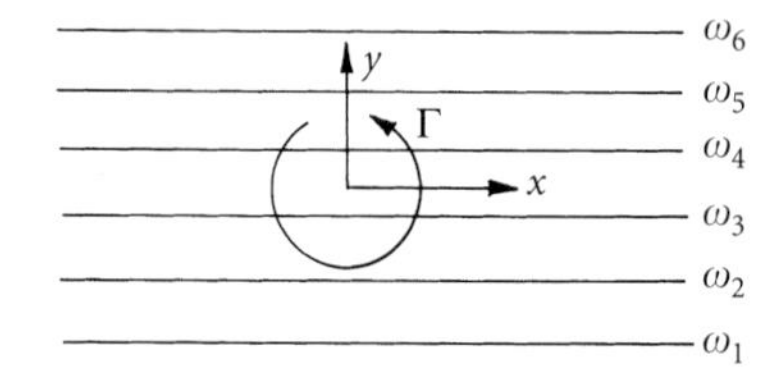

Figure 10.9 The wind up of a weak vorticity field by a point vortex. (See Example 10.5.)

according to

$$\frac{\partial \omega}{\partial t} + \frac{\Gamma}{r^2}\frac{\partial \omega}{\partial \theta} = \nu \nabla^2 \omega$$

and confirm that, if viscosity is neglected, the form of ω for $t > 0$ is, $\omega = Sr \sin[\theta - \Gamma t / r^2]$.

Sketch the distribution of ω and show that this represents a progressive spiralling of the vorticity field by the point vortex. Confirm that the solution above is valid provided that r is much greater than the diffusive length-scale $\sqrt{\nu t}$ and $r^2/\Gamma t$ is greater than, or of the order of, unity. Now show that the integral scale of the eddy associated with the spiralled vorticity field grows as $l \sim \sqrt{\Gamma t}$ (a result to which we shall return), while the characteristic length-scale for gradients in vorticity (at a given radius) continually decreases according to $\eta \sim r^3/\Gamma t$. Finally, confirm that the enstrophy, palinstrophy, and rate of generation of palinstrophy, averaged over the angle θ, are

$$\left\langle \tfrac{1}{2}\omega^2 \right\rangle_\theta = \tfrac{1}{4}S^2 r^2, \qquad \left\langle \tfrac{1}{2}(\nabla\omega)^2 \right\rangle_\theta = \tfrac{1}{2}S^2 + S^2\Gamma^2 t^2/r^4$$

$$-\left\langle S_{ij}(\nabla\omega)_i(\nabla\omega)_j \right\rangle_\theta = \left\langle \frac{2\Gamma}{r^3}\frac{\partial\omega}{\partial\theta}\frac{\partial\omega}{\partial r} \right\rangle_\theta = \frac{2S^2\Gamma^2 t}{r^4} \approx \frac{1}{t}\left\langle (\nabla\omega)^2 \right\rangle_\theta$$

10.2 Coherent vortices: a problem for the classical theory

Education is an admirable thing, but it is well to remember from time to time that nothing that is worth knowing can be taught. (Oscar Wilde)

The early theories of Batchelor and Kraichnan are beautiful and compelling. However, they lacked experimental confirmation. This is important because Batchelor's theory assumes that the turbulence remembers only u^2. This excludes the possibility of, say, long-lived, coherent vortices, an omission which turned out to be crucial. Here we consider the role of coherent patches of vorticity; that is, vortices whose life greatly exceeds the eddy turn-over time, l/u.

10.2.1 The evidence

In Batchelor's theory the vorticity is treated rather like a passive tracer which is teased out by the turbulent velocity field. However, as computers became more powerful, and it was possible to perform high-resolution numerical experiments, it soon became clear that this is not the whole story. While filamentation of vorticity does indeed occur, just as envisaged by Batchelor, the high-resolution computations revealed another, quite distinct process at work. It seems that small, intense patches of vorticity, embedded in the initial conditions, can survive the filamentation process and form long-lived, coherent vortices. These concentrated vortices behave quite differently to the filamentary vorticity, interacting with each other rather like a collection of point vortices. Occasionally two or more coherent vortices approach each other and merge, or perhaps a weaker vortex is destroyed by a larger one. Thus, as the flow evolves, the number of coherent vortices decreases, and their average size increases.

The picture we now have is one of two sets of dynamical processes coexisting in the same flow. The bulk of the vorticity is continually teased out into filaments, feeding the enstrophy cascade as suggested by Batchelor. However, within this sea of filamentary debris, bullets of coherent vorticity fly around, increasing in size and decreasing in number through a sequence of viscous mergers. For some initial conditions the coherent vortices are sufficiently strong and numerous to outlive the filamentary debris, and so after a period of many turn-over times, say 100 l/u, much of the residual vorticity is located within the coherent vortices. Such a case is shown schematically below. Note that the vortices on the right did not *emerge* as a result of the enstrophy cascade. Rather, they were latent in the initial conditions and have *survived* the cascade. Note also that Figure 10.10(b) represents a rather exhausted state, containing only a few per cent of the initial enstrophy. In effect, the enstrophy cascade has run its course and state (b) represents some sort of residual field of depleted vorticity. In order to show that the vortices on the right of Figure 10.10 were indeed embedded in the initial conditions, and did not emerge as part of the cascade process, it is necessary to track the evolution of the vortices backwards in time to $t = 0$. This has been done by several authors, most notably McWilliams (1990), who found that the coherent vortices evident in the final stages of the flow invariably had their origins in small blobs of intense vorticity in the initial flow field.

The ability of small patches of intense vorticity to survive the filamentation process is generally attributed to the inviscid equation

$$\frac{D^2}{Dt^2}[\nabla\omega] + \left[\frac{1}{4}\omega^2 - S_1^2 - S_2^2\right]\nabla\omega = \text{terms of order}\left[\frac{DS_i}{Dt}\nabla\omega, \frac{D\omega}{Dt}\nabla\omega\right] \tag{10.21}$$

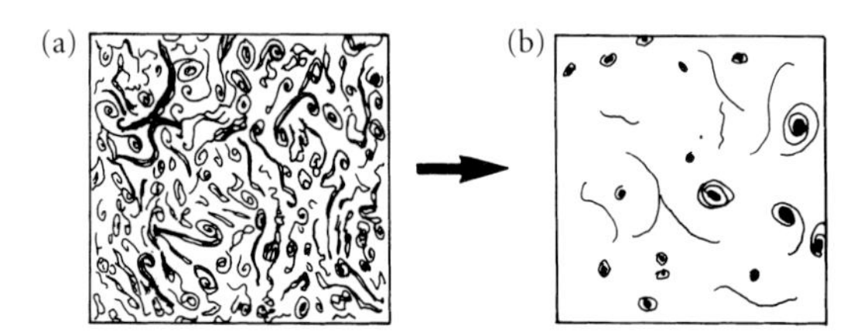

Figure 10.10 Schematic of vorticity contours for a turbulent flow. (a) The vorticity field near the beginning of the evolution. (b) The vorticity once the enstrophy cascade has largely run its course. The enstrophy in (b) is only a few per cent of that in (a).

where S_1 and S_2 are the strain fields $\partial u_x/\partial x$ and $\frac{1}{2}(\partial u_x/\partial y + \partial u_y/\partial x)$, respectively. This equation follows directly from $D\omega/Dt = 0$. Now it is observed in some (but not all) of the numerical simulations that the Lagrangian rates of change of ω, S_1, and S_2 are much smaller than the corresponding rate of change of $\nabla\omega$. If we accept this empirical statement then we have,

$$\frac{D^2}{Dt^2}[\nabla\omega] + \left[\frac{1}{4}\omega^2 - S_1^2 - S_2^2\right]\nabla\omega \approx 0. \tag{10.22}$$

It follows that vorticity gradients grow exponentially in regions where $S_1^2 + S_2^2 > \frac{1}{4}\omega^2$, and oscillate in regions where $\frac{1}{4}\omega^2 > S_1^2 + S_2^2$, the oscillations probably corresponding to a local rotation of the vorticity field. The former case leads to the filamentation of vorticity shown in Figure 10.5, while the latter could be indicative of coherent vortices, or at least that is the theory. In short, filamentation relies on a strain field which is strong enough to rip the vortices apart, and in the absence of such a strain field, the vortices survive. The necessarily condition for survival, $\frac{1}{4}\omega^2 > S_1^2 + S_2^2$, is known as the Okubo–Weiss criterion, and it has received some degree of support from numerical simulations (Brachet et al. 1988).

It should be stressed, however, that (10.22) is more of an interpretation of the simulations rather than a deductive theory. There is no real justification for neglecting the terms on the right of (10.21) and indeed regions where the criterion fails are readily found. Nevertheless, we have the empirical observation that, for certain initial conditions, the peaks in vorticity, say $\hat{\omega}$, survive the filamentation process and so are remembered by the flow. In such cases Batchelor's energy specturm (10.8) and (10.9) must be generalized to (Bartello and Warn 1996),

$$\left\langle [\Delta \mathrm{v}(r)]^2 \right\rangle = u^2 g(r/ut, \hat{\omega}t) \tag{10.23}$$

$$E(k) = u^3 t\ h(kut, \hat{\omega}t) \tag{10.24}$$

where u^2 and $\hat{\omega}$ are constants. This is not, in general, a self-similar spectrum.

Example 10.6 The Okubo–Weiss criterion in terms of pressure
Show that the Okubo–Weiss criterion can be written in terms of pressure as follows:

$$\nabla^2(p/\rho) = 2\left[\tfrac{1}{4}\omega^2 - S_1^2 - S_2^2\right] = \tfrac{1}{2}\omega^2 - S_{ij}S_{ij}.$$

Example 10.7 An alternative to the Okubo–Weiss criterion
Show that, in an Euler flow,

$$\frac{D^2}{Dt^2}\frac{\partial\omega}{\partial x_i} + P_{ij}\frac{\partial\omega}{\partial x_j} = -2\frac{\partial\omega}{\partial x_j}\frac{D}{Dt}\frac{\partial u_j}{\partial x_i}$$

where P_{ij} is the pressure Hessian,

$$P_{ij} = \frac{\partial^2 (p/\rho)}{\partial x_i \partial x_j}.$$

If we follow the logic of (10.21) and neglect the Lagrangian rate of change of $\partial u_i / \partial x_j$, by comparison with that of $\nabla\omega$, we obtain

$$\frac{D^2}{Dt^2}\frac{\partial \omega}{\partial x_i} + P_{ij}\frac{\partial \omega}{\partial x_j} \approx 0.$$

In this case positive eigenvalues of the pressure Hessian tend to give rise to coherent vortices. If λ_1 and λ_2 are the eigenvalues of P_{ij} then

$$\lambda_1 + \lambda_2 = P_{ii} = 2\left(\tfrac{1}{4}\omega^2 - S_1^2 - S_2^2\right).$$

Thus positive eigenvalues (coherent vortices) will tend to occur if P_{ii} is large and positive, which is similar to the Okubo–Weiss criterion.

10.2.2 *The significance*

Let us return to (10.23) and (10.24),

$$\left\langle [\Delta \mathrm{v}(r)]^2 \right\rangle = u^2 g(r/ut, \hat{\omega} t), \qquad E(k) = u^3 t\, h(kut, \hat{\omega} t)$$

where u^2 and $\hat{\omega}$ are constants. The significance of (10.23) and (10.24) is that they remove much of the justification for the estimate $l \sim ut$, and so all of the scalings based on Batchelor's self-similar spectrum become suspect. We can rewrite these in the form

$$\left\langle [\Delta \mathrm{v}(r)]^2 \right\rangle = u^2 g(r/l^+, \hat{\omega} t), \qquad E(k) = u^2 l^+ h(kl^+, \hat{\omega} t)$$

where l^+ is an integral scale defined by

$$l^+ = \sqrt{2u^2 / \langle \omega^2 \rangle} = utF(\hat{\omega} t).$$

(F is some dimensionless function of $\hat{\omega} t$.) The numerical simulations of Bartello and Warn (1996), Herring et al. 1999, and Lowe (2001) (see Figure 10.11) suggest that these expressions may be simplified to the self-similar form

$$\left\langle [\Delta \mathrm{v}(r)]^2 \right\rangle \approx u^2 g(r/l^+), \qquad E(k) \approx u^2 l^+ h(kl^+) \tag{10.25a}$$

Of course Batchelor's self-similar spectrum is a special case of this, in which $l^+ \sim ut$ and $\omega^2 \sim t^{-2}$. However the existence of the second invariant $\hat{\omega}$ means that there is no particular reason to suppose that Batchelor's scalings apply and indeed the simulations tend to suggest

$$l^+ \sim ut/\sqrt{\hat{\omega} t} \sim \sqrt{t}, \qquad \langle \omega^2 \rangle \sim \hat{\omega} t^{-1}, \qquad \beta \sim \hat{\omega} t^{-2}. \tag{10.25b}$$

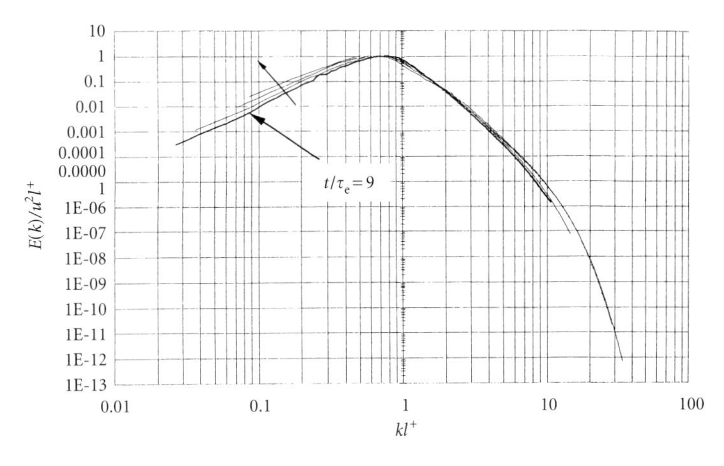

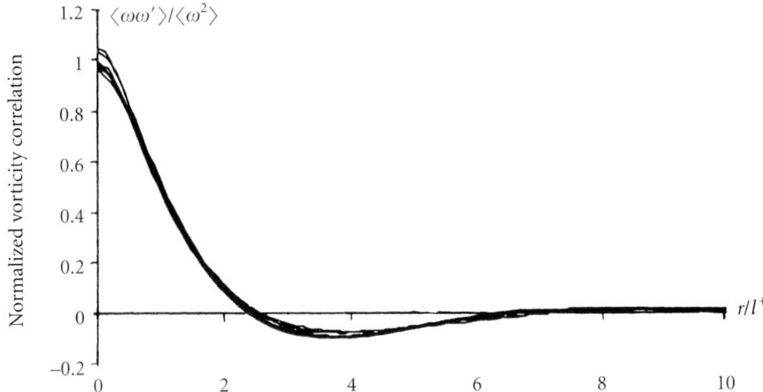

Figure 10.11 The form of $\langle\omega\omega'\rangle/\langle\omega^2\rangle$ and $E(k)$ obtained from DNS. E, k, and r are normalized using the self-similar scalings: E/u^2l^+, kl^+, and r/l^+. The data spans the range $t/\tau = 9$ to 140 where τ is the initial eddy turn-over time. The self-similar scaling works well (Lowe 2001).

Interestingly, this $\sqrt{t}$ scaling for l is exactly what emerged from the model problem (i.e. Example 10.5) given at the end of Section 10.1.8, in which a point vortex winds up the local vortex lines. This suggests that the growth in l might be related to the winding up of the filamentary vorticity by coherent vortices. Note that, with this form of self-similarity, the inertial-range spectrum, (10.19a), and structure-function, (10.20a), become

$$E \sim (\hat{\omega}/t)k^{-3}, \qquad \eta \ll k^{-1} \ll l \tag{10.25c}$$

$$\left\langle[\Delta v]^2\right\rangle \sim (\hat{\omega}/t)r^2, \qquad \eta \ll r \ll l \tag{10.25d}$$

These might be compared with the equivalent expressions for Batchelor's self-similar spectrum

$$E \sim t^{-2}k^{-3}, \qquad \left\langle[\Delta v]^2\right\rangle \sim t^{-2}r^2.$$

10.3 The governing equations in statistical form

In turbulent flow, the actual initial conditions cease to have any effect after sufficiently long intervals of time. This shows that the theory of turbulent flow must be a statistical theory. (Landau and Lifshitz 1959)

So far we have largely avoided translating the governing equations of two-dimensional turbulence into statistical form. However, if progress is to be made, then at some point we must establish these equations. In the following sections we do just that, developing the theory in a way which mirrors our description of three-dimensional turbulence in Section 6.2. For simplicity, we restrict ourselves to turbulence which is statistically homogeneous and isotropic.

10.3.1 *Correlation functions, structure functions, and the energy spectrum*

In this section we focus on the kinematics of two-dimensional turbulence. That is, we develop the language of statistical turbulence theory. We leave the question of dynamics to Section 10.3.2.

10.3.1.1 The second-order velocity correlation function

As in three dimensions, we start by introducing the second-order velocity correlation tensor

$$Q_{ij}(\mathbf{r}) = \langle u_i(\mathbf{x})u_j(\mathbf{x}+\mathbf{r})\rangle = \langle u_i u_j' \rangle$$

(Of course, Q_{ij} also depends on t, so strictly we should write Q_{ij} $(\mathbf{r}, t)$.) Note that, because of statistical homogeneity, Q_{ij} does not depend on $\mathbf{x}$ and has the geometric property that $Q_{ij}(\mathbf{r}) = Q_{ji}(-\mathbf{r})$. A special case of Q_{ij} is

$$Q_{xx}(r\hat{\mathbf{e}}_x) = u^2 f(r) = u^2 - \tfrac{1}{2}\langle(\Delta \mathrm{v})^2\rangle$$

where f is the usual longitudinal velocity correlation function. It turns out to be convenient to introduce a sequence of length-scales, l_n, defined in terms of $f(r)$:

$$(l_n)^n \sim \int_0^\infty r^{(n-1)} f(r)\,dr.$$

For low values of n, say $n = 1$ or 2, l_n provides a convenient measure of the extent of the region within which the velocities are appreciably correlated (the size of the big eddies). In three-dimensional turbulence it is conventional to use l_1 to define the integral scale. However, in two dimensions, we shall find it more convenient to define the integral scale through l_2, which turns out to be of the same order as l^+.

10.3.1.2 The relationship between the structure function and the energy spectrum

Let us now explore the relationship between $f(r)$, $E(k)$, and $\langle[\Delta \mathrm{v}]^2\rangle$. Following arguments similar to those of Chapter 8, we may show that, in isotropic turbulence, $Q_{ii}(r)$ and $E(k)$ are simply related by the Hankel transform pair,[16]

$$E(k) = \int_0^\infty \tfrac{1}{2}\langle \mathbf{u} \cdot \mathbf{u}' \rangle kr J_0(kr)\, dr$$

$$\tfrac{1}{2}\langle \mathbf{u} \cdot \mathbf{u}' \rangle = \int_0^\infty E(k) J_0(kr)\, dk.$$

Similarly, anticipating (10.27b), the relationship between $f(r)$ and $E(k)$ is readily shown to be,

$$E(k) = \tfrac{1}{2}u^2 \int_0^\infty f(r)[J_1(kr)k^2r^2]\, dr$$

$$u^2 f(r) = 2\int_0^\infty E(k)[J_1(kr)/(kr)]\, dk.$$

(Note that the first of these confirms that, for small k, $E = Ik^3/16\pi + \cdots$.) In terms of the second-order structure function the second of these equations becomes

$$\tfrac{1}{2}\langle[\Delta v(r)]^2\rangle = u^2(1-f) = \int_0^\infty E(k)H(kr)\, dk$$

where $H(kr)$ is the dimensionless function

$$H(\chi) = 1 - 2J_1(\chi)/\chi.$$

It turns out that a good approximation to $H(\chi)$ is,

$$H(\chi) = (\chi/\pi)^2, \quad \chi < \pi; \qquad H(\chi) = 1, \quad \chi > \pi$$

and so we have the approximate relationship

$$\tfrac{1}{2}\langle[\Delta v(r)]^2\rangle \approx \int_{\pi/r}^\infty E(k)\, dk + (r/\pi)^2 \int_0^{\pi/r} k^2 E(k)\, dk. \qquad (10.26)$$

In words this says that $\frac{1}{2}\langle[\Delta v(r)]^2\rangle$ is approximately equal to the kinetic energy held in eddies of size r or less, plus r^2/π^2 times the enstrophy held in eddies of size r or greater. This is reminiscent of the interpretation of $\langle[\Delta v(r)]^2\rangle$ in three-dimensional turbulence, and is more or less what we would expect from a crude physical picture of $\langle[\Delta v(r)]^2\rangle$. (See Section 3.2.5 or else Section 8.1.6.)

Notice that, if the inertial-range spectrum takes the form $E(k) \sim A(t)k^{-n}$, $n > 1$, then the corresponding form of $\langle[\Delta v(r)]^2\rangle$

[16] The energy spectrum $E(k)$ is defined as $\pi k \Phi_{ii}(k)$, where Φ_{ij} is the two-dimensional Fourier transform of Q_{ij}: $\Phi_{ij}(\mathbf{k}) = (2\pi)^{-2}\int Q_{ij}(\mathbf{r})\exp(-j\mathbf{k}\cdot\mathbf{r})\, d\mathbf{r}$. The Hankel transform pair relating $E(k)$ to Q_{ii} is obtained by integrating over the azimuthal angle in the two-dimensional transform integrals.

depends on the value of n. To see this let us assume that Re is very large, so that $E(k)$ is dominated by the inertial range, and rewrite (10.26) as

$$\tfrac{1}{2}\langle[\Delta v(r)]^2\rangle \approx \int_{\pi/r}^{\pi/\eta} E(k)\,dk + (r/\pi)^2 \int_{\pi/l}^{\pi/r} k^2 E(k)\,dk, \; E(k) = Ak^{-n}.$$

(Here l is the integral scale and η the microscale.) It is readily confirmed that,

$$\tfrac{1}{2}\langle[\Delta v(r)]^2\rangle \sim A(t)r^{n-1}, \quad \text{for } n < 3$$

$$\tfrac{1}{2}\langle[\Delta v(r)]^2\rangle \sim A(t)r^2 l^{n-3}, \quad \text{for } n > 3$$

$$\tfrac{1}{2}\langle[\Delta v(r)]^2\rangle \sim A(t)r^2 \ln(l/r), \quad \text{for } n = 3.$$

Thus the usual rule, that $E(k) \sim k^{-n}$ implies $\langle[\Delta v(r)]^2\rangle \sim r^{n-1}$, which holds in three-dimensional turbulence, does not generally apply in two dimensions. In fact numerical experiments indicate that n is greater than, or equal to, 3. Thus we expect $\langle[\Delta v(r)]^2\rangle \sim r^2$, possibly with a logarithmic correction, in the inertial range. The reason for this unexpected behaviour lies in the enstrophy integral on the right of (10.26), which measures the contribution to $\langle[\Delta v(r)]^2\rangle$ from eddies of size greater than r. When $n > 3$ this integral is dominated by the largest eddies, and these, in turn, make the largest contribution to the structure function. Thus the inertial-range scaling $\langle[\Delta v(r)]^2\rangle \sim A(t)r^2 l^{n-3}$ is a manifestation of the fact that $\langle[\Delta v(r)]^2\rangle$ is dominated by the *enstrophy* of eddies of size l. By contrast, in three dimensions, $\langle[\Delta v(r)]^2\rangle$ in the inertial range is dominated by the *energy* of eddies of size r.

So, in two dimensions we lose the useful rule of thumb that $\langle[\Delta v(r)]^2\rangle$ is a measure of the energy held in eddies of size r or less. One way around this difficulty is to use the signature function, $V(r, t)$, rather than the structure function, $\langle[\Delta v(r)]^2\rangle$, to represent the energy distribution in real space. We shall not pursue this in detail here, but the interested reader will find a starting point in Exercises 10.2 and 10.3. (See also Section 6.6.1 for a discussion of the signature function in three-dimensional turbulence.)

10.3.1.3 The general form of Q_{ij} and S_{ijk} in isotropic turbulence

Let us now return to Q_{ij} and its third-order counterpart. We have already seen that the symmetries implied by isotropy, and the requirement that $\nabla \cdot \mathbf{u} = 0$, allows the three-dimensional velocity correlation tensor, Q_{ij}, to be expressed purely in terms of $f(r)$. The same is true in two dimensions, although the end result is a little different. Following arguments similar to those laid out in Section 6.2, it is straightforward (though tedious) to show that,

$$Q_{ij}(\mathbf{r}) = u^2\left\{\frac{\partial}{\partial r}(rf)\delta_{ij} - \frac{r_i r_j}{r}f'(r)\right\}. \tag{10.27a}$$

We note in passing that,

$$Q_{ii} = \langle \mathbf{u}\cdot\mathbf{u}'\rangle = \frac{u^2}{r}\frac{\partial}{\partial r}(r^2 f). \tag{10.27b}$$

Next, we introduce the third-order (or triple) velocity correlation tensor,

$$S_{ijl}(\mathbf{r}) = \langle u_i(\mathbf{x})u_j(\mathbf{x})u_l(\mathbf{x}+\mathbf{r})\rangle = \langle u_i u_j u_l'\rangle.$$

It too can be written in terms of a single scalar function, $K(r)$, defined this time by,

$$u^3K(r) = \langle u_x^2(\mathbf{x})u_x(\mathbf{x}+r\hat{\mathbf{e}}_x)\rangle.$$

Of course, K is the usual longitudinal triple velocity correlation function. Again, symmetry and continuity arguments can be deployed, in a manner similar to those used in Section 6.2, to show that, for isotropic turbulence,

$$S_{ijl} = u^3\left\{\frac{r_i\delta_{jl} + r_j\delta_{il}}{2r}\frac{\partial}{\partial r}(rK) - \frac{r_i r_j r_l}{r}\frac{\partial}{\partial r}\left(\frac{K}{r}\right) - \frac{r_l\delta_{ij}K}{r}\right\}. \tag{10.28a}$$

10.3.1.4 The streamfunction and vorticity correlation functions

The tensors Q_{ij} and S_{ijk}, or equivalently $f(r)$ and $K(r)$, are the common currency of statistical turbulence theory. However, since we are in two dimensions, it seems appropriate to introduce two additional correlation functions: the streamfunction and vorticity correlations, $\langle\psi\psi'\rangle$ and $\langle\omega\omega'\rangle$. In Chapter 6 we saw that

$$\langle\boldsymbol{\omega}\cdot\boldsymbol{\omega}'\rangle = -\nabla^2\langle\mathbf{u}\cdot\mathbf{u}'\rangle$$

which, when combined with (10.27b), yields

$$\langle\omega\omega'\rangle = \frac{1}{2r}\frac{\partial}{\partial r}r\frac{\partial}{\partial r}\frac{1}{r}\frac{\partial}{\partial r}r^2\langle(\Delta \mathrm{v})^2\rangle.$$

For small r this tells us that,

$$\langle(\Delta \mathrm{v})^2\rangle = \langle\omega^2\rangle r^2/8 + 0(r^4)$$

and it is readily confirmed that the higher-order terms in the expansion are

$$\langle(\Delta \mathrm{v})^2\rangle = \frac{3\langle\omega^2\rangle r^2}{(4!)} - \frac{3\langle(\nabla\omega)^2\rangle r^4}{(4!)^2} + \frac{3\langle(\nabla^2\omega)^2\rangle r^6}{2(4!)^3} + \cdots. \tag{10.28b}$$

Turning now to $\langle \psi\psi' \rangle$ we have[17]

$$\langle \mathbf{u} \cdot \mathbf{u}' \rangle = -\nabla^2 \langle \psi\psi' \rangle \qquad (10.29a)$$

from which,

$$\frac{\partial}{\partial r} \langle \psi\psi' \rangle = -u^2 rf. \qquad (10.29b)$$

This suggests that a convenient definition of the integral scale, l, is:

$$l^2 = \frac{\langle \psi^2 \rangle}{u^2} = \int_0^\infty rf\, dr. \qquad (10.30)$$

Finally, we note that the Loitsyansky integral (10.17) can be rewritten in terms of $\langle \psi\psi' \rangle$. Substituting for $\langle \mathbf{u} \cdot \mathbf{u}' \rangle$ using (10.29a) we find,

$$I = -\int \mathbf{r}^2 \langle \mathbf{u} \cdot \mathbf{u}' \rangle\, d\mathbf{r} = 4 \int \langle \psi\psi' \rangle\, d\mathbf{r}. \qquad (10.31a)$$

Note that, since volume averages are equivalent to ensemble averages in homogeneous turbulence we can rewrite (10.31a) as

$$I = \left[\int_V 2\psi\, d\mathbf{x} \right]^2 \Big/ V \qquad (10.31b)$$

for some large volume V. Interestingly, the net angular momentum in a *closed* domain V is $H = \int 2\psi\, d\mathbf{x}$, since $(\mathbf{x} \times \mathbf{u})_z = 2\psi - \nabla \cdot (\psi \mathbf{x})$ and ψ is, by convention, zero on the boundary. This suggests that there may be a link between I and angular momentum, as is the case in three dimensions.

Example 10.8
Show that

$$\int_0^\infty f\, dr = \frac{2}{u^2} \int_0^\infty [E(k)/k]\, dk.$$

This provides an alternative measure of the integral scale.

10.3.2 *The two-dimensional Karman–Howarth equation and its consequences*

So far we have merely exploited some kinematic relationships. This provides us with the language of statistical turbulence theory, but does not, in itself, advance the theory in any way. If we are to say anything definite about the evolution of a turbulent flow then we must introduce some dynamics. Let us see where this leads.

[17] Remember that ψ is defined via $\mathbf{u} = \nabla \times (\psi \hat{\mathbf{e}}_z)$.

10.3.2.1 The derivation of the two-dimensional Karman-Howarth equation

Using the notation $\mathbf{u}' = \mathbf{u}(\mathbf{x}') = \mathbf{u}(\mathbf{x}+\mathbf{r})$, we have,

$$\frac{\partial u_i}{\partial t} = -\frac{\partial}{\partial x_l}(u_i u_l) - \frac{\partial}{\partial x_i}\left(\frac{p}{\rho}\right) + \nu\nabla^2 u_i$$

$$\frac{\partial u_j'}{\partial t} = -\frac{\partial}{\partial x_l'}(u_j' u_l') - \frac{\partial}{\partial x_j'}\left(\frac{p'}{\rho}\right) + \nu(\nabla')^2 u_j'.$$

On multiplying the first of these by u_j', and the second by u_i, adding the two and averaging, we obtain

$$\frac{\partial}{\partial t}\left\langle u_i u_j'\right\rangle = -\left\langle u_i \frac{\partial u_j' u_l'}{\partial x_l'} + u_j' \frac{\partial u_i u_l}{\partial x_l}\right\rangle - \frac{1}{\rho}\left\langle u_i \frac{\partial p'}{\partial x_j'} + u_j' \frac{\partial p}{\partial x_i}\right\rangle + \nu\left\langle u_i(\nabla')^2 u_j' + u_j'\nabla^2 u_i\right\rangle.$$

We now note that the operations of differentiation and taking averages commute, that u_i is independent of $\mathbf{x}'$ while u_j' is independent of $\mathbf{x}$, and that $\partial/\partial x_i$ and $\partial/\partial x_j'$ operating on averages may be replaced by $-\partial/\partial r_i$ and $\partial/\partial r_j$, respectively. The end result is,

$$\frac{\partial Q_{ij}}{\partial t} = \frac{\partial}{\partial r_l}\left[S_{jli} + S_{ilj}\right] + 2\nu\nabla^2 Q_{ij} \tag{10.32}$$

Note that we have dropped the terms involving pressure since it may be shown that, as in three dimensions, isotropy requires that $\left\langle p u_j'\right\rangle = 0$ (see Chapter 6). Finally, we substitute for Q_{ij} and S_{ijl} using (10.27a) and (10.28a). After a little algebra we obtain the two-dimensional analogue of the Karman–Howarth equation:

$$\frac{\partial}{\partial t}\left[u^2 r^3 f\right] = u^3 \frac{\partial}{\partial r}\left[r^3 K\right] + 2\nu u^2 \frac{\partial}{\partial r}\left[r^3 f'(r)\right]. \tag{10.33}$$

It is reassuring that such a simple equation should emerge from all of this algebra.

10.3.2.2 Four consequences of the two-dimensional Karman–Howarth equation

There are four immediate consequences of equation (10.33). These are listed below.

1. *The rate of growth of the integral scale.* First, we can use the Karman–Howarth equation to obtain an explicit expression for the rate of change of the integral scale defined through (10.30). Using (10.29b) to rewrite our dynamic equation in terms of $\langle\psi\psi'\rangle$, and noting that u^2 is conserved for $\mathrm{Re}\to\infty$, we obtain,

$$\frac{dl^2}{dt} = 2u\int_0^\infty K\,dr, \qquad \mathrm{Re}\to\infty. \tag{10.34a}$$

Since $K(r)$ is found to be positive in two-dimensional turbulence this confirms that l increases with time.

2. *The form of the third-order structure function in the dissipation range.* Second, substituting (10.28b) into (10.33) we find that, at first and second order in r, the Karman–Howarth equation yields,

$$\frac{du^2}{dt} = -\nu\langle\omega^2\rangle, \qquad \frac{d}{dt}\frac{1}{2}\langle\omega^2\rangle = -\nu\langle(\nabla\omega)^2\rangle,$$

which we knew already, while at next order,

$$6u^3K(r) = -\frac{r^5}{256}\left\langle S_{ij}(\nabla\omega)_i(\nabla\omega)_j\right\rangle + \cdots.$$

The term on the left, $-\langle S_{ij}(\nabla\omega)_i(\nabla\omega)_j\rangle$, is familiar as the rate of generation of palinstrophy. If we now introduce the third-order longitudinal structure function

$$\left\langle[\Delta v(r)]^3\right\rangle = \left\langle[u_x(\mathbf{x}+r\hat{\mathbf{e}}_x) - u_x(\mathbf{x})]^3\right\rangle$$

we can rewrite our expression as,

$$\left\langle(\Delta v)^3\right\rangle = 6u^3K(r) = -\frac{r^5}{256}\left\langle S_{ij}(\nabla\omega)_i(\nabla\omega)_j\right\rangle + \cdots. \qquad (10.34b)$$

Note that the fact that $\langle(\Delta v)^3\rangle$ scales as r^5 for small r tells us that $\langle(\partial u_x/\partial x)^3\rangle = 0$. Thus the skewness of $\partial u_x/\partial x$ is zero, which is not at all the case in three dimensions. Nevertheless, there is some similarity between two and three dimensions in as much as the form of $\langle(\Delta v)^3\rangle$ for small r is directly related to the rate of generation of enstrophy in three dimensions, and to the rate of generation of palinstrophy in two dimensions.

3. *The form of the third-order structure function in the inertial range.* The third consequence of (10.33) concerns the inertial range where it is possible to obtain expressions which are reminiscent of Kolmogorov's four-fifth's law. However, unlike the four-fifth's law these results are not exact. Rather, they rely on certain plausible, though ultimately heuristic, hypotheses.

Recall that, in three dimensions, the four-fifth's law rests on the existence of a range of eddies which are large enough not to be influenced by viscosity, yet small enough to be in a state of quasi-equilibrium (see Section 6.2.2). (Remember that, in three-dimensions, the small eddies have a very fast characteristic timescale, and so to them the large scales, and in particular the energy flux, appears to be almost steady.) In two dimensions, however, the inertial range is not in statistical equilibrium. That is to say, the turn-over time of the large eddies is the same as that of the small eddies (Table 10.2). So, unlike three-dimensional flow, we cannot picture the small eddies as being in statistical equilibrium. It is this lack of an equilibrium range in two dimensions which excludes an exact equivalent to the four-fifth's law. Nevertheless, it is possible to introduce additional hypotheses to compensate for the lack of an equilibrium range. When combined with the Karman–Howarth equation, these hypotheses yield predictions about the shape of $\langle[\Delta v(r)]^3\rangle$ in the inertial range.

Let us start by rewriting (10.33) in terms of structure functions,

$$\frac{\partial}{\partial t}\left[r^3\langle(\Delta v)^2\rangle\right] = -\frac{\partial}{\partial r}\left[\frac{1}{3}r^3\langle(\Delta v)^3\rangle\right].$$

Here we have taken u^2 to be constant and ignored the viscous term on the assumption that we are well removed from the dissipation range. In three dimensions we would proceed to the four-fifths law by ignoring the time derivative on the left (see Section 6.2.2). However, we are not at liberty to do this in two-dimensions. Instead, Lindborg (1999) has suggested that we substitute for $\langle[\Delta v(r)]^2\rangle$ using expansion (10.28b). He then assumed that the time derivative of $\langle[\Delta v(r)]^2\rangle$ is dominated by the first term in the expansion. (Actually, Lindborg's analysis is a little more involved than this as it deals with forced turbulence, but it reduces to the description above in the case of freely evolving turbulence.) The end result is

$$\langle(\Delta v)^3\rangle = 6u^3K(r) = \tfrac{1}{8}\beta r^3.$$

Comparing this with (10.20b) we see that $c_3 = 1/8$. Some support for this expression is given in Lindborg and Alvelius (2000), though it is by no means clear how justified the underlying assumptions are.

An alternative (and more defencible) strategy is to assume that the inertial-range scaling

$$\langle[\Delta v]^2\rangle = c_2\beta^{2/3}r^2$$

is correct, but to allow for the fact that c_2 may be time dependant. Two obvious cases are Batchelor's self-similar spectrum, where c_2 is constant and

$$\langle[\Delta v]^2\rangle \sim t^{-2}r^2$$

and the self-similar scaling (10.25a, b), in which c_2 is time dependant and

$$\langle[\Delta v]^2\rangle \sim (\hat{\omega}/t)r^2.$$

We now use the inviscid Karman–Howarth equation to find the relationship between $\langle[\Delta v(r)]^2\rangle$ and $\langle[\Delta v(r)]^3\rangle$. This yields,

$$\langle[\Delta v]^3\rangle = -\frac{r}{2}\frac{\partial}{\partial t}\langle[\Delta v]^2\rangle, \quad \eta \ll r \ll l. \tag{10.34c}$$

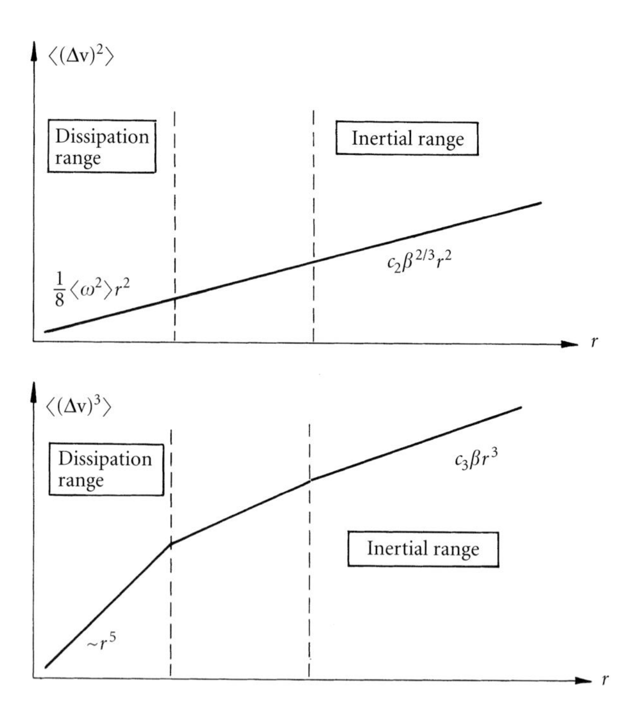

Figure 10.12 Schematic representation of $\langle(\Delta v)^2\rangle$ and $\langle(\Delta v)^3\rangle$ in two-dimensional turbulence.

We might think of this as the two-dimensional analogue of Kolmogorov's four-fifth law. In the case of Batchelor's self-similar spectrum this becomes

$$\langle[\Delta v]^3\rangle = \frac{r}{t}\langle[\Delta v]^2\rangle.$$

This is illustrated in Figure 10.12. For the case of the self-similar scaling (10.25b) we have

$$\langle[\Delta v]^3\rangle = \frac{r}{2t}\langle[\Delta v]^2\rangle.$$

4. *The rate of change of Loitsyansky's integral.* The fourth, and final, consequence of the Karman–Howarth equation comes from integrating (10.33) to give,

$$\frac{dI}{dt} = 4\pi[u^3r^3K]_\infty + 2\nu u^2[4\pi r^3 f'(r)]_\infty \tag{10.35}$$

where I is the two-dimensional Loitsyansky integral defined by (10.17). This is the analogue of the evolution equation for Loitsyansky's integral in three dimensions. If remote points in the flow happen to be statistically independent, in the sense that f and K are exponentially small at large r, then (10.35) predicts that the two-dimensional Loitsyansky integral is conserved. However, as in three dimensions, remote points are *not* statistically independent. We shall return to this issue shortly.

10.3.2.3 The two-dimensional Karman–Howarth equation in spectral space

We close this section by noting that the two-dimensional Karman–Howarth equation (10.33) may be recast in Fourier space as an equation for the rate of change of $E(k, t)$. Recall that

$$E(k) = \frac{1}{2}u^2k^2 \int_0^\infty J_1(kr)r^2 f(r)\, dr.$$

Thus, if we multiply (10.33) by $r^{-1}J_1(kr)$ and integrate from $r=0$ to $r=\infty$, we find, after a little algebra

$$\frac{\partial E}{\partial t} = T(k,t) - 2\nu k^2 E$$

$$T(k,t) = k^3 \int_0^\infty \frac{\partial}{\partial r}[r^3u^3K(r)]\frac{J_1(kr)}{2kr}\, dr.$$

Here $T(k)$ is the spectral kinetic energy transfer function (Figure 10.13), first introduced in Chapter 8. It can also be expressed as,

$$T = -\frac{\partial \Pi}{\partial k}, \qquad \Pi = -k^4 \int_0^\infty \frac{\partial}{\partial r}[r^3u^3K(r)]\frac{J_2(kr)}{2(kr)^2}\, dr$$

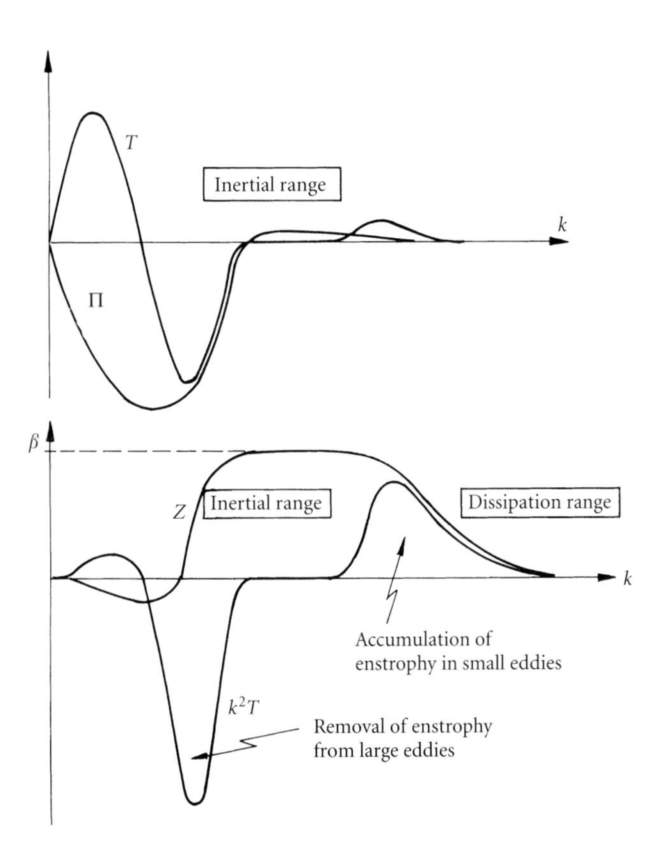

Figure 10.13 Schematic shape of $T(k)$, $\Pi(k)$, $k^2T(k)$, and $Z(k)$ in two-dimensional turbulence.

or

$$k^2 T = -\frac{\partial Z}{\partial k}, \qquad Z(k) = k^6 \int_0^\infty \frac{\partial}{\partial r}\left[r^3 u^3 K(r)\right]\left[\frac{J_3(kr)}{(kr)^3} - \frac{J_2(kr)}{2(kr)^2}\right] dr$$

where $\Pi(k)$ and $Z(k)$ are the spectral kinetic energy and enstrophy fluxes, respectively. Noting that $\int_0^\infty T\, dk = 0$ and $\int_0^\infty k^2 T\, dk = 0$ we have,

$$\frac{d}{dt}\int_0^\infty E\, dk = -2\nu \int_0^\infty k^2 E\, dk$$

$$\frac{d}{dt}\int_0^\infty k^2 E\, dk = -2\nu \int_0^\infty k^4 E\, dk$$

which are the spectral equivalents of (10.4) and (10.5).

10.3.3 *Loitsyansky's integral in two-dimensions*

We now return to Loitsyansky's integral. At first sight (10.35) is very appealing. It suggests that, if the long-range correlations are sufficiently weak, then I might be an invariant of two-dimensional turbulence:

$$I = -\int r^2 \langle \mathbf{u} \cdot \mathbf{u}' \rangle\, d\mathbf{r} = 4\pi u^2 \int_0^\infty r^3 f\, dr = \text{constant?}$$

However, our experience in three dimensions suggest that we must look carefully at the form of f and K as $r \to \infty$, since it is by no means obvious that the terms on the right of (10.35) should vanish. The purpose of this section is to estimate K_∞ and f_∞ and thus determine dI/dt.

As in three dimensions, the key to establishing K_∞ and f_∞ is to note that the pressure field is 'non-local' in space. That is to say, a fluctuation in $\mathbf{u}$ at one point in the flow sends out pressure waves, which travel infinitely fast in an incompressible fluid, and these produce pressure forces, and hence accelerations, which fall off algebraically with distance from the source. Thus, because of pressure, a fluctuation in $\mathbf{u}$ at one point is felt everywhere within the fluid and this can set up long-range velocity correlations which decline algebraically, rather than exponentially, with distance. In short, remote points in a turbulent flow need not be statistically independent.

In three dimensions we saw that the pressure waves decline as $p \sim r^{-3}$, which can lead to terms of order r^{-4} in K_∞, and r^{-6} in f_∞. This is sufficient to ensure that Loitsyansky's integral is time dependent, and not an invariant as had been traditionally assumed. However, we saw also that the pressure-induced, long-range

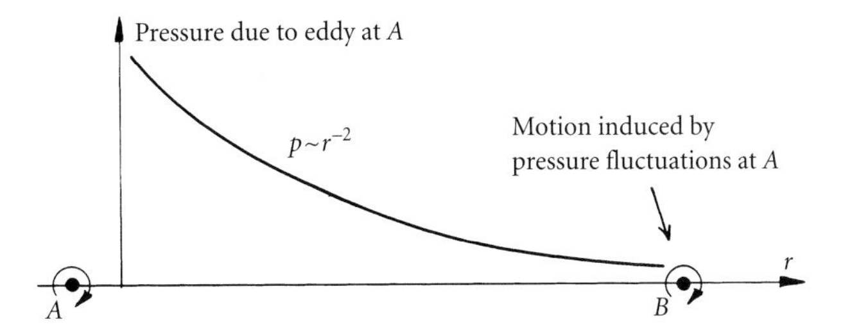

Figure 10.14 A schematic representation of long-range pressure forces.

correlations, though finite, are rather feeble in three dimensions. Thus, Loitsyansky's integral is, in fact, only a weak function of time in freely evolving, three-dimensional turbulence.

Let us now explore the analogous problem in two dimensions. We shall not reproduce all of the algebraic steps, since much of the analysis is a direct extension of that which appears in Chapter 6. It turns out that, in two dimensions, $p_\infty \sim r^{-2}$, $K_\infty \sim r^{-3}$, and $f_\infty \sim r^{-5}$. (See Figure 10.14.) The net result is that I is, in fact, time dependent, as in three dimensions. Of course, the key question is: how strong is that time dependence? Is it rather weak, as in three-dimensions, or do we have $I \sim t^4$, as suggested by Batchelor's self-similar spectrum?

Let us start by establishing the strength of the long-range velocity correlations. Taking the divergence of the Navier–Stokes equation we find

$$\nabla^2 p = -\rho \frac{\partial^2 u_i u_j}{\partial x_i \partial x_j}.$$

Next we use the Biot–Savart law to invert this equation, which yields

$$p' = p(\mathbf{x}') = -\frac{\rho}{4\pi} \int \frac{\partial^2 u_i u_j}{\partial x_i \partial x_j} \ln\left\{(x - x')^2 + (y - y')^2\right\} d\mathbf{x}.$$

The non-local nature of the pressure field is immediately apparent. The pressure at $\mathbf{x}'$ is influenced by the motion at all locations.

To illustrate the point, consider a vorticity field which is instantaneously localized in space, in the sense that $|\boldsymbol{\omega}|$ becomes exponentially small as $|\mathbf{x}| \to \infty$. The far-field pressure distribution set up by this localized disturbance can be found by expanding $\ln(\mathbf{x} - \mathbf{x}')^2$ as a Taylor series in $|\mathbf{x}'|^{-1}$. The first two terms in the expansion lead to integrals of $(\partial^2 u_i u_j / \partial x_i \partial x_j)$ and $(\partial^2 u_i u_j / \partial x_i \partial x_j) x_k$, respectively. Both integrands can be rewritten as divergences. These, in turn, can be converted into surface integrals which vanish when the surface recedes to infinity. The first non-zero term in the expansion is then,

$$4\pi p'/\rho \sim -\frac{\partial^2}{\partial x_i' \partial x_j'} \left[\ln(\mathbf{x}')^2\right] \int u_i u_j \, d\mathbf{x} + \text{HOT}$$

and so it follows that $p_\infty \sim r^{-2}$.

Now the long-range pressure forces cannot influence $f(r)$ directly since, because of isotropy, the pressure forces are absent from (10.33). However, when we derive a dynamic equation for S_{ijk}, which is analogous to (10.32), we find,

$$\rho \frac{\partial S_{ijk}}{\partial t} = \rho \langle uuuu \rangle - \frac{\partial}{\partial r_k} \langle u_i u_j p' \rangle - \left\langle u'_k \left[u_i \frac{\partial p}{\partial x_j} + u_j \frac{\partial p}{\partial x_i} \right] \right\rangle \tag{10.36}$$

where $\mathbf{r} = \mathbf{x}' - \mathbf{x}$ and $\langle uuuu \rangle$ is a symbolic representation of terms involving fourth-order correlations. (Again, the reader should consult Chapter 6 for the details.) This time, isotropy is not sufficient to dispense with the pressure terms. It turns out that, because of the algebraic decay of pressure, there is an algebraic decay in $\langle u_i u_j p' \rangle_\infty$:

$$\langle u_l u_m p' \rangle_\infty \sim -\frac{\rho}{4\pi} \frac{\partial^2}{\partial r_i \partial r_j} \ln(r^2) \int \left\langle u_l u_m \left[u''_i u''_j - \left\langle u''_i u''_j \right\rangle \right] \right\rangle d\mathbf{x}''$$

(Actually, in deriving this expression, we have to assume that the fourth-order cummulants are vanishingly small at $r \to \infty$, so that certain surface integrals vanish. However, following the arguments in Chapter 6, this seems a reasonable approximation for fully developed turbulence.) It follows, from (10.36), that for $r \to \infty$,

$$\rho \frac{\partial S_{ijk}}{\partial t} \sim r^{-3} \int \langle uuuu \rangle \, d\mathbf{r} + \text{HOT}.$$

In fact, when the details are followed though, we obtain,

$$\frac{d}{dt} [u^3 K]_\infty = \frac{1}{2\pi r^3} \int \langle ss' \rangle \, d\mathbf{r} \tag{10.37}$$

where $s = u_x^2 - u_y^2$. Thus it seems that the triple correlations decay as $K_\infty \sim r^{-3}$ and so, by virtue of (10.33), the double correlations fall off as $f_\infty \sim r^{-5}$.

Let us now return to (10.35). The viscous term on the right vanishes while the inertial term is finite. Combining (10.35) with our estimate of K_∞ yields, at last, an equation for the rate of change of I (Davidson 2000, 2001):

$$\frac{d^2 I}{dt^2} = 2 \int \langle ss' \rangle \, d\mathbf{r} \tag{10.38}$$

[Valid only for fully developed turbulence].

So I is, after all, time dependent. Moreover, since $\int \langle ss' \rangle d\mathbf{r}$ can be rewritten as

$$\int \langle ss' \rangle \, d\mathbf{r} = V^{-1} \left[\int_V s \, d\mathbf{x} \right]^2 > 0 \tag{10.39}$$

it follows that I must ultimately increase with time. We might compare this with the three-dimensional result,

$$\frac{d^2I}{dt^2} = 6\int \langle ss' \rangle \, d\mathbf{r}. \tag{10.40}$$

The next step is to try and evaluate the fourth-order correlation on the right of (10.38). However, at this point we become the victim of the *closure problem* of turbulence. That is, we can obtain an equation for $\partial\langle uuuu \rangle / \partial t$ only in terms of $\langle uuuuu \rangle$, which in turn depends on the sixth-order correlations, and so on. We are left with a problem. There is no rigorous means of evaluating the term on the right of (10.38). In three dimensions we know from experiments that it is rather small, so that I is only a weak function of time. Perhaps this is also true in two dimensions? On the other hand, Bachelor's self-similar spectrum predicts $I \sim t^4$. Of course, there is a vast difference between $I \sim t^0$ and $I \sim t^4$. We must turn to the numerical evidence to determine which, if either, prevails. It turns out that the limited numerical evidence available to us is closer to t^2 than either t^0 or t^4, although the numerical resolution in most direct numerical simulations is rather poor at the large scales. For example the numerical simulations of Bartello and Warn (1996), Chasnov (1997), and Ossai and Lesieur (2001), gave $I \sim t^{2.0}$, $I \sim t^{1.6}$, and $I \sim t^{2.5}$, respectively. This $I \sim t^2$ growth is consistent with the self-similar scaling (10.25a, b).

10.4 Variational principles for predicting the final state in confined domains

In this penultimate section we turn to freely evolving turbulence in *confined* domains. The approximate conservation of u^2, combined with the continual growth in l, means that two-dimensional turbulence in a confined domain will eventually evolve into a quasi-steady flow, with a length-scale comparable with the domain size. This is illustrated in Figure 10.15. The idea of order emerging from chaos in somewhat contrary to intuition. However, this is precisely what is observed in many numerical simulations. Once the quasi-steady state is reached, which occurs when $l \sim R$, R being the domain size, the flow settles down to a quasi-laminar motion which then decays relatively slowly due to friction on the boundary. The time taken to reach the quasi-steady state is $t \sim R/u$, while the characteristic timescale for the subsequent decay is R^2/ν. This is illustrated in Figure 10.16.

The idea that, at high Re, $\langle \omega^2 \rangle$ monotonically falls to some minimum, but that u^2 is (approximately) conserved in the process, immediately suggests the use of some variational principle to predict

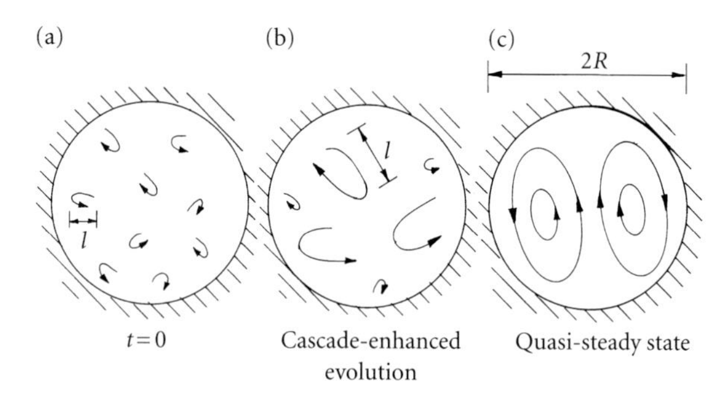

Figure 10.15 The evolution of two-dimensional turbulence in a confined domain.

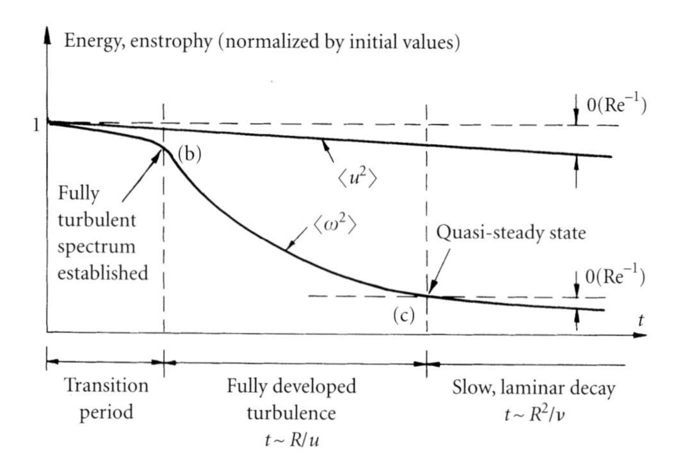

Figure 10.16 The variation of energy and enstrophy with time.

the quasi-steady state depicted in Figure 10.15(c). Actually there has been a number of hypotheses of this type, and often they work reasonably well. In general these theories identify the integral invariants of the flow, invoke some variational hypothesis (e.g. the minimization of $\langle\omega^2\rangle$ subject to the conservation of u^2), and then predict the quasi-steady state which emerges from the cascade-enhanced destruction of enstrophy. Like all variational principles, these theories have a seductive appeal, in that they are able to make specific predictions for a multitude of geometries based on some simple hypothesis. However, one must be cautious in applying such theories. Sometimes they work and sometimes they do not. This is because, although quite plausible, they are ultimately heuristic in nature.

There are two such theories in common use: *minimum enstrophy* and *maximum entropy*. We shall describe each in turn.

10.4.1 *Minimum enstrophy*

One of the more beautiful theories of two-dimensional turbulence is the *minimum enstrophy theory* (Leith 1984). The idea is that enstrophy falls monotonically during the cascade-enhanced evolution and that this occurs on the fast timescale of the eddy turn-over time. Once a quasi-steady state is established, the flow evolves on the slow timescale of R^2/ν. This suggests, but does not prove, that the flow shown in Figure 10.15(c) corresponds to a minimization of $\langle\omega^2\rangle$ subject to the conservation of u^2 (and any other invariant which may be relevant). To gain some idea of how this works consider flow in a circular domain, as shown in Figure 10.15. We would expect that, until stage (c) is reached, both u^2 and $H = 2\int\psi\, dV$ will decline on the slow viscous timescale R^2/ν, while $\langle\omega^2\rangle$ declines on the fast (cascade-enhanced) timescale of l/u. (Numerical experiments confirm that this is so.) Thus we might anticipate that the quasi-steady state corresponds to an absolute minimum of $\langle\omega^2\rangle$, subject to the conservation of u^2 and of angular momentum.[18] Minimizing $\langle\omega^2\rangle$ in this way is equivalent to minimizing the functional,

$$F = \int [R^2\omega^2 - \lambda^2(\nabla\psi)^2 + 2\lambda^2\Omega\psi]\, dV \tag{10.41}$$

where ψ is the streamfunction and λ and Ω are Lagrange multipliers (constants) which are determined by the initial conditions. Application of the calculus of variations shows that the absolute minimum of F, compatible with no slip boundary conditions, corresponds to,

$$\frac{\omega}{\Omega} = 1 - \frac{\lambda\, J_0(\lambda r/R)}{2J_1(\lambda)} \tag{10.42}$$

where J_0 and J_1 are Bessel functions. The Lagrange multipliers are fixed by the initial values of H and u^2. Specifically, if we define u^2 through $\pi R^2 u^2 = \frac{1}{2}\int \mathbf{u}^2\, dV$, we obtain, on integrating (10.42),

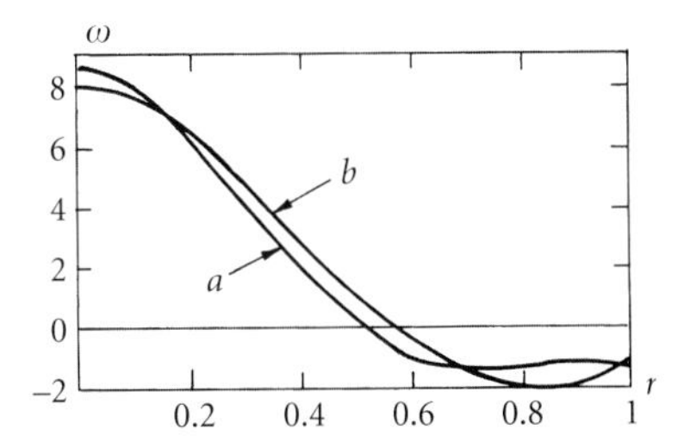

Figure 10.17 Comparison of the computed final state in a circular domain with that predicted by minimum enstrophy: (a) computed vorticity; (b) predicted vorticity. $(\pi R^3 u/H) \sim 1.5$.

$$H = -\left(\frac{\pi}{4}\right)\Omega R^4 \frac{J_3(\lambda)}{J_1(\lambda)} \tag{10.43}$$

$$\left[\frac{\pi R^3 u}{H}\right]^2 = \frac{[2J_2^2(\lambda) - 3J_1(\lambda)J_3(\lambda)]}{J_3^2(\lambda)}. \tag{10.44}$$

We may regard (10.44) as defining λ and (10.43) as fixing Ω. A comparison of (10.42) with a numerical experiment (performed by Li, Montgomergy, and Jones 1996) is shown in Figure 10.17. In this particular case, the comparison seems reasonably favourable.

[18] At first sight in may seem surprising that this final state is stable and does not revert back to turbulence. After all, it has a very high Re. Actually, it turns out that minimizing $\langle\omega^2\rangle$ subject to the conservation of $\langle\mathbf{u}^2\rangle$ produces a flow which is linearly stable to inviscid disturbances. The point is that minimizing $\langle\omega^2\rangle/\langle\mathbf{u}^2\rangle$ is equivalent to maximizing $\langle\mathbf{u}^2\rangle/\langle\omega^2\rangle$, and we know from the Kelvin–Arnold theorem that maximizing energy subject to the conservation of all of the invariants of the vorticity field yields an Euler flow which is linearly stable (e.g. see Davidson 2001).

It should be stressed, however, that the minimum enstrophy theory is not always this successful and indeed often goes badly wrong. In general, it suffers from three major drawbacks. First, while seeming plausible, it is ultimately heuristic. In particular, the quasi-steady state need not correspond to an absolute minimum in $\langle \omega^2 \rangle$. Second, the transition from a cascade-enhanced evolution to a slow diffusive evolution is not always clear-cut. Third, at finite Re, H and u^2 are not exactly conserved. This is particularly problematic when comparing the theory with numerical experiments, since the latter are invariably performed at rather modest values of Re.

10.4.2 Maximum entropy

Maximum entropy is another variational hypothesis designed to do much the same as minimum enstrophy. In effect, it defines some measure of mixing and then assumes that the turbulence maximizes this mixing (rather than minimizing enstrophy) subject to the conservation of u^2 and H. It has its roots in the early work of Joyce and Montgomery (1973), and has recently received renewed interest, partly as a result of the work of Robert and Sommeria (1991). Maximum entropy has the appearance of being more profound than the minimum enstrophy theory because it borrows some of the ideas and language of statistical mechanics. In practice, however, it is a heuristic model, exhibiting the same advantages and disadvantages of the simpler enstrophy theory.

Maximum entropy has its supporters and its detractors, and only the passage of time will tell who is correct. Those who regard it with suspicion point out that, for certain classes of initial conditions, its predictions coincide with those of minimum enstrophy, and so its apparent success in those cases may be simply a manifestation of the reliability of minimum enstrophy. Moreover, the theory involves a number of Lagrange multipliers, analogous to those in (10.41), which should be determined by the initial conditions. However, some practitioners of maximum entropy fail to fix the Lagrange multipliers in this way, but rather use them as 'free parameters' which are chosen to give a best fit for the final state (when compared to numerical experiments). Not unnaturally, this arouses suspicion amongst the unconverted. However, the theory also has its supporters. A positive review of the maximum entropy theory is given in Frisch (1995).

Example 10.9 A violation of Batchelor's similarity hypothesis?

Consider the freely-decaying turbulence described by (10.42)–(10.44). Suppose that the radius of the domain, R, is much greater than the initial integral scale, l_0, and that the turbulence is initiated by a force which is random in space but statistically isotropic and homogeneous. Then the bulk of the turbulence (that which is remote from the boundaries) will also be approximately isotropic, and it will remain so for many turn-over times. The success of equations (10.42)–(10.44) in predicting the final state of the flow shows that the turbulence must remember how much angular momentum it had at $t = 0$, since the initial value of H influences the final flow. Does this violate Batchelor's hypothesis that homogeneous turbulence remembers only its energy density, u^2?

10.5 Quasi-two-dimensional turbulence: bridging the gap with reality

Two-dimensional turbulence . . . is a consequence of the construction of large computers. (Pullin and Saffman 1998)

Strictly two-dimensional turbulence is of limited practical interest since, to some degree, all real flows are three-dimensional. In geophysics, however, there are many flows which are quasi-two-dimensional in the sense that certain aspects of the turbulence are governed by a form of two-dimensional dynamics. We close this chapter by introducing the simplest of two-dimensional models relevant to geophysics: that of a shallow layer of water in rapid rotation (Figure 10.18). In order to keep the analysis brief we shall ignore density stratification. We shall also assume that the rotation rate is independent of position.

10.5.1 *The governing equations for shallow-water, rapidly rotating flow*

We start with a brief derivation of the governing equations for shallow-water, rapidly rotating flow. Although these equations are two-dimensional, we shall see that they differ from the two-dimensional Navier–Stokes equation. Many of the more subtle issues are side stepped in our derivation, but the interested reader will find a full

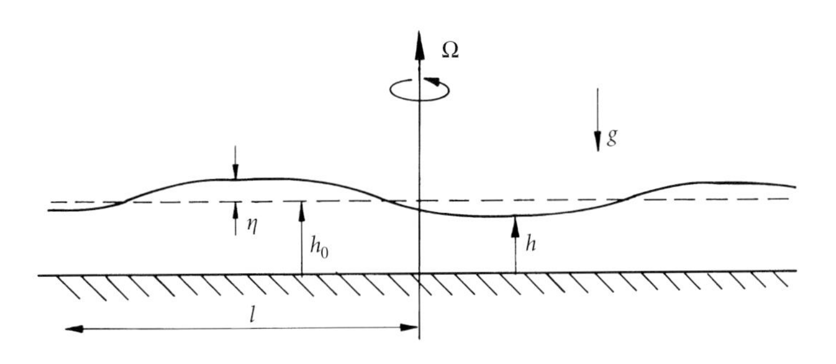

Figure 10.18 A shallow layer of water in rapid rotation.

account in Pedlosky (1979). Let $\boldsymbol{\Omega} = \Omega\hat{\mathbf{e}}_z$ be the global rotation rate, $h(x, y)$ the depth of fluid, $\eta(x, y)$ the height of the free-surface above the equilibrium depth h_0, and l the horizontal dimension of a typical motion in the fluid. We introduce coordinates x, y, z which rotate with the layer and use $\mathbf{u}$ to denote the velocity field measured in the rotating frame.

The terms shallow-water and rapidly rotating are intended to indicate that $\delta = h/l$ and $\varepsilon = u/\Omega l$ are both small. For simplicity we shall also assume that $\eta/h \ll 1$. The fact that δ is small allows us to adopt standard shallow-water theory in which the pressure distribution is hydrostatic to second order in δ. The shallow-water equations are

$$\mathbf{u} = \mathbf{u}_H(x, y) + u_z\hat{\mathbf{e}}_z + O(\delta^2)$$

$$\frac{D\mathbf{u}_H}{Dt} = -g\nabla\eta + 2\mathbf{u}_H \times \boldsymbol{\Omega} + \text{friction} + O(\delta^2)$$

$$\nabla \cdot (h\mathbf{u}_H) = -\frac{\partial h}{\partial t}$$

where $\mathbf{u}_H$ is the depth-averaged horizontal velocity, $2\mathbf{u}_H \times \boldsymbol{\Omega}$ is the Coriolis force arising from the rotation of the coordinate system, and $D/Dt = \partial/\partial t + \mathbf{u}_H \cdot \nabla$ is the convective derivative based on the mean horizontal velocity. The second equation above is the horizontal momentum equation with the pressure gradient evaluated using the hydrostatic approximation. The third equation represents conservation of mass. Let us temporarily ignore friction and invoke the condition $\eta/h \ll 1$. Our equations then simplify to

$$\frac{D\mathbf{u}_H}{Dt} = -g\nabla\eta + 2\mathbf{u}_H \times \boldsymbol{\Omega} \tag{10.45}$$

$$\frac{D\eta}{Dt} = -h_0\nabla \cdot \mathbf{u}_H. \tag{10.46}$$

We now make use of the fact that the Rossby number, $\varepsilon = u/\Omega l$, is small and expand $\mathbf{u}_H$ and η in the series

$$\mathbf{u}_H = \mathbf{u}_H^{(0)} + \varepsilon\mathbf{u}_H^{(1)} + \cdots, \qquad \eta = \eta^{(0)} + \varepsilon\eta^{(1)} + \cdots.$$

Substituting these expansions into (10.45) gives, to leading order in ε,

$$2\mathbf{u}_H^{(0)} \times \boldsymbol{\Omega} = g\nabla\eta^{(0)} \tag{10.47}$$

This is the so-called *geostrophic balance*, in which the Coriolis force is balanced by a pressure gradient. The curl of (10.47) tells us that $\mathbf{u}_H^{(0)}$ is solenoidal and so we can introduce a streamfunction for $\mathbf{u}_H^{(0)}$,

$$\mathbf{u}_H^{(0)} = \nabla \times (\psi\hat{\mathbf{e}}_z).$$

In fact (10.47) tells us that ψ and $\eta^{(0)}$ are proportional:

$$\psi = -\frac{g}{2\Omega}\eta^{(0)}. \tag{10.48}$$

Physically this says that the streamlines run parallel to the crests and valleys of the surface undulations. Notice that (10.47) does not contain any time derivatives. To get an evolution equation for $\mathbf{u}_H^{(0)}$ we must go to next order in ε in (10.45) and (10.46). That is, we consider weak departures from the geostrophic balance. This procedure yields

$$\frac{D\mathbf{u}_H^{(0)}}{Dt} = -g\nabla(\varepsilon\eta^{(1)}) + 2\varepsilon\mathbf{u}_H^{(1)} \times \mathbf{\Omega} \tag{10.49}$$

$$\frac{D\eta^{(0)}}{Dt} = -h_0\nabla\cdot(\varepsilon\mathbf{u}_H^{(1)}) \tag{10.50}$$

where we now assume that the convective derivative is based on $\mathbf{u}_H^{(0)}$, rather than on $\mathbf{u}_H$. Next, taking the curl of (10.49), and eliminating $\nabla\cdot\mathbf{u}_H^{(1)}$ using (10.50), gives us,

$$\frac{D\omega}{Dt} = \frac{2\Omega}{h_0}\frac{D\eta^{(0)}}{Dt}$$

where ω is the z-component of $\nabla\times\mathbf{u}_H^{(0)}$. The final step is to replace $\eta^{(0)}$ by ψ using (10.48). The end result is

$$\frac{D}{Dt}(\omega + \alpha^2\psi) = 0, \qquad \alpha^2 = 4\Omega^2/gh_0. \tag{10.51}$$

This is the governing equation we have been after. It is sometimes called the *quasi-geostrophic equation* for shallow-water flow, and it governs the evolution of the so-called *potential vorticity*

$$q = \omega + \alpha^2\psi = -\nabla^2\psi + \alpha^2\psi. \tag{10.52}$$

So in this type of motion it is the potential vorticity, rather than the vorticity itself, which is materially conserved. From now on we shall drop the subscript and superscript on $\mathbf{u}_H^{(0)}$, so that $\mathbf{u}(x, y)$ represents the leading-order contribution to the horizontal motion, with $\nabla\times\mathbf{u} = \omega\hat{\mathbf{e}}_z = -(\nabla^2\psi)\hat{\mathbf{e}}_z$. It is readily confirmed that (10.51) possesses the integral invariant

$$E = \tfrac{1}{2}\rho\int(\mathbf{u}^2 + \alpha^2\psi^2)\,dA = \tfrac{1}{2}\rho\int\psi q\,dA \tag{10.53}$$

which represents the combined kinetic and potential energy of the fluid. Also (10.52) may be inverted to give

$$\psi(\mathbf{x}) = (2\pi)^{-1}\int K_0(\alpha s)q(\mathbf{x}')\,d\mathbf{x}', \qquad \mathbf{s} = \mathbf{x}' - \mathbf{x}, \tag{10.54}$$

where K_0 is the usual modified Bessel function. This might be compared with

$$\psi(\mathbf{x}) = -(4\pi)^{-1}\int \ln(s^2)\omega(\mathbf{x}')\,d\mathbf{x}', \qquad \mathbf{s} = \mathbf{x}' - \mathbf{x}. \tag{10.55}$$

10.5.2 The Karman–Howarth equation for shallow-water, rapidly rotating turbulence

Let us now return to turbulence. The first thing to note is that the far-field velocity induced by a patch of potential vorticity in quasi-geostrophic turbulence is much weaker than the equivalent far-field velocity associated with a patch of vorticity in strictly two-dimensional flow. (In the quasi-geostrophic case $|\mathbf{u}|$ falls off exponentially with r, while in two-dimensional Navier–Stokes it falls as r^{-1}.) So, as an eddy moves around in our shallow layer, carrying its potential vorticity with it, its movement has little influence in the far field. We would expect, therefor, that the long-range correlations discovered by Batchelor and Proudman (see Section 6.3.4) are largely absent in quasi-geostrophic turbulence.

Next we observe that, for isotropic turbulence (10.51) yields the statistical equation

$$\frac{\partial}{\partial t}(\langle \mathbf{u} \cdot \mathbf{u}' \rangle + \alpha^2 \langle \psi \psi' \rangle) = \frac{1}{r}\frac{\partial}{\partial r}\frac{1}{r}\frac{\partial}{\partial r}(u^3 r^3 K) + (\text{friction term}) \tag{10.56}$$

where K is the usual triple correlation function. (This is left as an exercise for the reader.) When α is zero this reverts to the Karman–Howarth equation for strictly two-dimensional turbulence. Thus all of the rotational and free-surface effects are tied up in the second term on the left. The dynamic properties of this interesting equation are still poorly understood, though there is some discussion in Pedlosky (1979). Finally we note that (10.54) and (10.56) may be combined to yield a simple evolution equation for $\langle \psi \psi' \rangle$:

$$\frac{\partial}{\partial t}\langle \psi \psi'(r) \rangle = \frac{1}{2\pi}\int K_0(\alpha|\mathbf{r} - \mathbf{r}''|)\left[\frac{1}{r}\frac{\partial}{\partial r}\frac{1}{r}\frac{\partial}{\partial r}(u^3 r^3 K(r))\right]'' d\mathbf{r}''. \tag{10.57}$$

This may be integrated to show that Loitsyansky's integral is conserved in rapidly-rotating, shallow-water turbulence. (See Exercise 10.1.)

Exercises

10.1 Use the definite integral

$$\int_0^{\pi} K_0\left(\alpha\sqrt{a^2 + b^2 - 2ab\cos\theta}\right) d\theta = \pi I_0(\alpha a) K_0(\alpha b), \quad b > a$$

to show that (10.57) can be integrated over the polar angle θ to give

$$\frac{\partial}{\partial t}\langle\psi\psi'\rangle = \alpha^2 I_0(\alpha r)\int_r^\infty \frac{d}{dr}[u^3r^3K(r)]''\,\frac{K_1(\alpha r'')}{\alpha r''}\,dr'' - \alpha^2 K_0(\alpha r)\int_0^r \frac{d}{dr}[u^3r^3K(r)]''\,\frac{I_1(\alpha r'')}{\alpha r''}\,dr''$$

where I_n and K_n are the usual modified Bessel functions. Hence confirm that

$$\frac{dI}{dt} = \frac{d}{dt}4\int\langle\psi\psi'\rangle\,d\mathbf{r} = 8\pi[Q(\infty) - Q(0)]$$

where I is Loitsyansky's integral and

$$Q(r) = \alpha r I_1(\alpha r)\int_r^\infty \frac{d}{dr}[u^3r^3K(r)]''\,\frac{K_1(\alpha r'')}{\alpha r''}\,dr'' + \alpha r K_1(\alpha r)\int_0^r \frac{d}{dr}[u^3r^3K(r)]''\,\frac{I_1(\alpha r'')}{\alpha r''}\,dr''$$

Finally show that, provided K_∞ declines faster than r^{-3}, I is an invariant of shallow-water, rapidly rotating turbulence. (You may assume that K is positive, as illustrated in Figure 10.12.) In this respect quasi-geostrophic turbulence is no different from conventional two-dimensional turbulence. However, given the weak far-field influence of an eddy in shallow-water, quasi-geostrophic turbulence, it is likely that K_∞ declines exponentially fast, and so, unlike in two-dimensional turbulence, I is indeed an invariant.

10.2 Consider the *two-dimensional signature function* defined by

$$V(r,t) = -\frac{r^2}{4}\frac{\partial}{\partial r}\frac{1}{r}\frac{\partial}{\partial r}\langle(\Delta v)^2\rangle.$$

Show that it is related to $E(k, t)$ by the Hankel transform pair,

$$V(r) = \int_0^\infty E(k)J_3(kr)k\,dk, \qquad E(k) = \int_0^\infty V(r)J_3(kr)r\,dr,$$

and that it possesses the following integral properties:

$$\tfrac{1}{2}\langle\mathbf{u}^2\rangle = \int_0^\infty V(r)\,dr = \int_0^\infty E(k)\,dk$$

$$\tfrac{1}{2}\langle\omega^2\rangle = \int_0^\infty [8V(r)/r^2]\,dr = \int_0^\infty k^2E(k)\,dk$$

$$\tfrac{1}{2}\langle\psi^2\rangle = \int_0^\infty [r^2V(r)/8]\,dr = \int_0^\infty [E(k)/k^2]\,dk.$$

10.3 Consider a field of homogeneous, two-dimensional turbulence composed of a random distribution of circular eddies of fixed size l_e. Townsend (1976) has shown that, in such a situation, the longitudinal correlation function is $f(r) = \exp[-r^2/l_e^2]$. Show that, in such a case,

$$kE(k) = \tfrac{1}{8}u^2(kl_e)^4\exp[-(kl_e)^2/4]$$

$$rV(r) = 2u^2(r/l_e)^4\exp[-(r/l_e)^2]$$

where $V(r)$ is defined in Exercise 10.2 above. Note that $rV(r)$ peaks at $r = \sqrt{2}l_e$. Sketch $rV(r)$ and $[kE(k)]_{k=\pi/r}$ and confirm that they have similar shapes. Now consider the more general case of two-dimensional turbulence with the energy spectrum

$$E(k) = Ak^m \exp[-k^2\eta^2], \quad m \leq 3.$$

Show that $V(r)$ takes the form

$$V(r) = A\frac{r^3\Gamma(5/2 + m/2)}{2^4\Gamma(4)\eta^{5+m}}M(\tfrac{5}{2} + \tfrac{m}{2}, 4, -r^2/4\eta^2)$$

where M is Kummer's hypergeometric function and Γ is the gamma function. Confirm that $V(r) > 0$ for all r, and that, in the range $k\eta \ll 1$ we have $[kE(k)]_{k=\pi/r} = \alpha_m rV(r)$ where the constant of proportionality, α_m, is close to unity (e.g. $\alpha_m = 8/\pi^2$ for $m = -3$). Compare the properties of $V(r)$ with the three-dimensional signature function introduced in Section 6.6.1.

10.4 Use the Hankel transform pair of Exercise 10.2 to show that

$$\int_0^\infty (rV(r))^2 \frac{dr}{r} = \int_0^\infty (kE(k))^2_{k=\pi/r} \frac{dr}{r}.$$

Suggested reading

Ayrton, H. (1919) *Proc. Roy. Soc. A*, **xcvi**, 249–256.

Bartello, P. and Warn, T. (1996) *J. Fluid Mech.*, **326**, 357–372.

Batchelor, G.K. (1969) *Phys. Fluids*, **12** (Suppl. II) 233–239.

Brachet, M.E. et al. (1988) *J. Fluid Mech.*, **194**, 333–349.

Chasnov, J.R. (1997) *Phys. Fluids*, **9**(1), 171–190.

Davidson, P.A. (2000) *J. Turbulence*, **1**.

Davidson, P.A. (2001) *An Introduction to Magnetohydrodynamics*. CUP.

Frisch, U. (1995) *Turbulence*. Cambridge University Press.

Fujiwhara, (1921) *Q.J.R. Met. Soc.*, **xlvii**.

Fujiwhara, (1923) *Q.J.R. Met. Soc.*, **xlix**.

Herring J.R., Kimura Y., and Chasnov J. (1999) *Trends in Mathematics*. Birkhauser.

Herring J.R. et al. (1974) *J. Fluid Mech.*, **66**, 417–444.

Joyce, G. and Montgomery, D. (1973) *J. Plasma Phys.*, **10**, 107–121.

Kraichnan, R.H. (1967) *Phys. Fluids*, **10**, 1417–1423.

Leith, C.E. (1984) *Phys. Fluids*, **27**(6), 1388–1395.

Lesieur, M. (1990) *Turbulence in Fluids*. Kluwer Academic Publishers.

Li, S., Montgomery, D., and Jones W.B. (1996) *J. Plasma Phys.*, **56**(3), 615–639.

Lin, J.-T. (1971) *J. Atmos. Sci.*, **29**(2), 394–396.

Lindborg, E. (1999) *J. Fluid Mech.*, **338**, 259–288.

Lindborg, E. and Alvelius, K. (2000) *Phys. Fluids*, **12**(5), 945–947.

Lowe, A. (2001) The direct numerical simulation of isotropic two-dimensional turbulence in a periodic square. Ph.D. Thesis, University of Cambridge.

McWilliams, J.C. (1990) *J. Fluid Mech.*, **219**, 361–385.

Ossai, S. and Lesieur, M. (2001) *J. Turbulence*, August.

Pedlosky, J. (1979) *Geophysical Fluid Dynamics*. Springer-Verlag.

Pouquet, A. et al. (1975) *J. Fluid Mech.*, **72**, 305–319.

Robert, R. and Sommeria, J. (1991) *J. Fluid Mech.*, **229**, 291–310.

Townsend A.A. (1976) *The Structure of Turbulent Shear Flow*. Cambridge University Press.

Epilogue

> It must be admitted that the principal result of fifty years of turbulence research is the recognition of the profound difficulties of the subject.
>
> S.A. Orszag (1970)

Little has changed since 1970. The story of turbulence is a story without an ending. After a century of concerted effort there are still relatively few statements which can be made about turbulence which are non-trivial and have any degree of generality. And of these handful of theories—one hesitates to call them laws—even fewer are rigorous in the sense that they follow directly from the equations of motion (Newton's second law). Almost all involve additional hypotheses which, though often plausible, are ultimately empirical. It is sometimes said that the log-law of the wall and Kolmogorov's four-fifths law are rigorous in the limit of Re $\rightarrow \infty$, but this is not true. The log-law requires that there exists a region of space adjacent to the wall where the large eddies are free from both viscous effects and the influence of the remote, core eddies. While this seems intuitively reasonable, it cannot be defended on strictly dynamical grounds since the transmission of information by the pressure field ensures that every eddy in a field of turbulence 'feels' every other eddy. Thus, the near-wall eddies, which give rise to the log-law *are*, to some degree, aware of the existence of the larger, remote eddies in the core of the flow. It just so happens that this coupling is weak. A similar, though less severe, difficulty arises in the derivation of Kolmogorov's four-fifths law. This requires that there exists a range of eddy sizes which are too large to be influenced by viscosity yet too small to 'feel' directly the time dependence of the large eddies. Again, it seems likely that such a range of eddies exists, particularly if one believes in cascades and the associated loss of information. However, we cannot *prove* that these vortices of intermediate size feel neither the very large nor the very small eddies. We can only hypothesize that this is so.

On the positive side we have a number of hypotheses which lead to non-trivial and reasonably general predictions; predictions which are often (but not always) found to be in agreement with the experimental evidence. Kolmogorov's theory of the small scales, including the two-thirds and four-fifths laws, the log-law of the wall for temperature and momentum, and Richardson's law for two-particle dispersion, are just a few examples. Now all of these 'laws' are flawed, in the sense that they appear to have a limited range of applicability, and we often find it difficult to predict in advance exactly what that range should be. So, for example, there is a vast literature devoted to exposing the discrepancies in Kolmogorov's theory. There is also a growing literature which questions the validity of the log law. But perhaps the most remarkable thing about these laws is not that they can break down, but that they work at all. After all, the foundations on which they are built are somewhat tenuous, yet within limitations they do indeed work. So we might argue that there *is* a body of knowledge which, though ultimately empirical, has the hallmark of a scientific theory. That is, predictions of some generality can be made, based on only a minimum of assumptions, and those predictions are well supported by the experimental data.

Unfortunately, these laws are of limited help to the engineer. He or she needs answers to questions of a complexity well beyond the limits of the near-rigorous theories. One cannot design an aerofoil based on the log-law or predict the wind loading on a building using Kolmogorov's theory. So estimating the influence of turbulence on practical problems still requires the use of complex, semi-empirical models. These models are becoming increasingly ingenious and sophisticated, but it is unlikely that their fundamental nature will change. Experience teaches us that we have no right to expect that the complex turbulent flows which face the engineer will ever be encompassed by some grand theory of turbulence, or even that such a theory exists.

There is, however, one bright spot on the horizon. The growing field of Direct Numerical Simulations (DNS), though still in its infancy, offers much. Admittedly, at present, only trivial geometries and relatively low Reynolds numbers can be realized. However, year-on-year the scope of the simulations becomes ever more ambitious, and the extraordinarily detailed information which these numerical experiments yield has already changed our view of turbulence. Of course any experiment, numerical or otherwise, is no substitute for a theory. Nevertheless, DNS looks destined to yield a rich source of data which will provide an invaluable testing ground, allowing us to refine and modify those theories which we have, and perhaps even to propose new ideas. It seems that Von Neumann's assessment of the 'problems of turbulence' has been vindicated, albeit half a century after his diagnosis. His hope to: 'break the deadlock by extensive, but well planned, computational effort . . . [thus yielding] . . . a reasonable chance of effecting real penetrations into this complex of problems, and gradually developing a useful, intuitive relationship to it. . .' (Von Neumann 1949) seems to have been amazingly prophetic. It is probable that, as the computations become better, we will begin to develop a clearer picture as to what an eddy is, how eddies interact and whether or not there is a multistage cascade as pictured by Richardson. Certainly, we need this kind of intuition before we can hope to improve our predictive models.

But we should not expect any sudden breakthrough in turbulence. The closure problem has proven to be a formidable barrier, and if the towering intellects of Landau and Kolmogorov could make only minor inroads into the subject, it seems improbable that nature will suddenly offer up all of its secrets. We leave the last word to the physical chemist Peter Atkins, whose comment about the second law of thermodynamics also seem to characterize our understanding of turbulence: 'We are the children of chaos, and the deep structure of change is decay. At root, there is only corruption, and the unstemmable tide of chaos. Gone is purpose; all that is left is direction. This is the bleakness we have to accept as we peer deeply and dispassionately into the heart of the universe.' (Peter Atkins 1984)

Vector identities and an introduction to tensor notation

A1.1 Vector identities and theorems

(1) Vector triple product:

$$\mathbf{a} \times (\mathbf{b} \times \mathbf{c}) = (\mathbf{a} \cdot \mathbf{c})\mathbf{b} - (\mathbf{a} \cdot \mathbf{b})\mathbf{c}$$
$$\mathbf{a} \times (\mathbf{b} \times \mathbf{c}) + \mathbf{b} \times (\mathbf{c} \times \mathbf{a}) + \mathbf{c} \times (\mathbf{a} \times \mathbf{b}) = 0.$$

(2) Mixed vector–scalar product:

$$(\mathbf{a} \times \mathbf{b}) \cdot \mathbf{c} = (\mathbf{b} \times \mathbf{c}) \cdot \mathbf{a} = (\mathbf{c} \times \mathbf{a}) \cdot \mathbf{b}.$$

(3) Grad, div and curl in Cartesian coordinates:

$$\nabla\phi = \frac{\partial\phi}{\partial x}\mathbf{i} + \frac{\partial\phi}{\partial y}\mathbf{j} + \frac{\partial\phi}{\partial z}\mathbf{k}$$

$$\nabla \cdot \mathbf{F} = \frac{\partial F_x}{\partial x} + \frac{\partial F_y}{\partial y} + \frac{\partial F_z}{\partial z}$$

$$\nabla \times \mathbf{F} = \left[\frac{\partial F_z}{\partial y} - \frac{\partial F_y}{\partial z}\right]\mathbf{i} + \left[\frac{\partial F_x}{\partial z} - \frac{\partial F_z}{\partial x}\right]\mathbf{j} + \left[\frac{\partial F_y}{\partial x} - \frac{\partial F_x}{\partial y}\right]\mathbf{k}.$$

(4) Grad, div and curl in cylindrical polar coordinates (r, θ, z):

$$\nabla\phi = \frac{\partial\phi}{\partial r}\hat{\mathbf{e}}_r + \frac{1}{r}\frac{\partial\phi}{\partial\theta}\hat{\mathbf{e}}_\theta + \frac{\partial\phi}{\partial z}\hat{\mathbf{e}}_z$$

$$\nabla \cdot \mathbf{F} = \frac{1}{r}\frac{\partial}{\partial r}(rF_r) + \frac{1}{r}\frac{\partial F_\theta}{\partial\theta} + \frac{\partial F_z}{\partial z}$$

$$\nabla \times \mathbf{F} = \left[\frac{1}{r}\frac{\partial F_z}{\partial\theta} - \frac{\partial F_\theta}{\partial z}\right]\hat{\mathbf{e}}_r + \left[\frac{\partial F_r}{\partial z} - \frac{\partial F_z}{\partial r}\right]\hat{\mathbf{e}}_\theta + \left[\frac{1}{r}\frac{\partial}{\partial r}(rF_\theta) - \frac{1}{r}\frac{\partial F_r}{\partial\theta}\right]\hat{\mathbf{e}}_z$$

$$\nabla^2\phi = \frac{1}{r}\frac{\partial}{\partial r}\left(r\frac{\partial\phi}{\partial r}\right) + \frac{1}{r^2}\frac{\partial^2\phi}{\partial\theta^2} + \frac{\partial^2\phi}{\partial z^2}$$

$$\nabla^2\mathbf{F} = \left[\nabla^2 F_r - \frac{1}{r^2}F_r - \frac{2}{r^2}\frac{\partial F_\theta}{\partial\theta}\right]\hat{\mathbf{e}}_r + \left[\nabla^2 F_\theta - \frac{1}{r^2}F_\theta + \frac{2}{r^2}\frac{\partial F_r}{\partial\theta}\right]\hat{\mathbf{e}}_\theta$$
$$+ (\nabla^2 F_z)\hat{\mathbf{e}}_z.$$

(5) Vector identities:

$$\nabla(\phi\psi) = \phi\nabla\psi + \psi\nabla\phi$$
$$\nabla(\mathbf{F}\cdot\mathbf{G}) = (\mathbf{F}\cdot\nabla)\mathbf{G} + (\mathbf{G}\cdot\nabla)\mathbf{F} + \mathbf{F}\times(\nabla\times\mathbf{G}) + \mathbf{G}\times(\nabla\times\mathbf{F})$$

$$\nabla\cdot(\phi\mathbf{F}) = \phi\nabla\cdot\mathbf{F} + \mathbf{F}\cdot\nabla\phi$$
$$\nabla\cdot(\mathbf{F}\times\mathbf{G}) = \mathbf{G}\cdot(\nabla\times\mathbf{F}) - \mathbf{F}\cdot(\nabla\times\mathbf{G})$$

$$\nabla\times(\phi\mathbf{F}) = \phi(\nabla\times\mathbf{F}) + (\nabla\phi)\times\mathbf{F}$$
$$\nabla\times(\mathbf{F}\times\mathbf{G}) = \mathbf{F}(\nabla\cdot\mathbf{G}) - \mathbf{G}(\nabla\cdot\mathbf{F}) + (\mathbf{G}\cdot\nabla)\mathbf{F} - (\mathbf{F}\cdot\nabla)\mathbf{G}$$

$$\nabla\times(\nabla\phi) = 0$$
$$\nabla\cdot(\nabla\times\mathbf{F}) = 0$$
$$\nabla^2\mathbf{F} = \nabla(\nabla\cdot\mathbf{F}) - \nabla\times(\nabla\times\mathbf{F}).$$

(6) Integral theorems:

$$\int_V \nabla\cdot\mathbf{F}\,dV = \oint_S \mathbf{F}\cdot d\mathbf{S} \qquad \int_S \nabla\times\mathbf{F}\cdot d\mathbf{S} = \oint_C \mathbf{F}\cdot d\mathbf{l}$$
$$\int_V \nabla\phi\,dV = \oint_S \phi\,d\mathbf{S} \qquad \int_S \nabla\phi\times d\mathbf{S} = -\oint_C \phi d\mathbf{l}$$
$$\int_V \nabla\times\mathbf{F}\,dV = -\oint_S \mathbf{F}\times d\mathbf{S}$$

(7) Helmholtz's decomposition. Any vector field $\mathbf{F}$ may be written as the sum of an irrotational and a solenoidal field. The irrotational field may, in turn, be written in terms of a scalar potential, ϕ, and the solenoidal field in terms of a vector potential, $\mathbf{A}$:

$$\mathbf{F} = -\nabla\phi + \nabla\times\mathbf{A}, \quad \nabla\cdot\mathbf{A} = 0.$$

The two potentials are solutions of

$$\nabla^2\phi = -\nabla\cdot\mathbf{F}, \qquad \nabla^2\mathbf{A} = -\nabla\times\mathbf{F}.$$

(8) Two useful vector relationships. In the first $\boldsymbol{\Omega}$ is any constant vector and $\mathbf{u}$ is solenoidal. In the second $\boldsymbol{\omega} = \nabla\times\mathbf{u}$ and $\boldsymbol{\omega}$ is confined to the sphere V_R.

$$[2\mathbf{x}\times(\mathbf{u}\times\boldsymbol{\Omega})]_i = [(\mathbf{x}\times\mathbf{u})\times\boldsymbol{\Omega}]_i + \nabla\cdot[(\mathbf{x}\times(\mathbf{x}\times\boldsymbol{\Omega}))_i\mathbf{u}]$$
$$\int_{V_R}\mathbf{u}\,dV = \tfrac{1}{3}\int_{V_R}\mathbf{x}\times\boldsymbol{\omega}\,dV$$

(9) Navier–Stokes equation in cylindrical polar coordinates in terms of stresses:

$$\frac{\partial u_r}{\partial t}+\left[(\mathbf{u}\cdot\nabla)u_r-\frac{u_\theta^2}{r}\right]=-\frac{1}{\rho}\frac{\partial p}{\partial r}+\frac{1}{\rho}\left[\frac{1}{r}\frac{\partial}{\partial r}(r\tau_{rr})+\frac{1}{r}\frac{\partial}{\partial\theta}(\tau_{r\theta})+\frac{\partial\tau_{rz}}{\partial z}-\frac{\tau_{\theta\theta}}{r}\right]$$

$$\frac{\partial u_\theta}{\partial t}+\left[(\mathbf{u}\cdot\nabla)u_\theta+\frac{u_r u_\theta}{r}\right]=-\frac{1}{\rho r}\frac{\partial p}{\partial\theta}+\frac{1}{\rho}\left[\frac{1}{r^2}\frac{\partial}{\partial r}(r^2\tau_{\theta r})+\frac{1}{r}\frac{\partial}{\partial\theta}(\tau_{\theta\theta})+\frac{\partial\tau_{\theta z}}{\partial z}\right]$$

$$\frac{\partial u_z}{\partial t}+(\mathbf{u}\cdot\nabla)u_z=-\frac{1}{\rho}\frac{\partial p}{\partial z}+\frac{1}{\rho}\left[\frac{1}{r}\frac{\partial}{\partial r}(r\tau_{zr})+\frac{1}{r}\frac{\partial}{\partial\theta}(\tau_{z\theta})+\frac{\partial\tau_{zz}}{\partial z}\right]$$

$$\tau_{rr}=2\rho\nu\frac{\partial u_r}{\partial r},\quad \tau_{\theta\theta}=2\rho\nu\left[\frac{1}{r}\frac{\partial u_\theta}{\partial\theta}+\frac{u_r}{r}\right],\quad \tau_{zz}=2\rho\nu\frac{\partial u_z}{\partial z}$$

$$\tau_{r\theta}=\rho\nu\left[r\frac{\partial}{\partial r}\left(\frac{u_\theta}{r}\right)+\frac{1}{r}\frac{\partial u_r}{\partial\theta}\right],\quad \tau_{\theta z}=\rho\nu\left[\frac{1}{r}\frac{\partial u_z}{\partial\theta}+\frac{\partial u_\theta}{\partial z}\right],$$

$$\tau_{zr}=\rho\nu\left[\frac{\partial u_r}{\partial z}+\frac{\partial u_z}{\partial r}\right].$$

(10) Navier–Stokes equation in cylindrical polar coordinates in terms of velocity:

$$\frac{\partial u_r}{\partial t}+\left[(\mathbf{u}\cdot\nabla)u_r-\frac{u_\theta^2}{r}\right]=-\frac{1}{\rho}\frac{\partial p}{\partial r}+\nu\left[\nabla^2 u_r-\frac{u_r}{r^2}-\frac{2}{r^2}\frac{\partial u_\theta}{\partial\theta}\right]$$

$$\frac{\partial u_\theta}{\partial t}+\left[(\mathbf{u}\cdot\nabla)u_\theta+\frac{u_r u_\theta}{r}\right]=-\frac{1}{\rho r}\frac{\partial p}{\partial\theta}+\nu\left[\nabla^2 u_\theta-\frac{u_\theta}{r^2}+\frac{2}{r^2}\frac{\partial u_r}{\partial\theta}\right]$$

$$\frac{\partial u_z}{\partial t}+(\mathbf{u}\cdot\nabla)u_z=-\frac{1}{\rho}\frac{\partial p}{\partial z}+\nu[\nabla^2 u_z].$$

(11) The Reynolds averaged Navier–Stokes equation for steady-on-average flow in cylindrical polar coordinates ($\nu=0$):

$$\left[(\bar{\mathbf{u}}\cdot\nabla)\bar{u}_r-\frac{\bar{u}_\theta^2}{r}\right]=-\frac{1}{\rho}\frac{\partial\bar{p}}{\partial r}-\left[\frac{1}{r}\frac{\partial}{\partial r}\left(r\overline{u_r'u_r'}\right)+\frac{1}{r}\frac{\partial}{\partial\theta}\left(\overline{u_r'u_\theta'}\right)+\frac{\partial}{\partial z}\left(\overline{u_r'u_z'}\right)-\frac{\overline{u_\theta'u_\theta'}}{r}\right]$$

$$\left[(\bar{\mathbf{u}}\cdot\nabla)\bar{u}_\theta+\frac{\bar{u}_r\bar{u}_\theta}{r}\right]=-\frac{1}{\rho r}\frac{\partial\bar{p}}{\partial\theta}-\left[\frac{1}{r^2}\frac{\partial}{\partial r}\left(r^2\overline{u_\theta'u_r'}\right)+\frac{1}{r}\frac{\partial}{\partial\theta}\left(\overline{u_\theta'u_\theta'}\right)+\frac{\partial}{\partial z}\left(\overline{u_\theta'u_z'}\right)\right]$$

$$[(\bar{\mathbf{u}}\cdot\nabla)\bar{u}_z]=-\frac{1}{\rho}\frac{\partial\bar{p}}{\partial z}-\left[\frac{1}{r}\frac{\partial}{\partial r}\left(r\overline{u_z'u_r'}\right)+\frac{1}{r}\frac{\partial}{\partial\theta}\left(\overline{u_z'u_\theta'}\right)+\frac{\partial}{\partial z}\left(\overline{u_z'u_z'}\right)\right].$$

A1.2 An introduction to tensor notation

This appendix provides a brief introduction to tensor notation for those readers who have not met it before. We are interested primarily in exposing the utility of tensor notation (sometimes called suffix notation), rather than in expounding the formal theory of tensor quantities. Much of the notation can be introduced in the context of vectors, and so we start with vectors and with the so-called *summation convention*.

Consider the advection diffusion equation for temperature $T(x, y, z, t)$:

$$\frac{\partial T}{\partial t} + (\mathbf{u} \cdot \nabla)T = \alpha \nabla^2 T.$$

When written in tensor (or suffix) notation this becomes,

$$\frac{\partial T}{\partial t} + u_i \frac{\partial T}{\partial x_i} = \alpha \frac{\partial^2 T}{\partial x_i^2}.$$

The rule is, if a suffix is repeated in any term then there is an implied summation with respect to that suffix over the range 1, 2, 3. Thus,

$$u_i \frac{\partial T}{\partial x_i} = u_1 \frac{\partial T}{\partial x_1} + u_2 \frac{\partial T}{\partial x_2} + u_3 \frac{\partial T}{\partial x_3}$$

$$\frac{\partial^2 T}{\partial x_i^2} = \frac{\partial^2 T}{\partial x_i \partial x_i} = \frac{\partial^2 T}{\partial x_1^2} + \frac{\partial^2 T}{\partial x_2^2} + \frac{\partial^2 T}{\partial x_3^2}.$$

This is called the *implied summation convention*. When working with vector (rather than scalar) equations we need another rule, called the *range convention*. This states that, if a suffix occurs just once in a term, then it is assumed to take all of the values 1, 2, 3 in turn. For example, the Euler equation,

$$\frac{\partial \mathbf{u}}{\partial t} + (\mathbf{u} \cdot \nabla)\mathbf{u} = -\nabla(p/\rho)$$

is written

$$\frac{\partial u_i}{\partial t} + u_j \frac{\partial u_i}{\partial x_j} = -\frac{\partial}{\partial x_i}\left(\frac{p}{\rho}\right).$$

Here j is a dummy suffix, indicating an implied summation, while i is a *free suffix*, indicating that the equation above represents the three components of a vector equation. Thus, in suffix notation, a vector such as $\mathbf{u}$ is written u_i where it is understood that i takes all of the values 1, 2, 3. The quantities ∇f and $\nabla \cdot \mathbf{u}$ are, in tensor notation,

$$\frac{\partial f}{\partial x_i}, \qquad \frac{\partial u_i}{\partial x_i}.$$

To write $\nabla \times \mathbf{u}$ in suffix notation we need an additional quantity, called the Levi–Civita symbol, ε_{ijk}. This is defined as follows:

$$\varepsilon_{123} = \varepsilon_{312} = \varepsilon_{231} = 1$$

$$\varepsilon_{132} = \varepsilon_{213} = \varepsilon_{321} = -1$$

$$\text{all other } \varepsilon_{ijk} = 0.$$

Thus $\varepsilon_{ijk} = 1$ if (i, j, k) appear in a cyclic order (sometimes called an even permutation of 1, 2, 3), $\varepsilon_{ijk} = -1$ if (i, j, k) are anticyclic (an odd permutation of 1, 2, 3), and $\varepsilon_{ijk} = 0$ if any two suffixes are equal. It is readily confirmed that

$$(\mathbf{a} \times \mathbf{b})_i = \varepsilon_{ijk} a_j b_k.$$

For example, noting that the only non-zero elements of ε_{1jk} are ε_{123} and ε_{132} we have

$$(\mathbf{a} \times \mathbf{b})_1 = \varepsilon_{1jk} a_j b_k = \varepsilon_{123} a_2 b_3 + \varepsilon_{132} a_3 b_2 = a_2 b_3 - a_3 b_2.$$

So the symbol ε_{ijk} provides us with a means of handling cross-products in suffix notation. It follows that,

$$\omega_i = (\nabla \times \mathbf{u})_i = \varepsilon_{ijk} \frac{\partial u_k}{\partial x_j}.$$

Since j and k are dummy variables, and ε_{ijk} changes sign if two of its suffixes are exchanged, this can also be written as,

$$\omega_i = (\nabla \times \mathbf{u})_i = \varepsilon_{ikj} \frac{\partial u_j}{\partial x_k} = -\varepsilon_{ijk} \frac{\partial u_j}{\partial x_k}.$$

Combining these yields yet another expression for $\nabla \times \mathbf{u}$:

$$\omega_i = (\nabla \times \mathbf{u})_i = \frac{1}{2} \varepsilon_{ijk} \left[\frac{\partial u_k}{\partial x_j} - \frac{\partial u_j}{\partial x_k} \right].$$

It is left to the reader to confirm that an inverse relationship also holds in the form,

$$\frac{\partial u_k}{\partial x_j} - \frac{\partial u_j}{\partial x_k} = \varepsilon_{jki} \omega_i.$$

(Try putting $j = 2$ and $k = 3$. You will find this gives the correct expression for ω_1.) Finally we note that ε_{ijk} is related to the Kronecker delta δ_{ijk} by

$$\varepsilon_{imn} \varepsilon_{ijk} = \delta_{mj} \delta_{nk} - \delta_{mk} \delta_{nj}.$$

So far we have introduced some notation, which we have called suffix or tensor notation, without introducing the notion of a tensor. This is our next task.[1] Actually we have already been using tensors without making it explicit. A tensor of zero order (or zero rank) is a scalar, a tensor of first order is simply a vector, while a tensor of second order is a quantity which depends on two suffixes—in effect,

[1] We shall restrict ourselves to tensors defined in terms of Cartesian coordinates, so that we need not distinguish between so-called covariant and contravariant tensors. Those interested in this distinction should consult Arfken (1985).

a matrix. Thus p and ρ are zero-order tensors, $\mathbf{u}$ and $\boldsymbol{\omega}$ are first-order tensors,[2] and

$$S_{ij} = \frac{1}{2}\left(\frac{\partial u_i}{\partial x_j} + \frac{\partial u_j}{\partial x_i}\right), \qquad W_{ij} = \frac{1}{2}\left(\frac{\partial u_i}{\partial x_j} - \frac{\partial u_j}{\partial x_i}\right)$$

are examples of second-order tensors. To multiply a second-order tensor by a first- or second-order tensor we just follow the rules of matrix algebra. Thus $S_{ij}W_{jk}$ is another second-order tensor, C_{ik}, the elements of which are obtained from S_{ij} and W_{jk} by matrix multiplication. Similarly, $S_{ij}u_j$ is a first-order tensor, D_i, whose elements are again given by matrix algebra.

One must count suffixes with care, however. $S_{ij}W_{jk}$ is a quantity which depends on only two suffixes, i and k, since j is a dummy suffix whose only role is to remind us to perform a summation. Similarly $S_{ij}u_j$ is a first-order tensor. However, $S_{ij}W_{mn}$ is a quantity which depends on four indices and so is not a tensor of second-order.

Note, however, that not all 3×3 matrices are second-order tensors, just as a quantity composed of three numbers, (a, b, c) need not be a vector. Let us consider the case of vectors.

Recall that a vector is a quantity which has direction and magnitude, and obeys the law of vector addition. Vectors have the important physical characteristic that their existence is independent of the coordinate system used to evaluate its components. For example, a force of 2 Newtons applied in a particular direction will always have a magnitude of 2 Newtons and always point in the same direction, irrespective of the coordinate system chosen to analyse the problem. This implies that the components of a vector must transform in a very particular way as we move from one coordinate system to another. Consider two Cartesian coordinate systems, (x, y, z) and (x', y', z'), which have unit vectors $\mathbf{i}, \mathbf{j}, \mathbf{k}$ and $\mathbf{i}', \mathbf{j}', \mathbf{k}'$. The primed and unprimed coordinate axes have a common origin, but are rotated relative to each other. A vector, say $\mathbf{F}$, can be written as

$$\mathbf{F} = F_x\mathbf{i} + F_y\mathbf{j} + F_z\mathbf{k}$$

in the unprimed system, or

$$\mathbf{F} = F'_x\mathbf{i}' + F'_y\mathbf{j}' + F'_z\mathbf{k}'$$

in the primed coordinate system. Since $\mathbf{F}$ has an identity which is independent of the coordinate system used, we have

$$F_x\mathbf{i} + F_y\mathbf{j} + F_z\mathbf{k} = F'_x\mathbf{i}' + F'_y\mathbf{j}' + F'_z\mathbf{k}'.$$

Thus the components of $\mathbf{F}$ transform according to the rule,

$$F'_x = (\mathbf{i} \cdot \mathbf{i}')F_x + (\mathbf{j} \cdot \mathbf{i}')F_y + (\mathbf{k} \cdot \mathbf{i}')F_z$$

[2] Strictly $\boldsymbol{\omega}$ is a first-order pseudo tensor, as we shall see.

$$F'_y = (\mathbf{i}\cdot\mathbf{j}')F_x + (\mathbf{j}\cdot\mathbf{j}')F_y + (\mathbf{k}\cdot\mathbf{j}')F_z$$
$$F'_z = (\mathbf{i}\cdot\mathbf{k}')F_x + (\mathbf{j}\cdot\mathbf{k}')F_y + (\mathbf{k}\cdot\mathbf{k}')F_z.$$

In tensor notation we write this as, $F'_i = a_{ij}F_j$, where a_{ij} is a matrix of direction cosines. (Similarly, F_i is related to F'_j through the transpose of a_{ij}: $F_i = a_{ji}F'_j$.) This transformation property is so important that it is sometimes used to *define* a vector. That is to say, if an ordered triple (A_x, A_y, A_z) transforms according to the law, $A'_i = a_{ij}A_j$, under a rotation of the coordinates, then we say A_x, A_y, and A_z are the components of a vector $\mathbf{A}$. If the ordered triple does not transform in this way then it is not a vector. Unfortunately, such a definition sounds a little formal and so it is easy to lose sight of the fact that all we are saying is that the physical characteristics of a vector (i.e. magnitude and direction) must transcend the coordinate system used to evaluate its components.

In a similar way we would like the physical characteristics of any second-order tensor to be independent of the coordinate system used to evaluate its elements. As with vectors, this implies that the elements of a second-order tensor must transform in a particular way when we move from one coordinate system to another. Perhaps this is best illustrated by an example. Consider Ohm's law for an electrically conducting material. For an isotropic material this tells us that $\mathbf{J} = \sigma\mathbf{E}$ where $\mathbf{J}$ is the current density (a vector), $\mathbf{E}$ is the electric field (another vector), and σ is the electrical conductivity (a scalar). If the material is anisotropic, however, a component of $\mathbf{E}$ in one direction can cause current to flow in other directions, and so $\mathbf{J}$ and $\mathbf{E}$ are not, in general, parallel. Ohm's law generalizes to $J_i = \sigma_{ij}E_j$. The quantity σ_{ij} is an example of a second-order tensor. It has nine components and is associated with two directions (those of $\mathbf{J}$ and $\mathbf{E}$). It looks like a 3×3 matrix, and indeed it is. However, it is not just any old matrix, because its elements must, like those of $\mathbf{J}$ and $\mathbf{E}$, transform in a definite way when the coordinate axes are rotated. Let us see if we can get the transformation rule for the elements of σ_{ij}. As before, we consider two coordinate systems (x, y, z) and (x', y', z'). The expression $J_i = \sigma_{ij}E_j$ applies equally in both since it is this equation which defines σ_{ij}. So we have

$$\sigma'_{mn}E'_n = J'_m = a_{mi}J_i = a_{mi}(\sigma_{ij}E_j) = a_{mi}\sigma_{ij}(a_{nj}E'_n)$$

or

$$(\sigma'_{mn} - a_{mi}a_{nj}\sigma_{ij})E'_n = 0.$$

Since this is true for all choices of E'_n we have the transformation rule

$$\sigma'_{mn} = a_{mi}a_{nj}\sigma_{ij}.$$

This rule is a direct consequence of our insistence that the equation $J_i = \sigma_{ij}E_j$ applies equally in all coordinate systems. The convention is to use this transformation law to test for tensor character. If a 3×3 matrix A_{ij} obeys this law it is a second-order tensor, otherwise it is not. In effect, we define a second-order tensor by the transformation law: *a second order tensor A_{ij} is a set of quantities depending on two directions and specified by nine components which transform according to $A'_{mn} = a_{mi}a_{nj}A_{ij}$*. Higher-order tensors are defined in a similar way.

There is one final distinction to be made. Cartesian tensors can be divided into two groups: so-called true tensors and pseudo-tensors. We may illustrate this distinction by focussing on first-order tensors, that is vectors. So far we have considered the transformation properties of vectors or tensors under a *rotation* of the coordinate system. Suppose we now consider *reflections*. In particular, consider a reflection about the origin in which each coordinate axis is reversed. We have moved from a right-handed coordinate system to a left-handed one in which $x' = -x$, $y' = -y$, $z' = -z$, $\mathbf{i}' = -\mathbf{i}$, $\mathbf{j}' = -\mathbf{j}$, and $\mathbf{k}' = -\mathbf{k}$. A true vector, such as velocity, transforms like $u'_x = -u_x$ etc. This leaves the physical direction of the vector unchanged since

$$\mathbf{u} = u_x\mathbf{i} + u_y\mathbf{j} + u_z\mathbf{k} = (-u'_x)(-\mathbf{i}') + \left(-u'_y\right)(-\mathbf{j}') + (-u'_z)(-\mathbf{k}').$$

Thus, after a reflection of the coordinates, a true vector, such as force or velocity, has the same magnitude and direction as before, although the numerical value of its components change sign.

Now consider a vector such as the angular momentum of a particle of mass m, $\mathbf{H} = \mathbf{x} \times (m\mathbf{u})$. Both $\mathbf{x}$ and $\mathbf{u}$ are true vectors and so their components change sign under a reflection of the coordinates. However, it follows from the definition of $\mathbf{H}$ that its components do not change sign under this coordinate transformation. For example, $H'_z = m(x'u'_y - y'u'_x) = m(xu_y - yu_x) = H_z$. Thus $\mathbf{H}$ transforms according to $H_x = H'_x$ etc. By implication, the physical direction of $\mathbf{H}$ reverses when we invert the coordinate system! Such a vector is called a *pseudo-vector*.

There are many examples of vector quantities which do this: angular velocity, magnetic fields, and vorticity are three examples. In fact any vector defined as the cross-product of two real vectors is a pseudo-vector. Physically this means that such quantities have a definite magnitude and a line of action, but their sense along that line of action is somewhat arbitrary and reverses if the handedness of the coordinate system reverses.

The most obvious example of this is torque, $\mathbf{T} = \mathbf{x} \times \mathbf{F}$. Suppose we apply a torque, which rotates a body clockwise in a horizontal plane. The torque has a definite magnitude and line of action, but whether we chose to say that $\mathbf{T}$ points up or down is a matter of convention. (The right-handed screw convention will have it pointing one way, while the left-handed convention will have it in the opposite direction.)

To avoid this difficulty we usually agree to restrict ourselves to right-handed coordinate systems and to the right-hand screw convention. There is then no need to distinguish between true vectors (sometimes called polar vectors) and pseudo-vectors (sometimes called axial vectors). However this convention is really just a mathematical convenience. It disguises the fact that true vectors and pseudo vectors are really two different types of physical quantities.

The idea of pseudo-vectors leads to the concept of a pseudo-scalar. A pseudo-scalar is a sort of degenerate scalar whose value changes sign as we move from a right-handed to a left-handed coordinate system. Such quantities are easily constructed from pseudo-vectors. For example, the helicity, $\mathbf{u} \cdot \nabla \times \mathbf{u}$, is a pseudo scalar since the numerical value of the components of $\mathbf{u}$ change sign under a reflection of the coordinates while those of $\nabla \times \mathbf{u}$, do not.

A more complete discussion of pseudo-tensors is given in Feynman et al. (1964).

Suggested reading

Arfken G. (1985) *Mathematical Methods for Physicists*, 3rd edition. Academic Press.

Feynman R.P., Leighton R.B. and Sands M. (1964) *Lectures on Physics*, Vol.I Chs 20,52 and Vol.II Ch. 31. Addison–Wesley.

The properties of isolated vortices: invariants, far-field properties, and long-range interactions

We have defined an eddy as a coherent blob of vorticity. Such a blob retains its identity in a turbulent flow since vorticity can spread by material movement or by diffusion only. It is of interest to identify the kinematic and dynamic properties of an isolated eddy since they play an important role in the large-scale dynamics of homogeneous turbulence. For example, they are crucial to understanding the origins of Saffman's invariant and Loitsyansky's integral, as well as the distinction between the two (see Section 6.3). They are also central to understanding the nature of Batchelor's long-range pressure forces (see Section 6.3.4). The topics we shall discuss are:

- The far-field velocity induced by an isolated eddy.
- The pressure distribution in the far field.
- Integral invariants of an isolated eddy.
- Long-range interactions between eddies.

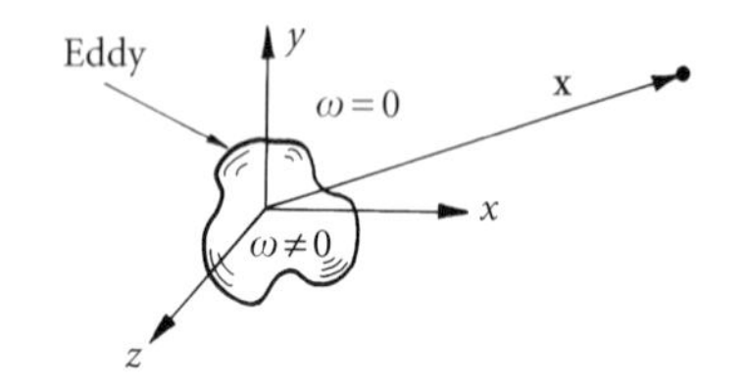

Figure A2.1 An isolated eddy located at $x=0$.

A2.1 The far-field velocity induced by an isolated eddy

Consider an isolated blob of vorticity sitting near $\mathbf{x}=0$ in an infinite fluid which is motionless at infinity (Figure A2.1). The vector potential, $\mathbf{A}$, for the velocity field is defined by

$$\nabla \times \mathbf{A} = \mathbf{u}, \qquad \nabla \cdot \mathbf{A} = 0 \tag{A2.1}$$

and is related to the vorticity by the expression,

$$\nabla^2 \mathbf{A} = -\boldsymbol{\omega}. \tag{A2.2}$$

This may be inverted using the Biot–Savart law to give,

$$\mathbf{A}(\mathbf{x}) = \frac{1}{4\pi}\int \boldsymbol{\omega}(\mathbf{x}')\frac{d\mathbf{x}'}{|\mathbf{x}' - \mathbf{x}|}. \tag{A2.3}$$

To find the vector potential some distance from the blob of vorticity we expand $|\mathbf{x}' - \mathbf{x}|^{-1}$ in a Taylor series in $|\mathbf{x}|^{-1}$. This yields,

$$\frac{1}{|\mathbf{x}' - \mathbf{x}|} = \frac{1}{r} - D_i(r)x'_i + B_{ij}(r)x'_i x'_j + \cdots \tag{A2.4}$$

where $r = |\mathbf{x}|$ and

$$D_i = \frac{\partial}{\partial x_i}\left(\frac{1}{r}\right), \qquad B_{ij} = \frac{1}{2}\frac{\partial^2}{\partial x_i \partial x_j}\left(\frac{1}{r}\right).$$

On substituting this expansion into integral (A2.3) we find,

$$4\pi \mathbf{A}(\mathbf{x}) = \frac{1}{r}\int \boldsymbol{\omega}'\, d\mathbf{x}' - D_i(r)\int x'_i \boldsymbol{\omega}'\, d\mathbf{x}' + B_{ij}(r)\int x'_i x'_j \boldsymbol{\omega}'\, d\mathbf{x}' + \cdots \tag{A2.5}$$

However, the first term on the right integrates to zero since

$$\int_V \omega_i\, d\mathbf{x} = \int_V \nabla \cdot [\boldsymbol{\omega} x_i]\, d\mathbf{x} = \oint_S x_i \boldsymbol{\omega} \cdot d\mathbf{S} = 0 \tag{A2.6}$$

the vorticity being confined to the region near $\mathbf{x} = 0$. Also, the second integral can be transformed using the relationship

$$\int_V \left[x_i \omega_j + x_j \omega_i\right] d\mathbf{x} = \int_V \nabla \cdot \left(x_i x_j \boldsymbol{\omega}\right) d\mathbf{x} = \oint_S x_i x_j \boldsymbol{\omega} \cdot d\mathbf{S} = 0 \tag{A2.7}$$

from which,

$$D_i \int x'_i \omega'_j\, d\mathbf{x}' = \frac{1}{2} D_i \int \left[x'_i \omega'_j - x'_j \omega'_i\right] d\mathbf{x}' = -\frac{1}{2}\mathbf{D} \times \int [\mathbf{x}' \times \boldsymbol{\omega}'] d\mathbf{x}'. \tag{A2.8}$$

Our expansion for the vector potential now simplifies to

$$4\pi \mathbf{A}(\mathbf{x}) = \mathbf{D}(r) \times \mathbf{L} + B_{ij}(r) \int x'_i x'_j \boldsymbol{\omega}'\, d\mathbf{x}' + \cdots \tag{A2.9}$$

where

$$\mathbf{L} = \frac{1}{2}\int (\mathbf{x} \times \boldsymbol{\omega})\, d\mathbf{x}. \tag{A2.10}$$

We shall see shortly that $\mathbf{L}$ is a measure of the linear momentum of the eddy and is an invariant of the motion. It is called the *linear impulse*. Substituting for $\mathbf{D}(r)$ and taking the curl gives the far-field velocity distribution

$$4\pi \mathbf{u}(\mathbf{x}) = (\mathbf{L} \cdot \nabla)\nabla(1/r) + \nabla\left(B_{ij}(r)\right) \times \int x'_i x'_j \boldsymbol{\omega}'\, d\mathbf{x}' + \cdots. \tag{A2.11}$$

Evidently the velocity in the far-field is $O(r^{-3})$ if $\mathbf{L}$ is finite and $O(r^{-4})$ if $\mathbf{L}$ happens to be zero. We can find the corresponding far-field scalar potential for $\mathbf{u}$ by rewriting (A2.11) as

$$4\pi\mathbf{u} = 4\pi\nabla\phi = \nabla[\mathbf{L}\cdot\nabla(1/r)] + \cdots$$

which tells us that the scalar potential for $\mathbf{u}$ in the far field has the form,

$$4\pi\phi = \mathbf{L}\cdot\nabla(1/r) + \cdots. \tag{A2.12}$$

The far-field angular momentum density, on the other hand, is

$$4\pi[\mathbf{x}\times\mathbf{u}(\mathbf{x})] = \mathbf{x}\times(\mathbf{L}\cdot\nabla)\nabla(1/r) + \mathbf{x}\times\left[\nabla(B_{ij})\times\int x_i'x_j'\boldsymbol{\omega}'\,d\mathbf{x}'\right] + \cdots.$$

Noting that $\mathbf{x}\times\nabla(1/r) = 0$ and expanding the vector triple product we can rearrange this to give,

$$4\pi[\mathbf{x}\times\mathbf{u}(\mathbf{x})] = \frac{\mathbf{L}\times\mathbf{x}}{r^3} + x_k\frac{\partial B_{ij}}{\partial x_m}\int x_i'x_j'\omega_k'\,d\mathbf{x}' - x_k\frac{\partial B_{ij}}{\partial x_k}\int x_i'x_j'\omega_m'\,d\mathbf{x}'. \tag{A2.13}$$

We shall return to these expressions shortly.

A2.2 The pressure distribution in the far field

The pressure field at large distances from the eddy may be found by taking the divergence of the Navier–Stokes equation to give

$$\nabla^2 p = -\rho\nabla\cdot[\mathbf{u}\cdot\nabla\mathbf{u}]$$

and then integrating this using the Biot–Savart law:

$$\frac{4\pi p}{\rho} = \int[\nabla\cdot(\mathbf{u}\cdot\nabla\mathbf{u})]'\frac{d\mathbf{x}'}{|\mathbf{x}'-\mathbf{x}|}.$$

Substituting for $|\mathbf{x}'-\mathbf{x}|$ using (A2.4) and noting that

$$x_i\nabla\cdot(\mathbf{u}\cdot\nabla\mathbf{u}) = \nabla\cdot(x_i\mathbf{u}\cdot\nabla\mathbf{u}) - \nabla\cdot[u_i\mathbf{u}]$$

we find,

$$\frac{4\pi p}{\rho} = \frac{1}{r}\int[\nabla\cdot(\mathbf{u}\cdot\nabla\mathbf{u})]'\,d\mathbf{x}' - D_i(r)\int[\nabla\cdot(x_i\mathbf{u}\cdot\nabla\mathbf{u} - u_i\mathbf{u})]'\,d\mathbf{x}'$$
$$+ B_{ij}(\mathbf{r})\int x_i'x_j'[\nabla\cdot(\mathbf{u}\cdot\nabla\mathbf{u})]'\,d\mathbf{x}' + \cdots.$$

The first two volume integrals convert to surface integrals which, since $|\mathbf{u}|\sim O(r^{-3})$, vanish on a sphere of infinite radius. The integrand of the third integral can be written as $2u_iu_j$ plus some divergences which also vanish, and so the leading-order contribution to the far-field pressure is simply,

$$\frac{4\pi p}{\rho} = \frac{\partial^2}{\partial x_i \partial x_j}\left(\frac{1}{r}\right)\int u_i' u_j' \, d\mathbf{x}' + \cdots . \tag{A2.14}$$

A2.3 Integral invariants of an isolated eddy

We now consider those integral invariants of an isolated eddy which are related to the principles of conservation of linear and angular momentum. These invariants are called the *linear impulse* and *angular impulse* of the eddy. They are both integrals of the vorticity field.

The first point to note is that a comparison of (A2.11), (A2.12), and (A2.14) yields,

$$\frac{d\mathbf{L}}{dt} = 0. \tag{A2.15}$$

The argument goes as follows. In a potential flow the viscous forces are zero since $\nu\nabla^2\mathbf{u} = -\nu\nabla \times \nabla \times \mathbf{u} = 0$, and so Bernoulli's theorem for unsteady potential flow demands,

$$\frac{\partial\phi}{\partial t} + \frac{p}{\rho} + \frac{\mathbf{u}^2}{2} = 0.$$

Since $p \sim O(r^{-3})$ and $u^2 \sim O(r^{-6})$ for large r this reduces to,

$$\frac{\partial\phi}{\partial t} = O(r^{-3}).$$

Equation (A2.15) follows immediately since $\phi \sim \mathbf{L}r^{-2} + \cdots$. Thus the linear impulse, $\mathbf{L}$, is an invariant.

The invariance of $\mathbf{L}$ is a direct consequence of the principle of conservation of linear momentum. This becomes clear if we note that,

$$\frac{\partial}{\partial t}\left[\frac{1}{2}(\mathbf{x} \times \boldsymbol{\omega})\right] = -\mathbf{u}\cdot\nabla\mathbf{u} + \nabla\left(\frac{u^2}{2}\right) + \frac{1}{2}\nabla\cdot[(\mathbf{x}\times\mathbf{u})\boldsymbol{\omega} - (\mathbf{x}\times\boldsymbol{\omega})\mathbf{u}]$$
$$-\frac{1}{2}\nu[\nabla(\mathbf{x}\cdot\nabla\times\boldsymbol{\omega}) + 2\nabla\times\boldsymbol{\omega} - \nabla\cdot((\nabla\times\boldsymbol{\omega})\mathbf{x})].$$

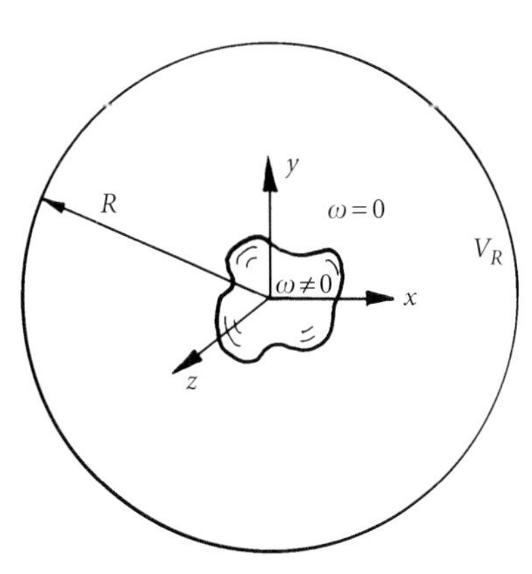

Figure A2.2 Control volume V_R used to evaluate the linear momentum.

(This follows from the evolution equation for $\boldsymbol{\omega}$.) Integrating over a large sphere of radius R which encloses all the vorticity (Figure A2.2), and using Bernoulli's theorem to evaluate $u^2/2$ in the far field, we find

$$\frac{d}{dt}\int_{V_R}\left[\frac{1}{2}(\mathbf{x}\times\boldsymbol{\omega})\right]dV = -\oint_{S_R}\mathbf{u}(\mathbf{u}\cdot d\mathbf{S}) - \oint_{S_R}\left(\frac{p}{\rho}\right)d\mathbf{S} - \oint_{S_R}\frac{\partial\phi}{\partial t}\,d\mathbf{S}.$$

Compare this with the linear momentum equation applied to V_R,

$$\frac{d}{dt}\int_{V_R} \mathbf{u}\, dV = -\oint_{S_R} \mathbf{u}(\mathbf{u}\cdot d\mathbf{S}) - \int_{S_R} (p/\rho)\, d\mathbf{S}.$$

Evidently,

$$\frac{d}{dt}\left[\int_{V_R} \mathbf{u}\, dV - \int_{V_R} \frac{1}{2}(\mathbf{x}\times\boldsymbol{\omega})\, dV\right] = \oint_{S_R} \frac{\partial\phi}{\partial t}\, d\mathbf{S} = O(R^{-1}).$$

In the limit $R\to\infty$ we have,

$$\mathbf{L} = \int_{V_R} \mathbf{u}\, dV + \text{constant}$$

which shows that, to within a constant, the linear impulse $\mathbf{L}$ is equal to the linear momentum in a sphere of large radius.

The constant may be found by direct evaluation of $\int \mathbf{u}\, dV$. The integration is complicated by the fact that $\int \mathbf{u}\, dV$ is only conditionally convergent, but the details are spelt out in Bachelor (1967). It turns out that, no matter how large we make V_R, there is always some linear momentum outside V_R. When the integrals are evaluated we find that the contribution to $\int \mathbf{u}\, dV$ from within V_R is $\frac{2}{3}\mathbf{L}$ while that from outside is $\frac{1}{3}\mathbf{L}$. Adding the two gives

$$\mathbf{L} = \int \mathbf{u}\, dV \tag{A2.16}$$

Linear impulse = Linear momentum

We now turn our attention to the angular momentum. Again we must be careful as it is by no means clear that $\int \mathbf{x}\times\mathbf{u}\, dV$ is a convergent integral, since $\mathbf{u}\sim O(r^{-3})$ at large r. This requires that we take a rather circuitous route to applying the principle of conservation of angular momentum. Let us introduce the angular momentum integral

$$\mathbf{H} = \int_{V_R} (\mathbf{x}\times\mathbf{u})\, d\mathbf{x}$$

where V_R is a sphere of radius R which completely encloses the vorticity field (Figure A2.2). We do not know what happens to $\mathbf{H}$ as $R\to\infty$, but it is certainly well defined for finite R. Now it is readily verified that

$$6(\mathbf{x}\times\mathbf{u}) = 2\mathbf{x}\times(\mathbf{x}\times\boldsymbol{\omega}) + 3\nabla\times(r^2\mathbf{u}) - \boldsymbol{\omega}\cdot\nabla(r^2\mathbf{x})$$

and so it follows that

$$\mathbf{H} = \frac{1}{3}\int_{V_R} \mathbf{x}\times(\mathbf{x}\times\boldsymbol{\omega})\, d\mathbf{x}. \tag{A2.17}$$

(The second term on the right integrates to give $3R^2 \int \boldsymbol{\omega}\, dV$, which is zero because of (A2.6), while the third term gives a surface integral which vanishes because $\boldsymbol{\omega}$ is zero on the surface of V_R.) This second measure of $\mathbf{H}$ is perfectly well behaved as $R \to \infty$ since there are no contributions to (A2.17) outside the vortex blob. Equation (A2.17) is called the *angular impulse* of the eddy. Now the principle of conservation of angular momentum tells us that,

$$\frac{d}{dt}\int_{V_R} (\mathbf{x} \times \mathbf{u})\, d\mathbf{x} = -\oint_{S_R} (\mathbf{x} \times \mathbf{u})\mathbf{u} \cdot d\mathbf{S}$$

there being no viscous torque on S_R because the flow outside S_R has no vorticity.[1] It follows from (A2.17) that

$$\frac{d}{dt}\left[\frac{1}{3}\int_{V_R} \mathbf{x} \times (\mathbf{x} \times \boldsymbol{\omega})\, d\mathbf{x}\right] = -\oint_{S_R} (\mathbf{x} \times \mathbf{u})\mathbf{u} \cdot d\mathbf{S}.$$

We can now safely take the limit $R \to \infty$ since both integrals are convergent. We find

$$\frac{d}{dt}\left[\frac{1}{3}\int_{V_R} \mathbf{x} \times (\mathbf{x} \times \boldsymbol{\omega})\, d\mathbf{x}\right] = O(R^{-3})$$

and so, in the limit $R \to \infty$, the angular impulse is conserved:

$$\frac{1}{3}\int_{V_R} \mathbf{x} \times (\mathbf{x} \times \boldsymbol{\omega})\, d\mathbf{x} = \text{constant}. \tag{A2.18}$$

In summary, then, an isolated eddy has two integral invariants, its linear impulse and angular impulse:

$$\underset{\text{(Linear impulse)}}{\mathbf{L} = \frac{1}{2}\int \mathbf{x} \times \boldsymbol{\omega}\, dV = \text{constant}}, \qquad \underset{\text{(Angular impulse)}}{\mathbf{H} = \frac{1}{3}\int \mathbf{x} \times (\mathbf{x} \times \boldsymbol{\omega})\, dV = \text{constant}}.$$

The invariance of these quantities corresponds to the conservation of linear momentum and angular momentum respectively. If the eddy has a finite linear impulse then $\mathbf{u}_\infty \sim O(r^{-3})$, whereas an eddy with finite angular impulse but zero linear impulse has $\mathbf{u}_\infty \sim O(r^{-4})$. Thus eddies which possess linear impulse cast a longer shadow than those that do not. It is this which lies behind the potent long-range correlations in Saffman's spectrum (see Section 6.3.5).

[1] The net viscous force on the surface S_R is $\oint \tau_{ij}\, dS_j$ where τ_{ij} is the viscous stress. This can be rewritten as the volume integral $\int \partial\tau_{ij}/\partial x_j\, dV = \rho\nu \int \nabla^2 \mathbf{u}\, dV = -\rho\nu \int \nabla \times \boldsymbol{\omega}\, dV$. Converting back to a surface integral we find that the net viscous force is, $\rho\nu \oint \boldsymbol{\omega} \times d\mathbf{S}$, which is zero since $\boldsymbol{\omega} = 0$ on S_R. In a similar way the net viscous torque on S_R can be written as $\rho\nu \oint \mathbf{x} \times (\boldsymbol{\omega} \times d\mathbf{S}) - 2\rho\nu \oint \mathbf{x}(\boldsymbol{\omega} \cdot d\mathbf{S})$, which is also zero.

A2.4 Long-range interactions between Eddies

We may use the results above to show how a Batchelor ($E \sim k^4$) or a Saffman ($E \sim k^2$) spectrum arises. To establish a Batchelor spectrum we assume that, at $t=0$, there are no long-range correlations. We then ask: can such correlations arise naturally by the dynamical equations? In particular, we shall derive a dynamic equation for the triple correlations of the form,

$$\frac{\partial}{\partial t}[u^3 K(r,t)]_\infty = \frac{\partial}{\partial t}\langle u_x^2 u_x' \rangle_\infty = \langle uuuu \rangle \tag{A2.19}$$

It turns out that the term involving fourth-order correlations decays as r^{-4} at large r. Thus, even if K_∞ is exponentially small at $t=0$, it acquires an algebraic tail for $t>0$. We can then use the Karman–Howarth equation to show that $K_\infty \sim r^{-4}$ implies $\langle uu' \rangle_\infty \sim r^{-6}$ in isotropic turbulence. (In anisotropic turbulence we find $\langle uu' \rangle_\infty \sim r^{-5}$). As shown in Chapter 6, an r^{-6} (or slower) decline in $\langle uu' \rangle_\infty$ ensures that the leading order term in $E(k)$ is $O(k^4)$. In Section 6.3.4 we established (A2.19) through a consideration of the long-range pressure forces. Here we shall show that it also follows directly from the far-field properties of an isolated vortex, that is, from the Biot–Savart law.

To establish a Saffman spectrum we must take a different route. We have seen in Section 6.3.5 that a $E \sim k^2$ spectrum can arise only if $\langle uu' \rangle_\infty \sim r^{-3}$. Such strong long-range correlations cannot arise spontaneously in homogeneous turbulence. The long-range pressure forces are not potent enough. Thus, to obtain a Saffman spectrum, we must ensure that $\langle uu' \rangle_\infty \sim r^{-3}$ at $t=0$. Again we can use the results above to ask: under what conditions is the assumption $\langle uu' \rangle_\infty \sim r^{-3}$ kinematically admissible? We shall see that the key distinction between Batchelor and Saffman spectra is that, in the former, the leading-order term in (A2.11) is zero (or near zero) for a typical eddy.

Let us start with $E \sim k^4$ spectra. Consider a turbulent flow in which a typical eddy has negligible linear impulse. In such cases we can use expansion (A2.11) to explain the origin of Batchelor's long-range correlations in a $E \sim k^4$ spectrum. We start by using (A2.11) to evaluate the x-component of $\mathbf{u}$ induced at $\mathbf{x}' = r\hat{\mathbf{e}}_x$ by a remote cloud of eddies located near $\mathbf{x}=0$ (Figure A2.3). It is not hard to show that,

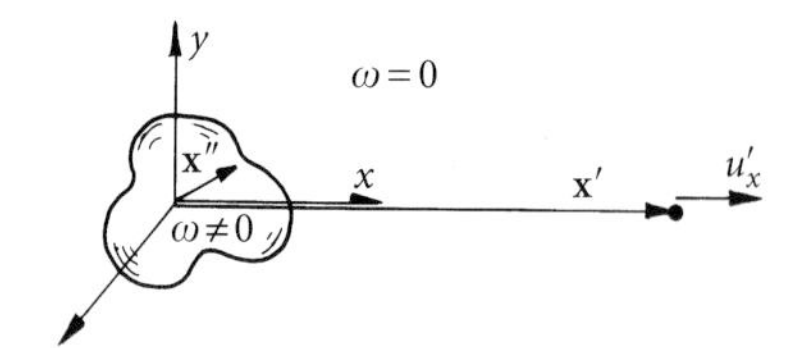

Figure A2.3 The velocity component u_x' induced at $\mathbf{x}' = r\hat{\mathbf{e}}_x$ a long distance from a cloud of eddies located near $x=0$.

$$4\pi u_x' = 4\pi u_x(r\hat{\mathbf{e}}_x) = \frac{3}{r^4}\int \left(x''y''\omega_z'' - x''z''\omega_y''\right) d\mathbf{x}'' + \cdots$$

Differentiating with respect to time we find, after a little algebra,

$$\frac{\partial u_x'}{\partial t} = \frac{3}{4\pi r^4}\int \left[2u_x^2 - u_y^2 - u_z^2\right]'' d\mathbf{x}'' + \cdots. \tag{A2.20}$$

As we shall see, it is this expression which lies behind the dynamic equation for the long-range triple correlations discussed in Section 6.3.4.

Now suppose we have a homogeneous sea of eddies (blobs of vorticity) including our cloud centred at $\mathbf{x}=0$. The Biot–Savart law then tells us that the velocity at $\mathbf{x}'$ has many contributions (from the different vortex blobs) but that at least part of $\partial u'_x/\partial t$ is given by (A2.20) and hence correlated to events in the vicinity of $\mathbf{x}=0$. We might surmise, correctly as it turns out, that

$$\frac{\partial}{\partial t}\left\langle u_x^2 u'_x \right\rangle = \frac{3}{4\pi r^4}\int \left\langle u_x^2 \left(2u_x^2 - u_y^2 - u_z^2\right)'' \right\rangle d\mathbf{x}'' \tag{A2.21}$$

where u_x is evaluated at $\mathbf{x}=0$. (Compare 6.96a with A2.21.) A rigorous derivation of (A2.21) is given in Appendix III using the more conventional pressure argument. For the present purposes the important point to note is that the $O(r^{-4})$ term in expansion (A2.11) gives rise directly to long-range triple correlations of order r^{-4}. These, in turn, lead to a $O(r^{-6})$ contribution to $\left\langle u_i u'_j \right\rangle$ in isotropic turbulence (r^{-5} in anisotropic turbulence) and to the time dependence of Loitsyansky's integral by equation (6.96b).

When the linear momentum of a typical eddy is non-zero, on the other hand, we have, to leading-order in r^{-1}, a larger velocity at $\mathbf{x}'=r\hat{\mathbf{e}}_x$:

$$u'_x = \frac{2L_x}{4\pi r^3} \tag{A2.22}$$

and this suggests the possibility of a r^{-3} contribution to $\left\langle u_x u'_x \right\rangle$. It is this which lies behind the Saffman spectrum

$$\left\langle u_x u'_x \right\rangle_\infty = u^2 f_\infty = \frac{\hat{L}}{4\pi r^3} + O(r^{-6}) \tag{A2.23}$$

$$E(k) - \frac{\hat{L}k^2}{4\pi^2} + O(k^4) \tag{A2.24}$$

where, $\hat{L}=\int\langle \mathbf{u}\cdot\mathbf{u}'\rangle\, d\mathbf{r}$. Note that $\hat{L}$ is an invariant of the Saffman spectrum so that, if $\hat{L}=0$ at $t=0$, then $E\sim k^4$ for all time. Thus, whether or not we have a Batchelor ($E\sim k^4$) or Saffman ($E\sim k^2$) spectrum depends on the initial value of $\hat{L}$. Equation (A2.22) suggests that a non-zero $\hat{L}$ requires a typical eddy to have a finite linear impulse $\mathbf{L}$. The central limit theorem then suggests that, if the turbulent eddies are randomly orientated,

$$\int_V \mathbf{u}\, dV \sim V^{1/2}.$$

This, in turn, tells us that integrals of the type

$$\frac{1}{V}\int_V \mathbf{u}\,dV \cdot \int_V \mathbf{u}'dV = \frac{1}{V}\iint \mathbf{u}\cdot\mathbf{u}'\,d\mathbf{r}dV \sim \int \mathbf{u}\cdot\mathbf{u}'\,d\mathbf{r}$$

are finite, which is consistent with the notion that $\hat{L}$ is non-zero in a Saffman spectrum.

In summary, then, whether we have a k^2 or k^4 spectrum depends on the mechanism which generates the turbulence. If enough linear momentum is injected into the fluid at $t = 0$, then we expect $E \sim k^2$ for all time. On the other hand, if the linear momentum is less than $O(V^{1/2})$ then a $E \sim k^4$ spectrum will emerge. Both situations are readily generated on the computer. However, researchers still cannot agree as to which category real turbulence, say grid turbulence, belongs. In 1967 Saffman suggested that grid turbulence is of the $E \sim k^4$ form, and this is consistent with the little experimental evidence we have.

Suggested reading

Batchelor, G.K. and Proudman, I. (1956) *Phil. Trans. A*, **248**, 369–405.

Batchelor, G.K. (1967) *An Introduction to Fluid Dynamics*. Cambridge Univ. Press.

Saffman, P.G. (1967) *J. Fluid Mech.*, **27**(3), 581–593.

Long-range pressure forces in isotropic turbulence

A3.1 A dynamic equation for the pressure-induced, long-range correlations

We provide here a proof of equations (6.96a, b) which govern the behaviour of the large scale (low-k) end of a $E \sim k^4$ spectrum in isotropic turbulence. We start by introducing

$$S_{i,j,k} = \langle u_i(\mathbf{x})u_j(\mathbf{x}')u_k(\mathbf{x}'')\rangle.$$

It is readily confirmed that, if viscous forces are neglected,

$$\begin{aligned}\frac{\partial S_{i,j,k}}{\partial t} = &-\frac{\partial}{\partial x_l}\left(\overline{u_i u_j' u_k'' u_l}\right) - \frac{\partial}{\partial x_i}\left(\overline{u_j' u_k'' p}\right)\\ &-\frac{\partial}{\partial x_l'}\left(\overline{u_i u_j' u_k'' u_l'}\right) - \frac{\partial}{\partial x_j'}\left(\overline{u_k'' u_i p'}\right)\\ &-\frac{\partial}{\partial x_l''}\left(\overline{u_i u_j' u_k'' u_l''}\right) - \frac{\partial}{\partial x_k''}\left(\overline{u_i u_j' p''}\right)\end{aligned} \tag{A3.1}$$

where, in the interests of brevity, we take $\rho = 1$ and use $\overline{(\sim)}$ rather than $\langle(\sim)\rangle$ to denote an average. Taking the divergence of (A3.1) yields,

$$\begin{aligned}\nabla_x^2\left(\overline{u_j' u_k'' p}\right) &= -\frac{\partial^2}{\partial x_i \partial x_l}\left[\left(\overline{u_i u_j' u_k'' u_l}\right) - \left(\overline{u_i u_l}\right)\left(\overline{u_j' u_k''}\right)\right]\\ \nabla_{x'}^2\left(\overline{u_k'' u_i p'}\right) &= -\frac{\partial^2}{\partial x_j' \partial x_l'}\left[\left(\overline{u_i u_j' u_k'' u_l'}\right) - \left(\overline{u_j' u_l'}\right)\left(\overline{u_i u_k''}\right)\right]\\ \nabla_{x''}^2\left(\overline{u_i u_j' p''}\right) &= -\frac{\partial^2}{\partial x_k'' \partial x_l''}\left[\left(\overline{u_i u_j' u_k'' u_l''}\right) - \left(\overline{u_k'' u_l''}\right)\left(\overline{u_i u_j'}\right)\right].\end{aligned}$$

It follows that we can eliminate pressure from the dynamic equation for $S_{i,j,k}$. In particular we find,

$$\begin{aligned}\frac{\partial S_{11,1}}{\partial t} + \overline{uuuu} = &\,2\frac{\partial}{\partial x_1}\nabla_x^{-2}\frac{\partial^2}{\partial x_m \partial x_n}\left[\left(\overline{u_m u_n u_1 u_1''}\right) - \left(\overline{u_m u_n}\right)\left(\overline{u_1 u_1''}\right)\right]\\ &+\frac{\partial}{\partial x_1''}\nabla_{x''}^{-2}\frac{\partial^2}{\partial x_m'' \partial x_n''}\left[\left(\overline{u_m'' u_n'' u_1^2}\right) - \left(\overline{u_m'' u_n''}\right)\left(\overline{u_1^2}\right)\right]\end{aligned} \tag{A3.2}$$

where, as in Appendix 2, $S_{11,1}(r\hat{\mathbf{e}}_1) = u^3K(r)$ and is $\overline{uuuu}$ a symbolic representation of terms involving the divergence of fourth-order correlations. We now adopt the Batchelor and Proudman (1956) position and assume that, at $t = 0$, well-separated points are statistically independent in the sense that cumulants of the form,

$$\left[u_i u_j' u_k'' u_l\right]_C = \left(\overline{u_i u_j' u_k'' u_l}\right) - \left(\overline{u_i u_j'}\right)\left(\overline{u_l u_k''}\right) - \left(\overline{u_i u_k''}\right)\left(\overline{u_l u_j'}\right) - \left(\overline{u_i u_l}\right)\left(\overline{u_j' u_k''}\right) \tag{A3.3}$$

are exponentially small for *well-separated points*. It follows that the terms on the right of (A3.2) of the form $[\overline{uuuu} - (\overline{uu})(\overline{uu})]$ will tend to zero as $|\mathbf{r}| = |\mathbf{x}'' - \mathbf{x}|$ becomes large. We can now use the Biot–Savart law and (A2.4) to evaluate the inverse Laplacians in (A3.2) for large r. Gauss' theorem ensures that the first two terms in the expansion are zero and so, for large r, we are left with

$$\begin{aligned}\frac{\partial S_{11,1}}{\partial t} + \overline{uuuu} = {} & \frac{1}{2\pi}\frac{\partial^3}{\partial r_1 \partial r_m \partial r_n}\left(\frac{1}{r}\right)\int\left[\left(\overline{u_m u_n u_1 u_1''}\right) - \left(\overline{u_m u_n}\right)\left(\overline{u_1 u_1''}\right)\right] d\mathbf{r} \\ & - \frac{1}{4\pi}\frac{\partial^3}{\partial r_1 \partial r_m \partial r_n}\left(\frac{1}{r}\right)\int\left[\left(\overline{u_m'' u_n'' u_1^2}\right) - \left(\overline{u_m'' u_n''}\right)\left(\overline{u_1^2}\right)\right] d\mathbf{r} \\ & + O(r^{-5}).\end{aligned}$$

However, the first integral on the right is zero, leaving us with

$$\frac{\partial S_{11,1}}{\partial t} + \overline{uuuu} = -\frac{1}{4\pi}\frac{\partial^3}{\partial r_1 \partial r_m \partial r_n}\left(\frac{1}{r}\right)\int\left[\left(\overline{u_m'' u_n'' u_1^2}\right) - \left(\overline{u_m'' u_n''}\right)\left(\overline{u_1^2}\right)\right] d\mathbf{r}. \tag{A3.4}$$

Now $S_{11,1}(r\hat{\mathbf{e}}_x) = u^3K(r)$ and so (A3.4) yields, at $t = 0$,

$$\frac{\partial}{\partial t}(u^3K)_\infty = \frac{3}{4\pi r^4}\int\left[\overline{u_x^2\left(2u_x''^2 - u_y''^2 - u_z''^2\right)}\right] d\mathbf{r}. \tag{A3.5}$$

Compare this with equation (A2.21) of Appendix 2, which was derived by a more heuristic (though physically more illuminating) route. Introducing the symbol s for $u_x^2 - u_y^2$, this may be rewritten as,

$$\frac{\partial}{\partial t}(u^3r^4K)_\infty = \frac{3}{4\pi}\int \overline{ss''}\, d\mathbf{r} = \frac{3}{4\pi}J \tag{A3.6}$$

where J is defined as the integral $\int \overline{ss''}\, d\mathbf{r}$. Finally, combining (A3.6) with (6.72) yields an explicit expression for the second-order derivative of Loitsyansky's integral (Davidson 2000):

$$\frac{d^2I}{dt^2} = 8\pi\frac{d}{dt}[u^3r^4K]_\infty = 6\int\langle ss'\rangle\, d\mathbf{r} = 6J. \tag{A3.7}$$

(In order to keep in line with the notation in the rest of the book we have reverted to the use of $\langle\sim\rangle$ for averages and to a single prime instead of a double prime.) Now the only assumption implicit in (A3.7) is that fourth-order cumulants for *well separated points* are exponentially small. There is some experimental evidence to suggest that, in fully developed turbulence, this might be so and consequently, following Davidson (2000), we might tentatively take (A3.7) to apply throughout the evolution of mature turbulence. Note that we have *not* used the quasi-normal (QN) hypothesis in deriving A3.7 as this would require the additional assumption that fourth-order cumulants are zero for *all* pairs of points, close or distant. In fact the QN equation of Proudman and Reid (1954) is a special case of (A3.7) since the QN estimate of $\langle ss' \rangle$ leads to

$$6J_{\text{QN}} = \frac{14}{5} \int \langle \mathbf{u} \cdot \mathbf{u}' \rangle^2 \, d\mathbf{r}$$

and Rayleigh's theorem yields,

$$(4\pi)^2 \int_0^\infty [E^2/k^2] \, dk = 2 \int \langle \mathbf{u} \cdot \mathbf{u}' \rangle^2 \, d\mathbf{r},$$

from which,

$$\frac{d^2 I}{dt^2} = 6J_{\text{QN}} = \frac{7}{5} (4\pi)^2 \int_0^\infty [E^2/k^2] \, dk \quad \text{(QN only)}.$$

We shall see shortly that this QN estimate of J overestimates its size by one or two orders of magnitude. The EDQNM model, on the other hand, arbitrarily removes a time derivative from (A3.7), uses the QN estimate of J, and introduces a free parameter, θ. It is,

$$dI/dt \sim \theta J_{\text{QN}} \quad \text{(EDQNM only)}.$$

It is hard to reconcile this equation with (A3.7).

A3.2 Experimental evidence for the strength of the long-range pressure forces

We have seen that, under certain conditions, the large scales in isotropic turbulence are governed by (6.96b):

$$\frac{d^2 I}{dt^2} = 8\pi \frac{d}{dt} \left[u^3 r^4 K \right]_\infty = 6 \int \langle ss' \rangle \, d\mathbf{r} = 6J \tag{A3.8}$$

where I is Loitsyansky's integral and $s = u_x^2 - u_y^2$. Now J is positive since

$$J = \int \langle ss' \rangle \, d\mathbf{r} = \left[\int s \, dV \right]^2 \Big/ V$$

for some large volume V. Thus we might write $J = \alpha u^4 l^3$, $\alpha \geq 0$ where l is an integral scale defined here as $l = (I/u^2)^{1/5}$. The positive parameter $\alpha(t)$ gives a dimensionless measure of the strength of the long-range effects in isotropic turbulence. We shall try to estimate its magnitude in a moment. First, however, let us see what the QN model would have us believe.

The QN prediction for α may be estimated as follows. Suppose that the longitudinal correlation function, $f(r)$, has the form, $f(r) = \exp[-r^2/L^2]$ for some L. Then Loitsyansky's integral becomes,

$$I = 8\pi u^2 L^5 \int_0^\infty \eta^4 e^{-\eta^2} d\eta = 3\pi^{3/2} u^2 L^5$$

and so we can relate L to the integral scale: $l = (9\pi^3)^{1/10} L$. The QN estimate of J is (Proudman and Reid 1954),

$$J_{QN} = \frac{7}{15} \int \langle \mathbf{u} \cdot \mathbf{u}' \rangle^2 d\mathbf{r}.$$

Substituting for $\langle \mathbf{u} \cdot \mathbf{u}' \rangle^2$ in terms of $f(r)$ and replacing L by l we find, $J_{QN} = 0.64 u^4 l^3$, and hence $\alpha_{QN} = 0.64$.

Let us now compare this with the experimental evidence. There are no direct measurements of J. However, we can infer a value for α if we examine the rate of decrease of kinetic energy or else the rate of change of I. Consider the equations

$$\frac{du^2}{dt} = -\beta \frac{u^3}{l} \qquad (A3.9)$$

$$\frac{d^2 I}{dt^2} = 6\alpha u^4 l^3 \qquad (A3.10)$$

where the integral scale l is defined by $I = u^2 l^5$. (The coefficient β is of the order of unity.) Now if we assume that α is small (we shall see later that this is so), and that it varies slowly with time, then these equations may be integrated to yield, to leading order in α,

$$u^2/u_0^2 = \left\{ 1 + \gamma\hat{t} - (6\alpha/5)\left[(\hat{t}^2/2) - \gamma^{-1} \int_0^{\hat{t}} \ln(1 + \gamma\hat{t}) d\hat{t} \right] \right\}^{-10/7} \qquad (A3.11)$$

Here u_0 and l_0 are the initial values of u and l, $\hat{t} = u_0 t / l_0$ and $\gamma = 7\beta/10$. We also find that,

$$I/I_0 = 1 + (6\alpha/\gamma)[\hat{t} - \gamma^{-1} \ln(1 + \gamma\hat{t})]. \qquad (A3.12)$$

Now most of the numerical and experimental data for variations of I/I_0 and u^2/u_0^2 are expressed in terms of polynomials of the form $\hat{t}^n$. However, we can relate α to n by choosing a best-fit polynomial to (A3.11) and (A3.12) for the appropriate range of $\hat{t}$.

Let us now consider the experimental data. There are six sets of data, which are relevant:

1. In the LES of Chasnov (1993) we find $I \sim \hat{t}^{0.25}$.
2. Lesieur (1990) reports similar simulation in which $u^2 \sim \hat{t}^{-1.4}$.
3. Lesieur et al. (1999, 2000) report LES in which, in the asymptotic state, $u^2 \sim \hat{t}^{-10/7}$.
4. In the final stages of decay wind-tunnel tests show an exponential, rather than algebraic, decay of $f_\infty(r)$. (e.g. see Batchelor and Proudman 1956.)
5. Comte–Bellot and Corrsin (1966) give $u^2 \sim \hat{t}^{-n}$ with n in the range $1.2 \rightarrow 1.4$, with an average of 1.26.
6. Warhaft and Lumley (1978) report $u^2 \sim \hat{t}^{-n}, n \sim 1.34$.

The estimates of α based on these measurements are shown below.

Data set	1	2	3	4	5	6
α	0.02	0.01	$\sim$0	$\sim$0	0.03	0.02

The difference between these estimates of α and α_{QN} is very striking, with $\alpha \sim 0.02$ yet $\alpha_{QN} \sim 0.64$. It is clear that there is considerable uncertainty in the precise value of α, but it seems that it is much less than unity. In short, the long-range pressure forces predicted by Batchelor and Proudman (1956) turn out to be weak.

As yet, there is no satisfactory explanation for the weakness of the long-range correlations. However, one naive interpretation is as follows. From (A2.20) and (A2.21) we see that the origin of the slow, algebraic decay in $\langle u_x^2 \, u'_x \rangle$ lies in the fact that the vorticity located near $\mathbf{x}$, which dominates u_x^2, also gives rise to a quadrupole far-field velocity, of order $u'_x \sim r^{-4}$. However, if the vorticity in the region of $\mathbf{x}$ is spatially convoluted, as suggested by Plate 8, then this quadrupole component will be small.

Suggested reading

See the reference list at the end of Chapter 6 for the references cited here.

APPENDIX 4

Hankel transforms and hypergeometric functions

A4.1 Hankel transforms

The table below refers to the Hankel transform pair

$$f(x) = \int_0^\infty k\hat{f}(k)J_\nu(kx)\,dk, \qquad \hat{f}(k) = \int_0^\infty xf(x)J_\nu(kx)\,dx.$$

$f(x)$	$\hat{f}(k)$
$x^n e^{-px^2}$ $(\nu + n + 2 > 0, p > 0)$	$\dfrac{k^\nu\Gamma(\frac{1}{2}\nu + \frac{1}{2}n + 1)}{2^{\nu+1}p^{1+(n+\nu)/2}\Gamma(\nu+1)} \times {}_1F_1(\frac{1}{2}\nu + \frac{1}{2}n + 1, \nu + 1, -k^2/4p)$
$x^\nu e^{-px^2}$ $(\nu > -1, p > 0)$	$\dfrac{k^\nu}{(2p)^{\nu+1}}\exp[-k^2/4p]$
$x^\nu e^{-px}$ $(\nu > -1, p > 0)$	$2p(2k)^\nu\ \Gamma(\nu + 3/2)(p^2 + k^2)^{-(\nu+3/2)}\pi^{-1/2}$
$x^{-1}e^{-px^2}$ $(\nu > -1, p > 0)$	$\dfrac{\pi^{1/2}}{2p^{1/2}}\exp[-k^2/8p]I_{\nu/2}(k^2/8p)$
$x^n e^{-px}$ $(\nu + n + 2 > 0, p > 0)$	$\dfrac{(k/2p)^\nu\Gamma(\nu + n + 2)}{p^{n+2}\Gamma(\nu+1)} \times F(\frac{1}{2}\nu + \frac{1}{2}n + 1,\ \frac{1}{2}\nu + \frac{1}{2}n + \frac{3}{2}, \nu + 1,\ -k^2/p^2)$
$x^{-1}e^{-px}$ $(\nu > -1, p > 0)$	$\dfrac{k^{-\nu}\left[(p^2 + k^2)^{1/2} - p\right]^\nu}{(p^2 + k^2)^{1/2}}$
$x^{-1}(x^2 + a^2)^{-1/2}$ $(a > 0, \nu > -1)$	$I_{\nu/2}\ (k\ a/2)\ K_{\nu/2}\ (k\ a/2)$
x^n $(-\nu < n + 2 < 3/2)$	$\dfrac{2^{n+1}\Gamma(\frac{1}{2}\nu + \frac{1}{2}n + 1)}{\Gamma(\frac{1}{2}\nu - \frac{1}{2}n)}k^{-(n+2)}$
$\dfrac{x^\nu}{(x^2 + a^2)^{1+\mu}}$ $(-1 < \nu < 2\mu + 3/2)$	$\dfrac{a^{\nu-\mu}k^\mu}{2^\mu\Gamma(1+\mu)}K_{\nu-\mu}(ka)$

Notation:

${}_1F_1$ Kummer's hypergeometric function, often written as M

${}_2F_1$ Gauss' hypergeometric function, often written as F

A4.2 Hypergeometric functions

Kummer's *confluent hypergeometric function*, $M(a, b, x)$, sometimes written as ${}_1F_1(a; b; x)$, is a solution of Kummer's equation,

$$x\frac{d^2f}{dx^2} + (b - x)\frac{df}{dx} - af = 0.$$

It is

$$M(a, b, x) = 1 + \frac{ax}{b} + \frac{a_2x^2}{b_2 2!} + \cdots + \frac{a_nx^n}{b_n n!} + \cdots$$

where

$$a_n = a(a + 1)(a + 2)\cdots(a + n - 1), \quad a_0 = 1.$$

Special cases of M are,

$$M(a, a, x) = e^x, \quad M(1, 2, 2x) = x^{-1}e^x \sinh x,$$
$$M(\tfrac{1}{2}, \tfrac{3}{2}, -x^2) = \sqrt{\pi}(2x)^{-1} \operatorname{erf} x,$$

and M satisfies the differential expressions

$$\frac{d^n}{dx^n}M(a, b, x) = \frac{a_n}{b_n}M(a + n, b + n, x),$$
$$\frac{x}{a}\frac{d}{dx}M(a, b, x) = M(a + 1, b, x) - M(a, b, x).$$

For $x \to \infty$ we have

$$M(a, b, x) \to \frac{\Gamma(b)}{\Gamma(a)}e^x x^{a-b}$$

and Kummer's transformation rule tells us that

$$M(a, b, x) = e^x M(b - a, b, -x).$$

Gauss' hypergeometric function, on the other hand, is

$$F(a, b, c, x) = {}_2F_1(a; b; c; x) = \sum_{n=0}^{\infty} \frac{a_n b_n}{c_n}\frac{x^n}{n!}.$$

Special cases of F are,

$$F(1, 1, 2, x) = x^{-1}\ln(1 - x), \quad F(\tfrac{1}{2}, \tfrac{1}{2}, \tfrac{3}{2}, x^2) = x^{-1}\arcsin x,$$
$$F(\tfrac{1}{2}, 1, \tfrac{3}{2}, -x^2) = x^{-1}\arctan x,$$

and F satisfies the differential expression

$$\frac{d^n}{dx^n} F(a, b, c, x) = \frac{a_n b_n}{c_n} F(a + n, b + n, c + n, x).$$

Suggested reading

Abramowitz, M. and Stegun, I.A. (1965) *Handbook of Mathematical Functions*. Dover.

The kinematics of homogeneous, axisymmetric turbulence

The assumption of isotropy is generally satisfied at the small scales of a turbulent flow, but only rarely at the large scales. A less restrictive assumption to make is that the turbulence possesses axial symmetry. That is to say, the statistical features of the turbulence are invariant for rotations about a preferred direction, say the z-axis, and for reflections in planes containing $\hat{\mathbf{e}}_z$ and perpendicular to $\hat{\mathbf{e}}_z$. (We shall not allow for mean helicity in this discussion, which would require a relaxation of the requirement for invariance under reflections.) This kind of symmetry is relevant to turbulence evolving in a uniform magnetic field, **B**, or subject to rapid rotation, where the rotation axis or **B** imparts a preferred direction to the turbulence.

The kinematics of axisymmetric turbulence was developed by Chandrasekhar (1950). We shall outline briefly some of the key results without offering any form of proof. Let $\mathbf{r} = \mathbf{x}' - \mathbf{x}$, and $\boldsymbol{\lambda}$ be a unit vector parallel to the axis of symmetry. We introduce two coordinates, r and μ, defined by

$$r = |\mathbf{r}|, \qquad \mu = (\mathbf{r} \cdot \boldsymbol{\lambda})/r \tag{A5.1}$$

and three differential operators,

$$D_r = \frac{1}{r}\frac{\partial}{\partial r} - \frac{\mu}{r^2}\frac{\partial}{\partial \mu}, \qquad D_\mu = \frac{1}{r}\frac{\partial}{\partial \mu}, \qquad D_{\mu\mu} = D_\mu D_\mu. \tag{A5.2}$$

The general form of Q_{ij} for homogeneous, axisymmetric turbulence in an incompressible fluid is,

$$Q_{ij} = Ar_i r_j + B\delta_{ij} + C\lambda_i\lambda_j + D(\lambda_i r_j + \lambda_j r_i) \tag{A5.3}$$

where A, B, C, and D are functions of r and μ which can, in turn, be derived from the two independent functions, Q_1 and Q_2, according to

$$A = (D_r - D_{\mu\mu})Q_1 + D_r Q_2, \tag{A5.4}$$

$$\begin{aligned} B = {} & \left[r^2(1-\mu^2)D_{\mu\mu} - r\mu D_\mu - (2 + r^2 D_r + r\mu D_\mu)\right]Q_1 \\ & - \left[r^2(1-\mu^2)D_r + 1\right]Q_2, \end{aligned} \tag{A5.5}$$

$$C = -r^2 D_{\mu\mu} Q_1 + (1 + r^2 D_r) Q_2, \tag{A5.6}$$

$$D = \left[\left(r\mu D_\mu + 1\right) D_\mu\right] Q_1 - r\mu D_r Q_2. \tag{A5.7}$$

Thus Q_{ij} is determined solely by Q_1 and Q_2, both of which turn out to be even functions of r and $r\mu$. In the special case of three-dimensional isotropic turbulence it turns out that $Q_2 = 0$ and $2Q_1 = -u^2 f(r)$ where f is the longitudinal correlation function. At the other extreme, two-dimensional isotropic turbulence corresponds to $Q_1 = 0$ and $Q_2 = -u^2 f(r)$. Expansions for Q_1 and Q_2 in r and μ are of the form,

$$Q_1 = \alpha_0 + r^2(\alpha_1 + \alpha_2 \mu^2) + O(r^4)$$
$$Q_2 = \beta_0 + r^2(\beta_1 + \beta_2 \mu^2) + O(r^4)$$

where the α's and β's are constants. The corresponding expansion for Q_{ij} is,

$$\begin{aligned} Q_{ij} = {} & [\alpha_1 - \alpha_2 + \beta_1] 2 r_i r_j \\ & + \left[r^2\left(2\alpha_2 - 4\alpha_1 - 3\beta_1\right)\right. \\ & \left. + r^2\mu^2\left(2\beta_1 - \beta_2 - 8\alpha_2\right) - \left(2\alpha_0 + \beta_0\right)\right]\delta_{ij} \\ & + \left[\beta_0 + r^2\left(3\beta_1 - 2\alpha_2 + \mu^2\beta_2\right)\right]\lambda_i\lambda_j \\ & + \left[2\alpha_2 - \beta_1\right] 2r\mu\left(\lambda_i r_j + \lambda_j r_i\right) + \text{H.O.T.} \end{aligned} \tag{A5.8}$$

In isotropic turbulence $\alpha_2 = \beta_0 = \beta_1 = \beta_2 = 0$, $Q_2 = 0$, and $6Q_1 = -\langle \mathbf{u}^2 \rangle f(r)$. The only non-zero coefficients are then,

$$2\alpha_0 = -\tfrac{1}{3}\langle \mathbf{u}^2 \rangle, \qquad 2\alpha_1 = \tfrac{1}{30}\langle \boldsymbol{\omega}^2 \rangle.$$

Perhaps the most important special cases of expansion (A5.8) are,

1. $\langle u_\parallel u_\parallel \rangle (r_\parallel, 0) = -2\alpha_0 - 2(\alpha_1 + \alpha_2) r_\parallel^2 + \cdots$
2. $\langle u_\parallel u_\parallel \rangle (0, r_\perp) = -2\alpha_0 - 4\alpha_1 r_\perp^2 + \cdots$
3. $\langle u_\perp u_\perp \rangle (r_\parallel, 0) = -(2\alpha_0 + \beta_0) - (4\alpha_1 + 6\alpha_2 + \beta_1 + \beta_2) r_\parallel^2 + \cdots$
4. $\langle u_\perp u_\perp \rangle (0, r_\perp) = -(2\alpha_0 + \beta_0) - (2\alpha_1 + \beta_1) r_\perp^2 + \cdots$

where $||$ and $\perp$ refer to directions parallel and perpendicular to $\boldsymbol{\lambda}$. If $\boldsymbol{\lambda}$ coincides with the z-axis we have,

$$\langle u_x^2 \rangle = \langle u_y^2 \rangle = -(2\alpha_0 + \beta_0), \quad \langle u_z^2 \rangle = -2\alpha_0,$$
$$\langle u_x u_y \rangle = \langle u_x u_z \rangle = \langle u_y u_z \rangle = 0.$$

The enstrophy can be shown to be,

$$\langle \boldsymbol{\omega}^2 \rangle = 4[15\alpha_1 + 5\alpha_2 + 5\beta_1 + \beta_2].$$

The spectral equivalent of (A5.3)–(A5.7) is

$$\Phi_{ij} = (F + G)[k^2\delta_{ij} - k_i k_j] - G[k_\parallel^2 \delta_{ij} + k^2 \lambda_i \lambda_j - k_\parallel (\lambda_i k_j + \lambda_j k_i)]$$

where F and G are even functions of k and μk. $G = 0$ in isotropic turbulence and $F = 0$ in two-dimensional turbulence.

Suggested reading

Chandrasekhar S. (1950) *Phil. Trans. Roy. Soc. London, A*, **242**, 557.

Subject Index